Lecture Notes in Computer Science　　　11164

Commenced Publication in 1973
Founding and Former Series Editors:
Gerhard Goos, Juris Hartmanis, and Jan van Leeuwen

More information about this series at http://www.springer.com/series/7409

Richang Hong · Wen-Huang Cheng
Toshihiko Yamasaki · Meng Wang
Chong-Wah Ngo (Eds.)

Advances in Multimedia Information Processing – PCM 2018

19th Pacific-Rim Conference on Multimedia
Hefei, China, September 21–22, 2018
Proceedings, Part I

 Springer

Editors
Richang Hong
Hefei University of Technology
Hefei
China

Meng Wang
Hefei University of Technology
Hefei
China

Wen-Huang Cheng
National Chiao Tung University
Hsinchu
Taiwan

Chong-Wah Ngo
City University of Hong Kong
Hong Kong
Hong Kong, China

Toshihiko Yamasaki
University of Tokyo
Tokyo
Japan

ISSN 0302-9743   ISSN 1611-3349   (electronic)
Lecture Notes in Computer Science
ISBN 978-3-030-00775-1   ISBN 978-3-030-00776-8   (eBook)
https://doi.org/10.1007/978-3-030-00776-8

Library of Congress Control Number: 2018954671

LNCS Sublibrary: SL3 – Information Systems and Applications, incl. Internet/Web, and HCI

This Springer imprint is published by the registered company Springer Nature Switzerland AG
The registered company address is: Gewerbestrasse 11, 6330 Cham, Switzerland

Preface

The 19th Pacific-Rim Conference on Multimedia (PCM 2018) was held in Hefei, China, during September 21–22, 2018, and hosted by the Hefei University of Technology (HFUT). PCM is a major annual international conference for multimedia researchers and practitioners across academia and industry to demonstrate their scientific achievements and industrial innovations in the field of multimedia.

It is a great honor for HFUT to host PCM 2018, one of the most longstanding multimedia conferences, in Hefei, China. Hefei University of Technology, located in the capital of Anhui province, is one of the key universities administrated by the Ministry of Education, China. Recently its multimedia-related research has attracted more and more attentions from local and international multimedia community. Hefei is the capital city of Anhui Province, and is located in the center of Anhui between the Yangtze and Huaihe rivers. Well known both as a historic site famous from the Three Kingdoms Period and the hometown of Lord Bao, Hefei is a city with a history of more than 2000 years. In modern times, as an important base for science and education in China, Hefei is the first and sole Science and Technology Innovation Pilot City in China, and a member city of WTA (World Technopolis Association). We hope that PCM 2018 is a memorable experience for all participants.

PCM 2018 featured a comprehensive program. We received 422 submissions to the main conference by authors from more than ten countries. These submissions included a large number of high-quality papers in multimedia content analysis, multimedia signal processing and communications, and multimedia applications and services. We thank our Technical Program Committee with 178 members, who spent much time reviewing papers and providing valuable feedback to the authors. From the total of 422 submissions, the program chairs decided to accept 209 regular papers (49.5%) based on at least three reviews per submission. In total, 30 papers were received for four special sessions, while 20 of them were accepted. The volumes of the conference proceedings contain all the regular and special session papers.

We are also heavily indebted to many individuals for their significant contributions. We wish to acknowledge and express our deepest appreciation to general chairs, Meng Wang and Chong-Wah Ngo; program chairs, Richang Hong, Wen-Huang Cheng and Toshihiko Yamasaki; organizing chairs, Xueliang Liu, Yun Tie and Hanwang Zhang; publicity chairs, Jingdong Wang, Min Xu, Wei-Ta Chu and Yi Yu, special session chairs, Zhengjun Zha and Liqiang Nie. Without their efforts and enthusiasm, PCM 2018 would not have become a reality. Moreover, we want to thank our sponsors: Springer, Anhui Association for Artificial Intelligence, Shandong Artificial Intelligence Institute, Kuaishou Co. Ltd., and Zhongke Leinao Co. Ltd. Finally, we wish to thank all committee members, reviewers, session chairs, student volunteers, and supporters. Their contributions are much appreciated.

September 2018

Richang Hong
Wen-Huang Cheng
Toshihiko Yamasaki
Meng Wang
Chong-Wah Ngo

Organization

General Chairs

Meng Wang — Hefei University of Technology, China
Chong-Wah Ngo — City University of Hong Kong, Hong Kong, China

Technical Program Chairs

Richang Hong — Hefei University of Technology, China
Wen-Huang Cheng — National Chiao Tung University, Taiwan, China
Toshihiko Yamasaki — University of Tokyo, Japan

Organizing Chairs

Xueliang Liu — Hefei University of Technology, China
Yun Tie — Zhengzhou University, China
Hanwang Zhang — Nanyang Technological University, Singapore

Publicity Chairs

Jingdong Wang — Microsoft Research Asia, China
Min Xu — University of Technology Sydney, Australia
Wei-Ta Chu — National Chung Cheng University, Taiwan, China
Yi Yu — National Institute of Informatics, Japan

Special Session Chairs

Zhengjun Zha — University of Science and Technology of China, China
Liqiang Nie — Shandong University, China

Contents – Part I

Special Session

Contents – Part II

Contents – Part III

Oral Session

CodedVision: Towards Joint Image Understanding and Compression via End-to-End Learning

Qiu Shen[1], Juanjuan Cai[2], Linfeng Liu[1], Haojie Liu[1], Tong Chen[1], Long Ye[2], and Zhan Ma[1(✉)]

[1] Nanjing University, Nanjing, China
{shenqiu,mazhan}@nju.edu.cn
[2] Communication University of China, Beijing, China
{caijuanjuan,yelong}@cuc.edu.cn

Abstract. We present a *CodedVision* framework to achieve image content understanding and compression jointly, leveraging the recent advances in deep neural networks. We have introduced an eight-layer deep residual network to extract image features for compression and understanding. For compression, a scalar quantizer and an entropy coder are utilized to remove redundancy. Rate-distortion optimization is integrated to improve the coding efficiency where rate is estimated via a piecewise linear approximation. A noticeable 7.8% BD-Rate (Bjontegaard delta rate) gain is presented against the state-of-the-art HEVC intra based image compression. For content understanding, we patch another residual network-based classifier to perform the classification, with reasonable accuracy at the current stage.

Keywords: Deep neural network · Content understanding
Image compression · Rate approximation · Classification

1 Introduction

Image, as an efficient and effective representation, illustrates the vivid natural world and events captured by either cameras or our human visual systems (HVS). It prevails at the entire Internet scale, with approximately billions of photos uploaded and shared daily. It's apparent that images require efficient compact representations to save storage and improve the exchange efficiency (e.g., social sharing). Towards this goal, images are often coded in JPEG, JPEG2000 [19], or even other file formats [30,33]. Meanwhile, images are used by human beings

This work is supported by the National Natural Science Foundation of China (Grant # 61422107, 61571215 and 61631016) and the Fundamental Research Funds for the Central Universities (Grant # 021014380053, 021014380091). Dr. Z. Ma is the corresponding author of this work.
Q. Shen, J. Cai – Authors contributed equally to this work.

R. Hong et al. (Eds.): PCM 2018, LNCS 11164, pp. 3–14, 2018.
https://doi.org/10.1007/978-3-030-00776-8_1

or machines intentionally for typical vision tasks [29], such as object detection [27], classification [12], retrieval [13] and so on. This undoubtedly demands an explicit understanding of the image content for these aforementioned activities. It naturally leads to a fundamental question: *can we have a compact and understandable presentation of images, which can benefit not only compression but also analysis simultaneously?*

Unfortunately, research efforts for image compression and understanding are often isolated for dominant applications. This is mainly because that image compression tries to convert the raw image signal in RGB color space into compressed binary stream, while image understanding extracts and analyzes the content mainly in feature space (e.g., gradients, SIFT [20], etc). Given that almost all images are stored and shared in compressed format, image understanding first needs to decode the binary stream into corresponding signal in raw domain (e.g., RGB or other color spaces), which brings extra computational burden. Furthermore, compressed images often incur noticeable quantization and filtering artifacts that might impair the efficiency when performing the vision tasks (i.e., a heavy compression would remove or smooth sharp edges of interested object).

There are intuitive advantages in combing the image compression and understanding jointly. For example, it allows the compression and understanding to be done in the edge, such as the front camera. Therefore, latency could be minimized when it does not require additional cycles to decode the compressed binary stream (often using remote servers) to corresponding RGB samples. This is a vital factor that could massively improve the emergency response efficiency for security surveillance scenarios. Moreover, leveraging the joint compression and understanding, we could mine effective features immediately for future applications, such as augmented reality (AR) [2], aesthetic tendency analysis [14,15], etc.

Because of these functional augments, numerous efforts have been devoted for decades to consider the vision task in compressed domain,such as retrieval [21], segmentation [26], analysis [3], saliency detection [11], and classification [36–38]. Even though, all these studies are still in very early phase and utilize the already compressed bit stream rather than performing the compression and understanding simultaneously during capturing. The limited and distorted information extracted from the compressed domain is not sufficient to guarantee the accuracy of the vision tasks. Furthermore, it has not yet presented any systematic model or framework that could generate a compact and efficient presentation to meet the requirements both of compression and analysis.

Fortunately, recent learning schemes, particularly those via deep neural networks (DNN) have shown brilliant potentials on image understanding [9,12,17,18] and compression [4,7,31,32]. Note that similar neural networks are used by aforementioned works, and features extracted from these networks also have much in common. Therefore, we expect that both image compression and understanding can be combined jointly via deep neural networks. In this work, we have developed a *CodedVision* framework, as shown in Fig. 1, to achieve the simultaneous image compression and understanding. More specifically, we have used an

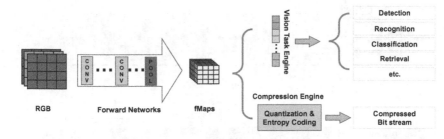

Fig. 1. Illustration of proposed *CodedVision* framework

8 layer ResNet [12] as the forward neural network to derive the compact feature maps (fMaps). Upon derived fMaps, we could add additional sub-system, such as Vision Task Engine (VTE), to perform the vision tasks; for instance, detection, recognition, classification, retrieval and so on. In the meantime, we could apply the Compression Engine (CE) composed by quantization and entropy coding to have densely compacted binary stream of the fMaps.

We test compression performance on a public Kodak (http://r0k.us/gra phics/kodak/) and CLIC (http://www.compression.cc/challenge/) test dataset, and achieve a noticeable BD-Rate [6] improvement (i.e., averaged 7.8% on Kodak images) against the JPEG2000 and BPG (a.k.a, a HEVC intra based image compression method). Meanwhile, classification is selected as the first attempt of vision task by concatenating a classifier network, and the experimental results prove that fMaps is useful for classification. In general, joint image understanding and compression via end-to-end learning is feasible, although it requires many further studies.

2 Visual Primitives

Light travels across the 3-D environment and finally reaches the sensor plane of the camera or the retina of our HVS to have imagery projection. To be honest, our HVS is the *most intelligent* camera by far. It senses the surrounding natural world visually, extracts the features of the interested object/event, and constructs the representation to be rendered and further understood in the brain. This actually coincides with the computational vision theory developed by David Marr decades ago [22], where our HVS first extracts low-level information (i.e, gradient, texture activity, color, etc) of the visible scene in the front (i.e., referred to as 2-D vision), followed by the connections among low-level features to depict the contours of the object (i.e., referred to as the 2.5D vision) and the final construction of the vivid 3-D vision with additional information augmented. With the procedure discussed so far, an image I, as the representation of the visual scene, can be simplified as the linear combinations of the visual features, i.e.,

$$I = \sum_{i=1}^{N} \alpha_i \cdot \mathbf{V}_f(i), \tag{1}$$

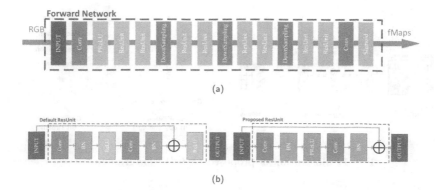

Fig. 2. Illustration of the forward network of *CodedVision*: (a) the architecture of forward network (b) residual network structure (left: default in [12]; right: proposed scheme)

with α_i as the weighting coefficient, $\mathbf{V}_f(i)$ as a typical visual stimuli (e.g., gradients, etc), and N as the total number of visual features that have contributed to the content representation. On the other hand, $\mathbf{V}_f(i)$ can be also seen as a i-th basis vector of the learned dictionaries (e.g., with either fixed or self-adaptive atoms), and associated α_i is the sparse representation of the weighting coefficients that can be derived conventionally via orthogonal match pursuit (OMP) [25]. This sparsity guarantees the efficiency of the compression. The same observations are also discussed in [34,35]. In practice, these features can be obtained via training, such as the K-SVD [1] or more recent various DNNs.

In the meantime, visual features, such as histogram of gradients (HOG) and SIFT, are the fundamental cues for various vision tasks (e.g., detection, classification, etc). Recently, we also have noticed that learned features via DNN are widespread in vision applications. It is natural to assume that the features used for vision tasks might be the same as these used for compression. This actually fits the discriminative intuition of our visual system. Our HVS extracts the salient (important and necessary) features. These features represent the meaningful data for object detection, classification, retrieval and so on. With these features well organized and compressed, they would offer identical accuracy for content understanding after reconstruction. Then, the fundamental question is *how to extract appropriate features that can be well utilized in both understanding and compression?*

3 The *CodedVision* Framework

Motivated by the recent advances in DNN, we have proposed to use DNN based learning for compact and salient feature (i.e., fMaps) extraction. Then, these features would go through different sub-systems to facilitate the compression and understanding, as shown in Fig. 1. For example, they are usually categorized, pooled, and fused for vision tasks. In practice, this could be realized via

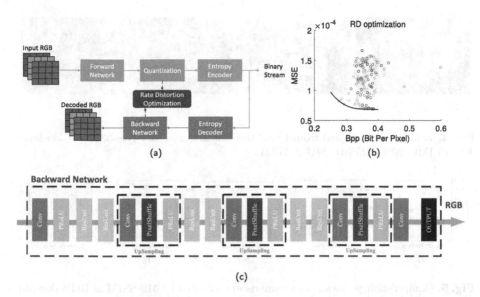

Fig. 3. Illustrations of (a) end-to-end learning framework for compression(b) various pairs of rate and distortion generated by different λ with a fixed amount of features. Rate-distortion optimization (RDO) is performed on these discrete points, getting the optimized red curve that covers all points on its up-right side. (c) the architecture of backward network

a designated vision task engine. But, as for the compression engine, features are typically quantized and entropy coded to maximally exploit the redundancy for better storage and exchange.

We have employed the deep residual network (ResNet) [12] to extract fMaps, because of its superior efficiency and fast convergence. Furthermore, to achieve a better convergence speed, we replace the default rectified linear units (ReLU) with the parametric rectified linear units (PReLU) and remove the nonlinear mapping after the short connection [23] for each residual unit, as shown in Fig. 2b. The forward network contains eight residual units as shown in Fig. 2a and all the down-sampled operations are using a stride-2 4×4 convolutional layer. The other convolutional layers all have 3×3 kernel size except the first and last layer, which uses 5×5.

3.1 Image Compression

Compression is achieved via exploiting the redundancy inside the output fMaps, as shown in Fig. 2a. Variational compression ratio could be achieved via adjusting the number of fMaps. A scalar quantizer is used to reduce the number of bits (e.g., we used fixed 6-bits in this work) for representing the exacted fMaps produced from the forward network, followed by an entropy encoder to further remove the statistical redundancy. This generally refers to as an compression *encoder*.

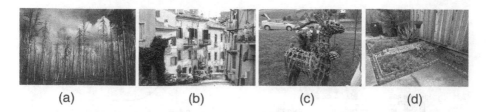

Fig. 4. Four images sampled from CLIC test dataset (a) casey-fyfe-3340 (b) lou-levit-369 (c) IMG-20161123 (d) IMG-20170211

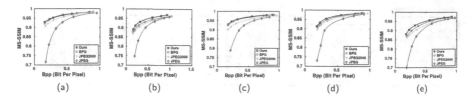

Fig. 5. Compression performance comparison measured by MS-SSIM in RGB domain, compared with JPEG, JPEG2000 and BPG (a) averaged BD-Rate on Kodak Dataset (b) casey-fyfe-3340 (c) lou-levit-369 (d) IMG-20161123 (e) IMG-20170211

Towards the optimal performance via an end-to-end learning framework presented in Fig. 3a, a corresponding *decoder* is implemented in the training phase. For instance, compressed binary stream is first entropy decoded to derive the fMaps, followed by a symmetrical backward network to reconstruct the signal from the reconstructed fMaps (shown in Fig. 3c). We choose the pixel-shuffle layers [28] as up-sampled operations considering their decent performance in super resolution.

In practice, rate-distortion optimization is often used to improve the image compression efficiency [24]. The actual bitrate depends on the entropy of the quantized feature maps. We propose to apply the Lagrangian optimization framework to jointly consider the rate loss L_R and l-2 distortion loss. Here the rate loss L_R is defined as

$$L_R = -\mathbf{E}[\log_2 P_q],\qquad(2)$$

where L_R is the entropy approximation of the fMaps at the bottleneck layer. Since the derivatives of the quantization function are almost zero, we apply a piecewise linear approximation of the discrete P_q to ensure it continuous and differentiable. Note that Ballé [5] also applied similar idea to perform joint rate and distortion optimization.

Network Training Here, we present more details on how to train CNNs used in the work. In practice, we use the open source data sets released by the Computer Vision Lab of ETH Zurich in CLIC competition. All the images in the training sets are split into 128×128 patches randomly with a data augmentation method

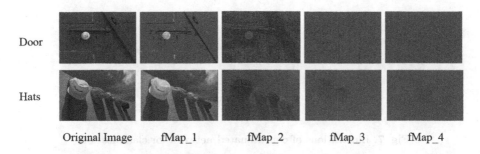

Fig. 6. Visualization of four-channel fMaps extracted from "door" and "hats" images

such as rotation or scaling, resulting in 80000 patches in total. The objective of training is to minimize the following loss function:

$$L = \frac{1}{N} \sum_{n=1}^{N} ||(Y_n|\theta) - X_n||^2 + \lambda L_R, \tag{3}$$

where X_n is the input image, Y_n is the decoded image, N represents the batch size (e.g. 64), and θ represents the parameters in neural network. We introduce the parameter λ to control the penalty of rate loss L_R which is generated from the rate estimation module as shown in Eq. (2). Inspired by transfer learning, we also apply an easy-to-hard learning method mentioned in the deblocking method named ARCNN [8] and first set λ to 0. Without any rate control, we use the optimizer Adam (an adaptive learning rate method) [16] with the learning rate 0.0001 to make fast convergence and generate the pre-training model first after 100 epochs. Then we increase the value of λ with an interval of 0.0001 from 0 to 0.002 every 5 epochs to progressively improve the level of rate constraint. Finally, it generates a lot of models with different rates and distortions to make a sophisticated RD optimization as shown in Fig. 3b.

Performance Evaluation. We evaluate our performance on the dataset released by CLIC and Kodak PhotoCD data set, and compare with existing codecs including JPEG, JPEG2000, and BPG. Figure 5 shows MS-SSIM performance over all 24 Kodak images and achieves an average 7.8% BD-Rate reduction over BPG[1]. Moreover, we have more impressive performance on CLIC test dataset. We select four typical images with different types from the dataset as test samples, as shown in Fig. 4. Finally, the BD-Rate has separately reduced by 33.54%, 9.65%, 13.31% and 19.96%, as illustrated in Fig. 5. As can be seen from the results, our approach outperforms BPG and JPEG2000 in both overall performance and separate comparison using individual test image.

---

[1] Given that BPG demonstrates the state-of-the-art coding efficiency, we mainly present the comparison against it.

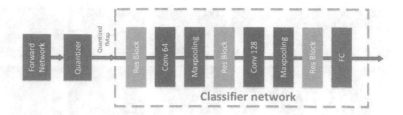

Fig. 7. Illustrations of concatenated network for classification

Table 1. mAP of classification with different sets of input fMaps

fMap_num	8	12	24
Quant	52.38%	55.45%	59.26%
no_Quant	57.03%	59.54%	60.39%

3.2 Image Understanding

As visualized in Fig. 6, we have found that the characteristics of image content (i.e., object, color, edge, texture, etc) are preserved among the ground-truth samples in RGB domain and fMaps extracted and quantized from the afore-mentioned network shown in Fig. 3a. Therefore, we expect that the quantized fMaps can be used for vision tasks. Towards this purpose, we have concatenated a *classifier* network with the *forward network + quantizer*, shown in Fig. 7. The classifier network adopts the same ResUnit used in forward network, while the quantized features are extended to 128 channels for classification task.

Network Training. 5k images in the training set of Pascal VOC 2012 dataset [10] are used as trainning data. The 256×256 crop is randomly sampled from the original image or its horizontal flip, and then compressed by the forward network. The output fMaps are fed into classifier network for training. The forward network for image compression is pre-trained as detailed in Sect. 3.1 and it is fixed while training the classifier network.

Performance Evaluation. The validation set of Pascal VOC 2012 dataset is used for testing. Every test image is randomly sampled by ten 256×256 crops, and the mean average precision (mAP) is used to measure the performance of classification. The result with different numbers of fMaps are presented in Table 1. The second line gives the result when fMaps are quantized with 6 bits, while the third line is the result with no quantization. It is evident that the classification result may be affected by the number and quantization degree of fMaps. More fMaps with bigger data size can achieve better performance. The performance loss caused by quantization may reduce with the increase of fMaps.

Thanks to the versatile framework of *CodedVsion*, receivers are able to perform the image classification in compressed domain without fully decoding. The

Table 2. Comparison of accuracy and processing time with other classification method

	mAP	Time for decoding	Time for classification
CodedVision	52.38%	0 s	7.46 s
Compressed with JPEG	64.24%	24.73 s	21.69 s
Uncompressed	69.30%	0 s	21.69 s

fMaps reconstructed from the binary stream are fed into the classifier network presented in Fig. 7 for classification. Table 2 compares the accuracy and processing time of *CodedVision* to the conventional method (e.g. images are compressed with JPEG), where the images are fully decoded into raw representation for classification. In the meantime, the results using uncompressed images are provided as a reference. To make a fair comparison, the network for the other two classification methods is a concatenation of forward network (in Fig. 2a) and classifier network (in Fig. 7), but trained with reconstructed and uncompressed raw images respectively. The compressed ratio is around 0.3Bpp. As seen from the table, the classification accuracy with compressed images are lower than that without compression, but with acceptable values. Moreover, the network of *CodedVision* is optimized for compression in this work, resulting in performance loss for classification. Consequently, joint optimization is required for further improvement. The processing time are collected on a server configured with an Intel Xeon E5-2620 CPU and a NVIDIA GTX 1080Ti GPU while processing 5822 images. It is apparently that *CodedVision* consumes the least processing time as it only needs to pass the coded bitstream without fully decoding. But for the others, we need to fully decode the image and perform the low-level feature extraction. Furthermore, the decoding part is required for images compressed with JPEG.

4   Concluding Remarks

A *CodedVision* system is developed to jointly consider the image understanding and compression, under an unified learning framework that leverages the advances in deep neural networks. We have demonstrated 7.8% BD-Rate gains over the state-of-the-art HEVC intra based image compression and reasonable accuracy in classification as mAP ranged from 52.38% to 59.26% in this preliminary study. Aforementioned results have shown the potentials of our proposed framework. There are many interesting avenues to explore to improve the performance. For instance, it is desirable to improve the architecture of forward network to provide more comprehensive features for vision task, investigate how the quantization of fMaps affect the accuracy of vision tasks, and explore the important factors of fMaps to realize more effective compression and understanding. Network models should be optimized for mobile platforms for low-power processing and massive adoption, such as innovative architectures, parameter compression, etc. Recently, deep network training mostly demands tremendous cost by using a large-scale database to guarantee the model generalization. Training using a small-size set or even federated learning is worthy of further in-depth study.

References

1. Aharon, M., Elad, M., Bruckstein, A.: k-svd: an algorithm for designing overcomplete dictionaries for sparse representation. IEEE Trans. Sig. Process. **54**(11), 4311–4322 (2006). https://doi.org/10.1109/TSP.2006.881199
2. Azuma, R.T.: A survey of augmented reality. Presence Teleop. Virt. Environ. **6**(4), 355–385 (1997)
3. Babu, R.V., Tom, M., Wadekar, P.: A survey on compressed domain video analysis techniques. Multimedia Tools Appl. **75**(2), 1043–1078 (2016)
4. Ballé, J., Laparra, V., Simoncelli, E.P.: End-to-end optimized image compression. CoRR abs/1611.01704 (2016). http://arxiv.org/abs/1611.01704
5. Ballé, J., Laparra, V., Simoncelli, E.P.: End-to-end optimized image compression. arXiv preprint (2016). arXiv:1611.01704
6. Bjontegaard, G.: Calculation of average PSNR differences between R-D curves. In: Document VCEG-M33, ITU-T VCEG 13th Meeting (2001)
7. Chen, T., Liu, H., Shen, Q., Yue, T., Cao, X., Ma, Z.: Deepcoder: a deep neural network based video compression. In: Visual Communications and Image Processing (VCIP), 2017, pp. 1–4. IEEE (2017)
8. Dong, C., Deng, Y., Change Loy, C., Tang, X.: Compression artifacts reduction by a deep convolutional network. In: Proceedings of the IEEE International Conference on Computer Vision, pp. 576–584 (2015)
9. Dong, C., Loy, C.C., He, K., Tang, X.: Learning a deep convolutional network for image super-resolution. In: Fleet, D., Pajdla, T., Schiele, B., Tuytelaars, T. (eds.) ECCV 2014. LNCS, vol. 8692, pp. 184–199. Springer, Cham (2014). https://doi.org/10.1007/978-3-319-10593-2_13
10. Everingham, M., Van Gool, L., Williams, C.K., Winn, J., Zisserman, A.: The pascal visual object classes (voc) challenge. Int. J. Comput. Vis. **88**(2), 303–338 (2010)
11. Fang, Y., Lin, W., Chen, Z., Tsai, C.M., Lin, C.W.: A video saliency detection model in compressed domain. IEEE Trans. Circuits Syst. Video Technol. **24**(1), 27–38 (2014)
12. He, K., Zhang, X., Ren, S., Sun, J.: Deep residual learning for image recognition. In: Proceedings of the IEEE Conference on Computer Vision and Pattern Recognition, pp. 770–778 (2016)
13. Hong, R., Hu, Z., Wang, R., Wang, M., Tao, D.: Multi-view object retrieval via multi-scale topic models. IEEE Trans. Image Process. **25**(12), 5814–5827 (2016)
14. Hong, R., Zhang, L., Tao, D.: Unified photo enhancement by discovering aesthetic communities from flickr. IEEE Trans. Image Process. **25**(3), 1124–1135 (2016)
15. Hong, R., Zhang, L., Zhang, C., Zimmermann, R.: Flickr circles: aesthetic tendency discovery by multi-view regularized topic modeling. IEEE Trans. Multimedia **18**(8), 1555–1567 (2016)
16. Kingma, D.P., Ba, J.: Adam: a method for stochastic optimization. arXiv preprint (2014). arXiv:1412.6980
17. Krizhevsky, A., Sutskever, I., Hinton, G.E.: Imagenet classification with deep convolutional neural networks. In: Advances in Neural Information Processing Systems, pp. 1097–1105 (2012)

18. LeCun, Y., Bengio, Y., Hinton, G.: Deep learning. Nature **521**(7553), 436 (2015)
19. Lee, D.T.: Jpeg 2000: retrospective and new developments. Proc. IEEE **93**(1), 32–41 (2005). https://doi.org/10.1109/JPROC.2004.839613
20. Lowe, D.G.: Object recognition from local scale-invariant features. In: Proceedings of the Seventh IEEE International Conference on Computer Vision, vol. 2, pp. 1150–1157 (1999). https://doi.org/10.1109/ICCV.1999.790410
21. Lu, Z.M., Li, S.Z., Burkhardt, H.: A content-based image retrieval scheme in jpeg compressed domain. Int. J. Innovative Comput. Inf. Control **2**(4), 831–839 (2006)
22. Marr, D.: Vision: A Computational Investigation into the Human Representation and Processing of Visual Information. Henry Holt and Co., Inc., New York (1982)
23. Nah, S., Kim, T.H., Lee, K.M.: Deep multi-scale convolutional neural network for dynamic scene deblurring. In: CVPR, vol. 1, p. 3 (2017)
24. Ortega, A., Ramchandran, K.: Rate-distortion methods for image and video compression. IEEE Sig. Process. Mag. **15**(6), 23–50 (1998)
25. Pati, Y.C., Rezaiifar, R., Krishnaprasad, P.S.: Orthogonal matching pursuit: recursive function approximation with applications to wavelet decomposition. In: Proceedings of 27th Asilomar Conference on Signals, Systems and Computers, vol. 1, pp. 40–44 (1993). https://doi.org/10.1109/ACSSC.1993.342465
26. Porikli, F., Bashir, F., Sun, H.: Compressed domain video object segmentation. IEEE Trans. Circ. Syst. Video Technol. **20**(1), 2–14 (2010)
27. Redmon, J., Divvala, S., Girshick, R., Farhadi, A.: You only look once: Unified, real-time object detection. In: Proceedings of the IEEE Conference on Computer Vision and Pattern Recognition, pp. 779–788 (2016)
28. Shi, W., Caballero, J., Huszár, F., Totz, J., Aitken, A.P., Bishop, R., Rueckert, D., Wang, Z.: Real-time single image and video super-resolution using an efficient sub-pixel convolutional neural network. In: Proceedings of the IEEE Conference on Computer Vision and Pattern Recognition, pp. 1874–1883 (2016)
29. Simpson, R.L.: Computer vision: an overview. IEEE Expert **6**(4), 11–15 (1991). https://doi.org/10.1109/64.85917
30. Sullivan, G.J., Ohm, J.R., Han, W.J., Wiegand, T.: Overview of the high efficiency video coding (HEVC) standard. IEEE Trans. Circuits Syst. Video Technol. **22**(12), 1649–1668 (2012). https://doi.org/10.1109/TCSVT.2012.2221191
31. Toderici, G., O'Malley, S.M., Hwang, S.J., Vincent, D., Minnen, D., Baluja, S., Covell, M., Sukthankar, R.: Variable rate image compression with recurrent neural networks. CoRR abs/1511.06085 (2015). http://arxiv.org/abs/1511.06085
32. Toderici, G., Vincent, D., Johnston, N., Hwang, S.J., Minnen, D., Shor, J., Covell, M.: Full resolution image compression with recurrent neural networks. CoRR abs/1608.05148 (2016). http://arxiv.org/abs/1608.05148
33. Wiegand, T., Sullivan, G.J., Bjontegaard, G., Luthra, A.: Overview of the h.264/avc video coding standard. IEEE Trans. Circuits Syst. Video Technol. **13**(7), 560–576 (2003)
34. Xue, Y., Wang, Y.: Video coding using a self-adaptive redundant dictionary consisting of spatial and temporal prediction candidates. In: Proceedings of the IEEE International Conference on Multimedia and Expo (ICME) (2014)
35. Zepeda, J., Guillemot, C., Kijak, E.: Image compression using sparse representations and the iteration-tuned and aligned dictionary. IEEE. J. Sel. Top. Sign. Process. **5**(5), 1061–1073 (2011)
36. Zhang, C., Cheng, J., Tian, Q.: Multiview label sharing for visual representations and classifications. IEEE Trans. Multimedia **20**(4), 903–913 (2018)

37. Zhang, C., Liu, J., Tian, Q., Xu, C., Lu, H., Ma, S.: Image classification by non-negative sparse coding, low-rank and sparse decomposition. In: 2011 IEEE Conference on Computer Vision and Pattern Recognition (CVPR), pp. 1673–1680. IEEE (2011)
38. Zhao, L., He, Z., Cao, W., Zhao, D.: Real-time moving object segmentation and classification from HEVC compressed surveillance video. IEEE Transactions on Circuits and Systems for Video Technology (2016)

Random Angular Projection for Fast Nearest Subspace Search

Binshuai Wang[1], Xianglong Liu[1(✉)], Ke Xia[1], Kotagiri Ramamohanarao[2], and Dacheng Tao[3]

[1] State Key Lab of Software Development Environment,
Beihang University, Beijing, China
xlliu@nlsde.buaa.edu.cn
[2] Department of CIS, The University of Melbourne, Melbourne, Australia
[3] UBTECH Sydney AI Centre, School of IT, FEIT,
University of Sydney, Sydney, Australia

Abstract. Finding the nearest subspace is a fundamental problem in many tasks such as recognition, retrieval and optimization. This hard topic has been seldom touched in the literature, except a very few studies that address it using the locality sensitive hashing for subspaces. The existing solutions severely suffer from poor scaling with expensive computational cost or unsatisfying accuracy, when the subspaces originally distribute with arbitrary dimensions. To address these problems, this paper proposes a new and efficient family of locality sensitive hash for linear subspaces of arbitrary dimension. It preserves the angular distances among subspaces by randomly projecting their orthonormal basis and further encoding them with binary codes. The proposed method enjoys fast computation and meanwhile keeps the strong collision probability. Moreover, it owns flexibility to easily balance the performance between the accuracy and efficiency. The experimental results in the tasks of face recognition and video de-duplication demonstrate that the proposed method significantly outperforms the state-of-the-art vector and subspace hashing methods, in terms of both accuracy and efficiency (up to 16× speedup).

Keywords: Nearest subspace search · Locality sensitive hash
Linear subspace · Subspace hashing

1 Introduction

Subspaces enjoy strong capability of describing the intrinsic structures of the data naturally. Therefore, they often offer convenient means for the data representation in the fields of pattern recognition, computer vision, and statistical learning. Typical applications include face recognition [11,23], video retrieval [26], person re-identification [24], and 2D/3D matching [17,19]. In these applications, usually finding the nearest subspaces (*i.e.*, subspace-to-subspace search)

© Springer Nature Switzerland AG 2018
R. Hong et al. (Eds.): PCM 2018, LNCS 11164, pp. 15–26, 2018.
https://doi.org/10.1007/978-3-030-00776-8_2

is a fundamental problem and inevitably involved, which means that when the amount of data explodes, searching the gigantic subspace database with high ambient dimensions will consume expensive storage and computation [8,9].

Locality-sensitive hashing (LSH) serves as one of the most successful techniques for the fast nearest neighbor search over the gigantic database, which achieves a good balance between the search performance and computational efficiency using small hash codes. [2] pioneered the LSH study for the vector data with high dimension (*i.e.*, vector-to-vector search), by introducing the linear projection paradigm to efficiently generate the hash codes for the specific similarity metric. In recent years, extensive research efforts have been devoted to developing learning based hashing methods for vector-to-vector search, and have been widely adopted in many tasks including large-scale visual search [7,22], object detection [3,4], and machine learning [14,15,18].

Despite the progress of hashing research for the classic vector-to-vector nearest neighbor search, rare studies have been done on developing LSH techniques for subspace-to-subspace search. To deal with this problem, it is desirable to represent subspaces with compact signatures without losing the locality information. Compared to the vector data, the subspace data is more complex and informative, mainly because that there are massive samples distributed in one subspace, and meanwhile there also exist the inclusion relations between different subspaces, even though their dimensions may vary largely. This certainly brings great difficulties to distinguish the similar and dissimilar subspaces, especially among the large-scale high dimensional database with large cardinality.

The most well-known pioneering approach for subspace search was proposed in [1]. It maps each subspace in $\mathbb{R}^n$ to its orthogonal projection matrix, and subsequently performs the nearest neighbor search equivalently over the projection matrix. [11] further turns this matrix into a high-dimensional vector, and then applies the random projection based LSH to generate binary codes. Such a solution can preserve the angular distance among subspaces with high probability. Besides, both [1,11] achieved fast search with encouraging performance. However, they heavily rely on the product of the orthogonal projection matrix, which inevitably takes expensive computational cost. [23] provided a new LSH solution by simply thresholding the angular distance for linear subspace. However, at querying stage the number of possible candidates often increases explosively, subsequently degenerating the sublinear search to the exhaustive search.

In this paper, we propose a new and efficient family of locality sensitive hash for fast nearest neighbor search over linear subspaces of arbitrary dimension. It adopts the simple random projection paradigm, named random angular projection (RAP), to efficiently capture the subspace distribution characteristics by the discriminative binary codes. Specifically, RAP can largely preserve the angle of subspaces, which is a fundamental subspace similarity metric. In Fig. 1 we demonstrate how RAP generates binary codes for the video sequences. To the best of our knowledge, our method is the fastest subspace LSH till now that enjoys strong locality sensitivity for linear subspaces. Theoretically and practically, we show that the proposed method can simultaneously promise discriminative and fast binary codes generation, compared to the state-of-the-arts.

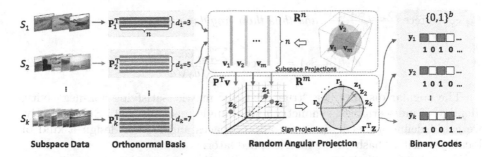

Fig. 1. Random angular projection for subspace hashing (*e.g.*, video sequences)

2 Subspace Angular Projection

Given a d-dimensional subspace $S \subset \mathbb{R}^n$, our goal is to find a locality sensitive hash $\mathcal{H} : \mathbb{R}^{n \times d} \mapsto \{0, 1\}^b$, *i.e.*, it can map an arbitrary d-dimensional subspace into a b-length binary code. Namely, if two subspaces are similar, $\mathcal{H}$ can encode them with similar hash codes.

2.1 Angular Distance

Before we elaborate on the detailed design of our subspace hash function, we first introduce the subspace distance metric commonly used in the literature.

Given any subspace S of d-dimension, it can be fully represented by its orthonormal basis $\mathbf{P} \in \mathbb{R}^{n \times d}$. The orthonormal basis has been proved to be a kind of good representation, which can fully characterize the feature of the subspace. Besides, it can be obtained efficiently by applying Gram-Schmidt process with unit-normalization on the given subspace.

Based on the orthonormal basis representation, the principal angle between subspaces can be defined formally. It reflects the correlations among the subspaces, and serves as an important and fundamental concept for subspaces in the literature. Given subspace S_1 and S_2 with dimension d_1 and d_2 ($d_1 \leq d_2$) in $\mathbb{R}^n$, and their orthonormal basis $\mathbf{P}_1$ and $\mathbf{P}_2$, their principal angles $\xi_1, \xi_2, \ldots,$ $\xi_{d_1} \in [0, \frac{\pi}{2}]$ can be defined recursively [12]:

$$\cos(\xi_i) = \max_{\mathbf{p}_i \in S_1} \max_{\mathbf{q}_i \in S_2} \left(\frac{\mathbf{p}_i^\top \mathbf{q}_i}{\|\mathbf{p}_i\|_2 \|\mathbf{q}_i\|_2} \right), \tag{1}$$

subject to

$$\mathbf{p}_i^\top \mathbf{p}_j = 0 \text{ and } \mathbf{q}_i^\top \mathbf{q}_j = 0, \text{ for } j = 1, 2, \ldots, i - 1.$$

Based on the principal angles, we can define the common angular distance:

Definition 1 Angular Distance. *Given subspace S_1 and S_2 in $\mathbb{R}^n$, with dimension d_1 and d_2 ($d_1 \leq d_2$) respectively, the angle between S_1 and S_2 is*

$\xi_{S_1,S_2} = \arccos \frac{\sum_{i=1}^{d_1} \cos^2(\xi_i)}{\sqrt{d_1}\sqrt{d_2}}$, *and thus their angular distance can be defined as*

$$dist(S_1, S_2) = \frac{1}{\pi} \arccos \frac{\sum_{i=1}^{d_1} \cos^2(\xi_i)}{\sqrt{d_1}\sqrt{d_2}}. \tag{2}$$

The angular distance is a "true" distance metric, satisfying non-negativity, identity of indiscernibles, symmetry and triangle inequality [11]. Subsequently, we could define a metric space for all subspaces and further design a kind of locality-sensitive hash on the metric space later.

Note that usually it is complicated to directly compute the angular distance based on (1). But fortunately we have the following fact [25]:

$$\|\mathbf{P}_1^\top \mathbf{P}_2\|_F^2 = \sum_{i=1}^{d_1} \cos^2(\xi_i). \tag{3}$$

Therefore, we can simplify the computation of the angle ξ_{S_1,S_2} by $\|\mathbf{P}_1^\top \mathbf{P}_2\|_F^2$.

2.2   Random Subspace Projection

It should be pointed out that the orthonormal basis is not unique for one subspace. To efficiently characterize a subspace, it is very critical to find an unique and compact representation for subspace data.

Suppose both $\mathbf{P}$ and $\bar{\mathbf{P}}$ are the orthonormal basis for subspace S of d-dimension, then given a unit vector $\mathbf{v} \in \mathbb{R}^n$, we have the following observation

$$\|\mathbf{P}^\top \cdot \mathbf{v}\|_F^2 = \|\bar{\mathbf{P}}^\top \cdot \mathbf{v}\|_F^2. \tag{4}$$

Inspired by this fact, we can introduce an α function that can uniquely characterize the orthonormal matrix form of subspace.

Definition 2 α **Function.** *A α function is defined based on a unit vector $\mathbf{v} \in \mathbb{R}^n$, mapping a d-dimensional subspace S in $\mathbb{R}^n$ to a real number in $[0,1]$*

$$\alpha_\mathbf{v}(S) = \left\|\mathbf{P}_S^\top \cdot \mathbf{v}\right\|_F^2 \tag{5}$$

where $\mathbf{P}_S$ is an orthonormal matrix of S and $\|\cdot\|_F$ denotes the Frobenius norm.

The α function is invariant to different orthonormal matrices for the same subspace, which means that it is able to capture the unique characteristic of the subspace data.

Example 1. *Let* $\mathbf{P}_1 = \begin{pmatrix} 1 & 0 \\ 0 & 1 \\ 0 & 0 \end{pmatrix}$ *and* $\bar{\mathbf{P}}_1 = \begin{pmatrix} \frac{1}{\sqrt{2}} & \frac{1}{\sqrt{2}} \\ \frac{1}{\sqrt{2}} & -\frac{1}{\sqrt{2}} \\ 0 & 0 \end{pmatrix}$ *be the orthonormal*

matrice of subspace S_1 in $\mathbb{R}^3$, $\mathbf{P}_2 = \begin{pmatrix} 0 & 0 \\ 1 & 0 \\ 0 & 1 \end{pmatrix}$ be the orthonormal matrix of another *subspace* S_2 *in* $\mathbb{R}^3$. *Then, for a unit vector* $\mathbf{v} = \left(\frac{1}{\sqrt{2}}, \frac{1}{\sqrt{2}}, 0\right)^\top$, *we have*

$$\|\mathbf{P}_1^\top \mathbf{P}_2\|_F^2 = \|\bar{\mathbf{P}}_1^\top \mathbf{P}_2\|_F^2 = 1, \ \textit{with} \ \xi_1 = 0, \xi_2 = \pi/2,$$

$$\alpha_\mathbf{v}(S_1) = \|\mathbf{P}_1^\top \cdot \mathbf{v}\|_F^2 = \|\bar{\mathbf{P}}_1^\top \cdot \mathbf{v}\|_F^2 = 1, \ \alpha_\mathbf{v}(S_2) = \|\mathbf{P}_2^\top \cdot \mathbf{v}\|_F^2 = \frac{1}{2}.$$

Based on α function, we can further define the angular projection function for subspace S of d dimension, by appending a proper basis $\alpha_0 d$,

$$z_\mathbf{v}(S) = \alpha_\mathbf{v}(S) + \alpha_0 d. \tag{6}$$

The projection function $z_\mathbf{v}$ serves as a useful and efficient representation of the subspace. Later we will show that such kind of projection induces a locality sensitive hash function for the angular distance, so that the similar subspaces will share the similar hash code. In the following theorem, we show that $z_\mathbf{v}$ has strong connections to the angular distance.

Theorem 1. *Given subspaces* S_1 *and* S_2 *in* $\mathbb{R}^n$, *with dimension* d_1 *and* d_2 *respectively, and*

$$\alpha_0 = \frac{\sqrt{2}}{\sqrt{n^3 + 2n^2}} - \frac{1}{n}, \tag{7}$$

if $\mathbf{v}$ *is a random variable vector uniformly distributed in the unit sphere of* $\mathbb{R}^n$, *then we have the following equation*

$$\mathbf{E}\left[z_\mathbf{v}(S_1) z_\mathbf{v}(S_2)\right] = \frac{2}{(n+2)n} \|\mathbf{P}_1^\top \mathbf{P}_2\|_F^2, \tag{8}$$

where $\mathbf{P}_1$ *and* $\mathbf{P}_2$ *are any orthonormal basis of subspace* S_1 *and* S_2.

Due to the space limit, we leave the proof to the supplementary material. As Eq. (3) states that the angular distance between two subspaces has a close relationship with the Frobenius norm of their orthonormal matrices product. The theoretical result here further indicates that the product of their projection values can be regarded equivalent to the Frobenius norm. Namely, we have the following fact

$$\mathbf{E}\left[z_\mathbf{v}(S_1) z_\mathbf{v}(S_2)\right] = \frac{2}{(n+2)n} \sum_{i=1}^{d_1} \cos^2(\xi_i). \tag{9}$$

We see that the projection function $z_\mathbf{v}$ owns the surprising property that it can preserve the angular distances.

Example 2. *Let* $\mathbf{P}_1 = (1,0)^\top$ *and* $\mathbf{P}_2 = (\cos\gamma, \sin\gamma)^\top$ (γ *is constant) respectively be the orthonormal matrix of subspace* S_1 *and* S_2 *in* $\mathbb{R}^2$, *with dimension* $d_1 = d_2 = 1$. *Then, for a unit vector* $\mathbf{v} = (\cos\tau, \sin\tau)^\top$, $\tau \sim U[0, 2\pi]$, *we have*

$$\mathbf{E}[\alpha_\mathbf{v}(S_1)\alpha_\mathbf{v}(S_2)] = \frac{\cos^2(\gamma)}{4} + \frac{1}{8}, \ \mathbf{E}[\alpha_\mathbf{v}(S_2)] = \frac{1}{2}, \ \mathbf{E}[\alpha_\mathbf{v}(S_1)] = \frac{1}{2}.$$

Thus, with $\alpha_0 = \frac{\sqrt{2}}{4} - \frac{1}{2}$ and $\xi_1 = \gamma$,

$$\mathbf{E}[z_\mathbf{v}(S_1)z_\mathbf{v}(S_2)] = \mathbf{E}[\alpha_\mathbf{v}(S_1)\alpha_\mathbf{v}(S_2)] + \alpha_0 \left(\mathbf{E}[\alpha_\mathbf{v}(S_1)] + \mathbf{E}[\alpha_\mathbf{v}(S_2)]\right) + \alpha_0^2 = \frac{\cos^2(\xi_1)}{4}.$$

Practically, to completely characterize the subspace with a large probability, we can employ a series of random projection function $\{z_{\mathbf{v}_1}, z_{\mathbf{v}_2}, \ldots, z_{\mathbf{v}_m}\}$ as one composite projection function $\mathcal{Z}$ and approximate the expectation in (9) by average. Specifically, for any subspace S, we can have an m-dimensional vectorial representation $\mathbf{z}_S$ using projection $\mathcal{Z}$:

$$\mathbf{z}_S = \mathcal{Z}(S) \doteq \left(z_{\mathbf{v}_1}(S), z_{\mathbf{v}_2}(S), \ldots, z_{\mathbf{v}_m}(S)\right)^\top. \tag{10}$$

And it is easy to see that $\lim_{m \to +\infty} \frac{1}{m}\mathbf{z}_{S_1}^\top \mathbf{z}_{S_2} = \mathbf{E}[z_\mathbf{v}(S_1)z_\mathbf{v}(S_2)]$. The projection vector enjoys both fast computation and good approximating power. When m is large enough, we can guarantee that the approximation nearly converges.

3   Locality Sensitive Subspace Hashing

The vectorial representation can be generated efficiently by random subspace projections. However, to promise a satisfying approximation to the subspace similarities, usually we have to rely on a large number of projections, forming a long vectorial representation. This inevitably brings expensive computation in the nearest subspace search. Fortunately, we have a new kind of locality sensitive hash (LSH) for subspace data, which can faithfully speed up the search by further encoding the vectorial representation into the binary codes, while preserving the locality among the subspaces.

Definition 3 Sign Random Projection. *A sign random projection $f_\mathbf{r}(\cdot)$, mapping a vector $\mathbf{x} \in \mathbb{R}^m$ to a binary bit using $\mathbf{r} \in \mathbb{R}^m$, is defined as*

$$f_\mathbf{r}(\mathbf{x}) = sgn(\mathbf{r}^\top \mathbf{x}). \tag{11}$$

For any two vectors $\mathbf{x}$ and $\mathbf{y}$, if $\mathbf{r}$ is a random variable vector, uniformly distributed in the unit sphere of $\mathbb{R}^m$, then we have the following fact [6]

$$\Pr[f_\mathbf{r}(\mathbf{x}) = f_\mathbf{r}(\mathbf{y})] = 1 - \frac{1}{\pi}\arccos\left(\frac{\mathbf{x}^\top \mathbf{y}}{\|\mathbf{x}\|_2\|\mathbf{y}\|_2}\right). \tag{12}$$

For the subspace search problem, we could further define a metric space (Ω, dist) for all subspaces of $\mathbb{R}^n$ based on the angular distance for subspaces in Eq. (2), where $\Omega = \{S \subset \mathbb{R}^n | S \text{ is a subspace of } \mathbb{R}^n\}$. Then, we find that the sign random projection combined with the random subspace projection function can serve as a new family of locality sensitive hash function in the metric space (Ω, dist). Namely, the following theorem holds:

Theorem 2. *When m is large enough, $\mathcal{H} = f_\mathbf{r} \circ \mathcal{Z}$ is a locality sensitive hash defined on (Ω, dist) approximately, i.e., $\mathcal{H}$ is $(\delta, \delta(1 + \epsilon), 1 - \delta, 1 - \delta(1 + \epsilon))$-sensitive with $\delta, \epsilon > 0$.*

Table 1. Comparison of collision probability p and the complexity t of encoding a subspace from orthonormal basis to a b-length binary code.

	RAP	BSS	GLH
p	$1 - \delta$	$1 - \delta$	-
t	$O(bm + dmn)$	$O(bn^2)$	$O(bdn)$

Now we have our random angular projection (RAP) based LSH function, and the theoretical result indicates that the composite function $\mathcal{H}$ satisfies the locality sensitivity to the angular distance, when using a large number of angular projections. The LSH function for subspace uses the following two steps: first projecting the subspace into a high-dimensional vector, and second encoding the vector into a binary code. Table 1 compares the locality sensitivity and time complexity of state-of-the-art solutions. The time complexity of RAP is linear to the space dimension n, which largely reduces the computation, compared to $O(n^2)$ of existing subspace hashing methods [11].

With the LSH property, following [10, 13] we can theoretically guarantee the sublinear search time based on the binary codes generated for all subspaces.

Theorem 3. *Given a subspace database $\mathcal{D}$ with N subspaces in $\mathbb{R}^n$, for a subspace query S, if there exists a subspace S^* such that $dist(S, S^*) \leq \delta$, then with $\rho = \frac{\ln p_1}{\ln p_2}$ (1) using N^ρ hash tables with $\log_{1/p_2} N$ hash bits, the random angular hash of an even order is able to return a database subspace $\bar{S}$ such that $dist(S, \bar{S}) \leq \delta(1 + \epsilon)$ with probability at least $1 - \frac{1}{c} - \frac{1}{e}$, $c \geq 2$; (2) the query time is sublinear to the entire data number N, with $N^\rho \log_{1/p_2} N$ bit generations and cN^ρ pairwise distances computation.*

4 Experiments

Subspace search has been widely involved in many applications, including face recognition, video search, action recognition, etc. To comprehensively evaluate our method RAP, here we choose the two popular tasks: face recognition and video search. For more experimental results on gesture recognition, please refer to the supplementary material. We compare our RAP method with the state-of-the-art subspace hashing methods, including BHZM [1], Binary Signatures for Subspaces (BSS) [11] and Grassmannian-based Locality Hashing (GLH) [23]. All these methods rely on the random way to generate hash codes. We also adopt several state-of-the-art unsupervised vector hashing methods as the baselines, including the random locality sensitive hashing (LSH) [2], and the learning based methods: iterative quantization (ITQ) [7], dubbed discrete proximal linearized minimization (DPLM) [20] and adaptive binary quantization (ABQ) [16].

4.1 Face Recognition

Face images with fixed pose under different illumination conditions lie near linear subspaces of low dimension [11]. Therefore, the face recognition in this scenario

(a) The Extended Yale Face Database B

(b) UQ_VIDEO database

Fig. 2. The examples of face recognition and video retrieval datasets

can be regarded as a nearest subspace search, where a set of face images belonging to the same person can be easily represented by a subspace.

Dataset: The two popular datasets are adopted: the Extended Yale Face Database B [5] containing 38 individuals each with about 64 near frontal images under different illuminations, and PIE database [21] consisting of 68 individuals each with 170 images under 5 near frontal poses, 43 different illumination conditions and 4 different expressions. On PIE database, we choose the frontal face images (C26 and C27). On the Extended Yale Face Database B, we randomly choose 30 images per individual as the database set, and 11 images from the rest images as the testing set. Similarly, on the PIE database the database and testing set contain 20 and 15 images respectively per individual (Fig. 2).

Experiment Setup: Each face image is represented by a $1,024$-length vector, and thus images of the same individual will constitute a subspace in $\mathbb{R}^{1024}$. For each individual in Yale Face, we extract a representative subspace of dimension $d = 9$ ($d = 14$ for PIE datasets) by taking the top principal components as BSS did. For the testing set we generate the query subspaces with different dimensions $d_q = 7, 9, 11$ on Yale Face and $d_q = 13, 14, 15$ on PIE respectively. For the vector hashing methods, without reduction we directly utilize the original face vectors.

Results: Tables 2 and 3 list the precision of the top-1 results and search time of different methods for the face recognition task. Here, the search time consists of two parts: the encoding time and the retrieval time per query subspace. In the table, the learning based vector hashing methods outperform the random LSH method. However, they heavily rely on the training process, which requires extra computational cost (see the train time). Moreover, we can see that the random subspace hashing methods RAP and BSS can perform much better than all vector hashing methods. This is because that the vector hashing, considering the vector data independently, can hardly capture the intrinsic distributions in the subspace. In all cases, our RAP achieves close, sometimes even better performance to BSS, which is consistent with our analysis in Table 1 that RAP and BSS share the same locality sensitivity power to the angular distance. As to the query time, RAP can get up to 16× speedup, compared to BSS.

Table 2. Face recognition performance on Yale Faces

	Methods	Precision (%), $b=512, d=9, m=10^4$			Search time (s)		
		$d_q=7$	$d_q=9$	$d_q=11$	$d_q=7$	$d_q=9$	$d_q=11$
Yale Face	LSH	$45.84_{\pm3.49}$			0.0025		
	ITQ	$47.89_{\pm3.86}$			0.0016 (train time 5.73)		
	DPLM	$46.84_{\pm5.10}$			0.0022 (train time 0.49)		
	ABQ	$58.42_{\pm1.05}$			0.0028 (train time 22.26)		
	BHZM	$14.21_{\pm5.41}$	$16.31_{\pm8.55}$	$16.84_{\pm3.93}$	0.0080	0.0080	0.0087
	BSS	$74.21_{\pm8.05}$	$72.63_{\pm4.27}$	$73.68_{\pm3.72}$	0.1349	0.1349	0.1219
	RAP	$72.10_{\pm6.13}$	$\mathbf{73.10}_{\pm8.58}$	$\mathbf{82.10}_{\pm6.09}$	0.0078	0.0078	0.0084

Table 3. Face recognition performance on PIE Faces.

	Methods	Precision (%), $b=1024, d=14, m=10^4$			Search time (s)		
		$d_q=13$	$d_q=14$	$d_q=15$	$d_q=13$	$d_q=14$	$d_q=15$
PIE Face	LSH	$57.64_{\pm2.35}$			0.0075		
	ITQ	$52.64_{\pm1.44}$			0.0077 (train time 13.48)		
	DPLM	$79.41_{\pm2.07}$			0.0079 (train time 6.82)		
	ABQ	$81.47_{\pm1.99}$			0.0137 (train time 3.17)		
	BHZM	$38.82_{\pm4.97}$	$41.76_{\pm2.72}$	$37.94_{\pm5.76}$	0.0166	0.0175	0.0186
	BSS	$97.05_{\pm2.63}$	$95.58_{\pm2.27}$	$95.29_{\pm2.35}$	0.1252	0.1274	0.1259
	RAP	$96.17_{\pm0.72}$	$\mathbf{96.47}_{\pm0.72}$	$\mathbf{97.35}_{\pm2.35}$	0.0089	0.0111	0.0114

Figure 3(a) and (b) further investigate the effect of the code length b on the precision performance and the speed acceleration (*i.e.*, speedup). Here, we attempt to simulate the real world case by randomly choosing d and d_q of each subspace from $\{3, 5, 7\}$. Here, since GLH works in a different way from BSS and RAP, its performance isn't affected by the code length. In the figures, we also show the results using exhaustive search ('Scan'). From the figures, we can see that all the subspace hashing methods can obviously improve the search efficiency, and our RAP gains close performance to BSS, due to the same collision probability. Moreover, with more hash bits, RAP is able to approximate or even outperform exhaustive search strategy, but completing the search at a much higher speed (more than $10\times$ speedup over BSS and Scan).

Figure 3(c) shows RAP performance with respect to the dimension m of the subspace projection vector (here we set $b = 2,000$). In general, a larger m promises better performance, which is consistent with our approximation analysis in Sect. 2.2. We can see that $m > 500$ is sufficiently large for a satisfying performance in practice.

4.2 Near-Duplicated Large-Scale Video Retrieval

Next, we evaluate the performance in the video retrieval task. Here, traditional vector hashing methods individually encode each video frame into a binary code,

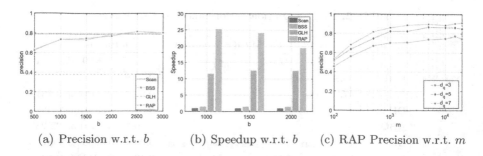

| (a) Precision w.r.t. b | (b) Speedup w.r.t. b | (c) RAP Precision w.r.t. m |

Fig. 3. The effect of different b and m on Yale Face. To simulate the real world scenarios, we adopt random d and d_q in (a) and (b), and random d in (c).

Table 4. Video retrieval performance on UQ_VIDEO.

Methods	MAP (%), $d = d_q = 5$			Search time (s)		
	$b = 128$	$b = 256$	$b = 512$	$b = 128$	$b = 256$	$b = 512$
LSH	30.73 ±1.01	30.20 ±0.34	31.53 ±0.06	5.5660	7.5578	16.7660
ITQ	28.92 ±0.30	–	–	5.3990	–	–
DPLM	34.09 ±1.60	37.24 ±0.95	39.15 ±0.91	5.0560	6.9249	16.8364
ABQ	25.39 ±0.29	24.72 ±0.04	25.92 ±0.23	9.6779	17.3563	29.2905
BSS	36.91 ±3.57	50.32 ±0.10	57.47 ±1.32	1.2842	1.2261	1.7796
RAP	**37.48** ±1.60	**51.50** ±0.04	**58.12** ±2.20	1.2688	1.2444	1.7593

but neglect the temporal relations among frames. The subspace hashing can address this problem by treating the consecutive frames as one subspace.

Dataset: The UQ_VIDEO [22] is one of largest databases for near duplicate video retrieval. It contains 169,952 videos in total and 3,305,525 key frames extracted from these videos. The database provides 24 selected query videos, each of which have a number of near duplicate videos, and other videos serve as background videos. We use the provided global color histogram feature with 162 dimensions as the input feature for each frame.

Experiment Setup: We first construct the database set with all the videos except those with very few frames, forming a database set with nearly 3,000,000 key frames. For each video, we use every $d = 5$ consecutive frames without overlap to form a subspace. At the search stage, we use the Hamming distance ranking strategy to rank all the video clips in the database set according to the averaged Hamming distances to the query.

Results: Table 4 shows the mean average precision (MAP) and search time using Hamming distance ranking in the video search task. It is easy to see that the subspace hashing methods perform much better than the state-of-the-art vector hashing methods. Besides, when using different number of hash bits, our RAP method can consistently achieve the best performance. Note that here we only adopt features of 162 dimensions, and thus RAP and BSS spend similar time cost on the search. But in practice, as the feature dimension increases RAP's speedup will become much more significant (see the speedup in Fig. 3(b) and the analysis in Table 1). Moreover, in this large-scale task, RAP obtains more than 5× faster than the vector hashing methods.

5   Conclusions

This paper mainly concentrated on the fast nearest neighbor search over massive subspace data, and proposed a new family of locality sensitive hashing method named random angular projection (RAP) for linear subspaces of arbitrary dimension. RAP encodes the subspace by random subspace projections and binary quantization, and theoretically enjoys fast computation and strong locality sensitivity. Our experiments in several popular tasks show that the proposed RAP method, obtaining state-of-the-art performance, can be a promising and practical solution to the nearest subspace search.

Acknowledgement. This work was supported by National Natural Science Foundation of China 61690202, 61872021, Australian Research Council Projects FL-170100117, DP-180103424, LP-150100671, IH180100002, and MSRA Collaborative Research Grant.

References

1. Basri, R., Hassner, T., Zelnik-Manor, L.: Approximate nearest subspace search. IEEE Trans. Pattern Anal. Mach. Intell. **33**(2), 266–278 (2011)
2. Charikar, M.S.: Similarity estimation techniques from rounding algorithms. In: ACM Symposium on Theory of Computing, pp. 380–388 (2002)
3. Cheng, J., Leng, C., Wu, J., Cui, H., Lu, H., et al.: Fast and accurate image matching with cascade hashing for 3d reconstruction. In: IEEE Conference on Computer Vision and Pattern Recognition, pp. 1–8 (2014)
4. Dean, T., Ruzon, M.A., Segal, M., Shlens, J., Vijayanarasimhan, S., Yagnik, J.: Fast, accurate detection of 100,000 object classes on a single machine. In: IEEE Conference on Computer Vision and Pattern Recognition, pp. 1814–1821 (2013)
5. Georghiades, A.S., Belhumeur, P.N., Kriegman, D.J.: From few to many: illumination cone models for face recognition under variable lighting and pose. IEEE Trans. Pattern Anal. Mach. Intell. **23**(6), 643–660 (2001)
6. Goemans, M.X., Williamson, D.P.: Improved approximation algorithms for maximum cut and satisfiability problems using semidefinite programming. J. ACM **42**(6), 1115–1145 (1995)
7. Gong, Y., Lazebnik, S., Gordo, A., Perronnin, F.: Iterative quantization: a procrustean approach to learning binary codes for large-scale image retrieval. IEEE Trans. Pattern Anal. Mach. Intell. **35**(12), 2916–2929 (2013)
8. Hong, R., Hu, Z., Wang, R., Wang, M., Tao, D.: Multi-view object retrieval via multi-scale topic models. IEEE Trans. Image Process. **25**(12), 5814–5827 (2016)

9. Hong, R., Zhang, L., Tao, D.: Unified photo enhancement by discovering aesthetic communities from flickr. IEEE Trans. Image Process. **25**(3), 1124–1135 (2016)
10. Jain, P., Vijayanarasimhan, S., Grauman, K.: Hashing hyperplane queries to near points with applications to large-scale active learning. In: Advances in Neural Information Processing Systems, pp. 928–936 (2010)
11. Ji, J., Li, J., Tian, Q., Yan, S., Zhang, B.: Angular-similarity-preserving binary signatures for linear subspaces. IEEE Trans. Image Process. **24**(11), 4372–4380 (2015)
12. Knyazev, A.V., Argentati, M.E.: Principal angles between subspaces in an a-based scalar product: algorithms and perturbation estimates. SIAM J. Sci. Comput. **23**(6), 2008–2040 (2002)
13. Liu, W., Wang, J., Mu, Y., Kumar, S., Chang, S.F.: Compact hyperplane hashing with bilinear functions. In: International Conference on Machine Learning, pp. 1–8 (2012)
14. Liu, X., Fan, X., Deng, C., Li, Z., Su, H., Tao, D.: Multilinear hyperplane hashing. In: IEEE Conference on Computer Vision and Pattern Recognition, pp. 5119–5127 (2016)
15. Liu, X., He, J., Deng, C., Lang, B.: Collaborative hashing. In: IEEE International Conference on Computer Vision, pp. 2139–2146 (2014)
16. Liu, X., Li, Z., Deng, C., Tao, D.: Distributed adaptive binary quantization for fast nearest neighbor search. IEEE Trans. Image Process. **26**(11), 5324–5336 (2017)
17. Marques, M., Stošić, M., Costeira, J.: Subspace matching: unique solution to point matching with geometric constraints. In: IEEE International Conference on Computer Vision, pp. 1288–1294 (2009)
18. Mu, Y., Hua, G., Fan, W., Chang, S.F.: Hash-svm: scalable kernel machines for large-scale visual classification. In: IEEE Conference on Computer Vision and Pattern Recognition, pp. 979–986 (2014)
19. Peng, C., Kang, Z., Cheng, Q.: Subspace clustering via variance regularized ridge regression. In: IEEE Conference on Computer Vision and Pattern Recognition, pp. 2931–2940 (2017)
20. Shen, F., Zhou, X., Yang, Y., Song, J., Shen, H.T., Tao, D.: A fast optimization method for general binary code learning. IEEE Trans. Image Process. **25**(12), 5610–5621 (2016)
21. Sim, T., Baker, S., Bsat, M.: The CMU pose, illumination, and expression (PIE) database. In: Proceedings of Fifth IEEE International Conference on Automatic Face and Gesture Recognition, pp. 53–58 (2002)
22. Song, J., Yang, Y., Huang, Z., Shen, H.T., Hong, R.: Multiple feature hashing for real-time large scale near-duplicate video retrieval. In: ACM on Multimedia Conference, pp. 423–432 (2011)
23. Wang, X., Atev, S., Wright, J., Lerman, G.: Fast subspace search via grassmannian based hashing. In: IEEE International Conference on Computer Vision, pp. 2776–2783 (2013)
24. Yang, X., Wang, M., Hong, R., Tian, Q., Rui, Y.: Enhancing person re-identification in a self-trained subspace. ACM Trans. Multimedia Comput. Commun. Appl. **13**(3), 27 (2017)
25. Zhang, F.: Matrix theory: basic results and techniques. Springer Science & Business Media, New York (2011)
26. Zhou, L., Bai, X., Liu, X., Zhou, J.: Binary coding by matrix classifier for efficient subspace retrieval. In: ACM International Conference on Multimedia Retrieval, pp. 1–8 (2018)

Image Denoising with Local Dense and Adaptive Global Residual Networks

Lulu Sun[1], Yongbing Zhang[1(✉)], Chenggang Yan[2], Xiangyang Ji[1,3],
Xinhong Hao[4], Yongdong Zhang[5], and Qionghai Dai[1,3]

[1] Graduate School at Shenzhen, Tsinghua University, Shenzhen 518055, China
zhang.yongbing@sz.tsinghua.edu.cn
[2] Institute of Information and Control,
Hangzhou Dianzi University, Hangzhou 310018, China
[3] Department of Automation, Tsinghua University, Beijing 100084, China
[4] Science and Technology on Mechatronic Dynamic Control Laboratory,
Beijing Institute of Technology, Beijing 100081, China
[5] School of Information Science and Technology,
University of Science and Technology of China, Hefei 230026, China

Abstract. Residual Networks (ResNet) and Dense Convolutional Networks (DenseNet) have shown great success in lots of high-level computer vision applications. In this paper, we propose a novel network with Local Dense and Adaptive Global Residual (LD+AGR) frameworks for fast and accurate image denoising. More precisely, we combine local residual/dense with global residual/dense to investigate the best performance dealing with image denoising problem. In particular, local/global residual/dense means the connection way of inner/outer recursive blocks. And residual/dense represents combining layers by summation/concatenation. Furthermore, when combining skip connections, we add some adaptive and trainable scaling parameters, which could adjust automatically during training to balance the importance of different layers. Numerous experiments demonstrate that the proposed network performs favorably against the state-of-the-art methods in terms of quality and speed.

Keywords: Adaptive global residual · Local dense · Image denoising

1 Introduction

Image denoising, which aims to recover a clear image from its degraded observation caused by noise contamination, is a classic and fundamental problem in computer vision [12–14]. Since image denoising is highly ill-posed, it is very challenging to achieve satisfactory results.

This work was partially supported by the National Science Foundations of China under Grant 61571254, Guangdong Natural Science Foundation 2017A030313353, and Shenzhen Fundamental Research fund under Grant JCYJ20170817161409809.

© Springer Nature Switzerland AG 2018
R. Hong et al. (Eds.): PCM 2018, LNCS 11164, pp. 27–37, 2018.
https://doi.org/10.1007/978-3-030-00776-8_3

Numerous image denoising methods have been proposed [1–4,9,21–23,26] in recent years with fantastic advancements. Most denoising methods are based on nonlocal self-similarity (NSS) priors [1,3,4,18,24]. NSS refers to the fact that a local patch often has many nonlocal similar patches across the image. Nonlocal means (NLM) [1] could be considered as a seminal work, bringing the new era of denoising by finding the NSS priors within a search window sliding across the image. It obtained a denoised patch by weighted averaging all other patches in the search window. Another famous benchmark, named block-matching and 3D filtering (BM3D) [4], remarkably combined NSS with an enhanced sparse representation in transform domain. It contained two general procedures: grouping and collaborative filtering. First, forming a 3D array by stacking together similar blocks. Second, obtaining 2D estimates of grouped blocks after performing collaborative filtering of the group. Instead of transforming images to other domains, low rank matrix approximation methods also attracted great attention in recent years. Representative and significant low-rank method was weighted nuclear norm minimization (WNNM) [9]. Based on the general prior knowledge that the larger singular values of the patch matrices of original image are more important than the smaller ones, WNNM achieved great success in image denoising.

Recently, methods based on neural networks [5–7,11,15,17,19] have shown significant success in many computer vision tasks, especially in image classification. Among these methods, Residual Networks (ResNet) [11,25] and Dense Convolutional Networks (DenseNet) [15] are attracting the most attention. Inspired by such achievements, we try to investigate the properties of the two architectures: residual and dense. In this paper, we not only combine the two elements in terms of local/global way, but also adding adaptive parameters to keep a good balance when combining various skip connections (Fig. 1).

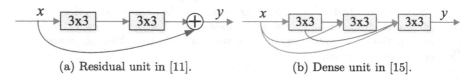

(a) Residual unit in [11].   (b) Dense unit in [15].

Fig. 1. Comparisons of residual/dense units. 3×3 represents a convolutional layer with kernel size of 3×3. $\oplus$ indicates the summation way of connecting. As shown in Fig. 3(b), many lines focusing on one point is the concatenation way for combination. Lines with the same color share the same value. The following symbols have the same meanings. (Color figure online)

2   Discussion of ResNet/DenseNet

ResNets are usually composed of lots of residual blocks, which only contains one skip connection and two convolutional layers. Such a simple residual architecture is easy to train. However, these units lack enough power to transmit sufficient information merely through cascading, leading to the lower ability of the whole network. Especially when dealing with image processing problems, these networks are not strong enough to extract features from massive data. In addition, it is likely to lose useful information during the process of deep-layers of delivery without any effective connection.

In contrast, DenseNets have plenty of skip connections in one dense block and the dense block is diverse to be able to simulate complex functions, which is beneficial to learn features. However, one big problem is that such powerful networks lack efficient contacts among outputs of each block. This will increase the time consumption of training. What is worse, no connections between blocks will cause some distortion when transmitting features.

Taking into account the shortcomings owned by single ResNet/DenseNet separately mentioned above, we are going to combine the two elements in two ways: local and global, which will be explained completely in the following sections.

3   Local Residual/Dense and Global Residual/Dense Networks

3.1   Local Residual and Global Residual Networks (LR+GR)

As shown in Fig. 2, both the local recursive block and global connecting way are the residual manner. So we name this style of framework as local residual and global residual networks (LR+GR). Normally, the first and last 3 × 3 convolutional layers are usually used for extracting features and reconstruction separately. In detail, this network is composed of three residual blocks, three inner and three outer identity skip connections, and two convolutional layers. In particular, we use parametric rectified linear unit (PReLU) [10] as activation function in all networks, which are omitted in the figures for simplicity.

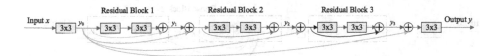

Fig. 2. Framework of local residual and global residual network.

3.2   Local Residual and Global Dense Networks (LR+GD)

From Fig. 3, we can see that the inner connecting way of each block is residual while the global manner is dense. Similarly, this kind of architecture is named as local residual and global dense networks (LR+GD). Particularly, there are three residual blocks and one summation skip connection in each unit. From the overall point of view, it uses dense style and there are six concatenating shortcuts.

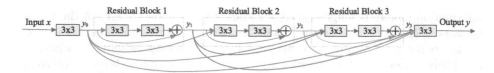

Fig. 3. Framework of local residual and global dense network.

3.3 Local Dense and Global Residual Networks (LD+GR)

If the recursive units are dense style while the global way is residual skip connection, we would call this framework as local dense with global residual network (LD+GR), as shown in Fig. 4. In particular, there are two dense blocks, two residual shortcuts and two convolutional layers in this network, and each block contains three convolutional layers and three dense skip connections.

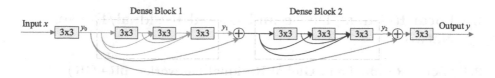

Fig. 4. Framework of local dense and global residual network.

3.4 Local Dense and Global Dense Networks (LD+GD)

Local dense with global dense networks (LD+GD) represent such frameworks that both inner and outer connections of blocks are dense, as shown in Fig. 5. To be specific, there are three concatenating lines in each dense unit and three skip connections in a global view.

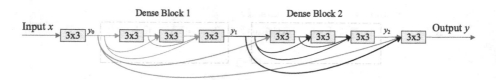

Fig. 5. Framework of local dense and global dense network.

4   Local Residual/Dense and Adaptive Global Residual Networks

4.1 Local Residual and Adaptive Global Residual Networks (LR+AGR)

Based on the framework of LR+GR, adding some trainable variables before summation, the network will become local residual and adaptive global residual network (LR+AGR). Seeing Fig. 6, there are three extra pairs of scaling parameters compared to the above LR+GR in Fig. 2.

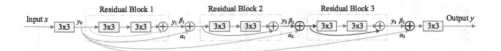

Fig. 6. Framework of local residual and adaptive global residual network.

4.2 Local Dense and Adaptive Global Residual Networks (LD+AGR)

Similarly, on the basis of LD+GR in Fig. 4, if we add some adaptive scaling parameters at the output of each dense block to balance the importance of each part automatically, the framework will become local dense and adaptive global residual network (LD+AGR), as shown in Fig. 7. We could see two pairs of scaling parameters after two dense blocks.

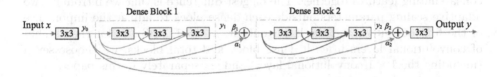

Fig. 7. Framework of local dense and adaptive global residual network.

4.3 Analysis and Discussions

In order to investigate more properties of the four basic frameworks and two adaptive ones mentioned above, we conducted the image denoising experiments using these networks. The training process has been recorded in Fig. 8(a). We controlled all the variables the same except the frameworks. As iteration increases, they are going to converge. Clearly, LD+AGR has the fastest convergence speed and achieves the best value at last. The following are LD+GD, LR+AGR, LD+GR, LR+GD, and LR+GR. Compared to LD+GR, LD+AGR has superior performance, which fully demonstrates the importance of introducing the adaptive and trainable scaling parameters.

5   The Proposed LD+AGR Networks for Image Denoising

5.1   Architecture

Referring to the framework of LD+AGR, we build the improved network, as shown in Fig. 9(b). It is composed of six dense blocks and six adaptive residual

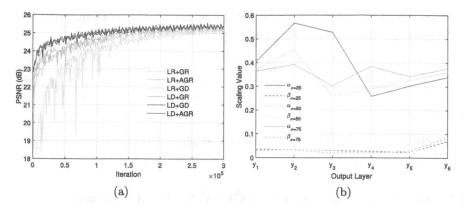

Fig. 8. (a) PSNR(dB) comparisons of six frameworks during training. (b) Adaptive α and β of different output layers in our LD+AGR.

skip connections. Focusing on one dense block (See Fig. 9(a)), there are six 3×3 convolutional layers for learning features continuously, one 1×1 convolutional layer for decreasing the dimension of feature mappings, and fifteen dense lines for concatenating features together. The biggest difference is that we introduce two adaptive scaling parameters outside each dense block to adjust the importance of the first output y_0 and the latter output $y_i (i = 1, ..., 6)$. As for the number of convolutional layers in each dense block and total blocks, we choose seven (including the 1×1 convolutional layer) and six separately in this paper.

5.2   Adaptive Parameters

We trained three models for image denoising with noise level $\sigma = 25$, 50, and 75 using our LD+AGR framework. The learned parameters α and β of different layers can be observed in Fig. 8(b). Intuitively, all αs are much bigger than βs, which means the original output y_0 plays a more important role than latter output layers. Moreover, all αs change rapidly while all βs shake slowly and softly. But the last output y_6 seems to be more important than the other five ones. We also conducted such experiments on the condition that all αs and βs are 0.5, but the denoising performance is far worse than the adaptive ones.

6   Experiments

In this section, we compare the proposed LD+AGR image denoising model with several state-of-the-art denoising methods, including BM3D [4], EPLL [26], WNNM [9], MLP [2], and PCLR [3]. The implementations are all from the publicly available codes provided by the authors.[1]

---

[1] The source code of the proposed method will be available after this paper is published.

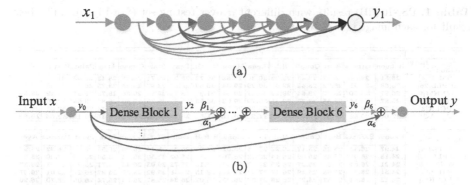

(a)

(b)

Fig. 9. (a) Dense block of our network. (b) Architecture of our LD+AGR networks. Grey circles represent 3×3 convolutional kernels, and white circle is 1×1 convolutional kernel.

Fig. 10. The 14 test images (grey, 256×256). From left to right: Baboon, Barbara, Boat, Couple, Hill, Lena, Monarch, R.R.Hood, Pentagon, Starfish, Cameraman, Man, Paint-full, Parrots.

6.1 Training Details

We use Berkeley Segmentation Dataset BSD500 [20] as the training set and 14 widely used test images as the testing set (It can be found in Fig. 10). To increase the training set, we segment these images to overlapping patches of size 50×50 with stride of 10. We use the deep learning library Tensorflow on an NVIDIA GTX TITAN X GPU with 3072 CUDA cores and 12 GB of RAM to implement all operations in our network. The filter weights are initialized using the "Xavier" strategy [8] and biases are generated by tf.constant initializer using Tensorflow. We use Adam [16] algorithm to optimize the loss function of Mean Square Error (MSE).

6.2 Quantitative Results

We record PSNR comparisons to other state-of-the-art algorithms on noise level $\sigma = 25$, 50, and 75 in Table 1. On the whole, our LD+AGR has the overwhelming superiority over the other methods on average, especially when $\sigma = 25$ and 50, the superiority can reach up to 0.33 dB and 0.36 dB over the second best methods on PSNR.

From Table 1, on average, we have the best results on three noise levels. Concretely, among 14 testing images, there are 13, 14, and 8 reconstructed images by our methods achieve the best performance. Hence, no matter on the whole or individuals, our LD+AGR shows tremendous advance over other methods in terms of PSNR.

Table 1. PSNR (dB) results with different σ over testing set (See Fig. 10). The best result for each image is **highlighted**.

Image / Method	Baboon	Barbara	Boat	Couple	Hill	Lena	Mona.	R.R.H.	Pent.	Star.	C.man	Man	Paint.	Parrots	Ave.
$\sigma = 25$															
BM3D[4]	26.51	29.54	28.42	28.44	29.06	30.36	28.88	30.73	28.21	28.71	29.08	28.43	29.03	31.55	29.07
EPLL[26]	26.68	28.91	28.72	28.42	29.03	30.29	29.34	30.63	28.07	28.97	29.25	28.73	29.28	31.59	29.14
WNNM[9]	26.81	**30.20**	28.77	28.64	29.29	30.87	29.84	31.22	28.72	29.54	29.64	28.73	29.51	32.12	29.56
MLP[2]	26.87	29.14	28.85	28.75	29.31	30.79	29.62	30.93	28.23	29.26	29.30	28.80	29.65	32.10	29.40
PCLR[3]	26.75	29.73	28.88	28.73	29.30	30.78	29.75	30.98	28.42	29.35	29.68	28.74	29.58	32.01	29.48
LD+AGR	**27.06**	29.52	**29.27**	**29.22**	**29.53**	**31.22**	30.16	31.55	**28.77**	**30.09**	29.95	**29.10**	30.18	**32.64**	**29.88**
$\sigma = 50$															
BM3D[4]	24.07	26.33	25.25	25.17	26.22	27.05	25.62	27.75	25.35	25.44	25.98	25.55	25.66	28.32	25.98
EPLL[26]	24.13	26.02	25.48	25.09	26.14	26.93	25.76	27.58	25.02	25.50	26.04	25.59	25.88	28.00	25.94
WNNM[9]	24.21	26.64	25.54	25.27	26.35	27.47	26.32	28.13	25.72	26.03	26.45	25.71	25.97	28.71	26.32
MLP[2]	24.31	26.41	25.68	25.49	26.47	27.54	26.24	28.15	25.48	25.92	26.37	25.87	26.21	29.01	26.37
PCLR[3]	24.12	26.44	25.57	25.27	26.26	27.38	26.25	28.02	25.34	25.87	26.60	25.74	26.05	28.70	26.26
LD+AGR	**24.52**	**26.80**	**26.07**	25.78	**26.67**	**27.81**	26.88	28.64	**25.83**	**26.48**	26.77	26.03	26.60	29.35	**26.73**
$\sigma = 75$															
BM3D[4]	23.05	24.39	23.61	23.56	24.64	25.20	23.69	25.96	23.85	23.71	24.17	23.99	23.81	26.33	24.28
EPLL[26]	22.94	24.04	23.77	23.44	24.50	24.97	23.64	25.84	23.45	23.76	24.21	24.07	23.89	26.11	24.19
WNNM[9]	23.18	24.70	23.87	23.65	24.74	25.72	24.31	26.19	**24.21**	24.13	24.60	24.10	24.07	26.69	24.58
MLP[2]	23.30	**24.76**	24.11	**23.92**	**24.97**	**25.79**	24.40	26.41	24.13	24.16	24.67	**24.39**	**24.40**	**27.08**	24.75
PCLR[3]	22.99	24.58	23.87	23.64	24.64	25.52	24.28	26.28	23.79	24.01	24.78	24.12	24.15	26.74	24.53
LD+AGR	**23.33**	**24.76**	**24.30**	23.83	24.94	25.76	**24.83**	26.69	**24.21**	**24.32**	24.98	24.29	24.39	27.05	**24.83**

6.3 Visual Quality

As shown in Fig. 11, similarly, our LD+AGR has the best visual quality compared to other methods. Especially, in the green and red windows, it is easy for us to recognize lines and shapes of the starfish in our result. Even with the noise level $\sigma = 75$, our method can still recover the most valuable information, which can be found in Fig. 12. In the green window, the head of butterfly in our

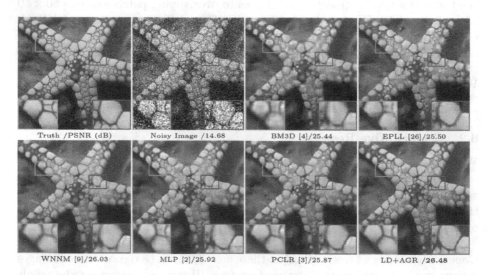

Fig. 11. Sample image denoised results on *Starfish* with state-of-the-art methods ($\sigma = 50$). (Color figure online)

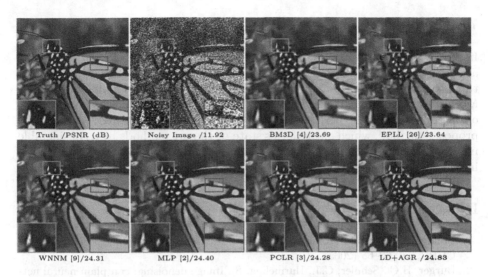

Fig. 12. Sample image denoised results on *Monarch* with state-of-the-art methods ($\sigma = 75$). (Color figure online)

recovered image is distinct from others. Likewise, in the red block, our pattern is also much sharper than the others. In a word, from the view of visual quality, our LD+AGR performs better than other state-of-the-art image denoising methods.

6.4 Running Time

We profile the time consumption of all the methods in a Matlab 2015b environment using the same machine (an NVIDIA GTX TITAN X GPU with 3072 CUDA cores and 12 GB of RAM) in Table 2. Obviously, based on the adaptive networks, our method has enormous advantage than all the traditional algorithms.

Table 2. Average running time (s) for one image with different noise level σ over testing set (See Fig. 10). The best result for each dataset is **highlighted**.

σ	BM3D [4]	EPLL [26]	WNNM [9]	MLP [2]	PCLR [3]	LD+AGR
$\sigma = 25$	41.78	46.29	160.71	2.22	69.68	**1.25**
$\sigma = 50$	44.08	45.17	114.96	2.18	124.71	**1.25**
$\sigma = 75$	56.47	45.59	180.36	2.17	130.79	**1.25**
Average	47.44	45.68	152.01	2.19	108.39	**1.25**

7   Conclusions

In this paper, we address the image denoising problem via a local dense and adaptive global residual (LD+AGR) network which learns high effective features to reconstruct the latent clean images from the corresponding noisy ones. Moreover, we introduce adaptive scaling parameters to balance the importance of different outputs. Experimental results fully illustrate the effectiveness of the proposed method, which outperforms state-of-the-art methods by a considerable margin in terms of PSNR. Noticeable improvements can also visually be found in the reconstruction results.

References

1. Buades, A., Coll, B., Morel, J.M.: A non-local algorithm for image denoising. In: CVPR, pp. 60–65 (2005)
2. Burger, H.C., Schuler, C.J., Harmeling, S.: Image denoising: can plain neural networks compete with BM3D? In: CVPR, pp. 2392–2399 (2012)
3. Chen, F., Zhang, L., Yu, H.: External patch prior guided internal clustering for image denoising. In: ICCV, pp. 603–611 (2015)
4. Dabov, K., Foi, A., Katkovnik, V., Egiazarian, K.: Image denoising by sparse 3-D transform-domain collaborative filtering. IEEE Trans. Image Process. **16**(8), 2080–2095 (2007)
5. Dong, C., Deng, Y., Change Loy, C., Tang, X.: Compression artifacts reduction by a deep convolutional network. In: ICCV, pp. 576–584 (2015)
6. Dong, C., Loy, C.C., He, K., Tang, X.: Learning a deep convolutional network for image super-resolution. In: Fleet, D., Pajdla, T., Schiele, B., Tuytelaars, T. (eds.) ECCV 2014. LNCS, vol. 8692, pp. 184–199. Springer, Cham (2014). https://doi.org/10.1007/978-3-319-10593-2_13
7. Girshick, R., Donahue, J., Darrell, T., Malik, J.: Rich feature hierarchies for accurate object detection and semantic segmentation. In: CVPR, pp. 580–587 (2014)
8. Glorot, X., Bengio, Y.: Understanding the difficulty of training deep feedforward neural networks. In: International Conference on Artificial Intelligence and Statistics, pp. 249–256 (2010)
9. Gu, S., Zhang, L., Zuo, W., Feng, X.: Weighted nuclear norm minimization with application to image denoising. In: CVPR, pp. 2862–2869 (2014)
10. He, K., Zhang, X., Ren, S., Sun, J.: Delving deep into rectifiers: surpassing human-level performance on imagenet classification. In: ICCV, pp. 1026–1034 (2015)
11. He, K., Zhang, X., Ren, S., Sun, J.: Identity mappings in deep residual networks (2016). arXiv preprint: arXiv:1603.05027
12. Hong, R., Hu, Z., Wang, R., Wang, M., Tao, D.: Multi-view object retrieval via multi-scale topic models. IEEE Trans. Image Process. **25**(12), 5814–5827 (2016)
13. Hong, R., Zhang, L., Tao, D.: Unified photo enhancement by discovering aesthetic communities from flickr. IEEE Trans. Image Process. **25**(3), 1124–1135 (2016)
14. Hong, R., Zhang, L., Zhang, C., Zimmermann, R.: Aesthetic tendency discovery by multi-view regularized topic modeling. IEEE Trans. Multimed. **18**(8), 1555–1567 (2016)
15. Huang, G., Liu, Z., Weinberger, K.Q., van der Maaten, L.: Densely connected convolutional networks (2016). arXiv preprint: arXiv:1608.06993

16. Kingma, D., Ba, J.: Adam: A method for stochastic optimization (2014). arXiv preprint: arXiv:1412.6980
17. Krizhevsky, A., Sutskever, I., Hinton, G.E.: Imagenet classification with deep convolutional neural networks. In: NIPS, pp. 1097–1105 (2012)
18. Liu, H., Xiong, R., Zhang, J., Gao, W.: Image denoising via adaptive soft-thresholding based on non-local samples. In: CVPR, pp. 484–492 (2015)
19. Long, J., Shelhamer, E., Darrell, T.: Fully convolutional networks for semantic segmentation. In: CVPR, pp. 3431–3440 (2015)
20. Martin, D., Fowlkes, C., Tal, D., Malik, J.: A database of human segmented natural images and its application to evaluating segmentation algorithms and measuring ecological statistics. In: ICCV, pp. 416–423 (2001)
21. Vincent, P., Larochelle, H., Bengio, Y., Manzagol, P.A.: Extracting and composing robust features with denoising autoencoders. In: ICML, pp. 1096–1103 (2008)
22. Mao, X.-J., Shen, C., Yang, Y.-B.: Image restoration using very deep fully convolutional encoder-decoder networks with symmetric skip connections. In: NIPS (2016)
23. Xie, J., Xu, L., Chen, E.: Image denoising and inpainting with deep neural networks. In: NIPS, pp. 341–349 (2012)
24. Xu, J., Zhang, L., Zuo, W., Zhang, D., Feng, X.: Patch group based nonlocal self-similarity prior learning for image denoising. In: ICCV, pp. 244–252 (2015)
25. Zhang, Y., Sun, L., Yan, C., Ji, X., Dai, Q.: Adaptive residual networks for high-quality image restoration. IEEE Trans. Image Process. **27**(7), 3150–3163 (2018). https://doi.org/10.1109/TIP.2018.2812081
26. Zoran, D., Weiss, Y.: From learning models of natural image patches to whole image restoration. In: ICCV, pp. 479–486 (2011)

Cross Diffusion on Multi-hypergraph
for Multi-modal 3D Object Recognition

Zizhao Zhang, Haojie Lin, Junjie Zhu, Xibin Zhao, and Yue Gao[✉]

Beijing National Research Center for Information Science and Technology KLISS,
School of Software, Tsinghua University, Beijing, China
zzz_14@126.com, haojie.lin@outlook.com, zhujj18@mails.tsinghua.edu.cn,
{zxb,gaoyue}@tsinghua.edu.cn

Abstract. 3D object recognition is a longstanding task in computer vision and has shown wide applications in computer aided design, virtual reality, etc. Current state-of-the-art methods mainly focus on 3D object representation for recognition. Concerning the multi-modal representations in practice, how to effectively combine such multi-modal information for recognition is still a challenging and urgent requirement. In this paper, we aim to conduct 3D object recognition using multi-modal information through a cross diffusion process on multi-hypergraph structure. Given multi-modal representations of 3D objects, the correlation among these objects is formulated using the multi-hypergraph structure each representation separately, which is able to model complex relationship among objects. To combine multi-modal representation, we propose a cross diffusion process on multi-hypergraph, in which the label information is propagated from multiple hypergraphs alternatively. In this way, the multi-modal information can be jointly combined through this cross diffusion process in multi-hypergraph structure. We have applied the proposed method in 3D object recognition using multiple representations. To evaluate the performance of the proposed cross diffusion method, we provide extensive experiments on two public 3D object datasets. Experimental results demonstrate that the proposed method can achieve satisfied multi-modal combination performance and outperform the current state-of-the-art methods.

Keywords: 3D Object Recognition · Cross diffusion
Hypergraph learning · Multi-modal combination

1 Introduction

The development of image processing, graphics hardware, and computing techniques has lead to wide applications of 3D objects, such as virtual/augment reality, 3D games/movies, and 3D architecture design. Such applications have lead to rapid increasing of 3D object data and urgent demand of 3D object understanding, such as 3D object recognition and retrieval [7,22]. Under such circumstances, effective 3D object recognition [13,18,23,30] has become an urgent requirement

© Springer Nature Switzerland AG 2018
R. Hong et al. (Eds.): PCM 2018, LNCS 11164, pp. 38–49, 2018.
https://doi.org/10.1007/978-3-030-00776-8_4

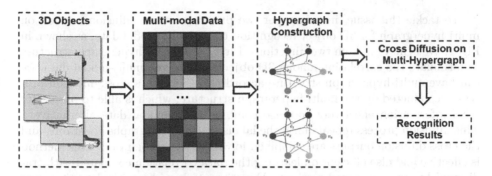

Fig. 1. The framework of the proposed 3D object recognition method using multi-modal data.

and also a fundamental challenge in computer vision, which has attracted much research attention in recent years.

Regarding the task of 3D object recognition, how to represent 3D object is an important task and has attracted much attention recently. Early methods [11,26] employ shape features to represent 3D content. Based on the volumetric data, 3D ShapeNet [28] and VoxNet [16] explored deep neural networks to generate 3D object representations. PointNet and PointNet++ [19,21] employed 3D point cloud data for deep representation. As volumetric data or point cloud could be not available in many cases, Some recent attempts employed multi-view data to represent 3D objects, and visual features were extracted accordingly. In Lighting Field descriptor [1], a set of 10 views captured from the vertices of a dodecahedron over a hemisphere are employed. In [17], a panoramic object representation for accurate model attributing was introduced to generate panoramic views to represent the surface and the orientation of 3D objects. Multi-View Convolutional Neural Networks [24] and Group-View Convolutional Neural Networks [3] introduced to employ deep neural networks for multi-view 3D object representation.

Although there have been many attempts on 3D object recognition, due to the varied circumstances in practice, existing methods for 3D representation have their own advantages and also specified limitations, and it is challenging to find the common best representation for all 3D objects. Under such circumstances, it is reasonable to consider different 3D object representation methods, and how to combine such multiple 3D representations towards better 3D object recognition performance has become an urgent and important requirement. This task is challenging from the following two aspects. On one hand, given the multi-modal data or multiple representations of 3D objects, it is needed to consider such multi-modal data simultaneously, which could be heterogeneous. On the other hand, the correlation among 3D objects should be deeply exploited during the multi-modal fusion process. We note that it is a common scenario of multiple data representations in different applications. Therefore, it is important to exploit the combination method for multi-modal representations.

To tackle this issue, in this paper, we propose a cross diffusion method on multi-hypergraph for 3D object recognition using multi-modal data, as shown in Fig. 1. We achieve it from two directions. First, we employ hypergraph structure to formulate the correlation among 3D objects, and given multi-modal data, we can have multi-hypergraph structure accordingly. Second, a cross diffusion process is conducted on the multi-hypergraph structure, which is able to propagate label information effectively and also combine multi-modal data alternatively. This diffusion process repeats until the labels on the hypergraphs are stable, and the cross diffusion outputs are combined for 3D object recognition. This method is effective and also efficient on learning the data correlation compared with traditional hypergraph based methods. Experiments have been conducted on two public datasets, i.e., the ModelNet40 dataset and National Taiwan University 3D dataset, to evaluate the performance of the proposed method. Experimental results and comparison with the state-of-the-art methods demonstrate that the proposed method can achieve better performance.

The rest of this paper is organized as follows. We briefly introduce related works on 3D object recognition and hypergraph learning in Sect. 2. The proposed method is provided in Sect. 3. Experiments and discussions are shown in Sect. 4. We finally conclude this paper in Sect. 5.

2   Related Works

In this section, we briefly introduce recent progress on 3D object recognition and hypergraph learning.

Recent successful 3D object recognition methods are mainly based on deep neural networks, which have shown superior performance compared with traditional methods. Using volumetric data, Wu et al. [28] lift 2.5D to 3D and proposed a 3D ShapeNets to represent 3D objects using a probability distribution of volumetric data, and a convolutional deep belief network was employed to generate the binary variable of 3D objects. Similarly, VoxNet [16] also utilized binary voxel grids and a 3D CNN architecture for 3D object representation. With point cloud data, Qi et al. [19] introduced PointNet, a deep neural networks, to process point clouds, in which both spatial transform network and a symmetry function were employed for the invariance to permutation. Qi et al. [21] further proposed PointNet++ to accumulates local features in a hierarchical architecture for 3D object representation. Kd-tree is exploited to subdivide the point clouds and aggregate features in Kd-network [12]. DGCNN [27] proposes Edge-Conv as a basic block in network, where the edge features of points and their neighbors are employed. In PointCNN [14], χ-Conv is proposed to aggregate features in each local pitches and a hierarchical network structure is applied as typical CNNs. Based on the multi-view representation of 3D objects, Su et al. [24] proposed a multi-view convolutional neural network (MVCNN). In this method, MVCNN first generated the view feature for each view using convolutional neural networks, which are pre-trained on ImageNet. Then, these view features were fused using a pooling procedure. Qi et al. [20] further employed multi-resolution views for 3D object representation, and the features were extracted from multi-resolution views. Xie et al. [29] proposed a deep nonlinear metric learning method to enforce the similarity between a pair of samples belonging to the same category to be small and vice versa. Guo et al. [6] proposed to a deep embedding network to deal with the complex intra-class and inter-class variations of 3D objects, and the deep convolutional network can be jointly supervised using classification loss and triplet loss. Considering the multi-modal 3D object data, Chen et al. [2] proposed a MV3D network to combine the point cloud data from LIDAR and the view data, which projected the point cloud data into images and then processed them together using CNNs. Hong et al. [8] proposed a multi-view regularized topic modeling, which fuses color, textural, and semantic channel features and adjusts the feature weights automatically.

Hypergraph learning has been widely used in many computer vision tasks [9,25,31,32], due to its superiority on high order correlation modeling. Huang et al. [10] proposed a probabilistic hypergraph to formulate the relevance relationship among images, which was used for image retrieval. Gao et al. [5] employed hypergraph structure to estimate the relations among images based on the visual and textual features for the task of social image search. Liu et al. [15] proposed an elastic net hypergraph learning for image clustering and classification, which employed the spare representation method to gain the robustness to data nose. Gao et al. [4] proposed a multi-hypergraph learning (MHL) method, which directly combined multiple hypergraphs during the learning process. In this method, multiple hypergraphs were constructed to formulate the object correlation. A learning process was conducted on the multi-hypergraph structure to jointly optimize the object label and the hypergraph weight. Noted that although multiple data can be used in this framework, these multi-modal data are still processed separately and they are only considered in the fused part. It is noted that it is still a challenging task to jointly consider different modal data together for 3D object recognition.

3   Cross Diffusion on Multi-Hypergraph for Multi-Modal 3D Object Recognition

In this section, we introduce our proposed cross diffusion method on multi-hypergraph for 3D object recognition. Figure 1 illustrates the main framework of the proposed method. Given the multiple representations of 3D objects, a set of hypergraphs are constructed to describe the complex correlation among these 3D objects. Then, the transition matrices for different hypergraphs are generated, and the label projection matrix is updated by conducting the diffusion on these hypergraphs. More specifically, considering the multiple representations, a cross diffusion process is repeated until convergence. The final label matrix can be used for 3D object recognition.

3.1   Multi-hypergraph Modeling

Given n 3D objects $\mathbf{X} = [x_1, \ldots, x_i, \ldots, x_n]$ and their corresponding n_m representations, e.g., n_m types of representations, the first task is to formulate the data correlation among these 3D objects. Considering the t-th type of representations, for these n 3D objects, a hypergraph $\mathcal{G}^t = (\mathcal{V}^t, \mathcal{E}^t, \mathbf{W}^t)$ is constructed to formulate the relationship among them, where $\mathcal{V}^t$ is vertex set, $\mathcal{E}$ is hyperedge set, and the diagonal entries of $\mathbf{W}^t$ denote the weights of hyperedges. Each vertex in $\mathcal{V}^t$ denotes one object in $\mathbf{X}$, and the hyperedges in $\mathcal{E}^t$ represent the connections among these 3D objects. There are many hyperedge generation methods, such as kNN, ε-ball distance and spare representation based methods. In this work, we generate hyperedges using the kNN method. More specifically, one vertex is chosen as the centroid each time, and its K nearest neighbors and the centroid itself are connected by a hyperedge $\mathbf{e}$. It is noted that other hypergraph generation methods can be also used here, and the hyperedges can be flexibly extended. For the total n_m modalities, n_m hypergraphs can be generated to represent the object correlations.

3.2   Transition Matrix Generation

The diffusion on hypergraph is based on the assumption that long-range similarities can be approximated by evaluating local similarities. In other words, local similarities (high values) are more important than far-away ones, and the diffusion process on hypergraph can be regarded as propagating local similarities to non-local points.

Here, we first generate the similarity matrix among vertices on the t-th hypergraph as follows:

$$\mathbf{\Lambda}^t(u, v) = \sum_{e \in \mathcal{E}^t} \frac{\mathbf{W}^t(e)\, \mathbf{H}^t(u, e)\, \mathbf{H}^t(v, e)}{\delta(e)}, \tag{1}$$

and

$$\mathbf{\Lambda}^t = \mathbf{H}^t \mathbf{W}^t \mathbf{D_e^{t}}^{-1} \mathbf{H}^{t\,\mathrm{T}}. \tag{2}$$

Then, the transition matrix $\mathbf{P}^t$ is generated by normalizing the similarity matrix $\mathbf{\Lambda}^t$ as

$$\mathbf{P}^t(i, j) = \frac{\mathbf{\Lambda}^t(i, j)}{\sum_{w \in \mathcal{V}^t} \mathbf{\Lambda}^t(i, w)}. \tag{3}$$

$\mathbf{P}^t$ can be written in the matrix form as $\mathbf{P}^t = \mathbf{D}^{t\,-1}\mathbf{\Lambda}^t$ where $\mathbf{D}^t$ is a diagonal matrix whose i-th diagonal element is $\mathbf{D}^t(i, i) = \sum_{j=1}^{|\mathcal{V}^t|} \mathbf{\Lambda}^t(i, j)$. $\mathbf{P}^t(i, j)$ indicates the transition probability of label from the i-th sample to the j-th sample.

In this way, $\mathbf{P}^t$ can be regarded as a generalization of Parzen window estimators on hypergraph structure, where the multiplication by $\mathbf{P}^t$ can be considered as a diffusion process on the t-th hypergraph.

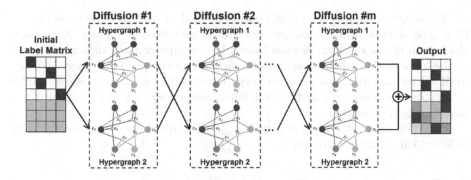

Fig. 2. Illustration of the cross diffusion process on multi-hypergraph.

3.3 Cross Diffusion Process on Multi-Hypergraph

Here we let $\mathbf{Y}_0 = \{\mathbf{y}_1, \mathbf{y}_2, \ldots, \mathbf{y}_n\}$ and $\mathbf{y}_i$ denotes the initial label vector of the i-th sample (vertex). For labeled samples, the j-th element is 1 if the i-th sample belongs to the j-th class, and it is 0 otherwise. For unlabeled samples, all elements are set to 0.5, indicating that there are no prior information about the category of these samples. Let $\mathbf{Y}_0^L$ denote the labeled part.

Here, we take two types of 3D object representations as an example. Given two types of data or features for representation, two hypergraphs can be constructed accordingly. First, we calculate the transition matrices $\{\mathbf{\Lambda}^1, \mathbf{\Lambda}^2\}$ from the two hypergraph structures using Eq. (2), respectively. Then, the according transition matrices $\{\mathbf{P}^1, \mathbf{P}^2\}$ can be obtained using Eq. (3).

During the cross diffusion process, the diffusion output from one hypergraph structure can be used as the input for the other hypergraph structure, and this process repeats until the output of each hypergraph convergence, as shown in Fig. 2. The cross diffusion process can be defined as:

$$\begin{aligned} \mathbf{Y}_{d+1}^1 &= \mathbf{P}^1 \times \mathbf{Y}_d^2 \\ \mathbf{Y}_{d+1}^{1L} &= \mathbf{Y}_0^L \end{aligned} \tag{4}$$

and

$$\begin{aligned} \mathbf{Y}_{d+1}^2 &= \mathbf{P}^2 \times \mathbf{Y}_d^1 \\ \mathbf{Y}_{d+1}^{2L} &= \mathbf{Y}_0^L \end{aligned} \tag{5}$$

where $\mathbf{Y}_d^k$ is the label matrix of the k-th hypergraph after d times diffusion. Considering that the labels of training data are fixed, we reset the labeled samples $\mathbf{Y}_{d+1}^{kL}$ to their initial values $\mathbf{Y}_0^L$ after each diffusion. After m steps, the cross diffusion procedure will converge and the overall label matrix can be computed as

$$\mathbf{Y}_{final} = \frac{1}{n_m} \sum_{i=1}^{n_m} \mathbf{Y}_d^i, \tag{6}$$

where $n_m = 2$ when two representations are available.

The above method can be easily extended to deal with three or more modalities. The alternative diffusion process can be described as follows. Each time, the output of one diffusion process from a hypergraph can be randomly selected as the input of the diffusion process of another hypergraph. This process repeats until all the diffusion processes on these hypergraphs converges. As the diffusion process contains only matrix multiplication process, it is very efficient when handling large scale data.

Based on the overall label matrix Y_{final}, 3D objects can be categorized into different categories accordingly.

4   Experiments and Discussions

4.1   Testing Datasets and Experimental Settings

To evaluate the performance of the proposed cross diffusion method on multi-hypergraph for multi-modal 3D object recognition, we have conducted experiments on two public benchmarks, including the Princeton ModelNet40 dataset [28] and the National Taiwan University (NTU) 3D model dataset [1]. Model-Net40 is composed of 12311 3D objects from 40 categories, such as *airplane, bookshelf, bottle, curtain, keyboard, lamp, laptop, piano, radio* and *toilet*. NTU contains 2012 3D objects from 67 categories, such as *boat, bomb, book, chip, car, fence, flashlight, helicopter, motorcycle* and *truck*.

In experiments, a set of training data are first selected. In the ModelNet40 dataset, we follow [28] to conduct the training/testing split. In the NTU dataset, 80% data for each category are used for training, and the other 20% data are used for testing. The classification accuracy is used as the evaluation metric.

For each 3D object, two recent view-based 3D features are used for representation, including Multi-View Convolutional Neural Networks feature (MVCNN) [24] and Group-View Convolutional Neural Networks (GVCNN) [3]. These two methods are selected as they have shown superior performance recently. First, a set of views are generated from each 3D object, and then the MVCNN and the GVCNN mehtods are performed to generated features, leading to a 4096-d feature for MVCNN and a 2048-d feature for GVCNN, in this case. We follow the instruction in [3,24] for feature extraction, and then each 3D object has two types of representations.

In experiments, the parameter K for hypergraph generation in the proposed method is empirically set as 10 and 5 on the ModelNet40 and NTU datasets, respectively.

To justify the effectiveness of the proposed method, the following methods are used for comparison.

- MVCNN [24] and GVCNN [3]: The methods of these two features.
- MVCNN+HL and GVCNN+HL: These two methods employ the hypergraph structure to formulate 3D object correlation using MVCNN and GVCNN feature individually. Then, the learning on hypergraph is conducted to conduct 3D object recognition using one type of features only.

- MVCNN+GVCNN+HL: In this method, the MVCNN feature and the GVCNN feature are concatenated directly. Then, a hypergraph is constructed using the concatenated feature and the hypergraph learning is conducted for 3D object recognition.
- MVCNN+GVCNN+MHL [4]: In this method, two hypergraphs are generated using MVCNN and GVCNN features respectively. Then, the joint learning on two hypergraphs are conducted to recognize 3D objects, where the weights for different hypergraphs are also optimized.
- Cross Diffusion on Multi-Hypergraph (CDMH), i.e., the proposed method.

4.2   Experimental Results and Discussions

Table 1. Experimental comparison of different methods in terms of recognition accuracy.

Method	ModelNet40	NTU
MVCNN	90.10%	79.89%
GVCNN	93.10%	82.30%
MVCNN+HL	90.68%	79.89%
GVCNN+HL	92.14%	82.84%
MVCNN+GVCNN+HL	93.23%	80.43%
MVCNN+GVCNN+MHL	96.19%	83.38%
CDMH	**96.76%**	**84.45%**

Experimental results on two datasets are shown in Table 1, and the confusion matrices of the proposed method on two datasets are provided in Fig. 3. As shown in these results, we can have the following observations:

1. The concatenating of multi-modal data may not always lead to better performance. For example, MVCNN+GVCNN+HL can achieve the improvement of recognition accuracy of 3.47% and 0.14% compared with MVCNN and GVCNN on the ModelNet40 dataset. While the reduce ration of recognition accuracy is 2.27% compared with GVCNN on the NTU dataset.
2. Compared with concatenating two features directly, jointly learning the correlations has shown better performance. MVCNN+GVCNN+MHL achieves much higher performance compared with MVCNN+GVCNN+HL on both dataset, where the gains are 3.17% and 3.67%, respectively.
3. Compared with traditional multi-hypergraph learning, the proposed cross diffusion method on multi-hypergraph achieves better performance, i.e., 96.76% and 84.45% on the ModelNet40 dataset and the NTU dataset, respectively.

The better performance of the proposed CDMH method can be dedicated to the following reasons. First, the hypergraph structure is able to model the

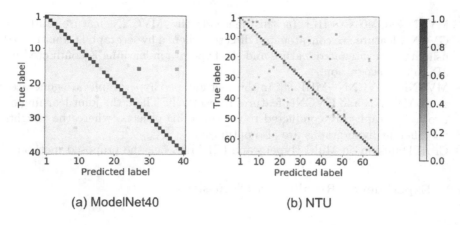

(a) ModelNet40   (b) NTU

Fig. 3. The confusion matrices on two datasets respectively.

complex 3D object correlation in the feature space. Regarding the multiple representations of 3D objects, the cross diffusion model can further jointly take multi-modal data into consideration during the label propagation process. The proposed cross diffusion method can better explore the interaction of different representations, rather than just simply combine them directly.

4.3   On Hypergraph Construction

In the proposed method, hypergraph construction is an important procedure, and parameter K is used to determine how many nearest neighbors of each centroid vertex are connected by one hyperedge, which controls the smoothness of the hypergraph structure. To evaluate the influence of different selections of parameter K on 3D object recognition, we have varied K from 3 to 20, and Fig. 4 provides the recognition accuracy performance curve with respect to the variation of parameter K on the ModelNet40 dataset and the NTU dataset, respectively.

As shown in these results, we can observe that the performance on Model-Net40 is stable, and there are only little difference with respect to different K. On the NTU dataset, performance is stable when K is not too large, such as less than 5. While when K is larger than 5, the performance degrades quickly. This is due to that many 3D object categories in the NTU dataset only have 10 to 15 objects. A large K will lead to very inaccurate hyperedges, and these K objects which are connected by one hyperedge may be not from the same category accordingly. Therefore, the performance will decrease rapidly if K is set too large.

4.4   On Diffusion Process

The number of iterations in the cross diffusion process is another important issue. In this part, we evaluate the influence of different times of iterations in the cross

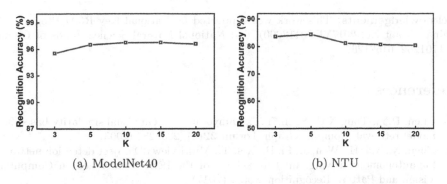

(a) ModelNet40   (b) NTU

Fig. 4. The performance comparison with respect to different K selections.

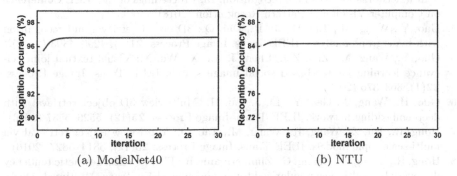

(a) ModelNet40   (b) NTU

Fig. 5. The performance comparison with respect to different iteration times in the cross diffusion process.

diffusion process. We have recorded the performance from 1 to 30 iterations and the results are demonstrated in Fig. 5

As shown in the results, the cross diffusion process converges very fast. The performance becomes stable after only 5 and 6 iterations on the ModelNet40 dataset and the NTU dataset, respectively, which demonstrates the efficiency of the proposed method.

5   Conclusion

In this paper, we propose a cross diffusion method on multi-hypergraph for multi-modal 3D object recognition. This method is able to jointly consider the multiple representations of 3D objects and exploit complex data correlation. Experiments on two public benchmarks and comparison with existing methods demonstrate the effectiveness of the proposed method. The proposed cross diffusion method on multi-hypergraph has the merit on conducting effective multi-modal information fusion. The proposed method is a general framework which can be used in other applications which contain multiple data representations.

Acknowledgements. This work was supported by National Key R&D Program of China (Grant No. 2017YFC0113000), and National Natural Science Funds of China (U1701262, 61671267).

References

1. Chen, D.Y., Tian, X.P., Shen, Y.T., Ouhyoung, M.: On visual similarity based 3D model retrieval. Comput. Graph. Forum **22**(3), 223–232 (2003)
2. Chen, X., Ma, H., Wan, J., Li, B., Xia, T.: Multi-view 3D object detection network for autonomous driving. In: Proceedings of the IEEE Conference on Computer Vision and Pattern Recognition, vol. 1 (2017)
3. Feng, Y., Zhang, Z., Zhao, X., Ji, R., Gao, Y.: GVCNN: group-view convolutional neural networks for 3D shape recognition. In: Proceedings of the IEEE Conference on Computer Vision and Pattern Recognition (2018)
4. Gao, Y., Wang, M., Tao, D., Ji, R., Dai, Q.: 3D object retrieval and recognition with hypergraph analysis. IEEE Trans. Image Process. **21**(9), 4290–4303 (2012)
5. Gao, Y., Wang, M., Zha, Z.J., Shen, J., Li, X., Wu, X.: Visual-textual joint relevance learning for tag-based social image search. IEEE Trans. Image Process. **22**(1), 363–376 (2013)
6. Guo, H., Wang, J., Gao, Y., Li, J., Lu, H.: Multi-view 3D object retrieval with deep embedding network. IEEE Trans. Image Process. **25**(12), 5526–5537 (2016)
7. Hong, R., Hu, Z., Wang, R., Wang, M., Tao, D.: Multi-view object retrieval via multi-scale topic models. IEEE Trans. Image Process. **25**(12), 5814–5827 (2016)
8. Hong, R., Zhang, L., Zhang, C., Zimmermann, R.: Flickr circles: aesthetic tendency discovery by multi-view regularized topic modeling. IEEE Trans. Multimed. **18**(8), 1555–1567 (2016)
9. Huang, Y., Liu, Q., Metaxas, D.: Video object segmentation by hypergraph cut. In: IEEE Conference on Computer Vision and Pattern Recognition, CVPR 2009, pp. 1738–1745. IEEE (2009)
10. Huang, Y., Liu, Q., Zhang, S., Metaxas, D.N.: Image Retrieval via Probabilistic Hypergraph Ranking. In: Proceedings of the IEEE Conference on Computer Vision and Pattern Recognition, pp. 3376–3383. IEEE (2010)
11. Klaser, A., Marszalek, M., Schmid, C.: A spatio-temporal descriptor based on 3D-gradients. In: Proceedings of the British Machine Vision Conference, pp. 275–1. British Machine Vision Association (2008)
12. Klokov, R., Lempitsky, V.: Escape from cells: deep KD-networks for the recognition of 3D point cloud models. In: 2017 IEEE International Conference on Computer Vision (ICCV), pp. 863–872. IEEE (2017)
13. Li, J., Chen, B.M., Lee, G.H.: SO-Net: self-organizing network for point cloud analysis. In: Proceedings of the IEEE Conference on Computer Vision and Pattern Recognition, pp. 9397–9406 (2018)
14. Li, Y., Bu, R., Sun, M., Chen, B.: PointCNN. arXiv preprint arXiv:1801.07791 (2018)
15. Liu, Q., Sun, Y., Wang, C., Liu, T., Tao, D.: Elastic net hypergraph learning for image clustering and semi-supervised classification. IEEE Trans. Image Process. **26**(1), 452–463 (2017)
16. Maturana, D., Scherer, S.: Voxnet: A 3D convolutional neural network for real-time object recognition. In: Proceedings of the Intelligent Robots and Systems, pp. 922–928. IEEE (2015)

17. Papadakis, P., Pratikakis, I., Theoharis, T., Perantonis, S.: PANORAMA: a 3D shape descriptor based on panoramic views for unsupervised 3D object retrieval. Int. J. Comput. Vis. **89**(2–3), 177–192 (2010)
18. Pontil, M., Verri, A.: Support vector machines for 3D object recognition. IEEE Trans. Pattern Anal. Mach. Intell. **20**(6), 637–646 (1998)
19. Qi, C.R., Su, H., Mo, K., Guibas, L.J.: Pointnet: deep learning on point sets for 3D classification and segmentation. In: Proceedings of the IEEE Conference on Computer Vision and Pattern Recognition, vol. 1, p. 4 (2017)
20. Qi, C.R., Su, H., Nießner, M., Dai, A., Yan, M., Guibas, L.J.: Volumetric and multi-view cnns for object classification on 3D data. In: Proceedings of the IEEE Conference on Computer Vision and Pattern Recognition, pp. 5648–5656 (2016)
21. Qi, C.R., Yi, L., Su, H., Guibas, L.J.: Pointnet++: deep hierarchical feature learning on point sets in a metric space. In: Advances in Neural Information Processing Systems, pp. 5105–5114 (2017)
22. Riegler, G., Ulusoy, A.O., Geiger, A.: OctNET: learning deep 3D representations at high resolutions. In: Proceedings of the IEEE Conference on Computer Vision and Pattern Recognition, vol. 3 (2017)
23. Shen, Y., Feng, C., Yang, Y., Tian, D.: Mining point cloud local structures by kernel correlation and graph pooling. In: Proceedings of the IEEE Conference on Computer Vision and Pattern Recognition, vol. 4 (2018)
24. Su, H., Maji, S., Kalogerakis, E., Learned-Miller, E.: Multi-view convolutional neural networks for 3D shape recognition. In: Proceedings of the IEEE International Conference on Computer Vision, pp. 945–953 (2015)
25. Tian, Z., Hwang, T., Kuang, R.: A hypergraph-based learning algorithm for classifying gene expression and ArrayCGH data with prior knowledge. Bioinformatics **25**(21), 2831–2838 (2009)
26. Vranic, D., Saupe, D.: 3D Shape Descriptor Based on 3D Fourier Transform. EURASIP, pp. 271–274 (2001)
27. Wang, Y., Sun, Y., Liu, Z., Sarma, S.E., Bronstein, M.M., Solomon, J.M.: Dynamic graph CNN for learning on point clouds. arXiv preprint arXiv:1801.07829 (2018)
28. Wu, Z., et al.: 3D ShapeNets: a deep representation for volumetric shapes. In: Proceedings of the IEEE Conference on Computer Vision and Pattern Recognition, pp. 1912–1920 (2015)
29. Xie, J., Dai, G., Zhu, F., Shao, L., Fang, Y.: Deep nonlinear metric learning for 3-D shape retrieval. IEEE Trans. Cybern. **48**(1), 412–422 (2016)
30. Xie, S., Liu, S., Chen, Z., Tu, Z.: Attentional ShapeContextNet for point cloud recognition. In: Proceedings of the IEEE Conference on Computer Vision and Pattern Recognition, pp. 4606–4615 (2018)
31. Zhang, Z., Bai, L., Liang, Y., Hancock, E.: Joint hypergraph learning and sparse regression for feature selection. Pattern Recogn. **63**, 291–309 (2017)
32. Zhu, L., Shen, J., Jin, H., Zheng, R., Xie, L.: Content-based visual landmark search via multimodal hypergraph learning. IEEE Trans. Cybern. **45**(12), 2756–2769 (2015)

View-Dependent Streaming of Dynamic Point Cloud over Hybrid Networks

Lanyi He, Wenjie Zhu, Ke Zhang, and Yiling Xu[✉]

Cooperative MediaNet Innovation Center,
Shanghai Jiao Tong University, Shanghai, China
yl.xu@sjtu.edu.cn

Abstract. Characterized by efficient and exquisite representation of the objects or scenarios in the real world, 3D point cloud has been widely applied in large amount of emerging applications such as virtual reality/augmented reality, automatic drive, gaming technologies or robotics. Each point of the data contains 3D geometry information and corresponding photometry information like color, intensity, normal or texture, leading to massive data capacity and severely influence the transmission quality with limited network resources. However, more than a half of the points in each point cloud frame are invisible as being occluded by others from the main viewpoint. To deal with the above issues, we propose a view-dependent streaming for dynamic point cloud based on the novel hybrid networks. We project the point cloud frame into six 2D frames and generate videos with different bitrates in consideration of various user interests. Therefore, differential transmission can be achieved such that the personalized contents like the current consumed viewpoint are transmitted via interactive broadband channel, while the less-attention contents can be pushed through general digital broadcasting channel. Therefore, benefit from existing hybrid transmission systems, reliable services with efficient utilization of limited transmission resources are achieved. Experimental results have shown considerable bandwidth saving based on the proposed scheme, maintaining satisfying reconstruction performance.

Keywords: Dynamic 3D point cloud · View-dependent presentation
Hybrid transmission

1 Introduction

The world has been witnessing a great transition in the multimedia service environments due to the rapid increase of hybrid transmission over both the broadband and broadcasting. As a developing media content, 3D point cloud has attracted significant attention in both academy and industry, and has been widely applied in emerging applications such as virtual reality/augmented reality, automatic drive, gaming technologies and robotics. Aimed at vivid rendering the objects or scenarios of the real world, millions of points integrated together to represent 3D models and each point consists of geometry information and corresponding photometry attribute information like color, intensity, normal or texture, leading to massive data capacity and break the balance with the limited network bandwidth and computation resources. Meanwhile,

© Springer Nature Switzerland AG 2018
R. Hong et al. (Eds.): PCM 2018, LNCS 11164, pp. 50–58, 2018.
https://doi.org/10.1007/978-3-030-00776-8_5

the existing traditional multimedia delivery systems have revealed massive inapplicability and weakness in the case of dynamic 3D point cloud streaming and in-time applications.

Much work has been done about the research of 3D point cloud transmission system. Yadvendar present an efficient transmission algorithm, which can facilitate real time interaction with a remote scene, and they detect and only transmit changes between the point cloud frames efficiently [1]. Chenguang proposed a transmission approach of the point cloud data of the object upper body, which the key inspiration is the unnecessary information is removed by using filters and the segmentation algorithm [2]. Carlos proposed using a dynamic coding approach to adjust the compression ratio in response to the network condition [3]. The resolution of the voxelized point cloud data is decided by the current throughput of the network, and it can be adjusted in time to maintain the stability of the frame rate. For teleoperation, which is the operation of a robot from a distance using information about the remote environment obtained from cameras and sensors, Ga-Ram presented a method that provides real-time point cloud streaming for teleoperation using H.264/AVC [4], during the processing of packing, the point cloud frame is converted to a disparity image before encoded using H.264/AVC.

Apparently, an essential feature of 3D vision is that users can consume specified media content from any viewpoint which surpass the immersive experience of tradition 2D contents. Nonetheless, no matter which the main viewpoint is, there always exists quite a few points are invisible as being occluded by others closer to the user. Thus, it is improper to treat consumed points and unconsumed ones equally, in other words, wasting unnecessary network resources on those invisible points. Meanwhile, the 6 Degree of Freedom (DoF) contents of VR/AR allow users to move freely without the limitation of cables, so the unreliability of wireless channel also challenge the delivery of large data volume of 3D point cloud sequences. Therefore, adaptive and reliable transmission control scheme should be designed to achieve comprehensive utilization of limited network resources and adorable immersive experience.

In this paper, we propose a view-dependent streaming architecture for dynamic point cloud over hybrid networks. IP streaming-based hybrid networks transmission protocols developed fast and widely applicable, such as MPEG Media Transport (MMT) [5–8], has catered for the emerging convergence of the broadband and digital broadcasting. Reference to those above well-developed frameworks, Fig. 1 shows our proposed streaming architecture that personalized and common media contents are streamed by broadband and broadcasting channels respectively. Moreover, unequal error protection is used to the geometry and attribute information since that the accurate coordinates of reconstructed point cloud are more important for guaranteeing the visual quality. This paper is organized as follows. Section 2 describes the preprocessing of the dynamic 3D point cloud data. In Sect. 3, we demonstrate the proposed streaming architecture. Experimental results are shown in Sect. 4, and the conclusion is provided in Sect. 5.

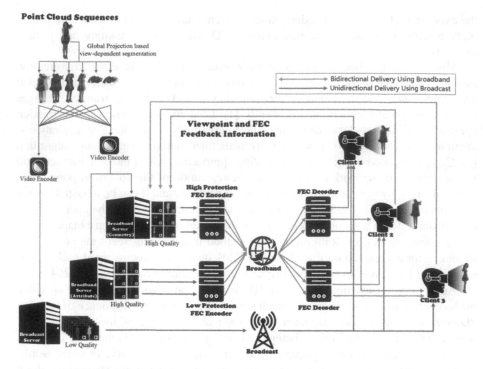

Fig. 1. Proposed view-dependent point cloud streaming architecture.

2   Point Cloud Preprocessing

To achieve view-dependent streaming, based on our previous works [9], we partition each point cloud frame into six segments and using efficient mapping algorithm to project 3D points into 2D image pixels, then the well-developed 2D compression approaches can be utilized for further compression.

In order to deal with the irregular 3D geometry of point cloud, we employ octree structure to decompose the 3D space and arrange the points on uniformed 3D grids called voxels and build a unique cubic bounding box which included all the points in each frame [10–12]. The unique resolution is determined based on such voxelized data. Next, we partition each point cloud frame into six fundamental view segments corresponding to the planes of the bounding box and create a mapping from each segment to a video frame. Color and depth based video frames are qualified to numerically preserve the attributes and geometry information of original point cloud frames. Meanwhile, the strong correlations of the color attribute associated with the nearby points in 3D space are well-preserved. Eventually, we acquire six complete views of each point cloud frame by storing color and depth video frames respectively, and the latest video coding standard H.265/MPEG-HEVC is employed to encode the output video. Consequently, for each point cloud sequences, several versions of video streams with different bitrates are stored in the server and each version includes six output videos corresponding to the fundamental views. Furthermore, any combination with different qualities, as long as all the six fundamental views included completely, is qualified to reconstruct the point cloud sequences successfully.

3 View-Dependent Based Hybrid Streaming

Our proposed streaming architecture base on hybrid networks, characterized by integrated the traditional digital broadcasting and the flexible broadband, that assumes an IP network with in-network intelligent caches close to the receiving entities, which not only actively cache the contents but also adaptively packetize and push the content to the receiving entities. Furthermore, our proposed streaming architecture jointly take the consumption features of 3D point cloud in different network conditions into consideration, in order to achieve the adaptive and reliable view-dependent streaming.

3.1 Hybrid Streaming Architecture

The diagram of our proposed streaming architecture is shown in Fig. 1. After point cloud preprocessing, multiple pairs of output video streams with different bitrates are generated and stored in the broadcast or broadband servers, and each pair including six fundamental video streams. The receiving media content of each client is made up of an entire point cloud frame with low bitrate and specific segments of the user's current viewpoint. Due to the analysis of consumption process, the entire frame with low bitrate is demanded by every node of the network terminal, and such common media content could be delivered by digital broadcasting to exploit its superiority for further increasing the transmission efficiency. Conversely, the content of current viewpoint of each user in a certain moment is probability different among the users, so that such personalized media contents, including the push-pull combination media distribution and interaction strategy, are more suitable to be delivered in the bidirectional broadband network to increase the system flexibility. Thus, the advantages of both broadcast and broadband have been utilized respectively to distribute the network bandwidth reasonably and optimize the streaming efficiency.

However, when it comes to User Datagram Protocol (UDP) based multimedia streaming system such as MMT or IPTV, or the emerging real-time applications such as holographic communication, especially in the wireless condition, the unreliability of network channel decrease the quality of experience observably. Fortunately, based on our view-dependent streaming, only the protection of consumed viewpoint is able to guarantee the users' experiences. Thus, our architecture provides an optional unequal Forward Error Correction (FEC) to maintain the transmission performance. Compared with color component, more accurate coordinates of a point cloud sequence gain better visual quality. Geometry information often plays a more significant role during point cloud rendering. So the redundancy distribution of the geometry and attribute packets also take the hierarchical priorities into consideration to jointly keep a relatively high user experience. Meanwhile, a point cloud frame generates one gray image represented coordinates and one color image represented RGB values, so the bitrate of geometry information after encoding is much smaller than that of attribute information.

Therefore, adding a small amount of coding bitrate to the geometry can bring considerable improvement of quality. Based on above analysis, we assign hierarchical transmission priorities to the geometry and attribute information of each viewpoint content. Concretely, in the limited IP network condition, we assign high enough bitrate to the geometry streams to guarantee the PSNR reaching the predefined threshold, and the remaining bandwidth is assigned to the attribute streams.

3.2 View-Dependent Dynamic Streaming

Supposed that a user is watching the front of a point cloud model via VR/AR headset, it does not accord with the common user behavior to skip watching sideward segments of the model and be transported instantaneously to the back of the model. In general, consumers are more likely to revolve round the model step by step, and the next segment that the user is going to watch may be always adjacent to the current one. Consequently, within a limited short time, it is reasonable to only deliver users' current consumed view and ignore other views, especially the opposite one. Based on the analysis above, although it is unlikely to watch the other unconsumed points within a short time, we still deliver a low quality version of the entire point cloud frame to avoid the exceptional circumstance.

We define two kinds of consumed views as shown in Fig. 2 corresponding to one or three fundamental views selected for streaming and 3D reconstruction. When user's viewpoint is one of the six projection angle exactly, it is the ideal situation shown in Fig. 2 (1), which only one high quality video is needed to reconstruct the consumed view, and other five views' reconstruction just use the low bitrate version. For the viewpoint randomly picked, for instance in Fig. 2 (2), it can be reconstructed by at most 3 fundamental views, so we utilize 3 high quality views combined with 3 low quality for streaming and reconstruction. The viewpoint information feed back to the broadband server from the clients, so the server is able to decide which views does the user need and send adaptively.

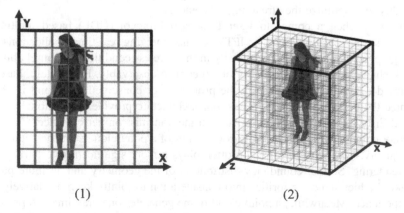

(1)   (2)

Fig. 2. Two kinds of consumed view: (1) can reconstructed by single fundamental view, and (2) need 3 fundamental views.

4 Experiments

In this section, we conduct the experiment to illustrate the performance of our proposed view-dependent streaming scheme. The datasets we employed are 360 elaborate point cloud sequences *Redandblack, Loot, Longdress* and *Soldier*, provided by 8i Labs and accepted by MPEG PCC group [15]. To evaluate the distortion during the point cloud transmission, we employ a full reference quality metric PSNR which combine the practices from video compression, and modified PSNR to survey the exceptional circumstance of switching viewpoint not in time. As for the original and the decoded point cloud V_{org}, V_{deg}, the calculation of the geometry and color PSNR is described by:

$$RMS(V_{deg}, V_{org}) = \sqrt{\frac{1}{K} \sum_{v_d \in V_{deg}} [[v_d - v_{org}]]^2} \tag{1}$$

$$PSNR = 10 log_{10} (Peak\_value^2 / RMS^2 (V_{deg}, V_{org})) \tag{2}$$

Furthermore, although there is little possibility for users to see the points of unconsumed views, we still define a modified PSNR to survey the exceptional circumstance of switching viewpoint not in time. When the consumed view is Fig. 3 (1), we take 0.1% RMS of low quality and 99.9% RMS of high quality to calculate the modified PSNR. For Fig. 3 (2), 3 high quality views are used for reconstruction so that the probability of seeing the low quality points is lower than (1). Thus, we take 0.01% RMS of low quality and 99.99% RMS of high quality for calculation. The coding performance is surveyed by bits per point (bpp), which is the ratio of the bitrate of the whole point cloud sequences and the number of points in the original model. Figure 3 shows reconstruction results and comparison with two kinds of viewpoints, the fundamental viewpoint and others selected randomly.

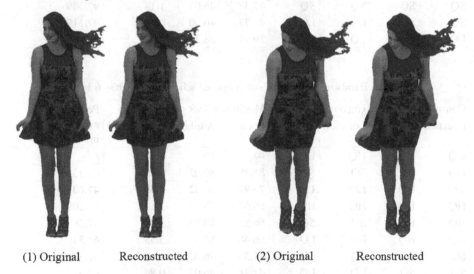

(1) Original   Reconstructed   (2) Original   Reconstructed

Fig. 3. Reconstruction results and comparison with two kinds of viewpoints: (1) one of the 6 fundamental viewpoints, and (2) viewpoints randomly picked.

We simulate the adaptive streaming of dynamic 3D point cloud over hybrid networks. The digital broadcasting takes charge of the basic layer, which contains the entire content of the point cloud frame, with low bitrate for visual quality assurance under the exceptional circumstance. As for the high quality viewpoint content and the feedback information, we utilize bidirectional broadband to transmit. Combined with the viewpoint feedback information, we provide 4 levels bitrate video streams for server decision, which are lossless coding ultra quality, lossy coding high quality, standard quality and low quality, marked as UQ, HQ, SQ and LQ.

Tables 1 and 2 shows the bandwidth saving of our proposed view-dependent streaming scheme, corresponding to two kinds of consumed viewpoints in Fig. 2. For instance, the combination of "UQ + LQ" own higher modified PSNR and lower bpp compared with full HQ. Therefore, the saving bandwidth from the unconsumed points can be spent meaningfully to improve the user experience by increasing the quality of the current consumed view.

Table 1. Bandwidth saving of our proposed scheme ("1 high + 6 low")

Consumed View		Unconsumed view		Modified PSNR		Total bpp (Geo + Att)	Percentage of average saving in bandwidth
Location	Attribute	Location	Attribute	Location	Attribute		
UQ	UQ	UQ	UQ	∞	∞	15.45	/
UQ	UQ	SQ	SQ	71.38	76.02	4.26	72.45%
UQ	**UQ**	**LQ**	**LQ**	**69.97**	**68.32**	**3.29**	**78.73%**
HQ	HQ	HQ	HQ	66.67	58.92	10.32	33.20%
HQ	HQ	SQ	SQ	65.41	58.84	3.25	78.96%
HQ	HQ	LQ	LQ	65.01	58.45	2.28	85.23%
SQ	SQ	SQ	SQ	41.38	46.02	1.18	92.38%
SQ	SQ	LQ	LQ	41.37	46.00	0.45	97.11%
LQ	LQ	LQ	LQ	39.97	38.32	0.21	98.66%

Table 2. Bandwidth saving of our proposed scheme ("3 high + 6 low")

Consumed View		Unconsumed view		Modified PSNR		Total bpp (Geo + Att)	Percentage of average saving in bandwidth
Location	Attribute	Location	Attribute	Location	Attribute		
UQ	UQ	UQ	UQ	∞	∞	15.45	/
UQ	UQ	SQ	SQ	81.38	86.02	8.52	40.73%
UQ	**UQ**	**LQ**	**LQ**	**79.97**	**78.32**	**8.08**	**47.00%**
HQ	HQ	HQ	HQ	66.67	58.92	10.32	33.20%
HQ	HQ	SQ	SQ	66.53	58.91	5.96	57.26%
HQ	HQ	LQ	LQ	66.48	58.87	5.53	63.54%
SQ	SQ	SQ	SQ	41.38	46.02	1.18	92.38%
SQ	SQ	LQ	LQ	41.38	46.02	0.84	94.52%
LQ	LQ	LQ	LQ	39.97	38.32	0.21	98.66%

5 Conclusion

In this paper, we propose a view-dependent streaming architecture for dynamic point cloud over hybrid networks, adaptive streaming can be achieved such that the feedback information and personalized contents like the current consumed viewpoint can be transmitted by flexible broadband channel, while the less-attention contents are transmitted through general digital broadcasting channel. Therefore, the advantages of both broadcast and broadband have been utilized respectively to distribute the network bandwidth reasonably and improve streaming efficiency of dynamic 3D point cloud.

In the future, we will concentrate on the dynamic compression algorithm of 3D point cloud, and the QoE based unequal error protection to the geometry and attribute information, to jointly improve the efficient point cloud streaming system.

Acknowledgment. This paper is supported in part by National Natural Science Foundation of China (61650101), Scientific Research Plan of the Science and Technology Commission of Shanghai Municipality (16511104203), in part by the 111 Program (B07022).

References

1. Champawat, Y., Kumar, S.: Online point-cloud transmission for tele-immersion. ACM SIGGRAPH International Conference on Virtual-Reality Continuum and ITS Applications in Industry ACM, pp. 79–82 (2012)
2. Yang, C., et al.: Development of a fast transmission method for 3D point cloud. In: Multimedia Tools & Applications, pp. 1–19 (2018)
3. Moreno, C., Chen, Y., Li, M.: A dynamic compression technique for streaming kinect-based Point Cloud data. In: International Conference on Computing, NETWORKING and Communications. IEEE, pp. 550–555 (2017)
4. Jang, G.R., et al.: Real-time point-cloud data transmission for teleoperation using H.264/AVC. In: IEEE International Symposium on Safety, Security, and Rescue Robotics. IEEE, pp. 1–6 (2015)
5. Information Technology—High Efficiency Coding and Media Delivery in Heterogeneous Environments—Part 1: MPEG Media Transport (MMT), document ISO/IEC 23008-1, Int. Org. Stand., Geneva, Switzerland, June 2014
6. Park, K., Lim, Y., Suh, D.Y.: Delivery of ATSC 3.0 services with MPEG media transport standard considering redistribution in MPEG-2 TS format. IEEE Trans. Broadcast. **62**(1), 338–351 (2016)
7. Lim, Y.: MMT, new alternative to MPEG-2 TS and RTP. In: Proceedings of the IEEE International Symposium on Broadband Multimedia System and Broadcasting. (BMSB), London, UK, pp. 1–5, June 2013
8. Lim, Y., et al.: MMT: an emerging MPEG standard for multimedia delivery over the Internet. IEEE Multimed. **20**(1), 80–85 (2013)
9. He, L., Zhu, W., Xu, Y.: Best-effort projection based attribute compression for 3D point cloud. In: Asia-Pacific Conference on Communications (APCC), Australia, Perth, December 2017
10. Schnabel, R., Klein, R.: Octree-based Point-Cloud Compression. Spbg **6**, 111–120 (2006)
11. Kammerl, J., et al.: Real-time compression of point cloud streams. In: IEEE International Conference on Robotics and Automation (ICRA), IEEE (2012)

12. Zhu, W., Xu, Y., et al.: Lossless point cloud geometry compression via binary tree partition and intra prediction. In: IEEE 19th International Workshop on Multimedia Signal Processing (MMSP). IEEE (2017)
13. Lim, Y., Park, K., Lee, J.Y., et al.: MMT: An emerging MPEG standard formultimedia delivery over the internet. IEEE Multimedia **20**(1), 80–85 (2013)
14. Information Technology—Dynamic Adaptive Streaming Over HTTP (DASH)—Part 1: Media Presentation Description and Segment Formats, document ISO/IEC 23009-1:2014, Int. Org. Stand., Geneva, Switzerland (2014)
15. d'Eon, E., Harrison, B., Myers, T., Chou, P.A.: Voxelized full bodies - A voxelized point cloud dataset. In: ISO/IEC JTC1/SC29 Joint WG11/WG1 (MPEG/JPEG) input document WG11M40059/WG1M74006, Geneva, January 2017

Video Captioning Based on the Spatial-Temporal Saliency Tracing

Yuanen Zhou[✉], Zhenzhen Hu, Xueliang Liu, and Meng Wang

Hefei University of Technology, Hefei, China
y.e.zhou.hb@gmail.com, huzhen.ice@gmail.com,
Liuxueliang1982@gmail.com, eric.mengwang@gmail.com

Abstract. Video captioning is a crucial task for video understanding and has attracted much attention recently. Regions-of-Interest (ROI) of video always contains the most interesting information for the audience. Different from the ROI of images, the ROI of videos has the property of temporally-continuity (e.g. a moving object, or an action in video clips), which is the focus of people's attention. Inspired by this insight we propose an approach to automatically trace the Spatial-Temporal Saliency content for video captioning by catching the temporal structure of ROI candidates. To this aim, we employ a set of modules named tracing LSTMs, each of which traces a single ROI candidate of feature maps across the entire video. The temporal structure of global features and ROI features are combined to obtain a rough understanding of video content and information of ROI, which is set as the initial states of the decoder to generate captions. We verify the effectiveness of our method on the public benchmark: the Microsoft Video Description Corpus (MSVD). The experimental results demonstrate that catching temporal ROI information by tracing LSTMs enhances the representation of input videos and achieves the state-of-the-art results.

Keywords: Video captioning · Regions-of-Interest (ROI)
Spatial-Temporal saliency · Tracing LSTM

1 Introduction

Visual captioning is to automatically generate natural language descriptions for visual information. This task, including image captioning and video captioning, is a crucial challenge in computer vision and has attracted much attention recently. It may benefit many practical applications, such as impaired people auxiliary, human-robot interaction, image retrieval and video retrieval.

There is complicated information contained in the visual content, such as scenes, motions, and objects, as well as relationships between them and their attributes. The earliest approaches [24–26, 30] attacked the task by firstly predicting semantic concepts or words (e.g., objects, subjects, and verbs) by the different classifier. Then a rigid sentence template is employed to generate a natural language description. This kind of method is intuitive. However, it is not flexible enough due to the limitation of the sentence template. Inspired by the recent advance of neural network in image

© Springer Nature Switzerland AG 2018
R. Hong et al. (Eds.): PCM 2018, LNCS 11164, pp. 59–70, 2018.
https://doi.org/10.1007/978-3-030-00776-8_6

recognition and machine translation, research [1, 3, 28, 29] has significantly improved the quality of caption generation based on the encoder-decoder framework by employing different neural networks as encoder to obtain visual representation as well as a recurrent neural network, especially Long Short-Term Memory (LSTM) [27], as decoder for variable-length captions generation.

A video, as consecutive visual content, contains richer spatial-temporal information. It makes video captioning to be a more challenge subtask of visual captioning than image captioning. Previous works [3, 4, 8, 22] have tried to exploit the temporal structure of video inputs, in order to enhance the visual representation of videos. Consecutive video inputs are treated in the same way and their different weights are not considered for understanding the video content. However, it would be better to automatically select significant or discriminative segments and regions for describing the video content. In [7], Yao et al. proposed a temporal attention mechanism to adaptively select the most relevant segments while generating the sentence. In [1], Yang et al. demonstrated the effectiveness of utilizing temporal cues of regions-of-interest.

However, the method of catching temporal regions-of-interest in [1] still needs to be improved. For example, a video about *a person dives into a pool*, as shown in Fig. 1. The global feature of each frame extracted by a CNN model pre-trained on ImageNet may be the representation of a swimming pool, because the swimming pool occupies a major area of each frame. In this case, it is difficult to select the diving process, which is the temporal regions-of-interest, by employing an attention model guided by the global feature, as done in [1]. To mitigate this problem, we utilize a module named tracing LSTM, as shown at the top of Fig. 1, which has also been used to detect concept words [2] in video clips. The intuition is that regions-of-interest are temporally-continuity in short video clips so that they can be traced over time from the local feature maps. We use a total of λ tracing LSTMs to capture out λ traces (temporal regions-of-interest), where λ is the number of spatial regions of the local feature maps. At the end of tracing LSTMs, we obtain λ temporal regions-of-interest candidates. For example, the diving process in the video, as shown in Fig. 1, should be one of the 16 candidates. Then we simply use a max-pooling operation over the λ temporal regions-of-interest candidates to acquire final temporal region-of-interest representation in this paper. We also use another LSTM to capture the temporal structure of global features. Then the temporal structure of global features and regions-of-interest features are combined to obtain a rough understanding of video content and particular information of regions-of-interest, which is set as the initial states of the decoder to generate captions. The decoder LSTM is also equipped with attention model as previous works [1, 6] to generate captions word by word. The overall framework of our model is illustrated in Fig. 1.

It is worthwhile to highlight the following aspects of the proposed model:

- We take the temporally-continuity property of regions-of-interest into consideration when catching the temporal structure of regions-of-interest for video captioning compared to most related work [1].
- To this aim, we reasonably revise the tracing LSTMs used to detect concept words [2] to propose the final temporal regions-of-interest features.
- Experiments on the benchmark dataset demonstrated that our method achieves comparable or even better results than the state-of-the-art methods for video captioning.

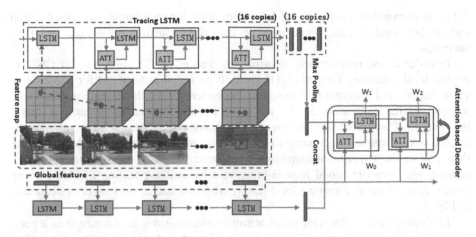

Fig. 1. The overall framework of our model. The concrete architecture of the attention module (ATT) is different, which will be formulated in the corresponding section.

2 Related Work

Inspired by the recent advance of deep neural network in image recognition and machine translation, research has significantly improved the quality of caption generation based on the encoder-decoder framework by employing different neural networks as encoder to obtain visual representation as well as a recurrent neural network, especially Long Short-Term Memory (LSTM), as decoder for variable-length captions generation. Venugopalan et al. [22] simply performed mean pooling over CNN features of frames and fed video representation into a two-layer LSTM for generating captions. Donahue et al. [21] explored video content representation by using CRFs to get semantic predictions of the object, subjects, and verbs. Then an LSTM layer was applied to generate sentences. Pan et al. [20] introduced visual-semantic embedding, which enforced the relationship between the entire sentence semantics and the visual content.

However, the above methods didn't take temporal dependencies between video frames into consideration and just regarded a video as a collection of images. Although Yao et al. [7] equipped decoder with a temporal attention mechanism to automatically select the most relevant temporal segments while generating the sentence, they merely used a weighted sum of segment-level CNN features. Therefore, later methods explored various encoders with the aim of incorporating better temporal structure into video representation. S2VT [8] firstly adopted an end-to-end sequence-to-sequence model, which was already applied to machine translation [17], to address both video encoding and sentence decoding. HRNE [4] leveraged a hierarchical recurrent neural encoder to capture video temporal structure in a long range and exploited temporal transitions with multiple granularities. In order to explore temporal discontinuity of video inputs, Baraldi et al. [3] proposed a novel LSTM cell which can identify discontinuity points between frames or segments and modify the temporal connections of the encoding layer accordingly. It incorporated a fine hierarchical structure of video.

All these approaches mostly take global information for each frame into consideration and neglect local details, which may be helpful for identifying objects and their attributes.

In order to incorporate local information, Yao et al. [7] utilized a 3D CNN to exploit local structure. Yang et al. [1] focused on regions-of-interest for consecutive video inputs and incorporated temporal dependencies of these regions. Although inspired by the motivation of [1], we adopt a different method to catch temporal information of interested regions. Specifically, Yang et al. [1] extracted the regions-of-interest in each frame by attending to local feature maps guided by the global feature, then caught the temporal information of interested regions by a recurrent model. Instead, we extract temporal regions-of-interest by utilizing tracing LSTMs starting from regions of the first frame's local feature map and tracing over time, as illustrated in Fig. 1.

Equipping the decoder with visual attention mechanism is also regarded as a good choice for visual captioning. Xu et al. [18] introduced two variants of attention mechanism for image captioning to learn a latent alignment while generating corresponding words. You et al. [19] proposed a model to selectively attend to semantic concept proposals and fuse them into hidden states and outputs of recurrent neural networks. Yu et al. [2] proposed the first semantic attention model for the video-to-language task. Further, to decide "when" and "what" to attend, adaptive attention has been proposed in [33, 34]. Our decoder is also equipped with attention mechanism to attend to frame-level global features as the most related work [1].

3   Our Method

Our method also targets to exploit temporal structures of both global and regions-of-interest features, which would obtain a rough understanding of video content and particular information of regions-of-interest respectively. The main difference between our work and the most related work [1] lies in the method of catching the temporal regions-of-interest. In this section, we first make an overall introduction to our approach. Then we elaborate on the tracing LSTMs module which used to propose temporal regions-of-interest candidates.

3.1   The Overall Framework

Given an input video with N frames and corresponding description sentences, our approach also adopts an encoder-decoder framework to model their mapping, which has been successfully applied to machine translation [17] and visual captioning [8, 16].

The encoding part of our model is composed of two modules: a plain LSTM exploited to catch temporal structure of global feature and tracing LSTMs used to catch temporal information of regions-of-interest. At the end of the encoding stage, we concatenate the last hidden states and memory cells of the two streams as an initialization of decoder. As shown in Fig. 1, we respectively extract global feature $x_t \in R^{1024}$ and local feature map $v_t \in R^{4*4*512}$ from the fully-connected layer and convolutional layer at time t by a pre-trained CNN. Hence we obtain a sequence of global features

$(x_1, x_2, \ldots, x_N)$ and local feature maps $(v_1, v_2, \ldots, v_N)$. We formulate the encoding process as the following equations:

$$h_t^G = \text{LSTM}\left(h_{t-1}^G, x_t\right), t = 1, \ldots, N \tag{1}$$

$$h_t^{Ri} = tracing\text{LSTM}\left(h_{t-1}^{Ri}, v_t\right), t = 1, \ldots, N \text{ and } i = 1, \ldots, 16 \tag{2}$$

$$h_N^R = MaxPooling\left(h_N^{R1}, h_N^{R2}, \ldots, h_N^{R16}\right) \tag{3}$$

$$h_0^D = Concat\left(h_N^R, h_N^G\right) \tag{4}$$

Where h_t^G and h_N^G are the hidden state of plain LSTM at time t and N. h_t^{Ri} is the hidden state of tracing LSTM at time t for the i-th trace. h_N^R denotes the final representation of temporal region-of-interest. h_0^D denotes the initial hidden state of decoder LSTM. The details of tracing LSTMs will be elaborated in Sect. 3.2. For a more detailed description of LSTM, please refer to [27].

For the decoder in our model, we equip a plain LSTM with attention mechanism and a word embedding layer as a language model according to related methods [1, 6, 7]. A sentence consisting of T words is represented as $(y_1, y_2, \ldots, y_T)$, encoded with one-hot vectors (1-of-D encoding, where D is the size of the vocabulary). As mentioned above, we initialize hidden state and memory cell of the decoder LSTM by two-stream encoder outputs at the last encoding step. Conditioned on the previous $t - 1$ words of the caption and a sequence of global features x, the decoder is trained to predict next word y_t. We model the conditional probability of next word with a softmax layer. The decoder is formulated by:

$$z_t = f_{att}\left(x, h_{t-1}^D\right) \tag{5}$$

$$h_t^D = \text{LSTM}\left(z_t, h_{t-1}^D, W_e y_{t-1}\right) \tag{6}$$

$$p(y_t | y_{t-1}, y_{t-2}, \ldots, y_0, x) \propto \left(y_t^T W_y h_t^D\right) \tag{7}$$

Where W_y denotes parameters of a linear transformation layer which maps the outputs of LSTM to word space, W_e denotes the word embedding matrices, z_t denotes the context vector, and f_{att} is the attention module in decoder. The internal computing process of this attention module is formulated by:

$$s_n^t = w^T \tanh\left(W_x x_n + W_h h_{t-1}^D + b_s\right) \tag{8}$$

$$\beta_n^t = softmax\left(s_n^t\right) \tag{9}$$

$$z_t = \sum_{n=1}^N \beta_n^t x_n \tag{10}$$

Where β_n^t is the attention weight at time t describing the relevance of the n-th feature in the input global feature sequences and s_n^t is the un-normalized relevance score.

We define the objective function of generating words as the negative logarithm of the likelihood:

$$\text{Loss} = -\sum_{t=1}^{T} logp(y_t|y_{t-1}, y_{t-2}, \ldots, y_0, \text{x}; \theta) \tag{11}$$

Where θ are parameters of the video captioning model and T is the length of the sentence. We minimize the objective function over all the training set.

3.2   Tracing LSTM

In [1], Yang et al. proposed a method to catch regions-of-interest and utilize their temporal cues. However, it is difficult to catch the region-of-interest by attending to local feature maps guided by the global feature when regions-of-interest occupy the relatively small area. So we propose to employ a set of tracing LSTMs to catch temporal regions-of-interest candidates and produce the final temporal regions-of-interest by the max-pooling operation. Tracing LSTMs has been used to detect concept words in [2]. We replace the last fully connection layer, which used to predict the concept confidence vector, with a max-pooling layer. Tracing LSTMs is composed of a plain LSTM and a specific spatial attention mechanism. Specifically, we use a total of λ tracing LSTMs to capture out λ traces (or temporal regions-of-interest candidates), where λ is the number of spatial regions of local feature maps (i.e. $\lambda = 4*4 = 16$ for $v \epsilon R^{4*4*512}$). By max-pooling these candidates, we obtain a representative temporal regions-of-interest. For each trace i, we maintain spatial attention weights $a_t^i \epsilon R^{4*4}$, indicating where to attend on (4*4) spatial grid locations of v_t, through video frames $t = 1, 2, \ldots, N$. The initial attention weight a_0^i at $t = 0$ is initialized with a one-hot matrix, for each of grid locations. We compute the hidden states $h_t^{Ri} \epsilon R^{512}$ of the tracing LSTM through $t = 1, 2, \ldots, N$ by:

$$c_t^i = a_t^i \otimes v_t \tag{12}$$

$$h_t^{Ri} = LSTM\left(c_t^i, h_{t-1}^{Ri}\right) \tag{13}$$

Where $A \otimes B = \sum_{j,k} A(j, k) * B(j, k, :)$. The input to LSTMs is the context vector $c_t^i \in R^{512}$, which is obtained by applying spatial attention weights a_t^i to the local feature maps v_t. It is worthwhile to note that the parameters of all the tracing LSTMs are shared. The attention weight vector a_t^i at time step t is updated as follows:

$$e_t^i(j, k) = v_t(j, k) \odot h_{t-1}^{Ri} \tag{14}$$

$$a_t^i = softmax\left(Conv\left(e_t^i\right)\right) \tag{15}$$

Where $\odot$ is element-wise product, and Conv(.) denotes two convolution operations before the softmax layer. For clarity, the computation process of spatial attention is illustrated in Fig. 2. The spatial attention weight a_t^i measure how each spatial grid location of local feature maps at time t is related to the trace (temporal regions-of-interest) being tracked through tracing LSTMs.

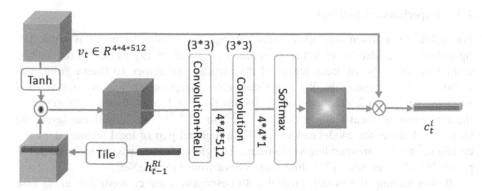

Fig. 2. Computation process of spatial attention (ATT) module in tracing LSTMs.

4 Experiments

Our proposed method is validated on Microsoft Video Description Corpus (MSVD) [31] which is widely used by most video captioning methods.

4.1 Dataset

The Microsoft Video Description Corpus is a popular benchmark for evaluating video captioning methods. It contains 1,970 video clips with about 80,000 English descriptions labeled by the Amazon Mechanical Turkers. The original dataset is composed of multi-lingual descriptions, but our experiments only consider English descriptions like many previous works [7, 8]. For fair comparisons with the state-of-the-art video captioning methods, we use the common split which divides the dataset into training, validation, and testing with 1,200 clips, 100 clips, and 670 clips respectively. This dataset mainly contains short video clips with a single action.

4.2 Evaluation Metrics

We also employ three popular metrics to evaluate our captioning results for the convenience of comparison: BLEU [15], CIDEr [14], and METEOR [13]. The BLEU measures the precision of n-grams between generated sentence and ground-truth descriptions. The METEOR measures the word correlation between candidate and reference sentences by generating an alignment based on exact token matching. As same as many previous captioning methods [4, 7, 8], we utilize the Microsoft COCO evaluation server [12] to compute all the scores in this paper for fair comparison. We also use a beam search with size 5 during sentence generation as done in [1]. We conduct experiments using the Torch Framework [11] based on the publicly available code[1]. Generally, the higher scores indicate the higher quality of the generated caption.

---

[1] https://github.com/ziweiyang/dualMemoryModel.

4.3   Experimental Settings

For a fair comparison, our experiment settings are similar to previous work [1]. Specifically, we also employ the pre-trained GoogLeNet [9] as our initial feature extractors. To unify the input length of the encoder, we gather 10 frames from each video at different intervals. We use GoogLeNet inception 5b feature as raw local features, then add a 2×2 max-pooling layer and a 1×1 convolution layer to reduce the dimension of local features from $R^{7*7*1024}$ to $R^{4*4*512}$. The additional layers are trained with the entire model end-to-end. Each regional part of local features depicts a certain region in corresponding video frame. The global feature we use is the output of pool5/7*7_s1 layer with 1024-dimension vector from GoogLeNet.

Before training the model, ground truth descriptions are converted to lower case and tokenized after having removed punctuation characters. The collection of word tokens composes the vocabulary of the corresponding dataset. The vocabulary of the MSVD dataset contains 12,593 words. During the training phase, a begin-of-sentence $<$ BOS $>$ tag is added at the beginning of each sentence and end-of-sentence $<$ EOS $>$ tag is added at its end. The size of hidden units of two LSTMs in the encoder is set to 512. The hidden size of decoder LSTM is 1024 and the word embedding size is set to 512. The attention size in the decoder is set to 100 as [1]. We also adopt the RMSPROP algorithm to update parameters, with learning rate $2 * 10^{-4}$ and other parameters using default values. The training batch size is set to 64. In order to alleviate the overfitting problem when training model, the strategy of dropout [10] is applied on the output of decoder LSTM with the rate of 0.5. The range of gradients is clipped to $[-5, 5]$ to prevent gradient explosion.

4.4   Experiment Results on MSVD

The Effect of Number of Traces
The λ is the number of spatial regions of the local feature map. We use a total of λ tracing LSTMs to capture out λ traces (temporal regions-of-interest candidates). On the one hand, if the value of λ is too small, it's difficult to capture temporal regions-of-interest in small size. On the other hand, if the value of λ is too large, the overhead of computation is too large. In this subsection, we study the performance variance with different λ values. The results are shown in Table 1. As same as previous work [1], we also consider METEOR as the main metric for model selection because METEOR shows good performance according to the consistency with human judgment. In the following experiments, we set $\lambda = 16$.

Comparison with the State-of-the-Art Methods
To validate the effectiveness of our proposed algorithm, we compare it with several state-of-the-art methods on video captioning.

The performance of baselines and our method on the MSVD dataset are listed in Table 2. Inspired by the motivation of utilizing temporal information of regions-of-interest for video captioning [1], so we are mainly interested to compare our method with the raw method of catching the temporal regions-of-interest in [1]. We choose DMRM + in5b(g) + DA, which is a model variant in [1], for comparison. It's worth to

Table 1. The effect of λ on the MSVD dataset.

λ	BLEU-4	METEOR	CIDEr
1	50.8	33.4	**75.5**
4	48.5	33.3	72.7
9	49.6	33.4	71.3
16	**51.0**	**33.9**	74.4

mention that in [1], the authors further employ the ground-truth sentences of training videos as semantic supervision [32] to improve their model's performance, and their final model performance is better than DMRM + in5b(g) + DA. Even so, the performance of our model is still comparative with theirs without using a specific Dual Memory Recurrent Model (DMRM) and additional supervision.

Table 2 shows that our method outperforms DMRM + in5b(g) + DA in all three metrics, which proves the effectiveness of our method for catching temporal regions-of-interest. By comparing our method and DMRM + in5b(g) + DA with the others, it demonstrates that video captioning can do benefit from catching temporal regions-of-interest.

Table 2. Experiment results on the MSVD dataset compared to the state-of-the-art methods. (–) indicates an unknown metric. (G) indicates the authors tried some feature extractors, and we choose the GoogLeNet version for fair comparison.

Methods	BLEU-4	METEOR	CIDEr
S2VT [8]	–	29.8	–
Temporal Attention [7]	41.9	29.6	51.7
aLSYMs(G) [6]	45.3	32.3	68.7
p-RNN [5]	49.9	32.6	65.8
HRNE with attention(G) [4]	43.8	33.1	–
Boundary-aware encoder [3]	42.5	32.4	63.5
DMRM + in5b(g) + DA [1]	50.0	33.2	73.2
Ours	**51.0**	**33.9**	**74.4**

Video Captioning Examples

To illustrate the performance of our caption results, we present some representative examples in Fig. 3, where contains sentences obtained from DMRM + in5b(g) + DA [1], our method and ground truth. It should be noted that the result of DMRM + in5b(g) + DA is obtained by running the released source code. The results demonstrate that catching temporal regions-of-interest via our method makes a contribution to obtaining more discriminative regional features to identify objects. The global features of each frame of the four video clips illustrated in Fig. 3 are dominated by another thing instead of regions-of-interest. It is to some extent difficult to capture temporal regions-of-interest by attending to local feature maps guided by global features, as done in [1]. By enumerating possible regions-of-interest at the beginning of video clips firstly, and then tracing this regions-of-interest over time, our method is able to enhance the representation of input video clips. For example, our method improves "man" to a more accurate word "dog" at the upper-left of Fig. 3. Similarly, our method corrects "UNK" to "deer", which is closer to the ground truth word "jackal" at the bottom-left of Fig. 3. To our surprise, our method corrects "monkey" to "man" and adds an object "ball" at the upper-right of Fig. 3. By finding details, our method produces a more meaningful verb "playing basketball" instead of "running" at the bottom-right of Fig. 3. These representative examples also give an intuitive motivation for our proposed method.

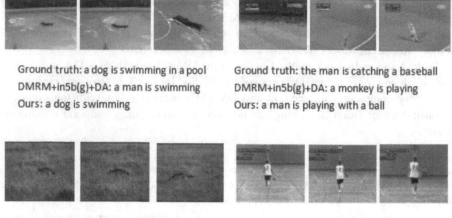

Ground truth: a dog is swimming in a pool
DMRM+in5b(g)+DA: a man is swimming
Ours: a dog is swimming

Ground truth: the man is catching a baseball
DMRM+in5b(g)+DA: a monkey is playing
Ours: a man is playing with a ball

Ground truth: a jackal is running over grass
DMRM+in5b(g)+DA: a UNK is running
Ours: a deer is running

Ground truth: a basketball player dribbling the ball
DMRM+in5b(g)+DA: a man is running
Ours: the boys are playing basketball

Fig. 3. Captions generated by DMRM + in5b(g) + DA, our method, and ground truth on MSVD test set.

5   Conclusions

In this paper, we propose to use tracing LSTMs to automatically catch temporal regions-of-interest for video captioning. We verify the effectiveness of our method for video captioning on public benchmark: the Microsoft Video Description Corpus (MSVD). The experimental results demonstrate that catching temporal regions-of-interest information by tracing LSTMs really enhances the representation of input videos and our approach obtains comparable or better results than state-of-the-art approaches on popular evaluation metrics.

However, our proposed method also has some issues that need to be solved. Firstly, how to pick out a meaningful temporal regions-of-interest from candidates instead of a simple max-pooling operation. Secondly, our proposed model would obtain confused video representation when encountering with long video clips with more than a single action, and how to extend our method to these long videos with a hierarchical structure is also needed to be explored. We leave these remainder problems as future work.

Acknowledgment. We would like to thank Ziwei Yang, who is one of the authors of [1], for providing us with the source code and preprocessed dataset.

References

1. Yang, Z., Han, Y., Wang, Z.: Catching the temporal regions-of-interest for video captioning. In: Proceedings of the 2017 ACM on Multimedia Conference. ACM (2017)
2. Yu, Y., et al.: End-to-end concept word detection for video captioning, retrieval, and question answering. In: Proceedings of the IEEE Conference on Computer Vision and Pattern Recognition (2017)
3. Baraldi, L., Grana, C., Cucchiara, R.: Hierarchical boundary-aware neural encoder for video captioning. In: CVPR (2017)
4. Pan, P., et al.: Hierarchical recurrent neural encoder for video representation with application to captioning. In: Proceedings of the IEEE Conference on Computer Vision and Pattern Recognition (2016)
5. Yu, H., et al.: Video paragraph captioning using hierarchical recurrent neural networks. In: Proceedings of the IEEE Conference on Computer Vision and Pattern Recognition (2016)
6. Gao, L., et al.: Video captioning with attention-based lstm and semantic consistency. IEEE Trans. Multimedia **19**(9), 2045–2055 (2017)
7. Yao, L., et al.: Describing videos by exploiting temporal structure. In: Proceedings of the IEEE International Conference on Computer Vision (2015)
8. Venugopalan, S., et al.: Sequence to sequence-video to text. In: Proceedings of the IEEE International Conference on Computer Vision (2015)
9. Szegedy, C., et al.: Going deeper with convolutions. In: Proceedings of the IEEE Conference on Computer Vision and Pattern Recognition (2015)
10. Srivastava, N., et al.: Dropout: a simple way to prevent neural networks from overfitting. J. Mach. Learn. Res. **15**(1), 1929–1958 (2014)
11. Collobert, R., Kavukcuoglu, K., Farabet, C.: Torch7: a matlab-like environment for machine learning. BigLearn, NIPS Workshop. No. EPFL-CONF-192376 (2011)
12. Chen, X., et al.: Microsoft COCO captions: data collection and evaluation server (2015). arXiv preprint arXiv:1504.00325
13. Denkowski, M., Lavie, A.: Meteor universal: language specific translation evaluation for any target language. In: Proceedings of the Ninth Workshop on Statistical Machine Translation (2014)
14. Vedantam, R., Lawrence Zitnick, C., Parikh, D.: Cider: consensus-based image description evaluation. In: Proceedings of the IEEE Conference on Computer Vision and Pattern Recognition (2015)
15. Papineni, K., et al.: BLEU: a method for automatic evaluation of machine translation. In: Proceedings of the 40th Annual Meeting on Association for Computational Linguistics. Association for Computational Linguistics (2002)
16. Vinyals, O., et al.: Show and tell: a neural image caption generator. In: Proceedings of the IEEE Conference on Computer Vision and Pattern Recognition (2015)
17. Sutskever, I., Vinyals, O., Le, Q.V.: Sequence to sequence learning with neural networks. In: Advances in Neural Information Processing Systems (2014)
18. Xu, K., et al.: Show, attend and tell: neural image caption generation with visual attention. In: International Conference on Machine Learning (2015)
19. You, Q., et al.: Image captioning with semantic attention. In: Proceedings of the IEEE Conference on Computer Vision and Pattern Recognition (2016)

20. Pan, Y., et al.: Jointly modeling embedding and translation to bridge video and language. In: Proceedings of the IEEE Conference on Computer Vision and Pattern Recognition (2016)
21. Donahue, J., et al.: Long-term recurrent convolutional networks for visual recognition and description. In: Proceedings of the IEEE Conference on Computer Vision and Pattern Recognition (2015)
22. Venugopalan, S., Xu, H., Donahue, J., Rohrbach, M., Mooney, R., Saenko, K.: Translating videos to natural language using deep recurrent neural networks. In: NAACL-HLT (2015)
23. Bahdanau, D., Cho, K., Bengio, Y.: Neural machine translation by jointly learning to align and translate. In: ICLR (2014)
24. Guadarrama, S., et al.: Youtube2text: recognizing and describing arbitrary activities using semantic hierarchies and zero-shot recognition. In: Proceedings of the IEEE International Conference on Computer Vision (2013)
25. Rohrbach, M., et al.: Translating video content to natural language descriptions. In: Proceedings of the IEEE International Conference on Computer Vision (2013)
26. Kulkarni, G., et al.: Baby talk: understanding and generating image descriptions. In: Proceedings of the 24th CVPR (2011)
27. Hochreiter, S., Schmidhuber, J.: Long short-term memory. Neural Comput. 9(8), 1735–1780 (1997)
28. Dong, J., et al.: Early embedding and late reranking for video captioning. In: Proceedings of the 2016 ACM on Multimedia Conference. ACM (2016)
29. Chen, L., et al.: SCA-CNN: spatial and channel-wise attention in convolutional networks for image captioning. In: CVPR (2017)
30. Fang, H., et al.: From captions to visual concepts and back. In: Proceedings of the IEEE Conference on Computer Vision and Pattern Recognition (2015)
31. Chen, D.L., Dolan, W.B.: Collecting highly parallel data for paraphrase evaluation. In: Proceedings of the 49th Annual Meeting of the Association for Computational Linguistics: Human Language Technologies, vol. 1, pp. 190–200. Association for Computational Linguistics (2011)
32. Yang, Z., et al. Review networks for caption generation. In: Advances in Neural Information Processing Systems (2016)
33. Lu, J., et al.: Knowing when to look: adaptive attention via a visual sentinel for image captioning. In: Proceedings of the IEEE Conference on Computer Vision and Pattern Recognition (CVPR), vol. 6 (2017)
34. Bin, Y., et al. Adaptively attending to visual attributes and linguistic knowledge for captioning. In: Proceedings of the 2017 ACM on Multimedia Conference. ACM (2017)

CS-DeCNN: Deconvolutional Neural Network for Reconstructing Images from Compressively Sensed Measurements

Wentao Wan, Guohui Li, and Peng Pan$^{(\boxtimes)}$

Huazhong University of Science and Technology, Wuhan 430074, Hubei, People's Republic of China
rode_wayne@163.com, {guohuili,panpeng}@hust.edu.cn

Abstract. One important research point of compressive sensing (CS) is to restore a high-dimensional signal as completely as possible from its compressed form, which has much lower dimensionality than the original. Several methods have been employed to this end, including traditional iterative methods as well as recurrent approaches based on deep learning. This paper proposes a novel architecture of a deconvolutional neural network (CS-DeCNN) to address the signal reconstruction problem in CS. Compared with state-of-the-art CS methods, our proposed method not only has the advantage of high running speed; but also achieves higher recovery precision. Besides, CS-DeCNN perform better with less space occupancy compared to other deep learning CS methods. All these advantages make CS-DeCNN more suitable for embedded systems. We demonstrate the efficacy of our method by carefully performing comparative experiments.

Keywords: Compressive sensing · Deconvolutional neural network Deep learning · Image reconstruction

1 Introduction

Compressive sensing (CS) is a technique used to highly compress and reconstruct a signal. CS can be formally expressed as follows: $\mathbf{y} = \Phi\mathbf{x}$, where $\mathbf{x} \in \mathbf{R}^N$ represents the original signal, $\mathbf{y} \in \mathbf{R}^N$ ($M \ll N$) represents the compressed signal, and Φ is the compressed matrix with an $M \times N$ basis. This is an "ill-posed" problem and generally has no unique solution. However, [1,5] showed that, if $\mathbf{x}$ is a sparse signal and Φ satisfies the RIP (restricted isometry property) condition, then a solution can be found by solving the following optimization problem:

$$\mathbf{x} = \arg\min_{\mathbf{x}} \|\mathbf{x}\|_0 \quad \text{s.t.} \quad \|\mathbf{y} - \Phi\mathbf{x}\|_2 \leq \varepsilon \tag{1}$$

where ε is a sufficiently small real number and $\|\mathbf{x}\|_0$ denotes the ℓ_0-norm of the vector $\mathbf{x}$. Usually $\mathbf{x}$ is not a sparse signal but a structured signal, which can be

© Springer Nature Switzerland AG 2018
R. Hong et al. (Eds.): PCM 2018, LNCS 11164, pp. 71–80, 2018.
https://doi.org/10.1007/978-3-030-00776-8_7

expressed as $\mathbf{x} = \Psi\mathbf{s}$, where Ψ is an $N \times N$ basis transformation matrix, and $\mathbf{s}$ is a k-sparse signal (The number of non-zero elements of $\mathbf{s}$ is not more than k). Then, the problem can be translated as follows:

$$\mathbf{S} = \arg\min_{\mathbf{s}} \|\mathbf{s}\|_0 \quad \text{s.t.} \quad \|\mathbf{y} - \Phi\Psi\mathbf{s}\|_2 \leq \varepsilon \tag{2}$$

This is an NP-hard problem. Terence Tao et al. proposed replace the ℓ_0-norm with the ℓ_1-norm [4] and transform the problem into a linear programming problem, which has mature solutions.

Many CS methods exist, including linear programming [4], the iterative greedy algorithms [2,3,8,20], and methods based on special models (e.g., Hidden Markov Tree Model [10] or Gaussian Scale Mixtures Model [12]). These methods are effective but still have two significant disadvantages: they both require a long runtime, and their results quickly degrade as the measurement rate (MR) declines, which causes them perform poorly under low measurement rates.

Since 2015, several deep learning methods have been developed in an attempt to overcome these challenges. The deep learning CS methods are usually much faster and produce more elegant results under low measurement rates. However, for these methods, **two new challenges must be addressed**. The first challenge is that many deep learning CS methods still cannot achieve higher reconstructed precision when compared with state-of-the-art non-deep learning CS methods under many measurement rates. The second challenge is well known: deep neural networks usually occupy a large amount of space, which limits them to space-constrained systems (such as embedded systems). Many scenarios of CS are embedded systems. Therefore, more lightweight deep learning models must be developed.

In this paper, we propose a novel class of multilayer deconvolutional neural networks (DeCNN) named CS-DeCNN to overcome these challenges. CS-DeCNN has the advantages of faster runtime compared to traditional non-deep learning methods and better performance under low measurement rates. Moreover, because DeCNN are used, CS-DeCNN achieves significantly improved reconstruction precision when compared with state-of-the-art deep learning CS methods (**addressing the first challenge**) and reduces the number of redundant connections. Therefore, our model becomes more lightweight (**addressing the second challenge**). We verify the above statements by carefully performing contrast experiments on a benchmark data set.

Innovations: We don't invent the DeCNN. The novelty of our paper comes from: **a)** This is the first time DeCNN be used in CS reconstructed process. **b)** Previous work, no matter SDA [19], DeepInverse [18] or ReconNet [14], they use one or several fully connected layer(s) to raise the dimension of the compressed signal. Most of the total parameters of these networks come from the fully connected layer(s). Using multi-layer deconvolutional layers to raise the dimension, CS-DeCNN reduces most of the parameters, so the model can be much more lighter and fit the application scenes of CS better. **c)** After abandoning many connections, CS-DeCNN achieves higher precision of reconstruction, which shows that previous models do encounter overfitting caused by redundant connections. In general, we get better CS reconstructed results in a lighter model.

2 Related Work

Ali Mousavi et al., who were the first to study compressed sensing via deep learning [19], used stacked denoising autoencoder (SDA) to reconstruct compressed signals. However, the SDA uses only the fully connected layers, which leads to overfitting and disaster of dimensionality when processing a large image. Their work also assumes that the probability distribution of each pixel in an image is conditional independence, which ignores the correlation between image pixels. A subsequent paper [14] proposed a method named ReconNet to use multi-layer convolutional neural networks (CNNs) followed a fully connected layer to restore signals block by block. By using CNNs, it performs better than the SDA method.

Another related work is DeepInverse [18]. It consists of a matrix multiplication operation followed by several convolutional layers. DeepInverse is not a blocky recovery method, therefore, it can recover an entire image all at once. The tradeoff for the convenience is a not very high recovery precision.

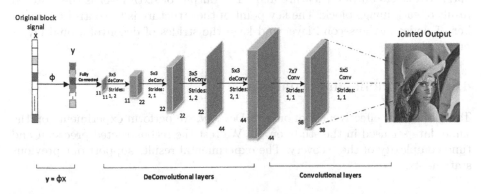

Fig. 1. Structure of CS-DeCNN at MR = 0.1. The size of recovered blocks outputed by CS-DeCNN is 34 * 34. Then these blocks will be stitched in place and fed into a denoiser to produce a full recovered image.

3 Structure of CS-DeCNN

In this section, we introduce the structure of CS-DeCNN which is a blocky recovery method. The method divides a picture into blocks of the same size, compresses each block and recovers them using CS-DeCNN. Then these recovered blocks will be stitched in place and fed into a denoiser to produce the final large reconstructed image. As Fig. 1 shows, $\mathbf{y}$ is used as the input ($\mathbf{y} = \Phi\mathbf{x}$, where

Φ is the compressed matrix with size $\mathbf{M} * \mathbf{N}$ and $\mathbf{x}$ is the vectorized original block signal). Then, the input is fully connected to the second layer. The number of neuron nodes in this layer is equal to the first squared number greater than the dimensionality of the input layer. For example: if $m = 12$ then the second layer has 16 nodes, if $m = 46$, then the second layer has 49 nodes. So, The number is $[17 * 17], [11 * 11], [7 * 7]$ and $[4 * 4]$ when $\mathbf{MR} = 0.25, 0.1, 0.04$ and 0.01. Then, we use multilayer deconvolutional networks followed the second layer to obtain the final reconstructed image block, which meets the size requirement.

We present here some details regarding our CS-DeCNN. As shown in Fig. 1, to increase the dimensionality of the input, the stride of each deconvolutional layer must be greater than 1. The size of kernels in the first and third deconvolutional layer is $3 * 5$, while that in the second and fourth deconvolutional layer is $5 * 3$. We use zero-padding to ensure that pixels at the edges of a block are not be ignored. After this dimensional increase, the model may output a block with a size larger than that of the original uncompressed image. The solution is to add two convolutional networks behind it to appropriately decrease the dimensionality to produce a result with the standard size. In CS-DeCNN, ReLU is applied to the output of each layer.

After the fully connected layer, the first and fourth layers both output 64 feature maps, the second and fifth layers output 32 feature maps, and the third and sixth layers output 1 feature map. The output of sixth layer is the restored result of each image block. The key point of the structure is to control the number of nodes in the second layer and keep the strides of deconvolutional layers bigger than 1.

4   Experiment and Analysis

To validate the effectiveness of our CS-DeCNN, we perform experiments on the same data set used in the study on [9]. We test the reconstructed precision and time complexity of the recovery. The experimental results support our previous statements.

4.1   Experimental Process

Data Set. We use the same set as in [9]. We convert the pictures in the training set into grayscale images and divide each image into $34 * 34$ sized blocks with a stride of 14 to form a set of about 21700 patches. We use a two-dimensional array of size $[34 * 34]$ to represent each image block. Each element of the array is the gray value of the corresponding pixel. There exist overlapping parts of the divided sub image blocks. This condition can allow the production of more samples in the training set. Then, we reshape each $[34 * 34]$ array to produce a one-dimensional vector of size 1156. This is our original block signal vector $\mathbf{x}$. We calculate the dimensionality of the compressed signal according to the sampling rate, denoted as $\mathbf{M}$. We then produce a Gaussian random matrix of size $\mathbf{M} * \mathbf{N}$ ($\mathbf{N}$ is the size of the original signal vector; in this case, it is 1156) and

Table 1. Quality of reconstruction(PSNR in dB) for 4 different images(chosen followed [14]) in the testing set using different methods under 4 measurement rates. The bottom line is the average result of all images in the testing set.

Image name	Algorithm	MR = 0.25		MR = 0.10		MR = 0.04		MR = 0.01	
		w/o BM3D	w/ BM3D	w/o BM3D	w/ BM3D	w/o BM3D	w/ BM3D	w/o BM3D	w/ BM3D
Barbara	TVAL3 [15]	24.19	24.20	21.88	22.21	18.98	18.98	11.94	11.96
	D-AMP [17]	**31.71**	**31.85**	20.52	20.99	11.17	11.40	5.80	5.80
	SDA [19]	23.19	23.20	22.07	22.39	20.49	20.86	18.59	18.76
	ReconNet [14]	23.25	23.46	21.89	22.20	20.39	20.68	18.61	18.75
	Ours	23.65	23.70	**22.40**	**22.58**	**21.11**	**21.27**	**19.00**	**19.14**
Fingerprint	TVAL3	22.70	22.71	18.69	18.70	16.04	16.05	10.35	10.37
	D-AMP	**27.57**	**27.56**	16.71	16.97	9.94	10.15	4.92	4.93
	SDA	24.28	23.45	20.29	20.31	16.87	16.83	14.83	14.82
	ReconNet	25.58	25.82	**20.76**	**21.00**	**16.93**	**16.97**	14.82	14.87
	Ours	25.75	25.60	20.72	20.90	16.61	16.63	**15.07**	**15.15**
Flinstones	TVAL3	24.05	24.07	18.88	18.92	14.88	14.91	9.75	9.77
	D-AMP	**28.20**	**28.32**	17.46	17.73	10.23	10.42	4.60	4.60
	SDA	20.88	20.21	18.40	18.21	16.19	16.18	13.90	13.95
	ReconNet	22.60	23.15	19.02	19.24	16.38	16.57	13.97	14.07
	Ours	24.08	24.29	**19.46**	**19.70**	**16.53**	**16.68**	**14.26**	**14.38**
Lena	TVAL3	28.67	28.71	24.16	24.18	19.46	19.47	11.87	11.89
	D-AMP	**30.62**	**30.63**	22.33	22.44	11.72	12.06	6.04	6.07
	SDA	25.89	25.70	23.81	24.15	21.18	21.55	17.84	17.95
	ReconNet	26.54	27.03	23.83	24.29	21.27	21.61	17.87	17.98
	Ours	28.21	28.28	**25.14**	**25.37**	**22.27**	**22.50**	**18.80**	**18.92**
Mean PSNR	TVAL3	27.84	27.87	22.84	22.86	18.39	18.40	11.31	11.34
	D-AMP	**30.99**	**30.97**	20.79	21.07	11.06	11.32	5.50	5.51
	SDA	24.72	24.55	22.43	22.68	19.96	20.21	17.29	17.40
	ReconNet	25.57	26.01	22.70	23.07	20.00	20.26	17.28	17.43
	Ours	27.14	27.28	**23.64**	**23.86**	**20.78**	**20.94**	**18.05**	**18.16**

Table 2. Average time complexity (in seconds) of different algorithms for reconstruct one picture in the testing set.

Algorithm	**MR = 0.25**	**MR = 0.10**	**MR = 0.04**	**MR = 0.01**
TVAL3 [15]	3.557	4.363	3.802	3.612
D-AMP [17]	33.152	42.455	36.197	35.137
SDA [19]	0.00873	0.00885	0.00921	0.00859
ReconNet [14]	0.0457	0.0472	0.0483	0.0474
CS-DeCNN	0.0503	0.0511	0.0524	0.0435

orthonormalize its rows to obtain our compressed matrix Φ. Using $\mathbf{y} = \Phi\mathbf{x}$ to produce the input of our network. Therefore, our input-label pair is $(\mathbf{y}, \mathbf{x})$. The size of our inputs are $\mathbf{y} = 289, 116, 46$ and 12 when $\mathbf{MR} = 0.25, 0.1, 0.04$ and 0.01. Then, The input y fully connects to a layer with size of $[17*17], [11*11], [7*7]$ and $[4 * 4]$ when $\mathbf{y} = 289, 116, 46$ and 12.

Training Methods. We conduct our experimental results in four measurement rates: **MR** $= 0.25, 0.10, 0.04$ and 0.01. Our experiment utilize the back propagation and the MSE loss. The loss function is as follows:

$$L = \frac{1}{S} \sum_{i}^{S} \| f(\mathbf{y}_i, \mathbf{W}) - \mathbf{x}_i \|^2 \tag{3}$$

where $f(\mathbf{y}_i, \mathbf{W})$ is the output of the network and $\mathbf{x}_i$ is the original image patch signal, also called the label. Here, we employ the stochastic batch gradient descent method. S denotes the batch size, which is set to 128. We divide all the input-label pairs into batches, select each batch as an input in turn, calculate the loss L, and adjust the network weights and biases. For all weights, we perform a random initialization. Our codes were constructed under the Tensorflow framework, and the AdamOptimizer [13] is applied to dynamically adjust the learning rate.

Obtain Results. After completing the training of CS-DeCNN, we divide the images in the testing set into non-overlapping patches of size $34 * 34$, produce the compressed signal using the same Gaussian random matrices we used to obtain the training set, and input them into the corresponding CS-DeCNN network. We obtain the output of each patch, stitch them into place, and then obtain the intermediate result. Because the picture is restored patch by patch, the transition between patches is not smooth. We used BM3D [7] to denoise the entire image to produce our final result.

4.2 Experimental Results

In this subsection, we design an experiment to compare the recovered accuracy and time complexity of different methods. We choose TVAL3 [15], D-AMP [17], SDA [19], and ReconNet [14] as baselines. They are state-of-the-art CS methods. our experiment shows that CS-DeCNN can achieve higher accuracy than these methods in most of the situations.

For the D-AMP method in our baselines, we used the method presented in [17], and according to the codes provided by the author, construct the experiment. We intended to use the default parameters, but found, in our experiments on D-AMP, that when $iter = 30$ is selected, the quality of recovery declines too quickly as the measurement rate decreases. Therefore, we set iter equal to 20, while the rest of the parameters were set to the default values.

We obtained the restored images (including the intermediate results that were not denoised by BM3D, indicated by 'w/o BM3D', and the final results, indicated by 'w/ BM3D') corresponding to the testing set images using different methods under different measurement rates. We calculated the PSNR value (in dB) between each restored image and their corresponding original image. The experimental results are shown in Table 1 (The results of TVAL3 [15] and SDA [19] were cited from [14], because we used the same data set to study the same problem). Table 1 shows that our CS-DeCNN achieved a higher average reconstruction precision than that of ReconNet and SDA under all four measurement rates. Under low measurement rates (mr = 0.10, 0.04, and 0.01), our method can achieve higher restoration precision compared to the traditional CS restoration methods (D-AMP and TVAL3). Moreover, when the measurement rate decreased from 0.25 to 0.01, the decrease of the mean PSNR for the classical reconstruction methods D-AMP and TVAL3 was 25.49 dB and 16.53 dB, respectively, while for CS-DeCNN, the decrease was only 9.09 dB. In addition, the mean PSNR for CS-DeCNN could still be 18.05 dB at **MR** = 0.01, an outcome that the traditional methods could not achieve.

Figure 2 shows the visual contrast of pictures corresponding to the recovered results obtained by the three methods under $MR = 0.10$. From top to bottom, we show the restoration results for the "Foreman" and "Parrots" images. These pictures, show that our method is superior to D-AMP and ReconNet in terms of the smoothness of image restoration and the average quality of detail recovery.

We compared the time complexity of these different methods by using each of them to recover all images (scales: $256 * 256$ and $512 * 512$) of the testing set and calculating the mean time required to recover one picture (in seconds). The results are shown in Table 2. For the traditional iterative CS algorithms (non-deep learning methods): D-AMP and TVAL3, we do these on the Intel(R) Core(TM) i7-5930K CPU with 64GB memory, using MATLAB 2014b on Windows 7 operating system; for deep learning methods: SDA, ReconNet and CS-DeCNN, we do these on NVIDIA GeForce GTX 1080 with 16 GB memory using Tensorflow 1.1.0 on Ubuntu 16.04 operating system. Our experiment shows that CS-DeCNN is about 2–3 order of magnitude faster than traditional CS methods like TVAL3 or D-AMP at all measurement rates. What should be noted that the speedup is not solely come from the utilization of the GPU. It is mainly because CS-DeCNN is not a iterative but a deep learning method utilized a simple structure - DeCNN, which can be parallelized and has very fast implementation. Other two deep learning methods are also running much faster than D-AMP and TVAL3. Similar results can also be found in [14,19].

5   Space Complexity Comparison of Three Deep Learning CS Methods

Another advantage of our network, compared to existed deep learning CS methods, is that our model is lighter. We elaborate on this by comparing the numbers of parameters of SDA [19], ReconNet [14] and CS-DeCNN. The results are shown in Table 3. Although the experiments in this paper use the compressed signal from single-channel grayscale images as inputs, to show the general situation, we compare the numbers of parameters in the situation with 3-channel color images. The results of SDA is calculated from a 3-layer stacked denoised autoencoder network (Fig. 1 in [19]).

From Table 3, we can clearly see that: under all 4 measurement rates, the number of our parameters is approximately 1/10 that of SDA and 1/4 that of ReconNet. Specifically, when MR=0.25, the number of parameters in SDA is

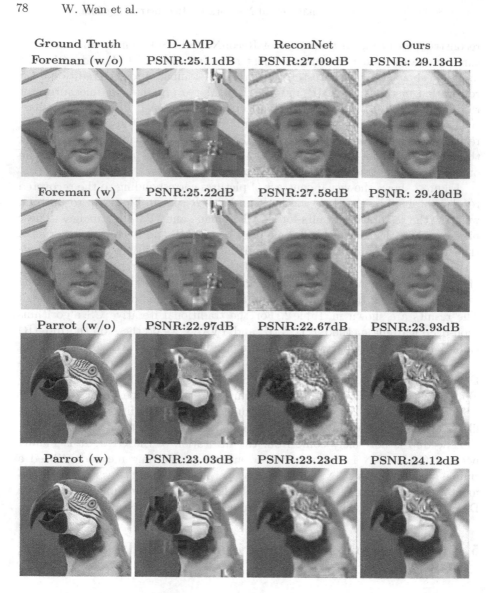

Fig. 2. Reconstruction results for the "Foreman" and "Parrot" images when MR is 0.10.

Table 3. The numbers of parameters in three deep learning CS methods under different measurement rates (1 **K** = 1000, 1 **M** = 1000000).

Neural network	**MR** = 0.25	**MR** = 0.10	**MR** = 0.04	**MR** = 0.01
SDA	9.0 M	3.7 M	1.4 M	360.8 K
ReconNet	2.7 M	1.1 M	468.2 K	147.9 K
CS-DeCNN	793.6 K	258.2 K	109.0 K	58.6 K

approximately 8226 K more than that in CS-DeCNN, and the total number of parameters in CS-DeCNN is only 794 K. The reason why the parameters of these networks decrease as the MR decreases is as follows: when MR decreases, the input dimensionality decreases. So, the number of connections of the fully connected layer(s) decrease(s).

What Should Be Emphasized is That: although CS-DeCNN has fewer parameters than state-of-the-art deep learning CS methods, it achieves higher recovery precision (see Table 1 and Fig. 2). In previous deep learning CS methods, the fully connected layer is utilized to raise the dimensionality of input signals. But we use multilayer deconvolutional neural networks to attain it. We believe that this kind of design has two advantages: first, it reduces the number of parameters in CS-DeCNN, makes CS-DeCNN more suitable for CS application scenarios; second, it makes the network lighter but not shallower just like the pruning which helps CS-DeCNN alleviate overfitting and get better recovered results. There are many prior works have showed that a sparser deep learning structure may usually perform better [6, 11, 16].

6 Conclusions

In this paper, we propose a non-iterative deep learning network framework called CS-DeCNN to recover the images compressed during CS. Compared to traditional state-of-the-art CS methods, CS-DeCNN has the advantages of high efficiency, high speed, and high reconstruction quality under low measurement rates. When compared to previous deep learning CS methods, CS-DeCNN performs better under all experimental measurement rates with much less space occupancy. All these advantages help CS-DeCNN become a better CS method especially for embedded systems.

References

1. Baron, D., Wakin, M.B., Duarte, M.F., Sarvotham, S., Baraniuk, R.G.: Distributed compressed sensing. Preprint **22**(10), 2729–2732 (2005)
2. Beck, A., Teboulle, M.: A fast iterative shrinkage-thresholding algorithm for linear inverse problems. Siam J. Imaging Sci. **2**(1), 183–202 (2009)
3. Blumensath, T., Davies, M.E.: Iterative hard thresholding for compressed sensing. Appl. Comput. Harmon. Anal. **27**(3), 265–274 (2009)
4. Candes, E.J., Terence, T.: Near-optimal signal recovery from random projections: universal encoding strategies? IEEE Trans. Inf. Theory **52**(12), 5406–5425 (2006)
5. Candès, E.J.: The restricted isometry property and its implications for compressed sensing. Comptes Rendus Math. **346**(9), 589–592 (2008)
6. LeCun, Y., Denker, J.S., Solla, S.A.: Optimal brain damage. In: International Conference on Neural Information Processing Systems, pp. 598–605 (1989)
7. Dabov, K., Foi, A., Katkovnik, V., Egiazarian, K.: Image denoising by sparse 3-d transform-domain collaborative filtering. IEEE Trans. Image Process. **16**(8), 2080–2095 (2007)

8. Donoho, D.L., Maleki, A., Montanari, A.: Message-passing algorithms for compressed sensing. In: Information Theory, pp. 1–5 (2010)
9. Dong, C., Loy, C.C., Kaiming, H., Xiaoou, T.: Learning a deep convolutional network for image super-resolution. In: Fleet, D., Pajdla, T., Schiele, B., Tuytelaars, T. (eds.) ECCV 2014. LNCS, vol. 8692, pp. 184–199. Springer, Cham (2014). https://doi.org/10.1007/978-3-319-10593-2_13
10. Duarte, M.F., Wakin, M.B., Baraniuk, R.G.: Wavelet-domain compressive signal reconstruction using a hidden markov tree model. In: IEEE International Conference on Acoustics, Speech and Signal Processing, pp. 5137–5140 (2008)
11. Han, S., Mao, H., Dally, W.J.: Deep compression: compressing deep neural networks with pruning, trained quantization and huffman coding. Fiber **56**(4), 3–7 (2015)
12. Kim, Y., Nadar, M.S., Bilgin, A.: Compressed sensing using a gaussian scale mixtures model in wavelet domain. In: IEEE International Conference on Image Processing, pp. 3365–3368 (2010)
13. Kingma, D.P., Ba, J.: Comput. Sci. A method for stochastic optimization, Adam (2014)
14. Kulkarni, K., Lohit, S., Turaga, P., Kerviche, R., Ashok, A.: ReconNet: non-iterative reconstruction of images from compressively sensed measurements. In: Proceedings of the IEEE Conference on Computer Vision and Pattern Recognition, pp. 449–458 (2016)
15. Chengbo, L., Wotao, Y., Hong, J., Yin, Z.: An efficient augmented lagrangian method with applications to total variation minimization. CAAM technical report. Comput. Optim. Appl. **56**(3), 507–530 (2013)
16. Li, H., Kadav, A., Durdanovic, I., Samet, H., Graf, H.P.: Pruning filters for efficient convnets. arXiv preprint arXiv:1608.08710 (2016)
17. Metzler, C.A., Maleki, A., Baraniuk, R.G.: From denoising to compressed sensing. IEEE Trans. Inf. Theory **62**(9), 5117–5144 (2016)
18. Mousavi, A., Baraniuk, R.G.: Learning to invert: signal recovery via deep convolutional networks. In: IEEE International Conference on Acoustics, Speech and Signal Processing, pp. 2272–2276 (2017)
19. Mousavi, A., Patel, A.B., Baraniuk, R.G.: A deep learning approach to structured signal recovery. In: 53rd Annual Allerton Conference on Communication, Control, and Computing (Allerton), pp. 1336–1343. IEEE (2015)
20. Needell, D., Tropp, J.A.: CoSaMP: iterative signal recovery from incomplete and inaccurate samples. Appl. Comput. Harmon. Anal. **26**(3), 301–321 (2008)

Semantic Correspondence Guided Deep Photo Style Transfer

Zhijiao Xiao, Xiaole Zhang, and Xiaoyan Zhang$^{(\boxtimes)}$

College of Computer Science and Software Engineering, Shenzhen University,
Shenzhen 518000, People's Republic of China
{cindyxzj,xyzhang15}@szu.edu.cn, zhangxiaole2016@email.szu.edu.cn

Abstract. The objective of this paper is to develop an effective photographic transfer method while preserving the semantic correspondence between the style and content images. A semantic correspondence guided deep photo style transfer algorithm is developed, which is to ensure that the semantic structure of the content image has not been changed while the color of the style images is being migrated. The semantic correspondence is constructed in large scale regions based on image segmentation and also in local scale patches using deep image analogy. Based on the semantic correspondence, a matting optimization is utilized to optimize the style transfer result to ensure the semantic accuracy and transfer faithfulness. The proposed style transfer method is further extended to automatically retrieve the style images from a database to make style transfer more-friendly. The experimental results show that our method could successfully conduct the style transfer while preserving semantic correspondence between diversity of scenes. A user study also shows that our method outperforms state-of-the-art photographic style transfer methods.

Keywords: Style transfer · Semantic correspondence · Deep learning

1 Introduction

Nowadays, more and more research focuses on photographic transfer. Unlike common style transfer, photographic transfer aims to change the color and contrast of the content images, but at the same time ensure the texture of the content images. The application of photographic transfer in our daily life has also become more attractive. Especially on social media, we can more easily modify our photographs to get the effect we want. However, image stylization is still a long-standing issue as it is a semantic related task. So far, existing techniques are either limited in the diversity of scenes or in the faithfulness of the stylistic match they achieve.

For style transfer, the main problem is finding the semantically meaningful correspondence between the style and content images. Some low-level matching methods, such as PatchMatch [1], are designed to match local intensities that

© Springer Nature Switzerland AG 2018
R. Hong et al. (Eds.): PCM 2018, LNCS 11164, pp. 81–93, 2018.
https://doi.org/10.1007/978-3-030-00776-8_8

they can not match under large visual variations. The latest method of deep image analogy [6] extends the PatchMatch [1] method. It searches the local patch correspondence in the feature domain instead of the image domain, which serves to guide semantically-meaningful visual attribute transfer. The deep image analogy [6] method could conduct style transfer for a pair of images those may be very different in appearance. However, the local patches in flat object regions with low frequency features are more probably wrongly matched, such as the examples in Fig. 1. In the first example image, the color of the mountain should be green instead of the color of the cloud (white) (the first row in Fig. 1(c)). The color of the sheet in the second image and the color of the fence and ground in the lower left corner in the third image are also wrongly transferred (see images in the second and third rows of Fig. 1(c)). In state-of-the-art deep photo style transfer [10] method, image region segmentation and Matting Laplacian [7] are utilized in their network to ensure the semantic accuracy and transfer faithfulness. The deep photo style transfer [10] method could preserve the image semantic structure after style transfer. However, this method does not produce dense correspondences between images, which could produce color inconsistence in an object region, such as the color of the sheet in the second image of Fig. 1(d). It also changes the texture of the content image and makes the style transfer result image distorted, such as the left side of the mountain in the first image and the sky in the third image of Fig. 1(d).

The objective of this paper is to develop an effective photographic transfer method while preserves the semantic correspondence between the style and content images. The semantic correspondence is constructed both in large scale regions based on image segmentation and local scale patches using deep image analogy. Based on the semantic correspondence, a deep matting optimization is utilized to optimize the style transfer result to ensure the semantic accuracy and transfer faithfulness. As shown in Fig. 1, compared to the original deep image analogy [6] and deep photo style transfer [10] methods, our method could more effectively preserve the semantic correspondence in the style transfer. Our results have better color and texture consistence in object regions. Our main contribution is to combine the segmentation images and the Nearest-neighbor Field Search (NNF) to find the semantic correspondence between the style and content images, and use a matting optimization to make sure the transfer faithfulness.

2   Related Work

We can divide existing style transfer algorithms into traditional style transfer algorithms and style transfer using deep network algorithms.

For traditional style transfer algorithms, Piti et al. [11] described an algorithm to transfer the full 3D color histogram using a series of 1D histograms. Image Analogies [4] was proposed to determine the relationship between a pair of images and then applied it to stylize other images. As this method is based on finding dense correspondence, it often requires that the image pair depicts the same type of scene. Hwang et al. [5] cast the problem of image enhancement as searching for the best transformation for each pixel in the given image

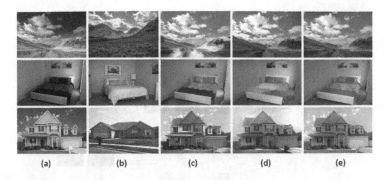

Fig. 1. Comparison of our method against other methods in [6] and [10]. (a) Content images; (b) Style images; (c) The style transfer results produced by the deep image analogy method [6]; (d) The style transfer results produced by the deep photo style transfer method [10]; (e) The style transfer results of our method.

The pixel-wise transformation was learned from the original and enhanced image pairs of the example. The enhanced version of the example image was generated by professionals using global transformations. It is challenging to get such image pairs for arbitrary style.

The first neural network approach to style transfer is the work by Gatys et al. [3] which has demonstrated impressive results in example-based image stylisation. This algorithm was developed for arbitrary stylization based on matching the feature correlations (Gram matrix) between deep features extracted by a trained network classifier within an iterative optimization framework. More recent alternative methods, semantic style transfer [2] introduced a novel concept to augment such generative architectures with semantic annotations, either by manually authoring pixel labels or using existing solutions for semantic segmentation. A patch-based approach named CNNMRF [8] operated on the patch of deep features to allow the algorithm to understand the pattern in the image locally, which improved the overall accuracy of the style transfer. Luan et al. [10] proposed a style transfer algorithm with image segmentation and matting to ensure the semantic accuracy and transfer faithfulness. The deep image analogy [6] method constructed the dense correspondence between images in the feature domain for semantically-meaningful visual attribute transfer. Most of the methods using VGG-19 deep neural network to extract the feature map for content images and style images. In order to effectively perform photographic transfer method while preserves the semantic correspondence between the style and content images, we propose to construct the semantic correspondence in large scale regions based on image segmentation and also in local scale patches using deep image analogy. We also employed VGG-19 deep neural network to extract the feature map for content images and style images.

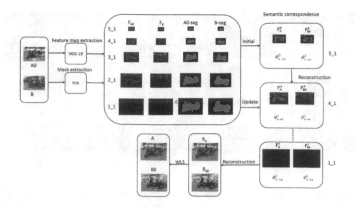

Fig. 2. Semantic correspondence network. A0 is the content image; B is the style image.

3   The Proposed Approach

For an image pair including a content image and a style image, the feature maps of them are extracted using a pre-trained deep neural network. Meanwhile, the segmentation maps of them are also generated. Then we use the semantic correspondence network (shown in Fig. 2) to combine the semantic structure and texture information of the content image with the color information of the style image to get the resulting style transfer image. As the image segmentation can not be always fully accurate, so in the final step, a matting optimization network is used to optimize the final style transfer result to ensure the semantic accuracy and transfer faithfulness.

3.1   Semantic Correspondence

The framework of our semantic correspondence algorithm is shown in Fig. 2. For the content image A0 and style image B, the deep feature maps F_{A0}^l and F_B^l are extracted by using a VGG-19 network [12]. A FCN network [9] is used to get segmentation images A0-seg and B-seg for the content image A0 and style image B, respectively. The FCN network could identify a variety of objects and segment the images based on the recognition result.

Then, the main task of the semantic correspondence is to construct the correspondence mapping functions $\Phi_{a\rightarrow b}$ and $\Phi_{b\rightarrow a}$ between images A0 and B according to four images (content image feature map F_{A0}, style image feature map F_B, content image segmentation A0-seg, and style image segmentation B-seg). $\Phi_{a\rightarrow b}$ is to map a pixel from A0 to B, and $\Phi_{b\rightarrow a}$ is to map a pixel from B to A0.

Based on the correspondence mapping functions, two feature maps F_A and F_{B0} are constructed, where F_A is the feature map for an image A which has the same spatial structure with the image A0 and has similar color feature to the image B, and F_{B0} is the feature map for an image B0 which has the same spatial structure with the image B and has similar color feature to the image A0.

Given the correspondence mapping functions $\Phi_{a \to b}$ and $\Phi_{b \to a}$ between images A0 and B, the style transfer could be described as,

$$A(p) = B(\Phi_{a \to b}(p)), and, B0(p) = A0(\Phi_{b \to a}(p)) \tag{1}$$

To further achieve symmetric and consistent mappings, bidirectional constraint is considered in the correspondence mapping, which is $\Phi_{a \to b}(\Phi_{b \to a}(p)) = p$ and $\Phi_{b \to a}(\Phi_{a \to b}(p)) = p$. p is a pixel index.

Nearest-neighbor Field Search is used to map A0 to A and B to B0. Through nearest neighbor patch randomly searches the better semantic-level correspondences between the two images, the search is two-way at the same time, $\Phi_{a \to b}^l$ and $\Phi_{b \to a}^l$ are constructed hierachically at each feature layer l, which is

$$\Phi_{a \to b}^l(p) = \arg\min_q \sum_{x \in N(p), y \in N(q)} (\|F_A^l(x) - F_B^l(y)\|^2 + \|F_{A0}^l(x) - F_{B0}^l(y)\|^2) \tag{2}$$

where $N(p)$ is the patch around pixel p. For each patch around pixel p in the source A (or A0), we find its nearest neighbor position $q = \Phi_{a \to b}^l(p)$ in the target B (or B0). The same for $\Phi_{b \to a}^l(p)$. Simultaneously, for reconstruction, F_A^l on the previous layer are rebuilt on the next layer in the network (the same for F_{B0}^l). The reconstruction of F_A^l is

$$F_A^{l-1} = F_{A0}^{l-1} \circ W_{A0}^{l-1} + R_B^{l-1} \circ (1 - W_{A0}^{l-1}) \tag{3}$$

where $\circ$ is element-wise multiplication on each channel of feature map, and W_{A0}^{l-1} is a 2D weight map. R_B^{l-1} is the result of the $F_B^l(\Phi_{a \to b}^l)$ after deconvolution in layer l.

The segmentation images are used to initialize the feature maps F_A^5 and F_{B0}^5, and the correspondence mapping functions $\Phi_{a \to b}^5$ and $\Phi_{b \to a}^5$. Only pixels in the same type of object in the content image and the style image can be defined as the corresponding position relationship. When updating feature maps F_A^l and F_{B0}^l, and the mapping function $\Phi_{a \to b}^l$, $\Phi_{b \to a}^l$ of the next layer, the segmentation images are also used to constrain the deep patch match result when updating F_A^l, F_{B0}^l, $\Phi_{a \to b}^l$, $\Phi_{b \to a}^l$.

After semantic correspondence, we get the correspondence functions $\Phi_{a \to b}^1$, $\Phi_{b \to a}^1$ which represent the corresponding position relationship of the image A0 to image B. Based on these two correspondence functions, two new images R_A and R_{B0} are constructed by replacing each patch in A0 and B by its corresponding patch in B and A0, respectively. Image R_A has the color of the image B and the spastical structure of the image A0. Image R_{B0} has the color of the image A0 and the spastical structure of the image B. Because images R_A and R_{B0} are constructed based on patch, they are not smooth. Finally, two smooth images A and B0 are optimized from the images R_A and R_{B0} through the weighted least square (WLS) [6].

3.2 Image Reconstruction

Deep Matting. The matting Laplacian matrices proposed by Levin et al. [7] is used to refine the style transfer. It expresses a grayscale matte as a locally affine combination of the input RGB channels. It only depends on the input content image. The matting loss function is expressed as

$$\mathcal{L}_m = V[R]^T M_I V[R] \tag{4}$$

where $\mathcal{L}_m$ is the photorealism regularization, M_I is a symmetric matrix calculated based on the original content image using the method in [7], $V[R]$ is the vectorized version of the result image R (R_A or R_{B0}) from the semantic correspondence network.

Loss Function. From the semantic correspondence network, we get four result images A, B0, R_A, R_{B0}. They are used as inputs for the matting optimization, which has loss function as

$$\mathcal{L}_{total} = \sum_{l=1}^{L} \alpha_l \mathcal{L}_c^l + \Gamma \sum_{l=1}^{L} \beta_l \mathcal{L}_s^l + \lambda \mathcal{L}_m \tag{5}$$

where $\mathcal{L}_c^l$ is the content loss. $\mathcal{L}_s^l$ is the style loss. $\mathcal{L}_m$ is the is the photorealism regularization (Eq. 4). α_l and β_l are the weights to configure layer preferences. L is the total number of convolutional layers and l indicates the l-th convolutional layer of the deep neural network. Γ is a weight that controls the style loss, λ is a weight that controls the photorealism regularization.

The content loss is

$$\mathcal{L}_c^l = \sum_{k=1}^{K} \frac{1}{N_l M_l} \sum_{ij} (F_l[O] - F_l[R])_{ij}^2 \tag{6}$$

where K is the number of the segmentation regions, N_l and M_l are the dimensionality of these feature maps in layer l, F_l is the feature matrix with (i, j) indicating its index. O is the content image, R is the image R_A (or R_{B0}).

The style loss is

$$\mathcal{L}_s^l = \sum_{k=1}^{K} \frac{1}{N_l^2 M_l^2} \sum_{ij} (G_l[S] - G_l[R])_{ij}^2 \tag{7}$$

where G_l is the Gram matrix and is defined as the inner product between the vectorized feature maps. S is the image A (or B0).

4 Experiments

In this section, we analyze the performance of the proposed algorithm. The experimental data we use comes from the work in [10] and VOC2012 dataset.

For the semantic correspondence network, we use the pre-trained VGG-19 as the feature extractor, and employed conv1_1, conv2_1, conv3_1, conv4_1 and conv5_1 as the content representation and style representation to extract the feature maps. We use the VOC2012 dataset to training a FCN network to segmented the two input images. The segmentation masks were resized according to the size of the feature maps in each layer.

In the matting optimization network, we also use the pre-trained VGG-19 as the feature extractor, and choose conv4_2 as the content representation, and conv1_1, conv2_1, conv3_1, conv4_1 and conv5_1 as the style representation. Reference to [10] we set the parameters $\Gamma = 100$. Since the two input images have the same spatial information, in order to better maintain the color information of the resulting image, we set $\alpha = 1/2$, and $\beta = 1/3$ in order to prevent the texture of the result image lost.

About the value of λ, we set $\lambda = 10000$. As shown in the Fig. 3, it is easy to find that while $\lambda = 100$, the result becomes too stylized, and while $\lambda = 1000000$, the color of the result image does not transfer well while setting $\lambda = 10000$ it can keep the texture and maintain the right color while preventing images from being too stylized.

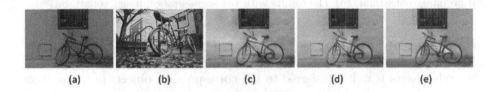

Fig. 3. Comparison the result images of different value of λ (a) Content images; (b) Style images; (c) $\lambda = 100$; (d) $\lambda = 10000$; (e) $\lambda = 1000000$.

4.1 The Effect of Segmentation Constraint

Using the semantic correspondence construction with segmentation constraint, the style transfer performs better than that without the segmentation image constraint. Some style transfer results are shown in Fig. 4. Some regions of the road in the first example image, the mountain in the third example image, and the trees in the fifth example image in Fig. 4 are matched to the wrong regions while without segmentation constraint. It is because that the patches in these regions are matched to those of wrong type of objects while without guidance of the segmentation images, This results in color inconsistence in these objects. In the second example image, the faucet does not match the target faucet well. The same problem exists in fourth example, where the tree is matched to background color. However, while using segmentation constraint, the color of these wrongly matched regions is corrected. By comparing the style transfers between methods with and without segmentation constraint, we find that objects in the content and style images may easily form the wrong correspondence in the style transfer

Fig. 4. Comparison of our method against the method without the segmentation image constraint. (a) Content images; (b) Style images; (c) The results without the segmentation image constraint; (d) The results with the segmentation image constraint.

while without the segmentation constraint. In contrast, the segmentation constraint could effectively solve this problem, so that the color of the objects in the style images is well transferred to the corresponding objects in the content images.

4.2   The Effect of Matting Optimization

In order to study the effect of matting optimization in the style transfer, the final results generated by our method are compared with those without matting optimization which are from the semantic correspondence network in Fig. 2. Some experimental results for the comparison are shown in Fig. 5. As shown in Fig. 5(c), in images generated without matting optimization, the textures of the wall and the floor in the first example are not well preserved. While using matting optimization, in the first image in Fig. 5(d), the textures are restored to be consistent with those of the original image. In the second example, the top right of the table is very bright in the result without matting optimization, which is different from the light distribution in the original content image. But this is corrected after using matting optimization (see Fig. 5(d)). In the third example, the light distribution in the result generated using matting optimization is also more consistent with the original content image comparing than that generated without matting optimization. The texture and light problems in the results generated without matting optimization are mainly from that the result image from the semantic correspondence network has some texture changed. Additionally, the segmentation accuracy also effects the style transfer effect in

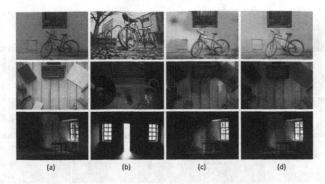

Fig. 5. Comparison between the final results of our method against the results without matting optimization. (a) Content images; (b) Style images; (c) The results without matting optimization; (d) The final results of our method.

the semantic correspondence network. While using matting optimization, the texture change and effect of inaccurate segmentation are refined to solve these problems. From Fig. 5, we can clearly see that the style transfer with matting optimization performs well for semantic photographic style transfer.

4.3   Compare with Related Methods

In order to study the advantage of our proposed style transfer method, it is compared to four state-of-the-art methods, which are the deep image analogy [6], deep photo style transfer [10], Neural transfer [3] and CNNMRF [8]. Some example results for the comparison are shown in Fig. 6. In the results generated by the deep image analogy method, some regions may be wrongly matched, such as the road is matched to the color of the grass in the fourth example and the trees are matched to the color of the cloud in the ninth example (in Fig. 6(c)). It is because that without large scale correspondence constraint the dense correspondence could be wrong, especially for flat region with low frequency features. The deep photo style transfer could preserve large scale region color correspondence, but it does not guarantee the image texture information, which results in image color distortion (Fig. 6(d)). The Neural transfer method transfers both

Table 1. User study result.

Methods	Selection rate
Deep image analogy	14.8%
Deep photo style transfer	11.9%
Neural transfer	8.2%
CNNMRF	4%
Our method	61.1%

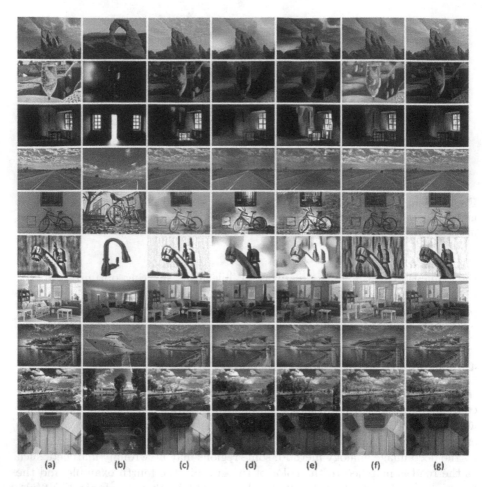

(a)   (b)   (c)   (d)   (e)   (f)   (g)

Fig. 6. Comparison of our method against related methods. (a) Content images; (b) Style images; (c) The results of the Deep image analogy [6]; (d) The result of the Deep Photo Style Transfer [10]; (e) The results of Neural transfer [3]; (f) The results of CNNMRF [8]; (g) The results of our method.

the color and texture in the style transfer. It could not preserve the semantic correspondence in the style transfer, such as the sky in the first example, the road in the fourth example and the faucet in the sixth example of Fig. 6(e). The results generated by the CNNMRF method have both color and texture change, and they also might have wrongly matched color, such as the wine glass of the second example and the background of the sixth example of Fig. 6(f). Our results outperforms those generated by the other four methods, as they have successfully preserved the semantic correspondence while conducting the photographic style transfer. Our method can preserve the texture and spatial structure of the content images unchanged while merging with the color of the style images well. Method [10] also use the same semantic mask.

Fig. 7. Our method with automatic style image retrieval. Top: Content images; Mid: Style images; Bottom: result images.

In order to significantly evaluate the effectiveness and advantage of the proposed approach, a user study was conducted to compare style transfer effect by our proposed method with those four related methods. The results obtained by our proposed method and the four related methods were shown side by side in a row. The ordering of the images in each row was randomly generated. The five results in each row are numbered from a to e. Users were required to choose the one they considered as having the best color migration and the best texture preservation from each set of results. The experimental results of 30 test photographs were used for this user study. There were a total of 40 participants aged from 20 to 50. A total of 1200 responses were collected. The user study result is summarized in Table 1. It shows that our results are preferred by a significantly higher percentage of responses (61.1%) than those by other methods. The other four methods all get a selection rate less than 15%. The CNNMRF method gets the smallest selection rate in this user study. This is not surprised as that the CNNMRF method changes the texture of the original images which is not preferred in the photographic style transfer.

4.4   Automatic Style Image Retrieval

In addition to the above contribution, we also tested to conduct automatic style image retrieval based on semantic similarity to make the style transfer to be automatic and more user-friendly. Users only need to give a content image. The content image is segmented with object recognition by using the pre-trained FCN network. Then, a set of style images which have similar semantic content with the input content photograph are retrieved from an image dataset. The image retrieval is conducted based on the similarity of recognized object species and the region size of objects. The user could use the first ranked image or select one retrieved image from the recommended top ranked images as the style image to conduct the style transfer using our proposed method.

In Fig. 7, we show some results of our method through automatic style image retrieval. We can easily find that the motorcycle, car, bus, and airplane all inherits the color of the corresponding target in the style image while maintaining the original texture. Meanwhile, the background also has a well color transfer. Our method could effectively perform style transfer between diversity of scenes.

5   Conclusion

This paper develops a semantic correspondence-guided deep photo style transfer algorithm, which is to ensure that the image's semantic structure of the content image has not been changed while the color of the style images is being migrated. The semantic correspondence is constructed in large scale regions based on image segmentation and also in local scale patches using deep image analogy. Based on the semantic correspondence, the matting optimization is performed to optimize the style transfer result to ensure the semantic accuracy and transfer faithfulness. We also extended the proposed method to conduct automatic style image retrieval based on semantic similarity to make the style transfer to be automatic and more user-friendly. The experimental results and user study shows that our method outperforms the current state-of-the-art photographic style transfer methods.

Acknowledgement. The authors wish to acknowledge the financial support from: (i) Chinese Natural Science Foundation under the Grant No. 61602313, 61620106008; (ii) Shenzhen Commission of Scientific Research and Innovations under the Grant No. JCYJ20170302153632883, JCYJ20160422151736824; (iii) Tencent "Rhinoceros Birds" - Scientific Research Foundation for Young Teachers of Shenzhen University; (iv) Startup Foundation for Advanced Talents, Shenzhen; (v) The Natural Science Foundation of Guangdong Province No. 2016A030310053, 2017A030310521.

References

1. Barnes, C., Shechtman, E., Finkelstein, A., Goldman, D.B.: Patchmatch: a randomized correspondence algorithm for structural image editing. ACM Trans. Graph. **28**, 3 (2009). (Proc. of SIGGRAPH)
2. Champandard, A.J.: Semantic style transfer and turning two-bit doodles into fine artworks. arXiv preprint arXiv:1603.01768 (2016)
3. Gatys, L.A., Ecker, A.S., Bethge, M.: A neural algorithm of artistic style. In: Computer Science, Computer Vision and Pattern Recognition (2015)
4. Hertzmann, A., Jacobs, C.E., Oliver, N., Curless, B., Salesin, D.H.: Image analogies. In: Conference on Computer Graphics and Interactive Techniques, pp. 327–340 (2001)
5. Hwang, S.J., Kapoor, A., Kang, S.B.: Context-based automatic local image enhancement. In: European Conference on Computer Vision, pp. 569–582 (2012)
6. Kang, S.B.: Visual attribute transfer through deep image analogy. ACM Trans. Graph. **36**(4), 120 (2017)
7. Levin, A., Lischinski, D., Weiss, Y.: A closed-form solution to natural image matting. IEEE Trans. Pattern Anal. Mach. Intell. **30**(2), 228–242 (2008)
8. Li, C., Wand, M.: Combining Markov random fields and convolutional neural networks for image synthesis. In: Computer Vision and Pattern Recognition, pp. 2479–2486 (2016)

9. Long, J., Shelhamer, E., Darrell, T.: Fully convolutional networks for semantic segmentation. In: Computer Vision and Pattern Recognition, pp. 3431–3440 (2015)
10. Luan, F., Paris, S., Shechtman, E., Bala, K.: Deep photo style transfer. In: Computer Vision and Pattern Recognition, pp. 6997–7005 (2017)
11. Pitie, F., Kokaram, A.C., Dahyot, R.: N-dimensional probability density function transfer and its application to color transfer. In: Tenth IEEE International Conference on Computer Vision, pp. 1434–1439 (2005)
12. Simonyan, K., Zisserman, A.: Very deep convolutional networks for large-scale image recognition. arXiv preprint arXiv:1409.1556 (2014)

Multiple-Level Feature-Based Network for Image Captioning

Kaidi Zheng[1(✉)], Chen Zhu[2], Shaopeng Lu[1], and Yonggang Liu[1]

[1] School of Software, Beihang University, Beijing, China
kaidi_zheng@163.com, lsp57263@163.com, liuyonggang@buaa.edu.cn
[2] Institute of Computing Technology, Chinese Academy of Sciences, Beijing, China
zhuchen01@ict.ac.cn

Abstract. Image captioning, which automatically describes the content of an image, has attracted interests recently. Due to the need for both fine-grained visual understanding and meaningful natural language expression, image captioning is a challenging task. Existing methods predominantly take one kind of image feature to generate the description while neglecting other useful features. This strategy leads to unsatisfied captioning result. To deal with this problem, we propose a multiple-level feature-based network for image captioning. In our method, three kinds of features are extracted from the image, representing analysis of different level of the image. Attention mechanism in our network is adopted to selectively attend to salient region or attribute of each feature when predicting each word of the caption. Experimental results show that our model can outperform the state-of-the-art methods on MS-COCO dataset. Compared with other methods, our network can lead to more accurate subject prediction and vivid description of sentences.

Keywords: Image captioning · Multilevel features · Attention

1 Introduction

Image captioning refers to the task that automatically describes the content of an image using properly formed sentences of natural language. Compared with image classification or machine translation, image captioning is a very challenging task, as it not only requires visual understanding of the image, but also needs the knowledge in Natural Language Processing (NLP).

Today, almost all image captioning methods take inspiration from the success of end-to-end training formulation in machine translation, i.e. the encoder-decoder architecture. Early methods, such as [20,21], use deep convolution neural network (CNN) as the encoder, which produce a rich visual representation of the input image. Then the visual features are fed to the decoder, which is usually implemented as recurrent neural network (RNN), for descriptive sentence generation. Recently, researchers found that fine-grained analysis of the input image is necessary for high quality of captioning. Therefore, high-level information

R. Hong et al. (Eds.): PCM 2018, LNCS 11164, pp. 94–103, 2018.
https://doi.org/10.1007/978-3-030-00776-8_9

extracted from the image is used frequently in state-of-the-art image captioning methods. For example, [7,22] extracted semantic concepts from the image, while Anderson et al. [1] detected image regions. These methods fuse their high-level information through attention mechanism, which lets captioning model attend to a semantically important concept or a salient region in the input image.

However, all the above methods only use one or two types of the image features, which restrict the improvement to be small. In fact, different kind of features can capture different levels of image information and complement each other. Specifically, visual features are global image-level description. Regional features focusing on salient regions or objects are more fine-grained than visual features. Semantic features represent higher-level knowledge based on the content of the image, rather than describing the object of image directly. These features usually play different roles in image captioning tasks. For example, regional features can usually lead to more accurate quantifiers, as they retain spatial information of the region. Contrasted, because semantic features are not limited to the entity in the image, they can give some verb tags that can not be detected by regional features.

Based on the above observation, we propose a multiple-level feature-based network (MLFN) for image captioning. We use different levels of features, i.e. visual, regional and semantic features, to encode the input image. These features are fused and decoded by a two-layer LSTM [9] network equipped with attention mechanism. When predicting each word of the caption sentence, the network can selectively attend to semantic concept or salient region of each feature, which plays a great role in performance improvement. To evaluate our model, we compare it with some current methods on MS-COCO dataset. Experimental results show that our model significantly outperforms these state-of-the-art methods.

The remaining of this paper is organized as follows. We review related previous works in Sect. 2 and introduce our model in Sect. 3. Section 4 gives the experimental results with the paper concluded in Sect. 5.

2   Related Work

Automatically generating a meaningful natural language description of an image, known as image captioning, has received a lot of attention of researchers. It has many important practical applications, such as in human computer interaction or helping visually impaired people. Currently, image captioning is inspired by the success of encoder-decoder frameworks in machine translation [2,3]. For encoding, CNNs are used to extract visual features, which are fed into the captioning model [10,14,20]. For decoding, RNNs are used for caption generation. In [14], vanilla RNN is used as decoder to predict the next word given the image and previous word, while in [20] vanilla RNN is replaced by LSTM. As a powerful tools, attention mechanism is widely used in image captioning recently [16,24]. This mechanism can guide the captioning model to attend to salient image region corresponding to each predicted word. Thus more correct caption can be generated. The work in [21] uses soft attention to dynamically select image regions, which most relevant to a target word. [13] constructs adaptive attention mechanism, which automatically decides when to look at the image and when to rely on the language model to generate the next word.

There are also works that use high level features to improve the performance of image captioning. One type of features is semantic concepts or tags detected from the image. The method in [22] learns to selectively attend to semantic concept proposals and fuse them into hidden states and outputs of recurrent neural networks. Gan et al. [7] use the probability of semantic tags to compose the parameters in a long short-term memory (LSTM) network. Another type of features are local region information of image. Li et al. [11] integrate global representation of image-level with local object-level informations. Anderson et al. [1] proposed a combined bottom-up and top-down attention mechanism, where bottom-up mechanism is based on the image regions detected by Faster R-CNN.

3   Approach

In this section, we first introduce the general image captioning framework, then we describe how to extract three different levels of features. Finally we give the architecture of proposed captioning model and the implementation details.

3.1   Review of Image Captioning

Before introducing our method, we first describe the general image captioning framework [20], i.e. the encoder-decoder architecture.

Given the input image I and its corresponding transcription $Y = \{y_1, \cdots, y_T\}$, with T words, the encoder-decoder architecture directly maximizes the following formulation:

$$\theta^* = \arg\max_{\theta} \sum_{(I,Y)} \log p(Y|I; \theta), \tag{1}$$

where θ is the parameters of captioning model. The joint probability distribution of Y can be modeled by chain rule as following:

$$\log p(Y|I; \theta) = \sum_{t=1}^{T} \log p(y_t|y_1, \cdots, y_{t-1}, I; \theta). \tag{2}$$

For model $p(y_t|y_1, \cdots, y_{t-1})$, the general image captioning framework uses a CNN to extract visual features of input image (encoding), then these features are fed into the RNN that generates sentences (decoding). That is to say:

$$p(y_t|y_1, \cdots, y_{t-1}) = f(RNN(y_{t-1}, CNN(I), h_{t-1})), \tag{3}$$

where f is non-linear function that outputs the probability of y_t and h_{t-1} is the output of RNN at time step $t - 1$.

Compared with the general framework, our captioning model extracts different features from the image, which will be introduced in the next section.

3.2 Feature Extraction

In our captioning model, we extract three different levels of image features: visual feature V, regional feature R and semantic feature S.

Visual Feature. As a specialized kind of deep neural network, convolutional neural networks are widely used in computer vision applications, such as image classification and object detection. Thanks to the residual learning framework designed to deal with the degradation problem, ResNet [8] has made great improvement in image classification task. In this paper, we use Resnet-101, pre-trained on ImageNet, to extract the visual feature. The output of $res5c$ layer is used as the visual feature V, whose dimension is $7 \times 7 \times 2048$.

Regional Feature. The regional feature $R = \{r_1, \cdots, r_{k_r}\}$ encodes the visual information of salient objects in the image. In this paper, we use Faster R-CNN [17] in conjunction with the ResNet-101 to detect objects and localize them with bounding boxes. For each bounding box i, we use the output of the last mean-pooling layer as the feature $r_i \in \mathbf{R}^{2048}$. In our implementation, the top $k_r = 20$ ranked bounding boxes are selected. Thus the dimension of regional feature R is 20×2048.

Semantic Feature. We use the network in [5] to extract the semantic feature $S = \{s_1, \cdots, s_{k_s}\}$. This model tries to train a CNN that outputs the semantic words, including noun, verb and adjective, that appear in the ground truth caption. Because we do not know the image bounding boxes corresponding to the words, the weakly-supervised approach of Multiple Instance Learning (MIL) [23] is adopted. In training, the first 1000 entries in captioning dictionary with hight probabilities are used as the target words. Using the network, we can get a 1000-dimensional vector, reflecting the probability distribution of these words in the input image. The top k words with the highest probability are selected as the final output. In our implementation, we set $k_s = 5$ and we can get 5 semantic words for each image. For the convenience of use in the next section, these words are represented as one-hot encoding.

3.3 Captioning Model

Our captioning model use attention mechanism to fuse different features. In this section, we first introduce the network architecture of our captioning model, then we describe the attention mechanism, finally we show how to generate the caption using our model.

Network Architecture. The overall captioning model of our method is composed of two LSTM layers as illustrated in Fig. 1. The input vector to the network

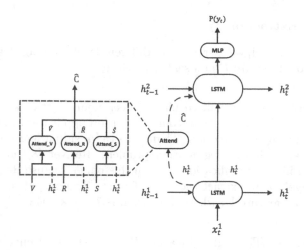

Fig. 1. The proposed captioning model

at each time step t consists of an encoding of the previously generated word and the mean-pooled features, i.e.

$$x_t^1 = [\bar{v}; \bar{r}; \bar{s}; W_e y_{t-1}], \tag{4}$$

where $\bar{v} = \frac{1}{k_v} \sum_{i=1}^{k_v} v_i$ is the mean-pooled visual feature, $\bar{r} = \frac{1}{k_r} \sum_{i=1}^{k_r} r_i$ is the mean-pooled regional feature and $\bar{s} = \frac{1}{k_s} \sum_{i=1}^{k_s} W_s s_i$ is the mean-pooled semantic feature. The vector y_{t-1} is one-hot encoding of the input word at time $t-1$. The matrices $W_e \in \mathbf{R}^{E \times N}$ and $W_s \in \mathbf{R}^{E \times N_s}$ are word embedding matrix, where E is the embedding dimension, N and N_s are the size of caption vocabulary and semantic tag vocabulary.

After obtaining the output h_t^1 of the first LSTM, we fuse the different features through attention mechanism. The combined feature $\hat{C}$ is fed into the second LSTM, whose ouput is used to predict the word of the caption.

Attention Module. We combine the visual feature V, regional feature R and semantic feature S into $\hat{C}$ through the attention module *Attend*. So that we can selectively attend to the important components of features and fuse them together. Because the computation of attended three features are similar, we introduce one of them for simply. Taken regional feature as example, given the output h_t^1 of the first LSTM, we first compute the normalized attention weight as

$$z_t = \mathbf{w}_z^T \tanh(W_R R + W_h h_t^1), \tag{5}$$

$$a_t = softmax(z_t), \tag{6}$$

where $W_R \in \mathbf{R}^{H \times D_R}$, $W_h \in \mathbf{R}^{H \times M}$ and $W_z \in \mathbf{R}^H$. M is the number of hidden units in LSTM and H is the the number of hidden units in attention module.

The attended regional feature is

$$\hat{R} = \sum_{i=1}^{k_r} a_{t,i} r_i. \tag{7}$$

Then we can get the combined feature $\hat{C} = [\hat{V}; \hat{R}; \hat{S}]$.

Word Prediction. After getting the output h_t^2 of caption LSTM at each time step t, the distribution over possible output words is given by

$$p(y_t) = softmax(W_p h_t^2 + b_p), \tag{8}$$

where $W_p \in \mathbf{R}^{N \times M}$ and $b_p \in \mathbf{R}^N$. The distribution over complete output sequences is calculated as

$$\log p(Y|I; \theta) = \sum_{t=1}^{T} \log p(y_t). \tag{9}$$

For inference, greedy search or beam search [20] can be used to generate the caption given an image.

3.4   Implementation Details

For regional features, we use Faster R-CNN to extract region proposals, which parameters of the model are the same as the publicly code[1]. To enable more object classes to be detected, we use Visual Genome dataset [19] as the training dataset, which contains 108 K images with dense annotations.

In the captioning model, we set the dimension of the input word embedding E to 512, the number of hidden units M in each LSTM to 1024 and the number of hidden units in attention module H to 1024. When training the network, we use ADAM algorithm with minibatch size of 10. The learning rate is set as 5×10^{-4}. The learning was stopped after 25 epochs. In inference, we use beam search with beam size equals 5.

4   Experimental Result

4.1   Dataset and Metrics

To evaluate the performance of the proposed captioning model, we use the MS-COCO 2014 caption dataset [12]. This dataset has 123, 287 images, each of which is given at least five captions by different AMT workers. For fair comparison, we use Karpathy splits [10] that have been extensively used in previous image captioning methods. This split contains 113, 287 images for training, 5, 000 images

---

[1] https://github.com/rbgirshick/py-faster-rcnn.

for validation and 5,000 images for testing. We use BLEU [15], CIDEr [18], METEOR [4] and ROUGE-L [6] as our evaluation metrics. All the metrics are computed by using the code[2] released by the COCO evaluation server. For pre-processing of caption, we also follow the publicly available method [10] to pre-process, yielding vocabulary sizes of 8791 for COCO.

4.2   Results on MS-COCO

Table 1 gives the performance of our proposed captioning model compared with other state-of-the-art methods on MS-COCO dataset. Among these methods, NIC [20] and soft-Attention [21] are the Classic methods in image captioning, ATT-FCN [22] and SCN-LSTM [7] solve the image captioning problem with semantic information. While GLA [11] and Up-Down [1] use Faster R-CNN to extract image regions, whose features are fed into LSTM. From the table, we find that our method can get the best performance compared with the state-of-the-art models, which improves the score of CIDEr from 1.127 to 1.160 on COCO. This result illustrates the effectiveness of multi-level features.

Table 1. Performance of our method compared with other state-of-the-art methods on MS-COCO dataset. The numbers in bold face are the best results and (-) indicates unknown scores.

Model	BLUE-1	BLUE-2	BLUE-3	BLUE-4	METEOR	ROUGE-L	CIDEr
NIC [20]	0.666	0.461	0.329	0.246	-	-	-
soft-Attention [21]	0.707	0.492	0.344	0.243	0.239	-	-
ATT-FCN [22]	0.709	0.537	0.402	0.304	0.243	-	-
GLA [11]	0.725	0.556	0.417	0.312	0.249	0.533	0.964
SCN-LSTM [7]	0.728	0.566	0.433	0.330	0.257	-	1.012
Up-Down [1]	0.772	-	-	0.362	0.270	0.564	1.135
MLFN(ours)	**0.779**	**0.619**	**0.482**	**0.371**	**0.279**	**0.577**	**1.160**

4.3   Analysis

To evaluate the effect of different features, we test four variants of our captioning model. The first variant only uses visual feature extracted by CNN. Besides of the visual feature, semantic feature is also used by the second variant. The semantic feature is replaced with regional feature in the third variant. The last variant uses all these features. From Table 2 we can see that using only visual feature gives the worst result. This is because the visual features are hard to capture all the details of image. The result of the third variant is better than that of the second one. This may because the regional feature is more discriminative than semantic feature, which semantic feature are more high-level. The last variant achieves

---

[2] https://github.com/tylin/coco-caption.

the best result. This shows that these three features can capture different level of informations of image and let us understand the image more deeply. Figure 2 gives some exemplar images and the generated captions by different variants. From figure we can see the fourth variant can give more accurate quantifier, adjective and noun for the caption sentences.

Table 2. Performance of the variants of our proposed method on MS-COCO dataset. From the first row to the last row are represents the first, the second, the third and the fourth variant respectively.

	BLUE-1	BLUE-2	BLUE-3	BLUE-4	METEOR	ROUGE-L	CIDEr
Vfeat	0.731	0.569	0.434	0.330	0.253	0.540	1.001
Vfeat+Sfeat	0.738	0.577	0.441	0.336	0.259	0.547	1.023
Vfeat+Rfeat	0.764	0.607	0.472	0.364	0.274	0.566	1.127
Vfeat+Sfeat+Rfeat	**0.779**	**0.619**	**0.482**	**0.371**	**0.279**	**0.577**	**1.160**

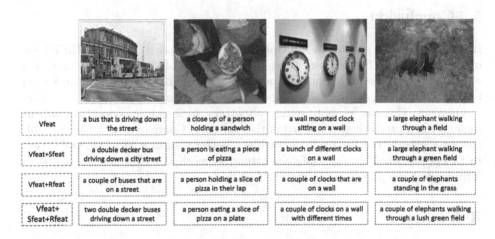

Vfeat	a bus that is driving down the street	a close up of a person holding a sandwich	a wall mounted clock sitting on a wall	a large elephant walking through a field
Vfeat+Sfeat	a double decker bus driving down a city street	a person is eating a piece of pizza	a bunch of different clocks on a wall	a large elephant walking through a green field
Vfeat+Rfeat	a couple of buses that are on a street	a person holding a slice of pizza in their lap	a couple of clocks that are on a wall	a couple of elephants standing in the grass
Vfeat+ Sfeat+Rfeat	two double decker buses driving down a street	a person eating a slice of pizza on a plate	a couple of clocks on a wall with different times	a couple of elephants walking through a lush green field

Fig. 2. The captioning results of different variants of our method. The first column indicates which variant is used to generate the caption.

We further qualitatively access the effect of semantic and regional features in Fig. 3. In this figure, we give some exemplar images, their semantic tags, the detected regions and the generated captions. For the semantic features, top-5 semantic words are listed. The semantic tags and regions can capture most of the information needed to generate the captions as illustrate by the left four columns of the figure. However if there are mistakes, the accuracy of captioning would also be ruined. For example, in the last column of the figure, semantic features mistakenly treats the cat in last image as a dog, result in the generated caption is incorrect.

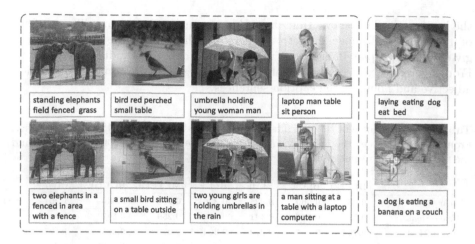

Fig. 3. Qualitative analysis on impact of different features. From the first row to last row are source images, semantic tags, detected regions and captions.

5 Conclusion

In this paper, we propose a multiple-level feature-based network(MLFN) for image captioning. The network takes different levels of image features as input, which has greater representational power than the single type of features used in existing method. Our caption model fuses these features through attention mechanism. This can let us to selectively attend to semantic concepts or salient regions when predicting each word of the sentence. Compared with the state-of-the-art methods, our approach can achieve better result on MS-COCO dataset. In future, we would design different attention mechanism for these features to fully exploit the potential of different features.

References

1. Anderson, P., et al.: Bottom-up and top-down attention for image captioning and visual question answering. arXiv: 1707.07998v2 (2017)
2. Bahdanau, D., Cho, K., Bengio, Y.: Neural machine translation by jointly learning to align and translate. In: ICLR (2015)
3. Cho, K., et al.: Learning phrase representations using RNN encoder-decoder for statistical machine translation. In: Proceedings of the 2014 Conference on Empirical Methods in Natural Language Processing (EMNLP), pp. 1724–1734 (2014)
4. Denkowski, M., Lavie, A.: Meteor universal: language specific translation evaluation for any target language. In: The Workshop on Statistical Machine Translation, pp. 376–380 (2014)
5. Fang, H., et al.: From captions to visual concepts and back. In: 2015 IEEE Conference on Computer Vision and Pattern Recognition (CVPR), pp. 1473–1482 (2015)
6. Flick, C.: ROUGE: a package for automatic evaluation of summaries. In: The Workshop on Text Summarization Branches Out, p. 10 (2004)

7. Gan, Z., et al.: Semantic compositional networks for visual captioning. In: 2017 IEEE Conference on Computer Vision and Pattern Recognition (CVPR), pp. 1141–1150 (2017)
8. He, K., Zhang, X., Ren, S., Sun, J.: Deep residual learning for image recognition. In: 2016 IEEE Conference on Computer Vision and Pattern Recognition (CVPR), pp. 770–778 (2016)
9. Hochreiter, S., Schmidhuber, J.: Long short-term memory. Neural Comput. **9**(8), 1735–1780 (1997)
10. Karpathy, A., Fei-Fei, L.: Deep visual-semantic alignments for generating image descriptions. IEEE Trans. Pattern Anal. Mach. Intell. **39**(4), 664–676 (2017)
11. Li, L., Tang, S., Deng, L., Zhang, Y., Tian, Q.: Image caption with global-local attention (2017)
12. Lin, T.-Y., et al.: Microsoft COCO: common objects in context. In: Fleet, D., Pajdla, T., Schiele, B., Tuytelaars, T. (eds.) ECCV 2014. LNCS, vol. 8693, pp. 740–755. Springer, Cham (2014). https://doi.org/10.1007/978-3-319-10602-1_48
13. Lu, J., Xiong, C., Parikh, D., Socher, R.: Knowing when to look: adaptive attention via a visual sentinel for image captioning. In: 2017 IEEE Conference on Computer Vision and Pattern Recognition (CVPR), pp. 3242–3250 (2017)
14. Mao, J., Xu, W., Yang, Y., Wang, J., Huang, Z., Yuille, A.: Deep captioning with multimodal recurrent neural networks (m-RNN). In: ICLR (2015)
15. Papineni, K., Roukos, S., Ward, T., Zhu, W.J.: BLEU: a method for automatic evaluation of machine translation. In: Proceedings of the 40th Annual Meeting on Association for Computational Linguistics, pp. 311–318 (2002)
16. Pedersoli, M., Lucas, T., Schmid, C., Verbeek, J.: Areas of attention for image captioning. In: ICCV-International Conference on Computer Vision (2017)
17. Ren, S., He, K., Girshick, R., Sun, J.: Faster R-CNN: towards real-time object detection with region proposal networks. Adv. Neural Inf. Process. Syst. **28**, 91–99 (2015)
18. Vedantam, R., Zitnick, C.L., Parikh, D.: CIDEr: consensus-based image description evaluation. In: Computer Vision and Pattern Recognition, pp. 4566–4575 (2015)
19. Vendrov, I., Kiros, R., Fidler, S., Urtasun, R.: Order-embeddings of images and language. arXiv preprint arXiv:1511.06361 (2015)
20. Vinyals, O., Toshev, A., Bengio, S., Erhan, D.: Show and tell: a neural image caption generator. In: 2015 IEEE Conference on Computer Vision and Pattern Recognition (CVPR), pp. 3156–3164 (2015)
21. Xu, K., et al.: Show, attend and tell: neural image caption generation with visual attention. In: International Conference on Machine Learning, pp. 2048–2057 (2015)
22. You, Q., Jin, H., Wang, Z., Fang, C., Luo, J.: Image captioning with semantic attention. In: 2016 IEEE Conference on Computer Vision and Pattern Recognition (CVPR), pp. 4651–4659 (2016)
23. Zhang, C., Platt, J.C., Viola, P.A.: Multiple instance boosting for object detection. In: Advances in Neural Information Processing Systems, pp. 1417–1424 (2006)
24. Zhou, L., Xu, C., Koch, P., Corso, J.J.: Watch what you just said: image captioning with text-conditional attention. In: Proceedings of the on Thematic Workshops of ACM Multimedia 2017, pp. 305–313. ACM (2017)

Collaborative Detection and Caption Network

Tianyi Wang[1], Jiang Zhang[2](✉), and Zheng-Jun Zha[3]

[1] iFlytek Co., Ltd., Hefei, China
tywang3@iflytek.com
[2] Hangzhou Dianzi University, Hangzhou, China
jiangzhang@hdu.edu.cn
[3] University of Science and Technology of China, Hefei, China
zhazj@ustc.edu.cn

Abstract. Recently it has been shown that deep recurrent neural network can be utilized to train video captioning systems. However, existing approaches often are perplexed by vagueness among videos, which often lead to grammatical correct but less germane results. In this paper, we propose an effective end-to-end network, called a Collaborative Detection and Caption network, which takes a video caption network as video-to-sentence sub-network and principle syntactic components detector as video-to-words sub-network. Our detector and caption network warp spatial-correlated attributes with temporal attention model and are optimized jointly which could facilitate each other. Experiments on the YouTube2Text, MPII movie description datasets and MVAD datasets consistently show that our proposed network can generate crucial contents needed for describing videos and thus enhance caption and detection performance simultaneously. Also, metric scores reported on those benchmarks have outperforms the state-of-the-art methods.

Keywords: Video caption · LSTM · Neural network

1 Introduction

With the application of Recurrent Neural Networks, modeling sequence task has shown great improvements in recent years especially in image/video captioning [2,6]. Video caption or video to text (VTT), could be regarded as a final core problem in computer vision domain and also be applied at many aspects, such as human-machine interaction and robotics. Compared to image caption, VTT suffers more difficulties, such as ensemble of objects, long periods of time and even the transition of scenes which thus brings more challenges to it.

Traditional VTT methods basically are established on encoder-decoder (ED) scheme to generate sentences sequentially. However, due to the restricted and vague hallmarks of image features, conventional works generate less germane results. When dealing to this problem, researchers combined very complicated

© Springer Nature Switzerland AG 2018
R. Hong et al. (Eds.): PCM 2018, LNCS 11164, pp. 104–114, 2018.
https://doi.org/10.1007/978-3-030-00776-8_10

deep CNN features, such as ResNet features [9], motion [5] and semantic features [28,30], to augment the performance level. These measures could be helpful because new features enhance the representative ability of network and diminish the vagueness. But those advantages could be sabotaged by the mismatch between visual input and textual output.

Further research of attention mechanism [13,20,29] help system cover more spatial and temporal relationship between frames. Spatial attention [29], however, would substantially increase the calculation burden because of the increasing dimension of visual feature to make sure it contain enough resolution.

In this paper, we incorporate temporal attention on space-correlated features instead of spatial-temporal attention, and design an syntactic components detector network to generate crucial textual words and solve the irrelevant problems. We also propose a novel collaborative end-to-end network for video caption. In all, our work mainly contributes on following two aspects:

(a) we design an network called Relation of Instances (RI) Network (RIN) to explore the objects among videos and association between them. In this paper, we detect subject, object and verb, namely SVO triplet as RI.
(b) we propose a Collaborative Detection and Caption Network (CDCN), taking a video caption network as video-to-sentence sub-network and RIN detector as video-to-words sub-network, where two sub-networks are trained jointly and enhanced mutually.
(c) we experiment on three popular video captioning benchmark, Microsoft Research Video Description Corpus (MSVDC) [8], the MPII Movie Description dataset (MPII-MD) [18] and Montreal Video Annotation Dataset (MVAD) [21]. Results demonstrate that our proposed approach outperform several state-of-art methods. For details, please refer to Sect. 3.3 Table 1.

2 Method

2.1 LSTM and Attention Model

In this work, we utilize LSTM [10] network and attention mechanism [3,20,26,27] to modeling our VTT network. We would not elaborate the principle of LSTM and Attention in details. Assume h_t is the hidden state, a_i is context information and x_t is the input. For conciseness, the LSTM unit and soft-attention mechanism in temporal sequence could be concisely defined as below,

$$h_t = LSTM(x_t, h_{t-1}). \tag{1}$$

$$\mathbf{c}_t = ATTN(\{\mathbf{a}_i\}_{i=1}^{M}, h_{t-1}). \tag{2}$$

2.2 Relationship of Instances Network

We have attenuate the reason why we add syntactic information into caption system. In this paper, we propose our RIN, a sequential method to dominantly generate syntactic components, Subject (S), Verb (V) and Object (O). In order to get labeled training data of SVO, we use third-party tools[1] to parse the captions and find out their subject, object and verb triplets. In encoding stage of RIN, we incorporate RGB features and 3DCNN features into our network to express static attributes and also the motion features. Distinguished from approaches for dense object detection in proposal way [17,22], we add a dense object-detection module, the multi-instance learning (MIL) [15,28] which proposed firstly in [6] to get 1000 way objects distribution, to generate spatial-correlated features. This features can be defined as $\mathbf{p}_i = \{p_i^w\}_{w=1}^{\mathcal{Y}}$. Here, we use CS to abbreviate $\{\mathbf{p}_i\}_{i=1}^{M}$, representing the ith frame's MIL. Finally, we incorporated the MIL features with temporal attention mechanism in order to attend not only spatially (MIL is spatial correlated) and temporally as well.

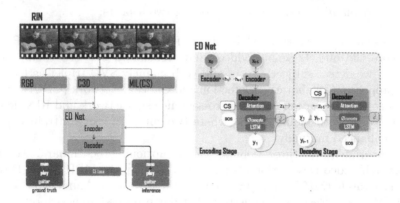

Fig. 1. Our RIN consist of encoder and decoder. Former part condense multi-modal visual features, and the MIL are integrated in via attention mechanism, while the latter one inference a precise SC output.

As shown in Fig. 1, we encoded the visual inputs by a LSTM layer, then decode out the RI (SVO). Suppose a video's visual input representation vector is $\{x^t\}_{t=1}^{T}$, which contains RGB, C3D and also MIL features. The dictionary is $\tilde{\mathcal{Y}}$, the $h_t \in \mathbb{R}^{D_h}$ is the hidden state of encoder LSTM and $z_t \in \mathbb{R}^{D_h}$ is the hidden state of decoder LSTM at time t, the t-th word in Y_{t-1} in caption is denoted by the one-hot representation in $\mathbb{R}^{|\tilde{\mathcal{Y}}|}$ space. The generating process of RI could be depicted as following:

$$h_t = LSTM(x_t, h_{t-1}) \tag{3}$$

$$\hat{c}_t^{(1)} = ATTN(CS, z_{t-1}^{(1)}) \tag{4}$$

$$z_t^{(1)} = LSTM(\phi(\hat{c}_t^{(1)}, \boldsymbol{E}_d d_{t-1}), z_{t-1}^{(1)}) \tag{5}$$

$$d_t = \arg\max(softmax(\boldsymbol{W}_{z^{(1)}} z_t^{(1)})) \tag{6}$$

---

[1] We use the Stanford Parser [12] to parse captions and choose the noun, verb and obj of the tokenization results as their subject-verb-object triplet for each caption.

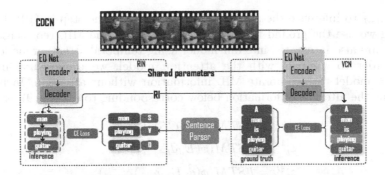

Fig. 2. CDCN yields RI via RIN and injects them into VCN to synthesize sentences. Encoder layers are shared between RIN and VCN in CDCN.

where $ATTN(\cdot)$ denotes the soft-attention mechanism described above in Sect. 2.1. So each $\hat{c}_t$ is jointly decided by SC context vectors and z_{t-1}. Because what the SA attend is temporal dimension, $M = T$. $\boldsymbol{E} \in \mathbb{R}^{|\tilde{\mathcal{Y}}| \times D_e}$ is the linear embedding matrix. For conciseness, we omit all the bias terms. ϕ is the concatenating function. The loss function of RIN is softmax loss, denoted by L_{rin}.

2.3 Collaborative Detection and Caption Network

As what have been argued before, our proposed CDCN take a video caption network (VCN) as video-to-sentence sub-network and RIN detector as video-to-words sub-network. Both sub-networks are trained jointly and enhanced mutually. Particularly, RIN outputs SVO triplet of a sentence, which define the basic content of a video, while VCN are used for sentence generation. The CDCN network have been depicted in Fig. 2, from which we can imply that CDCN quite like a multi-task architecture. In Fig. 2, VCN and RIN share the encoder layers and diverge before the decoding stage. As the encoding process has been illustrated in Sect. 2.2 (3), we would give the decoding approach below. Following equations depict the VCN's decoding process:

$$\hat{d} = \boldsymbol{E}_0 d_0 + \boldsymbol{E}_1 d_1 + \boldsymbol{E}_2 d_2 \tag{7}$$

$$\hat{c}_t^{(2)} = ATTN(CS, z_{t-1}^{(2)}) \tag{8}$$

$$z_t^{(2)} = LSTM(\phi(\hat{d}, \hat{c}_t^{(2)}, \boldsymbol{E}_y y_{t-1}), z_{t-1}^{(2)}) \tag{9}$$

$$y_t = \arg\max(softmax(\boldsymbol{W}_{z^{(2)}} z_t^{(2)})) \tag{10}$$

$$t \in [1, T], i \in [1, M]$$

$\hat{d}$ is build by d_0, d_1, d_2, coming from (6) which represent the SVO triplet. Equation (7) denotes the processing method when dealing with $\hat{d}$. Embedding transformations of each term in S-V-O $\boldsymbol{E}_0, \boldsymbol{E}_1, \boldsymbol{E}_2$ are independent. We combine the $\hat{d}$ with hidden state z_t of last LSTM model and last time step's word embedding vector y_{t-1} to inference the possible word at current time step in (10). During training we use the ground target word as the y_{t-1}. The MIL representations $\{\mathbf{p}_i\}_{i=1}^M$ are also integrated into our model using temporal attention model.

In order to compare with the attention model, we proposed a another sequence model to incorporate MIL information without attention. In briefly, we define the alternative formation below corresponding to (3) (5) (9) as

$$h_t = LSTM(\phi(x_t, CS), h_{t-1}) \tag{11}$$

$$z_t^{(1)} = LSTM(\phi(\boldsymbol{E}_d d_{t-1}), z_{t-1}^{(1)}) \tag{12}$$

$$z_t^{(2)} = LSTM(\phi(\hat{d}, \boldsymbol{E}_y y_{t-1}), z_{t-1}^{(2)}) \tag{13}$$

where the $M = T$. The experiment results would be shown in Sect. 3.3. The training loss of VCN is also softmax loss which denoted via L_{VCN}. Because the CDCN is consisted of VCN and RIN, then L_{CDCN} can be defined as,

$$L_{CDCN} = (1 - \alpha^n)L_{VCN} + \alpha^n L_{RIN}, \tag{14}$$

where $\alpha \leq 1$, n is the function of iteration steps. As the RIN is easier to converge than VCN, the initial weight of RIN would be large and progressively decay when the training step grow. When the α^n approximate zero, the L_{RIN} would nearly stop optimizing while only predict svo output for VCN.

3 Experiments

Experiments are implemented on three benchmarks MSVDC, MPII and MVAD. First, we explore difference of diverse structure in Table 2(a). Also, we evaluate the precision of SVO respectively. Second, we compare our results with several state-of-the-art results on MSVD in Table 1. Eventually, we report our scores on challenging MPII-MD and MVAD dataset in Table 3(a), (b).

3.1 Datasets

MSVDC. This corpus consists of 1970 video clips, maintained more than 70,000 captions, lasted 5.3 h. The total number of clip-sentence pairs is about 80,000. We split them into train (1200), test (670), val (100) respectively [27].

MPII. This dataset was proposed by Xu et al. [18] in 2016. This dataset last a duration of 73.6 h, consists of 68,000 clips and the vocabulary size is 24,000+. In all, there is 68,337 clips and sentence pairs. We utilize the standard splits, same as the settings in [11,24,30].

MVAD. This movie description dataset was created by MILA, UdeM, and it contains about 49,000 movie snippets that accompanied with just one caption and the vocabulary size is 18,000+. It last a duration of 84.6 h which is the most among all datasets we evaluated. We followed the official splits [21].

3.2   Implementation Details

Model Training. The 3D-CNN features is 4096-way fc7-1 which are extracted from model pre-trained through the SPORT1M [5]. The frame features are ResNet-152 [9]'s pool5, 2048 way, and VGG-16 fc7, 4096 way. The CS features are generated from VGG-16 MIL network's final 1000-way. The word embedding size is set as 900, empirically, while the dimension of RI is 300 for each item in svo. LSTM layers' hidden size are 1000. We sample one frame by each ten of video, drop all frames more than 60 and pad zero frames for videos less than 60f. We utilize 0.5 dropout rate after each embedding layer and clip gradients at 10. α of (14), is 0.5 and we use the step-function to control the decay rate of α^n in which the step width is 3 epochs.

Model Testing. During the test stage, the golden y_{t-1} and d_{t-1} would be supplanted by last time step's output of decoder in VCN and RIN. We implement decoding process with the Beam Search width equals to 5.

Compared Algorithms. We compared our metrics with several previous methods to verify the performance of our designs. (1) Long Shot-Term Memory (LSTM) [25] (2) S2VT [24] (3) Temporal Attention (TA) [27] (4) LSTM embedding (LSTM-E) [14] (5) Paragraph RNN decoder (p-RNN) [29] (6) Hierarchical Recurrent Neural Encoder (HRNE) [13] (7) Hierarchical LSTM with Adjusted Temporal Attention (hLSTMat) [20] (8) LSTM with Multi-Faceted Attention (TM) (9) Supervised Tessellation [11].

Evaluation Metric. We use BLEU [16], ROUGE-L [7], METEOR [4], CIDEr [23] as the performance metrics, which are the most frequently used metrics and most correlated with human-beings. We evaluate the metrics via employing the open-source code[2] released with MSCOCO Evaluation Server [1].

3.3   Experimental Results

We run several configurations of our proposed method and evaluate them respectively. We use these abbreviation below to denote the system dependencies. (1) **ResNet(R)** [9]: we use the ResNet152's output pool5 feature as a basic visual input. (2) **ResNet-C3D(RC)** [5,14]: we add 3D-CNN features fc7-1, generated through the network in [5] and combine them with pool5. (3) **ResNet-C3D-CSA(RCSA)**: upon the setting (2), CS features is warped in by attention model. (4) **CDCN using VGG16-C3D(CDCN-VC)**: in order to remove the positive effects coming from advanced features in ResNet, we also implemented an ablated experiment that used VGG16's fc7 4096-way features as input. (5) **CDCN using VGG16-C3D-CSA(CDCN-VCSA)**: (6) **CDCN-ResNet(CDCN-R)**: implementation of our CDCN baseline with ResNet features only. (7) **CDCN-C3D(CDCN-RC)**: embedding C3D features in network's encoder. (8) **CDCN-ResNet-C3D-CS(CDCN-RCS)**:

---

[2] https://github.com/tylin/coco-caption.

Table 1. CIDEr, BLEU@N, ROUGEL and METEOR scores on MSVD

Method	CIDEr	BLEU1	BLEU2	BLEU3	BLEU4	ROUGEL	METEOR
(1) LSTM	–	–	–	–	33.3	–	29.1
(2) TA	51.7	80.0	64.7	52.6	41.9	–	29.6
(3) S2VT	–	–	–	–	–	–	29.8
(4) LSTM-E	–	78.8	66.0	55.4	45.3	–	31.0
(5) HRNE	–	81.1	68.6	57.8	46.7	–	**33.9**
(6) p-RNN	65.8	81.5	70.4	60.4	49.9	–	32.6
(7) hLSTMat	68.9	82.9	72.0	62.7	52.8	70.5	33.4
(8) TM	73.8	82.9	72.2	63.0	53.0	–	33.6
(9) CDCN-VC	68.3	81.8	70.8	60.2	50.0	69.1	32.0
(10) CDCN-VCSA	71.1	82.7	71.6	62.0	51.2	69.9	33.0
(11) CDCN-R	70.5	81.5	69.7	60.1	50.3	69.4	32.8
(12) CDCN-RCS	74.4	83.0	71.5	62.0	52.3	70.7	33.4
(13) CDCN-RCSA	**75.2**	**83.3**	**72.4**	**63.3**	**53.3**	**70.8**	33.6

Scores are reported as percentage (%)

adding C3D features and CS into CDCN network's encoder (11)(12)(13). (9) **CDCN-ResNet-C3D-CSA(CDCN-RCSA)**: combining the CDCN-RC with CS-attention model. This setting could lead to the proposed best results. **MSVD** Table 2(a) shows our performances of different model structures on MSVD dataset. It can be seen that high-qualitative features could enhance the caption performance. In Table 2(a), result (3) obviously outperform the result (1)–(2), which explain the advantages of CS-Attention model. Result (4) demonstrate our model's robustness even without such fine features. Further, the CDCN always outperform general ED network with equal features inputs, which could be proved via comparing the result (4) (5) and (7) to (1)–(3) respectively. Overall, our CDCN could generate better scores with same visual features than conventional ED network.

Quantitative Analysis: (A) MSVDC In Table 1, our CDCN-RCSA (CR) achieves the best scores in ROUGE-L and CIDEr and also on BLEU@N. Our baseline (11) could also achieves a improvement in many metric. While TM (8) got approximate scores at BLEU@N and METEOR, but our (13) outperform it in each metric especially the CIDEr. We could compare (10) with (5) and (6), which all scores are calculated based on VGG features. Even though our results lack of spatial attention paradigm, scores indicate that our results are still better than methods in [13,29]. Although HRNE (5) achieved the highest score on METEOR, but the obvious much lower values on other metrics make us hardly to judge its performance objectively.

Table 2. Auto metric scores on MPII-MD and M-VAD

(a) CDCN Vs. Baseline on MSVD

Model	Cr	B4	M
(1) R	67.0	49.9	31.8
(2) RC	71.6	50.7	33.0
(3) RCSA	73.0	51.6	33.3
(4) CDCN-R	70.5	50.3	32.8
(5) CDCN-RC	74.3	53.0	33.3
(6) CDCN-RCS	74.4	52.3	33.4
(7) CDCN-RCSA	**75.2**	**53.3**	**33.6**

(b) Precision of SVO on MSVD

Pr	S	V	O
(1)RCSA	86.38	45.99	24.36
(2)CDCN-RCSA	86.19	51.80	27.33
(3)RIN in (2)	88.51	54.18	40.75
(4)RIN Only	86.12	52.84	39.55

*Scores are reported as percentage(%)

(B) Precision of SVO. In order to evaluate the effect of RIN, we test the precision of S, V, O in RIN and VCN of CDCN. In Table 2(b), we list the precision of our detected results on SVO triplet. From the results of (1) and (2), precision achieve a great improvements in Verb and Object. Because most of the captions' subject in test set are "PERSON", the precision rates in S are so similar. Result (3) is the precision of RIN output in CR, while (4) is from an independent RIN network. Through the difference between (3) and (4), (1) and (2), we can infer that RIN's and VCN's performance have definitely been mutually improved via the CDCN strategy.

(C) MPII and MVAD. Scores reported in Table 3(a) and Table 3(b) indicate that the performance are relatively much lower than those in MSVD. This is mostly caused by its high diversity in visual content and textual expression. As a result, making captions be at a nonplus. Some of those metrics are close to 0. So in order to evaluate our approaches more objectively, we take more consideration on CIDEr and METEOR. Specifically, CR network made a improvement over the Visual Labels [19] on METEOR and CIDEr on both datasets. Results in Table 3(a), (b) imply that even in harder circumstances, our approach perform better than previous results.

Qualitive Analysis: In Fig. 3, we show several results from MSVD. Column A of Fig. 3 illustrate three results with correct RI and precise caption comparing

Table 3. Auto metric scores on MPII-MD and M-VAD

(a) MPII dataset

Method	CIDEr	METEOR
(1)LSTM	-	6.7
(2)S2VT	8.8	7.1
(3)LSTM-E	-	7.3
(4)Visual Labels	10.0	7.0
(5)RIN-C	8.3	7.4
(6)RIN-CLA	**10.3**	**7.6**

Values are reported as percentage(%)

(b) M-VAD dataset

Method	CIDEr	METEOR
(1)LSTM	-	6.1
(2)S2VT	-	6.7
(3)LSTM-E	-	6.7
(4)Visual Labels	-	6.4
(5)HRNE	-	6.8
(6)RIN-C	7.9	6.9
(7)RIN-CLA	**9.8**	**7.0**

Values are reported as percentage(%)

to targets. What's more, some results without utilizing RIN, fail to predict the correct or relative content. But after implementation of CR, results can be transformed into more relative ones. Take the row three for example, the action of this video was misclassified as "cooking", however, the CR could detect "pouring" VCN, correct the CDCN sentence finally. Sometimes the RIN could not generate relative svo triplet of a caption, but final output sentence of our CR network still could synthesize a more convincible result. In picture (2) of column B, although dog are not relative to this image, CR network still use the verb, "walk" to turn the action from "Sitting" into "Walking" successfully. This advantage prove that RIN could highly improved the decoding accuracy in VCN. But there are still some circumstances that CR fail to fix feasibly. From column B again, picture (1)'s mistake may come from the rarity of underlined word ferret, which may be neglected by models. Picture (3) indicates that inaccurate svo could cause some troubles. Comparing to RCSA's result, the subject in CR sentence is wrong. The error may be caused by the biases of datasets which accommodate more corpus about human-swimming pair.

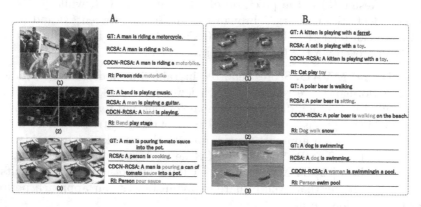

Fig. 3. Partial results on MSVD, column (A) are correct and relevant results while (B) are results with different type of mistakes. For details please refer to Sect. 3.3

4   Conclusion

In this paper, we propose a novel framework, named Collaborative Detection and Caption Network (CDCN) for video caption. It generates syntactic information from the proposed Relationship of Instance Network (RIN) and then combines these priors into a Video Caption Network (VNC) to generate the target sentences. By warping spatial-correlated attributes with temporal attention model together and being optimized jointly, it achieves superior performance on several benchmark datasets, i.e., MSVD, MPII-MD and M-VAD, which show that it can resist to the vagueness perplexity among videos and generate words more related to the ground truth than other end-to-end methods. The objective SVO precision also proves the effectiveness of the proposed CDCN.

Acknowledgment. This work was supported by the National Natural Science Foundation of China (NSFC) under Grants 61622211, 61472392, 61751304 and 61620106009.

References

1. Chen, X., et al.: Microsoft coco captions: data collection and evaluation server. arXiv preprint arXiv:1504.00325 (2015)
2. Chen, X., Zitnick, C.L.: Mind's eye: a recurrent visual representation for image caption generation, pp. 2422–2431 (2014)
3. Cho, K., Courville, A., Bengio, Y.: Describing multimedia content using attention-based encoder-decoder networks. IEEE Trans. Multimed. **17**(11), 1875–1886 (2015)
4. Denkowski, M., Lavie, A.: Meteor universal: Language specific translation evaluation for any target language. In: The Workshop on Statistical Machine Translation, pp. 376–380 (2014)
5. Du, T., Bourdev, L., Fergus, R., Torresani, L., Paluri, M.: C3d: generic features for video analysis. Eprint Arxiv (2014)
6. Fang, H., et al: From captions to visual concepts and back. In: 2015 IEEE Conference on Computer Vision and Pattern Recognition (CVPR), pp. 1473–1482 (2015). https://doi.org/10.1109/CVPR.2015.7298754
7. Flick, C.: ROUGE: a package for automatic evaluation of summaries. In: The Workshop on Text Summarization Branches Out, p. 10 (2004)
8. Guadarrama, S., et al.: YouTube2Text: recognizing and describing arbitrary activities using semantic hierarchies and zero-shot recognition. In: IEEE International Conference on Computer Vision, pp. 2712–2719 (2013)
9. He, K., Zhang, X., Ren, S., Sun, J.: Deep residual learning for image recognition, pp. 770–778 (2015)
10. Hochreiter, S., Schmidhuber, J.: Long short-term memory. Neural Comput. **9**(8), 1735–1780 (1997)
11. Kaufman, D., Levi, G., Hassner, T., Wolf, L.: Temporal tessellation: a unified approach for video analysis (2017). 2(9), II-II
12. Klein, D., Manning, C.D.: Accurate unlexicalized parsing. In: Proceedings of the 41st Annual Meeting on Association for Computational Linguistics-Volume 1, pp. 423–430. Association for Computational Linguistics (2003)
13. Pan, P., Xu, Z., Yang, Y., Wu, F., Zhuang, Y.: Hierarchical recurrent neural encoder for video representation with application to captioning, no. 6, pp. 1029–1038 (2015)
14. Pan, Y., Mei, T., Yao, T., Li, H., Rui, Y.: Jointly modeling embedding and translation to bridge video and language. In: 2016 IEEE Conference on Computer Vision and Pattern Recognition (CVPR), pp. 4594–4602, June 2016. https://doi.org/10.1109/CVPR.2016.497
15. Pan, Y., Yao, T., Li, H., Mei, T.: Video captioning with transferred semantic attributes (2016)
16. Papineni, K., Roukos, S., Ward, T., Zhu, W.J.: BLEU: a method for automatic evaluation of machine translation. In: Meeting on Association for Computational Linguistics, pp. 311–318 (2002)
17. Ren, S., He, K., Girshick, R., Sun, J.: Faster R-CNN: towards real-time object detection with region proposal networks. In: International Conference on Neural Information Processing Systems, pp. 91–99 (2015)
18. Rohrbach, A., Rohrbach, M., Tandon, N., Schiele, B.: A dataset for movie description. In: Proceedings of the IEEE Conference on Computer Vision and Pattern Recognition (CVPR) (2015)

19. Rohrbach, A., et al.: Movie description. Int. J. Comput. Vis. (2017). http://resources.mpi-inf.mpg.de/publications/D1/2016/2310198.pdf
20. Song, J., Guo, Z., Gao, L., Liu, W., Zhang, D., Shen, H.T.: Hierarchical LSTM with adjusted temporal attention for video captioning (2017)
21. Torabi, A., Pal, C., Larochelle, H., Courville, A.: Using descriptive video services to create a large data source for video annotation research. Computer Science (2015)
22. Uijlings, J.R., Van De Sande, K.E., Gevers, T., Smeulders, A.W.: Selective search for object recognition. Int. J. Comput. Vis. **104**(2), 154–171 (2013)
23. Vedantam, R., Zitnick, C.L., Parikh, D.: Cider: consensus-based image description evaluation. In: IEEE Conference on Computer Vision and Pattern Recognition, pp. 4566–4575 (2015)
24. Venugopalan, S., Rohrbach, M., Donahue, J., Mooney, R., Darrell, T., Saenko, K.: Sequence to sequence - video to text (2015)
25. Venugopalan, S., Xu, H., Donahue, J., Rohrbach, M., Mooney, R.J., Saenko, K.: Translating videos to natural language using deep recurrent neural networks. CoRR abs/1412.4729 (2014). http://arxiv.org/abs/1412.4729
26. Xu, K., et al.: Show, attend and tell: neural image caption generation with visual attention. In: Computer Science, pp. 2048–2057 (2015)
27. Yao, L., et al.: Describing videos by exploiting temporal structure. In: IEEE International Conference on Computer Vision, pp. 4507–4515 (2015)
28. You, Q., Jin, H., Wang, Z., Fang, C., Luo, J.: Image captioning with semantic attention, pp. 4651–4659 (2016)
29. Yu, H., Wang, J., Huang, Z., Yang, Y., Xu, W.: Video paragraph captioning using hierarchical recurrent neural networks. In: Computer Vision and Pattern Recognition, pp. 4584–4593 (2016)
30. Yu, Y., Ko, H., Choi, J., Kim, G.: Video captioning and retrieval models with semantic attention. arXiv preprint arXiv:1610.02947 (2016)

Video-Based Person Re-identification with Adaptive Multi-part Features Learning

Jingjing Wu$^{(\boxtimes)}$, Jianguo Jiang , Meibin Qi , Hao Liu ,
and Meng Wang

Hefei University of Technology, Hefei 230009, Anhui, China
hfutwujingjing@mail.hfut.edu.cn, jgjiang@hfut.edu.cn, qimeibin@163.com,
hfut.haoliu@gmail.com, eric.mengwang@gmail.com

Abstract. Video-based person re-identification plays a significant role in the video surveillance, which can automatically judge whether two non-overlapping video sequences of the pedestrian belong to the same class or not. However, many factors make it challenging, such as different viewpoints and illumination among different cameras, the occlusion, *etc.* Aiming at increasing the robustness to the occlusion, this paper extracts multi-part appearance features and the feature weight of each part is learned according to its importance. Besides, in order to fully utilize the information included in the video sequences, this paper combines the appearance features and spatial-temporal features of pedestrian by learning several independent metric kernels and fusing the learned metric distances. Extensive experiments on two public benchmark datasets, *i.e.*, the iLIDS-VID and PRID-2011 datasets, demonstrate the effectiveness of the proposed method.

Keywords: Video-based person re-identification
Multi-part appearance features · Adaptive feature weight learning
Distance fusion

1 Introduction

Person Re-identification (Re-ID) aims to match people across multiple surveillance cameras with non-overlapping views, which has many applications in the video surveillance, such as cross-view object tracking and object retrieval, *etc.* Generally, Re-ID can be divided into image-based and video-based Re-ID. Video-based Re-ID is close to real-world application requirements, so this paper focuses on it. However, many factors make video-based Re-ID difficult to tackle, including the variance of viewpoints and illumination among different cameras, the

Supported by organization National Natural Science Foundation of China Grant 61876056, 61632007 and Key Research and Development Project of Anhui Province, China (1704d0802183).

© Springer Nature Switzerland AG 2018
R. Hong et al. (Eds.): PCM 2018, LNCS 11164, pp. 115–125, 2018.
https://doi.org/10.1007/978-3-030-00776-8_11

occlusion, *etc.* To solve these problems, video-based Re-ID has been extensively studied with methods generally falling into two main aspects: feature extraction and metric learning.

Methods of feature extraction are designed to extract the hand-crafted features [1,3,13,21,22,26] or the deep features [10–12,14,15,19,23–25], which should be both discriminative and robust to the variance among different cameras. The hand-crafted features include the appearance feature and the spatial-temporal feature, such as color, texture, edge, optical flow, *etc.* Literature [22] exploits the periodicity exhibited by a walking person to generate representation, which aligns the spatial-temporal appearance of a pedestrian globally. The deep features usually include the spatial features extracted from Convolutional Neural Network (CNN) and the temporal features between video frames extracted from Recurrent Neural Network (RNN). Literature [19] designs a network whose spatial pooling layer can select regions from each frame and attention temporal pooling can select informative frames over the sequence. In [8], the final human descriptors are composed of deep features and hand-crafted local features.

Although the deep features have achieved nice results on large datasets, the research on the hand-crafted features is still significative when the system computing power is weak and training samples are insufficient. Thus, this paper pays attention to the hand-crafted features.

In general, some methods [2,16,20] divide the images into several horizontal stripes to obtain the local and detailed features. These methods have achieved promising performance. While this paper divides the pedestrian image into grids to obtain more detailed features. As we know, when identifying pedestrians, the attention to different parts of pedestrian, such as the upper body or the lower body or the whole body, is different. Inspired by the human visual mechanism, we extract the appearance feature from the grids to construct multi-part appearance features, including the whole part, the upper part and the lower part appearance features. Then the importance of each part appearance feature can be obtained by training these appearance features. Thus, the effect of the useful part can be enlarged and the effect of poor part, such as occlusion region, can be reduced.

Distance metric learning algorithms [7,18,21,26,27] usually map the features into the metric space, where the intra-class variations are smaller and the inter-class variations are larger, so we can calculate the distances between the probe and the gallery in the metric space. At last, the gallery videos are ranked according to the calculated distances. Zhu *et al.* [27] propose a method that learns intra-video and inter-video distance metrics simultaneously to make the representation more compact and discriminative. In [21], You *et al.* exploit a feature representation constituted by Histogram of Oriented Gradient 3D (HOG3D) [6] and the average pooling of color histograms and Local Binary Pattern (LBP) [5]. And a discriminative distance model is proposed to optimize towards the realization of the top-push distance constraint. Literature [26] jointly learns a pair of feature projection matrices and a pair of dictionaries by integrating the information contained in labeled and unlabeled pedestrian videos.

The effective combination of the appearance and the spatial-temporal features is beneficial to describe the pedestrians. In some algorithms [8,21], in order to fully exploit the information in the video and achieve the functional complementarity of these two features, the appearance feature is concatenated with the spatial-temporal feature. However, it's unreasonable to concatenate these two features when the dimensions of the two features are not in the same order of magnitude. Therefore, in this paper, several metric kernels can be obtained by feeding the appearance and the spatial-temporal features into the metric learning respectively, then these features are combined by fusing different learned metric distances.

To sum up, this paper presents an adaptive multi-part appearance features weight learning and distance fusion method, named AMFL, to address the Re-ID issue. The main contributions of this paper are as follows: (1) We divide the image and extract the multi-part appearance features. Then we obtain the importance of each part appearance feature by training. Therefore, it can reduce the effect of the poor region and increase robustness to the occlusion. (2) We combine the multi-part appearance features and the spatial-temporal feature with distance fusion to achieve better complementarity of these features. (3) This paper achieves the state-of-the-art performance.

2   The Proposed Method

2.1   Framework

The procedure of the proposed method consists of two parts: training stage and testing stage depicted in Fig. 1. In detail, the framework includes five steps:

(a) The image is divided into 6 grids, and the Local Maximal Occurrence (LOMO) feature [9] is extracted from each grid to generate the upper part, lower part and whole part appearance features.
(b) The HOG3D feature is extracted from each pedestrian video as the spatial-temporal feature.
(c) Top-push Distance Learning (TDL) [21] is performed with the spatial-temporal HOG3D feature.
(d) The distance metric learning and feature weight learning are conducted in turn with the multi-part appearance features.
(e) In the testing stage, we calculate the multi-part appearance features distances and spatial-temporal feature distances between the samples respectively, then sum them to obtain the final distance between the samples.

2.2   Features Extraction

Multi-part Appearance Features Extraction. To get more detailed local information, we equally divide the original image into 3 horizontal stripes and 2 vertical columns, thus we get 3×2 grid-based division image.

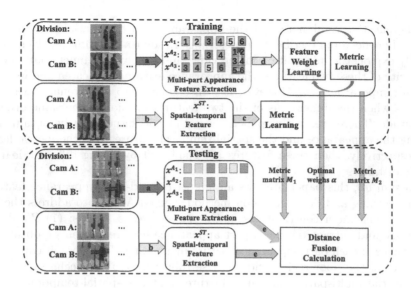

Fig. 1. An overview of the proposed method. The red box shows the extraction method of multi-part appearance features: Each color block represents the appearance feature of the grid with the same number in the pedestrian image. Best viewed in color.

Then the LOMO feature which consists of the color and texture information is extracted from each grid as the appearance feature. The whole part appearance feature is obtained by concatenating the LOMO features of 6 grids in order, which is denoted as x^{A_1}. In most cases, top (bottom) 4 grids approximately include the upper (lower) body of the pedestrian. Based on this observation, the upper (lower) part appearance feature is defined as the concatenation of LOMO features which are extracted from top (bottom) 4 grids, and we denote it as x^{A_2} (x^{A_3}). The Principal Component Analysis (PCA) is utilized to reduce the x^{A_1} dimension to the same dimension as x^{A_2} and x^{A_3}. Thus, we obtain appearance features of three parts named multi-part appearance features, and denote them as $x^A = [x^{A_1}, x^{A_2}, x^{A_3}]$. The specific orders of concatenation are illustrated in red box of the Fig. 1.

The multi-part appearance features are extracted from each image in the video. Given that person p has N_p images, we denote the appearance feature column vector of person p for part m ($m \in [1, 2, 3]$) as $x_{p,n}^{A_m}$. Among them, n means that the features are extracted from the n-th ($n \in [1, 2, \ldots, N_p]$) image of video. The final appearance feature of person p for part m ($\mathbf{x}_p^{A_m}$) is obtained by putting N_p feature vectors $x_{p,n}^{A_m}$ through the average pooling:

$$x_p^{A_m} = \frac{1}{N_p} \sum_{n=1}^{N_p} x_{p,n}^{A_m} \tag{1}$$

Spatial-Temporal Feature Extraction. The video contains not only rich appearance features but also abundant spatial-temporal clues. The combination of the appearance and the spatial-temporal features is useful to describe the pedestrians. Thus, the 1200-dimensional HOG3D feature is extracted from the pedestrian video as spatial-temporal feature. Furthermore, the HOG3D feature has accumulated global information of the video, while the multi-part appearance features contain local information of the video. Therefore, the combination of these features is also the combination of the global and the local information of the video. We denote the HOG3D feature column vector of person p as x_p^{ST}.

2.3   Metric Learning

Distance Fusion. The general tasks of distance metric learning are to pull the positive feature pairs closer and push the negative feature pairs further by learning the metric space of identified samples. Not only that, but this paper also integrates the multi-part appearance features and the spatial-temporal features with the distance fusion. Concretely, we conduct distance metric learning twice with the multi-part appearance features and the spatial-temporal features respectively. After that, the distances between samples of the appearance features and the spatial-temporal features are calculated separately and then added as the final distances of samples.

Training Strategy. (1) Training with the spatial-temporal HOG3D features: This paper performs TDL with the HOG3D feature, which could keep the feature intra-class distances smaller than the related minimum inter-class distances. The most commonly used distance, Mahalanobis distance, is applied as the distance between the sample HOG3D features x_i^{ST}, x_j^{ST}, which is denoted $D(x_i^{ST}, x_j^{ST})$:

$$D(x_i^{ST}, x_j^{ST}) = (x_i^{ST} - x_j^{ST})^T M_1 (x_i^{ST} - x_j^{ST}), \tag{2}$$

where M_1 is a semi-definite matrix. TDL minimizes the following objective loss function in terms with the metric matrix M_1:

$$f(M_1) = (1 - \beta) \sum_{x_i^{ST}, x_j^{ST}, y_i = y_j} D(x_i^{ST}, x_j^{ST})$$

$$+ \beta \sum_{x_i^{ST}, x_j^{ST}, y_i = y_j} [D(x_i^{ST}, x_j^{ST}) - \min_{y_k \neq y_i} D(x_i^{ST}, x_k^{ST}) + \rho]_+, \tag{3}$$

where ρ is a slack parameter and $\beta \in [0, 1]$ refers to a weighting parameter that balances the two terms and y_i, y_j, y_k are the labels of the sample HOG3D features $x_i^{ST}, x_j^{ST}, x_k^{ST}$ respectively. Then the gradient descent method is exploited to optimize the loss function and obtain the metric matrix M_1 for the spatial-temporal feature.

(2) Training with the multi-part appearance features: In order to weigh the contributions of multi-part appearance features $x^A = [x^{A_1}, x^{A_2}, x^{A_3}]$, we distribute the weights $\alpha = [\alpha_1, \alpha_2, \alpha_3]^T$ to appearance features of three parts,

and $\alpha_1, \alpha_2, \alpha_3 \geqslant 0$, $\sum_{q=1}^{3} \alpha_q = 1$. So the distance between sample appearance features $\boldsymbol{x}_i^A, \boldsymbol{x}_j^A$ further becomes:

$$D_{\boldsymbol{\alpha},\mathbf{M}_2}(\boldsymbol{x}_i^A, \boldsymbol{x}_j^A) = [\sum_{q=1}^{3} \alpha_q (\boldsymbol{x}_i^{A_q} - \boldsymbol{x}_j^{A_q})^T]\mathbf{M}_2[\sum_{q=1}^{3} \alpha_q (\boldsymbol{x}_i^{A_q} - \boldsymbol{x}_j^{A_q})]. \tag{4}$$

Based on the new distance function, we can get the new objective loss function by substituting the Eq. (4) into Eq. (3):

$$\begin{aligned}
f'(\mathbf{M}_2, \boldsymbol{\alpha}) = (1-\beta) \sum_{\boldsymbol{x}_i^A, \boldsymbol{x}_j^A, y_i=y_j} D_{\boldsymbol{\alpha},\mathbf{M}_2}(\boldsymbol{x}_i^A, \boldsymbol{x}_j^A) \\
+ \beta \sum_{\boldsymbol{x}_i^A, \boldsymbol{x}_j^A, y_i=y_j} [D_{\boldsymbol{\alpha},\mathbf{M}_2}(\boldsymbol{x}_i^A, \boldsymbol{x}_j^A) - \min_{y_k \neq y_i} D_{\boldsymbol{\alpha},\mathbf{M}_2}(\boldsymbol{x}_i^A, \boldsymbol{x}_k^A) + \rho]_+,
\end{aligned} \tag{5}$$

To simplify our notations, $D_{\boldsymbol{\alpha},\mathbf{M}_2}(\boldsymbol{x}_i^A, \boldsymbol{x}_j^A)$ can be reformulated as:

$$D_{\boldsymbol{\alpha},\mathbf{M}_2}(\boldsymbol{x}_i^A, \boldsymbol{x}_j^A) = \boldsymbol{\alpha}^T \mathbf{C}_{i,j} \mathbf{M}_2 \mathbf{C}_{i,j}^T \boldsymbol{\alpha} = tr(\mathbf{C}_{i,j}\mathbf{M}_2\mathbf{C}_{i,j}^T\mathbf{W}), \tag{6}$$

where $\mathbf{C}_{i,j} = [(\boldsymbol{x}_i^{A_1} - \boldsymbol{x}_j^{A_1}), (\boldsymbol{x}_i^{A_2} - \boldsymbol{x}_j^{A_2}), (\boldsymbol{x}_i^{A_3} - \boldsymbol{x}_j^{A_3})]^T$, $\mathbf{W} = \boldsymbol{\alpha}\boldsymbol{\alpha}^T$, $tr(.)$ is the trace of matrix. Therefore, the objective loss function can be reformulated as:

$$\begin{aligned}
f'(\mathbf{M}_2, \mathbf{W}) = (1-\beta) \sum_{\boldsymbol{x}_i, \boldsymbol{x}_j, y_i=y_j} tr(\mathbf{C}_{i,j}\mathbf{M}_2\mathbf{C}_{i,j}^T\mathbf{W}) \\
+ \beta \sum_{\boldsymbol{x}_i, \boldsymbol{x}_j, y_i=y_j} [tr(\mathbf{C}_{i,j}\mathbf{M}_2\mathbf{C}_{i,j}^T\mathbf{W}) - \min_{y_k \neq y_i} tr(\mathbf{C}_{i,k}\mathbf{M}_2\mathbf{C}_{i,k}^T\mathbf{W}) + \rho]_+.
\end{aligned} \tag{7}$$

For the objective loss function in Eq. (7), our algorithm minimizes it iteratively in terms with $\mathbf{M}_2$ and $\mathbf{W}$. The stochastic gradient descent method is employed as well to optimize parameters $\mathbf{W}$ and $\mathbf{M}_2$ in turn, until the objective loss function converges. We define a set of indices $(i, j, k) \in \mathcal{N}(\mathbf{M}_2^l, \mathbf{W}^l)$, if and only if the indices (i,j,k) trigger the second term of hinge loss (Eq. (7)) at step l. And the gradient $\mathbf{GW}^l$ at step l is computed by:

$$\begin{aligned}
\mathbf{GW}^l = \frac{\partial f'}{\mathbf{W}}|_{\mathbf{W}=\mathbf{W}^l, \mathbf{M}_2=\mathbf{M}_2^l} \\
= (1-\beta) \sum_{i,j}(\mathbf{C}_{i,j}\mathbf{M}_2\mathbf{C}_{i,j}^T) + \beta \sum_{(i,j,k)\in\mathcal{N}(\mathbf{M}_2^l,\mathbf{W}^l)}(\mathbf{C}_{i,j}\mathbf{M}_2\mathbf{C}_{i,j}^T - \mathbf{C}_{i,k}\mathbf{M}_2\mathbf{C}_{i,k}^T),
\end{aligned} \tag{8}$$

Similarly, we can calculate $\mathbf{GM}_2^l$ in this way:

$$\begin{aligned}
\mathbf{GM}_2^l = \frac{\partial f'}{\mathbf{M}_2}|_{\mathbf{M}_2=\mathbf{M}_2^l, \mathbf{W}=\mathbf{W}^l} \\
= (1-\beta) \sum_{i,j}(\mathbf{C}_{i,j}^T\mathbf{W}\mathbf{C}_{i,j}) + \beta \sum_{(i,j,k)\in\mathcal{N}(\mathbf{M}_2^l,\mathbf{W}^l)}(\mathbf{C}_{i,j}^T\mathbf{W}\mathbf{C}_{i,j} - \mathbf{C}_{i,k}^T\mathbf{W}\mathbf{C}_{i,k}).
\end{aligned} \tag{9}$$

Testing Stage. In the testing stage, the learned metric $\mathbf{M}_1, \mathbf{M}_2$ and the optimal weights $\boldsymbol{\alpha}$ can be exploited to perform Re-ID by matching a probe person video $\boldsymbol{x}_i$ against a gallery set $\{\boldsymbol{x}_j\}$ in another camera view. The final distance between a probe video $\boldsymbol{x}_i$ and a gallery video $\boldsymbol{x}_j$ is computed by summing normalized distance $D(\boldsymbol{x}_i^{ST}, \boldsymbol{x}_j^{ST})$ and normalized distance $D_{\boldsymbol{\alpha}, \mathbf{M}_2}(\boldsymbol{x}_i^A, \boldsymbol{x}_j^A)$:

$$D(\boldsymbol{x}_i, \boldsymbol{x}_j) = D(\boldsymbol{x}_i^{ST}, \boldsymbol{x}_j^{ST}) + D_{\boldsymbol{\alpha}, \mathbf{M}_2}(\boldsymbol{x}_i^A, \boldsymbol{x}_j^A). \tag{10}$$

3 Experiment Results

3.1 Datasets and Settings

PRID-2011 Dataset: The PRID-2011 dataset [4] consists of video pairs recorded from two different cameras in uncrowded outdoor scenes. 385 persons are recorded in camera view A, and 749 persons in camera view B. Among all persons, 200 persons are recorded in both camera views, which are the samples this paper employs to evaluate. We randomly select 100 persons to form the training set, while the remaining 100 persons are used to constitute the testing set. The statistics of the dataset are listed in Table 1.

Table 1. The statistics of the PRID-2011 and iLIDS-VID datasets. Symbol "#" means "the number of".

Dataset	# ID	# Video	# Camera	# Training Sample	#Testing Sample
PRID-2011	934	1134	2	100	100
iLIDS-VID	300	600	2	150	150

iLIDS-VID Dataset: The iLIDS-VID dataset [17] contains 600 video of 300 randomly sampled people. Each person has one pair of video from two camera views. It is captured in an airport arrival hall. Compared with the PRID-2011 dataset, it is more challenging due to its heavier occlusion. We randomly divide the dataset into training set and testing set by half, namely, there both are 150 video pairs in the training set and testing set. The statistics of the dataset are shown in Table 1.

Implementation Details: All the experiments are under a single query setting. When training with the multi-part appearance features and the spatial-temporal HOG3D feature, parameters are all the same with the literature [21], which include ρ, β, the maximum number of iterations, the initialization metric matrix and the gradient step. And we initialize the weights $\boldsymbol{\alpha}^0 = [\frac{1}{3}, \frac{1}{3}, \frac{1}{3}]$.

Evaluation Metrics: We adopt Cumulated Matching Characteristics (CMC) curve to evaluate the performance of Re-ID methods for all datasets in this paper. And we only report the cumulated matching accuracies at selected ranks in tables instead of plotting the actual curves. The experiments are repeated 10 times to get average results.

3.2   Performance Comparison

Results on iLIDS-VID: We first evaluate AMFL against the state of the arts on the iLIDS-VID dataset. The results of comparisons are shown in Table 2. It can be concluded that AMFL can beat the compared methods on the iLIDS-VID dataset by a large margin. More specifically, AMFL achieves 82.6% Rank-1 matching rate, which is 20.6% higher than that of the best compared method ASTPN [19]. Compared with the baseline method TDL, our result increases by 26.3% at Rank-1, and the brief introduction of the baseline method is described in the literature [21] of the Sect. 1. Moreover, our method has better performance than the deep learning methods, including ASTPN, DFGP [8] and literature [25]. To be specific, our Rank-1 matching rate is over twice that of the DFGP.

Table 2. The results of comparisons with the state of the arts on iLIDS-VID and PRID-2011 datasets. Results are shown as matching rates (%) at Rank-n, $n \in [1, 5, 10, 20]$.

Method	iLIDS-VID Dataset				PRID-2011 Dataset			
	Rank-1	Rank-5	Rank-10	Rank-20	Rank-1	Rank-5	Rank-10	Rank-20
AMFL (Ours)	**82.6**	**94.7**	**97.1**	**98.8**	**80.9**	**95.3**	97.8	**99.4**
TDL[21]	56.3	87.6	95.6	98.3	56.7	80.0	87.6	93.6
Zhang et al. [23]	55.3	85.0	91.7	95.1	72.8	92.0	95.1	97.6
SCPDL[26]	56.8	86.3	94.2	96.6	74.5	92.1	94.3	96.6
ASTPN[19]	62.0	86.0	94.0	98.0	77.0	95.0	**99.0**	99.0
Zhang et al. [22]	44.3	71.4	83.7	91.7	64.1	87.3	89.9	92.0
Zhou et al. [25]	55.2	86.5	-	97.0	79.4	94.4	-	99.3
DFGP[8]	34.5	63.3	74.5	84.4	51.6	83.1	91.0	95.5

Results on PRID-2011: As shown in Table 2, it can be observed that AMFL achieves the best Rank-1 matching rate among the competing methods on the PRID-2011 dataset. Specifically, AMFL achieves 80.9% Rank-1 matching rate, which improves the Rank-1 matching rate by 24.2% compared with the TDL. Besides, AMFL outperforms the deep learning methods ASTPN, DFGP and literature [25].

Furthermore, it is obvious that the results of iLIDS-VID dataset with serious occlusion problem are superior to those of PRID-2011 dataset, which may testify that the proposed method reduces the effect of the poor region and has an advantage of the occlusion problem.

3.3   Effect of Major Components

We perform concrete analysis of AMFL on the PRID-2011 dataset to prove the validity of major components.

Effect of Distance Fusion: To verify the effectiveness of feature fusion method based on distance fusion, we concatenate the multi-part appearance features and the spatial-temporal feature, both of these features are the same as those in the AMFL, and then perform the TDL. The results are listed in method CM1 of Table 3. It can be observed that AMFL performs much better than the CM1, which could demonstrate the better validity of the distance fusion.

Table 3. The results of comparative methods on PRID-2011 dataset. Results are shown as matching rates (%) at Rank-n, $n \in [1, 5, 10, 20]$.

Method	Rank-1	Rank-5	Rank-10	Rank-20
AMFL (Ours)	**80.9**	**95.3**	**97.8**	**99.4**
CM1	60.6	83.4	89.6	95.1
CM2	70.6	94.7	97.5	99.1
CM3	76.2	93.7	97.0	99.3

Effect of Multi-part Appearance Features: In order to better illustrate the role of multi-part appearance features, we design a comparative method named CM2 without the multi-part appearance features. Specifically, the LOMO feature is directly extracted from the original image and then combined with the HOG3D feature through distance fusion. Experimental results in Table 3 demonstrate that notable improvement on rank-1 matching rate can be obtained by multi-part appearance features.

Effect of Adaptive Feature Weight Learning: We perform the experiments without the adaptive feature weight learning to prove the effectiveness of this component. To be specific, we manually set the feature weights $\alpha = [\frac{1}{3}, \frac{1}{3}, \frac{1}{3}]$, the results are shown in Table 3 method CM3. It can be observed that our method with adaptive feature weights has better performance than the CM3 with manual feature weights, which can illustrate the adaptive feature weight learning can effectively improve the matching rates of Re-ID.

4 Conclusion

In this paper, multi-part appearance features are extracted, and the proportion of the appearance feature in distinctive local part is increased by training. Moreover, we combine the multi-part appearance features and the spatial-temporal feature by the distance fusion. The novel combination method of these features could fully employ the information in the video. Experiments on two datasets can demonstrate that the proposed method outperforms the state of the arts.

References

1. Cho, Y.J., Yoon, K.J.: Improving person re-identification via pose-aware multi-shot matching. In: 2016 IEEE Conference on Computer Vision and Pattern Recognition (CVPR), pp. 1354–1362, June 2016. https://doi.org/10.1109/CVPR.2016.151
2. Chu, H., Qi, M., Liu, H., Jiang, J.: Local region partition for person re-identification. Multimedia Tools Appl. **7**, 1–17 (2017)
3. Gao, C., Wang, J., Liu, L., Yu, J.G., Sang, N.: Temporally aligned pooling representation for video-based person re-identification. In: 2016 IEEE International Conference on Image Processing (ICIP), pp. 4284–4288, September 2016. https://doi.org/10.1109/ICIP.2016.7533168
4. Hirzer, M., Beleznai, C., Roth, P.M., Bischof, H.: Person re-identification by descriptive and discriminative classification. In: Heyden, A., Kahl, F. (eds.) SCIA 2011. LNCS, vol. 6688, pp. 91–102. Springer, Heidelberg (2011). https://doi.org/10.1007/978-3-642-21227-7_9
5. Hirzer, M., Roth, P.M., Köstinger, M., Bischof, H.: Relaxed pairwise learned metric for person re-identification. In: Fitzgibbon, A., Lazebnik, S., Perona, P., Sato, Y., Schmid, C. (eds.) ECCV 2012. LNCS, vol. 7577, pp. 780–793. Springer, Heidelberg (2012). https://doi.org/10.1007/978-3-642-33783-3_56
6. Klser, A., Marszalek, M., Schmid, C.: A spatio-temporal descriptor based on 3D-gradients. In: British Machine Vision Conference 2008, Leeds, September 2008
7. Li, W., Wang, X.: Locally aligned feature transforms across views. In: Computer Vision and Pattern Recognition, pp. 3594–3601 (2013)
8. Li, Y., Zhuo, L., Li, J., Zhang, J., Liang, X., Tian, Q.: Video-based person re-identification by deep feature guided pooling. In: 2017 IEEE Conference on Computer Vision and Pattern Recognition Workshops (CVPRW), pp. 1454–1461, July 2017. https://doi.org/10.1109/CVPRW.2017.188
9. Liao, S., Hu, Y., Zhu, X., Li, S.Z.: Person re-identification by local maximal occurrence representation and metric learning. In: 2015 IEEE Conference on Computer Vision and Pattern Recognition (CVPR), pp. 2197–2206, June 2015. https://doi.org/10.1109/CVPR.2015.7298832
10. Liu, H., Jie, Z., Jayashree, K., Qi, M., Jiang, J., Yan, S., Feng, J.: Video-based person re-identification with accumulative motion context. IEEE Trans. Circuits Syst. Video Technol. **PP**(99), 1 (2017). https://doi.org/10.1109/TCSVT.2017.2715499
11. Liu, W., Mei, T., Zhang, Y., Che, C., Luo, J.: Multi-task deep visual-semantic embedding for video thumbnail selection. In: Computer Vision and Pattern Recognition, pp. 3707–3715 (2015)
12. Liu, X., Liu, W., Mei, T., Ma, H.: Provid: progressive and multi-modal vehicle re-identification for large-scale urban surveillance. IEEE Trans. Multimedia **20**(3), 645–658 (2017)
13. Liu, Z., Chen, J., Wang, Y.: A fast adaptive spatio-temporal 3D feature for video-based person re-identification. In: 2016 IEEE International Conference on Image Processing (ICIP), pp. 4294–4298, September 2016. https://doi.org/10.1109/ICIP.2016.7533170
14. Ma, H., Liu, W.: A progressive search paradigm for the internet of things, pp. 76–86 (2018)
15. McLaughlin, N., Rincon, J.M.d., Miller, P.: Recurrent convolutional network for video-based person re-identification. In: 2016 IEEE Conference on Computer Vision and Pattern Recognition (CVPR), pp. 1325–1334, June 2016. https://doi.org/10.1109/CVPR.2016.148

16. Song, Z., Cai, X., Chen, Y., Zeng, Y., Lv, L., Shu, H.: Deep convolutional neural networks with adaptive spatial feature for person re-identification. In: IEEE Advanced Information Technology, Electronic and Automation Control Conference, pp. 2020–2023 (2017)
17. Wang, T., Gong, S., Zhu, X., Wang, S.: Person re-identification by discriminative selection in video ranking. IEEE Trans. Pattern Anal. Mach. Intell. **38**(12), 2501–2514 (2016). https://doi.org/10.1109/TPAMI.2016.2522418
18. Xiang, T., Gong, S., Zheng, W.S.: Transfer re-identification: from person to set-based verification. In: IEEE Conference on Computer Vision and Pattern Recognition, pp. 2650–2657 (2012)
19. Xu, S., Cheng, Y., Gu, K., Yang, Y., Chang, S., Zhou, P.: Jointly attentive spatial-temporal pooling networks for video-based person re-identification, pp. 4743–4752 (2017)
20. Yi, D., Lei, Z., Liao, S., Li, S.Z.: Deep metric learning for person re-identification. In: 2014 22nd International Conference on Pattern Recognition, pp. 34–39, August 2014. https://doi.org/10.1109/ICPR.2014.16
21. You, J., Wu, A., Li, X., Zheng, W.S.: Top-push video-based person re-identification. In: 2016 IEEE Conference on Computer Vision and Pattern Recognition (CVPR), pp. 1345–1353, June 2016. https://doi.org/10.1109/CVPR.2016.150
22. Zhang, W., Ma, B., Liu, K., Huang, R.: Video-based pedestrian re-identification by adaptive spatio-temporal appearance model. IEEE Trans. Image Process. **26**(4), 2042–2054 (2017)
23. Zhang, W., Yu, X., He, X.: Learning bidirectional temporal cues for video-based person re-identification. IEEE Trans. Circuits Syst. Video Technol. **PP**(99), 1 (2017). https://doi.org/10.1109/TCSVT.2017.2718188
24. Zheng, S., Li, X., Men, A., Guo, X., Yang, B.: Integration of deep features and hand-crafted features for person re-identification. In: 2017 IEEE International Conference on Multimedia Expo Workshops (ICMEW), pp. 674–679, July 2017. https://doi.org/10.1109/ICMEW.2017.8026267
25. Zhou, Z., Huang, Y., Wang, W., Wang, L., Tan, T.: See the forest for the trees: joint spatial and temporal recurrent neural networks for video-based person re-identification. In: IEEE Conference on Computer Vision and Pattern Recognition, pp. 6776–6785 (2017)
26. Zhu, X., et al.: Semi-supervised cross-view projection-based dictionary learning for video-based person re-identification. IEEE Trans. Circuits Syst. Video Technol. **PP**(99), 1 (2017). https://doi.org/10.1109/TCSVT.2017.2718036
27. Zhu, X., Jing, X.Y., Wu, F., Feng, H.: Video-based person re-identification by simultaneously learning intra-video and inter-video distance metrics. In: International Joint Conference on Artificial Intelligence, pp. 3552–3558 (2016)

Visual-SLIM: Integrated Sparse Linear Model with Visual Features for Personalized Recommendation

Siyang Chen$^{(\boxtimes)}$, Feng Xue, and Haobo Zhang

Hefei University of Technology, Hefei 230601, Anhui, China
{siyang,hbzhang2017}@mail.hfut.edu.cn, feng.xue@hfut.edu.cn

Abstract. With the increasingly complexity and dynamically of information, recommendation system has been a key solution to alleviate the problem of information overloaded. Most recommender system models users' preferences toward items based on users' historical implicit feedback with item (*e.g.*, product purchase history, browsing logs, *etc.*). They typically make recommendation for a target user based on her profiles only (*e.g.*, the user's previous activities), ignoring the existence of other valuable information on items such as the visual features of images corresponding to the items. As a downside, it may limit the performance of recommender systems to some extent. This paper proposes a joint prediction model (visual-SLIM) which extends SLIM method with visual information to predict people's preference. The proposed approach automatically generates the missing items scores for a target user by aggregating observed user-item interaction matrix and learning linear regression model with items' visual information. It would not only improve performance of the model, but also do help to better analysis of the effects of visual information on user's opinions. Extensive experiments conducted on the real-world dataset of the Amazon have demonstrated the effectiveness of our proposed model.

Keywords: Recommender system · Sparse linear methods
Implicit feedback

1 Introduction

Along with the rapid development of the Internet, e-commerce is an emerging industry in the Internet. For online shopping users, the biggest problem they have encountered is that how to pick up products that meet their real needs from billions of products. One way to address this problem is to apply personalized

---

This work is supported by the National Natural Science Foundation of China (No. 61772170, 61472115), the National Key Research and Development Program of China (No. 2017YFB0803301) and the Fundamental Research Funds for the Central Universities (No. JZ2017YYPY0234).

© Springer Nature Switzerland AG 2018
R. Hong et al. (Eds.): PCM 2018, LNCS 11164, pp. 126–135, 2018.
https://doi.org/10.1007/978-3-030-00776-8_12

recommendation technologies to automatically generate a unique recommendation list for each user.

In the past few years, various algorithms in the literature have been developed to build top-K recommender systems [26]. These algorithms can be divided into collaborative filtering methods and content-based methods. Collaborative Filtering (CF) methods do not require the prior acquisition of product or user characteristics. They only rely on the user's past behavior (*e.g.*, browse histories) [33] to make recommendation. Researchers have developed a lot of these algorithms, For example, clustering models [19,23], Bayesian models [27,28] and factor models [14,32] and so on. Matrix factorization (MF) [7] is the most effective and popular technique that encode users and items into latent vectors (*aka.* embeddings) in a shared latent space and the preference of users towards items can be modeled as the inner product of their corresponding embeddings. The content-based filtering (CBF) method [24] is based on the extracted user and item features to realize recommendation. The CBF uses the characteristics of users and items to estimate the degree of match (*i.e.*, similarity) between them and finally recommends the best matched products to corresponding users. However, these aforementioned algorithms produce ranked list for each user based on implicit feedback or explicit feedback only and do not take any auxiliary information into consideration.

It is quite easy for content providers to collect side information about items from Internet. These information can be integrated with top-K recommender systems. Rearchers in [20] used titles, reviews, and comments of books as item features; Bao *et al.* [1] leveraged users' physical location to enrich the sparse rating matrix. Nevertheless, these approaches either only focus on the content of items which ignoring users' feedback or simply consider the content of items as a supplement to the rating matrix so as to generate minor improvements. A recent work [22] attempts to incorporate side information about items into top-K recommender systems to further enhance the accuracy of recommendation, which will limit the performance of recommendation by assuming the side information is reproduced in the same way with users' feedback. Here we are interested in the visual information of items which is usually ignored by existing CF approaches and integrate it with a joint discriminative model. If we don't like the appearance of this dress, we probably won't buy it. The work in [10] also demonstrates the importance of visual features. Therefore we believe that visual features are not negligible for recommendation system.

In general, we propose a joint prediction approach for implicit feedback which integrates the visual features of items with user-item interaction matrix in different model. The proposed model takes the effect of item visual features on user preferences modeling. To be more specific, it uncovers the recommendation score for unobserved items and complete the recommendation matrix for all items by performing low-rank aggregation over the users' observed items, on top of which we build a regularized linear regression with visual features to model the preference of users discriminatively. Extensive experiments conducted on the real-world dataset of Amazon which demonstrate the effectiveness of our proposed model and outperforms other compared methods significantly.

2   Related Work

The core of a personalized recommender system is in collaborative filtering, which infers the preference of users by their past behaviors and the interactions from similar users on items [31]. Generally,there are two types of memory-based models and model-based methods in CF approaches [3].

Memory-based models generate recommendation item lists for a target user through leveraging the neighbor information of users or items. We need to estimate the similarities between items and then obtain prediction of the corresponding user on a new item by taking users' specific past recommendation activities for similar items into account. Researchers in [21] propose a Sparse Linear Method (SLIM) model which learns a sparse coefficient matrix on user-item feedbacks by optimizing a regression-based objective function. Different from models based on memory, model-based approaches model predictive scores with different object function to generate a final list of recommendations. Varieties of model-based algorithms have been proposed, matrix factorization (MF) is the most successful implementation of model-based methods which learn embedding vector of users and items [13]. For instance, Cremonesi *et al.* [4] capture the individualized preferences of users on items by performing singular value decomposition (SVD) over item-user interaction matrix. The problem of overfitting is easy to occur when learning the SVD model. Therefore, relevant regularized learning methods have been proposed. In [11,29], the authors propose a weighted regularized matrix factorization model(WRMF) which learn the latent factors by assuming all the missing data as negative feedback. Rendle *et al.* [25] consider item recommendation as ranking problem and present BPR-Opt criterion which based on maximum posterior estimator to solve it.

In contrast to focusing on user-item interaction matrix only, some researchers pay their close attention on incorporating visual features [9] into user modeling for enhancing the performance of recommendation. For example, McAuley *et al.* [17] incorporate visual information with users' implicit feedback by learning the similarity between items to make recommendation. He *et al.* [6] proposed a visual personalized ranking model (VBPR) which incorporates the visual features of images into the framework of BPR [25]. In [8], the authors propose a multi-view object retrieval method using multi-scale topic models. For visual features, researchers in [16] propose a multi-scale triplet CNN network. More recently, the work in [15] developed a triple-net deep network, which mapping into the same latent space together with images and user preferences and outperforms CF methods.

3   Proposed Approach

In this section, we introduce the details of our visual-SLIM model. We use the symbol m and n to denote the number of items and users, the item-user interaction matrix $R \in \mathbb{R}^{m \times n}$ which each entry r_{ij} is 1 or 0, denoting whether user j has consumed item i or not. The problem of top-k item recommendation is to select k items for a user that are most likely to be consumed by the user. Besides the user-item interaction matrix R, given the visual-feature matrix $F \in \mathbb{R}^{m \times d}$, where d denotes the dimension of image features. In special, the column vector of all 1s is denoted by the symbol of $\mathbf{1}$. The problem is tackled by predicting the missing entries (*i.e.*, zero entries) of R, and rank items by score for recommendation.

3.1   Visual-SLIM

We aim to generate recommendation list for users where the impact of visual features on users' preference is taken into account. Therefore, the visual-SLIM can be splited into the following components, a self-completion component which use SLIM [21] to fill the missing entries of matrix R and the regression model based on the visual feature of items to generate the final recommendation list. The structure of the proposed visual-SLIM can be briefly describe as Fig. 1.

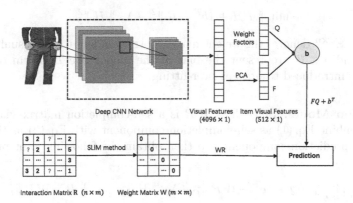

Fig. 1. Diagram of our preference predictor

Self-Completion Component Using SLIM. Due to the sparsity of the item-user interaction matrix R, we first resume the interaction matrix $\hat{R}$ without missing values by using Sparse LInear Method (SLIM) [21]. In the SLIM method, we assume the full interaction matrix can be recovered by a sparse coefficient matrix as weight matrix. In other words, it is actually a linear way to learn the similarity between items. Hence we can recover the interaction matrix by the formulation $\hat{R} = WY$ with $W \geq 0$, where W is $m \times m$ weight matrix for all users. We view pre-given item-user interaction matrix Y as the ground-truth matrix and learn the sparse coefficient matrix W as the minimizer optimization problem by the following object function:

$$\min_{W} \parallel R - WR \parallel_F^2 \; s.t. \; rank(W) \leq \lambda_{\theta}, \tag{1}$$

where $\| \bullet \|_F^2$ denotes the Frobenius norm of matrix. In order to guarantee the sparsity of matrix W, we add constraint on the rank of matrix W by the hyperparameter λ_θ and do help to accelerate calculatio since the sparsity of matrix W.

Linear Regression Model with the Visual Feature. Given the visual feature matrix F of items and the recovered item-user interaction matrix $\hat{R}$, we aim to build a prediction model which takes the visual-feature matrix F into consider to generate ranked recommendation list. In the real world, different people may have different regions of interest in a item image. We hence assume that the different parts of the visual feature may have different effects on the users' appearance, that is, the user's purchase or click behaviors may depend on different subset of the visual feature of items. In particular, we can simply build linear regression model to capture the importance of different parts of visual features to users' purchase or click activities when make recommendation. We build linear regression model with the following formulation by minimizing the following regularized least squares loss function for all users:

$$\min \| FH + 1b^T - \hat{R} \|_F^2 + \lambda_\beta \| H \|_F^2, \tag{2}$$

where $H \in \mathbb{R}^{d \times n}$ denotes the weight matrix corresponding to the visual features of items and $b \in \mathbb{R}^n$ represent the bias of the model, the L_2 norm of H is a regularizer introduced to prevent over-fitting.

Integration Model. Since $\hat{R} = WR$ is a self-completion matrix via Eq. (1), we can combine Eq. (1) as self-completion component with Eq. (2) as the linear regression prediction component and then obtain the following joint prediction model:

$$\min_{W,H,b} \| FQ + 1b^T - WR \|_F^2 + \lambda_\beta \| H \|_F^2 + \lambda_\gamma \| R - WR \|_F^2 \tag{3}$$

where λ_γ is the trade-off parameter for balancing the proportion of the self-completion component and the visual component. This joint model integrates visual features into predictors of people's opinions, and is helpful to enhance the performance of top-K recommendation system.

3.2 Learning Algorithm

Since there are two set of parameters in the joint minimization Eq. (3), we first fix the weight matrix W in self-completion component, which has the following closed-form solutions for H and b by performing minimization over the joint model,

$$H = (F^T GF + \lambda_\beta I_d)^{-1} F^T GWR \tag{4}$$

$$b = \frac{1}{n}(WR - FH)^T \tag{5}$$

where G is the symbol of centering matrix, I_n denotes the identity matrix. By combining Eq. (4) with Eq. (3), The minimization problem in (3) can be reformulated into the following problem,

$$\min_{W} \parallel YWR \parallel_F^2 +\lambda_\beta \parallel XWR \parallel_F^2 +\lambda \parallel R - WR \parallel_F^2 \tag{6}$$

where $X = (F^T G F + \lambda_\beta I_d)^{-1} F^T G$ and $Y = G(FX - I_n)$. In order to further simplify Eq. (6), we define the following auxiliary matrices,

$$U = \begin{bmatrix} Y \\ \sqrt{\lambda_\beta} X \\ \sqrt{\lambda_\gamma} I_n \end{bmatrix}, V = \begin{bmatrix} O_{(m+d) \times n} \\ \sqrt{\lambda_\gamma} R \end{bmatrix} \tag{7}$$

Here we use $O_{(m+d) \times n}$ to denote a $(m + d) \times n$ matrix in which each item is 0, thus we can reformulate Eq. (6) as

$$\min_{W} \parallel UWR - V \parallel_F^2 \tag{8}$$

For matrix U, we denote ite SVD [2] as $U = A_U \sum_U B_U^T$. We also get $\mathcal{P}_{U,\mathcal{L}} = A_U A_U^T$, $\mathcal{P}_{U,\mathcal{R}} = B_U B_U^T$ by performing projection operation. In the same way, we can get $\mathcal{P}_{R,\mathcal{L}}$ and $\mathcal{P}_{R,\mathcal{R}}$ by performing projection on matrix R. Sondermann [30] and Friedland&Torokhti [5] derives an convenient theorem for solving Eq. (8):

$$W^* = U^\dagger Q_{(\lambda)} R^\dagger \tag{9}$$

where $Q = \mathcal{P}_{F,\mathcal{L}} V \mathcal{P}_{Y,\mathcal{R}}$, $\parallel \bullet \parallel^\dagger$ represents the pseudo-inverse of matrix. The symbol of $Q_{(\lambda)}$ means the truncation of Q at the position λ.

After obtaining the closed-form solution for W^* in Eq. (9), we can recover the full interaction matrix $\hat{R}^*$ in which the scores denotes the preference of all items. Then we can rank the scores for non-purchased items in a descending order and generate the top-K recommendation list for users.

4   Experiment

In this section, we introduce the details about experiment datasets and present our experimental results and discussions.

4.1   Datasets

We conduct experiments on two publicly Amazon datasets [18]: Amazon Men and Amazon Women. We consider users' purchase histories as implicit feedback and crawl images for all items and use deep CNN network to extract their visual features. We process each Amazon dataset by converting user's rating scores to 1s and discard the users and items with less than 5 interaction. Table 1 shows statistics of our datasets.

Table 1. Statistics of the Amazon datasets.

Datasets	Users	Items	Feedback
Amazon men	3967	5000	25351
Amazon women	9859	8000	65393

For the detailed information about visual features, we use pre-trained deep CNN network [12] with five layers of convolution layers and three layers of full-connected layers to extract visual features. The output of network produced by full-connected layer is a 4096 dimensional vector. Due to the high sparsity of visual features we extracted, We use PCA to reduce the dimension of visual features we extracted to 512 for subsequent optimization.

4.2   Comparison Approache and Evaluation Criteria

We compare with the following methods:

(1) Sparse Linear Model (SLIM) [21]. This is a linear model to learn a sparse coefficient matrix between items by SGD. Since the high sparsity of the coefficient matrix, it has lower time complexity than other MF methods and can be calculated in parallel;
(2) Bayesian Personalized Ranking (BPR-MF) [25]. This pair-wise method that optimizes the position between the positive and negative samples by SGD and is the state-of-the-art methods for implicit feedback.

For each dataset, we adopt the leave-one-out strategy to split dataset into training set and testing set according to timestamp of the review for each user j. We use the training set to train a model and evaluate the performance on testing set. We adopt AUC (Area Under the ROC curve) to measure the performance of above baseline methods.

4.3   Experimental Results

Table 2 presents the best performance of our proposed method and two baseline methods in the term of AUC on Amazon dataset. We use the same latent dimensions for all baseline methods.

Table 2. Comparison results on Amazon datasets.

Datasets	SLIM	BPR-MF	visual-SLIM	Improvement
Amazon women	0.5734	0.5836	0.6644	13%
Amazon men	0.5665	0.5757	0.6345	10%

From Table 2, we can find that the baseline BPR-MF method works better than SLIM method for all the datasets. However, our visual-SLIM which incorporates visual information into SLIM produces better result than BPR-MF. This demonstrates that the visual information is essential for improving the performance of recommender systems. On the other hand, our visual-SLIM method produces best results among all the comparison methods on the Amazon datasets. This shows that our proposed method provides an effective model to integrate visual information for top-K recommender systems. Moreover, we also show the results of our visual-SLIM method with different parameter setting. Figure 2 shows the fluctuation trend of AUC with different parameters setting. We can find that the AUC increases as λ increases or λ_γ decreases where λ controls the strength of truncation and λ_γ decides the proportion of SLIM component. As shown in Fig. 2(a), our visual-SLIM method produces better result as the value of λ increases, which demonstrates. In Fig. 2(b), we show the trend of AUC with increasing the value of λ_γ. In general, the decline of λ_γ that the increase of the proportion of visual component can produces better result. This also demonstrate the importance of visual information. We also find that the visual-SLIM method works better on Amazon women dataset than Amazon men dataset. Presumably this is because women are more susceptible to visual information when making decisions. To summarize, our visual-SLIM approach is effective and outperforms all baseline methods.

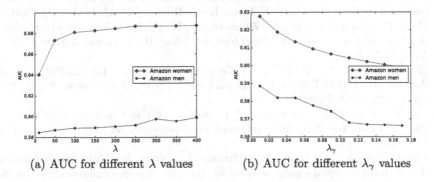

(a) AUC for different λ values   (b) AUC for different λ_γ values

Fig. 2. AUC for different parameters settings

5   Conclusions and Future Work

In this paper, we proposed a new joint discriminative model called visual-SLIM model to exploit both the item-user implicit feedback and the visual information of items to produce top-K recommendation for users. Our experiments conducted on multiple Amazon datasets which demonstrate that visual information is very important for improving the quality of traditional recommendation systems based on implicit feedback. For future work, we plan to extend our model to the recommender systems based on explicit feedback.

References

1. Bao, Y., Fang, H., Zhang, J.: TopicMF: simultaneously exploiting ratings and reviews for recommendation. In: AAAI, vol. 14, pp. 2–8 (2014)
2. Brand, M.: Fast low-rank modifications of the thin singular value decomposition. Linear Algebra Appl. **415**(1), 20–30 (2006)
3. Breese, J.S., Heckerman, D., Kadie, C.: Empirical analysis of predictive algorithms for collaborative filtering. In: Proceedings of the Fourteenth Conference on Uncertainty in Artificial Intelligence, pp. 43–52. Morgan Kaufmann Publishers Inc. (1998)
4. Cremonesi, P., Koren, Y., Turrin, R.: Performance of recommender algorithms on top-n recommendation tasks. In: Proceedings of the Fourth ACM Conference on Recommender Systems, pp. 39–46. ACM (2010)
5. Friedland, S., Torokhti, A.: Generalized rank-constrained matrix approximations. SIAM J. Matrix Anal. Appl. **29**(2), 656–659 (2007)
6. He, R., McAuley, J.: VBPR: visual Bayesian personalized ranking from implicit feedback. In: AAAI, pp. 144–150 (2016)
7. He, X., Zhang, H., Kan, M.Y., Chua, T.S.: Fast matrix factorization for online recommendation with implicit feedback. In: Proceedings of the 39th International ACM SIGIR Conference on Research and Development in Information Retrieval, pp. 549–558. ACM (2016)
8. Hong, R., Hu, Z., Wang, R., Wang, M., Tao, D.: Multi-view object retrieval via multi-scale topic models. IEEE Trans. Image Process. **25**(12), 5814–5827 (2016)
9. Hong, R., Zhang, L., Tao, D.: Unified photo enhancement by discovering aesthetic communities from Flickr. IEEE Trans. Image Process. **25**(3), 1124–1135 (2016)
10. Hong, R., Zhang, L., Zhang, C., Zimmermann, R.: Flickr circles: aesthetic tendency discovery by multi-view regularized topic modeling. IEEE Trans. Multimed. **18**(8), 1555–1567 (2016)
11. Hu, Y., Koren, Y., Volinsky, C.: Collaborative filtering for implicit feedback datasets. In: Eighth IEEE International Conference on Data Mining, ICDM 2008, pp. 263–272. IEEE (2008)
12. Jia, Y., et al.: Caffe: convolutional architecture for fast feature embedding. In: Proceedings of the 22nd ACM International Conference on Multimedia, pp. 675–678. ACM (2014)
13. Karatzoglou, A., Amatriain, X., Baltrunas, L., Oliver, N.: Multiverse recommendation: n-dimensional tensor factorization for context-aware collaborative filtering. In: Proceedings of the Fourth ACM Conference on Recommender Systems, pp. 79–86. ACM (2010)
14. Koren, Y., Bell, R., Volinsky, C.: Matrix factorization techniques for recommender systems. Computer **42**(8), 30–37 (2009)
15. Lei, C., Liu, D., Li, W., Zha, Z.J., Li, H.: Comparative deep learning of hybrid representations for image recommendations. In: Proceedings of the IEEE Conference on Computer Vision and Pattern Recognition, pp. 2545–2553 (2016)
16. Liu, J., et al.: Multi-scale triplet CNN for person re-identification. In: Proceedings of the 2016 ACM on Multimedia Conference, pp. 192–196. ACM (2016)
17. McAuley, J., Leskovec, J.: Image labeling on a network: using social-network metadata for image classification. In: Fitzgibbon, A., Lazebnik, S., Perona, P., Sato, Y., Schmid, C. (eds.) ECCV 2012. LNCS, vol. 7575, pp. 828–841. Springer, Heidelberg (2012). https://doi.org/10.1007/978-3-642-33765-9_59

18. McAuley, J., Targett, C., Shi, Q., Van Den Hengel, A.: Image-based recommendations on styles and substitutes. In: Proceedings of the 38th International ACM SIGIR Conference on Research and Development in Information Retrieval, pp. 43–52. ACM (2015)
19. Merialdo, A.K.B.: Clustering for collaborative filtering applications. Intell. Image Process. Data Anal. Inf. Retr. **3**, 199 (1999)
20. Mooney, R.J., Roy, L.: Content-based book recommending using learning for text categorization. In: Proceedings of the Fifth ACM Conference on Digital Libraries, pp. 195–204. ACM (2000)
21. Ning, X., Karypis, G.: Slim: Sparse linear methods for top-n recommender systems. In: IEEE 11th International Conference on Data Mining (ICDM), pp. 497–506. IEEE (2011)
22. Ning, X., Karypis, G.: Sparse linear methods with side information for top-n recommendations. In: Proceedings of the Sixth ACM Conference on Recommender Systems, pp. 155–162. ACM (2012)
23. OâĂŽConnor, M., Herlocker, J.: Clustering items for collaborative filtering. In: Proceedings of the ACM SIGIR Workshop on Recommender Systems, vol. 128. UC Berkeley (1999)
24. Pazzani, M.J., Billsus, D.: Content-based recommendation systems. In: Brusilovsky, P., Kobsa, A., Nejdl, W. (eds.) The Adaptive Web. LNCS, vol. 4321, pp. 325–341. Springer, Heidelberg (2007). https://doi.org/10.1007/978-3-540-72079-9_10
25. Rendle, S., Freudenthaler, C., Gantner, Z., Schmidt-Thieme, L.: BPR: Bayesian personalized ranking from implicit feedback. In: Proceedings of the Twenty-Fifth Conference on Uncertainty in Artificial Intelligence, pp. 452–461. AUAI Press (2009)
26. Ricci, F., Rokach, L., Shapira, B., Kantor, P.B.: Recommender Systems Handbook, 1st edn. Springer, Heidelberg (2010). https://doi.org/10.1007/978-0-387-85820-3
27. Salakhutdinov, R., Mnih, A.: Bayesian probabilistic matrix factorization using Markov chain Monte Carlo. In: Proceedings of the 25th International Conference on Machine Learning, pp. 880–887. ACM (2008)
28. Savia, E., Puolamaki, K., Sinkkonen, J., Kaski, S.: Two-way latent grouping model for user preference prediction. arXiv preprint arXiv:1207.1414 (2012)
29. Sindhwani, V., Bucak, S.S., Hu, J., Mojsilovic, A.: One-class matrix completion with low-density factorizations. In: IEEE 10th International Conference on Data Mining (ICDM), pp. 1055–1060. IEEE (2010)
30. Sondermann, D.: Best approximate solutions to matrix equations under rank restrictions. Statistische Hefte **27**(1), 57 (1986)
31. Tan, Z., He, L.: An efficient similarity measure for user-based collaborative filtering recommender systems inspired by the physical resonance principle. IEEE Access **5**, 27211–27228 (2017)
32. Wu, J.: Binomial matrix factorization for discrete collaborative filtering. In: Ninth IEEE International Conference on Data Mining, ICDM 2009, pp. 1046–1051. IEEE (2009)
33. Yi, X., Hong, L., Zhong, E., Liu, N.N., Rajan, S.: Beyond clicks: dwell time for personalization. In: Proceedings of the 8th ACM Conference on Recommender Systems, pp. 113–120. ACM (2014)

Residual Learning Dehazing Net

Yili Gu and Xinguang Xiang[✉]

School of Computer Science and Engineering,
Nanjing University of Science and Technology,
Nanjing 210094, People's Republic of China
YiliGu94@163.com, xgxiang@mail.njust.edu.cn

Abstract. Single haze removal is a challenging ill-posed problem. Most existing methods solving this dilemma depend on atmospheric physical scattering model. In other words, they recover haze-free images by estimating the atmospheric transmission. In this paper, we proposed a new recovery model called Residual Adding model, which takes dehazing procedure as a hazy image adding a loss image. Based on this new model, we proposed a single image dehazing network built with Conditional Generative Adversarial Nets (CGAN), called Residual Learning Dehazing Network (RLD-Net). Benefiting from the new model, the RLD-Net is designed as not only an end-to-end dehazing network but also a point-to-point mapping network. That means RLD-Net can take a hazy image as input and a corresponding clear image as output without any extract calculation like inversing atmospheric physical scattering model. Experimental results on both synthesized hazy images and real-world hazy images demonstrate our outstanding performance.

Keywords: Dehazing · Image restoration · Residual adding model
Residual Learning Dehazing Network

1 Introduction

1.1 Prior Work

Haze is a common atmospheric phenomenon which atmospheric light is absorbed and scattered by small particles and water droplets floating in the air. This phenomenon will cover images' details, reduce images' saturation and contrast. Moreover, the existence of haze will influence follow-up computer vision research like classification, recognition and detection. Therefore, image dehazing technology is an essential and indispensable task.

However, dehazing is an ill-posed problem because the haze is relying on the unknown depth. To solve the problem, lots of methods have been proposed. One kind of solution is using addition information or multiple images to estimate the depth information like [1–5]. Another kind of method is using only one image to remove haze which is convenience and having more development potential [6, 7]. As mentioned above, this paper will introduce a novel dehazing method with single image.

© Springer Nature Switzerland AG 2018
R. Hong et al. (Eds.): PCM 2018, LNCS 11164, pp. 136–145, 2018.
https://doi.org/10.1007/978-3-030-00776-8_13

1.2   The Key of Dehazing

Dehazing is an important work, not only for the clear vision but also for the advanced computer vision research [24, 25, 27, 28]. Dehazing methods are divided into two types, one is the image restoration and the other is image enhancement. The image enhancement methods achieve dehazing by enhancing the image information like contrast, which is easy to implement but lacking the scientific basis. In contrast, image restoration methods are much more reliable because they are relying on the scientific theory—atmospheric physical scattering model. The atmospheric physical scattering model is wildly used to computer graphics and computer vision [5, 22, 23]. This is a model describes the relationships between hazy image and haze-free image:

$$I(x) = J(x)t(x) + A(1 - t(x)) \tag{1}$$

In Eq. (1), I represents the hazy image, J is the corresponding haze-free image and A is global atmospheric light of J, t is medium transmission which depends on the depth inform of J. t can be calculate by Eq. (2):

$$t(x) = e^{-\beta d(x)} \tag{2}$$

In Eq. (2), β is the scattering coefficient about atmosphere and d is the image depth information.

From the Eqs. (1) and (2) we can know that both traditional methods [6, 7] and deep learning methods of dehazing [8, 9] are sharing the same opinion, that the key to recover a clean image is to restructure the transmission map t. And A will be calculated by the traditional methods.

However, in AOD-Net [10], authors have proved that estimate A and t respectively will extend their errors. Thus, they proposed a novel method that calculating A and t at the same time using an expression K. Although this method has made progress, it has still not deviated from the atmospheric physical scattering model and that's means we still need to calculate the transmission and global atmospheric light.

1.3   Innovation

In summary, there are two major shortcomings in prior works. The first one is the error of estimating t and A. And the second one is the lack of end-to-end systems, which is quite mature in other areas of computer vision [19–21].

In order to avoid the influence of t and A, we propose a new recover model called Residual Adding Model. This model takes the dehazed image as the sum of hazy image and the residual image. And the Residual Learning Dehazing Network (RLD-Net), proposed by us, is also depending on this new model. These two innovations bring two main advantages.

First of all, it is obviously that the innovations avoid the influence of error about t and A.

Secondly, benefiting from the new recover model, the RLD-Net is designed as not only an end-to-end dehazing network but also a point-to-point mapping network. That means RLD-Net can take a hazy image as input and a corresponding clear image as output without any extract calculation like inversing atmospheric physical scattering model.

This paper is organized as follows. In Sect. 2, we will introduce our networks in details, including the new dehazing model and the network design. In Sect. 3, we will display our dehazing results on both synthetic images and natural images. In last section we will summarize the advantages and disadvantages of our method and look to the future of application.

2 Residual Learning Dehaze Network

In this section, firstly, we will give the novel understanding about hazy image and propose the new recovery model. Secondly, the proposed RLD-Net will be explained in details.

2.1 Residual Adding Model

Usually, researchers consider haze as a scattering phenomenon forming by natural. Thus, dehazing must comply with physical guidelines as using atmospheric physical scattering model. In other words, the key to dehaze is to calculate the transmission and the global atmospheric light. In this paper, we propose a new recovery model that can be used without calculating transmission and the global atmospheric light.

Now, let's ignore the principle of haze formation and just pay attention on the influence of haze. The existence of haze will cover details of the images, not only the edge information but also the color contrast, saturation and brightness. We define these losses as R called residual image, and the model should be:

$$J(x) = I(x) + R(x) \tag{3}$$

Where J(x) is haze-free image, I(x) is hazy image and x represent the pixel.

Now, we can clearly see that, if we can learn the R from I, we can restore the J from I.

2.2 Network Design

The RLD-Net is designed on CGAN [11] model which has two parts, the generator G and the discriminator D. Both of these parts can be replaced by learning networks as illustrated in Figs. 1 and 2.

The generator G has a symmetrical structure. The first half of the network is used to extract features, and the second half of the network is used to restore residual images. The dotted lines in Fig. 1 represent the replication and connection. The role of these operations is to supplement the details which lost after convolution and deconvolution. Each convolution and deconvolution operation are followed by a batch norm and nonlinear mapping which except the last layer is tanh, the rest are relu. The most important layer is the last deconvolution layer which restores a residual image with the same size of input hazy image. Adding these two images will give you the final, clear image.

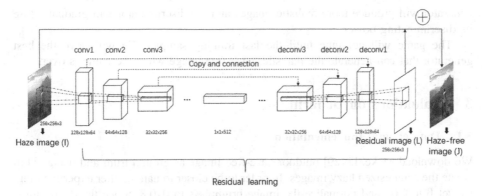

Fig. 1. Structure of generator (G)

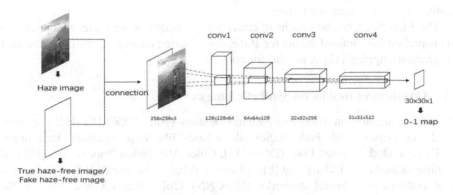

Fig. 2. Structure of discriminator (D)

The discriminator D has a similar structure with the first half of G which only composed of convolution structures and their accompanying operations. Putting a hazy image and a true haze-free image or a fake haze-free image which generated by G into D, D will give us a result graph with only of 0, 1 values, describing the reliability of this image as a true haze-free image.

2.3 Training of CGAN

The objective function of CGAN [11] is:

$$\min_{G} \min_{D} V(D, G) = E_{x \sim p_{data}^{(x)}} \left[logD(x|y) \right] + E_{z \sim p_z^{(z)}} \left[log(1 - D(G(z|y))) \right] \quad (4)$$

The training process of CGAN is a game with two subnetworks alternately optimized.

First, we fixed the generator G and optimized the discriminator D. This is a process that discriminator becomes smarter and generator becomes clumsy. Second, after an epoch, we fixed the discriminator D and optimized the generator G. In this time, the generator will produce more realistic images and the discriminator will gradually lose its discriminating power.

The game will continue until the last training sample. We will have the best generator that can generate the realistic, haze-free image when the game is over.

3 Evaluations on Dehazing

3.1 Datasets and Implementation

We download Make3D [29] outdoor haze-free image as ground-truth and using (1) to create the synthesized hazy images. In order to be closer to nature, after experimenting, we set β to 0.02 and normalize the image brightness to 0–0.8. Since the depth information given in dataset is limited, we use bilinear interpolation to obtain a complement depth information map. We resize the image size to 256×256 and use 400 images for training and 134 images for testing.

The RLD-Net takes 80 epochs to converge and performs well after 80 epochs. In this paper, we have trained the net for 100 epochs. During the experiments, we also find the gradient clipping [12] is helpful for convergence.

3.2 Quantitative Results on Synthetic Images

We test our method on Middlebury Stereo Datasets (2001–2006) [13–15]. We compared our method with both traditional methods like Fast Visibility Restoration (FVR) [16], Dark Channel Prior (DCP) [17], Color Attenuation Prior (CAP) [18] and learning methods like Dehaze-Net [8], All-in-one dehazing network (AOD) [10], multiscale convolutional neural networks (MS-CNN) [26]. Figures 3 and 4 show the dehazing effect of different methods. In comparison, we can obviously see the advantages of RLD-Net that the recovered images are bright and clear.

To quantitatively access these methods, we use peak signal-to-noise ratio (PSNR), structure similarity (SSIM), and minimum mean square error(MSE). The results are showed in Tables 1 and 2.

3.3 Quantitative Results on Natural Images

Figures 5 and 6 shows the dehazing results of our method and compared with other methods, including [8, 10, 16–18, 26].

The methods of dehazing by restoring transmission usually have defect on white objects because the transmission is close to zero when the color is similar to atmosphere. Figure 7 shows the effect of RLD-Net on white object. What's more, the RLD-Net has outstanding retention of details.

In addition to a good ability of dehazing, RLD-Net can also reduce the halo effect. Figure 8 shows the result.

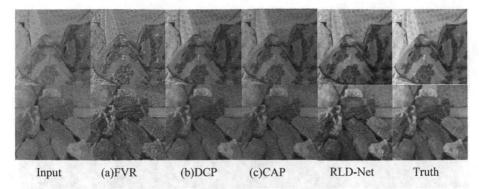

| Input | (a)FVR | (b)DCP | (c)CAP | RLD-Net | Truth |

Fig. 3. The test results of traditional methods and RLD-Net. The first image is the hazy image and the last image is the ground-truth.

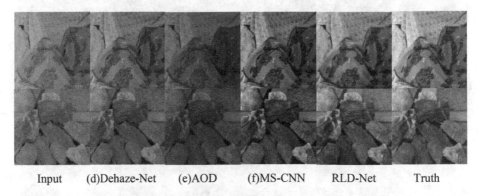

| Input | (d)Dehaze-Net | (e)AOD | (f)MS-CNN | RLD-Net | Truth |

Fig. 4. The test results of learning methods and RLD-Net. The first image is the hazy image and the last image is the ground-truth.

Table 1. The average results of PSNR, SSIM, MSE on Middlebury Stereo Datasets (2001–2006), compared with traditional methods

Methods	FVR [16]	DCP [17]	CAP [18]	**RLD-Net**
PSNR	12.0244	10.9457	17.0117	**18.4132**
SSIM	0.4581	0.4716	0.8878	**0.9208**
MSE	0.0688	0.0894	0.0243	**0.0163**

Table 2. The average results of PSNR, SSIM, MSE on Middlebury Stereo Datasets (2001–2006), compared with learning methods

Methods	Dehaze-Net [8]	AOD [10]	MS-CNN [26]	**RLD-Net**
PSNR	11.9251	11.7396	15.2118	**18.4132**
SSIM	0.4460	0.6867	0.8159	**0.9208**
MSE	0.0710	0.0730	0.0353	**0.0163**

Input   (a)FVR   (b)DCP   (c)CAP   RLD-Net

Fig. 5. Dehazing results of traditional methods on natural images. We can see the proposed method has more natural result.

Input   (d)Dehaze-Net   (e)AOD   (f)MS-CNN   RLD-Net

Fig. 6. Dehazing results of learning methods on natural images. We can see the proposed method has higher brightness and clearer details

(a)input   (b)output

Fig. 7. Dehazing on white object and the details are still clear.

(a)input   (b)output

Fig. 8. RLD-Net can reduce halo effect

4 Conclusion

The paper proposed residual adding model and RLD-Net. The residual adding model innovatively considers the dehazing process as the sum of two images. This is a new recovery model that can recover hazy image without calculating the transmission. RLD-Net is designed on this new model and it is not only an end-to-end dehazing network, but also a point-to-point mapping dehazing network. We compared our method with both traditional methods and learning methods, using criterion of SSIM, PSNR and MSE. The experiments prove that RLD-Net is excellent and has great development potential.

References

1. Schechner, Y.Y., Narasimhan, S.G., Nayar, S.K.: Instant dehazing of images using polarization. In: Proceedings of the IEEE Conference on Computer Vision and Pattern Recognition, vol. 1, pp. 325–332 (2001)
2. Shwartz, S., Namer, E., Schechner, Y.Y.: Blind haze separation. In: Proceedings of the IEEE Conference on Computer Vision and Pattern Recognition, vol. 2, pp. 1984–1991 (2006)
3. Narasimhan, S.G., Nayar, S.K.: Chromatic framework for vision in bad weather. In: Proceedings of the IEEE Conference on Computer Vision and Pattern Recognition, vol. 1, pp. 598–605 (2000)
4. Nayar, S.K., Narasimhan, S.G.: Vision in bad weather. In: Proceedings of the Seventh IEEE International Conference on Computer Vision, vol. 2, pp. 820–827 (1999)
5. Narasimhan, S.G., Nayar, S.K.: Contrast restoration of weather degraded images. IEEE Trans. Pattern Anal. Mach. Intell. 25(6), 713–725 (2003)
6. Fattal, R.: Single image dehazing. In: Proceedings of ACM SIGGRAPH (2008)
7. Tan, R.: Visibility in bad weather from a single image. In: Proceedings of the IEEE Conference on Computer Vision and Pattern Recognition (2008)
8. Cai, B., Xu, X., Jia, K., Qing, C., Tao, D.: Dehaze-Net: an end-to-end system for single image haze removal. IEEE Trans. Image Process. 25(11), 5187–5198 (2016)
9. Ren, W., Liu, S., Zhang, H., Pan, J., Cao, X., Yang, M.-H.: Single image dehazing via multi-scale convolutional neural networks. In: Leibe, B., Matas, J., Sebe, N., Welling, M. (eds.) ECCV 2016. LNCS, vol. 9906, pp. 154–169. Springer, Cham (2016). https://doi.org/10.1007/978-3-319-46475-6_10
10. Li, B., Peng, X., Wang, Z., Xu, J., Feng, D.: AOD-Net: all-in-one dehazing network. In: IEEE International Conference on Computer Vision, pp. 4780–4788. IEEE Computer Society (2017)
11. Mirza, M., Osindero, S.: Conditional generative adversarial nets. Computer Science, pp. 2672–2680 (2014)
12. Pascanu, R., Mikolov, T., Bengio, Y.: On the difficulty of training recurrent neural networks. ICML 3(28), 1310–1318 (2013)
13. Scharstein, D., Szeliski, R.: A taxonomy and evaluation of dense two-frame stereo correspondence algorithms. Int. J. Comput. Vis. 47(1–3), 7–42 (2002)
14. Scharstein, D., Szeliski, R.: High-accuracy stereo depth maps using structured light. In: IEEE Computer Society Conference on Computer Vision and Pattern Recognition, vol. 1, pp. I–195. IEEE (2003)
15. Scharstein, D., Pal, C.: Learning conditional random fields for stereo. In: IEEE Conference on Computer Vision and Pattern Recognition, pp. 1–8. IEEE (2007)
16. Tarel, J.-P., Hautiere, N.: Fast visibility restoration from a single color or gray level image. In: IEEE 12th International Conference on Computer Vision, pp. 2201–2208. IEEE (2009)
17. He, K., Sun, J., Tang, X.: Single image haze removal using dark channel prior. IEEE Trans. Pattern Anal. Mach. Intell. 33(12), 2341–2353 (2011)
18. Zhu, Q., Mai, J., Shao, L.: A fast single image haze removal algorithm using color attenuation prior. IEEE Trans. Image Process. 24(11), 3522–3533 (2015)
19. Wang, Z., et al.: Self-tuned deep super resolution. In: Proceedings of the IEEE Conference on Computer Vision and Pattern Recognition Workshops, pp. 1–8 (2015)
20. Schuler, C.J., Hirsch, M., Harmeling, S., Scholkopf, B.: Learning to deblur. IEEE Trans. Pattern Anal. Mach. Intell. 38(7), 1439–1451 (2016)
21. Xie, J., Xu, L., Chen, E.: Image denoising and inpainting with deep neural networks. In: Advances in Neural Information Processing Systems, pp. 341–349 (2012)

22. McCartney, E.J.: Optics of the Atmosphere: Scattering by Molecules and Particles, 421 p. Wiley, New York (1976)
23. Meng, G., Wang, Y., Duan, J., Xiang, S., Pan, C.: Efficient image dehazing with boundary constraint and contextual regularization. In: Proceedings of the IEEE International Conference on Computer Vision, pp. 617–624 (2013)
24. Wang, M., Gao, Y., Ke, L., Rui, Y.: View-based discriminative probabilistic modeling for 3D object retrieval and recognition. IEEE Trans. Image Process. 22(4), 1395–1407 (2013)
25. Hong, R., Zhenzhen, H., Wang, R., Wang, M., Tao, D.: Multi-view object retrieval via multi-scale topic models. IEEE Trans. Image Process. 25(12), 5814–5827 (2016)
26. Zeng, L., Xu, X., Cai, B., et al.: Multi-scale convolutional neural networks for crowd counting (2017)
27. Li, Z., Tang, J., He, X.: Robust structured nonnegative matrix factorization for image representation. IEEE Trans. Neural Netw. Learn. Syst. 29(5), 1947–1960 (2018)
28. Li, Z., Tang, J.: Unsupervised feature selection via nonnegative spectral analysis and redundancy control. IEEE Trans. Image Process. 24(12), 5343–5355 (2015)
29. Make3D. http://make3d.cs.cornell.edu/data.html#make3d

Temporal-Contextual Attention Network for Video-Based Person Re-identification

Di Chen, Zheng-Jun Zha$^{(\boxtimes)}$, Jiawei Liu, Hongtao Xie, and Yongdong Zhang

University of Science and Technology of China, Hefei, China
zhazj@ustc.edu.cn

Abstract. Video-based person re-identification aims to identify a specific person in surveillance videos from different cameras. This paper presents a new Temporal-Contextual Attention Network (TCA-Net) for person re-identification in videos. The TCA-Net exploits temporally local context among consecutive frames to concentrate selectively on crucial frames within a video sequence. Specifically, the network consists of a Convolutional Neural Network (CNN) module and a temporal-contextual attention block. The CNN module embeds each video frame into a convolutional representation, and the temporal-contextual attention block learns the importance of a video frame for re-identification by exploiting the local context among the frame and its neighboring frames. The feature of a video sequence is then obtained by aggregating frame-level features weighted by frame importance. We evaluate the proposed TCA-Net on a challenging dataset MARS. The experimental results have demonstrated the effectiveness of the proposed approach.

Keywords: Person re-identification · Temporal context
Visual attention

1 Introduction

Person re-identification aims to search for the person-of-interest from no overlapping camera views. It has attracted significant attention in recent years due to its importance for many practical applications, such as intelligent surveillance, activity analysis and criminal investigation [14,15,19]. Despite recent remarkable progress on person re-identification, it remains an challenging task due to occlusion, background clutters, variations in illumination and camera views, as well as changes in human poses *etc.*

Recent years have witnessed numerous research on re-identifying person-of-interest within surveillance images. Existing approaches mainly focus on learning discriminative feature of person appearance [9,26,27] and/or deriving distance metric for feature matching [10,14,25]. Despite remarkable progress on person re-identification in images, little attention has been paid for re-identifying person-of-interest in videos. Video-based person re-identification is to search for a target person from video sequences captured by non-overlapping cameras give a query

© Springer Nature Switzerland AG 2018
R. Hong et al. (Eds.): PCM 2018, LNCS 11164, pp. 146–157, 2018.
https://doi.org/10.1007/978-3-030-00776-8_14

Time

Fig. 1. Example video sequence of pedestrian with pose changes, occlusion and background clutters.

sequence of the person. Compared to an image with person appearance, a video sequence presents much more content related to person, such as motion patterns, person gait *etc.* Moreover, consecutive frames within a video sequence present person appearance with different body poses and from different viewpoints, providing valuable information towards addressing the challenge of pose variation, occlusion, and viewpoint change *etc.* One crucial task of video-based person re-identification is to learn discriminative video representation which is able to identify the same person and distinguish different ones.

Some traditional approaches have been developed as a straightforward extension of image-based re-identification solution. They learn a feature vector for each frame and aggregate frame-level features across time by a pooling function (e.g., average or max pooling) to form a feature for a video sequence [19,20]. These approaches treat all the frames within a sequence as equally important during pooling, resulting sub-optimal video representation for person re-identification. A video sequence presents pose changes, occlusion, background clutters *etc.* within different frames as shown in Fig. 1. Correspondingly, the different frames offers non-equally important cues for re-identifying person, even partial frames are useless for re-identification. Hence, there is a demand to estimate the importance of each frame and concentrate on the informative frames.

To this end, visual attention mechanism have been applied into video-based person re-identification [13,21,30]. A few of preliminary works design a temporal attention module to attend to relevant video frames [17,30]. In [30], the importance of a frame is inferred by considering its own visual content and its correlation to the previous frame. However, the efficacy of the attention is sensitive to the quality of the previous frame and suffers from error accumulation from previous low-quality frames. Liu *et al.* [17] proposed a Quality Aware Network for quality scores prediction. However, the quality score of a frame is generate by a convolution neural network, which completely ignores the temporal context information in the video. On the other hand, a video sequence possesses temporal consistency on content across neighboring frames. A frame is correlated to not only the frames previous to it but also subsequent frames. The exploration of such temporally local context among consecutive frames is valuable for learning accurate attention and can also alleviate the influence of low-quality frames.

In this paper, we propose a new temporal-contextual attention network (TCA-Net) to learn effective video representation for person re-identification by exploiting temporally local context among consecutive frames. The TCA-Net learns to concentrate selectively on crucial frames within a video sequence. It essentially infers which frames are informative for identifying person and attaches higher importance to them. As illustrated in Fig. 2, the proposed TCA-Net consists of a base Convolutional Neural Network (CNN) and a Temporal-Contextual Attention block. The CNN extracts convolutional features from video frames, then the convolutional features are fed into LSTM to capture temporal information with hidden states. The attention block perceives the convolutional feature of current time-step and its neighboring hidden states, and generates the importance score of current frame. Features of frames in the whole video sequence are then aggregated by the final weight assignment operation according to their importance. By concentrating on informative frames with robust and accurate temporal attentions, TCA-Net is able to learn robust and discriminative video representation, leading to satisfactory results of person re-identification in videos. We conduct experiments on a real-world dataset of video-based person re-identification, i.e., MARS, to evaluate the proposed TCA-Net. The experimental results have demonstrated the effectiveness of TCA-Net.

2   Method

2.1   Convolutional Feature

Given a video sequence $V = \{v_1, v_2, ..., v_T\}$, where T represents the total number of video frames, we extract features from the frames through an embedding function:

$$X = \{\phi(v_1), \phi(v_2), ..., \phi(v_T)\} = \{x_1, x_2, ..., x_T\}, \tag{1}$$

Here $\phi(\cdot)$ is the embedding function and could be a Convolutional Neural Network (CNN) model. Here, we adopt the DenseNet-121 [11] as the function $\phi(\cdot)$, where the dense blocks strengthen feature propagation and alleviate the problem of vanishing-gradient. DenseNet-121 consists of a convolutional layer, four dense blocks and a 7×7 average pooling layer. We use the output of the pooling layer as frame-level feature.

2.2   Temporal-Contextual Attention Module

The Temporal-Contextual Attention Module consists of LSTM cells and an attention block. The LSTM encodes the convolutional features of video frames

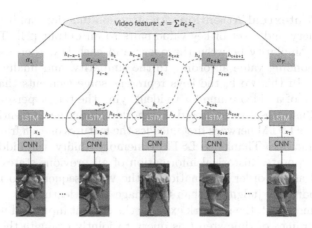

Fig. 2. The architecture of the proposed Temporal-Contextual Attention Network, which consists of a base CNN and a temporal-contextual attention block. The attention block learns the importance of each frame, which is represented as $\{\alpha_t\}_{t=1}^T$. At each time-step t, the attention block computes the importance score α_t based on the feature of the frame at hand and the hidden representation of the neighboring frames. The final video representation $\overline{x}$ is the weighted aggregation as $\{x_t\}_{t=1}^T$.

sequentially. The attention block learns to predict importance score for each frame and leads to better representation for video-based person re-identification.

Long Short-Term Memory. Long Short-Term Memory (LSTM) has strong power in processing sequential data [8,23]. It has been successfully applied to video analysis due to its improvement of generation of descriptions from intermediate visual representations derived from conventional visual models [4]. At each step the model is auto-regressive, consuming the previously generated symbols as additional input when generating the next [7]. Specifically, at each time step t, a new sample x_t (i.e., the t-th frame of the video sequence) is fed to LSTM cell together with the hidden representation h_{t-1} of the last frame. A standard LSTM consists of an input gate i_t, an output gate o_t, together with a forget gate f_t. Its formulations are illustrated as Eqs. (2) (4):

$$\begin{pmatrix} i_t \\ o_t \\ f_t \\ g_t \end{pmatrix} = \begin{pmatrix} \sigma \\ \sigma \\ \sigma \\ \tanh \end{pmatrix} \left(W \begin{pmatrix} x_t \\ h_{t-1} \end{pmatrix} \right) \tag{2}$$

$$c_t = f_t \odot c_{t-1} + i_t \odot g_t \tag{3}$$

$$h_t = o_t \odot \tanh(c_t) \tag{4}$$

where i_t, o_t, f_t, c_t and h_t denote the input gate, output gate, forget gate, cell state and hidden representation respectively, $\odot$ denotes Hadamard product, W is learnable weight parameters inside the gates.

Temporal-Contextual Attention. An attention function can be described as mapping a query and a set of key-value pairs to an output [24]. This function calculates the similarity between the query and each key, then assigns weight to the corresponding value according to the similarity, and finally outputs the weighted sum. In this work, the keys represent some elements that determine the importance of a video frame for re-identifying the target person.

Figure 2 illustrates the proposed Temporal-Contextual Attention Network (TCA-Net). The LSTM network first encodes the features of each frame extracted by the CNN module. Thanks to LSTM's memory ability, the hidden representation h_t incorporates historical information of all previous states and current input, as well as the order information of the video sequence, so it can better assist the importance judgment than the image-based feature x_t.

At each time step t, the network exploits the current input and nearest neighboring video frames of time step t as query to jointly evaluate the importance score e_t of the current input as follows:

$$e_t = W_{value} \left(\tanh \left(W_{key} \begin{pmatrix} h_{t-k} \\ \cdots \\ x_t \\ \cdots \\ h_{t+k} \end{pmatrix} \right) \right) \tag{5}$$

where h_{t-k} and h_{t+k}, $k \in (1, K)$ denotes the 2K-nearest hidden representations of time step t, W_{key} and W_{value} denote key and value matrix, respectively. The importance score of each frame is then computed as the response of a softmax layer for normalization:

$$\alpha_t = \frac{\exp(e_t)}{\sum_{u=1}^{T} \exp(e_u)} \tag{6}$$

The importance score α_t is then used to aggregate the feature vectors of each frame into a final video representation, which is formulated as follows:

$$\overline{x} = \sum_{t=1}^{T} \alpha_t x_t \tag{7}$$

2.3   Loss Function and Optimization

Person re-identification is often viewed as a multi-classification problem, and the network is trained using the loss function (mostly softmax loss) of multi-classification problems. Although softmax loss is widely used for its effectiveness and simplicity, it does not explicitly optimize the features to have higher similarity score for positive pairs and lower similarity score for negative pairs. In order to solve this problem, multiple loss functions have been proposed to learn highly discriminative features, such as triplet loss [22], large-margin softmax loss [16]

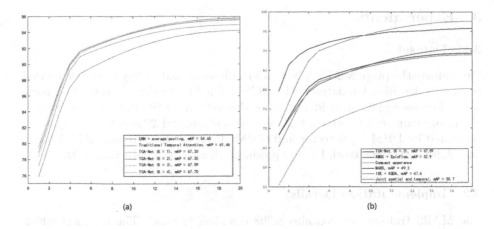

Fig. 3. Performance comparison in terms of CMC curves on the MARS dataset.

and SphereFace loss [18]. In this paper, we adopt ArcFace loss [3], which achieves excellent result in large-scale face recognition, instead of softmax loss to train the CNN module because of its excellent geometrical interpretation, so that it can learn highly discriminative robust features.

For the same purpose, we propose a batch-similarity loss to explicitly minimize intra-class distance and maximize inter-class distance. Due to softmax loss is shown its excellent performance in optimizing the probability of each class, we exploit both softmax loss and batch-similarity loss to train the Temporal-Contextual Attention block.

In each mini-batch iteration, the similarity between each sample pair are calculated by cosine metric, and then we compute the gap between the result and ground-truth:

$$cos_{ij} = \frac{\overline{x}_i \cdot \overline{x}_j}{\|\overline{x}_i\| \|\overline{x}_j\|} \tag{8}$$

$$s_{ij} = \frac{cos_{ij} + 1}{2} \tag{9}$$

$$L_{sim} = \sum_{i=1}^{m} \sum_{j=1}^{m} (s_{ij} - gt_{ij})^2 \tag{10}$$

where s_{ij} denotes the cosine similarity which is normalized between 0 and 1, $gt_{ij} = 1$ if i and j have the same identity, otherwise $gt_{ij} = 0$, m is the size of mini-batch, and L_{sim} represents the batch-similarity loss.

To optimize both intra-class and inter-class distance, we combine the softmax loss L_{soft} and batch-similarity loss L_{sim} to train the whole network. The final loss function is:

$$Loss = L_{soft} + \lambda L_{sim} \tag{11}$$

where the hyper-parameter λ is set to 0.1 in our experiment.

3 Experiments

3.1 Dataset

We evaluate the proposed TCA-Net on a challenging and widely used video-based person re-identification dataset: MARS [28]. The MARS dataset is the first large scale video based person re-identification dataset with 1,261 different pedestrians whom are captured by at least 2 cameras and around 20,000 video sequences generated by DPM [5] detector and GMMCP [2] tracker. Among 20,715 track-lets, 3,248 distractor track-lets are produced due to false detection or tracking.

3.2 Implementation Details

The MARS training set contains 8,298 tracklets in total. The tracklets which have 8 video frames or more are around 97%. Only 260 tracklets contain 8 frames or less. Hence, we select 8 consecutive video frames as an input unit in the experiments, and the size of the input sequence is $8 \times 224 \times 224 \times 3$. A tracklet with less than 8 frames is extended by duplicating its last frame. We first pre-train the DenseNet-121 model on ImageNet [12] dataset, and then fine-tune it on MARS training sets. Once finished, we freeze the CNN model and train the TCA-Net. The input image is resized to 224×224, the dimension of the hidden representation of LSTM is set to 256. We adopt the batched stochastic gradient descent with momentum 0.9 to optimize the parameters with an initial learning rate set to 0.01 and then dropped to 0.001. In label estimation and evaluation steps, feature representations belonging to the same video will form a single feature that represents the video through average pooling. Finally, the resulting features are l_2 regularized for subsequent metric learning and distance calculation.

To evaluate the performance, the Cumulative Matching Characteristic (CMC) curve [1] and the mean average precision (mAP) [29] are used. The CMC curve reflects the matching accuracy of the algorithm in top-k samples, in this paper, we approximately describe the CMC curve with Rank-1, Rank-5, and Rank-20 scores. The mAP considers both precision and the order, it calculates the precision of each sample in the corresponding position in the ranked sequence. Roughly speaking, the CMC curve reflects the precision of a model, while mAP represents the model recall.

3.3 Component Analysis

We conduct specific experiments to investigate the effect of each component of our network by conducting several analytic experiments. Baseline is the Densenet-121 which pretrained on Imagenet dataset and fine-tuned on MARS training sets with ArcFace loss, then using average pooling instead of temporal aggregation.

All models are jointly trained with softmax loss and batch-similarity loss, we further enhance the features from all models through metric learning, in our

Table 1. Performance comparison in terms of rank-k accuracy and mAP on the MARS dataset.

Methods	Rank-1	Rank-5	Rank-20	mAP
Baseline(CNN + avg pooling)	75.86	88.79	94.24	54.65
Traditional Temporal Attention	77.17	90.96	94.95	64.16
TCA-Net($K = 1$)	78.33	91.41	95.71	67.30
TCA-Net($K = 2$)	78.89	91.57	95.56	67.35
TCA-Net($K = 3$)	**79.55**	91.41	**95.81**	67.59
TCA-Net($K = 4$)	79.44	**91.67**	95.76	**67.70**

Table 2. Performance comparison of TCA-Net and the state-of-the-art methods on MARS dataset.

Methods	Rank-1	Rank-5	Rank-20	mAP
Compact Appearance [27]	55.5	70.2	80.2	-
IDE + XQDA [28]	65.3	82.0	89.0	47.6
MARS [28]	68.3	82.6	89.4	49.3
AMOC + EpicFlow [19]	68.3	81.4	90.6	52.9
Joint Spatial and Temporal [30]	70.6	90.0	**97.6**	50.7
TCA-Net($K = 3$)	**79.55**	**91.41**	95.81	**67.6**

experiments we used Cross-view Quadratic Discriminant Analysis (XQDA) [14]. Figure 3(a) plots the CMC curves of our model under different K values, and compares the curves of the baseline and the traditional temporal attention model. Detailed numerical results are recorded in Table 1. Traditional temporal attention model uses the current feature and the hidden representation of last step as input. Figure 3(a) plots the CMC curves of our model under different K values, and compares the curves of the baseline and the traditional temporal attention model. Detailed numerical results are recorded in Table 1. Compared with baseline, traditional temporal attention model has brought 1.31% and 9.51% improvement to Rank-1 and mAP, respectively. This shows the effectiveness of traditional temporal attention on aggregated temporal information, which can adaptively calculate the importance of video frames and assigns weights according to their importance, so it can improve the performance of video-based person re-identification.

Different Numbers of Neighboring Frames. We also explored the effect of different K values on performance. When $K = 1$, the model uses the current input together with its last frame and next frame for importance decision. With the introduction of the information of next frame, the model can analyze the importance more accurately, so the performance is higher than the traditional temporal attention. As K increases, the network uses more distant video frames to assist in importance judgment. Intuitively, the more information involved in the judgment, the more accurate the result will be. However, we noticed that when K increased from 3 to 4, the performance has dropped to some extent. This may be because when the distance is too far, the correlation between these video frames and the input frame becomes very weak or even disappears. Therefore, the information of these video frames does not provide much help for the importance analysis. Sometimes it even misleads the judgment and has a negative impact.

Table 3. The performance of using different loss functions.

Methods	Rank-1	Rank-5	Rank-20	mAP
Softmax Loss	77.42	91.11	95.40	64.48
Batch-similarity Loss	78.84	90.96	95.30	66.96
Softmax Loss and Batch-similarity Loss	**79.55**	**91.41**	**95.81**	**67.59**

Ablation Study of Loss Function. As mentioned in Sect. 2.1, softmax only focuses on how to distinguish samples of different identities, and does not care whether the distance of same identity samples are close in the latent space. This may lead to situations where the intra-class distance is larger than the inter-class distance. Since the distribution of training data is very different from the distribution of real-world data, when the network faces a new set of samples that have never been seen before, this embedding mode will produce a great error.

In this paper, we propose a batch-similarity loss to alleviate this problem. The optimization criterion of the loss is to maximize the similarity of features of the same identity while minimizing the similarity of different identity features. The result of ablation study of the Temporal-Contextual Attention Network is well reflected in Table 3. Both methods have K = 3, and all the other components are identical except training loss. We can observe from Table 3 that, jointly using the softmax loss and batch-similarity loss, leads to improvements in the performance than using only one type loss. This confirms the importance of using these two loss simultaneously.

LSTM Versus CNN. We utilize the hidden representation of LSTM as context in our model instead of utilizing CNN feature directly. At time step t, for LSTM, the historical information of all previous states is available. Benefit from the gate mechanism, LSTM can selectively decide whether to accept input information and whether to discard historical information, so noise and overlapping information can be well relieved. Meanwhile, CNN features do not contain the order information of the video sequence, the attention model which utilizes context CNN features as input could not capture temporal information of the video sequence. Experimental results are listed in Table 4 and all methods have $K = 3$. We can observe that, using the hidden representation of LSTM as context, leads to improvements in the performance than using CNN feature directly. This confirms our aforementioned statement.

We also experimented with using RNN or BLSTM [6] as context embedding module instead. Different from LSTM, RNN does not have the ability to filter historical information and input information, so it may introduce many overlapping information or noisy information. And we found that BLSTM produced nearly identical results as LSTM.

Table 4. Performance comparison to other context embedding modules.

Methods	Rank-1	Rank-5	Rank-20	mAP
CNN	78.54	91.52	95.71	67.57
RNN	78.79	**91.62**	95.76	67.09
BiLSTM	79.44	91.31	95.45	67.59
LSTM	**79.55**	91.41	**95.81**	**67.60**

3.4 Comparison with the State-of-the-Art Methods

Table 2 shows the results of our TCA-Net and other state-of-the-art methods, while Fig. 3(b) shows the corresponding CMC curve. Prior to this, the best results on the MARS dataset came from Joint Spatial and Temporal [30], which uses the traditional temporal attention model to aggregate temporal information, using six RNNs to sweep feature map from six directions to get spatial contextual feature, and finally concatenate temporal and spatial feature as final feature. The structure of this model is more complex and it is not easy to implement. And our method is more intuitive, concise and easy to train. Our method is 8.95% and 16.9% higher on Rank-1 and mAP than it respectively, which is a considerable increase. This result demonstrates the effectiveness of our method in video-based person re-identification tasks.

4   Conclusion

In this paper we developed a new Temporal-Contextual Attention Network (TCA-Net) for person re-identification in videos. The proposed TCA-Net learns to focus on crucial frames for identifying person-of-interest in a video sequence by exploiting temporally local context among consecutive frames. Its essentially learns the importance of each frame for re-identification and then aggregates frame-level convolutional features by the corresponding importance scores. We conducted evaluation on a challenging and widely used person re-identification dataset, i.e., MARS. The experimental results have shown that TCA-Net is able to learn robust and discriminative video representation, leading to satisfactory results of person re-identification in videos.

Acknowledgement. This work was supported by the National Key R&D Program of China under Grant 2017YFB1300201, the National Natural Science Foundation of China (NSFC) under Grants 61622211, 61472392 and 61620106009 as well as the Fundamental Research Funds for the Central Universities under Grant WK2100100030.

References

1. Bolle, R.M., Connell, J.H., Pankanti, S., Ratha, N.K., Senior, A.W.: The relation between the ROC curve and the CMC. In: 2005 IEEE Workshop on Automatic Identification Advanced Technologies, pp. 15–20. IEEE (2005)
2. Dehghan, A., Modiri Assari, S., Shah, M.: GMMCP tracker: globally optimal generalized maximum multi clique problem for multiple object tracking. In: 2015 IEEE Conference on Computer Vision and Pattern Recognition (CVPR), pp. 4091–4099. IEEE (2015)
3. Deng, J., Guo, J., Zafeiriou, S.: Arcface: additive angular margin loss for deep face recognition (2018). arXiv preprint: arXiv:1801.07698
4. Donahue, J., et al.: Long-term recurrent convolutional networks for visual recognition and description. In: 2015 IEEE Conference on Computer Vision and Pattern Recognition (CVPR), pp. 2625–2634. IEEE (2015)
5. Felzenszwalb, P.F., Girshick, R.B., McAllester, D., Ramanan, D.: Object detection with discriminatively trained part-based models. IEEE Trans. Pattern Anal. Mach. Intell. **32**(9), 1627–1645 (2010)
6. Graves, A., Schmidhuber, J.: Framewise phoneme classification with bidirectional LSTM and other neural network architectures. Neural Netw. **18**(5–6), 602–610 (2005)
7. Graves, A.: Generating sequences with recurrent neural networks (2013). arXiv preprint: arXiv:1308.0850
8. Hochreiter, S., Schmidhuber, J.: Long short-term memory. Neural Comput. **9**(8), 1735–1780 (1997)
9. Hong, R., Zhang, L., Zhang, C., Zimmermann, R.: Flickr circles: aesthetic tendency discovery by multi-view regularized topic modeling. IEEE Trans. Multimed. **18**(8), 1555–1567 (2016)
10. Hong, R., Hu, Z., Wang, R., Wang, M., Tao, D.: Multi-view object retrieval via multi-scale topic models. IEEE Trans. Image Process. **25**(12), 5814–5827 (2016)
11. Huang, G., Liu, Z., Weinberger, K.Q., van der Maaten, L.: Densely connected convolutional networks. In: 2017 IEEE conference on Computer Vision and Pattern Recognition (CVPR), pp. 4700–4708. IEEE (2017)
12. Krizhevsky, A., Sutskever, I., Hinton, G.E.: Imagenet classification with deep convolutional neural networks. In: Advances in Neural Information Processing Systems (NIPS), pp. 1097–1105 (2012)
13. Li, S., Bak, S., Carr, P., Wang, X.: Diversity regularized spatiotemporal attention for video-based person re-identification (2018). arXiv preprint
14. Liao, S., Hu, Y., Zhu, X., Li, S.Z.: Person re-identification by local maximal occurrence representation and metric learning. In: 2015 IEEE Conference on Computer Vision and Pattern Recognition (CVPR), pp. 2197–2206. IEEE (2015)
15. Liu, H., Feng, J., Qi, M., Jiang, J., Yan, S.: End-to-end comparative attention networks for person re-identification. IEEE Trans. Image Process. **26**(7), 3492–3506 (2017)

16. Liu, W., Wen, Y., Yu, Z., Yang, M.: Large-Margin Softmax Loss for Convolutional Neural Networks. In: 2016 International Conference on Machine Learning (ICML), pp. 507–516 (2016)
17. Liu, Y., Yan, J., Ouyang, W.: Quality aware network for set to set recognition. In: 2017 IEEE Conference on Computer Vision and Pattern Recognition (CVPR), pp. 4694–4703. IEEE (2017)
18. Liu, W., Wen, Y., Yu, Z., Li, M., Raj, B., Song, L.: Sphereface: deep hypersphere embedding for face recognition. In: 2017 IEEE Conference on Computer Vision and Pattern Recognition (CVPR), pp. 212–220. IEEE (2017)
19. Liu, H., et al.: Video-based person re-identification with accumulative motion context. IEEE Trans. Circuits Syst. Video Technol. $\mathbf{PP(99)}$, 1 (2017)
20. Mclaughlin, N., Rincon, J.M.D., Miller, P.: Recurrent convolutional network for video-based person re-identification. In: 2016 IEEE Conference on Computer Vision and Pattern Recognition (CVPR), pp. 1325–1334. IEEE (2016)
21. Ouyang, D., Zhang, Y., Shao, J.: Video-based person re-identification via spatio-temporal attentional and two-stream fusion convolutional networks. Pattern Recognit. Lett. (2018)
22. Schroff, F., Kalenichenko, D., Philbin, J.: Facenet: a unified embedding for face recognition and clustering. In: 2015 IEEE Conference on Computer Vision and Pattern Recognition (CVPR), pp. 815–823. IEEE (2015)
23. Sundermeyer, M., Schlüter, R., Ney, H.: LSTM neural networks for language modeling. In: 2012 Annual Conference of the International Speech Communication Association (ISCA) (2012)
24. Vaswani, A., et al.: Attention is all you need. In: Advances in Neural Information Processing Systems (NIPS), pp. 5998–6008. NIPS (2017)
25. You, J., Wu, A., Li, X., Zheng, W.S.: Top-push video-based person re-identification. In: 2016 IEEE Conference on Computer Vision and Pattern Recognition (CVPR), pp. 1345–1353. IEEE (2016)
26. Xiao, T., Li, S., Wang, B., Lin, L., Wang, X.: End-to-end deep learning for person search (2017). arXiv preprint
27. Zhang, W., Hu, S., Liu, K.: Learning compact appearance representation for video-based person re-identification (2017). arXiv preprint: arXiv:1702.06294
28. Zheng, L., et al.: MARS: a video benchmark for large-scale person re-identification. In: Leibe, B., Matas, J., Sebe, N., Welling, M. (eds.) ECCV 2016, Part VI. LNCS, vol. 9910, pp. 868–884. Springer, Cham (2016). https://doi.org/10.1007/978-3-319-46466-4_52
29. Zheng, L., Shen, L., Tian, L., Wang, S., Wang, J., Tian, Q.: Scalable person re-identification: a benchmark. In: 2015 IEEE International Conference on Computer Vision (ICCV), pp. 1116–1124. IEEE (2015)
30. Zhou, Z., Huang, Y., Wang, W., Wang, L., Tan, T.: See the forest for the trees: joint spatial and temporal recurrent neural networks for video-based person re-identification. In: 2017 IEEE Conference on Computer Vision and Pattern Recognition (CVPR), pp. 6776–6785. IEEE (2017)

Synthetic Aperture Based on Plenoptic Camera for Seeing Through Occlusions

Heng Zhang[✉], Xin Jin, and Qionghai Dai

Shenzhen Key Laboratory of Broadband Network and Multimedia,
Graduate School at Shenzhen, Tsinghua University, Shenzhen 518055, China
18930861549@163.com, jin.xin@sz.tsinghua.edu.cn,
qhdai@tsinghua.edu.cn

Abstract. De-occlusion is a classical challenge in computer vision to many research fields. In this paper, a synthetic aperture algorithm is proposed to see through occlusions based on plenoptic camera (PLC). The sheltered object is located onto the zero-disparity plane by deriving the geometric relationship between the focus plane and the object plane using the virtual camera array structure of PLC. Pixel selection relying on depth and motion clustering is proposed to pick out the pixels coming from sheltered object. Finally, a synthetic aperture image (SAI) is obtained by averaging the selected pixels so that the sheltered objects can be seen. Compared with the existing work, the proposed method provides much better visual quality in achieving clearer sheltered objects and much more blurry occlusions.

Keywords: Synthetic aperture · De-occlusion · Zero-disparity plane relocation
Depth+motion guided pixel selection · Plenoptic camera

1 Introduction

De-occlusion is a classical challenge in computer vision to many research fields. Fortunately, the technology named synthetic aperture imaging [1] has the ability to achieve de-occlusion by fusing the spatial information from captured light field. By warping and fusing the multiple view images, synthetic aperture imaging can simulate a virtual camera with a large convex lens and focus on different frontal-parallel planes so that the sheltered object on the virtual focal plane can be seen. However, the camera array or camera gantry [1] on which the technology is implemented has a poor performance in the portability and operation. Therefore, the hand-held plenoptic camera [2] (PLC) is employed to achieve de-occlusion in our research. As one type of the most advanced devices in recording the light field, PLC attracts great attentions owing to its superior advantages in single shot acquisition and portability. A variety of PLCs' applications, like refocusing, VR, AR, have been developed, within which seeing through occlusions can be investigated.

However, performing synthetic aperture for PLCs is challenging. The PLC inserts a micro-lens array (MLA) between the main lens and an imaging sensor to record the light field [2]. The limited number of pixels on the sensor and the aperture of the main lens force a constrained angular resolution in the acquired light field. Small disparities between the nearby sub-aperture images (SIs) reduce not only accuracies of warping but also the size of simulated virtual lens, which makes synthetic aperture shows limited efficiency in de-occluding for PLCs.

Synthetic aperture algorithms usually consist two steps: alignment and synthesizing. After alignment, the sheltered objects are aligned well in all views. The "well aligned" means two things: on the one hand, the sheltered objects are sharp in all views; on the other hand, there are no disparities for sheltered objects. Furthermore, those pixels from sheltered objects are selected and averaged in synthesizing stage. Finally, a SAI is obtained where the sheltered objects can be seen.

For alignment, the existing methods can be mainly classified into two categories: parallax-based methods and homography-based methods. In the researches of Wang et al. [3] and Levoy et al. [4], parallaxes among the views were computed by block matching, using which the view images were warped and the transformed images were obtained. To make the parallax more accurate, Shrestha et al. [5] and Dong et al. [6] transformed the spatial parallax into the phase difference in the frequency domain. But, considering the blur of side SIs made by the pixels at edges in macro-pixels, the block matching methods will lead notable errors into SAI. The homography-based methods, like those proposed by Xu et al. [7], Barron et al. [8] and Robert et al. [9], derived the homography matrices to warp viewpoints to a target view-point. However, the errors introduced by feature detecting methods are on the same order of magnitude as the disparities between nearby SIs.

For synthesizing, the existing methods can also be classified into two categories: averaging methods and pixel selection. In the research of Vaish et al. [10, 11] and Zhang et al. [12], the well-aligned images are averaged directly in pixels. However, since the pixels on the occlusions are only averaged, look blurred, the visual quality of the processing results is not acceptable. To improve the visual quality, pixel selection methods were proposed by developing criteria in selecting correct pixels from the views. Yang et al. [13], Xiao et al. [14] and Yatziv et al. [15] picked out pixels from all views by minimizing the energy function or maximizing the confidence of pixels used to make up SAI. While, prior information, like the depth of sheltered object, is needed, which results in low robustness to errors of depth estimation. Yang et al. [16] attached labels to all pixels according to the depth map, picked out pixels which are deemed from the sheltered objects and averaged them. But, the accuracy of the depth map, which is mainly determined by the efficiency of depth estimation, may directly affect the quality of SAI.

Thus, to overcome above problems, in this paper, a synthetic aperture algorithm is proposed for seeing through occlusions using PLCs. We propose to relocate the sheltered objects onto the zero-disparity plane by deriving the geometric relationship between the focus plane and the object plane using the virtual camera array structure of PLC. All the SIs are well aligned, based on which pixel selection relying on depth and motion clustering is proposed to pick out the pixels coming from sheltered object. Finally, a SAI is obtained by averaging the selected pixels. Compared with the existing methods, the processing results generated by the proposed method provide much better visual quality in achieving clearer sheltered objects and much more blurry occlusions.

The rest of this paper is organized as follows. Section 2 describes the proposed algorithm in detail. The experimental results and analysis are provided in Sect. 3. Section 4 concludes the paper with future work.

2 Proposed Algorithm

2.1 The Overall Diagram of the Proposed Algorithm

The diagram of proposed algorithm is shown in Fig. 1 where the overstriking blocks are unique. It mainly consists of two steps: *Zero-disparity Plane Relocation* and *Depth +motion guided Pixel Selection*. *Zero-disparity Plane Relocation* is to align SIs by relocating the sheltered objects onto the zero-disparity plane after deriving geometric relationship between the focus plane and the object plane based on the virtual camera array structure of PLC. *Depth+motion guided Pixel Selection* is to pick out the pixels coming from sheltered object by pixel selection relying on depth and motion clustering.

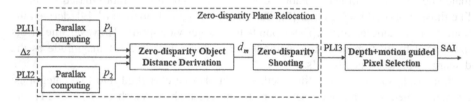

Fig. 1. The diagram of proposed algorithm.

First, for sheltered objects, two plenoptic images (PLIs) are taken along the axis at two object distances randomly, where the distance difference, Δz, can be measured by range finder. Then, *Zero-disparity Plane Relocation* is executed, within which *Parallax computing* is first applied to retrieve the parallaxes of the sheltered objects in the two PLIs. SIs are extracted from a PLI [17], based on which depth maps are estimated [18]. Considering the occlusions are closer to the camera than the sheltered objects, pixels in the occlusion region in a SI are clustered into two groups according to the depth values or module of motion vectors and the group of pixels that is farther to the camera is selected. Treating the group of pixels as the pixels from the sheltered objects, parallaxes between center SI and its neighborhoods are derived by block matching algorithm [19]. Thus, p_1 and p_2 are retrieved for the sheltered objects in PLI1 and PLI2, respectively. Then, *Zero-disparity Object Distance Derivation* is executed to derive the focus distance of the sheltered object relative to the position where the PLI1 is taken, d_m, using geometric relationship among the virtual cameras. The details of the algorithm will be introduced in Subsect. 2.2.

Using d_m as the displacement of PLC, *Zero-disparity Shooting* is performed by moving the PLC along the axis and taking a new plenoptic image, *PLI3* in which the sheltered objects stand on the zero-disparity plane so that the corresponding pixels in all the SIs are well aligned.

Then, *Depth+motion guided Pixel Selection* is applied to the SIs of *PLI3* to select pixels from the sheltered objects according to the fused map generated by depth and motion clustering and average them to generate a SAI. The details of the algorithm will be introduced in Subsect. 2.3.

2.2 Zero-Disparity Object Distance Derivation

In order to align the sheltered objects accurately for sharper results in the synthetic aperture image, we propose to relocate the sheltered objects by moving the PLC so that the focus plane, i.e. zero-disparity plane, coincides with the object plane. However, how to derive the distance between the focus plane of a PLC and the object plane, i.e. the zero-disparity object distance, denoted by d_m in Fig. 1, is the key problem.

Considering a PLC can be treated as a camera array consisting of many small virtual cameras capturing corresponding SIs [20], the relative distance between focus plane and object plane of a PLC can be derived from the geometry of the virtual camera array. Every small virtual camera is made up by virtual lens and virtual sensor plane. According to work of Mignard *et al.* [21], the virtual lenses are arranged at the focal plane while the virtual sensor planes coincide with the focus plane of PLC, which is shown in Fig. 2(a). In addition, as shown in Fig. 2(b), this equivalent structure is employed to compute the disparities between nearby SIs. By the aid of this structure and parameters of virtual camera, the equations can be listed to denote the relation between disparity, focus length and object length. Furthermore, the relative distance can be gained by solving simultaneous formulas. As one of the prerequisites, the coefficients of the formulas should be consistent. Therefore, the parameters of PLC should be kept in the whole proposed algorithm.

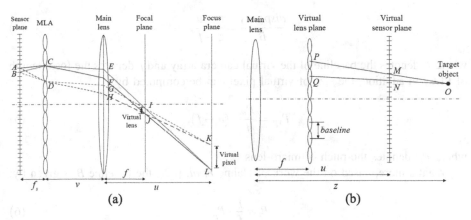

Fig. 2. The equivalent structure of PLC. (a): the positions of virtual lenses and virtual sensor plenas in the equivalent structure of PLC [20, 21]; (b): the equivalent structure is utilized to compute the disparities between nearby SIs.

Denoting the focus distance of the PLC and the object distance of the sheltered objects by u and z when the PLI1 is taken, respectively, d_m is given by:

$$d_m = z - u. \tag{1}$$

Generally, the object distance z and focus length u are unknown. While, by capturing two PLIs under the same parameters of PLC, the distance difference Δz is known. Using two PLIs, z and u can be obtained by solving the following simultaneous formulas:

$$\begin{cases} p_1 = f(u, z) \\ p_2 = f(u, z + \Delta z) \end{cases}, \tag{2}$$

where $f(u, z)$ denotes the relation between parallax p that is obtained by block matching algorithm [19] and u, z.

The parallax is usually denoted in pixels. Thus, the parallax for the target object, like MN in Fig. 2(b), can be written as:

$$p = \frac{disparity}{P_{vp}}, \tag{3}$$

where $disparity$ denotes the parallax between nearby SIs and P_{vp} denotes the size of virtual pixels.

According to the equivalent structure [21], the virtual sensor planes share the same space at the focus plane of PLC but have no interaction each other. As shown in Fig. 2 (b), by similar triangles, $disparity$ is given by:

$$\frac{disparity}{B} = \frac{z - u}{z - f}, \tag{4}$$

where B denotes the baseline of the virtual camera array and f denotes the focal length of PLC. In addition, the size of virtual pixel can be computed by:

$$P_{vp} = \frac{P_m}{f} \cdot (u - f), \tag{5}$$

where P_m denotes the pitch of micro-lens.

What's more, based on the work of Hahne et al. [22], the baseline B is given by:

$$B = \frac{f}{f_s} \cdot P_p, \tag{6}$$

where f_s denotes the focal length of micro-lens and P_p denotes the size of pixel on sensor plane of PLC.

By Eqs. (3)–(6), the relation can be written as follows:

$$f(u,z) = \frac{z - uf}{z - ff_s} \frac{f}{u - f} \frac{P_p}{P_m}. \tag{7}$$

Based on Eqs. (1), (2) and (7), the values of d_m can be obtained. Furthermore, we can move PLC until the focus plane coincides with the object plane and take another PLI, denoted as *PLI3* in Fig. 1. Therefore, the sheltered object not only is sharp but also has no parallax in all SIs which are extracted from *PLI3*.

2.3   Depth+Motion Guided Pixel Selection

Using the SIs in *PLI3*, the SAI is generated by the proposed *depth+motion guided pixel selection*. It is well known that the occlusions bring great difficulties into depth estimating and motion vector computing. Considering the sheltered objects are farther from the camera, which generally presents larger depth and smaller relative motion among the SIs compared with the occlusions in well aligned SIs, we propose to fuse these two cues to cluster the pixels.

Considering the structure of PLC, the edge SIs are darker than those middle SIs because they are made up by edge pixels in all macro-pixels. This phenomenon makes it difficult to estimate motion vectors accurately. Therefore, we prefer to fuse depth and motion vectors in middle SIs and use depth in edge SIs to select pixels. A map is generated for clustering the pixels to pick out the pixels from sheltered objects. The intensity at pixel (i,j) is given by:

$$I_{s,t}^{mer}(i,j) = \begin{cases} I_{s,t}^{dep}(i,j) + k \cdot I_{s,t}^{MV}(i,j), & if \ bright(I_{s,t}) > T \cdot \frac{1}{w}\sum_i\sum_j (bright(I_{i,j})); \\ I_{s,t}^{dep}(i,j), & if \ bright(I_{s,t}) \leq T \cdot \frac{1}{w}\sum_i\sum_j (bright(I_{i,j})). \end{cases} \tag{8}$$

where $I_{s,t}^{dep}$ and $I_{s,t}^{MV}$ are the view image $I_{s,t}$'s normalized depth map and normalized module of motion vector where values represent the length of motion vector, respectively; $bright(I_{s,t})$ denotes the arithmetical average brightness of this view image; w denotes the number of SIs; k is the weight and T is the threshold.

Then, FCM (Fuzzy C-means) [23] is performed on the map to cluster the pixels into two groups:

$$\Omega_1 = \{p(i,j) \in I^{mer} || I^{mer}(i,j) - I_1| > |I^{mer}(i,j) - I_2|\}, \tag{9}$$

$$\Omega_2 = \{p(i,j) \in I^{mer} || I^{mer}(i,j) - I_1| < |I^{mer}(i,j) - I_2|\}, \tag{10}$$

where I_1 and I_2 are two cluster centers from FCM.

In general, the depth values of deeper object are larger than those of shallower ones. Besides, the sheltered object stands on the zero-disparity plane while occlusions don't for the SIs extracted from *PLI3*. In other words, the pixels inside occlusions have larger motion vectors than those of pixels from sheltered object. Thus, the k in Eq. (8) is always negative. Furthermore, group with larger I^{mer} is selected as pixels from the sheltered objects. The selected pixels' set Ω_3 can be written as:

$$\Omega_3 = \begin{cases} \Omega_1, avgmer(\Omega_1) > avgmer(\Omega_2); \\ \Omega_2, avgmer(\Omega_1) < avgmer(\Omega_2). \end{cases} \quad (11)$$

$$avgmer(\Omega_k) = \frac{1}{N_k} \sum_{(u,v)\in\Omega_k} I^{mer}(u,v), \quad (12)$$

where $avgmer(\Omega_k)$ denotes the mean of merged values in set Ω_k and N_k denotes the number of elements in set Ω_k.

Finally, according to Ω_3, the corresponding pixels in each SI are selected as:

$$\Omega_4 = \{q(u,v) \in SI | p(u,v) \in \Omega_3\}, \quad (13)$$

where Ω_4 denotes the set of pixels picked out from SI. And, they are averaged by:

$$I_{SAI}(i,j) = \frac{1}{M(i,j)} \sum_{I_{s,t}\in SIs} I_{s,t}(i,j), \quad (14)$$

where $I_{s,t}(i,j) \in \Omega_4(s,t)$ and $M(i,j)$ denotes the number of selected pixels at position (i,j).

Considering small size of simulated virtual lens may cause holes in SAI, inpainting [24] is employed to fill the holes by propagating structure to fill the missing regions after sampling the other parts.

3   Experiments and Results

To demonstrate the effectiveness of the proposed algorithm, three scenes as shown in Fig. 3 are tested. As sample images shown in the figure, grid, grass and battledore serve as occlusions. The distance between the occlusion and main lens is around 9 cm and distance between the occlusion and sheltered object is around 10 cm for all scenes. The grid and battledore are regularized and sparse while the grass is irregular and dense. The plenoptic images are captured by Lytro Illum [25] in the resolution of 5368 by 7728.

Four state-of-the-art methods are compared: Vaish's method [10], where homography matrices were used to warp view images and the aligned images were averaged directly; Zhang's method [12], which used improved matching algorithm to warp view images and averaged the warped view images directly; Yatziv's method [15], which selected pixels with largest confidence in each macro-pixel to make up a SAI; and Yang's method [16], which used homography metrices to warp view images and weighted averaged the pixels based on depth map which is captured by Kinect.

The processing results are shown in Fig. 4. One basic reason for blur in results (d)–(o) is that the SIs are warped by homography matrices which lead notable errors into the process of aligning SIs. For example, the green lines in (d), (g), (j) and (m) are blurrier than those in (a). In addition, results in (d)–(i) have a poor performance in visual effect as the pixels from occlusions are kept. Considering the bad performance of robustness, results in (j)–(l) are influenced by the basis of clustering, like the depth map. The proposed results have the better performance in the de-occlusion and visual effect than other results, like the white lines at the top of toy in Fig. 4(q) almost disappear.

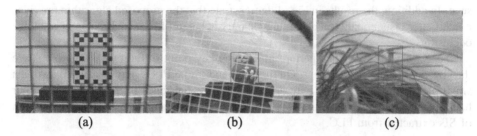

(a)   (b)   (c)

Fig. 3. Three scenes to test the effectiveness of proposed algorithm.

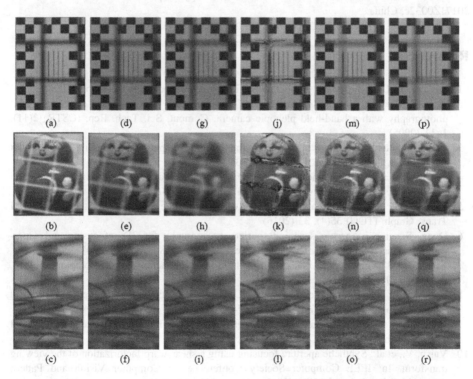

Fig. 4. The results of different methods. (a)–(c) are reference image (In red block in Fig. 3); (d)–(f) are the results of Vaish's method [10]; (g)–(i) are the results of Zhang's method [12]; (j)–(l) are the results of Yatziv's method [15]; (m)–(o) denote the results of Yang's method [16] and (p)–(r) denote the results of proposed method. (Color figure online)

4 Conclusions

In this paper, a synthetic aperture algorithm is proposed to see through occlusions based on PLC. The sheltered object is located onto the zero-disparity plane by deriving the geometric relationship between the focus plane and the object plane using the virtual camera array structure of PLC. Pixel selection relying on depth and motion clustering is proposed to pick out the pixels coming from sheltered object. Finally, a SAI is obtained by averaging the selected pixels so that the sheltered objects can be seen. In addition, the results in Fig. 4 demonstrate that the proposed algorithm provides much better visual quality in achieving clearer sheltered objects and much more blurry occlusions.

To get clearer synthetic aperture image, the key point of future work is to enlarge the size of simulated virtual lens. On the one hand, more PLCs may be needed so that a small plenoptic camera array with dense and sparse baselines is created. On the other hand, we are going to try the methods proposed to extend angular and spatial resolution of SIs extracted from PLC.

Acknowledgement. This work was supported in part by Shenzhen project JCYJ20170307 153135771 and Foundation of Science and Technology Department of Sichuan Province 2017JZ0032c, China.

References

1. Levoy, M.: Light fields and computational imaging. Computer **39**(8), 46–55 (2006)
2. Ren, N., Levoy, M., Bredif, M., Duval, G., Horowitz, M., Hanrahan, P.: Light field photography with a hand-held plenopic camera. Comput. Sci. Tech. Rep. (CSTR) **2**(11), 1–11 (2005)
3. Wang, J., et al.: Synthetic aperture integral imaging display with moving array lenslet technique. J. Disp. Technol. **11**(10), 827–833 (2015)
4. Levoy, M., et al.: Burst photography for high dynamic range and low-light imaging on mobile cameras. ACM Trans. Graph. (TOG) **35**(6), 192 (2016)
5. Shrestha, S., et al.: Computational imaging with multi-camera time-of-flight systems. ACM Trans. Graph. (TOG) **35**(4), 33 (2016)
6. Dong, S., et al.: Aperture-scanning fourier ptychography for 3D refocusing and super-resolution macroscopic imaging. Opt. Express **22**(11), 13586–13599 (2014)
7. Xu, Y., et al.: Camera array calibration for light field acquisition. Front. Comput. Sci. **9**(5), 691–702 (2015)
8. Barron, J.T., et al.: Fast bilateral-space stereo for synthetic defocus. In: Computer Vision and Pattern Recognition (CVPR), pp. 4466–4474 (2015)
9. Robert, M., et al.: PiCam: an ultra-thin high performance monolithic camera array. ACM Trans. Graph. (TOG) **32**(6), 166 (2013)
10. Vaish, V., et al.: Synthetic aperture focusing using a shear-warp factorization of the viewing transform. In: IEEE Computer Society Conference on Computer Vision and Pattern Recognition-Workshops, CVPR Workshops, p. 129 (2005)
11. Vaish, V., et al.: Using plane + parallax for calibrating dense camera arrays. In: Computer Vision and Pattern Recognition (CVPR), vol. 1, p. 1 (2004)

12. Zhang, H., Jin, X., Dai, Q.: Synthetic aperture based on plenoptic cameras for seeing behind occlusion. In: Intelligent Signal Processing and Communication Systems (ISPACS), pp. 801–806 (2017)
13. Yang, T., et al.: High performance imaging through occlusion via energy minimization-based optimal camera selection. Int. J. Adv. Robot. Syst. 10(11), 393 (2013)
14. Xiao, Z., Lipeng, S., Guoqing, Z.: Seeing beyond foreground occlusion: a joint framework for SAP-based scene depth and appearance reconstruction. IEEE J. Sel. Top. Signal Process. 11(7), 979–991 (2017)
15. Yatziv, L., Guillermo, S., Marc, L.: Lightfield completion. In: 2004 International Conference on Image Processing ICIP 2004, vol. 3, pp. 1787–1790 (2004)
16. Yang, T., et al.: Kinect based real-time synthetic aperture imaging through occlusion. Multimed. Tools Appl. 75(12), 6925–6943 (2016)
17. Dansereau, D.G., Oscar, P., Stefan, B.W.: Decoding, calibration and rectification for lenselet-based plenoptic cameras. In: Computer Vision and Pattern Recognition (CVPR), pp. 1027–1034 (2013)
18. Wang, T.C., Efros, A.A., Ramamoorthi, R.: Occlusion-aware depth estimation using light-field cameras. In: IEEE International Conference on Computer Vision (ICCV), pp. 3487–3495 (2015)
19. Mahmoudi, M., Guillermo, S.: Fast image and video denoising via nonlocal means of similar neighborhoods. IEEE Signal Process. Lett. 12(12), 839–842 (2005)
20. Hahne, C., et al.: Baseline and triangulation geometry in a standard plenoptic camera. Int. J. Comput. Vis. 126(1), 21–35 (2018)
21. Mignard-Debise, L., John, R., Ivo, I.: A unifying first-order model for light-field cameras: the equivalent camera array. IEEE Trans. Comput. Imaging 3(4), 798–810 (2017)
22. Hahne, C., et al.: Baseline of virtual cameras acquired by a standard plenoptic camera setup. In: 3DTV-Conference: The True Vision-Capture, Transmission and Display of 3D Video (3DTV-CON), pp. 1–3 (2014)
23. Bezdek, J.C.: Cluster validity with fuzzy sets. pp. 58–73 (1973)
24. Liu, J., et al.: Tensor completion for estimating missing values in visual data. IEEE Trans. Pattern Anal. Mach. Intell. 35(1), 208–220 (2013)
25. https://support.lytro.com/hc/zh-tw

Neutrosophic C-means Clustering with Local Information and Noise Distance-Based Kernel Metric Image Segmentation

Zhenyu Lu[1,2], Yunan Qiu[1], and Tianming Zhan[3(✉)]

[1] School of Electronic and Information Engineering,
Nanjing University of Information Science
and Technology, Nanjing, Jiangsu, China
[2] Jiangsu Collaborative Innovation Center on Atmospheric
Environment and Equipment, Nanjing, Jiangsu, China
[3] School of Information and Engineering,
Nanjing Audit University, Nanjing, Jiangsu, China
ztm@nau.edu.cn

Abstract. The traditional FCM algorithm is developed on the basis of classical fuzzy theory, though the classical fuzzy theory has its own limitations. The lack of expressive ability of uncertain information makes it hard for FCM algorithm to handle clustered boundary pixels and outliers. This paper proposes a Neutrosophic C-means Clustering with Local Information and Noise Distance-based Kernel Metric for Image Segmentation (NKWNLICM). The concept of local fuzzy information and noise distance in the Neutrosophic C-means Clustering Algorithm (NCM) is introduced in the paper. The algorithm improves the efficiency by leaving out parameter setting for different noises when segmenting pictures, and it also improves the robustness. Simulation results show that the algorithm has better segmentation results for noisy images.

Keywords: Image segmentation · Noise clustering · Fuzzy clustering
Neutrosophic clustering

1 Introduction

The traditional FCM algorithm is developed on the basis of fuzzy theory and is widely applied on computer vision [1–3]. However, the fuzzy theory has certain limitations [4]. The lack of expressive ability of uncertain information makes the FCM algorithm unable to handle clustered boundary pixels and outliers when segmenting the image. The traditional FCM algorithm only considers the pixel's gray information when performing image segmentation, and ignores the spatial neighborhood information of the image, which makes the algorithm very sensitive to noise and isolated points [5]. When it is used to process images with noises, the obtained results will suffer from a lower quality [6]. Due to above drawbacks of the FCM algorithm, Dave [7] proposed a noise clustering algorithm (NC), which uses a subset of parameters to represent noise class based on the FCM algorithm. The algorithm reduces the side-effect of noises on the final clustering results to some extent. Krinidis et al. [8] proposed a fuzzy local

© Springer Nature Switzerland AG 2018
R. Hong et al. (Eds.): PCM 2018, LNCS 11164, pp. 168–178, 2018.
https://doi.org/10.1007/978-3-030-00776-8_16

information C-means algorithm (FLICM), which used a fuzzy local information to associate local spatial information with local gray information. The algorithm improves the robustness in the algorithm to noise data processing. In order to improve the segmentation performance of the FLICM algorithm, Guo [9] improved the fuzzy local information by using pixel neighborhood variance information and replace the Euclidean distance with kernel distance. Fuzzy C-means clustering with local information and kernel metric for image segmentation (KWFLICM) is obtained with strong robustness and noise immunity. Guo et al. [10] improved the FCM on the basis of the Neutrosophic theory, and proposed the Neutrosophic c-means clustering algorithm (NCM). The algorithm not only includes the degree of membership, but also contains the uncertainty and opposition. The FCM algorithm makes the classification of the boundary region more obvious, and optimize the performance of denoising process. Jian et al. [11] propose a novel framework for underwater image saliency detection by exploiting Quaternionic Distance Based Weber Descriptor (QDWD), pattern distinctness, and local contrast. The algorithm incorporates quaternion number system and principal components analysis (PCA) simultaneously, so as to achieve superior performance.

In this paper, we propose a new clustering algorithm, which is the Neutrosophic C-means clustering with local information and noise distance-based kernel metric for image segmentation (NKWNLICM). The algorithm is applied for the segmentation study of noisy images. Introducing the local fuzzy information and noise distance in the kernel space, which makes the algorithm gain better denoising performance.

2 Noise Clusters Algorithm

The noise clustering algorithm (NC) [7] considers noises as an independent class. It regards the noise distance δ, representing the distance between the sample point and the center of the noise cluster, as a constant. It is a key parameter that is critical to the performance of noise clustering. Based on this argument, a simplified statistical average is used to calculate δ [7];

$$\delta^2 = \lambda \frac{\sum_{i=1}^{N} \sum_{k-1}^{c} d_{ik}^2}{Nc}, \tag{1}$$

Where λ is a noise multiplier used to adjust the effect of noise distance on the algorithm; N indicates the total number of sample points; c represents the number of sample clusters; d_{ik} represents the Euclidean distance between the sample x_i and the cluster center v_k.

In the NC algorithm, u_{*k} is used to indicate the degree of membership of the pixel existing in the noise class. The mathematical expression is shown as follows:

$$u_{*k} = 1 - \sum_{k=1}^{c} u_{ik}, \quad \forall i \in \{1, 2, 3, \ldots, N\}, \tag{2}$$

The NC algorithm changes the membership degree constraints based on the FCM algorithm and introduces the noise distance. Its clustering target expression [12] is expressed as follows:

$$J(U, V) = \sum_{i=1}^{N} \sum_{k=1}^{c} u_{ik}^m d^2(x_i, v_k) + \sum_{i=1}^{N} \delta^2 \left(1 - \sum_{k=1}^{c} u_{ik}\right)^m, \quad (3)$$

where $U = \{u_{ik}\}_{c \times N}$ denotes a fuzzy membership matrix; $V = \{v_k\}_{c \times 1}$ denotes a cluster center matrix; N denotes the total number of sample points; c denotes a number of sample clusters; $x_i(i = 1, 2, 3, \ldots, N)$ denotes a sample set; u_{ik} indicates a degree of membership of the ith sample x_i belonging to the kth class area; $v_k(k = 1, 2, 3, \ldots, c)$ denotes the kth cluster center; $d(x_i, v_k)$ is the Euclidean distance between the sample x_i and the cluster center v_k; $m \in [1, +\infty]$ is the fuzzy weighted index, which is usually specified as two; and, δ^2 is the noise distance.

Due to the effect of noise distance in the algorithm, the noise clustering algorithm is robust and can get better results when dealing with noisy data. Therefore, the concept of noise distance can be combined with other clustering algorithms to improve the robustness of the algorithm.

3 Neutrosophic C-means Cluster

3.1 Neutrosophic Theory

In order to address those limitations of the classical fuzzy theory [4] and improve its capability of processing and expressing uncertain information, Smarandache [13] proposed the Neutrosophic theory, which is a generalization of other extended theories. The Neutrosophic theory can not only represent non-deterministic issues in a better way, but also work out the unsolved problems when applying the fuzzy theory.

The basic idea of the Neutrosophic theory is that any viewpoint has a degree of truth, uncertainty, and falsity. Hence, T, I, and F have been introduced as Neutrosophic Components, which represent the authenticity, uncertainty, and absurdity of events respectively. These neutral elements are named true, indeterminate and false values.

3.2 Neutrosophic C-means Clustering Algorithm

In cluster analysis, traditional fuzzy clustering methods can only describe the degree of every group. In fact, especial for the samples on the boundary region between different groups, it is difficult to determine which group they belong to and what partitions they join in. In order to solve these problems, Guo et al. [10] improved the FCM on the basis of the Neutrosophic theory, and proposed the Neutrosophic c-means clustering algorithm (NCM) [10]. A new unique set A has been proposed, which regards as the union of the determinant clusters and indeterminate clusters. Let $A = C_j \cup B \cup R, j = 1, 2, \ldots, c$ where C_j is an indeterminate cluster, B regards the clusters in boundary regions, R is associated with noisy data and $\cup$ is the union operation. B and R are two kinds of indeterminate clusters. T is defined as the degree to determinant clusters, I is the degree to the boundary clusters, and F is the degree belonging to the noisy data set. Considering the clustering with indeterminacy, a new objective function and membership are defined as:

$$
J(T, I, F, C) = \sum_{i=1}^{N} \sum_{k=1}^{C} (w_1 T_{ik})^m \|x_i - v_k\|^2 + \sum_{i=1}^{N} \sum_{k=1}^{\binom{c}{2}} (w_2 I_{2ik})^m \|x_i - \overline{v_{2k}}\|^2
$$

$$
+ \sum_{i=1}^{N} \sum_{k=1}^{\binom{c}{3}} (w_3 I_{3ik})^m \|x_i - \overline{v_{3k}}\|^2 + \sum_{i=1}^{N} \sum_{k=1}^{\binom{c}{4}} (w_4 I_{4ik})^m \|x_i - \overline{v_{4k}}\|^2 + \ldots, \tag{4}
$$

$$
+ \sum_{i=1}^{N} \sum_{k=1}^{\binom{c}{c}} (w_c I_{cik})^m \|x_i - \overline{c_{ck}}\|^2 + \sum_{i=1}^{N} (\overline{w_{c+1}} F_i)^m
$$

where w_i is the weight factor. δ is used to control the number of objects considered as outliers. When the clustering number C is greater than 3, the objective function is very complex and time consuming. After simplification, the objective function is rewritten as:

$$
J(T, I, F, C) = \sum_{i=1}^{N} \sum_{k=1}^{C} (w_1 T_{ik})^m \|x_i - v_k\|^2 + \sum_{i=1}^{N} (w_2 I_i)^m \|x_i - \overline{v_{imax}}\|^2 + \sum_{i=1}^{N} (\overline{w_{c+1}} F_i)^m \delta^2, \tag{5}
$$

in which

$$
\overline{v}_{imax} = \frac{v_{p_i} + v_{q_i}}{2},
$$
$$
p_i = \underset{k=1,2,\ldots,C}{\arg\max}\, (T_{ik}),
$$
$$
q_i = \underset{k \neq p_i \cap k=1,2,\ldots,C}{\arg\max}\, (T_{ik}).
$$

In above equations, m is a constant, and p_i and q_i are the cluster numbers with the biggest and second biggest value. When the p_i and q_i are identified, the $\overline{v}_{imax}$ is calculated and its value is a constant number for each data point i, and will not change any more. T_{ik}, I_i and F_i are the membership values belonging to the determinate clusters, boundary regions and noisy data set, $0 < T_{ik}, I_i, F_i < 1$ which satisfy with the following formula:

$$
\sum_{k=1}^{C} T_{ik} + I_i + F_i = 1, \tag{6}
$$

The partitioning is carried out through an iterative optimization of the objective function, and the membership T_{ik}, I_i, F_i and the cluster centers v_k are updated in each iteration. The v_{imax} is calculated according to indexes of the largest and second largest value of T_{ik} of each iteration. The iteration will sustain until $\max\left\{\left|T_{ik}^{(h+1)} - T_{ik}^{(h)}\right|\right\} < \varepsilon$ or $h \geq H_{\max}$, in which ε a termination criterion between 0 and 1 is, k is the iteration step, and h is the number of iterations.

4 Proposed Method

In NCM algorithm, since the objective function does not involve any spatial information, if it is directly used for image segmentation, the ideal segmentation result can not be achieved. In addition, pixel tags identified by the principle of maximum membership may produce segmentation errors. Therefore, spatial neighborhood information should be added to the objective function to reduce the influence of undesired factors on the final determination of the membership function.

Based on the existing problems of NCM algorithm, we propose a new clustering algorithm, Neutrosophic C-means clustering with Local Information and Kernel Metric noise distance-based for Image Segmentation (NKWNLICM). The objective function is

$$
\begin{aligned}
J(T,I,F,c) = &\sum_{i=1}^{N}\sum_{k=1}^{c}(w_1 T_{ik})^m(\|\Phi(x_i) - \Phi(v_k)\|^2 + G_{ik}) \\
&+ \sum_{i=1}^{N}\sum_{l=1}^{\binom{2}{c}}(w_2 I_{il})^m(\|\Phi(x_i) - \Phi(\overline{v_l})\|^2 + \overline{G_{il}}) + \sum_{i=1}^{N}(w_3 F_i)^m\delta^2,
\end{aligned}
\tag{7}
$$

where T_{ik} denotes the extent to which element i belongs to cluster k, I_{il} denotes the degree to which element i belongs to two cluster boundaries in cluster c, F_i denotes the degree to which element i belongs to noise; $\overline{v_l}$ is the average of any two classes value. w_i is the weighting factor, δ is the noise distance, $\overline{G_{il}}$ and G_{ik} are local information.

When the clustering number C is greater than 3, the objective function in Eq. (7) is very complex and time consuming. In this situation, if we only consider the two closest determinate clusters which have the top two largest membership values, the objective function will be simplified. Meanwhile, computation cost will be reduced without decreasing the clustering accuracy greatly.

$$
\delta^2 = \lambda\left[\frac{\sum_{i=1}^{N}\sum_{k=1}^{c}\|\Phi(x_i) - \Phi(v_k)\|^2}{Nc}\right],
\tag{8}
$$

$$
G_{ik} = \sum_{j\in N_i, i\neq j} w_{ij}(1 - T_{jk})^m\|\Phi(x_i) - \Phi(v_k)\|^2,
\tag{9}
$$

where T_{ik}, I_i and F_i are the membership values belonging to the determinate clusters, boundary regions and noisy data set, $0 < T_{ik}, I_i, F_i < 1$ which satisfy with the following formula:

$$\sum_{k=1}^{C} T_{ik} + \sum_{l=1}^{\binom{2}{c}} I_{il} + F_i = 1, \tag{10}$$

According to the above formula, the Lagrange objective function is constructed as

$$L(T, I, F, c, \lambda) = \sum_{i=1}^{N} \sum_{k=1}^{c} (w_1 T_{ik})^m (\|\Phi(x_i) - \Phi(v_k)\|^2 + G_{ik})$$

$$+ \sum_{i=1}^{N} \sum_{l=1}^{\binom{2}{c}} (w_2 I_{il})^m (\|\Phi(x_i) - \Phi(\overline{v_l})\|^2 + \overline{G_{il}}) + \sum_{i=1}^{N} (w_3 F_i)^m \delta^2 \tag{11}$$

$$+ \sum_{i=1}^{N} \lambda_i \left(\sum_{k=1}^{c} T_{ik} + \sum_{l=1}^{\binom{2}{c}} I_{il} + F_i - 1 \right) = 0,$$

To minimize the Lagrange objective function, we use the following operations:

$$\frac{\partial L}{\partial T_{ik}} = m(w_1 T_{ik})^{m-1} \left(\|\Phi(x_i) - \Phi(v_k)\|^2 + G_{ik} \right) - \lambda_i, \tag{12}$$

$$\frac{\partial L}{\partial I_{il}} = m(w_2 I_{il})^{m-1} \left(\|\Phi(x_i) - \Phi(\overline{v_l})\|^2 + \overline{G_{il}} \right) - \lambda_i, \tag{13}$$

$$\frac{\partial L}{\partial F_i} = m(w_3 F_i)^{m-1} \delta^2 - \lambda_i, \tag{14}$$

$$\frac{\partial L}{\partial v_k} = -2 \sum_{i=1}^{N} (w_1 T_{ik})^2 (\Phi(x_i) - \Phi(v_k)), \tag{15}$$

The norm is specified as the Euclidean norm.
Let $\frac{\partial L}{\partial T_{ik}} = 0$, $\frac{\partial L}{\partial I_i} = 0$, $\frac{\partial L}{\partial F_i} = 0$, $\frac{\partial L}{\partial v_k} = 0$, then

$$T_{ij} = \frac{1}{w_1} \left(\frac{\lambda_i}{m} \right)^{\frac{1}{m-1}} \left(\|\Phi(x_i) - \Phi(v_k)\|^2 + G_{ik} \right)^{-\frac{1}{m-1}}, \tag{16}$$

$$I_{il} = \frac{1}{w_2}\left(\frac{\lambda_i}{m}\right)^{\frac{1}{m-1}}\left(\|\Phi(x_i) - \Phi(\overline{v_l})\|^2 + \overline{G_{il}}\right)^{-\frac{1}{m-1}}, \tag{17}$$

$$F_i = \frac{1}{w_3}\left(\frac{\lambda_i}{m}\right)^{\frac{1}{m-1}}\delta^{-\frac{2}{m-1}}, \tag{18}$$

$$v_k = \frac{\sum\limits_{i=1}^{N}(w_1 T_{ik})^m T_{ik}x_i}{\sum\limits_{i=1}^{N}(w_1 T_{ik})^m T_{ik}}, \tag{19}$$

Let $\left(\frac{\lambda_i}{m}\right)^{\frac{1}{m-1}} = K_i$,

$$\begin{aligned}
1 = &\sum_{j=1}^{c}\frac{K_i}{w_1}\left(\|\Phi(x_i) - \Phi(v_k)\|^2 + G_{ik}\right)^{-\frac{1}{m-1}} \\
&+ \sum_{l=1}^{\binom{2}{c}}\frac{K_i}{w_2}\left(\|\Phi(x_i) - \Phi(\overline{v_l})\|^2 + \overline{G_{il}}\right)^{-\frac{1}{m-1}} + \frac{K_i}{w_3}\delta^{-\frac{2}{m-1}},
\end{aligned} \tag{20}$$

$$\begin{aligned}
K_i = \Bigg[&\frac{1}{w_1}\sum_{k=1}^{c}\left(\|\Phi(x_i) - \Phi(v_k)\|^2 + G_{ik}\right)^{-\frac{1}{m-1}} \\
&+ \sum_{l=1}^{\binom{2}{c}}\frac{1}{w_2}\left(\|\Phi(x_i) - \Phi(\overline{v_l})\|^2 + \overline{G_{il}}\right)^{-\frac{1}{m-1}} + \frac{1}{w_3}\delta^{-\frac{2}{m-1}}\Bigg]^{-1},
\end{aligned} \tag{21}$$

Therefore,

$$T_{ik} = \frac{K_i}{w_1}\left(\|\Phi(x_i) - \Phi(v_k)\|^2 + G_{ik}\right)^{-\frac{1}{m-1}}, \tag{22}$$

$$I_{il} = \frac{K_i}{w_2}\left(\|\Phi(x_i) - \Phi(\overline{v_l})\|^2 + \overline{G_{il}}\right)^{-\frac{1}{m-1}}, \tag{23}$$

$$F_i = \frac{K_i}{w_3}\delta^{-\frac{2}{m-1}}, \tag{24}$$

$$\overline{v_l} = \frac{1}{2}(v_q + v_p),\ p, q \in \{1, 2, \ldots, c\}, p \neq q, \tag{25}$$

The partitioning is carried out through an iterative optimization of the objective function, and the membership T_{ik}, I_i, F_i and the cluster centers v_k are updated in each iteration. The v_{imax} is calculated according to indexes of the top two largest value of T_{ik} in each iteration. The iteration will continuous until $\max\left\{\left|T_{ik}^{(h+1)} - T_{ik}^{(h)}\right|\right\} < \varepsilon$ or $h \geq H_{max}$, where ε a termination criterion between 0 and 1, k is the iteration step, and h is the number of iterations.

The above equations allow the formulation of NKWNLICM algorithm. It can be summarized in the following steps:

Step 1 Initialize the $m, \varepsilon, w_1, w_2, w_3, H_{max}$ and w_{ij};

Step 2 Initialize $T^{(0)}, I^{(0)}$ and $F^{(0)}$, let $h = 0$;

Step 3 Calculate the centers vectors $v^{(h)}$ and cluster boundary $\overline{v_{il}}$ at h step using Eqs. (19) and (25);

Step 4 Calculate noise distance using Eq. (8);

Step 5 Update $T^{(h)}$ to $T^{(h+1)}$, $I^{(h)}$ to $I^{(h+1)}$, and $F^{(h)}$ to $F^{(h+1)}$ using Eqs. (22), (23) and (24);

Step 6 If $\left|T^{(h+1)} - T^{(h)}\right| < \varepsilon$ or $h \geq H_{max}$ then stop; otherwise return to **Step 3**, let $h = h + 1$;

Step 7 Assign each data into the class with the largest $TM = [T, I, F]$ value: $x(i) \in h^{th}$ class if $h = \arg \max_{j=1,\ldots,c+2} (TM_{ij})$.

5   Experimental Results

In this section, we compare the proposed method with other approaches by FCM, FGFCM, FLICM, NCM and KWFLICM to demonstrate their performance in clustering.

In the experiments, the parameter m has the same meaning to the fuzzification constant in the fuzzy clustering algorithm, and its value usually selected as 2. We selected $N_R = 8, \lambda_s = 3$, and $\lambda_g = 0.5$. The noise type is 20% salt and pepper noise.

By comparison, it can be clearly seen that the NKWNLICM method has the best segmentation effect, and noise is significantly less than other methods.

Table 1 shows the evaluation indicators based on entropy obtained by running the various methods on various images, where the minimum values are marked in bold. It can be seen from the table that the evaluation index value corresponding to the segmentation result obtained by the NKWNLICM method is the smallest, and these results fully show that the NKWNLICM method has a superior effect on noise pictures.

An entropy-based evaluation function(E), which combines both the layout entropy (Hr(L)) and the expected region entropy(Hl(L)), is often used in measuring the effectiveness of a segmentation method [14].

(a) The original image (b) Noise picture   (c) FCM   (d) FGFCM

(e) FLICM   (f) NCM   (g) KWFLICM   (h) NKWNLICM

Fig. 1. Comparison of camera image segmentation results

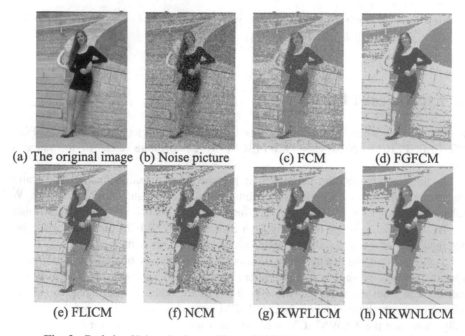

(a) The original image (b) Noise picture   (c) FCM   (d) FGFCM

(e) FLICM   (f) NCM   (g) KWFLICM   (h) NKWNLICM

Fig. 2. Berkeley University image library 388016 image segmentation results

Table 1. The E of the segmentation result obtained by different methods on each image

Figure	Metric	FCM	FGFCM	FLICM	NCM	KWFLICM	NKWNLICM
Figure 1	Hr(L)	1.7306	1.7076	1.7248	1.7249	1.7072	1.7026
	Hl(L)	0.4637	0.4576	0.4625	0.4536	0.4566	0.4548
	E	2.1943	2.1652	2.1873	2.1785	2.1638	**2.1574**
Figure 2	Hr(L)	1.8981	1.8813	1.8799	1.8932	1.8792	1.8698
	Hl(L)	0.5176	0.5057	0.5064	0.5184	0.5062	0.4985
	E	2.4157	2.3870	2.3863	2.4116	2.3854	**2.3683**

6 Conclusions

In this paper, we proposed a new clustering algorithm, Neutrosophic C-means clustering with Local Information and noise distance-based Kernel Metric for Image segmentation (NKWNLICM), and applied it to the segmentation study of noise images. By introducing the concept of local fuzzy information and noise distance in the NCM algorithm, the algorithm does not need parameters setting for different noises when segmenting pictures, which makes the algorithm gain better denoising performance. The efficiency of the proposed method is evaluated on grayscale image segmentation applications. Experimental results show that the algorithm has better segmentation results for noisy images. In addition, we plan to apply the method to the more complex data in our future works.

Acknowledgments. This work has been supported in part by the National Natural Science Foundation of China (Grant No. 61773220, 61502206), the Nature Science Foundation of Jiangsu Province under Grant (No. BK20150523)

References

1. Hong, R., Zhang, L., Zhang, C., et al.: Flickr circles: aesthetic tendency discovery by multi-view regularized topic modeling. IEEE Trans. Multimed. **18**(8), 1555–1567 (2016)
2. Hong, R., Zhang, L., Tao, D.: Unified photo enhancement by discovering aesthetic communities from Flickr. IEEE Trans. Image Process. **25**(3), 1124–1135 (2016)
3. Hong, R., Hu, Z., Wang, R., et al.: Multi-view object retrieval via multi-scale topic models. IEEE Trans. Image Process. **25**(12), 5814–5827 (2016)
4. Miyamoto, S., Ichihashi, H., Honda, K.: Algorithms for fuzzy clustering. Stud. Fuzziness Soft Comput., vol. 229 (2008). https://doi.org/10.1007/978-3-540-78737-2
5. Bezdek, J., Hathaway, R., Sobin, M.: Convergence theory for fuzzy c-means: counterexamples and repairs. IEEE Trans. Syst. Man Cybern. **17**(5), 873–877 (1987)
6. Chuang, K.S., Tzeng, H.L., Chen, S.: Fuzzy c-means clustering with spatial information for image segmentation. Comput. Med. Imaging Graph. **30**(1), 9 (2006)
7. Dave, R.N.: Characterization and detection of noise in clustering. Pattern Recognit. **12**(11), 657–664 (1991)
8. Krinidis, S., Chatzis, V.: A robust fuzzy local information C-means clustering algorithm. IEEE Trans. Image Process. Publ. IEEE Signal Process. Soc. **19**(5), 1328–1337 (2010)

9. Gong, M., Liang, Y., Shi, J.: Fuzzy C-means clustering with local information and kernel metric for image segmentation. IEEE Trans. Image Process. **22**(2), 573–584 (2013)
10. Cuo, Y.H., Sengur, A.: NCM: neutrosophic C-means clustering algorithm. Pattern Recognit. **48**(8), 2710–2724 (2015)
11. Jian, M., Qi, Q., Dong, J., Yin, Y., Lam, K.M.: Integrating QDWD with pattern distinctness and local contrast for underwater saliency detection. J. Vis. Commun. Image Represent. **53**, 31–41 (2018)
12. Kim, S., Chang, D.Y., Nowozin, S.: Image segmentation using higher-order correlation clustering. IEEE Trans. Pattern Anal. Mach. Intell. **36**(9), 1761–1774 (2014)
13. Smarandache, F.: Neutrosophy: Neutrosophic Probability, Set, and Logic: Analytic Synthesis and Synthetic Analysis. Philosophy, Cambridge (1998)
14. Zhang, H., Fritts, J.E.: Entropy-based objective evaluation method for image segmentation. In: Proceedings of Spie, vol. 5307, pp. 38–49 (2003)

Reflection Separation Using Patch-Wise Sparse and Low-Rank Decomposition

Jie Guo[1]([⊠]), Chunyou Li[1], Zuojian Zhou[2], and Jingui Pan[1]

[1] State Key Lab for Novel Software Technology, Nanjing University, Nanjing, China
guojie@nju.edu.cn
[2] School of Information Technology,
Nanjing University of Chinese Medicine, Nanjing, China

Abstract. This paper introduces a robust method for removing objectionable reflection interference in photographs captured through a piece of transparent medium. We exploit the fact that a group of image patches extracted from multiple correlated images with similar transmission lie in a very low-dimensional subspace, leading to a low-rank matrix after patch assembly. This allows us to formulate reflection separation as a per-patch sparse and low-rank decomposition problem which can be well solved by the ALM-ADM strategy. To eliminate the influence of unwanted reflection in patch searching and ensure that the extracted patches has a high similarity regarding their transmission layers, we introduce a new patch similarity metric based on both image intensities and gradients. This improves the performance of reflection separation. In addition, since our method does not require image reconstruction from gradient, color-shifting artifacts can be significantly ameliorated and more scene details can be preserved. Experimental results on both synthetic images and various real-world examples demonstrate that the proposed method achieves high quality reflection separation and performs favorably against many existing techniques.

Keywords: Reflection separation · Sparse · Low-rank · Patch

1 Introduction

Photographs taken in front of a transparent medium (e.g., a glass) would inevitably contain both the transmitted background scene and reflections of the objects on the same side of the camera. Generally, the reflections are undesirable and should be removed as they can obstruct the original scene, severely impacting the performance of many computer vision tasks. As a consequence, decomposing an observed image into a transmission layer and a reflection layer, i.e., the problem of image reflection separation, is a natural need and becomes an active research topic in computer vision [22].

Unfortunately, the reflection separation problem is an inherent ill-posed problem with more unknowns than equations to solve. To condense the space of

© Springer Nature Switzerland AG 2018
R. Hong et al. (Eds.): PCM 2018, LNCS 11164, pp. 179–188, 2018.
https://doi.org/10.1007/978-3-030-00776-8_17

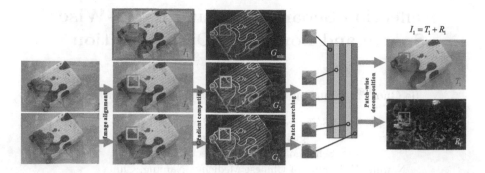

Fig. 1. The proposed reflection separation pipeline. Note that the separated reflection layer R_1 of the reference image I_1 has been magnified by a factor of four for better visualization. (Color figure online)

candidate solutions, additional information or assumptions are usually required. Existing techniques that use only a single input image usually rely on hand-crafted priors such as ghosting effects [18] and relative smoothness [2,11,17,21] to formulate objective functions. Although this category of methods are simple in data capture and have attracted increasing attention, their performance is quite limited in real-world scenarios due to the strong ambiguity involved and that the aforementioned assumptions are not always hold [2].

Currently, more effective and reliable techniques still adopt a set of several images to make the problem tractable. Besides employing some special facilities such as polarizers [9,16] and flashes [1], a practical strategy that is accessible to the everyday user acquires those images by simply varying the camera positions [5–8,10,15,19,20,23]. These methods exploit the motion of the camera as a cue for reflection separation. Given that gradient represents an important cue in addressing the reflection separation problem, most approaches try to obtain the gradient of the transmission layer first based on the correlation among those images [5–8]. Though effective, reconstructing an image from its gradient may exhibit very noticeable color shifts inside the processed region.

In this paper, we propose a patch-wise approach to recover the transmission layers of superimposed images, utilizing multiple images captured by a consumer-level camera at slightly different viewpoints. The key observation underlying our method is that similar patches from multiple correlated images approximately lie in a low-dimensional subspace. Therefore, if we collect each patch as a column of a matrix, the transmission layers tend to have low rank while the residual reflections can be viewed as gross pixel corruptions which are usually sparse in the spatial support. This prior is considered more general in practice than either ghosting cues or relative smoothness.

Based on the above observation, we formulate the reflection separation as a sparse and low-rank matrix decomposition problem [4,24] for each patch sequence. As depicted in Fig. 1, we first use SIFT-flow [14] to align all corre-lated images to a selected reference image (highlighted in the red box). Then, we extract patches from the reference image using a moving window and search for the similar patches from the other warped images. The similarity between a pair of patches are measured based on both their pixel intensities and gradients. After that, similar patches are assembled into a matrix and the matrix is further decomposed into a low-rank part (i.e., the transmission layer) and a sparse part (i.e., the reflection layer) with nuclear norm and l_1-norm minimization. We show that this constrained optimization problem can be efficiently solved by the augmented Lagrange multiplier (ALM) method with alternating direction minimizing (ADM) strategy [4,13]. Compared with previous work, our method has a number of desired properties: (1) our method is robust and fully automatic without any user assistance; (2) no professional photographing tools and skills are required; (3) no additional constraint is imposed on the targeting scene. Moreover, since we perform optimization with pixel intensities directly, color shifts can be significantly avoided.

2   Motivation

Our reflection separation method relies on K $(K > 1)$ images of the same target scene taken in front of a glass with slightly different camera locations. Similar to most previous work, we assume that the pixel value of each observed image $I_k(\mathbf{p})$ $(k = 1, 2, ..., K)$ can be decomposed into a transmission component $T_k(\mathbf{p})$ and a reflection component $R_k(\mathbf{p})$, satisfying

$$I_k(\mathbf{p}) = T_k(\mathbf{p}) + R_k(\mathbf{p}). \tag{1}$$

Unlike the reflection layers R_k that would vary significantly across multiple images, the transmission layers T_k keep almost unchanged if the target scene behind the glass is static. As a consequence, when these observed images are warped into the same coordinate system, the matrix composed of vectorized transmission layers can be naturally modeled as a low-rank matrix (ideally a rank one matrix). Moreover, due to the Fresnel reflectance, the transmitted signals usually dominate the observed image compared to the reflected ones. This makes the reflected signals sparse and low-intensity.

3   The Proposed Method

To better utilize the low-rank and sparse structures in multiple superimposed images, we perform reflection separation in a patch-wise manner. This section explains our method in details.

3.1   Image Alignment

The first step of our method is to automatically warp all input images to a selected reference image. In the proposed method, we estimate the warping function using SIFT-flow since it is proved to work well in most cases [8,10]. Built

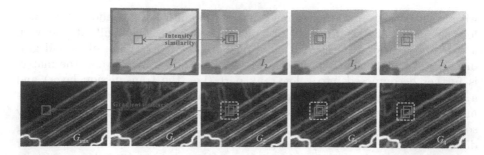

Fig. 2. Patch searching from multiple aligned images. The green box indicates a given image patch from the reference image while the blue boxes represent the most similar patches obtained from other warped images. Benefiting by image alignment, we can restrict the searching area to the local window (dashed yellow boxes) around the selected patch. Note that patch similarity is evaluated based on both pixel intensities and gradients. (Color figure online)

upon dense SIFT features, SIFT-flow aligns two images by minimizing matching cost and keeping the flow field smooth. Some other methods adopt 2D homographs to align multiple images [7]. However, 2D homographs are more suitable for handling flat scenes. After image warping with SIFT-flow, some image areas may contain invalid values due to misalignment. To alleviate this problem, we simply crop all the images in the sequence using the same size in our current implementation.

3.2   Patch Searching

Since our method works in a patch-wise manner, a robust and fast patch searching procedure is quite important. With image alignment, we can quickly find similar patches from multiple warped images by only conducting patch comparisons in a local window surrounding a given image patch extracted from the reference image.

As demonstrated in Fig. 2, for each image patch in the reference image (e.g., c_{ref}), we search for $2(K-1)$ most similar patches from the other $K-1$ warped images based on some distance metric. Since the images have already been properly aligned, we can restrict the searching area to the local window (dashed yellow boxes) around the selected patch. This will significantly accelerate the searching procedure.

It is well known that the image gradient provides an important cue in reflection separation [22]. Therefore, it is prevalent to treat the reflection separation problem in the gradient domain first, and then reconstruct the final results through Poisson integration [5–8]. Unfortunately, the reconstruction procedure may introduce color shifts and blur some heavily textured regions. To avoid these problems and still use the important gradient information in reflection separation, we leverage image gradient in patch similarity measure. Since the reflection layer is usually smooth and has low pixel intensities, the salient edges are more likely to belong to the transmission layer (see Fig. 2). Moreover, due to the large variations of the reflection, the reflected edges rarely appear at the same pixel locations for multiple images. Therefore, we can simply estimate the gradient of the reference image's transmission layer as

$$G_{min}(\mathbf{p}) = \arg \min_{G_k(\mathbf{p})} \{|G_k(\mathbf{p})|\} \tag{2}$$

in which $G_k = \nabla I_k$ is the gradient of the kth warped image.

With both image intensities and gradients, we can define the similarity measure between a pair of patches as

$$d(\mathbf{c}_{ref}, \mathbf{c}_{i,k}) = w_c \|\mathbf{c}_{ref} - \mathbf{c}_{i,k}\| + w_g \|\mathbf{g}_{ref} - \mathbf{g}_{i,k}\| \tag{3}$$

where $\mathbf{c}_{i,k}$ is the ith patch from the kth warped image, $\mathbf{g}_{i,k}$ is its gradient, and $\| \cdot \|$ denotes the l_2 norm of the vectorized patch. w_c and w_g are two weights controlling the relative importance of the two terms. Currently, we set w_c and w_g both as 1. Note that $\mathbf{g}_{ref}$ is extracted from G_{min} at the same spatial location of $\mathbf{c}_{ref}$ instead of the reference image's gradient map, as shown in Fig. 2. This can eliminate the influence of the gradient from the reflection layer, making similarity measure more accuracy.

3.3 Sparse and Low-Rank Decomposition

These $2K - 1$ similar patches extracted from multiple correlated images are then stacked as the columns of a large matrix $\mathbf{I}$. According to the low-rankness and the sparsity of these image patches mentioned above, our goal is to decompose the matrix $\mathbf{I}$ into a low-rank matrix $\mathbf{T}$ representing the transmission layer and a sparse matrix $\mathbf{R}$ representing the reflection layer. The objective function has the following compact form with matrix-vector notations:

$$\min_{\mathbf{T},\mathbf{R}} \|\mathbf{T}\|_* + \lambda\|\mathbf{R}\|_1 \\ \text{s.t.} \ \mathbf{I} = \mathbf{T} + \mathbf{R} \tag{4}$$

in which $\|\mathbf{T}\|_*$ is the nuclear norm of $\mathbf{T}$ defined by the sum of its singular values and $\|\mathbf{R}\|_1$ is the l_1 norm of $\mathbf{R}$ defined by its component-wise sum of absolute values. λ is a weighting parameter providing a trade-off between the sparse and low-rank components.

This is a standard sparse and low-rank matrix decomposition problem which can be well solved by the ALM-ADM strategy [13]. The augmented Lagrange function of the above optimization problem is

$$\mathcal{L}(\mathbf{T}, \mathbf{R}, \mathbf{Y}, \mu) = \|\mathbf{T}\|_* + \lambda\|\mathbf{R}\|_1 + <\mathbf{Y}, \mathbf{I} - \mathbf{T} - \mathbf{R}> + \frac{\mu}{2}(\|\mathbf{I} - \mathbf{T} - \mathbf{R}\|_F^2) \tag{5}$$

in which $\mathbf{Y}$ is the Lagrange multiplier while μ is the penalty parameter for the violation of the equality constraint. Equation 5 is solved by alternately minimizing over $\mathbf{T}$, $\mathbf{R}$ and $\mathbf{Y}$, while keeping the other fixed. The update rules are given by

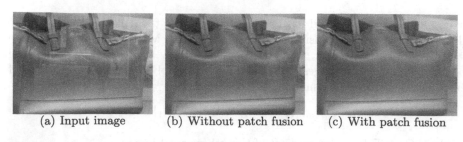

    (a) Input image   (b) Without patch fusion   (c) With patch fusion

Fig. 3. Visual comparison between patch-wise reflection removal without and with patch fusion.

$$\mathbf{T}^* = \mathcal{D}_{1/\mu}[\mathbf{I} - \mathbf{R} + \mathbf{Y}/\mu] \tag{6}$$

$$\mathbf{R}^* = \mathcal{S}_{\lambda/\mu}[\mathbf{I} - \mathbf{T} + \mathbf{Y}/\mu] \tag{7}$$

$$\mathbf{Y}^* = \mathbf{Y} + \mu(\mathbf{I} - \mathbf{T} - \mathbf{R}). \tag{8}$$

Here $\mathcal{D}_{1/\mu}$ is an SVT operator [3] defined as $\mathcal{D}_{1/\mu}(\mathbf{A}) = \mathbf{U}[\mathrm{sgn}(\mathbf{\Sigma})\max(|\mathbf{\Sigma}| - \mu^{-1}, 0)]\mathbf{V}^T$, in which $\mathbf{U\Sigma V}^T$ is a singular value decomposition of $\mathbf{A}$. $\mathcal{S}_\tau(x) = \mathrm{sgn}(x)\max(|x| - \tau, 0)$ is a soft-thresholding operator.

3.4 Patch Fusion

After per-patch treatment, adjacent patches may create annoying blocking arti-facts as shown in Fig. 3(b). To alleviate this problem, we extract patches from the reference image using a moving window of size W with a stride D, satisfying $0 < D < W$. This ensures that a single pixel is covered by multiple patches. In our method, the pixels in overlapping patches are averaged to produce the final output. Figure 3(c) shows the result after patch fusion. As seen, blocking artifacts are reduced significantly and the final image becomes very smooth.

4 Experiments

In this section, we evaluate our method on both synthetic and real image col-lection data and compare its performance against several state-of-the-art tech-niques. The experiments are conducted on a PC with Intel Core i7-6900 CPU and 16G RAM. Throughout this paper, we set the patch size $W = 32$ and the stride of moving window $D = \lfloor \frac{W}{4} \rfloor$. As suggested by [4], we set $\lambda = 1/\sqrt{2K - 1}$.

We first test our method on some synthetic image sequences. Each sequence is generated as follows. We randomly select a pair of images from the MSCOCO dataset [12]. One image is served as the transmission layer which is the ground truth of this sequence, and the other image is served as the reflection layer. As shown in Fig. 4, we slightly increase the size of the reflection image and randomly slide the transmission image on it. The overlapping area is blended with different weights (0.7 for transmission and 0.3 for reflection) to generate one image in the sequence. Current, each sequence contains five images with identical

Fig. 4. Illustration of synthetic images generation.

transmission in which the first image is selected as the reference image. Figure 5 presents quantitative evaluation of different reflection separation methods on four synthetic sequences. The error metrics used for evaluation are peak signal-to-noise ration (PSNR) and structural similarity (SSIM) with respect to the ground truth. Since both methods of [8, 10] perform reflection separation in the gradient domain and require image reconstruction with Poisson equations, scene details such as heavy textures (e.g., the fur on the dog) and hue (e.g., the hue of the sofa) may no longer preserved. On the contrary, our method successfully removes the reflections while preserving those importance scene details as much as possible. Out of all the methods we evaluated, our method achieves the highest PSNR value and the highest SSIM value in each case.

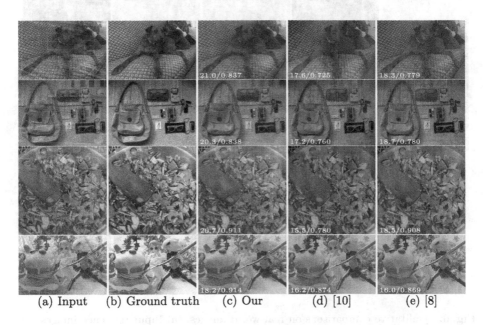

(a) Input   (b) Ground truth   (c) Our   (d) [10]   (e) [8]

Fig. 5. Quantitative comparisons on four synthetic cases for which ground truths of transmissions are given. The PSNR/SSIM errors are provided for each method in the lower-left corners. Please zoom in to view the details.

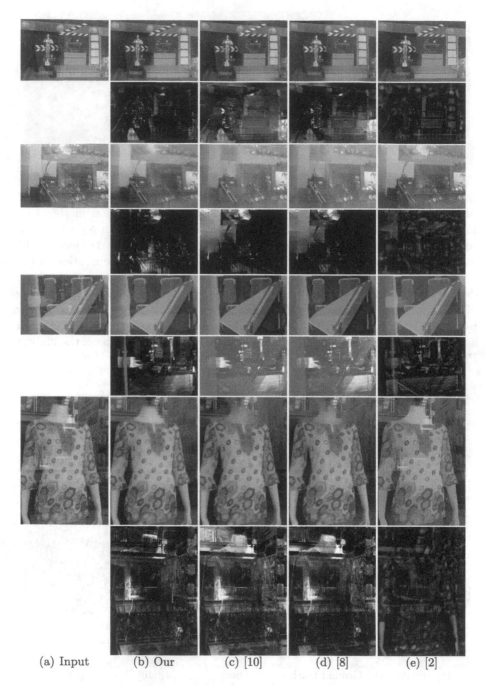

(a) Input   (b) Our   (c) [10]   (d) [8]   (e) [2]

Fig. 6. Qualitative comparisons on real-world images. (a) Input reference images. (b–e) Separated transmission and reflection layers by the methods of ours, Li and Brown [10], Han and Sim [8] and Arvanitopoulos et al. [2].

Figure 6 showcases the results of reflection separation on real-world inputs. In this figure, we compare our method to two effective multiple-image methods (i.e., [8,10]) and a state-of-the-art single-image method (i.e., [2]). As seen, the single-image method generally achieves insufficient separation since strong ambiguities are involved. Multi-image methods tend to achieve high-quality results with least traces of reflection. Again, since the method of [8,10] rely on Poisson image reconstruction from image gradient, some artifacts such as color shifts (see the third case) and image blurring (see the second and fourth cases) may occur. As expected, our method clearly outperforms other approaches and preserve much of the scene details.

5   Conclusions

Separation an image taken through glass windows into a transmission layer and a reflection layer is a well-known ill-posed problem that has long been an open challenge. By utilizing a sequence of images, we have proposed a patch-based approach for automatic reflection separation. Our method works by first wrapping multiple input images to a selected reference image using SIFT-flow to assist in the searching of similar patches. Based on a novel patch similarity metric, we form a large matrix with several patches which are expected to have similar transmission and enforce the low-rank subspace constraint to achieve per-patch reflection separation. Experimental results on various synthetic and real-world images demonstrate the effectiveness and versatility of the proposed method.

References

1. Agrawal, A., Raskar, R., Nayar, S.K., Li, Y.: Removing photography artifacts using gradient projection and flash-exposure sampling. ACM Trans. Graph. **24**(3), 828–835 (2005)
2. Arvanitopoulos, N., Achanta, R., Susstrunk, S.: Single image reflection suppression. In: The IEEE Conference on Computer Vision and Pattern Recognition (CVPR), July 2017
3. Cai, J.F., Cands, E.J., Shen, Z.: A singular value thresholding algorithm for matrix completion. SIAM J. Optim. **20**(4), 1956–1982 (2010)
4. Candès, E.J., Li, X., Ma, Y., Wright, J.: Robust principal component analysis? J. ACM **58**(3), 11:1–11:37 (2011)
5. Gai, K., Shi, Z., Zhang, C.: Blindly separating mixtures of multiple layers with spatial shifts. In: 2008 IEEE Conference on Computer Vision and Pattern Recognition, pp. 1–8 (2008)
6. Gai, K., Shi, Z., Zhang, C.: Blind separation of superimposed moving images using image statistics. IEEE Trans. Pattern Anal. Mach. Intell. **34**(1), 19–32 (2012)
7. Guo, X., Cao, X., Ma, Y.: Robust separation of reflection from multiple images. In: The IEEE Conference on Computer Vision and Pattern Recognition (CVPR), June 2014
8. Han, B.J., Sim, J.Y.: Reflection removal using low-rank matrix completion. In: The IEEE Conference on Computer Vision and Pattern Recognition (CVPR), July 2017

9. Kong, N., Tai, Y.W., Shin, J.S.: A physically-based approach to reflection separation: From physical modeling to constrained optimization. IEEE Trans. Pattern Anal. Mach. Intell. **36**(2), 209–221 (2014)

10. Li, Y., Brown, M.S.: Exploiting reflection change for automatic reflection removal. In: The IEEE International Conference on Computer Vision (ICCV), December 2013

11. Li, Y., Brown, M.S.: Single image layer separation using relative smoothness. In: The IEEE Conference on Computer Vision and Pattern Recognition (CVPR), June 2014

12. Lin, T.-Y., et al.: Microsoft COCO: common objects in context. In: Fleet, D., Pajdla, T., Schiele, B., Tuytelaars, T. (eds.) ECCV 2014. LNCS, vol. 8693, pp. 740–755. Springer, Cham (2014). https://doi.org/10.1007/978-3-319-10602-1_48

13. Lin, Z., Chen, M., Ma, Y.: The augmented Lagrange multiplier method for exact recovery of corrupted low-rank matrices. Technical report, UIUC (2009)

14. Liu, C., Yuen, J., Torralba, A.: SIFT flow: dense correspondence across scenes and its applications. IEEE Trans. Pattern Anal. Mach. Intell. **33**(5), 978–994 (2011)

15. Nandoriya, A., Elgharib, M., Kim, C., Hefeeda, M., Matusik, W.: Video reflection removal through spatio-temporal optimization. In: The IEEE International Conference on Computer Vision (ICCV), October 2017

16. Sarel, B., Irani, M.: Separating transparent layers through layer information exchange. In: Pajdla, T., Matas, J. (eds.) ECCV 2004. LNCS, vol. 3024, pp. 328–341. Springer, Heidelberg (2004). https://doi.org/10.1007/978-3-540-24673-2_27

17. Schechner, Y.Y., Kiryati, N., Basri, R.: Separation of transparent layers using focus. Int. J. Comput. Vision **39**(1), 25–39 (2000)

18. Shih, Y., Krishnan, D., Durand, F., Freeman, W.T.: Reflection removal using ghosting cues. In: The IEEE Conference on Computer Vision and Pattern Recognition (CVPR), June 2015

19. Simon, C., Kyu Park, I.: Reflection removal for in-vehicle black box videos. In: The IEEE Conference on Computer Vision and Pattern Recognition (CVPR), June 2015

20. Sun, C., Liu, S., Yang, T., Zeng, B., Wang, Z., Liu, G.: Automatic reflection removal using gradient intensity and motion cues. In: Proceedings of the 2016 ACM on Multimedia Conference, pp. 466–470. ACM (2016)

21. Wan, R., Shi, B., Hwee, T.A., Kot, A.C.: Depth of field guided reflection removal. In: 2016 IEEE International Conference on Image Processing (ICIP), pp. 21–25 (2016)

22. Wan, R., Shi, B., Duan, L.Y., Tan, A.H., Kot, A.C.: Benchmarking single-image reflection removal algorithms. In: International Conference on Computer Vision (ICCV) (2017)

23. Xue, T., Rubinstein, M., Liu, C., Freeman, W.T.: A computational approach for obstruction-free photography. ACM Trans. Graph. **34**(4), 79:1–79:11 (2015)

24. Zhou, X., Yang, C., Zhao, H., Yu, W.: Low-rank modeling and its applications in image analysis. ACM Comput. Surv. **47**(2), 36:1–36:33 (2015)

Conditional Feature Coupling Network for Multi-persons Clothing Parsing

Jiaming Guo, Zhuo Su$^{(\boxtimes)}$, Xianghui Luo, Gengwei Zhang, and Xiwen Liang

School of Data and Computer Science, National Engineering Research
Center of Digital Life, Sun Yat-sen University, Guangzhou, China
suzhuo3@mail.sysu.edu.cn

Abstract. Clothing parsing provides some significant cues to analyze the dressing collocation and occasion. In this paper, we propose a novel clothing parsing framework with deep end-to-end conditional feature coupling network for the photographic multi-persons in the fashion scene, and annotate a multi-persons clothing dataset for the effectiveness demonstration. Our parsing framework has three sub-networks, including the coarse parsing network (CPN), the multi-pose feature network (MFN) and the coupling residual network (CRN). CPN and MFN generate a coarse segmentation intermediary and 28 pose-indicated heat maps, respectively. CRN receives these auxiliary information and generates the fine-tuning clothing parsing result. To verify the generality and effectiveness of our parsing framework, we compare our method with the state-of-the-art parsing and segmentation methods such as Deeplab [2] and Co-CNN [7] on our multi-persons clothing dataset and some fashion clothing benchmarks. Experimental evaluations on these datasets demonstrate that our framework has a superior performance in the parsing task. In particular, our CFCN achieves 88.74% accuracy on the multi-persons clothing dataset, which is significantly higher than 86.50% by Deeplab. The project is available at https://github.com/suzhuoi/CFCNet.

Keywords: Clothing parsing · Deep learning · Coupling network

1 Introduction

Clothing parsing aims to recognize all dressing items and label them in the pixel-level. It opens the door of many fashion-oriented applications, such as fashion analysis, and clothes retrieval. Previous work of clothing parsing mainly emphasizes on single-person scenes [6,7,14,16]. Although some researches on multi-objects segmentation have been proposed recently [2], they are not specially

This research is supported by the National Natural Science Foundation of China (61502541, 61772140, 61502546), the Natural Science Foundation of Guangdong Province (2016A030310202), the Science and Technology Planning Project of Zhongshan (2016A1044), and the Fundamental Research Funds for the Central Universities (Sun Yat-sen University, 16lgpy39).

R. Hong et al. (Eds.): PCM 2018, LNCS 11164, pp. 189–200, 2018.
https://doi.org/10.1007/978-3-030-00776-8_18

designed for clothing parsing. Simply extend these methods to multi-persons clothing parsing problem could not obtain the optimal performance. Therefore, we propose a novel clothing parsing framework for multi-persons photos and establish a multi-person annotation dataset to verify the effectiveness of our solution.

Human poses have strong association with clothing. Therefore, many researchers make attempts in parsing clothing with the assistance of the pose information. Yamaguchi [14] combined the pose information with some specific features and used conditional random fields (CRFs) to generate an expected clothing parsing result. Although some achievements had been made in human parsing by adding pose information into deep neural network [12], directly applying these methods into clothing parsing data set could not obtain the expected results because the annotations different from clothing parsing to human parsing. Especially in multi-persons scenes, we need to further explore the connection between clothing segments and multiple poses under the deep framework.

In this paper, we propose an end-to-end clothing parsing framework for multi-persons photos and establish a multi-persons clothing dataset which contains thousands of multi-persons images with no less than 18 semantic labels. In our framework, we initially obtain coarse segmentation and multi-persons poses intermediary by the coarse parsing network (CPN) and the multi-pose feature network (MFN). The multi-persons poses heat maps would firstly enter the residual kernel in the coupling residual network (CRN) and then add to the coarse segmentation result, forming the mixing segmentation intermediary. The final parsing result is obtained by fine-tuning this mixing intermediate in the CRFs models. We respectively conduct experiments on the multi-persons and some solo clothing dataset to prove the effectiveness and generality of our framework.

In summary, our main contributions are as follows.

- Propose an deep end-to-end conditional feature coupling network for the multi-persons clothing parsing.
- Design three sub-networks, including a coarse parsing network for the primary semantic description, a multi-persons pose network for the joint skeleton estimation, and a coupling residual network in conditional association for the refining parsing results.
- Establish a multi-persons clothing data set, which contains thousands of fashion multi-persons annotations with no less than 18 clothing labels.

2   Related Work

Current clothing parsing methods could be divided into three categories, including AND/OR graph models, conditional random fields parsing, and deep parsing framework. The AND/OR graph model for clothing parsing builds AND/OR pairs over the parselets [4,5,13]. By the classification, combination and selection of the parselets, the parsing result is generated at the root of this graph.

Conditional random fields show powerful expansibility in clothing parsing. Prior information such as super-pixel and pose estimation could be added into

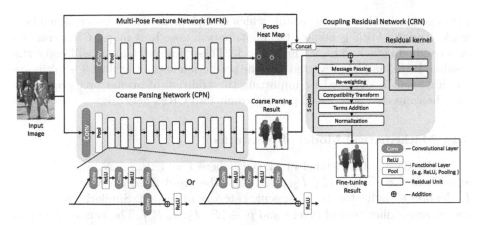

Fig. 1. Overview of multi-persons clothing parsing network, which is constructed by three sub-networks. Instead of directly assemble with the coarse parsing result, the pose heat maps would enter the residual kernel and add with the coarse parsing result.

CRFs models and show promising result [11]. Yamaguchi et al. [14] proposed pose prior. And later, he proposed a three-stage parsing pipeline which retrieved similar images to assist parsing [15]. Liu et al. [9] and Yang et al. [16] proposed co-parsing pipelines for videos and images, respectively.

Deep parsing framework could be divided into multi-stage and pixel-level parsing. For the multi-stage parsing, Liu et al. [10] proposed the matching-CNN scheme, which could parse clothing by mapping the similar items to the input image. And Liang et al. [6] proposed the active template scheme, which could generate the shape of each label by active template network and active shape network. For the pixel-level parsing, Liang et al. [7] employed the hidden layers of a clothing classification network to the parsing network and achieve state-of-the-art performance in solo photos.

However, these deep learning based clothing parsing methods did not make full use of pose information or local color information. The CRFs parsing only used primary feature representation instead of the deep neural network. Zheng et al. [17] tried combining deep network with CRF model, and outperformed the state-of-the-art in Pascal VOC 2012 dataset. In this paper, we propose an end-to-end multi-persons images parsing network based on pose estimation and CRFs. Distinguished with other methods, we employ a residual kernel to describe pose-clothing relationship.

3   Technical Details

Our proposed framework mainly consists of three components, including coarse parsing network (CPN), multi-pose feature network (MFN), and coupling residual network (CRN), as illustrated in Fig. 1. More specifically, the CPN generates the primary clothing parsing result, which is implemented as a deep residual network. And the MFN generates the multi-person pose heat maps. CRN consists of a residual kernel and a CRFs network. These pose heat maps will enter the residual kernel and be converted into the prediction of the clothing. The ultimate parsing result is obtained by importing the addition of primary parsing result and the converted pose-based parsing result into CRFs network.

3.1   Mathematical Model of Networks

Let l_i be the predicted label of pixel i, where $i = 1, 2, ..., N$, and N is the amount of pixels. $L = \{L_1, L_2, ..., L_x\}$ are the pre-defined clothing categories. L_i denotes an indicator from a specific category in this set. Similarly, p_i denotes the corresponding pose of pixel i and $p_i \in \{P_1, P_2, ..., P_y\}$. The parsing problem could be formulated as an energy function with deep networks approximation, which jointly considers clothing and poses features by adding extra potential function in the following,

$$E(l) = \sum_i \psi_l(l_i) + \sum_i \psi_P(l_i) + \sum_{i<j} \psi_w(l_i, l_j), \tag{1}$$

where $E(l)$ is the energy function of the whole network. The first term $\psi_l(l_i)$ estimates cost at the location i with the label l_i. And the pose potential term $\psi_P(l_i)$ measures the cost of assigning label l_i to pixel i under the pose p_i. The third term $\psi_w(l_i, l_j)$ is the pairwise potential function, measuring the cost between labels. ψ_l and ψ_P are both unary potential function, which could be expanded as follows,

$$\psi_l(l_i) = \psi_l(l_i|x_i), \psi_p(l_i) = \psi_p(l_i|p_i, x_i), \tag{2}$$

where $\psi_l(l_i)$ could be obtained from the coarse parsing network. And the $\psi_P(l_i)$ is obtained from the residual kernel. The third potential function ψ_w is defined as

$$\psi_w(l_i, l_j) = \mu(l_i, l_j) \sum_{m=1}^{M} \omega^m k_G^m(f_i, f_j). \tag{3}$$

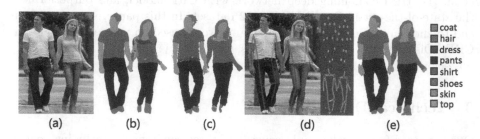

(a)   (b)   (c)   (d)   (e)

Fig. 2. Intermediate results of our framework. (a) Original image. (b) Primary parsing result. (c) Groundtruth. (d) Multi-person pose estimation. (e) Final parsing result.

where k_G^m is the mth Gaussian kernel with weight ω^m. The features f_i and f_j are extracted from the corresponding pixels i, j with their RGB channels and positions. $\mu(l_i, l_j)$ is a compatibility function, which indicates the probability of the co-occurrence between the pose p and the label l.

3.2 Coarse Parsing Network

Multivariate potential function is designed to fine-tuning the initial segmentation according to color and relative position among pixels, rather than segments the images directly by itself. Therefore, a coarse segmentation term is established in our clothing potential function (Eq. 1). Since the segmentation could be regarded as a region representation problem, a deep convolutional neural network (DCNN) solution provides a feasible way in practices. A coarse parsing network (CPN) is proposed to provides some coarse initial regions for further parsing work.

The details of our CPN are demonstrated in Table 1. CPN mainly consists of 11 residual units. Each residual unit contains 3 or 4 convolutional layers and 3 ReLU activation functions, as illustrated at the extensive frame with dotted line in Fig. 1. The input and output blob sizes of residual units may be different, therefore in some circumstances, we set a convolutional layer in bypass to modify the shortcut output. Before the forth residual unit, feature maps output by residual units or convolutional layer gradually reduce from 300×300 to 38×38, while the channels gradually increase from 64 to 512. This design may capture high dimensional local features of the image. CPN begins upsampling after seventh residual unit. We adopt three-stages upsampling scheme, where the feature maps are upsampled from 75×75, 150×150 to 300×300. Meanwhile, we keep the channel size to 256 during upsampleing stage. Examples of coarse parsing are shown in Fig. 2.

Table 1. Details of our CPN. Each ResGroup contains several residual units and the residual units in the same group have the same channel size and output size.

Network	Image	Conv.	Pool	ResG.1	ResG.2	ResG.3	ResG.4	Conv.
Output size	300×300	150×150	75×75	38×38	75×75	150×150	300×300	300×300
Channel size	3	64	64	256	512	256	256	18

3.3 Multi-pose Feature Network

Note that our framework is an end-to-end parsing network, MFN will be trained and tested together with other part of the framework. The existing multi-person pose estimation networks [1] are too large to be selected as the MFN. Therefore, we propose a resnet-based MFN with proper scale to generate the pose prior information. Similar to the CPN structure, MFN consists of 9 residual units. Instead of setting three residual units after the first upsampling, MFN only set one residual unit in this stage. The reason is that more residual units show limit improvement and costing more space and time. The pose configuration we adopted is similar to COCO keypoint dataset. We only keep the nose keypoint to represent the head. And we add the neck keypoint. Therefore, we have 14 keypoints and 14 limbs in total. The limbs configuration is demonstrated in Fig. 2.

Since the amount of people is unknown, the network would output the probability score of each pixel by the heat map, to indicate if the pixel belongs to a certain joint or body part. Examples of multi-persons pose estimation are shown in Fig. 2. Let T_i be the output, where $i \in [1, 2, 3, \ldots\ldots, n]$, and n is the sum of the amount of joints and body parts. T_i is a 2-dimensional matrix with the same size as the input image. Each element of T_i denotes the probability of the pixel belonging to the pose i. This design could convert the pose estimation problem into an image parsing problem, so that the results could be more easily accepted by the residual kernel.

3.4 Coupling Residual Network

Our proposed framework has a deep relationship with the CRFs parsing model [17], which could obtain a sound parsing accuracy by establishing the association between labels-to-labels and pixels-to-labels. As illustrated in Fig. 3, after adding the CRFs layer, the boundaries of the parsing results become more precise and acceptable than CPN results. By adding the multi-pose prior information, some error prediction could be suppressed and the little items could be accurately obtained.

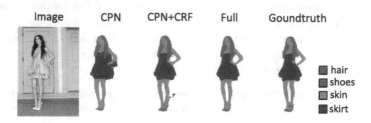

Fig. 3. Results by CPN, CRFs, our full model and ground truth annotations. Our full model improves the parsing result by suppressing some error areas and accurately parsing the little items.

Given the pose heat maps and RGB values, we expect to predict the probability of each clothing label. Since deep networks have the ability to approximate any function, we adopt a residual kernel to represent this distribution. The residual kernel consists of two residual units, which accept image and pose heat maps as input and produce the prediction on clothing. This prediction will add with the coarse parsing results, and then input to CRFs layer. The final parsing result could be obtained after fine-tuning by CRFs layer. For the loss function $L(\theta)$, a softmax loss is applied for the clothing parsing result,

$$l_1(\theta) = -\frac{1}{M} \sum_{m=1}^{M} \sum_{k=1}^{K} \hat{G}_k^{(m)} \cdot log \left[G_k^{(m)'} / \sum_{l=1}^{K} G_l^{(m)'} \right], \qquad (4)$$

where $\widehat{G}_k^{(m)}$ denotes the truth distribution map of label k in the mth image, and $G_k^{(m)'}$ denotes the corresponding label map generated by our CFCN. Besides, we set the euclidean loss for pose estimation, to ensure the MFN could generate the correct pose heat maps,

$$l_2\left(\theta\right) = \frac{1}{M} \sum_{m=1}^{M} \sum_{p=1}^{P} \parallel T_p^{(m)'} - \widehat{T}_p^{(m)'} \parallel_2^2, \tag{5}$$

where $\widehat{T}_p^{(m)}$ denotes the truth distribution map of pose p in the mth image. The weighted sum of the two loss functions is treated as the total loss function,

$$L(\theta) = \alpha l_1(\theta) + (1 - \alpha)l_2(\theta) + \beta\|\theta\|^2. \tag{6}$$

The coefficient α is the weight of loss functions and the coefficient β is the weight decay ratio. In our experiments, the parameter settings are $\alpha = 0.5$, $\beta = 1.6 \times e{-}3$.

4 Experiments

Although it is possible to train the entire framework once for the final result, we develop a two-phases scheme to train our networks because the over-complicated scale of our networks. CPN and MFN are trained firstly, ensuring them to generate expected coarse segmentation and pose estimation results. CRN is added later then the whole framework is trained. Since CRN only need to fine-tune the coarse results, the convergence procedure is determinate and easy. For the ATR dataset and fashion clothing dataset, our network parameters are initialized by Gaussian distribution with standard deviation $\sigma = 0.01$. And the model trained on the ATR dataset is chosen as the pre-trained model for the multi-persons clothing dataset. The learning rate is set to 10e−8 for CPN as well as MFN implementation. While fine-tuning the whole framework, we divide the learning rate by 10 to obtain a better result. In order to avoid over fitting, the weight decay parameter is set to $1.6 \times 10e{-}3$, and the momentum is set to 0.99. Moreover, the time τ of CRF iteration is $\tau = 5$. More iterations limit the improvement and also increasing the risk of the gradient vanishing. All experiments are conducted on NVIDIA GeForce GTX 1070 GPU and Intel i7-6700 CPU.

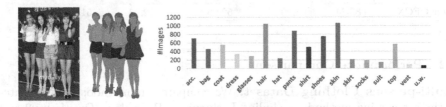

Fig. 4. Examples in our multi-persons dataset (Left). And the distribution of our 17 semantic labels in multi-persons dataset (Right).

4.1 Datasets

In this paper, we emphasize on the multi-persons clothing parsing problem. Therefore, we propose a multi-persons clothing parsing dataset to evaluate our parsing framework. However, in order to prove the versatility of the proposed method, we also conduct experiments on solo human parsing and clothing parsing datasets—ATR dataset and fashion clothing dataset.

Multi-persons Clothing Dataset. Multi-persons clothing dataset is a collection of real-life multi-persons photos, which are crawled through the search engines. We carefully annotate the crawled images with 18 standard labels. More details of our multi-persons clothing dataset are demonstrated in Fig. 4. The dataset contains 1060 photos in total, in which 106 images are used for testing, 106 for validating, and 848 for training.

Fashion Clothing Dataset. This dataset consists of three open-source fashion datasets, including Clothing Co-Parsing (CCP) [16], Fashionista [14] and Colorful Fashion Parsing Data (CFPD) [8]. These three dataset have 1,004, 685 and 2,682 photos respectively, up to 4,371 photos in total. We standardize the annotations into the 18 semantic labels and split all the photos into a training set of 3,933 photos and a test set of 473 photos.

ATR Dataset. Active Template Regression (ATR) dataset is a large and widely used solo human parsing dataset. It consists of the original ATR dataset and the supplementary Chictopia10k dataset. Original ATR dataset contains 7703 images while the Chictopia10k dataset has a total of 10,003 images. According to the experimental settings of Co-CNN [7], we take 1,000 images for testing, 700 for validation and the rest of the images for training.

Table 2. Comparison of multi-persons parsing performances with our frameworks and some state-of-the-art parsing and segmentation methods.

Method	Accuracy	F.g. accuracy	Avg. precision	Avg. recall	Avg. F-1 score
Yamaguchi's [14]	70.18	13.68	24.59	13.45	17.39
Paperdoll [15]	74.13	23.91	31.11	19.02	23.06
Attentaion [3]	86.16	67.40	40.89	46.32	43.46
Deeplab v2 [2]	86.50	66.90	49.46	50.07	49.76
Our CPN	86.16	63.52	51.57	47.87	46.70
Our CPN+CRF	87.78	65.28	51.76	49.47	49.81
Our CFCN	88.74	68.20	52.95	52.73	51.97

4.2 Performance Comparisons

Multi-persons Clothing Dataset. We compare our methods with 4 state-of-the-art parsing methods, including Fashionista, Paperdool, DeepLab v2 and Attention, illustrated in Table 2. For DeepLab v2 and Attention, we train them directly on our dataset without additional steps. However, Fashionista and

Table 3. Results on fashion clothing dataset. Our method outperforms other methods in all five indices. Our full model (CFCN) achieves the best results.

Method	Acc.	F.g.acc.	Avg.prec.	Avg.recall	Avg.F-1
Yamaguchi's	81.32	32.24	23.74	23.68	22.67
Paperdoll	87.17	50.59	45.80	34.20	35.13
Our CPN	88.59	55.26	45.57	42.24	43.02
Our CPN+CRF	89.84	61.35	49.06	46.72	46.18
Our CFCN	90.35	60.35	51.48	47.31	48.23

Table 4. Versatility experiments on ATR dataset. Our method remains top performance among all parsing and segmentation methods.

Method	Acc.	F.g.acc.	Method	Acc.	F.g.acc.
Yamaguchi's [14]	84.38	55.59	DeepLab v2 [2]	91.54	73.32
Paperdoll [15]	88.96	62.18	Attention [3]	92.68	78.10
M-CNN [10]	89.57	73.98	Our CFCN	95.45	82.10
ATR [6]	91.11	71.04	Co-CNN [7]	96.02	83.57

Paperdool are only designed for single-person photos, we have to adopt a human detector to extract every person in the images and parse them one by one.

As the basic network, CPN could outperform the paperdoll parsing over 12.03% in accuracy and 23.64% in F-1 score. By using CRFs layer as the post processor, the noise labels could be suppressed and edges could be refined, hence the CPN+CRF could exceed Deeplab v2. Our full model CFCN achieve the best performance on multi-persons clothing dataset, which indicates that the pose prior still have positive effect for multi-persons parsing task.

Fashion Clothing Dataset. The evaluation values of our frameworks under the fashion clothing set are demonstrated in Table 3. All three versions of our method surpass the paperdoll parsing in the F-1 score. The CRFs and the pose prior could effectively improve the network performance. In addition, the CRFs improves the F.g. accuracy over 6.09%, which indicates that adding the CRFs layer could assist the network in identifying the non-background parts. Having considering the pose prior information, the network performance could be further improved. However, compare to the performance in the multi-persons clothing dataset, the improvement is less significant because of the simple pose configuration and no occlusion (Table 4).

ATR Dataset. We compare our conditional feature coupling network with 7 parsing and segmentation methods on ATR dataset. The CFCN exceeds Attention [3] and achieves top 2 performance in ATR dataset, which proves the versatility of our CFCN. The ablation study is illustrated in Fig. 5. With the involvement of CRFs and pose prior, the network performance is gradually improving.

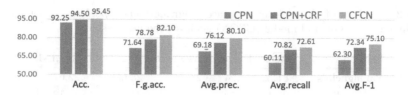

Fig. 5. Ablation study on ATR dataset. Coupling multiple features could improve the parsing performance effectively.

By effectively suppressing the noise label and refining the edges, the improvement brought by the CRFs is more significant than the pose prior.

4.3   Visual Evaluation

As illustrated in Fig. 6, our framework is robust enough to parse the foreground clothing from multi-persons photos with the occlusion, deficiency of parts, and cluttered background. The human shapes are perfectly obtained and the labels with strong association with human bodies are nicely parsed, e.g. hair and skin. However, our framework may fail in parsing small items like glasses, bags and accessories. Unbalance data ratio may be the main reason. Small items take up little proportion in our datasets. The framework may neglect these small data, being insensitive to small items. Therefore, we would add smoothing algorithms and small object detection terms to our framework in the future work.

Fig. 6. Multi-persons clothing parsing results of our CFCN framework. It could correctly parse the indoor and outdoor multi-persons fashion photos with complicated background.

5   Discussion and Conclusion

In this paper, we propose an end-to-end multi-persons clothing parsing framework. This framework adopts a coarse parsing network and a multi-poses feature network to generate two conditional terms, and uses a coupling residual network to establish a conditional random fields model over these conditional terms. Instead of directly assembling the pose feature and the coarse parsing result [12], we adopt the residual kernel to map the pose feature into the clothing label space and then add to the coarse parsing result. We prove the effectiveness of this implementation by the experiment conducted on several fashion dataset.

In the future work, we would enlarge the real-life multi-persons dataset for further experiments. Additionally, we would explore more conditional terms like object detection to help recognize tiny objects, and utilize some edge smoothing methods to improve the visual effect.

References

1. Cao, Z., Simon, T., Wei, S.E., Sheikh, Y.: Realtime multi-person 2D pose estimation using part affinity fields. In: IEEE International Conference on Computer Vision, vol. 1, p. 7 (2017)
2. Chen, L.C., Papandreou, G., Kokkinos, I., Murphy, K., Yuille, A.L.: DeepLab: semantic image segmentation with deep convolutional nets, atrous convolution, and fully connected CRFs. IEEE Trans. Pattern Anal. Mach. Intell. **40**(4), 834–848 (2016)
3. Chen, L.C., Yang, Y., Wang, J., Xu, W., Yuille, A.L.: Attention to scale: scale-aware semantic image segmentation. In: IEEE Computer Vision and Pattern Recognition, pp. 3640–3649 (2016)
4. Dong, J., Chen, Q., Huang, Z., Yang, J., Yan, S.: Parsing based on parselets: a unified deformable mixture model for human parsing. IEEE Trans. Pattern Anal. Mach. Intell. **38**(1), 88–101 (2016)
5. Dong, J., Chen, Q., Shen, X., Yang, J., Yan, S.: Towards unified human parsing and pose estimation. In: IEEE Computer Vision and Pattern Recognition, pp. 843–850 (2014)
6. Liang, X., et al.: Deep human parsing with active template regression. IEEE Trans. Pattern Anal. Mach. Intell. **37**(12), 2402–2414 (2015)
7. Liang, X., et al.: Human parsing with contextualized convolutional neural network. In: IEEE International Conference on Computer Vision, pp. 1386–1394 (2015)
8. Liu, S., et al.: Fashion parsing with weak color-category labels. IEEE Trans. Multimedia **16**(1), 253–265 (2014)
9. Liu, S., et al.: Fashion parsing with video context. IEEE Trans. Multimedia **17**(8), 1347–1358 (2015)
10. Liu, S., et al.: Matching-CNN meets KNN: quasi-parametric human parsing. In: IEEE Computer Vision and Pattern Recognition, pp. 1419–1427 (2015)
11. Wu, Q., Boulanger, P.: Enhanced reweighted MRFs for efficient fashion image parsing. ACM Trans. Multimedia Comput. Commun. Appl. **12**(3), 42 (2016)
12. Xia, F., Wang, P., Chen, X., Yuille, A.L.: Joint multi-person pose estimation and semantic part segmentation. In: IEEE Computer Vision and Pattern Recognition, pp. 6769–6778 (2017)
13. Xia, F., Zhu, J., Wang, P., Yuille, A.L.: Pose-guided human parsing by an and/or graph using pose-context features. In: AAAI, pp. 3632–3640 (2016)
14. Yamaguchi, K., Kiapour, M.H., Ortiz, L.E., Berg, T.L.: Parsing clothing in fashion photographs. In: IEEE International Conference on Computer Vision, pp. 3570–3577 (2012)

15. Yamaguchi, K., Kiapour, M.H., Ortiz, L.E., Berg, T.L.: Retrieving similar styles to parse clothing. IEEE Trans. Pattern Anal. Mach. Intell. **37**(5), 1028–1040 (2015)
16. Yang, W., Luo, P., Lin, L.: Clothing co-parsing by joint image segmentation and labeling. In: IEEE Conference on Computer Vision and Pattern Recognition, pp. 3182–3189 (2014)
17. Zheng, S., et al.: Conditional random fields as recurrent neural networks. In: IEEE International Conference on Computer Vision, pp. 1529–1537 (2015)

Intra-view and Inter-view Attention for Multi-view Network Embedding

Yueyang Wang[1]($\boxtimes$), Liang Hu[2], Yueting Zhuang[1], and Fei Wu[1]

[1] College of Computer Science and Technology,
Zhejiang University, Hangzhou, China
{yueyangw,yzhuang,wufei}@zju.edu.cn
[2] School of Mechanical Engineering, Zhejiang University, Hangzhou, China
hl111325003@zju.edu.cn

Abstract. Network Embedding, which represents nodes in networks with efficient low-dimensional vectors, has been proved useful in a variety of applications. However, most existing approaches study single-view networks but not the multi-view networks with multiple types of relationships between nodes. Meanwhile, they ignore the rich features associated with the nodes, which is common in real world. In this paper, we propose a novel network embedding method, Intra-view and Inter-view attention for Multi-view Network Embedding (I^2MNE), which leverages both the multi-view network structure and the node features to efficiently generate node representations. Specially, we introduce the intra-view attention when aggregating node features from neighbors for each single view and the inter-view attention when integrating representations across different views. Experiments on two real-world networks show that our approach outperforms other counterpart network embedding methods.

Keywords: Multi-view network · Network embedding
Network representation learning · Attention

1 Introduction

With the development of Internet, information networks are ubiquitous in the real world (e.g. academic networks [21,22], movie recommendation networks [18]). These information networks have attracted deep insight and one of the most popular analysis methods is the network embedding, i.e., network representation learning. Network embedding embeds the networks into the low-dimensional space to directly measure the neighborhood similarities between different nodes and has been proved to be highly effective when solving the traditional tasks such as node classification [2], clustering [3], link prediction [1].

Traditional network embedding approaches (e.g. DeepWalk [15], node2vec [6], LINE [24]) mainly focus on the representation learning for networks with a single type of relationship. Later, researchers further propose some convolutional neural network embedding methods (e.g. GCN [8], graphSAGE [7]) to encode

© Springer Nature Switzerland AG 2018
R. Hong et al. (Eds.): PCM 2018, LNCS 11164, pp. 201–211, 2018.
https://doi.org/10.1007/978-3-030-00776-8_19

both network structure and features of nodes. However, the majority of practical information networks are *de facto* multi-view networks, involving multiple types of relationships and node features. Figure 1 shows an example of multi-view network. In this academic network, relationships of authors may include co-author relationship which means whether two authors have been collaborated on a paper and citing relationships which means whether one author cited the papers written by the other one. Features of authors may include the research interests and titles of papers they write.

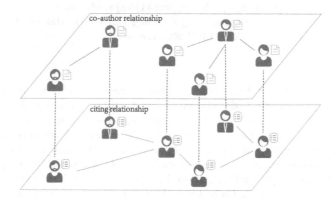

Fig. 1. An example of multi-view network which contains multiple types of relationships between nodes and different node features in each single view.

Recently, the multi-view network embedding problems has received more attention. Various methods proposed for learning representations with multiple views perform well on many applications [4]. Nevertheless, they suffer from specific limitations as well. Methods [5,10,20] based on matrix factorization are faced with the expensive computational cost, thus are not suitable for large-scale data. Clustering methods [12,27,28] neglect the differcnt importance of different vies with a lack of consideration about weight learning. And more recent methods like [16,19] based on deep learning miss the utilization of node features which promote the embedding performance besides the network structure.

In this paper, we propose Intra-view and Inter-view attention for Multi-view Network Embedding (I^2MNE), a novel method to overcome the limitations mentioned above. The intra-view attention is introduced to specify the different importance of neighbors when aggregating neighbor features of each node for single view. Similarly, the inter-view attention is also introduced to assign the different significance of views when integrating representations across different views. The attention weights and the representations can be efficiently trained through the back propagation algorithm.

We conduct experiments on two real-world datasets of different domains. The effectiveness of our approach is evaluated on the classification task. The experimental results show that I^2MNE outperforms other state-of-the-art methods. To summarize, we make the following contributions:

- We propose I²MNE to study multi-view network embedding, which aims to learn node representations by leveraging structure and feature information from multiple views.
- We introduce intra-view attention when aggregating the node features from neighbors and inter-view attention when integrating representations across different views to learn robust node representations.
- We conduct experiments on two real-world multi-view networks. Experimental results demonstrate the effectiveness and efficiency of our proposed approach over many competitive methods.

The rest of this paper is organized as follows. In Sect. 2, we describe the problem definition. In Sect. 3, we present the I²MNE algorithm for multi-view network embedding in detail. In Sect. 4, we analyze the learned node representation and compare the proposed model with the existing network embedding approaches on two real world datasets. Conclusions are given at the end.

2 Problem Definition

In this section, we formally define the problem of network embedding in multi-view network. Firstly, the multi-view network is defined as follows:

Definition 1 (Multi-view Network). *A multi-view network is defined as* $G = (V, E, C)$, *where* V *is the set of nodes representing objects;* $E = \cup_{k=1}^{K} E_k$, K *is the number of views,* $E_k \subseteq V \times V$ *is the set of edges representing relationships between two nodes in view* k *and* $C = \cup_{k=1}^{K} C_k$, $C_k = \{f_i, \forall n_i \in V\}$ *denotes the features of objects in view* k, $f_i \in \mathbb{R}^F$ *denotes features of node* n_i *and* F *is the dimension of features.*

Our goal is to learn the node representations in the multi-view network. We define this problem as follows:

Definition 2 (Multi-view Network Embedding). *Given a multi-view network, denoted as* $G = (V, E, C)$, *the aim of multi-view network embedding is to learn low-dimensional representations* $O \in \mathbb{R}^{|V| \times d}$, *where* $d \ll |V|$ *is the number of embedding dimensions.*

3 Our Approach

In this section, we present our approach, I²MNE, which embeds the nodes with features in multi-view network into a common space. We first aggregate the features of each node's neighbors for single view to encode the node proximities and node features. Then we integrate node representations across different views. Inspired by the recent progress of the attention mechanism [11], we introduce the intra-view attention (shown in Fig. 2) to automatically specify different weights to nodes within neighbors and inter-view attention (shown in Fig. 3) to assign different weights to views.

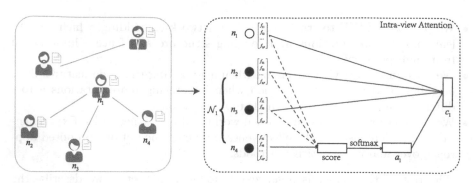

Fig. 2. The Intra-view Attention. It specifies different importance to neighbors (n_2, $n_3, n_4 \in \mathcal{N}_1$ in this example) of each node (n_1 in this example).

3.1   Embedding Generation

Intra-view Attention. In each single view, for each node n_i with its feature $\boldsymbol{f}_i$ and its neighbors' features $\boldsymbol{F}_i \in \mathbb{R}^{|\mathcal{N}_i| \times F}$, where $\mathcal{N}_i$ denotes the neighbors of n_i, we first introduce a content-based score function as follow:

$$\boldsymbol{s}_i = \text{score}(\boldsymbol{f}_i, \boldsymbol{F}_i) = \boldsymbol{f}_i W_a \boldsymbol{F}_i^{\text{T}} \tag{1}$$

where $W_a \in \mathbb{R}^{F \times F}$ is the intra-view attention weight matrix. This function indicates the importance of neighbors' features to node n_i. To make the scores comparable across different nodes, we use a softmax function to normalize them as follow:

$$[\boldsymbol{a}_i]_j = \text{softmax}_j(\boldsymbol{s}_i) = \frac{\exp([\boldsymbol{s}_i]_j)}{\sum_{j'=1}^{|\mathcal{N}_i|} \exp([\boldsymbol{s}_i]_{j'})} \tag{2}$$

where $[\cdot]_j$ means the j-th value of the vector and $j \in \{1, \cdots, |\mathcal{N}_i|\}$.

Then, we introduce the context-vector $\boldsymbol{c}_i$ which captures the relevance of features between n_i and its neighbors $\mathcal{N}_i$ using the normalized attention scores as follow:

$$\boldsymbol{c}_i = \boldsymbol{a}_i \boldsymbol{F}_i \tag{3}$$

Next, we use a weight matrix $W \in \mathbb{R}^{F \times F'}$, where F' is the dimension of hidden layer, to get the aggregated vector as follow:

$$\boldsymbol{f'}_i = \sigma(W[\boldsymbol{f}_i \oplus \boldsymbol{f}_{\mathcal{N}_i} \oplus \boldsymbol{c}_i]) \tag{4}$$

where $\boldsymbol{f}_{\mathcal{N}_i}$ is the feature vector aggregated by the neighbors of n_i, σ is the activation function and $\oplus$ is the concatenation operation. And the aggregation strategy of $\boldsymbol{f}_{\mathcal{N}_i}$ can be chosen as [7].

Finally, we apply the aggregated vector with a normalization to get the hidden representation $\boldsymbol{z}_i$ and search P depth for neighbor sampling [7]. The details can be found in Algorithm 1 step 1–12.

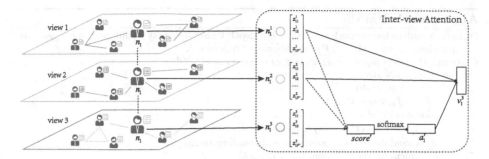

Fig. 3. Inter-view Attention. It assigns different importance to hidden representations from different views.

Inter-view Attention. For each z_i^k of view k, where $k \in \{1, 2, \cdots, K\}$, we introduce inter-view attention as follow:

$$s'^k_i = \mathrm{score}'(z_i^k, \bar{Z}_i^k) = z_i^k W_{a'}^k \bar{Z}_i^{k\,\mathrm{T}} \tag{5}$$

$$[a'_k]_j = \mathrm{softmax_j}(s'^k_i) = \frac{\exp([s'^k_i]_j)}{\sum_{j'=1}^{K-1} \exp([s'^k_i]_{j'})} \tag{6}$$

where $\bar{Z}_i^k$ is the set of all the hidden representations for node n_i except view k, $j \in \{1, \cdots, K-1\}$ and $W_{a'}^k \in \mathbb{R}^{F' \times F'}$ is the inter-view attention weight matrix.

Then, the view-context vector v_i^k is defined to integrate the hidden representations from different views as follow:

$$v_i^k = a'_k \bar{Z}_i^k \tag{7}$$

Finally, the final node representation is defined as follow:

$$o_i = \sigma(W'[Z_i \oplus V_i]) \tag{8}$$

where $Z_i = [z_i^1 \oplus z_i^2 \oplus \cdots \oplus z_i^K]$, $V_i = [v_i^1 \oplus v_i^2 \oplus \cdots \oplus v_i^K]$ and $W' \in \mathbb{R}^{2KF' \times d}$ is the weight matrix. The details can be found in Algorithm 1 step 13–14.

3.2 Parameters Learning of I²MNE

For each single view, we apply a graph-based loss function with negative sampling techniques [13,14] to $z_i, \forall n_i \in V$ as follow:

$$L_s(z_i) = -\log\left(\sigma(z_i^T z_j)\right) - \sum_{n=1}^{N} E_{v_n \sim P_n(n_j)} \log\left(\sigma(-z_i^T z_{v_n})\right) \tag{9}$$

where n_j is a node that co-occurs near n_i on fixed-length random walk, v_n is a negative sample, $\sigma(x) = 1/(1 - \exp(-x))$ is the sigmoid function, N is the number of negative samples and P_n denotes the negative sampling distribution.

Algorithm 1. I^2MNE

Input: A multi-view network $G = (V, E, C)$; depth P; non-linearity σ; differentiable aggregate
   functions $g_p, \forall p \in \{1, \cdots, P\}$; neighborhood function $\mathcal{N}$
Output: the node representations $o_i \in O$ where $O \in \mathbb{R}^{|V| \times d}$
1: **for** $k = 1$ to K **do**
2:      **for** $p = 1$ to P **do**
3:          $f_i^0 \leftarrow f_i, \forall n_i \in V$
4:          **for** $n_i \in V$ **do**
5:              $f_{\mathcal{N}_i}^p \leftarrow g_p(\{f_j^{p-1}, \forall n_j \in \mathcal{N}_i\})$
6:              calculate the context-vector c_i^p according to Eq.(3)
7:              calculate the $f_i'^p$ according to Eq.(4)
8:          **end for**
9:          $f_i^p \leftarrow f_i^p / \|f_i^p\|_2, \forall n_i \in V$
10:     **end for**
11:     $z_i^k \leftarrow f_i^P, \forall n_i \in V$
12: **end for**
13: calculate the view-context vector $v_i^k, \forall k \in \{1, \cdots, K\}, \forall n_i \in V$ according to Eq.(7)
14: calculate the final node representation $o_i, \forall n_i \in V$ according to Eq.(8)

This loss function encourages that the embeddings of connected nodes are similar
to each other, while enforcing that the embeddings of disparate nodes are highly
distinct.

For multi-views integration, we apply cross-entropy loss for the classification
task as follow:

$$L_m = \sum_{n_i \in S} L(z_i, y_i) \tag{10}$$

where S is the set of labeled nodes, y_i is the label of node n_i, and L is the
cross-entropy loss function.

With the above definitions, the overall loss function is defined as follows:

$$L = \sum_{n_i \in V} L_s(z_i) + L_m \tag{11}$$

Our objective is to minimize the overall loss function and it can be efficiently
optimized with the back propagation algorithm [17]. Following the suggestion of
[16], in each iteration of the optimization, we first optimize graph-based loss on
each single view, learn the hidden representations and tune the parameters of
intra-view attention. Then update the parameters of inter-view attention with
the labeled data by optimizing the cross-entropy loss.

4   Experiment

4.1   Datsets

The detailed statistics of the datasets are shown in Table 1.

Table 1. Statistics of datasets

Dataset	# of nodes	# of edges in each view		Features in each view		# of labels
AMiner	16,604	62,115 (co-author)	266,802 (citing)	interest (co-author)	title (citing)	8
Flickr	35,314	3,017,530 (friendship)	3,496,495 (tag-sim)	- (friendship)	tag (tag-sim)	171

AMiner. We use AMiner dataset [25][1] to analyze the research fields of authors. There are two views in the multi-view author network including the co-author relationship view and the citing relationship view. The former is constructed by authors who publish the same paper and the latter is constructed by authors who cite others' paper. The features of authors in the co-author view are the research interests. We use word2vec [13] pre-trained by English Wikipedia Skip-Gram[2] to learn the textual features. And the titles of all the papers published by each author are treated as the node features in the citing view. we use Doc2vec [9] pre-trained by English Wikipedia DBOW[4] to learn the textual features. We choose all the papers from the most popular venue[3] in eight research fields defined by [23] and select all the relative authors who publish these papers. There are 16604 authors with labels in the filtered dataset.

Flickr. We use Flickr dataset [26][4] to analyze the community membership of users. The multi-view user network includes the friendship view and the tag-similarity view. For the latter one, we first calculate the user similarity by their tags using TF-IDF. Then we construct the tag-similarity relationship using 100-nearest neighbors. And the textual features are the tags of users learned by word2vec.

4.2 Compared Algorithms

In this section, our proposed approach is compared with the following methods for performance analysis:

Single-view Methods:

- LINE [24]: A network embedding method without node features.
- Deepwalk/node2vec [6,15]: We find that the results of them have no significant differences, so we use $p = 1$ and $q = 1$ [6] in node2vec for comparison.
- GraphSAGE [7]: A network embedding method with node features.
- I²MNE-Intra: A variant of our proposed method only use the intra-view attention for single views.

---

[1] https://aminer.org/aminernetwork.
[2] https://github.com/jhlau/doc2vec.
[3] 1. IEEE Trans. Parallel Distrib. Syst. 2. STOC 3. IEEE Communications Magazine 4. ACM Trans. Graph. 5. CHI 6. ACL 7. CVPR 8. WWW.
[4] http://dmml.asu.edu/users/xufei/datasets.html.

Multi-view Methods

- *-concat: We concatenate the embeddings learned from all the single views by the node2vec, GraphSAGE and I^2MNE, respectively.
- *-mean: We calculate the mean average of the embeddings learned from all the single views by the node2vec and GraphSAGE, respectively.
- MVE [16]: A multi-view network embedding method with attention mechanism, which can also apply to our problem but cannot utilize the features of nodes and the attention is different from ours.
- I^2MNE-NoInter: A variant of our proposed method without the inter-view attention and calculate the mean average embeddings of all the single views.
- I^2MNE: our proposed method for multi-view network embedding with node features.

4.3 Parameter Settings

For all the methods except *-concat, the dimension is set as 128 by default. For the concatenation methods, the dimension is set as $128K$, where K is the number of views. The dimension of features is set as 300. The number of negative samples is set as 5, and the learning rate is set as 0.001. For node2vec, we set the walk length as 40, the window size as 10. All the embedding vectors are finally normalized.

4.4 Results

We evaluate the network embeddings on the classification task. A logistic regression classifier fed by the embeddings of all labeled nodes is employed. We set 75% nodes as training data and the rest of nodes are used for testing. The classification experiments are repeated independently for 10 times and the averaged Macro-F1 and Micro-F1 measures are reported in Table 2. Note that, for single-view methods, the best results on single views are reported. From this table, we have the following observations:

(1) For the single view, I^2MNE-Intra achieves the best performance than the other single-view methods. In addition, I^2MNE-concat also outperforms the other concatenation methods for the multi-view network. These indicate that the intra-view attention can capture the impact of neighbors' features on the nodes and improve the performance.
(2) On the Flickr dataset, I^2MNE achieves significant improvement comparing to all the methods on both measurements. I^2MNE-NoInter already outperforms all the baselines in terms of Macro-F1 due to the importance of intra-view attention. And I^2MNE gains further improvement over I^2MNE-NoInter with 0.0534 Macro-F1 score. It shows that the inter-view attention plays an important role in our methods.

(3) Especially for the AMiner dataset, I^2MNE achieves significant improvements than other methods with $0.0768 \sim 0.1125$ gains in terms of Macro-F1. In addition, we observe that I^2MNE achieves the best performance in terms of Macro-F1 while I^2MNE-NoInter gets the best results in terms of Micro-F1. It is perhaps that the classes in the AMiner dataset are imbalance, so the Macro-F1 which treats all the classes equally is more reasonable than the Micro-F1 which equally treats all the instances. From the result, we also see that the inter-view attention can preserve more distinction between classes compared to the intra-view attention.

Table 2. Results of classification on both datasets.

Category	Algorithm	AMiner		Flickr	
		Macro-F1	Micro-F1	Macro-F1	Micro-F1
Single view	LINE	0.4219	0.5804	0.5108	0.8438
	Deepwalk/node2vec	0.5494	0.6878	0.4937	0.8524
	GraphSAGE	0.6134	0.7533	0.5462	0.8971
	I^2MNE-Intra	**0.6380**	**0.7608**	**0.5650**	**0.8996**
Multi view	node2vec-concat	0.5739	0.7601	0.4981	0.8342
	GraphSAGE-concat	0.6095	0.8036	0.5602	0.8942
	I^2MNE-concat	**0.6116**	**0.8055**	**0.5748**	**0.8973**
	node2vec-mean	0.5884	0.7860	0.4938	0.8507
	GraphSAGE-mean	0.5786	0.7950	0.5134	0.8923
	MVE	0.5527	0.7544	0.4919	0.8836
	I^2MNE-NoInter	0.5735	**0.7977**	0.5340	0.8896
	I^2MNE	**0.6652**	0.7769	**0.5874**	**0.8962**

5   Conclusion

In this paper, we propose two effective attentions to learn representations of multi-view networks associated with node features, named I^2MNE. We introduce the intra-view attention to leverage the feature information from neighbors and the inter-view attention to make full use of the information from different views. Experiments on two real-world datasets demonstrate that our proposed model is effective and efficient for multi-view network embedding. In the future, we plan to apply our model for more tasks (e.g. link prediction). What's more, we also plan to investigate the embedding of networks with edge features.

Acknowledgments. This work is partially supported by the National Natural Science Foundation of China (Nos. U1509206, 61625107, U1611461), the Key Program of Zhejiang Province, China (No. 2015C01027).

References

1. Backstrom, L., Leskovec, J.: Supervised random walks: predicting and recommending links in social networks. In: ACM International Conference on Web Search and Data Mining, pp. 635–644 (2011)
2. Bhagat, S., Cormode, G., Muthukrishnan, S.: Node classification in social networks. Comput. Sci. **16**(3), 115–148 (2012)
3. Ding, C.H.Q., He, X., Zha, H., Gu, M., Simon, H.D.: A min-max cut algorithm for graph partitioning and data clustering. In: IEEE International Conference on Data Mining, pp. 107–114 (2001)
4. Elkahky, A.M., Song, Y., He, X.: A multi-view deep learning approach for cross domain user modeling in recommendation systems. In: International Conference on World Wide Web, pp. 278–288 (2015)
5. Greene, D.: A matrix factorization approach for integrating multiple data views. In: European Conference on Machine Learning and Knowledge Discovery in Databases, pp. 423–438 (2009)
6. Grover, A., Leskovec, J.: node2vec: scalable feature learning for networks. In: ACM SIGKDD International Conference on Knowledge Discovery and Data Mining, p. 855 (2016)
7. Hamilton, W., Ying, Z., Leskovec, J.: Inductive representation learning on large graphs. In: Advances in Neural Information Processing Systems, pp. 1024–1034 (2017)
8. Kipf, T.N., Welling, M.: Semi-supervised classification with graph convolutional networks (2016)
9. Lau, J.H., Baldwin, T.: An empirical evaluation of doc2vec with practical insights into document embedding generation (2016)
10. Liu, J., Wang, C., Gao, J., Han, J.: Multi-view clustering via joint nonnegative matrix factorization (2013)
11. Luong, T., Pham, H., Manning, C.D.: Effective approaches to attention-based neural machine translation. In: Proceedings of the 2015 Conference on Empirical Methods in Natural Language Processing, pp. 1412–1421 (2015)
12. Ma, G., et al.: Multi-view clustering with graph embedding for connectome analysis. In: ACM on Conference on Information and Knowledge Management, pp. 127–136 (2017)
13. Mikolov, T., Sutskever, I., Chen, K., Corrado, G., Dean, J.: Distributed representations of words and phrases and their compositionality. In: International Conference on Neural Information Processing Systems, pp. 3111–3119 (2013)
14. Mnih, A., Teh, Y.W.: A fast and simple algorithm for training neural probabilistic language models. In: International Conference on International Conference on Machine Learning, pp. 419–426 (2012)
15. Perozzi, B., Alrfou, R., Skiena, S.: DeepWalk: online learning of social representations, pp. 701–710 (2014)
16. Qu, M., Tang, J., Shang, J., Ren, X., Zhang, M., Han, J.: An attention-based collaboration framework for multi-view network representation learning, pp. 1767–1776 (2017)
17. Rumelhart, D.E., Hinton, G.E., Williams, R.J.: Learning representations by back-propagating errors. Nature **323**(6088), 399–421 (1986)
18. Shi, C., Zhang, Z., Luo, P., Yue, Y., Yue, Y., Wu, B.: Semantic path based personalized recommendation on weighted heterogeneous information networks. In: ACM International on Conference on Information and Knowledge Management, pp. 453–462 (2015)

19. Shi, Y., Han, F., He, X., Yang, C., Luo, J., Han, J.: mvn2vec: preservation and collaboration in multi-view network embedding (2018)
20. Singh, A.P., Gordon, G.J.: Relational learning via collective matrix factorization. In: ACM SIGKDD International Conference on Knowledge Discovery and Data Mining, pp. 650–658 (2008)
21. Sun, Y., Barber, R., Gupta, M., Aggarwal, C.C., Han, J.: Co-author relationship prediction in heterogeneous bibliographic networks. In: International Conference on Advances in Social Networks Analysis and Mining, pp. 121–128 (2011)
22. Sun, Y., Han, J., Aggarwal, C.C., Chawla, N.V.: When will it happen?: relationship prediction in heterogeneous information networks, pp. 663–672 (2012)
23. Swami, A., Swami, A., Swami, A.: metapath2vec: scalable representation learning for heterogeneous networks. In: ACM SIGKDD International Conference on Knowledge Discovery and Data Mining, pp. 135–144 (2017)
24. Tang, J., Qu, M., Wang, M., Zhang, M., Yan, J., Mei, Q.: LINE: large-scale information network embedding. In: International Conference on World Wide Web (2015)
25. Tang, J., Zhang, J., Yao, L., Li, J., Zhang, L., Su, Z.: ArnetMiner: extraction and mining of academic social networks. In: ACM SIGKDD International Conference on Knowledge Discovery and Data Mining, pp. 990–998 (2008)
26. Wang, X., Tang, L., Liu, H., Wang, L.: Learning with multi-resolution overlapping communities. Knowl. Inf. Syst. **36**(2), 517–535 (2013)
27. Xia, T., Tao, D., Mei, T., Zhang, Y.: Multiview spectral embedding. IEEE Trans. Syst. Man Cybern. Part B **40**(6), 1438–1446 (2010)
28. Zhou, D., Burges, C.J.C.: Spectral clustering and transductive learning with multiple views. In: Proceedings of the Twenty-Fourth International Conference on Machine Learning, pp. 1159–1166 (2007)

Multi-modal Sequence to Sequence Learning with Content Attention for Hotspot Traffic Speed Prediction

Binbing Liao[1], Siliang Tang[1], Shengwen Yang[2], Wenwu Zhu[3], and Fei Wu[1(✉)]

[1] College of Computer Science and Technology,
Zhejiang University, Hangzhou, China
{bbliao,siliang,wufei}@zju.edu.cn
[2] Baidu Inc., Beijing, China
yangshengwen@baidu.com
[3] Department of Computer Science and Technology,
Tsinghua University, Beijing, China
wwzhu@tsinghua.edu.cn

Abstract. Traffic speed prediction is a crucial and fundamental task of the intelligent transportation systems (ITS). Due to the dynamic and non-linear nature of the traffic, this task is difficult. Nonetheless, the collection of crowd map queries data brings new ways to solve this problem. Generally speaking, in a short period of time, a large amount of crowd map queries aiming at the same destination may lead to traffic congestion. For instance, large queries for Family Restaurant during the dinner time lead to traffic jams around it. However, traffic speed prediction with crowd map queries is challenging due to the complexity and scale of the map queries, as well as their modalities. To bridge the gap, we propose Multi-Seq2Seq-Att for hotspot traffic speed prediction. Multi-Seq2Seq-Att is a multi-modal sequence learning model that deals with two sequences in different modalities, namely, the query sequence and the traffic speed sequence. The main idea of Multi-Seq2Seq-Att is to learn to fuse the multi-modal sequence with content attention. With this method, Multi-Seq2Seq-Att addresses the modality gap between queries and the traffic speed. Experiments on real-world datasets from Baidu Map demonstrates a 24% relative boost over other state-of-the-art methods.

Keywords: Traffic speed prediction · Map query · Content attention

1 Introduction

With the ever-increasing urbanization process, traffic congestion has become a common urban problem around the world. As a crucial task of the ITS, accurate and real-time traffic prediction is particularly useful for many applications like route guidance, traffic network planning, and congestion avoidance [17]. However,

© Springer Nature Switzerland AG 2018
R. Hong et al. (Eds.): PCM 2018, LNCS 11164, pp. 212–222, 2018.
https://doi.org/10.1007/978-3-030-00776-8_20

this task is challenging due to the complexity and dynamic property of the traffic environment in urban cities.

Previous methods for traffic prediction can be classified into traffic prediction with unimodal and multi-modal data. On one hand, autoregressive integrated moving average (ARIMA) [1] and its family [15, 16] are widely used for traffic prediction. However, ARIMA and its family require high computational resources which make them not suitable for large-scale problems. Taking the stochastic and nonlinear nature of traffic into account, some non-parametric methods are proposed for traffic prediction, such as k-NN [3], RF [9] and SVR [8]. Recently, some deep learning methods are proposed for traffic prediction such as deep belief network [7] and stacked autoencoders (SAEs) [13]. Most of the aforementioned works consider the traffic prediction for highways, whose traffic condition are relatively stable and simple. While in urban cities, the traffic is highly dynamic and varies greatly due to diverse and complicated factors (e.g., crowd activities). Thus methods which ignore the multi-modal factors may no longer work.

On the other hand, a few researchers attempted to predict the traffic with related multi-modal data. [5, 11] proposed an optimization framework to extract traffic indicators based on location-based social media and incorporate them into traffic prediction via linear regression. However, using a linear regression may be insufficient when the traffic is a nonlinear system. [14] proposed an LSTM neural network for traffic prediction using microwave detector data. Nonetheless, in real-world road systems, only a small fraction of the road segments are deployed with sensors. For those road segments without sensors, previous methods may no longer work.

With the rapid growth of mobile technology, map applications (e.g., Baidu Map and Google Map) provide a rich source of information for traffic prediction. Figure 1 shows the average traffic speed and crowd query counts around the Capital Gym, Beijing on April 8, 2017. Query counts at time t are the numbers of queries issued for the Capital Gym, with the estimated time of arrival for these queries to be t. We can tell that the number of query counts (in red) are higher more than that of average historical query counts (in blue) around 18 PM, which predicts a sudden drop in the traffic speed. Note that several queries targeting the same destination in a short duration could lead to traffic jams at their designated destination after a while. Therefore, crowd map queries can provide early warnings (The number shown in [10] is 46 min) of traffic jams, which is good for many applications in ITS like route guidance, traffic control, and especially the congestion avoidance. Interestingly, the burst of map queries in a short duration normally indicates a "hotspot" being held at the targeting destination (the "hotspot" is "Fish Leong Concert" in Fig. 1). Therefore, the exploration of the hotspot from lots of crowd map queries brings a new way to explain the traffic jams.

We intend to make full use of crowd map queries in traffic speed prediction problems. However, there are two challenges to the integration of query and road traffic speed data: (1) **Spatiotemporal variation.** The raw crowd map queries targeting the same destination, like the Capital Gym, can be initiated

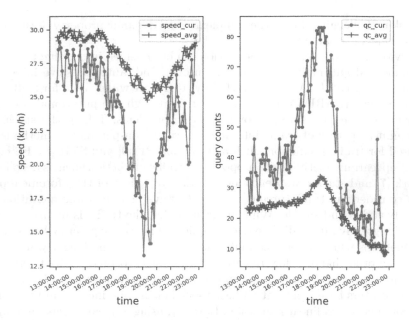

Fig. 1. The traffic speed (left) and crowd query counts (right) around the Capital Gym, Beijing on April 8, 2017. The red "dot" denotes the current traffic speed (query counts) while the blue "plus" represents the average historical traffic speed (query counts). At 19:00, there is the Fish Leong Concert in the Capital Gym. (Color figure online)

at different origins, at the different time and by individual users; (2) **Modality Gap.** The queries of traffic speed are from different modalities and have different distributions.

Enlighted by the idea that performance improvement can be achieved by properly integrating multi-modal information from various sources, this paper aims to predict future traffic speed with appropriate integration of current road traffic speed and crowd map queries. Technically, the contributions of this paper can be described in two aspects.

1. This paper proposes Multi-Seq2Seq-Att for hotspot traffic speed prediction. Multi-Seq2Seq-Att is a learning framework that deals with sequences of different modalities, (i.e., the query sequence and the traffic speed sequence). The main idea of Multi-Seq2Seq-Att attempts is to learn to fuse the multi-modal sequence with content attention. As a result, Multi-Seq2Seq-Att addresses the modality gap between the queries and the data of traffic speed.
2. Furthermore, the generality of Multi-Seq2Seq-Att makes it promising for many sequential multi-modal learning applications, such as the application of text and speech.

This paper is organized as follows: Sect. 2 presents the problem definition. Following that, in Sect. 3, we describe the proposed Multi-Seq2Seq-Att in detail.

Sect. 4 presents qualitative and quantitative results of different models. Finally, we make the conclusion in Sect. 5.

2   Problem Definition

Assume $\mathcal{L} = \{l^i | i = 1, 2, \ldots, K\}$ is a collection of K road segments, where a road segment is a section of a road. Let $\boldsymbol{v}^l = (v_1^l, v_2^l, \ldots, v_t^l)$ be the traffic speed of the road segment $l \in \mathcal{L}$, where v_t^l is a scalar that denotes the traffic speed of one specific road segment l at time t. Assume $\mathcal{Q} = \{q^i | i = 1, 2, \ldots, N\}$ is a corpus of N users' map query records. Each map query record q^i is defined by a triple $q^i = (t_s^i, s^i, d^i)$, which satisfies: (1) t_s^i is the starting time of query q^i; (2) s^i is the origin (source location) of q^i; (3) d^i is the destination. To simplify the problem, the superscript is removed without confusion in the remaining part of this paper.

Specifically, for the road segment l, given the previous traffic speed $V^p = (v_1^l, v_2^l, \ldots, v_t^l)$ and the query records $\mathcal{Q}$, our object is to maximize the conditional probability of observing the future traffic speed $V^f = (v_{t+1}^l, v_{t+2}^l, \ldots, v_{t+w}^l)$:

$$p_\theta(V^f | V^p, \mathcal{Q}) = \prod_{m=1}^{w} p_\theta(v_{t+m} | v_1, v_2, \ldots, v_{t+m-1}, \mathcal{Q}_{<=t}) \qquad (1)$$

In the equation above, $\mathcal{Q}_{<=t} = \{q^i | t_s^i <= t\}$ and w is the prediction horizon and θ is a parameter. Given the map query records $\mathcal{Q}$, the previous traffic speed slot V^p and the future traffic speed V^f of K road segments, our training objective is to maximize the following log likelihood w.r.t. the model parameter θ:

$$\underset{\theta}{\arg\min} \quad -\frac{1}{K} \sum_{k=1}^{K} \log \; p_\theta(V^f | V^p, \mathcal{Q}) \qquad (2)$$

3   Methods

Followed by our previous work [10], we address the spatiotemporal variation with hotspot discovery. In this section, we introduce the sequence to sequence model (Seq2Seq) and our multi-modal sequence to sequence learning with attention (Multi-Seq2Seq-Att) for hotspot traffic speed prediction.

3.1   Seq2Seq

For each hotspot, given the historical traffic speed slot $\boldsymbol{V}^c = (\boldsymbol{v}_{t-w+1}, \boldsymbol{v}_{t-w+2}, \ldots, \boldsymbol{v}_t)$ of selected k road segments, we aim to forecast their future traffic speed slot $\widehat{\boldsymbol{V}}^f = (\hat{\boldsymbol{v}}_{t+1}, \hat{\boldsymbol{v}}_{t+2}, \ldots, \hat{\boldsymbol{v}}_{t+w})$, where $\boldsymbol{v}_t = (v_t^1, v_t^2, \ldots, v_t^k)^T$ is the traffic speed of k road segments at time t, w is the prediction horizon.

As shown in Fig. 2, a sequence to sequence (Seq2Seq) network is applied to model the traffic speed. It consists of two LSTM [6] layers. The bottom LSTM layer (colored red) encodes information in the current traffic speed slot $\boldsymbol{V}^c$ while the second LSTM layer (colored green) decodes the encoding information of $\boldsymbol{V}^c$ to predict the future traffic speed slot $\widehat{\boldsymbol{V}}^f$.

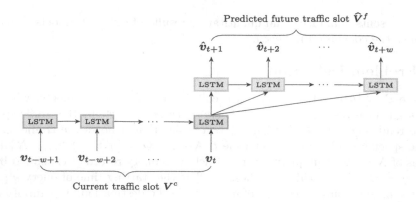

Fig. 2. The structure of the Seq2Seq network. (Color figure online)

3.2 Multi-Seq2Seq-Att

The map queries issued by users at a certain time can be utilized to foresee the traffic speed around the queried destination after a while. Considering a set of queries triggered earlier than 15:00 whose destinations are all around the Capital Gym, the arrival time towards the destinations can be estimated as [10]. The number of queries (ϕ_t) implies the information of how many individuals would arrive at the Capital Gym at the estimated arrival time, which would be beneficial to traffic speed prediction around the Capital Gym. Note that here only the map queries for the Capital Gym issued earlier than 15:00 is utilized to guarantee the foreseeable characteristic of map queries.

However, unleashing the power of the crowd map queries for traffic prediction is challenging due to the modality gap. To bridge the gap, we propose a multi-modal sequence to sequence framework with content attention as shown in Fig. 3. Motivated by [12], the idea of Multi-Seq2Seq-Att is to consider all the hidden states of the traffic encoder (self attention) and query encoder (multi-modal attention) when decoding the current target hidden state h_{t+1} through two content attention. Specifically, for self attention, given the current target hidden state h_{t+1}, the traffic source hidden state $\bar{h}_s^t$, the traffic context vector c_{t+1}^t is derived as follows:

$$\text{score}(h_{t+1}, \bar{h}_s^t) = h_{t+1}^T W_a^t \bar{h}_s^t \tag{3}$$

$$a_{t+1}^t = \text{self\_align}(h_{t+1}, \bar{h}_s^t) = \frac{\exp(\text{score}(h_{t+1}, \bar{h}_s^t))}{\sum_{s'} \exp(\text{score}(h_{t+1}, \bar{h}_{s'}^t))} \tag{4}$$

$$c_{t+1}^t = (a_{t+1}^t)^T \bar{h}_s^t. \tag{5}$$

where W_a^t is the traffic attention weight matrix, a_{t+1}^t is the traffic alignment vector and c_{t+1}^t is the traffic context vector.

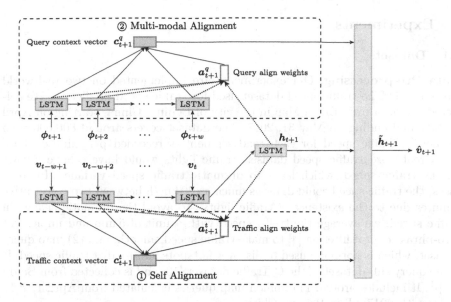

Fig. 3. The structure of the Multi-Seq2Seq-Att network. At time step t+1, the bottom (top) self (multi-modal) alignment infers an alignment weight vector a_{t+1}^t (a_{t+1}^q) based on the current target state h_{t+1} and all source states $\bar{h}_s^t$ ($\bar{h}_s^q$). A global context vector c_{t+1}^t (c_{t+1}^q) is then computed as the weighted average, according to a_{t+1}^t (a_{t+1}^q), over all the source states.

For multi-modal attention, given the current target hidden state h_{t+1}, the query source hidden state $\bar{h}_s^q$, the query context vector c_{t+1}^q is derived as follows:

$$\text{score}(h_{t+1}, \bar{h}_s^q) = h_{t+1}^T W_a^q \bar{h}_s^q \tag{6}$$

$$a_{t+1}^q = \text{multi-modal\_align}(h_{t+1}, \bar{h}_s^q) = \frac{\exp(\text{score}(h_{t+1}, \bar{h}_s^q))}{\sum_{s'} \exp(\text{score}(h_{t+1}, \bar{h}_{s'}^q))} \tag{7}$$

$$c_{t+1}^q = (a_{t+1}^q)^T \bar{h}_s^q. \tag{8}$$

where W_a^q is the query attention weight matrix, a_{t+1}^q is the query alignment vector and c_{t+1}^q is the query context vector.

Given the target hidden state h_{t+1}, the traffic context vector c_{t+1}^t, and the query context vector c_{t+1}^q, we utilize a simple concatenation layer to combine the information from all the vectors to produce an attentional hidden state as follows:

$$\tilde{h}_{t+1} = \tanh(W_c[h_{t+1}; c_{t+1}^t; c_{t+1}^q]) \tag{9}$$

The attentional vector $\tilde{h}_{t+1}$ is then fed into the linear-regression layer to produce the predicted traffic speed $\hat{v}_{t+1}$ as:

$$\hat{v}_{t+1} = W_s \tilde{h}_{t+1} \tag{10}$$

4   Experiments

4.1   Datasets

Data Pre-processing. Our experiments are implemented on two real-world datasets: (1) The traffic speed dataset used for traffic speed prediction. We collected the data from Baidu Map in the city of Beijing, China, starting on April 1, 2017 and ending on May 31, 2017. The dataset covers around 800,000 road segments. Traffic speed for each road segment is recorded per minute. Since it's a real word traffic speed dataset, traffic lights would have a huge impact on the traffic speed, which leads to dramatic traffic speed variance. In some cases, the traffic speed could differ as much as 20 km/h between two consecutive minutes due to the existence of traffic lights. To avoid the dramatic change in traffic speed, we average the traffic speed with a unit of 5 min and implement zero-phrase digital filtering [4] to make traffic speed smooth; and (2) map query dataset, which is normally used to discover hotspots and predict traffic speed. It is the query sub-dataset of the Q-Traffic dataset [10] and is collected from Baidu Map[1]. It includes around 114 million map queries calculated from April 1, 2017 to May 31, 2017, all in Beijing, China.

Correlation Analysis. The relation between traffic speed and query counts in the same spot is tested. For each spot, the average traffic speed of k $(k = 5)$ adjacent road segments with a window of 5 min is used. And We also collect traffic speed together with map query counts data for all selected spots(hotspots). Rank correlation coefficient from Spearman is used in our test and the result $\rho = -0.57$ with a P-value$= 4.64 \times e^{-14}$ shows that a strong negative correlation exists between the query counts and the average traffic speed, which forecasts the potential to predict traffic speed statistics with our map query data.

4.2   Baselines

We compare our proposed model with the following methods.

- Random forests regression (RF) [9]: RF is a widely-used machine learning method for prediction and regression, which constructs a multitude of decision tree at training time and outputs the mean prediction of the individual trees;
- Support vector regression (SVR) [8]: SVR is a regression version of SVM, which is widely used in the area of traffic prediction;
- Gated recurrent unit (GRU) [2]: GRU is a variant version of RNN which has been widely used in regression and prediction tasks. We compare our model with the Seq2Seq model which consists of two GRUs;
- Seq2Seq: We compare our model with the basic Seq2Seq model which consists of two LSTMs;
- Seq2Seq_Att: Seq2Seq_Att is the Seq2Seq model with content attention;
- Init(Q): We use the encoding of the query to initialize the encoder of the Seq2Seq;

---

[1] http://map.baidu.com/.

- T$\star$_Q$\star$_OP: We compare our model with 11 $(2*2*3-1)$ variants T$\star$_Q$\star$_OP, where $\star = \{B, A\}$, $OP = \{Add, Stack, Att\}$ represents the fusion (add, stack or attention) of traffic (before or after encoding) and query (before or after encoding).

4.3 Evaluation Metrics

We implement two metrics to properly evaluate the performance of models proposed, which are mean absolute error and mean square error.

$$MSE = \frac{1}{T} \sum_{t=1}^{T} (v_t - \hat{v}_t)^2 \tag{11}$$

$$MAE = \frac{1}{T} \sum_{t=1}^{T} |v_t - \hat{v}_t| \tag{12}$$

where v_t and $\hat{v}_t$ are the traffic speed ground truth and predicted speed at time t, respectively.

4.4 Results

Firstly, We show the effectiveness of our proposed Multi-Seq2Seq-Att for traffic prediction. Figure 4 demonstrates the performance, measured by MSE and MAE, of different methods. We can observe from the image that deep learning based methods (e.g. Seq2Seq and Multi-Seq2Seq-Att) outcompete many traditional methods like RF and SVR. It is observed that Seq2Seq model is slightly better than GRU, the major difference which separates these two is the sequence layer (one is LSTM, the other is GRU), based on that, we select LSTM as sequence

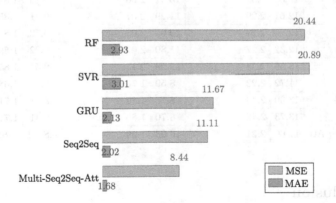

Fig. 4. Best performance of different methods for traffic speed prediction. Lower MSE (MAE) means better performance. (Best viewed in the electronic version)

layer in our final Seq2Seq network. Among all deep learning methods, our Multi-Seq2Seq-Att has the best performance. Compared with Seq2Seq, Multi-Seq2Seq-Att achieves 24% and 17% relative performance increase measured by MAE and MSE, respectively. This improvement may be explained in a way that our model better leverages the query features and the attention model.

Secondly, We show the effectiveness of the query and the attention mechanism. Table 1 compares the performance of all variants with the different number of hidden units. The performances of Seq2Seq_Att and T⋆_Q⋆_Att are better than that of Seq2Seq, which demonstrate the effectiveness of the attention mechanism and the query. In addition, T⋆_Q⋆_Att model outperforms all the T⋆_Q⋆_Add and T⋆_Q⋆_Stack, showing the power of attention. Moreover, the fusion after encoding is better than the fusion before encoding (raw input). As a result, our Multi-Seq2Seq-Att network which considers both the attention mechanism and the map query achieves the best performance. Furthermore, as we increase the number of hidden units, the performance will improve accordingly. However, as the hidden units number exceeds 32, the performance decreases.

Table 1. Comparison of all variants and our Multi-Seq2Seq-Att with different number of hidden units. The results with the best performance are marked in bold.

Methods	Num hidden units = 8		Num hidden units = 16		Num hidden units = 32	
	MSE	MAE	MSE	MAE	MSE	MAE
Seq2Seq	14.34	2.41	11.11	2.02	11.32	2.02
Seq2Seq_Att	15.10	2.51	10.14	2.02	9.83	1.87
Init(Q)	12.87	2.36	9.89	1.88	9.82	1.71
TB_QB_Stack	12.83	2.32	11.27	2.05	9.92	1.93
TB_QB_Add	14.20	2.51	13.69	2.43	12.92	2.28
TB_QB_Att	12.34	2.27	10.33	2.03	9.62	1.86
TB_QA_Stack	12.55	2.31	10.51	2.06	9.15	1.84
TB_QA_Add	12.61	2.29	11.29	2.07	9.41	1.85
TB_QA_Att	12.24	2.27	10.09	2.02	9.72	1.89
TA_QB_Stack	12.32	2.27	11.80	2.02	9.92	1.89
TA_QB_Add	13.50	2.38	12.28	2.17	9.54	1.85
TA_QB_Att	11.72	2.22	8.59	1.76	8.84	1.68
TA_QA_Stack	13.90	2.46	8.76	1.82	8.87	1.71
TA_QA_Add	13.73	2.42	8.70	1.85	9.41	1.71
Multi-Seq2Seq-Att	11.60	2.21	10.03	1.99	**8.44**	**1.68**

5   Conclusion

We study the problem of how to model map queries and to fully utilize them to assist traffic speed prediction. In this paper, we propose Multi-Seq2Seq-Att which integrates the sequence learning from different modalities. The attention part of Multi-Seq2Seq-Att fuses the information of map queries and traffic speed, addressing the modality gap between the map queries and traffic speed. As a result, Multi-Seq2Seq-Att achieves 24% relative improvement over other state-of-the-art methods on our datasets fetched from Baidu Map.

Acknowledgments. This work was supported in part by 973 program (No. 2015CB352302, 2015CB352300), the National Natural Science Foundation of China (Nos. 61625107, 61751209, U1611461), the Key Program of Zhejiang Province, China (No. 2015C01027) and Chinese Knowledge Center of Engineering Science and Technology (CKCEST).

References

1. Ahmed, M.S., Cook, A.R.: Analysis of freeway traffic time-series data by using Box-Jenkins techniques. No. 722 (1979)
2. Chung, J., Gulcehre, C., Cho, K., Bengio, Y.: Empirical evaluation of gated recurrent neural networks on sequence modeling. arXiv preprint arXiv:1412.3555 (2014)
3. Davis, G.A., Nihan, N.L.: Nonparametric regression and short-term freeway traffic forecasting. J. Transp. Eng. **117**(2), 178–188 (1991)
4. Gustafsson, F.: Determining the initial states in forward-backward filtering. IEEE Trans. Sig. Process. **44**(4), 988–992 (1996)
5. He, J., Shen, W., Divakaruni, P., Wynter, L., Lawrence, R.: Improving traffic prediction with tweet semantics. In: IJCAI, pp. 1387–1393 (2013)
6. Hochreiter, S., Schmidhuber, J.: Long short-term memory. Neural Comput. **9**(8), 1735–1780 (1997)
7. Huang, W., Song, G., Hong, H., Xie, K.: Deep architecture for traffic flow prediction: deep belief networks with multitask learning. IEEE Trans. Intell. Transp. Syst. **15**(5), 2191–2201 (2014)
8. Jin, X., Zhang, Y., Yao, D.: Simultaneously prediction of network traffic flow based on PCA-SVR. In: Liu, D., Fei, S., Hou, Z., Zhang, H., Sun, C. (eds.) ISNN 2007. LNCS, vol. 4492, pp. 1022–1031. Springer, Heidelberg (2007). https://doi.org/10.1007/978-3-540-72393-6_121
9. Leshem, G., Ritov, Y.: Traffic flow prediction using adaboost algorithm with random forests as a weak learner. Proc. World Acad. Sci. Eng. Technol. **19**, 193–198 (2007)
10. Liao, B., Zhang, J., Wu, C., McIlwraith, D., Chen, T., Yang, S., Guo, Y., Wu, F.: Deep sequence learning with auxiliary information for traffic prediction. In: Proceedings of the 24th ACM SIGKDD International Conference on Knowledge Discovery and Data Mining. ACM (2018)
11. Liu, X., Kong, X., Li, Y.: Collective traffic prediction with partially observed traffic history using location-based social media. In: Proceedings of the 25th ACM International on Conference on Information and Knowledge Management. pp. 2179–2184. ACM (2016)
12. Luong, T., Pham, H., Manning, C.D.: Effective approaches to attention-based neural machine translation. In: Proceedings of the 2015 Conference on Empirical Methods in Natural Language Processing, pp. 1412–1421 (2015)
13. Lv, Y., Duan, Y., Kang, W., Li, Z., Wang, F.Y.: Traffic flow prediction with big data: a deep learning approach. IEEE Trans. Intell. Transp. Syst. **16**(2), 865–873 (2015)

14. Ma, X., Tao, Z., Wang, Y., Yu, H., Wang, Y.: Long short-term memory neural network for traffic speed prediction using remote microwave sensor data. Transp. Res. Part C: Emerg. Technol. **54**, 187–197 (2015)
15. Tran, Q.T., Ma, Z., Li, H., Hao, L., Trinh, Q.K.: A multiplicative seasonal arima/garch model in evn traffic prediction. Int. J. Commun. Network Syst. Sci. **8**(04), 43 (2015)
16. Williams, B.M., Hoel, L.A.: Modeling and forecasting vehicular traffic flow as a seasonal arima process: theoretical basis and empirical results. J. Transp. Eng. **129**(6), 664–672 (2003)
17. Zhang, J., Wang, F.Y., Wang, K., Lin, W.H., Xu, X., Chen, C.: Data-driven intelligent transportation systems: a survey. IEEE Trans. Intell. Transp. Syst. **12**(4), 1624–1639 (2011)

Modeling Text with Graph Convolutional Network for Cross-Modal Information Retrieval

Jing Yu[1,2], Yuhang Lu[1,2], Zengchang Qin[3(✉)], Weifeng Zhang[4,5],
Yanbing Liu[1], Jianlong Tan[1], and Li Guo[1]

[1] Institute of Information Engineering, Chinese Academy of Sciences, Beijing, China
{yujing02,luyuhang,liuyanbing,tanjianlong,guoli}@iie.ac.cn
[2] School of Cyber Security,
University of Chinese Academy of Sciences, Beijing, China
[3] Intelligent Computing and Machine Learning Lab,
School of ASEE, Beihang University, Beijing, China
zcqin@buaa.edu.cn
[4] Hangzhou Dianzi University, Hangzhou, China
zwf.zhang@gmail.com
[5] Zhejiang Future Technology Institute, Jiaxing, China

Abstract. Cross-modal information retrieval aims to find heterogeneous data of various modalities from a given query of one modality. The main challenge is to map different modalities into a common semantic space, in which distance between concepts in different modalities can be well modeled. For cross-modal information retrieval between images and texts, existing work mostly uses off-the-shelf Convolutional Neural Network (CNN) for image feature extraction. For texts, word-level features such as bag-of-words or word2vec are employed to build deep learning models to represent texts. Besides word-level semantics, the semantic relations between words are also informative but less explored. In this paper, we model texts by graphs using similarity measure based on word2vec. A dual-path neural network model is proposed for couple feature learning in cross-modal information retrieval. One path utilizes Graph Convolutional Network (GCN) for text modeling based on graph representations. The other path uses a neural network with layers of non-linearities for image modeling based on off-the-shelf features. The model is trained by a pairwise similarity loss function to maximize the similarity of relevant text-image pairs and minimize the similarity of irrelevant pairs. Experimental results show that the proposed model outperforms the state-of-the-art methods significantly, with 17% improvement on accuracy for the best case.

1 Introduction

For past a few decades, online multimedia information in different modalities, such as image, text, video and audio, has been increasing and accumulated explosively. Information related to the same content or topic may exist in various

© Springer Nature Switzerland AG 2018
R. Hong et al. (Eds.): PCM 2018, LNCS 11164, pp. 223–234, 2018.
https://doi.org/10.1007/978-3-030-00776-8_21

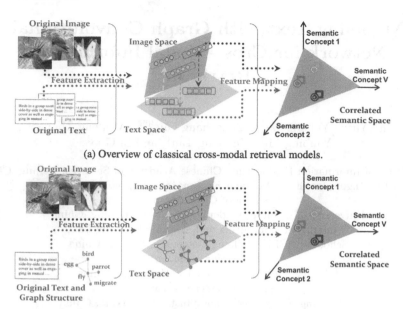

(a) Overview of classical cross-modal retrieval models.

(b) Overveiw of our proposed cross-modal retrieval model.

Fig. 1. Comparison of classical cross-modal retrieval models to our model. (a) Classical models adopt feature vectors to represent grid-structured multimodal data; (b) Our model can handle both irregular graph-structured data and regular grid-structured data simultaneously.

modalities and has heterogeneous properties, that makes it difficult for traditional uni-modal information retrieval systems to acquire comprehensive information. There is a growing demand for effective and efficient search in the data across different modalities. Cross-modal information retrieval [13,17,20] enables users to take a query of one modality to retrieve data in relevant content in other modalities.

The mainstream solution for cross-modal retrieval is to project the features of different modalities into a common semantic space and measure their similarity directly. Thus, feature representation is the footstone for cross-modal information retrieval. Existing work treats the irregular-structured data (i.e. text, protein network) as "flat" features in a similar way as modeling grid-structured data (i.e. image, audio, video). Take text-image retrieval for example. Recent works [19,22] extract the image features by pre-trained Convolutional Neural Network (CNN) [8], which can leverage the local information in the grid-structured data to represent the visual semantics. For text representation, deep models are also widely applied to extract high-level semantics based on the sequential word embeddings. CNN-based methods yield competitive results in image-sentence retrieval. Meanwhile, Recurrent Neural Networks (RNN) gains remarkable multimodal retrieval accuracy. However, these vector-space models treat the input words as "flat" embeddings for the downstream task. More specifically, they only consider the context relations in the text modeling regardless of other important relations.

Recent research has found that the global semantic relations among words can provide rich semantics and can effectively promote the text classification performance [16]. Inspired by their work, we aim to combine deep models to explore the global word relations in representing the irregular-structured text data. Such relations are leveraged for enhancing the generalization ability of text in cross-modal retrieval tasks. In this paper, we propose one of the possible solutions, that is, representing a text by a structured and featured graph and learning text features by a graph-based deep model, i.e. Graph Convolutional Network (GCN) [1,5]. Such a graph can well capture the semantic relations among words. The GCN model has a great ability to learn local and stationary features on graphs. Figure 1 shows the comparison of our model to classical cross-modal retrieval models. Based on this graph representation for texts, we propose a dual-path neural network, called **Graph-In-Network (GIN)**, for cross-modal information retrieval.

The main contributions can be summarized as follows:(1)We propose to model text by graphs using similarity measure based on word2vec, which realizes cross-modal retrieval between irregular-structured and regular grid-structured data; (2) The model can jointly learn the textual and visual representations as well as similarity metric, providing an end-to-end training mode; (3) Experimental results show the superior performance of our model over the state-of-the-art methods.

2   Related Work

Cross-Modal Information Retrieval. The generic solution for cross-modal retrieval is to learn a common semantic space for different modalities of data and measure their similarity directly. Traditional statistical correlation analysis methods, typically like Canonical Correlation Analysis (CCA) [13], aim to maximize the pairwise correlations between two sets the data of different modalities. In order to leverage the semantic information, semi-supervised methods [17,22] and supervised methods [14,18] are proposed to explore the label information and achieve great progress. With the advances of deep learning in multimedia applications, DNN-based cross-modal methods are in the ascendant. This kind of methods generally construct two subnetworks for modeling data of different modalities and learn their correlations by a joint layer. Wang et al. [19] uses two branches of neural networks for learning textual-visual embeddings and realize effective end-to-end fine-tuning. In this work, we also follow the DNN-based routine to model the matched and mismatched text-image pairs.

Graph Convolutional Network (GCN). To render the extension of CNN to irregular graphs, [1] proposes graph convolutional network, which allows convolutions on the graphs to be solved as multiplications in the graph spectral domain. Besides, [5] further simplifies GCN by a first-order approximation of graph spectral convolutions, resulting in more efficient filtering operations. Based on GCN, recent work [6] proposes a novel method for learning a similarity metric between irregular graphs. A siamese graph convolutional network is introduced for similarity matching. Different from our work, the two branches of the model come from the same image modality and the two branches share weights. Their model can only handle graph-structured modal data, which can been seen as a special case of our framework.

3   Methodology

In this section, we introduce a novel dual-path neural network to simultaneously learn multi-modal representations and similarity metric in an end-to-end mode. In the text modeling path (top in Fig. 2, that the convolution part is referred to the blog of GCNs[1]) contains two key steps: *graph construction* and *GCN modeling*.

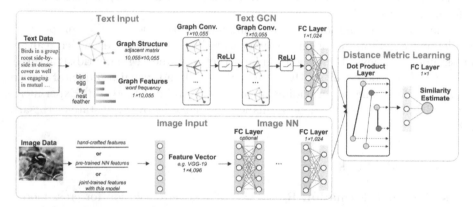

Fig. 2. The structure of the proposed model is a dual-path neural network: i.e., text Graph Convolutional Network (text GCN) (top) and image Neural Network (image NN) (bottom). The text GCN for learning text representation contains two layers of graph convolution on the top of constructed featured graph. The image NN for learning image representation contains layers of non-linearities initialized by off-the-shelf features. They have the same dimension in the last fully connected layers. The objective is a global pairwise similarity loss function.

3.1   Text Modeling

Graph Construction: In this work, we represent a text by a featured graph to combine the strengths of structural information with semantic information together. Given a set of texts, we extract the most common words, denoted as $W = [w_1, w_2, ..., w_N]$, from all the unique words in this corpus and represent each word by a pre-trained *word2vec* embedding. For the graph structure, we construct a k-nearest neighbor graph, denoted as $G = (V, E)$. Each vertex $v_i \in V$

---
[1] http://tkipf.github.io/graph-convolutional-networks/.

is corresponding to a unique word and each edge $e_{ij} \in E$ is defined by the *word2vec* similarity between two words:

$$e_{ij} = \begin{cases} 1 & \text{if } w_i \in N_k(w_j) \text{ or } w_j \in N_k(w_i) \\ 0 & \text{otherwise} \end{cases} \tag{1}$$

where $N_k(\cdot)$ denotes the set of k-nearest neighbors by computing the cosine similarity between word *word2vec* embeddings. k is the parameter of neighbor numbers (set to 8 in our following experiments). The graph structure is stored by an adjacent matrix $A \in \mathbb{R}^{N \times N}$. For the graph features, each text document is represented by a *bag-of-words* vector and the frequency value of word w_i serves as the 1-dimensional feature on vertex v_i. In this way, we combine structural information of word similarity relations and semantic information of word vector representation in a featured graph. Note that the graph structure is identical for a corpus and we use different graph features to represent each text in a corpus.

GCN Modeling: Deep network models have become increasingly popular and achieved breakthroughs in many text analysis tasks. However, classical deep network models are defined for grid-structured data and can not be easily extended to graphs. It's challenging to define the local neighborhood structures and the vertex orders for graph operations. Recently, Graph Convolutional Network (GCN) is proposed to generalize Convolutional Neural Network (CNN) to irregular-structured graphs. In this paper, the text features are learnt by GCN given the graph representation of a text document.

Given a text, we define its input graph feature vector by F_{in} and denote the output feature vector after graph convolution by F_{out}. In order to keep the filter K-localized in space and computationally efficient, [1] proposes a approximated polynomial filter defined as $g_\theta = \sum_{k=0}^{K-1} \theta_k T_k(\widetilde{L})$, where $T_k(x) = 2x T_{k-1}(x) - T_{k-2}(x)$ with $T_0(x) = 1$ and $T_1(x) = x$, $\widetilde{L} = \frac{2}{\lambda_{max}} L - I_N$ and λ_{max} denotes the largest eigenvalue of L. L is the normalized graph Laplacian for the input graph structure. The filtering operation can then be written as $F_{out} = g_\theta F_{in}$. In our model, we use the same filter as in [1]. For the graph representation of a text document, the ith input graph feature $f_{in,i} \in F_{in}$ is the word frequency of vertex v_i. Then the ith output feature $f_{out,i} \in F_{out}$ is given by:

$$f_{out,i} = \sum_{k=0}^{K-1} \theta_k T_k(\widetilde{L}) f_{in,i} \tag{2}$$

where we set $K = 3$ in the experiments to keep each convolution at most 3-steps away from a center vertex. Our text GCN contains two layers of graph convolutions, each followed by Rectified Linear Unit (ReLU) activation to increase non-linearity. A fully connected layer is successive with the last convolution layer to map the text features to the common latent semantic space. Given a text document T, the text representation f_t learnt by the text GCN model $H_t(\cdot)$ is denoted by $f_t = H_t(T)$.

3.2   Image Modeling

For modeling images, we adopt a neural network (NN) containing a set of fully connected layers (bottom in Fig. 2). We have three options of initializing inputs by hand-crafted feature descriptors, pre-trained neural networks, or jointly trained end-to-end neural networks. In this paper, the first two kinds of features are used for fair comparison with other models. The input visual features are followed by a set of fully connected layers for fine-tuning the visual features. Similar to text modeling, the last fully connected layer of image NN maps the visual features to the common latent semantic space with the same dimension as text. In experimental studies, we tune the number of layers and find that only keeping the last semantic mapping layer without feature fine-tuning layers can obtain satisfactory results. Given an image I, the image representation f_{img} learnt by the model from image NN $H_{img}(\cdot)$ is represented by $f_{img} = H_{img}(I)$.

3.3   Objective Function

Distance metric learning is applied to estimate the relevance of features learned from the dual-path model. The outputs of the two paths, i.e. f_t and f_{img}, are in the same dimension and combined by an inner product layer. The successive layer is a fully connected layer with one output $score(T, I)$, denoting the similarity score function between a text-image pair. The training objective is a pairwise similarity loss function proposed in [7], which outperforms existing works in the problem of learning local image features. In our research, we maximize the mean similarity score u^+ between text-image pairs of the same semantic concept and minimize the mean similarity score u^- between pairs of different semantic concepts. Meanwhile, we also minimises the variance of pairwise similarity score for both matching σ^{2+} and non-matching σ^{2-} pairs. The loss function is formally by:

$$Loss = (\sigma^{2+} + \sigma^{2-}) + \lambda \max(0, m - (u^+ - u^-)) \tag{3}$$

where λ is used to balance the weight of the mean and variance, and m is the margin between the mean distributions of matching similarity and non-matching similarity. $u^+ = \sum_{i=1}^{Q_1} \frac{score(T_i, I_i)}{Q_1}$ and $\sigma^{2+} = \sum_{i=1}^{Q_1} \frac{(score(T_i, I_i) - u^+)^2}{Q_1}$ when text T_i and image I_i are in the same class. While $u^- = \sum_{j=1}^{Q_2} \frac{score(T_j, I_j)}{Q_2}$ and $\sigma^{2-} = \sum_{j=1}^{Q_2} \frac{(score(T_j, I_j) - u^-)^2}{Q_2}$ when T_j and I_j are in different classes. We sequentially select $Q_1 + Q_2 = 200$ text-image pairs from the training set for each mini-batch in the experiments.

4   Experimental Studies

4.1   Datasets

Experiments are conducted on four widely used benchmark datasets. Each dataset contains a set of text-image pairs. dataset (Eng-Wiki for short) [13] contains 2,866 image-text pairs divided into 10 classes. Each image is represented by a 4,096-dimensional vector extracted from the last fully connected layer of VGG-19 model [15]. Each text is represented by a graph with 10,055 vertices. **NUS-WIDE** dataset consists of 269,648 image-tag pairs We select samples in the 10 largest classes as adopted in [22]. For images, we use 500-dimensional bag-of-features. For tags, we construct a graph with 5,018 vertices. **Pascal VOC** dataset consists of 9,963 image-tag pairs belonging to 20 classes. The images containing only one object are selected in our experiments as [14,18] For the features, 512-dimensional Gist features are adopted for the images and a graph with 598 vertices is used for the tags. **TVGraz** dataset contains 2,594 image-text pairs [10]. We choose the texts that have more than 10 words. Each image is represented by a 4,096-dimensional VGG-19 feature and each text is represented by a graph with 8,172 vertices.

4.2 Evaluation and Implementation

To evaluate the performance of our model, we conduct experiments for cross-modal retrieval tasks, i.e. text-query-images and image-query-texts. The mean average precision (MAP) and precision-recall (PR) curves [13] are used to evaluate the performance of all the algorithms on the four datasets. For all the datasets, we randomly select matched and non-matched text-image pairs and form 40,000 positive samples and 40,000 negative samples for training. The ground truth labels are binary denoting whether the pairs are from the same class or not. We train the model for 50 epochs with mini-batch size 200. We adopt the dropout ratio of 0.2, learning rate 0.001 with an Adam optimisation, and regularisation 0.005. m and λ are set to 0.6 and 0.35, respectively. In the semantic mapping layers of both text and image paths, the reduced dimensions are set to 1,024, 500, 256, 1,024, 1,024 for Eng-Wiki, NUS-WIDE, Pascal, and TVGraz, respectively.

4.3 Experimental Results

(1) Comparison with State-of-the-Art Methods. We compare our proposed GIN with a number of state-of-the-art models. The MAP scores of all the methods on the five benchmark datasets are shown in Table 1. All the other models are well cited work in this field. Since not all the papers have tested these four datasets, for fair comparison, we compare our model to methods on their reported datasets with the same preprocessing conditions. From Table 1, we can have the following observations:

First, GIN outperforms all the compared methods over the four datasets for the text-query-image task. On the Eng-Wiki, Pascal, NUS-WIDE, and TVGRaz datasets, the MAP scores of GIN are about 35.70%, 17.14%, 12.9%, and 1.3% higher than the second best results, respectively. It's obvious that no matter for the rich text or for the sparse tags, our model gains superior performance than other models. The reason is that the proposed model effectively keeps the inter-word semantic relations by representing the texts with graphs, which has

Table 1. MAP score comparison of text-image retrieval on four given benchmark datasets.

Method	Text query	Image query	Average	Dataset
CCA [13]	0.1872	0.2160	0.2016	Eng-Wiki
SCM [13]	0.2336	0.2759	0.2548	
TCM [11]	0.2930	0.2320	0.2660	
LCFS [18]	0.2043	0.2711	0.2377	
LGCFL [3]	0.3160	0.3775	0.3467	
ml-CCA [12]	0.2873	0.3527	0.3120	
GMLDA [14]	0.2885	0.3159	0.3022	
GMMFA [14]	0.2964	0.3155	0.3060	
AUSL [22]	0.3321	0.3965	0.3643	
JFSSL [17]	0.4102	**0.4670**	0.4386	
GIN (ours)	**0.7672**	0.4526	**0.6099**	
CCA [13]	0.2667	0.2869	0.2768	NUS-WIDE
LCFS [18]	0.3363	0.4742	0.4053	
LGFCL [3]	0.3907	0.4972	0.4440	
ml-CCA [12]	0.3908	0.4689	0.4299	
AUSL [22]	0.4128	**0.5690**	0.4909	
JFSSL [17]	0.3747	0.4035	0.3891	
GIN (ours)	**0.5418**	0.5236	**0.5327**	
CCA [13]	0.2215	0.2655	0.2435	Pascal
CDFE [9]	0.2211	0.2928	0.2569	
BLM [14]	0.2408	0.2667	0.2538	
GMLDA [14]	0.2448	0.3094	0.2771	
GMMFA [14]	0.2308	0.3090	0.2699	
CCA3V [2]	0.2562	0.3146	0.2854	
LCFS [18]	0.2674	0.3438	0.3056	
JFSSL [17]	0.2801	**0.3607**	0.3204	
GIN (ours)	**0.4515**	0.3170	**0.3842**	
CM [10]	0.4500	0.4600	0.4550	TVGraz
SM [10]	0.5850	0.6190	0.6020	
SCM [13]	0.6960	0.6930	0.6945	
TCM [11]	0.7060	0.6940	0.6950	
GIN (ours)	**0.7196**	**0.8188**	**0.7692**	

been ignored by other methods that only word frequency or context information. Such inter-word relations are enhanced and more semantically relevant words are activated with the successive layers of graph convolutions, resulting in better generalization ability for un-seen text data.

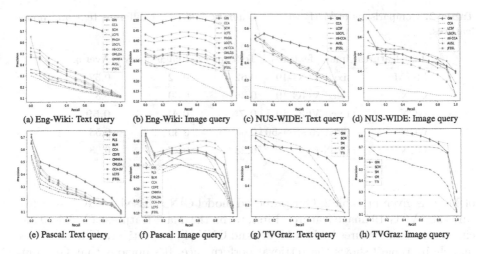

(a) Eng-Wiki: Text query   (b) Eng-Wiki: Image query   (c) NUS-WIDE: Text query   (d) NUS-WIDE: Image query

(e) Pascal: Text query   (f) Pascal: Image query   (g) TVGraz: Text query   (h) TVGraz: Image query

Fig. 3. Precision-recall curves on the four datasets.

Second, the MAP score of GIN for the image-query-text task is superior to most of the compared methods. GIN ranks the second best on Eng-Wiki and NUS-WIDE, the third best on Pascal and the best on TVGraz and Chi-Wiki. Since GIN uses off-the-shelf feature vectors for image view, it's normal that the performance is comparable with state-of-the-art results. Different from the observations on other datasets, the improvement for image-query-text is greater than that for text-query-image. The reason is that, for the image view, the compared algorithms represent images by bag-of-features with SIFT descriptors while we utilize 4096-dimensional CNN features, which are proved to be much more powerful than the hand-crafted feature descriptors. GIN achieves the best average MAP over all the competitors, especially outperforming the second best method JFSSL by 17.13% on Eng-Wiki.

The precision-recall (PR) curves of image-query-text and text-query-image are plotted in Fig. 3. For JFSSL, we show its best MAP after feature selection (see Table 7 in [17]). Since JFSSL hasn't reported the PR curves corresponding to the best MAP, we use its reported PR curves in [17]. For the text-query-image task, it's obvious that GIN achieves the highest precision than the compared methods with almost all the recall rate on the four benchmark datasets. For the image-query-text task, GIN outperforms other competitors with almost all the recall rate on Eng-Wiki. For NUS-WIDE dataset, GIN is only inferior to AUSL and LGCFL. For Pascal dataset, GIN is just slightly inferior to JFSSL. On the whole, GIN is comparable with state-of-the-art methods for the image-query-text task.

(2) Comparison with Baseline Models. Besides our proposed model, we implement another four baseline models to evaluate the influence of the variation in text features and image features on the retrieval performance. All the experiments are conducted on the Eng-Wiki dataset. The retrieval performance

Table 2. Comparisons of MAP with baseline methods w.r.t different text and image features.

Text features	Image features	Text query	Image query	Average
LSTM	fixed VGG-19	0.62	0.42	0.52
CNN	fixed VGG-19	0.36	0.30	0.33
GCN	fixed VGG-19	**0.75**	**0.43**	**0.59**
GCN	fixed ResNet-50	0.66	0.39	0.53
GCN	CNN-5	0.28	0.27	0.28

of MAP is given in Table 2. Our proposed model GIN is based on GCN text features and VGG-19 image features. First, we fix the image features of VGG-19 and change the text features by LSTM [21] and CNN [4], respectively. The first three models in Table 2 shows the retrieval performance. It's obvious that GIN outperforms other models especially for the text retrieval task, which indicates the power of GCN in semantic representation of texts. The MAP of LSTM is inferior to GCN while CNN performs the worst. Then we fix the text features of GCN and change the image features by ResNet-50 and CNN with five convolution layers (CNN-5), respectively. Particularly, CNN-5 is trained end-to-end with our proposed model. We obtain the same conclusion that GIN performs the best. The model using ResNet-50 is slightly worse than using VGG-19. CNN-5 performs the worst because that shallow convolutional networks are detrimental to high-level image feature representation. What's more, the training process of GIN is 5 times faster than CNN+VGG-19 and 8 times faster than LSTM+VGG-19.

4.4 Parameters Analysis

We conduct several experiments on the Eng-Wiki datasets to explore how parameters, i.e. m and λ in the loss function, affect the cross-modal retrieval performance. In Table 3, we range the value of m from 0.4 to 0.6 and range λ from 0.25

Table 3. Experiments on the influence of the parameters m and λ.

m	λ	Text Query	Image Query	Average
0.40	0.35	0.553	0.384	0.469
0.50	0.35	0.622	0.463	0.543
0.60	0.35	**0.808**	0.460	**0.634**
0.70	0.35	0.643	**0.473**	0.558
0.80	0.35	0.606	0.448	0.527
0.60	0.25	0.788	0.441	0.615
0.60	0.30	0.795	0.450	0.623
0.60	0.40	0.791	0.452	0.621

to 0.4 and show the model's MAP scores. From the results we can see that the model is not much sensitive to λ in the range of 0.25 to 0.4. On the contrary, the range of m has obvious impact on the final cross-modal retrieval performance. The average MAP scores range from 0.47 to 0.63 when varying the value of λ. In general, 0.35 for λ and 0.6 for m are the relative best settings for our model.

5 Conclusion

In this paper, we propose a novel cross-modal retrieval model named GIN that takes both irregular graph-structured textual representations and regular vector-structured visual representaions into consideration to jointly learn coupled feature and common latent semantic space. A dual path neural network with graph convolutional networks and layers of nonlinearities is trained using a pairwise similarity loss function. Extensive experiments on five benchmark datasets demonstrate that our model considerably outperform the state-of-the-art models. Besides, our model can be widely used in analyzing heterogeneous data lying on irregular or non-Euclidean domains.

Acknowledgments. This work is supported by the National Key Research and Development Program (Grant No. 2017YFC0820700) and the Fundamental Theory and Cutting Edge Technology Research Program of Institute of Information Engineering, CAS (Grant No. Y7Z0351101).

References

1. Defferrard, M., Bresson, X., Vandergheynst, P.: Convolutional neural networks on graphs with fast localized spectral filtering. In: NIPS, pp. 3837–3845 (2016)
2. Gong, Y., Ke, Q., Isard, M., Lazebnik, S.: A multi-view embedding space for internet images, tags, and their semantics. TPAMI **106**(2), 210–233 (2014)
3. Kang, C., Xiang, S., Liao, S., Xu, C., Pan, C.: Learning consistent feature representation for cross-modal multimedia retrieval. TMM **17**(3), 276–288 (2017)
4. Kim, Y.: Convolutional neural networks for sentence classification (2014). arXiv preprint arXiv:1408.5882
5. Kipf, T.N., Welling, M.: Semi-supervised classification with graph convolutional networks. In: ICLR (2017)
6. Ktena, S.I., Parisot, S., Ferrante, E., Rajchl, M., Lee, M., Glocker, B., Rueckert, D.: Distance metric learning using graph convolutional networks: Application to functional brain networks (2017). arXiv. 1703.02161
7. Kumar, B.G.V., Carneiro, G., Reid, I.: Learning local image descriptors with deep siamese and triplet convolutional networks by minimizing global loss functions. In: CVPR, pp. 5385–5394 (2016)
8. Lecun, Y., Bottou, L., Bengio, Y., Haffner, P.: Gradient-based learning applied to document recognition. IEEE **86**(11), 2278–2324 (1998)
9. Lin, D., Tang, X.: Inter-modality face recognition. In: ECCV, pp. 13–26 (2006)
10. Pereira, J.C., Coviello, E., Doyle, G., Rasiwasia, N., Lanckriet, G.R., Levy, R., Vasconcelos, N.: On the role of correlation and abstraction in cross-modal multimedia retrieval. TPAMI **36**(3), 521–535 (2014)

11. Qin, Z., Yu, J., Cong, Y., Wan, T.: Topic correlation model for cross-modal multimedia information retrieval. Pattern Anal. Appl. **19**(4), 1007–1022 (2016)
12. Ranjan, V., Rasiwasia, N., Jawahar, C.V.: Multi-label cross-modal retrieval. In: ICCV, pp. 4094–4102 (2015)
13. Rasiwasia, N., et al.: A new approach to cross-modal multimedia retrieval. In: ACM-MM, pp. 251–260 (2010)
14. Sharma, A., Kumar, A., Daume, H., Jacobs, D.W.: Generalized multiview analysis: a discriminative latent space. In: CVPR, pp. 2160–2167 (2012)
15. Simonyan, K., Zisserman, A.: Very deep convolutional networks for large-scale image recognition. In: ICLR (2015)
16. Wang, C., Song, Y., Li, H., Zhang, M., Han, J.: Text classification with heterogeneous information network kernels. In: AAAI, pp. 2130–2136 (2016)
17. Wang, K., He, R., Wang, L., Wang, W., Tan, T.: Joint feature selection and subspace learning for cross-modal retrieval. TPAMI **38**(10), 2010–2023 (2016)
18. Wang, K., He, R., Wang, W., Wang, L., Tan, T.: Learning coupled feature spaces for cross-modal matching. In: ICCV, pp. 2088–2095 (2013)
19. Wang, L., Li, Y., Lazebnik, S.: Learning deep structure-preserving image-text embeddings. In: CVPR, pp. 5005–5013 (2016)
20. Yu, J., Cong, Y., Qin, Z., Wan, T.: Cross-modal topic correlations for multimedia retrieval. In: ICPR, pp. 246–249 (2012)
21. Zaremba, W., Sutskever, I., Vinyals, O.: Recurrent neural network regularization (2014). arXiv preprint arXiv:1409.2329
22. Zhang, L., Ma, B., He, J., Li, G., Huang, Q., Tian, Q.: Adaptively unified semi-supervised learning for cross-modal retrieval. In: IJCAI, pp. 3406–3412 (2017)

Smoothness Assisted Interactive Face Annotation via Neural Network

Jin Sun, Hao He, Hengli Luo, and Liyan Zhang[✉]

College of Computer Science and Technology,
Nanjing University of Aeronautics and Astronautics,
Nanjing, People's Republic of China
gabbysunjin@gmail.com, {hehao,herryluo}@nuaa.edu.cn,
zhangly84@126.com

Abstract. Face annotation aiming to tag faces with identities, is an essential tool for image retrieval and management of character-centered photo albums. Conventional face annotation systems have a demand for fully labeled training data, which is hard to get in that manual annotation is a tedious and high-cost work. The aim of our work is to reduce as much as possible manual workload in face annotation. Toward that end, we proposed a smoothness-based model for semi-automatic face annotation, which first applies smoothness constraint and context information such as co-occurrence relationship(*e.g.* two faces extracted from one photo must have different identities) to a neural network and then determines a candidate list of faces that need to be annotated manually on the basis of active learning strategies. Experimental evaluations on two real photo albums show the effectiveness of our proposed model.

Keywords: Face annotation · Smoothness constraint · Neural network

1 Introduction

Recently, owing to the popularization of mobile phones and digital cameras, people tend to share their photos or videos on social networking services such as *Instagram* and *YouTube*, which causes the generation of tremendous image data. With the explosion of massive image data, image retrieval and management become the desirable needs. Face annotation that tags faces with identities becomes an essential tool in that most photos depict human, especially on social networks.

A straightforward idea for face annotation is to train a classifier and then annotate faces according to the prediction of classifier. However, when faced with the deficiency of labeled training data, the system suffers from the regressive performance. Thus some researches attempt to utilize active learning strategies, which can exploit users' feedback. However, manual annotation is a high-cost and tedious work. Due to the bigger number of unlabeled image data, some

© Springer Nature Switzerland AG 2018
R. Hong et al. (Eds.): PCM 2018, LNCS 11164, pp. 235–245, 2018.
https://doi.org/10.1007/978-3-030-00776-8_22

researches utilize clustering algorithms. However, once the clusters are determined the remaining work becomes all manual and if there are some mistakes in clusters it needs users to correct it.

Faced with these problems, we propose a semi-automatic face annotation system. We first train a classifier with a small amount of labeled faces, which doesn't achieve desirable performance due to the lack of training data. Then, we exploit information of unlabeled images by utilizing clustering algorithm to measure the compactness among images which we call smoothness constraint and add these constraints to our classifier. Finally, according to the classifier, we ask for users' feedback on the basis of active learning strategies. This model outperforms other models in that it can exploit unsupervised information in the case of the deficiency of labeled training data.

The rest of this paper is organized as follows. The related researches in face annotation are listed in Sect. 2. The detail description of smoothness constraint is presented in Sect. 3. Section 4 contains the architecture of our framework and the process of active annotation. The performance of our proposed model is shown in Sect. 5. The conclusion of this paper and future work are presented in Sect. 6.

2   Related Work

With the rapid development of research in computer vision, recent years have witnessed the importance of labeled images as training data. Besides, face annotation is a helpful tool for the management of photo albums. Thus face annotation that aims to tag person with identities has gained more attention and lots of achievements have been made in the last two decades.

These researches can be roughly divided into two main categories, classification based and clustering based. Besides, there are some strategies applied to the face annotation systems. For an instance, some researches apply active learning strategies which can ask for users' feedback to their models.

Classification based methods [16][13] use labeled faces as training dataset to train a classifier that can be used to predict the posterior distribution of unlabeled facial images. Kumar *et al.* [6] proposed a two-stage framework to correctly detect and recognize faces in an image collection by using labeled faces as training dataset.

Recently, with deep learning is extensively studied in computer vision, some literature tend to use neural networks in face annotation systems. Schroff *et al.* [9] proposed a unified system which is based on learning a Euclidean embedding using a deep convolutional network for a face recognition task.

The main advantage of classification based methods is that it can predict the posterior distribution of unlabeled facial images. However, face recognition is still a challenging topic in that it relies on the fully labeled training data and most algorithms perform well under a controlled environment due to the lighting conditions and large pose variations.

Clustering based methods [11][8][20][18], first use clustering algorithms to divide facial images into different clusters and then annotate faces with their cluster information. Zhang *et al.* [20] proposed a framework to cluster facial images, and then annotate unlabeled images by cluster labels. Adams *et al.* [3] leveraged density-based clustering to consolidate annotations by crowd workers. The advantage of these methods is that they can exploit some useful information from unlabeled images. However, once the clusters are determined the remaining work becomes all manual and if there are some mistakes in clusters it need users to correct it, which is a big burden on users. Besides, it can only obtain the cluster information of unlabeled facial images without detail posterior distribution information.

Recently, active learning that can address the problem of the deficiency of labeled data is widely used in the field of image annotation. Many query strategies such as uncertainty and information gain are proposed in the research of active learning. Due to the conventional face annotation systems need fully labeled image data, many literature would like to add active learning in their frameworks. Ye *et al.* [15] introduced an active annotation and learning framework for the face recognition task. Tian *et al.* [11] proposed a interactive face annotation framework combining unsupervised and interactive learning.

Faced with these problems, we propose a semi-automatic face annotation framework, which utilizes clustering information and users' feedback based on a neural network.

3   Smoothness Constraint

Smoothness is used to measure compactness among samples. Due to the large variation of faces in rotation angle, two faces from one person may have smaller similarity. However, if there are many other similar images between these two faces, they are inclined to have the same identity. As shown in Fig. 1, all data can be divided into three clusters. Although the distance between P_1 and P_2 is smaller than P_1 and P_3, the density of the path between P_1 and P_3 is higher, namely, the path between P_1 and P_3 is more smooth. So, P_1 and P_3 have the same label. Integrate smoothness in our classifier can effectively take this situation into account. Besides, we can exploit some context information.

3.1   Context Information

Given a photo album P that contains k images $\{ x_1, x_2, ..., x_k \}$. n facial images $\{f_1, f_2, ..., f_n\}$ are detected in P album. $f_i^{x_k}$ indicates that face f_i is extracted from image x_k. The set of facial images F can be divided into two subsets, labeled set F^l and unlabeled set F^u, where $l + u = n$.

Co-occurrence: We can extract faces regions by using SeetaFace [1]. If two faces extracted from one image, i.e $(f_i^{x_k} = f_j^{x_k}) \land (f_i \neq f_j)$, we know that these two faces must have different identities, then we can add co-occurrence constraint $co(f_i, f_j) = -1$ between these two faces, which indicates that these two faces must have different identities. Otherwise, $co(f_i, f_j) = 0$ indicates that we are not sure the faces having the same identity.

Fig. 1. An example of smoothness

Appearance Information: According to [19], given the detected faces we can extract clothes region. Besides, we can take advantage of images' meta data such as picture taken-time. In a short period of time, people with different identity have a great possibility wearing different clothes. [14] provided a appearance feature extraction method. Hence, we can get appearance similarity $s^a(f_i, f_j)$. Besides, we implement time decay factor to control the effect of appearance in identifying people [17]. Consequently, appearance similarity can be formulated as $s^{ap}(f_i, f_j) = s^a(f_i, f_j) \times e^{-|t_i - t_j|/2\triangle t^2}$, where $\triangle t$ denotes the threshold of the time interval, t_i denotes the picture taken-time.

3.2 Smoothness Calculation

Based on feature and context information of faces, we can calculate similarity between two faces. According to similarity matrix, we can divide facial images into different clusters and then calculate smoothness of each cluster. This process has three stages.

Stage 1 (Compute similarity). For each face f_i, $h(f_i)$ is its representation in feature space. s^h indicates the similarity between two facial images in feature space.

$$s^h(f_i, f_j) = exp(-\mu \parallel h(f_i) - h(f_j) \parallel^2) \tag{1}$$

The similarity between two faces can be formulated as:

$$s(f_i, f_j) = \begin{cases} -1 & co(f_i, f_j) = -1 \\ P(\Omega_I | s^h, s^{ap}) & co(f_i, f_j) = 0 \end{cases} \tag{2}$$

Here, Ω_I denotes that face f_i and f_j have the same identity. s^{ap} indicates the appearance similarity between two facial images which can be calculated in last section.

According to Bayesian rules, the equal (2) can be rewritten as:

$$P(\Omega_I | s^h, s^{ap}) = \frac{P(s^h | \Omega_I) \times P(\Omega_I) \times P(s^{ap} | \Omega_I)}{P(s^h) \times P(s^{ap} | s^h)} \tag{3}$$

In equal (3), s^{ap} and s^h are independent for each other, so, $P(s^{ap}|s^h) = P(s^{ap})$. Hence, equal (3) can be rewritten as:

$$P(\Omega_I|s^h, s^{ap}) = \frac{P(s^h|\Omega_I) \times P(\Omega_I) \times P(s^{ap}|\Omega_I)}{P(s^h) \times P(s^{ap})} \tag{4}$$

In equal(4):

$$P(s^h) = P(s^h|\Omega_I) \times P(\Omega_I) + P(s^h|\Omega_E) \times P(\Omega_E) \tag{5}$$

$$P(s^{ap}) = P(s^{ap}|\Omega_I) \times P(\Omega_I) + P(s^{ap}|\Omega_E) \times P(\Omega_E) \tag{6}$$

In equals (5)(6), Ω_E denotes that two faces f_i and f_j are from different people. $P(s^h|\Omega_I)$, $P(\Omega_I)$, $P(s^{ap}|\Omega_I)$, $P(s^h|\Omega_E)$, $P(s^{ap}|\Omega_E)$, $P(\Omega_E)$ can be calculated using training set [19]. By calculating the similarity among facial images, we can get a similarity matrix $S^* \in \mathbb{R}^{n \times n}$, where $S_{ij}^* = s(f_i, f_j)$.

Stage 2 (Clustering). Different from conventional clustering process, our method constructs graph among faces(adds edges between faces) to analyze the density of each cluster. The procedure of constructing graph among faces is shown in Algorithm 1. The threshold is e. If the weight between two faces is higher than e, we add a edge between these two faces. Here, the edge weight is proportional to similarity between these two faces. Therefore, through the process we can divide facial images into different clusters.

Algorithm 1. *Construct graph*

Input: S^*, $f_i, i \in 1, 2, 3..., n$
Output: Constructed graph among facial images.
1 start index k;
2 initial queue $Q = k$;
3 initial graph $g = \emptyset$;
4 threshold sim ε;
5 **while** Q *is not empty* **do**
6   pop q in Q;
7   insert q into g.vertex;
8   **for** $j < n$ *and* $q! = j$ **do**
9   **if** $S_{qj}^* < \varepsilon$ **then**
10   insert j into g.vertex;
11   insert $edge(q, j)$ into g.cdgco;
12   **else**
13   continue;

14 **return** g;

Stage 3 (Compute smoothness). For each graph, we can calculate smoothness according to the labels of samples and edge weight between samples. The smoothness of graph g can be formulated as follows:

$$S_g = \sum_i^E e_i(y^{e_i^1} - y^{e_i^2})^2 \tag{7}$$

where E indicates the number of edges in graph g, e_i indicates the i-th edge weight, $y^{e_i^1}$ and $y^{e_i^2}$ indicates the labels of the two vertex of the edge e_i.

Based on the ideal faces clustering, samples in each cluster should have the same identity, so the smoothness of each cluster is smaller in that the value of $(y^{e_i^1} - y^{e_i^2})$ is close to 0.

4   Interactive Face Annotation

Based on the smoothness constraint that mentioned in the last section, we can apply this constraint to our classifier. Next, based on the model, we determine a candidate list of faces for manual annotation according to active learning strategies.

4.1   Framework

Figure 2 shows the framework of the proposed face annotation method. The architecture of the framework can be roughly divided into three parts.

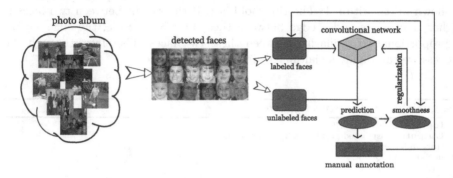

Fig. 2. Architecture of the proposed face annotation framework

Firstly, we establish the convolutional neural network which is used as a face classifier. Based on this classifier, we can train the labeled images and predict the posterior distribution of unlabeled images, which we can use in the following two parts.

Secondly, we cluster all faces and then calculate the smoothness among all facial images, which we can use to retrain the convolutional neural network. The loss function can be rewritten as follows:

$$L = \sum_{i}^{l} \iota(y^{f_i}, y^*) + \sum_{g} \theta S_g \tag{8}$$

Where $\iota(.)$ indicates the loss of labeled training data. y^{f_i} represents the prediction of face f_i and y^* represents the true label of f_i, S_g is the smoothness of each graph that we gain in the last section and θ is regularization coefficient.

Finally, according to the posterior distribution of unlabeled facial images, we can determine a candidate list of faces that need to be annotated manually.

4.2   Active Annotation

There are two main standard selection strategies in active learning [7][12], uncertainty and information gain.

uncertainty:

$$f^* = argmax(-\sum_{i\in classes} P_i log P_i) \tag{9}$$

where P_i indicates the probability of the unlabeled facial image belonging to class i.

information gain:

$$f^* = argmax\, UN(f) * (\frac{1}{U} * \sum_{u=1}^{U} sim(f, f_u)^\beta) \tag{10}$$

where $UN(f)$ indicates the uncertainty of every sample, which can be estimated by equal (9). U represents the number of unlabeled images. $sim(f_i, f_j)$ denotes the similarity between two images. β control the importance of data density.

The uncertainty strategy aims to choose the sample with most uncertainty, whereas the information gain strategy tends to implement exploration on feature correlation among samples. The information gain strategy is much more expensive in that we need to compute the similarity between every two image samples. In this work, we would like to choose the uncertainty strategy as the primary active learning criterion.

5   Experiments

In this section, the proposed method are evaluated on two photo albums. Before we carry out experiments using our proposed approach, we need to preprocess images. The preprocessing pipeline is: (1)detect faces in images (2) face alignment (3) extract features.

An open source C++ face recognition engine *seetaface*[1] provides three modules for face detection, alignment and feature extraction. The identification rate on LFW [2] dataset is up to 97.1%. In this paper, we use *SeetaFace* recognition engine to preprocess images.

5.1   Datasets

Experiments are carried out on two family photo albums. As shown in Table 1, The descriptions of these two datasets are shown as follows. The Fig. 3 denotes the distribution of faces in these two datasets.

Table 1. Description of dataset

Dataset	CMU	Wedding
Images	591	530
Faces	1064	1433
Identities	37	31

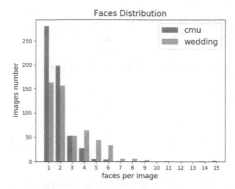

Fig. 3. Faces distribution of two datasets

CMU dataset[4]: is a public family photo album containing 591 images, 1064 faces and 37 identities, including five family members and their friends.

Wedding: is a dataset downloaded from Web Picasa. It is a photo album captured in a wedding ceremony, containing 530 images, 1433 faces and 31 identities including a couple, their relatives and friends.

5.2   Clustering Evaluation

The evaluation consists of two aspects, the accuracy of edges added between faces and the proportion of faces added into graphs.

Figure 4 shows the clustering results of two datasets. As shown in Fig. 4(a), the clustering accuracy rises as the threshold rises. On the contrary, the proportion declines, as shown in Fig. 4(b).

The clustering accuracy has direct impact on the accuracy of the classifier. On the other hand, the number of faces added into graphs decide the amount of information that we can exploit. In order to balance accuracy and proportion, we choose 0.72 as the threshold when constructing graphs.

5.3   Performance

On each dataset, we carry out five contrast experiments to show the performance of our proposed framework.

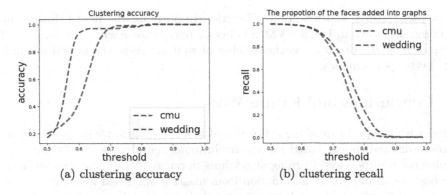

(a) clustering accuracy   (b) clustering recall

Fig. 4. Clustering evaluation on two datasets

We utilize two convolutional neural networks, VGG [10] that has 16 layers and ResNet [5] that has 50 layers. We apply smoothness on these two neural networks and compare the results with the model that smoothness is not applied to. Besides, we utilize a conventional image recognition model, SVM [12].

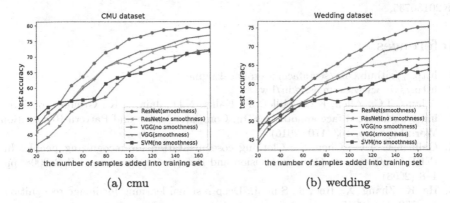

(a) cmu   (b) wedding

Fig. 5. Classification performance on two datasets

Figure 5 shows the results of different models on two datatsets. The results indicate that the convolutional networks that smoothness is applied to have better performance. The classification accuracy of two networks, VGG and ResNet without smoothness on CMU dataset are about 70% and 75%. The accuracy of our proposed model are up to 77% and 80%. The classification accuracy of VGG and ResNet without smoothness on Wedding dataset are 65% and 67%. Comparatively, the accuracy of our proposed method are up to 70% and 75%.

Besides, we compare the performance of neural networks with a conventional recognition model, SVM. The SVM model needs to extract the features of faces in advance. Different from SVM, our model uses a new convolutional network to extract features and modify loss function to optimize the model during the training. As shown in Fig. 5, SVM gets better performance when the number of samples is small. However, as the number of samples rises, the neural network gets better performance.

6   Conclusions and Future Work

Faced with the problem of the deficiency of labeled images, we propose a novel semi-automated model based on a convolutional neural network. Besides, we implement unsupervised learning algorithms in our model so that we can make the best use of important information from massive unlabeled image data. The experiment results on several datasets demonstrate that our model has better performance. However, we implement the proposed model in family photo album, which has smaller number of images compared to social networks. From a scalability perspective, with the need of the image management on social networking services, the idea can be used under these application scenarios.

Acknowledgment. This work was supported in part by the National Natural Science Foundation of China under Grant 61572252, Grant 61772268 and Grant 61720106006, and in part by the Natural Science Foundation of Jiangsu Province under Grant BK20150755.

References

1. https://github.com/seetaface/SeetaFaceEngine
2. http://vis-www.cs.umass.edu/lfw/
3. Adams, J.C., Allen, K.C., Miller, T., Kalka, N.D., Jain, A.K.: Grouper: optimizing crowdsourced face annotations. In: Computer Vision and Pattern Recognition Workshops, pp. 163–170 (2016)
4. Gallagher, A.C., Chen, T.: Clothing cosegmentation for recognizing people. In: IEEE Conference on Computer Vision and Pattern Recognition, CVPR 2008, pp. 1–8 (2008)
5. He, K., Zhang, X., Ren, S., Sun, J.: Deep residual learning for image recognition. pp. 770–778 (2015)
6. Kumar, V., Namboodiri, A., Jawahar, C.V.: Semi-supervised annotation of faces in image collection. In: Signal Image & Video Processing, pp. 1–9 (2017)
7. Mackay, D.J.C.: Information-based objective functions for active data selection. Neural Comput. 4(4), 590–604 (1992)
8. Otto, C., Wang, D., Jain, A.: Clustering millions of faces by identity. IEEE Trans. Pattern Anal. Mach. Intell. 40(2), 289–303 (2017)
9. Schroff, F., Kalenichenko, D., Philbin, J.: Facenet: a unified embedding for face recognition and clustering. In: Proceedings of the IEEE conference on computer vision and pattern recognition, pp. 815–823 (2015)
10. Simonyan, K., Zisserman, A.: Very deep convolutional networks for large-scale image recognition. Computer Science (2014)
11. Tian, Y., Liu, W., Xiao, R., Wen, F., Tang, X.: A face annotation framework with partial clustering and interactive labeling. In: IEEE Conference on Computer Vision and Pattern Recognition, CVPR 2007, pp. 1–8 (2007)

12. Tong, S., Koller, D.: Support vector machine active learning with applications to text classification. J. Mach. Learn. Res. **2**(1), 45–66 (2002)
13. Wang, D., Hoi, S.C., He, Y., Zhu, J., Mei, T., Luo, J.: Retrieval-based face annotation by weak label regularized local coordinate coding. IEEE Trans. Pattern Anal. Mach. Intell. **36**(3), 550–563 (2014)
14. Song, Y., Leung, T.: Context-aided human recognition – clustering. In: Leonardis, A., Bischof, H., Pinz, A. (eds.) ECCV 2006. LNCS, vol. 3953, pp. 382–395. Springer, Heidelberg (2006). https://doi.org/10.1007/11744078_30
15. Ye, H., et al.: Face recognition via active annotation and learning. In: ACM on Multimedia Conference, pp. 1058–1062 (2016)
16. Zhang, L., Chen, L., Li, M., Zhang, H.: Automated annotation of human faces in family albums. In: Eleventh ACM International Conference on Multimedia, pp. 355–358 (2003)
17. Zhang, L., Kalashnikov, D.V., Mehrotra, S.: A unified framework for context assisted face clustering. In: ACM Conference on International Conference on Multimedia Retrieval, pp. 9–16 (2013)
18. Zhang, L., Wang, X., Kalashnikov, D.V., Mehrotra, S., Ramanan, D.: Query-driven approach to face clustering and tagging. IEEE Trans. Image Process. **25**(10), 4504–4513 (2016)
19. Zhang, W., Zhang, T., Tretter, D.: Beyond face: improving person clustering in consumer photos by exploring contextual information. In: IEEE International Conference on Multimedia and Expo, pp. 1540–1545 (2010)
20. Zhang, Y., Tang, Z., Zhang, C., Liu, J., Lu, H.: Automatic face annotation in TV series by video/script alignment. Neurocomputing **152**, 316–321 (2015)

An End-to-End Real-Time 3D System for Integral Photography Display

Shenghao Zhang, Zhenyu Wang, Mingtong Zhu, and Ronggang Wang[✉]

School of Electronic and Computer Engineering,
Shenzhen Graduate School, Peking University, Shenzhen, China
1601214034@sz.pku.edu.cn, imailming@gmail.com,
{wangzhenyu,rgwang}@pkusz.edu.cn

Abstract. Integrated photography display is a commonly used technology to provide users 3D videos with horizontal and vertical parallaxes. However, 3D video contents under natural scenes are very scarce and many existing methods can only provide parallaxes in one direction using binocular cameras. In this work, we set up an end-to-end real-time virtual view synthesis system for integrated photography display, in which 64 views (4 real views and 60 virtual views) can be synthesized using only 4 cameras placed in a rectangle. This paper mainly makes two contributions. Firstly, we propose a new method of polar alignment for two-dimensional camera placement (2 × 2), which aligns horizontal and vertical polar lines at the same time and preserves the effective area of the final image as large as possible after clipping. Secondly, in the process of virtual view synthesis, we introduce an accelerated strategy by combining CPU thread-level parallelism and GPU data-level parallelism to make full use of the computing resources. Experimental results show that our system can capture 4-way 480P video sequences under natural scenes, synthesize 64-way views and render the resulting 4K video combined from 64-way views on a 3D monitor in real-time.

Keywords: Image-Domain-Warping (IDW)
Graphics Processing Unit (GPU) · Polar alignment · Real-time
Virtual view synthesis

1 Introduction and Related Work

Integrated photography display is a true 3D display technology that uses microlens arrays to reproduce 3D space scenes and shows users the 3D effect by providing both horizontal and vertical parallaxes at the same time. However, most of the existing 3D videos are post-synthesized, and the videos under natural scenes are very scarce. The conventional way of generating a 3D video requires N cameras to shoot the desired N views directly, which has several limitations, such as the unacceptable computational overhead of data compression and transmission [1]. A better solution is to acquire M viewpoints through

© Springer Nature Switzerland AG 2018
R. Hong et al. (Eds.): PCM 2018, LNCS 11164, pp. 246–256, 2018.
https://doi.org/10.1007/978-3-030-00776-8_23

M cameras ($M < N$) and synthesize the remaining $N - M$ virtual views by means of a key technology called virtual view synthesis [2].

For virtual view synthesis, one of the commonly used traditional algorithms is depth-image-based rendering (DIBR) [3]. However, building the required high-quality dense depth maps is difficult and dealing with the annoying holes caused by occlusion is also a great challenge. In [4], Image Domain Warping (IDW) algorithm was first proposed for Stereoscopic 3D (S3D) content processing. Farre et al. [5] established an IDW-based system that automatically generates high-quality virtual views. To reduce the computational complexity of IDW, Yao et al. [6] built a real-time multi-view conversion system with the aid of GPU platform. Experimental results showed that IDW can automatically estimate sparse disparity information and synthesize higher-quality virtual viewpoint images through a non-linear transformation in the image spaces.

Most of the related methods, including our previous work [7], can only produce videos with 3D effect in one direction on a binocular camera. Instead, we use 4 cameras and place them in a rectangle (2×2) to get a quad camera platform. The horizontal and vertical camera pairs in the quad camera can be used to generate the horizontal and vertical parallaxes respectively. Considering the speed and accuracy of feature search, the polar lines in both directions are aligned in a robust and efficient way. On the other hand, to generate 3D videos under natural scenes in real-time, an IDW-based end-to-end 3D system is designed for the quad camera through some acceleration strategies at the algorithm and platform levels.

The rest of the paper is structured as follows. In Sect. 2, we describe the proposed two-dimensional polar alignment algorithm, followed by the implementation and detailed acceleration strategies of the virtual view synthesis algorithm based on IDW. Experimental results tested in real scenes are shown in Sect. 3 and Sect. 4 concludes the paper.

2   Our Approach

The proposed end-to-end real-time 3D system consists of two components: (1) two-dimensional polar alignment; (2) virtual view synthesis algorithm and acceleration. The details are illustrated as follows.

2.1   Two-Dimensional Polar Alignment

Polar alignment is a commonly used optimization strategy for global search, which reduces the search range for feature matching while improving the accuracy. When the cameras are placed in two-dimension, as shown in Fig. 1, camera #1 should align with camera #2 horizontally and camera #3 vertically. Our proposed algorithm cannot only achieve two-dimensional polar alignment but also make the effective area of the image after clipping as large as possible. Next, we take the polar alignment between camera #1, #2, and #3 as an example.

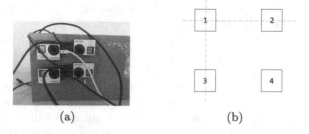

(a)   (b)

Fig. 1. The quad camera and polar alignment model. 1, 2, 3 and 4 correspond to four cameras and dotted lines indicate polar lines.

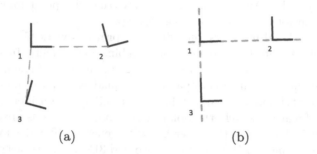

(a)   (b)

Fig. 2. Polar alignment process. The coordinate system represents a camera and the origin of the coordinate system represents the optical center of a camera.

Figure 2 (a) shows the original poses of the three cameras. It can be seen that each camera's orientation varies, the two dotted lines are not vertical, and the three optical centers are not actually coplanar.

Figure 2 (b) represents the polar alignment result. Experiments show that the distance moved by the optical center is the most important factor affecting the size of a final clipped image. In theory, regardless of other operations, the two-dimensional polar alignment can be achieved by moving only one camera's optical center. Our algorithm minimizes the total movement distance of the optical centers of three cameras to preserve the effective area of an image. The final conversion results between the three cameras we need are:

$$\begin{pmatrix} R_1 & T_1 \\ 0 & 1 \end{pmatrix} * P_1 = \begin{pmatrix} I & d_x \\ 0 & 1 \end{pmatrix} * \begin{pmatrix} R_2 & T_2 \\ 0 & 1 \end{pmatrix} * P_2, \tag{1}$$

$$\begin{pmatrix} R_1 & T_1 \\ 0 & 1 \end{pmatrix} * P_1 = \begin{pmatrix} I & d_y \\ 0 & 1 \end{pmatrix} * \begin{pmatrix} R_3 & T_3 \\ 0 & 1 \end{pmatrix} * P_3, \tag{2}$$

where P_i is a 3D point in the i-th camera coordinate system, R_i and T_i are the rotation and translation information needed by the i-th camera when aligning the polar lines, and vector d_x: $(d\ 0\ 0)'$ and d_y: $(0\ d\ 0)'$ represent the spacing between the optical centers of i-th and j-th cameras on the horizontal or vertical axis.

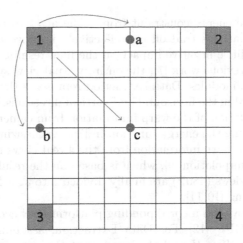

Fig. 3. Virtual views placement and warp relationship. a and b represent the middle position in the horizontal and vertical directions, and c is the midpoint of the rectangle. The blue boxes represent four real views and the white boxes represent 60 virtual views. (Color figure online)

To obtain R_i and T_i above, the first step is fixing the optical center of camera #1 (set T_1 to a zero vector). We start by assuming that T_2 and T_3 are all zero matrices, then R_1 can be obtained through *svd* decomposition of Eq. 3, where R_{ij} and T_{ij} are the rotation and translation information from i-th camera to j-th camera respectively, and they can be obtained using VisualSFM in [8]. Based on R_1, all the variables except T_1 in Eqs. 1 and 2 can be derived.

$$R_1 * \begin{pmatrix} T_{12} \\ T_{13} \end{pmatrix} = \begin{pmatrix} T_2 \\ T_3 \end{pmatrix} + \begin{pmatrix} d_x \\ d_y \end{pmatrix}. \tag{3}$$

Next step is solving T_1. The moving distances of optical centers of camera #1, #2, and #3 are denoted as x, y, and z, respectively, and the relationships of x, y, and z can be represented as $y - x = T_2$ and $z - x = T_3$. Then we get:

$$y^T * y + z^T * z + x^T * x = (x + T_2)^T * (x + T_2) + (x + T_3)^T * (x + T_3) + x^T * x. \tag{4}$$

By finding the partial derivative of x in Eq. 4 and setting it to 0, we can get the final T_1, T_2, and T_3 in Eqs. 1 and 2.

Through the algorithm above, the two-dimensional polar alignment can be achieved and an effective image after clipping is as large as possible.

2.2 Virtual Views Synthesis Algorithm and Acceleration

With polar aligned frame images, we design an IDW-based algorithm to synthesize virtual views based on the main ideas of [7], where a binocular camera becomes a quad camera, and 8 horizontal views become $8 \times 8 = 64$ views (60 virtual views). As shown in Fig. 3, due to a large amount of data to be processed, we only consider the energy constraint from camera #1 to the three red dots. The results show that this tradeoff can effectively improve the computational efficiency and have little negative impact on the final results.

Similar to our previous work [7], the entire virtual view synthesis algorithm is divided into three modules. Data Extraction represents the feature matching (disparity extraction) in the horizontal and vertical directions. Calculation Warp represents the calculation of the warp information from camera #1 to the three red dots in Fig. 3 using the energy function in [7]. Synthesizing View represents the calculation of the warp information from the 4 real views to other 60 virtual views by bilinear interpolation [9], which is based on the results of module Calculation Warp. All views (480P) are finally divided into $32 \times 24 = 768$ triangles for affine transforming [10,11].

Each module above has a corresponding platform-level acceleration strategy using OpenCL on GPU [12]. For Data Extraction, assuming the resolutions of input images are $W \times H$ and the number of the extracted features is N, the FAST feature detection [13] needs $W \times H$ work-items in GPU, and each work-item corresponds to one pixel. Similarly, the calculation of the BRIEF descriptor [14] needs N work-items and the feature matching needs H work-items in GPU. For Calculating Warp, successive over-relaxing (SOR) is an efficient method, which is accelerated using the thread-level parallelism. For Synthesizing View, the calculation of the affine transformation between each mesh triangles is independent, so that we can allocate one work-item corresponding to a mesh.

However, for our scenes, there are two other factors that seriously affect the efficiency of the real-time system: the frame preprocessing of the quad camera and the huge amount of data to be processed when synthesizing the 60 virtual views. We also put forward two targeted strategies as follows.

The first strategy is the parallel frame preprocessing of the quad camera. After time-consuming testing, the sequential frame capture including polar alignment of the four cameras takes almost 40 ms, which is unacceptable. Figure 4 shows our acceleration strategy, in which four separate threads are used to capture frames from four cameras, align the polar lines (the polar alignment rules are obtained offline) and add the final polar aligned four frames to one list. Simultaneously, our core algorithm constantly fetches frames from the list to synthesize virtual views. The CPU-based multi-thread parallelism above can reduce the frame preprocessing time to less than 20 ms and the frame capture operations of the four cameras are almost synchronous, which means normal movement of the object is allowed.

The second strategy is the virtual view synthesis based on the CPU + GPU platform. Experiments show that even if one GPU is used, Synthesizing View module still occupies almost 60% of the execution time in the core algorithm. Because the GTX-980 GPU we use only has 32768 work-items, the total required number of work-items for this module equals to the number of the triangle meshes in 60 virtual views, which is $32 * 24 * 60 = 46080$. Therefore, we adopt a CPU+GPU parallelism scheme [15] and use two GPUs to synthesize virtual views. From Fig. 5, two separate threads are created for virtual view synthesis.

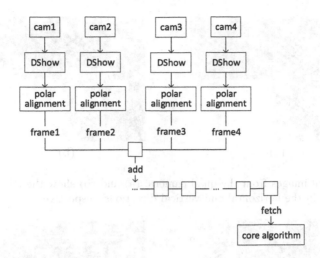

Fig. 4. CPU-based multi-thread parallelism for frames capture.

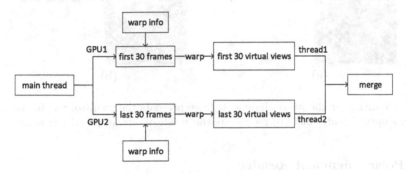

Fig. 5. CPU+2GPUs parallel parallelism.

GPU1 in thread1 is used to synthesize the first 30 virtual views and GPU2 in thread2 is used to synthesize the last 30 virtual views.

3 Experimental Result

Based on the two-dimensional polar alignment algorithm and the accelerated virtual view synthesis algorithm, we set up an end-to-end 3D system for integrated photography display. The system uses four GoPro Hero4 black cameras as input devices, synthesizes 64 views with a server containing one CPU (Intel Core I7 7800x) and two GPUs (GTX 980), and renders a 4K image combined from the 64 views on a 3D display in real-time. This section describes the experimental results in three parts, which are the polar alignment of the quad camera, the parallax effect of the 64-way virtual view images, and the speed of the accelerated 3D system. Since there is no standard dataset for the quad camera, we mainly use the data collected in real scenes in our experiments.

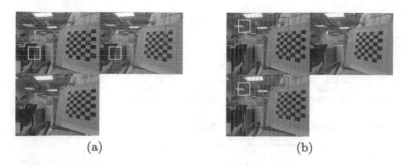

(a)   (b)

Fig. 6. Original images from the quad camera. (a) and (b) show the misalignment of the polar lines in the horizontal and vertical directions, respectively.

(a)   (b)

Fig. 7. Results after the polar alignment. (a) represents the result of moving only one camera's optical center and (b) represents the result of our proposed algorithm.

3.1 Polar Alignment Result

We test the polar alignment algorithm in terms of the accuracy of alignment and the size of the effective area after clipping. As can be seen from Fig. 6, the original three images' polar lines are misaligned in both directions. Figure 7 shows that the horizontal and vertical polar lines are aligned at the same time with our proposed algorithm, and more important, the size of the effective area after clipping is near the original size of the image compared with the scheme moving only one camera's optical center.

3.2 Disparity Information for 64-Way Views

The disparity test of 64-way views consists of two parts. The first is that the deformation of the virtual view cannot be too large, and the second is that the final synthesized 64 views should have parallaxes in both the horizontal and vertical directions. Figure 8 shows the final 64-way views synthesized by our algorithm. It can be seen that our method yields no holes and scarcely visible artifacts (e.g., boundary effects, halos). Taking the top row views and the left column views as examples. As can be seen from Figs. 9 and 10, the horizontal and vertical parallaxes exist at the same time.

Fig. 8. The image merged with 64-way views. The red boxes indicate the four views from the quad camera and in order to avoid parallax jumping, the box in the lower right corner can also be a virtual view. The positions of 8 virtual views in the horizontal direction are 0.0, 0.14, 0.28, 0.42, 0.57, 0.72, 0.86 and 1.0 (the interval in the vertical direction is the same). (Color figure online)

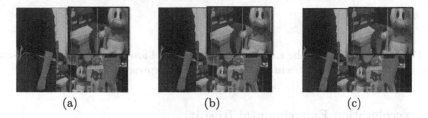

(a)   (b)   (c)

Fig. 9. Output views in positions 0.28, 0.57, 0.86 (left to right), corresponding to (a), (b) and (c), respectively. The green line is located exactly 310 pixels from the left boundary of an image, and the red line is located at the right edge of the person. The distance between the two lines increases from left to right, as seen in close-ups, and indicates the change in parallax. (Color figure online)

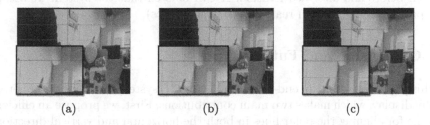

(a)   (b)   (c)

Fig. 10. Output views in positions 0.28, 0.57, 0.86 (top to bottom), corresponding to (a), (b) and (c), respectively. The green line is located exactly 180 pixels from the top boundary of an image, and the red line is located at the top edge of the door. The distance between the two lines decreases from left to right, as seen in close-ups, and indicates the change in parallax. (Color figure online)

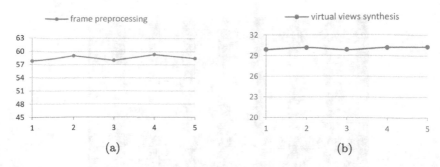

Fig. 11. The frame rate of frame preprocessing and synthesis of 60 virtual views. The x-axis represents the number of experiments and the y-axis represents the frame rate.

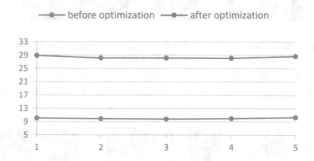

Fig. 12. The frame rate of the entire system before and after acceleration. The x-axis represents the number of experiments and the y-axis represents the frame rate.

3.3   Acceleration Experimental Result

The input of the end-to-end real-time 3D system is four 480P video sequences from the quad camera and the output is a 4K video sequence combined from the 64 views. Figure 11 shows the frame rate of frame preprocessing and synthesis of 60 virtual views after acceleration. Figure 12 shows the final speed of the entire system before and after acceleration. It can be seen that the system we set up can produce a 3D video in real-time (beyond 28 fps).

4   Conclusion and Future Work

In this paper, we set up an end-to-end real-time 3D system for integrated photography display, which makes two main contributions. First, we propose an efficient scheme for aligning the polar lines in both the horizontal and vertical directions at the same time and preserving the effective area of the final image as large as possible after clipping. Second, an accelerated optimization strategy combining CPU thread-level parallelism and GPU data-level parallelism (two GPUs) is introduced to synthesize 60 virtual views. Extensive experimental results show that the proposed system for integrated photography display can generate 3D videos under natural scenes with both horizontal and vertical parallaxes in real-time. For future work, we suggest that expanding the reference views for the warp into four cameras can improve the final 3D effect.

Acknowledgement. Thanks to National Natural Science Foundation of China 61672063 Shenzhen Peacock Plan, Shenzhen Research Projects of JCYJ2016050617222 7337 and GGFW2017041215130858, Shenzhen Key Laboratory for Intelligent Multimedia and Virtual Reality ZDSY201703031405467.

References

1. Smolic, A., Kauff, P., Knorr, S., Hornung, A., Kunter, M., Muller, M., Lang, M.: Three-dimensional video postproduction and processing. Proc. IEEE **99**(4), 607–625 (2011)
2. Smolic, A., Muller, K., Dix, K., Merkle, P., Kauff, P., Wiegand, T.: Intermediate view interpolation based on multiview video plus depth for advanced 3D video systems. In: 15th IEEE International Conference on Image Processing (ICIP). IEEE, San Diego, pp. 2448–2451 (2008)
3. Fehn, C.: Depth-image-based rendering (DIBR), compression, and transmission for a new approach on 3D-TV. In: Stereoscopic Displays and Virtual Reality Systems XI, San Jose, pp. 93–105, May 2004
4. Lang, M., Hornung, A., Wang, O., Poulakos, S., Smolic, A., Gross, M.: Nonlinear disparity mapping for stereoscopic 3D. In: ACM, Los Angeles, vol. 29, No. 4, p. 75, July 2010
5. Farre, M., Wang, O., Lang, M., Stefanoski, N., Hornung, A., Smolic, A.: Automatic content creation for multiview autostereoscopic displays using image domain warping. In: ICME, pp. 1–6, IEEE. Barcelona, July 2011
6. Yao, S.J., Wang, L.H., Lin, C.L., Zhang, M.: Real-time stereo to multi-view conversion system based on adaptive meshing. J. Real-Time Image Process. **14**(2), 481–499 (2018)
7. Wang, R., Luo, J., Jiang, X., Wang, Z., Wang, W., Li, G., Gao, W.: Accelerating Image-Domain-Warping Virtual View Synthesis on GPGPU. IEEE Trans. Multimed. **19**(6), 1392–1400 (2017)
8. Zheng, E., Wu, C.: Structure from motion using structure-less resection. In: Proceedings of the IEEE International Conference on Computer Vision, pp. 2075–2083. IEEE, Santiago, Chile (2015)
9. Scarpino, M.: OpenCL in Action: How to Accelerate Graphics and Computation. Manning Publication, Shelter Island (2011)
10. Chaurasia, G., Sorkine, O., Drettakis, G.: Silhouette-aware warping for image-based rendering. In: Computer Graphics Forum, vol. 30, No. 4, pp. 1223–1232, June 2011
11. Liu, F., Gleicher, M., Jin, H., Agarwala, A.: Content-preserving warps for 3D video stabilization. In: ACM Transactions on Graphics (TOG), Vol. 28, No. 3, p. 44, New Orleans, Louisiana July, 2009
12. Fang, J., Varbanescu, A. L., Sips, H.: A comprehensive performance comparison of CUDA and OpenCL. In: International Conference Parallel Processing (ICPP), pp. 216–225, Taipei City, Taiwan, September, 2011

13. Rosten, E., Drummond, T.: Machine learning for high-speed corner detection. In: Leonardis, A., Bischof, H., Pinz, A. (eds.) ECCV 2006. LNCS, vol. 3951, pp. 430–443. Springer, Heidelberg (2006). https://doi.org/10.1007/11744023_34
14. Calonder, M., Lepetit, V., Ozuysal, M., Trzcinski, T., Strecha, C., Fua, P.: BRIEF: computing a local binary descriptor very fast. IEEE Trans. Pattern Anal. Mach. Intell. 34(7), 1281–1298 (2012)
15. Momcilovic, S., Ilic, A., Roma, N., Sousa, L.: Dynamic load balancing for real-time video encoding on heterogeneous CPU + GPU systems. IEEE Trans. Multimed. 16(1), 108–121 (2014)

Deep Discriminative Quantization Hashing for Image Retrieval

Jingbo Fan, Chuanchuan Chen, and Yuesheng Zhu[✉]

Communication and Information Security Lab, Shenzhen Graduate School,
Peking University, Shenzhen, China
{fjb,chenchuanchuan}@pku.edu.cn, zhuys@pkusz.edu.cn

Abstract. In this paper, we present an efficient deep supervised hashing method to learn robust hash codes for content-based image retrieval on large-scale datasets. Deep hashing methods have achieved some good results in image retrieval by training the network with classification loss and constructing hash functions as a latent layer. However, the classification loss does not impose a sufficient constraint on the network to make sure that similar images can be encoded to similar binary codes. As a supplement to classification loss, a new loss is delicately designed in our method. After trained with the joint objective functions, the network can generate more discriminative hash codes, which will increase the performance of retrieval. Our method outperforms the state-of-the-art methods by an obvious margin on three datasets CIFAR-10, CIFAR-100 and MNIST. Especially, the improvement is more impressive when the code length is short and the category number is large.

Keywords: Image retrieval · Deep hashing
Discriminative quantization hashing

1 Introduction

With the explosive growth of image data in recent years, content-based image retrieval (CBIR) has become a more and more challenging field. A commonly used retrieval method is nearest-neighbor search. To find relevant images from datasets, we need to generate semantically based features and rank the feature distance between the query and datasets. However, the searching time cost is typically high for the large-scale image datasets. So the hashing methods based on approximate nearest neighbor are widely used in CBIR [1,2,16]. The target of these methods is to map similar semantic images to similar binary codes, which means that we can calculate Hamming distance with reduced time and memory costs.

Hashing methods can be divided into two strategies: data-independent hashing and data-dependent hashing. A representative data-independent hashing method Locality Sensitive Hashing (LSH) [1] was presented in early years, which

© Springer Nature Switzerland AG 2018
R. Hong et al. (Eds.): PCM 2018, LNCS 11164, pp. 257–266, 2018.
https://doi.org/10.1007/978-3-030-00776-8_24

aims to map similar images to similar binary codes through random projections. Compared with data-independent hashing, data-dependent methods can reduce the length of binary codes and achieve better retrieval performance. Data-dependent methods consist of unsupervised hashing and supervised hashing. Unsupervised methods learn binary codes without supervised information. Spectral hashing (SH) [17], Iterative Quantization (ITQ) [2] and Discrete Graph Hashing (DGH) [11] are representative unsupervised hashing methods. Supervised hashing takes advantage of label information to learn the hash function, such as Binary Reconstructive Embedding (BRE) [7], Iterative Quantization with Canonical Correlation Analysis (CCA-ITQ) [2] and Supervised discrete hashing (SDH) [14].

With the rapid development of convolutional neural networks (CNN) [6], deep hashing methods based on CNN are proposed to learn more efficient binary representations, i.e. CNNH [18], DNNH [8], DRSCH [20], DHN [21], DPSH [10]. A typical point-wise deep hashing method supervised semantics-preserving deep hashing (SSDH) [19] trains the network with classification loss and two additional constraints to generate hash code by quantizing the output of hidden layer. Though achieving state-of-the-art results, SSDH learns dispersedly distributed features. So after quantization, the set of binary codes from the same category is large and similar binary codes are not always projected from similar images, which will be harmful to the retrieval performance.

To solve the problem, we need to learn more discriminative features from the network. In this paper, a new deep hashing method Deep Discriminative Quantization Hashing (DDQH) is proposed. To make up the deficiency of feature constraints from classification loss, we design a discriminative quantization loss. For input images, our new loss aims to increase the cosine similarity between the learned features and the feature center of the corresponding category. Trained with the joint classification loss and discriminative quantization loss, the distributions of the features are well optimized: the features from the same category are closer, and those from different categories are farther. So the quantized hash codes are more discriminative.

We conduct experiments on CIFAR-10 [5], CIFAR-100 [5] and MNIST datasets respectively. With our method on CIFAR-10 dataset, mAP and mean precision of Hamming radius 2 for the code length 48 are 94.28% and 92.20%, which are 2.94% and 7.09% higher than those in [19]. When dealing with datasets with more categories (i.e., CIFAR-100) or generating shorter hash codes (i.e., 6-bit), DDQH improves the margin more, verifying the effectiveness of our method.

2   Our Approach

2.1   Overall Framework

We have two goals to achieve: (i) to conveniently learn real-valued features and quantize the features to binary codes in a point-wise neural network, and (ii) to specifically design the network to make it more adaptable to hash retrieval, i.e., the hash codes of the same category are projected to similar Hamming spaces.

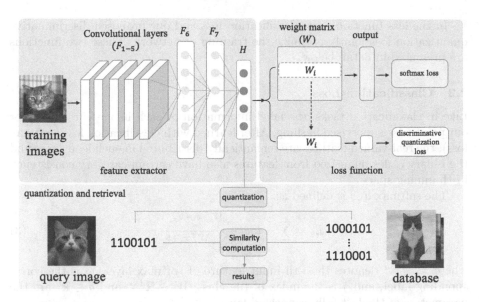

Fig. 1. Overview of our deep discriminative quantization hashing (DDQH) method.

Our proposed DDQH method satisfies the two goals simultaneously with high image retrieval performance.

As shown in Fig. 1, our method can be divided into three parts: feature extractor, loss function, quantization and retrieval. In the feature extractor part, we utilize convolutional neural network (CNN) to extract features. Our method is capable of fitting all the CNN architecture proposed in recent years (AlexNet [6], VGGNet [15], ResNet [3]). Without loss of generality, we choose AlexNet as our basic network. AlexNet includes five convolutional layers (F_{1-5}), two fully connected layers (F_{6-7}) and the output layer. To generate binary codes with multiple lengths, a hidden layer H with the nodes number of corresponding length N is added before the output layer. Let a_n be the F_{1-7} output of the n-th image. W^H and b^H are the weight and bias of the hidden layer, and h_n is the corresponding output of binary codes. The activation function of hidden layer is the hyperbolic tangent ($tanh$) function, which is defined as:

$$\varphi(x) = \frac{e^x - e^{-x}}{e^x + e^{-x}} \tag{1}$$

In the quantization and retrieval part, we quantize the output of hidden layer and then do the retrieval based on Hamming distance between query and database images. The binary quantization function can be written as:

$$h_n = \delta(\varphi((W^H)^T a_n + b^H) \geq 0) \tag{2}$$

where $\delta(condition) = 1$ when $condition$ is satisfied, and $\delta(condition) = 0$ if not.

In the loss function part, classification loss and our proposed discriminative quantization loss jointly optimize the training of network. These two functions are presented in detail below.

2.2   Classification Loss

Like in classification tasks, the label information of each image is requisite for our point-wise supervised hashing. After learned with classification loss (softmax loss), the real-valued features are semantically distributed in euclidean space and the binary codes quantized from features also have certain category consistency in Hamming space.

The softmax loss is defined as:

$$L_s = \sum_i -log(\frac{e^{W_{y_i}^T x_i + b_{y_i}}}{\sum_j e^{W_j^T x_i + b_j}}) \tag{3}$$

where $x_i \in \mathbb{R}^d$ denotes the i-th input feature of softmax layer, y_i is the corresponding label, and j is the index of the class. $W \in \mathbb{R}^{d \times c}$ and $b \in \mathbb{R}^c$ are the parameters of the last fully connected layer.

However, the softmax loss is not sufficient for deep hashing. To illustrate it, we make a toy example and visualize the distribution of features. In the example, we choose LeNet [9] as our network structure and the first 4 classes of MNIST as our training dataset. Also, we edit the nodes number of the last latent layer to 2 for visualization. Since a zero-centered activation function is used in our framework, every quadrant represents different Hamming subspace, and features located in each quadrant are quantized to different hash codes. In Fig. 2(a), with only softmax loss, while the decision boundary for classification is clear, the feature points are diffused and a few points of the same category are distributed in different quadrant, which will cause the performance of hashing retrieval down. In other words, such bad feature distribution for retrieval cannot be optimized by softmax loss since the classification accuracy is unaffected.

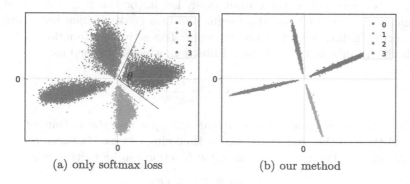

(a) only softmax loss   (b) our method

Fig. 2. The 2-d visualization of the feature distribution. In (a), only softmax loss is used in network. In (b), our novelly designed discriminative quantization loss is added together with softmax loss.

2.3 Discriminative Quantization Loss

To make the features more discriminative, we add a discriminative quantization loss in the last layer. Like in Fig. 2(a), we need to keep the features of the same class lie in the same quadrant. So decreasing the angle θ is a good choice. We can minimize the following objective function to make θ small:

$$-\frac{c_{y_i}^T x_i}{\|c_{y_i}^T\| \|x_i\|} \tag{4}$$

The $c_{y_i} \in \mathbb{R}^d$ denotes the feature center of the y_i-th class, which we should make the y_i-th features approach. To avoid the influence of the scale, we normalize the features and the feature center.

However, to calculate the feature center c, we should take all features of the same category into consideration and average them, which is inefficient and will cost a lot of time and memory. So we choose the weight matrix W of the last fully connected layer as the approximate feature center matrix, and each column W_j represents the feature center vector of the j-th class. To keep the efficiency of W in classification, W is not updated by the discriminative quantization loss and is only changed by the softmax loss in the backpropagation. In other words, by optimizing the discriminative quantization loss, only the parameters of layer 1 to layer H (i.e. $W^{F_{1-7},H}$) are updated.

The overall loss function can be written as:

$$L = L_s + \lambda L_d = \sum_i (-log(\frac{e^{W_{y_i}^T x_i + b_{y_i}}}{\sum_j e^{W_j^T x_i + b_j}}) - \lambda \frac{W_{y_i}^T x_i}{\|W_{y_i}^T\| \|x_i\|}) \tag{5}$$

where L_d is the discriminative quantization loss and the hyper parameter λ is used for balancing the two losses.

3 Experiments

In this section, we evaluate our method on three public datasets, i.e., CIFAR-10, CIFAR-100 and MNIST.

3.1 Datasets

CIFAR-10 [5] consists of 60,000 color images, which belong to ten classes. Each class contains 6,000 images. There are 50,000 training images and 10,000 test images, which all have the size of 32×32.

CIFAR-100 [5] contains 60,000 color images from 100 categories (600 images per category). The 100 classes in the CIFAR-100 are grouped into 20 superclasses, and each superclass is divided to 5 subclasses. The size is the same as CIFAR-10.

MNIST contains 70,000 grayscale images, which are split up to 60,000 training images and 10,000 testing images. The dataset includes 10 classes of handwriting digits from 0 to 9. The size of image is 28 × 28.

3.2   Evaluation Protocols

Following [19], we adopt three widely used evaluation protocols: mean Average Precision (mAP) for different code lengths, precision at k samples for 48 bits and mean precision within Hamming radius 2 for different code lengths.

For the three datasets, we use the officially provided train/test split for our training. Following the settings in [8,18], we sample 100 random images from each class in test set to form a query set with 1000 (10,000 for CIFAR-100) images, and use it for performance evaluation.

3.3   Implementation Details

Our method is implemented using PyTorch [13]. Following the settings of [19], all of our input images are resized to 256 × 256 and then cropped to the size of 227 × 227 randomly, which is the input size of AlexNet. Our network is trained by SGD with 0.9 momentum for 128 epochs. The learning rate is 0.001 and is reduced by a factor of ten after 64 epochs. The weight decay parameter is 0.0005.

3.4   Experiments on the Parameter λ

The hyper parameter λ influences the weight of our discriminative quantization loss in the overall objective function Eq. (5). The retrieval results will degrade when the hyper parameter λ is not appropriate. So we conduct an experiment to validate the effect of λ on results.

Without loss of generality, we choose CIFAR-10 as our training set and 48 as our bit length. A reasonable range 0 to 0.3 is exploited to learn different models. The model results of mean precision within Hamming radius 2 are shown in Fig. 3.

As shown in Fig. 3, when $\lambda = 0.1$, we get the best performance. Thus, we set the λ value as 0.1 in the following experiments.

3.5   Comparisons of Activation Functions

It is worth noting that we can choose different activation functions in our hidden layer. In this section, we pick four activation functions (tanh, sigmoid, ReLU and identity function) and conduct a comparison between our method and the method with only softmax loss. To demonstrate the influence from activation function better, we compare the performance between two tasks: classification and retrieval.

As shown in Fig. 4, our method outperforms the contrast one in both classification and retrieval tasks, and we get the best performance when choosing

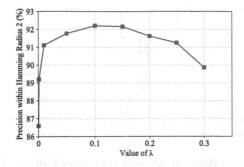

Fig. 3. The performance of mean precisions within Hamming radius 2 on CIFAR-10 with different value of λ

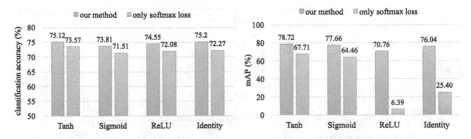

Fig. 4. The classification and retrieval performance of mAP on CIFAR-100 with 48 code length when using different activation functions

the tanh activation function. Interestingly, when using ReLU and identity function, the retrieval mAP with only softmax loss is 6.39% and 25.40%, while it improves to around 70% when using DDQH. So our method has enough generalization capability to fit different kinds of activation functions.

3.6 Comparisons of Retrieval Results

We compare DDQH with some hashing approaches, including two state-of-the-art deep hashing methods (SSDH [19], ADSH [4]) and some representative non-deep hashing methods (LSH [1], ITQ [2], CCA-ITQ [2], SDH [14]). We use the source codes provided by authors to implement these methods. For fair comparison, all the deep hashing methods, including ADSH, SSDH and DDQH, use AlexNet with ImageNet pretrained weights. Following the settings in [8], we represent each image with a 512-dimensional GIST [12] vector in non-deep hashing methods.

The comparisons of our methods against the others are shown in Figs. 5, 6, 7, and Table 1. Referring to the results, DDQH provides the best performance for different code lengths. When the code length is 48, the mAP is improved by a margin of 2.94% compared with the state-of-the-arts for CIFAR-10, and the improvement is 5.40% for CIFAR-100. Especially when the code length is short,

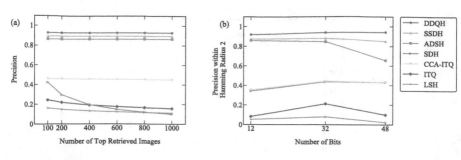

Fig. 5. Comparative evaluation of retrieval performance of our DDQH method and other methods on CIFAR-10. (a) Precision at k samples. (b) Mean precision within Hamming radius 2

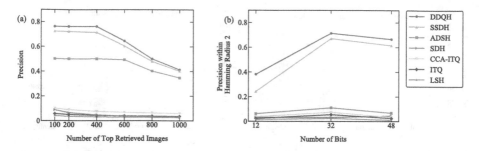

Fig. 6. Comparative evaluation of retrieval performance of our DDQH method and other methods on CIFAR-100. (a) Precision at k samples. (b) Mean precision within Hamming radius 2

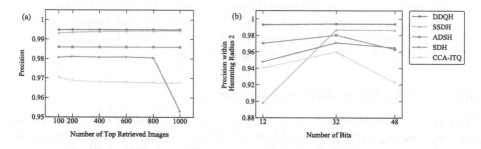

Fig. 7. Comparative evaluation of retrieval performance of our DDQH method and other methods on MNIST. (a) Precision at k samples. (b) Mean precision within Hamming radius 2

i.e., 6, the improvements are larger. For example, the mAP improves 9.76% for MNIST when the code length is 6, which is far more than the improvements of the other length.

The results prove that our loss function is capable to make the quantization more discriminative and the overall method is robust enough to handle hard situations, such as when the code length is short and the category number is large.

Table 1. Comparisons of mAP (%) on CIFAR-10, CIFAR-100 and MNIST with the state-of-the-arts.

Method	CIFAR-10				CIFAR-100				MNIST			
	6-bit	12-bit	32-bit	48-bit	6-bit	12-bit	32-bit	48-bit	6-bit	12-bit	32-bit	48-bit
LSH	3.35	7.42	8.92	15.33	1.19	1.24	1.43	1.66	1.07	5.70	30.67	24.62
ITQ	8.35	16.86	28.15	30.82	1.44	1.68	1.85	2.01	1.14	5.16	29.71	37.12
CCA-ITQ	25.60	29.23	31.94	32.85	1.96	2.61	3.40	3.75	38.39	94.03	95.99	92.26
SDH	25.04	33.38	40.45	35.86	1.33	1.84	3.84	4.93	37.12	94.80	97.11	96.50
ADSH	82.83	91.67	91.13	90.61	5.10	10.60	37.36	56.67	91.89	98.35	99.07	98.91
SSDH	88.24	90.45	91.47	91.34	12.38	57.37	72.02	73.32	88.93	99.09	99.15	99.50
DDQH	**92.20**	**93.31**	**93.93**	**94.28**	**41.07**	**71.21**	**78.08**	**78.72**	**98.69**	**99.29**	**99.51**	**99.54**

4   Conclusion

In this paper, we propose a novel Deep Discriminative Quantization Hashing (DDQH) method. In our method, the discriminative quantization loss together with softmax loss is utilized to learn more discriminative feature representation, which is more appropriate for hashing retrieval. Combined with proper quantization choice, we have achieved the state-of-the-art performance on the CIFAR-10, CIFAR-100 and MNIST datasets. Moreover, our method outperforms the others with larger margin when the category number is large or the code length is short.

Acknowledgment. This work was supported in part by the Shenzhen Municipal Development and Reform Commission (Disciplinary Development Program for Data Science and Intelligent Computing); and by Shenzhen International cooperative research projects GJHZ20170313150021171.

References

1. Andoni, A., Indyk, P.: Near-optimal hashing algorithms for approximate nearest neighbor in high dimensions. In: 2006 47th Annual IEEE Symposium on Foundations of Computer Science (FOCS 2006), pp. 459–468 (2006)
2. Gong, Y., Lazebnik, S.: Iterative quantization: a procrustean approach to learning binary codes. In: CVPR 2011, pp. 817–824 (2011)
3. He, K., Zhang, X., Ren, S., Sun, J.: Deep residual learning for image recognition. In: 2016 IEEE Conference on Computer Vision and Pattern Recognition (CVPR), pp. 770–778 (2016)
4. Jiang, Q.Y., Li, W.J.: Asymmetric deep supervised hashing. In: AAAI Conference on Artificial Intelligence (2018)
5. Krizhevsky, A., Hinton, G.: Learning multiple layers of features from tiny images. Master's thesis, Department of Computer Science, University of Toronto (2009)
6. Krizhevsky, A., Sutskever, I., Hinton, G.E.: ImageNet classification with deep convolutional neural networks. In: Pereira, F., Burges, C.J.C., Bottou, L., Weinberger, K.Q. (eds.) Advances in Neural Information Processing Systems 25, pp. 1097–1105. Curran Associates Inc., New York (2012)

7. Kulis, B., Darrell, T.: Learning to hash with binary reconstructive embeddings. In: Bengio, Y., Schuurmans, D., Lafferty, J.D., Williams, C.K.I., Culotta, A. (eds.) Advances in Neural Information Processing Systems 22, pp. 1042–1050. Curran Associates Inc., New York (2009)
8. Lai, H., Pan, Y., Liu, Y., Yan, S.: Simultaneous feature learning and hash coding with deep neural networks. In: 2015 IEEE Conference on Computer Vision and Pattern Recognition (CVPR), pp. 3270–3278 (2015)
9. Lecun, Y., Bottou, L., Bengio, Y., Haffner, P.: Gradient-based learning applied to document recognition. Proc. IEEE **86**(11), 2278–2324 (1998)
10. Li, W., Wang, S., Kang, W.: Feature learning based deep supervised hashing with pairwise labels. In: Proceedings of the Twenty-Fifth International Joint Conference on Artificial Intelligence, IJCAI 2016, New York, NY, USA, 9–15 July 2016, pp. 1711–1717 (2016)
11. Liu, W., Mu, C., Kumar, S., Chang, S.F.: Discrete graph hashing. In: Ghahramani, Z., Welling, M., Cortes, C., Lawrence, N.D., Weinberger, K.Q. (eds.) Advances in Neural Information Processing Systems 27, pp. 3419–3427. Curran Associates Inc., New York (2014)
12. Oliva, A., Torralba, A.: Modeling the shape of the scene: a holistic representation of the spatial envelope. Int. J. Comput. Vis. **42**(3), 145–175 (2001)
13. Paszke, A., et al.: Automatic differentiation in PyTorch. In: NIPS-W (2017)
14. Shen, F., Shen, C., Liu, W., Shen, H.T.: Supervised discrete hashing. In: 2015 IEEE Conference on Computer Vision and Pattern Recognition (CVPR), pp. 37–45 (2015)
15. Simonyan, K., Zisserman, A.: Very deep convolutional networks for large-scale image recognition. CoRR abs/1409.1556 (2014)
16. Wang, J., Zhang, T., Song, J., Sebe, N., Shen, H.T.: A survey on learning to hash. IEEE Trans. Pattern Anal. Mach. Intell. **40**(4), 769–790 (2018)
17. Weiss, Y., Torralba, A., Fergus, R.: Spectral hashing. In: Koller, D., Schuurmans, D., Bengio, Y., Bottou, L. (eds.) Advances in Neural Information Processing Systems 21, pp. 1753–1760. Curran Associates Inc., New York (2009)
18. Xia, R., Pan, Y., Lai, H., Liu, C., Yan, S.: Supervised hashing for image retrieval via image representation learning. In: AAAI Conference on Artificial Intelligence (2014)
19. Yang, H.F., Lin, K., Chen, C.S.: Supervised learning of semantics-preserving hash via deep convolutional neural networks. IEEE Trans. Pattern Anal. Mach. Intell. **40**(2), 437–451 (2018)
20. Zhang, R., Lin, L., Zhang, R., Zuo, W., Zhang, L.: Bit-scalable deep hashing with regularized similarity learning for image retrieval and person re-identification. IEEE Trans. Image Process. **24**(12), 4766–4779 (2015)
21. Zhu, H., Long, M., Wang, J., Cao, Y.: Deep hashing network for efficient similarity retrieval. In: AAAI Conference on Artificial Intelligence, pp. 2415–2421 (2016)

Automatic 3D Garment Fitting Based on Skeleton Driving

Haozhong Cai[1], Guangyuan Shi[1], Chengying Gao[1(✉)], and Dong Wang[2]

[1] School of Data and Computer Science, Sun Yat-sen University, Guangzhou, China
mcsgcy@mail.sysu.edu.cn
[2] College of Mathematics and Informatics,
South China Agricultural University, Guangzhou, China

Abstract. In this paper, we propose an automated 3D garment fitting method. The proposed method can automatically position a 3D garment model onto a human model without any user manual intervention. Given a garment model and a human model with various postures and shapes, we firstly segment the garment and the human model separately into mutually corresponding parts by our improved segmentation method. Secondly, we extract the skeletons of these parts accurately according to the segmentation. According to the skeletal information, we analyze the posture difference between the human and garment models. Thirdly, we change the posture of the garment model according to the difference of skeletal posture. Fourthly, we position the garment model onto the human model roughly with feature points of skeletons. Lastly, we constrain the garment mesh to resolve the penetration between human and garment models. With the purpose of achieving more realistic dressing result, we use mass-spring system to simulate garment. The experimental results show that the proposed method is stable and efficient.

Keywords: Garment deformation · Mesh segmentation
Skeleton extraction · Mass-spring system

1 Introduction

Fitting 3D garment models onto virtual human models with various postures and shapes is necessary in many fields such as virtual try-on application, film and video games. It is a popular research field in computer graphics and there are many relevant works have been done.

Most existing automatic dressing methods require that garment models should be first fitted on a reference human model. Given a target human model, these methods [4, 9, 11–14] transfer the garment from the reference human model to the target human model according to the corresponding relationship between the two human models. A representative method, which is proposed in [11], firstly, transforms the reference human model into a tetrahedral mesh. Then it deforms the reference human body according to the posture of the target human

© Springer Nature Switzerland AG 2018
R. Hong et al. (Eds.): PCM 2018, LNCS 11164, pp. 267–277, 2018.
https://doi.org/10.1007/978-3-030-00776-8_25

body. Finally it separates the garment from the reference body to complete the deformation of the garment. However, if there is not an available human model, these methods will be unable to complete the task of dressing.

Another type of methods to solve 3D garment positioning problem is garment fitting, which only takes a single human model and a garment model as input. A 3D clothing fitting method based on the geometric feature matching is proposed in [8]. However, it requires users to manually select the feature points and can only apply to garment models that exactly match the human model in posture and size. Li et al. [10] use the skeletal-driving method to change the posture of the garment and then adjust the garment locally to get a more realistic result. However, it is not fully automatic due to the manually skeleton adjusting process. The machine learning based method [7] trains a clothing model for a specific garment by carrying out physical simulation of garment on human bodies with various postures and shapes. The clothing model can be used to fit the garment on any human body. However, for different garment models, the tedious training process needs to be repeated, which is time-consuming. In addition, an automatic 3D garment positioning method based on surface metric is proposed in [17]. It defines a surface energy and manages to minimize the energy by animating the humanoid avatar. However, it is time-consuming to obtain an accurate result. And because of lacking garment structural information, the initial positioning step is not accurate enough, which increases the difficulty to calculate the minimum surfaces energy.

Another novel approach of dressing is proposed in [5]. A set of primitive actions is identified according to the vast majority of motions observed in human dressing. Then human character is trained to put on simulated clothing by himself. However, the time required to train a human character is unsuitable for real-time application.

In this paper, we propose an automatic 3D garment fitting method. Our method has the following advantages: First, it is fully automatic without any user intervention. Second, it does not need a reference human model. Third, it can be applied for human models with various postures and shapes. Fourth, it builds up the correspondence between human models and garment models without predefined information. The paper has the following contributions.

- we propose an efficient and stable method for automatically fitting garment models onto human models.
- we make appropriate improvement to the Reeb Graph based segmentation method and get more accurate segmentation results.
- we propose an novel approach to adjust the posture of garment models automatically based on the structural difference between the human skeleton and the garment skeleton.

2   Automatic 3D Garment Fitting Method

An overview of the proposed automatic 3D garment fitting method is depicted in Fig. 1. First of all, the human model and garment model need to be segmented

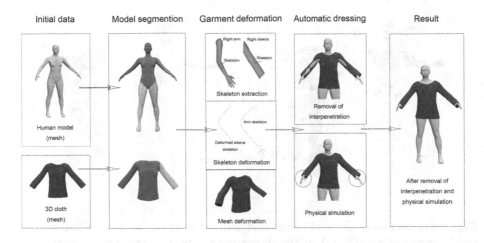

Fig. 1. Pipeline overview. Steps of our automatic 3D garment fitting method are presented here.

with an improved discrete Reeb graph based method. In general, a human model is divided into five parts: the torso, left arm, right arm, left leg and right leg. Then the garment should be deformed properly to fit the human body. After the garment deformation, the garment can be positioned according to the feature points of skeletons extracted. Finally, to generate natural and realistic shape, the penetration between garment model and human model is removed and a physical simulation based on physical force is carried out. In this section, detailed steps of the proposed automatic 3D garment fitting method are presented.

2.1 Model Segmentation

Discrete Reeb Graph Based Method. A discrete Reeb graph based segmentation method, which is quite robust against noise, posture variation and frame change, is proposed in [18]. The method is based on a morse function, namely geodesic distance.

The results of segmenting the human and garment models by adopting the discrete Reeb graph based method [18] are presented in Fig. 2(a). It is obvious that the tops of the upper arm and tight are classified as parts of torso. Though the problem is not critical in some scenarios, it will lead to a distortion deformed garment model in our fitting algorithm, as shown in Fig. 3(b). Therefore, we adjust the region boundaries to promote the accuracy of segmentation and make it more suitable for garment fitting process. For comparison, the deformation result of garment model which is segmented by our improved segmentation method is presented in Fig. 3(c).

Adjustment of Upper Arm and Upper Thigh. A more reasonable segmentation should refer to the location of the human joints as shown in Fig. 2(b).

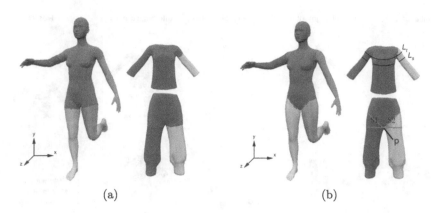

(a)   (b)

Fig. 2. Segmentation results of human and garment models. (a) By the discrete Reeb graph based method [18]. (b) By the improved discrete Reeb graph based method.

(a)   (b)   (c)

Fig. 3. Garment deformation results after segmenting garment. (a) Human model. (b) Segmented by the discrete Reeb graph based method [18]. (c) Segmented by our improved segmentation method.

Therefore, we should adjust the upper arm and upper thigh after segmenting models by the discrete Reeb graph based method [18].

The adjustment of the upper arm is based on the fact that a vertex on the mesh must belong to segment S if it is only adjacent to the vertices in S. For the first level set L_s with three connective sets, as shown in Fig. 2(b), let the connective sets of the left arm, the torso and the right arm be $S_l^{L_s}$, $S_t^{L_s}$, and $S_r^{L_s}$, respectively. Given a level set L_t such that L_s is adjacent to L_t and $f_{ss}(L_t) < f_{ss}(L_s)$, where $f_{ss}(L)$ denotes the Morse function value of level set L. Obviously, there is only one connective set in L_t, as shown in Fig. 2(b). For every vertex $\mathbf{v}$ in L_t, $\mathbf{v}$ belongs to the set of right arm, if it satisfies the following proposition.

$$\exists \mathbf{v}' \in S_r^{L_s}, \mathbf{v} \leftrightarrow \mathbf{v}' \text{ and } \forall \mathbf{v}'' \in (S_t^{L_s} \cup S_l^{L_s}), \mathbf{v}'' \not\leftrightarrow \mathbf{v}, \tag{1}$$

In addition the vertex $\mathbf{v}$ belongs to the set of left arm, if it satisfies the following proposition.

$$\exists \mathbf{v}' \in S_l^{L_s}, \mathbf{v} \leftrightarrow \mathbf{v}' \text{ and } \forall \mathbf{v}'' \in (S_r^{L_s} \cup S_t^{L_s}), \mathbf{v}'' \not\leftrightarrow \mathbf{v}, \tag{2}$$

where $\leftrightarrow$ represents that the vertices are adjacent, while $\nleftrightarrow$ represents that the vertices are not adjacent. Most of the vertices are divided into the correct segments after adjustment, as shown in Fig. 2(b).

For adjusting upper thigh, we define two boundary surfaces $S1$ and $S2$ between the human thighs and the torso, as shown in Fig. 2(b), and classify the vertices below the surfaces as parts of thighs. The intersection point of the left and right boundary surfaces is $P(P_x, P_y, P_z)$, as shown in Fig. 2(b).

S_{leftleg}, S_{rightleg} and S_{torse} denote the vertices sets of the left leg, right leg and torso respectively after applying the method [18], and M denotes the number of vertices of S_{torse}. Therefore, the point P can be calculated as follows:

$$P_x = \frac{1}{M} \sum_{v \in S_{\text{torse}}} v_x, \quad P_y = max(v_y) \text{ for } v \in S_{\text{leftleg}} \cup S_{\text{rightleg}}. \quad (3)$$

Let k_l be the slope of the left straight line, then the surfaces $S1$ and $S2$ can be given as follows:

$$
\begin{aligned}
S1: \quad & -k_l x + y = -k_l P_x + P_y. \\
S2: \quad & k_l x + y = k_l P_x + P_y.
\end{aligned}
\quad (4)
$$

2.2 Garment Deformation

In this section, a method of deforming garment model to fit the human body is proposed and the result of garment deformation is presented in Fig. 4. There are three steps of this method. Firstly, we use mean curvature flow (MCF) to computing curve skeletons of the input human model and garment model [16]. Secondly, the skeleton of garment model is deformed to fit the skeleton of human model. Finally, the garment mesh is deformed according to the information of the deformed garment skeleton.

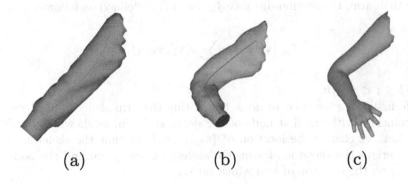

(a)   (b)   (c)

Fig. 4. Deformation of a sleeve. (a) Original sleeve. (b) Sleeve after deformation. (c) Human arm.

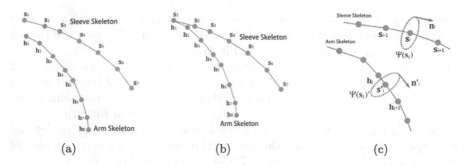

(a)   (b)   (c)

Fig. 5. Abstract representation of mesh skeleton. (a) Arm skeleton and sleeve skeleton. (b) Making first nodes of the two skeletons coincide. (c) Deformation of garment vertices according to garment skeleton nodes.

Skeleton Extraction. The curve-skeleton is a data structure contained points connected by edges in three-dimensional space [3]. It is easier to deal with this structure than the original mesh data. The mean curvature flow (MCF) based method [16], which is adopted, requires the mesh to be watertight. Therefore, the garment mesh and the segment meshes need to be preprocessed to obtain closed meshes, which should be approximated to the original meshes in shape. Finally, we adopt ear clipping method [6] to fill the holes.

Deformation of Garment. Before the deformation of garment mesh, we should adjust the garment skeleton according to the structural difference between the human skeleton and the garment skeleton. Suppose that $S = \{s_1, \ldots, s_n\}$ represents the set of sleeve skeleton nodes, and $H = \{h_1, \ldots, h_m\}$ represents the set of arm skeleton nodes, as shown in Fig. 5(a). Let n, m be the sizes of S and H respectively, and $d(v_1, v_2)$ denotes the Euclidean distance between points v_1 and v_2 in three-dimensional space. On the basis of the hypothesis that sleeve is longer than arm, the section-distance $f_{sec}(v_i, v_j)$ is defined as follows:

$$f_{sec}(v_i, v_j) = \sum_{k=i}^{j-1} d(v_k, v_{k+1}), \tag{5}$$

where $1 \leq i < j \leq m$.

The first thing we need to do is translating the arm skeleton until its first node coincides with the first node of the sleeve skeleton, as shown in Fig. 5(b). After that, we change the location of $\{s_2, \ldots, s_n\}$, so that the skeleton of the sleeve overlaps the skeleton of arm. For skeleton node s_i, we find the nodes h_j and h_{j+1} on the skeleton of arm which satisfy

$$f_{sec}(h_0, h_j) \leq f_{sec}(s_0, s_i) \leq f_{sec}(h_0, h_{j+1}). \tag{6}$$

Now the target position of s_i is located on the line segment of h_j and h_{j+1}, which is denoted by s_i', then s_i' can be linearly interpolated by h_j and h_{j+1}:

$$s_i' = h_j + \frac{f_{sec}(s_0, s_i) - f_{sec}(h_0, h_j)}{d(h_j, h_{j+1})} \cdot (h_{j+1} - h_j). \tag{7}$$

After the deformation of skeleton, the garment mesh should be deformed properly. For a skeleton node s, let $\Psi(s)$ denote the set of mesh vertices corresponding to s. Suppose that the directions of the skeleton nodes at s_i and s_i' are n_i and n_i' respectively. The transformation of $\Psi(s_i)$ includes rotation and translation. Firstly, the set of mesh vertices $\Psi(s_i)$ should be rotated until $n_i = n_i'$. Then a translation should be performed with the translation vector $v_i = s_i' - s_i$. Finally we use Laplacian surface deformation algorithm [15] to obtain global deformation of garment model.

2.3   Automatic Dressing

After deformation of garment model, the garment model and human model have similar posture. On the basis of these models, an automatic dressing system is proposed in this section. The system consists of three steps. Firstly, the garment model is positioned according to the feature points of human and garment skeletons. Then the penetration between garment model and human model is removed. Finally, physical simulation is carried out to generate a natural and realistic shape.

Garment model is positioned according to the feature points presented in Fig. 6. According to the segmentation results in Sect. 2.1 and skeletons extracted in Sect. 2.2, the feature points f_1, f_2, f_3, f_4, c_1, c_2, c_3 and c_4 can be obtained directly. The feature points f_0, f_5, c_0 and c_5 can be calculated as follows:

$$f_0 = (f_1 + f_2)/2, \quad f_5 = (f_3 + f_4)/2,$$
$$c_0 = (c_1 + c_2)/2, \quad c_5 = (c_3 + c_4)/2. \tag{8}$$

After obtaining the feature points f_0, f_5, c_0 and c_5, the translation vectors of a shirt v_{shirt} and a trouser $v_{trouser}$ can be given as follows:

$$v_{shirt} = f_0 - c_0, \quad v_{trouser} = f_5 - c_5. \tag{9}$$

Our penetration removing algorithm is similar to the algorithm in [7]. For a vertex c_i on the garment model, we find a human vertex h_j nearest by c_i. Let n_{h_j} denote the normal of the vertex h_j. We use the projection of $(c_i - h_j)$ onto n_{h_j} to determine whether c_i moves through the human mesh. Then we minimize the following two equations to resolve the penetration of the garment, and make the deformation of garment more natural.

$$\eta(C) = \sum_{(i,j)\in\Omega\wedge i\in P} (-\epsilon + n_{h_j}^{\mathrm{T}}(c_i - h_j)), \tag{10}$$

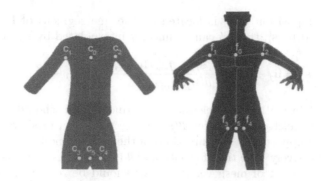

Fig. 6. Skeleton feature points of garment and human models

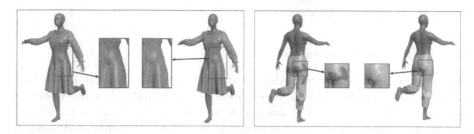

Fig. 7. Penetration handling results.

$$\lambda(C) = \sum_{i \in P} \parallel (\mathbf{c}_i - \hat{\mathbf{c}}_i) - \frac{1}{\mid N_i \mid} \sum_{j \in N_i} (\mathbf{c}_j - \hat{\mathbf{c}}_j) \parallel^2 . \tag{11}$$

where ϵ denotes the distance between the garment and the surface of the human; Ω is a set of correspondences between garment vertices and human vertices; P is a set of clothing vertices penetrated through the human mesh; $\hat{\mathbf{c}}_i$ are the vertices of the garment before adjusting; N_i is the set of vertices adjacent to vertex i. The results after removing the penetration are presented in Fig. 7, which reflect the fact that the penetration method has a great performance.

To generate natural and realistic shape, the deformed garment model is simulated using a mass-spring system based on physical forces, such as gravity, friction force and internal force. The edges of the garment mesh represent the tension-springs, and the rest length of the tension-spring is the initial edge length.

3   Experiments

We have tested our algorithm with human models in different postures, which are derived from MakeHuman [2]. In addition, the different types of garment models we used are derived from 3D modelling software [1]. The application has been developed in C++ and the rendering has been performed by OpenGL.

The results of automatic dressing for different types of garment and human models are presented in Fig. 8. The garment models are deformed reasonably

(a)   (b)

(c)   (d)

Fig. 8. Simulation results of different types of garment and human models. (a) Results of a dress on human models. (b) Results of a pant on human models. (c) Results of a shirt on human models. (d) Results of a shirt and a pant on human models.

and the automatic dressing system also has a good performance. In addition, the final simulation results are realistic and natural for various types of garment and human models. For example, as shown in the right most figure of Fig. 8(a), the dress is fit well though the shins separate.

To measure the efficiency of our method. The calculation time of the main steps is given in Tables 1, 2 and 3. It is obvious that the processes of skeleton extraction, Laplacian mesh deformation and removal of penetration contribute to most of the computational cost. The running time of the algorithm is basically within the acceptable range. With the increase of the number of garment vertices, the running time of the algorithm increases obviously, while the number of human vertices has little impact on the performance. The primary reason lies in the fact that the Laplacian mesh deformation algorithm, which is more time-consuming, only deals with garment models.

Compared with other automatic 3D garment fitting methods, our method is more efficient. For example, given a garment model with nearly six thousand vertices, the 3D garment fitting method based on surface metric [17] takes about 25 s to fit the garment, while our method only takes about 13 s.

Table 1. Computation time of our proposed automatic 3D garment fitting method

Garment	Human	Garment vertices	Human vertices	Model segmentation (ms)	Garment deformation (ms)	Automatic dressing (ms)	Total (ms)
Dress	Human 1	6134	8797	1510	6048	5371	12929
Shirt 1	Human 1	3106	8797	911	3207	661	4779
Shirt 1	Human 3	3106	13380	1648	3864	495	6007
Shirt 2	Human 2	3550	13380	1882	4329	1456	7667
Pant 1	Human 4	4200	13380	2188	6751	2342	11281
Pant 2	Human 5	3062	13380	2582	5019	1182	8783

Note: The calculation time has been computed with a 4 × 2.40 GHz processor.

Table 2. Detailed computation time of model segmentation

Model	Vertices	Computation of level sets (ms)	Classification of level sets (ms)	Coarse segmentation (ms)	Segmentation refinement (ms)	Total (ms)
Dress	6134	43	746	33	4	834
Shirt 1	3106	18	193	11	4	233
Shirt 2	3550	34	301	29	2	374
Pant 1	4200	56	247	45	11	368
Pant 2	3062	43	155	36	7	248
Human 1	8797	55	580	31	3	676
Human 2	13380	96	1337	61	4	1505
Human 3	13380	94	1247	61	4	1412
Human 4	13380	121	1606	76	10	1819
Human 5	13380	137	1980	73	12	2206

Note: The calculation time has been computed with a 4×2.40 GHz processor.

Table 3. Detailed computation time of garment deformation

Garment	Human	Garment vertices	Human vertices	Hole filling (ms)	Skeleton extraction (ms)	Laplacian deformation (ms)	Total (ms)
Dress	Human 1	6134	8797	241	2115	3321	6048
Shirt 1	Human 1	3106	8797	235	1646	1104	3207
Shirt 1	Human 3	3106	13380	694	1903	1090	3864
Shirt 2	Human 2	3550	13380	645	1924	1581	4329
Pant 1	Human 4	4200	13380	134	4806	1733	6751
Pant 2	Human 5	3062	13380	122	3575	1250	5019

Note: The calculation time has been computed with a 4×2.40 GHz processor.

4   Conclusion

In this paper, we present a method of dressing 3D human characters automatically. Given a 3D human model and garment model, the method fits the garment model onto the human body by global garment posture deformation and local garment adjustment. A more natural and authentic fitting result is obtained by physical simulation of garment model. Our method is feasible, stable and efficient, which can be used in a variety of applications such as CAD tools, virtual fitting room, games, and film.

There are still some shortcomings. For example, the Laplacian mesh deformation is relative time-consuming during the fitting process. We believe that alternative deformation algorithms or parallel computing methods can be used to improve the performance.

Acknowledgement. This work was supported, in part, by National Natural Science Foundation of China under Grant 61472455, Guangzhou Science Technology and Innovation Commission (GZSTI16EG14/201704030079).

References

1. 3dsmax (2018). http://www.autodesk.fr/products/3ds-max/overview
2. Makehuman (2018). http://www.makehuman.org
3. Au, O.K.C., Tai, C.L., Chu, H.K., Cohen-Or, D., Lee, T.Y.: Skeleton extraction by mesh contraction. ACM Trans. Graph. **27**(3), 44 (2008)
4. Brouet, R.: Design preserving garment transfer. ACM Trans. Graph. **31**(4), 36 (2012)
5. Clegg, A.: Animating human dressing. ACM Trans. Graph. **34**, 116:1–116:9 (2015)
6. Eberly, D.: Triangulation by ear clipping. Geometric Tools (2008)
7. Guan, P.: DRAPE: DRessing Any Person. ACM Trans. Graph. **31**(4), 35 (2012)
8. Igarashi, T., Hughes, J.F.: Clothing manipulation. ACM Trans. Graph. **22**(3)
9. Lee, Y.: Automatic pose-independent 3D garment fitting. Comput. Graph. **37**(7), 911–922 (2013)
10. Li, J.: Fitting 3D garment models onto individual human models. Comput. Graph. **34**(6), 742–755 (2010)
11. Li, J., Lu, G.: Customizing 3D garments based on volumetric deformation. Comput. Ind. **62**(7), 693–707 (2011)
12. Mingmin, Z.: Topology-independent 3D garment fitting for virtual clothing. Multimed. Tools Appl. **74**(9), 3137–3153 (2015)
13. Narita, F.: Texture preserving garment transfer. In: ACM SIGGRAPH 2015 Posters, p. 91. ACM (2015)
14. Narita, F., Saito, S., Kato, T., Fukusato, T., Morishima, S.: Pose-independent garment transfer. In: SIGGRAPH Asia 2014 Posters, p. 12. ACM (2014)
15. Sorkine, O.: Laplacian surface editing. ACM Int. Conf. Proceeding Ser. **71**, 175–184 (2004)
16. Tagliasacchi, A., Alhashim, I., Olson, M., Zhang, H.: Mean curvature skeletons. In: Computer Graphics Forum, vol. 31, pp. 1735–1744. Wiley Online Library (2012)
17. Tisserand, Y.: Automatic 3D garment positioning based on surface metric. Comput. Animat. Virtual Worlds **28**(3–4), e1770 (2017)
18. Werghi, N.: A functional-based segmentation of human body scans in arbitrary postures. IEEE Trans. Syst. Man Cybern. Part B (Cybern.) **36**(1), 153–165 (2006)

Real-Time RGBD Reconstruction Using Structural Constraint for Indoor AR

Chen Wang[1] and Yue Qi[1,2(✉)]

[1] State Key Laboratory of Virtual Reality Technology and Systems,
Beihang University, Beijing 100191, China
qy@buaa.edu.cn
[2] Qingdao Research Institute of Beihang University, Qingdao 266100, China

Abstract. RGBD-based 3D indoor scene reconstruction has been paid much attention due to the advantage of consumer depth camera. It is significant for many interactive application, especially in augmented reality. At present, the AR system mainly focus on the issue of the instabilities in the registration without any marker (i.e. error accumulation in camera pose estimate). Current methods generally consider isolate point cloud pairwise as the argument in the registration and ignore the prior correlation of geometric structures in the indoor scene. In our work, we focus on the issue and propose a novel, structural-based AR framework. Specifically, we use a two-pass scheme strategy to execute the system. The first pass tracks camera and analyze scene structure timely at video rate. We apply structural constraint to the iterative-closest-point algorithm and generate a new pose optimization strategy. We also incorporate the structure information into the global model integration and improve the reconstruction quality. Comparing with other state-of-the-art online reconstruction methods, our approach significantly reduces pose drift. The second pass simultaneously processing occlusion between virtual objects and real scene with the advent of prior structure analysis to improve the realism in AR.

Keywords: 3D reconstruction · Camera tracking
Structural constraint · Augmented reality

1 Introduction

With the advent of the consumer depth camera (such as Microsoft Kinect, Asus Xtion or Structure sensor), real-time 3D indoor scene reconstruction receives much attention. It is significant for many interactive application, especially in augmented reality. RGBD-based 3D reconstruction method aims to estimate the camera's motion trajectory in real time and generate a complete scene model. At

This work was supported in part by the National Natural Science Foundation of China under Grant 61572054, in part by the Applied Basic Research Program of Qingdao under Grant 16-10-1-3-xx.

R. Hong et al. (Eds.): PCM 2018, LNCS 11164, pp. 278–289, 2018.
https://doi.org/10.1007/978-3-030-00776-8_26

present, various reconstruction methods have acquired compelling improvement in real-time performance [1,2], stability [3,4] and model quality [5–7]. However, the main issue still existing is error accumulation in camera pose estimate, which result in the real-virtual registration instabilities in AR. Recently several online methods [8–10,21] have been proposed to improve the drift-accumulation. Current methods usually consider isolate point cloud pairwise as the argument in the registration and ignore the prior correlation of geometric structures in the indoor scene. By analyzing the scenario structure, the scene usually contains many planar surfaces (e.g., ceils), which are compact representations of dense points and reflect a stable structure associations across frames. In addition, some objects always contain various curved surfaces (e.g., cups), which can represent the details and make it more real.

In our work, we focus on the surface structure as a reliable constrained information. We first classify all the initial point cloud data in the scene, divide it into different plane regions and local curved regions, and parameterize the region. In the meanwhile, analytic scene structure throughout the whole reconstruction. In the registration step, the structural constraint can filter the data association, so that the matching degree can be enhanced and the speed of convergence can be reduced. In the scene construction step, the surface structure can be used to refining the local details, so that we can obtain more real results. In the fusion step, the continuous and common scene planar structure has been used for positioning the virtual object without any marker. Finally, we can form a real-virtual fusion system with geometric reality and generate a global 3d indoor scene model.

Overall, our approach comprises the following contributions:

- A novel real-time 3D reconstruction framework for marker-less AR system, which parameterizing the local surface of the input vertex map and considering different geometric consistency to the online process.
- A structural constraint iterative-closest-point incorporating geometric primitives correspondence relationships, which can better constrain the continuous camera tracking problem.
- A scene construction refined method including two aspects: weighted integration based on point geometrical features, and implicit isosurface extraction by calculating the intersection point of the line and surface.
- A realistic augmented reality application combining the continuous scene planar structure to virtual object positioning without any marker.

2   Related Work

Camera Registration. The registration issue (i.e., camera pose estimation) is essential to simultaneous localization and mapping (SLAM) system. The core problem is integrating the frame to the half-baked scene model accumulated from previous frames. Unknown scenario structure and noise interference make drift occur at any moment. The early research was focused on monocular RGB sensor, using either pose graph optimization [11] or bundle adjustment [12] to minimizing the error of image re-projection and pose graph distribution. Given the depth measurement is inferred from RGB image feature description, the detail of final model is inadequately expressed.

Real-time dense localisation from consecutive depth map make hand-held online scanning possible. For small-scale scenes, classical PTAM [13], DTAM [14] has shown compelling result. It should be noticed that the ICP algorithm play an important role at the tracking stage. Given the RGB image can provide priori constraint, improved ICP framework considering the RGB priori like color-based weighting [15], edge feature-based weighting [16], has been introduced. These methods still do not solve the fundamental problem from the geometric level. In our work, we naturally incorporate structural constraint to the ICP framework, which is necessary to find the correspondences for valid pairwise removing the redundancy in the point cloud.

Geometric Structure Analyze. The structure in indoor scene usually contain a series of commonalities, such as some existing planar regions like walls, floors, ceiling, etc., and some objects with smoothing surface like cup, vase and chair. This prior structure constraint can be exploited to high-quality 3D reconstruction. Structural information has been proved useful in other work [7,17,18] for 3D scanning reconstruction. These method either only developed based on point cloud representations, or used merely for the camera tracking procedure in the current frame. We not only naturally incorporate the structural constraint to the ICP process, but also see it as a global data structure to improve the quality of the model reconstruction. Beyond this simple geometric construction, the high-curvature regions structure that can reflect local details is also widely concerned, [5,19] has proved valuable.

3   Method Overview

A schematic overview of our pipeline is shown in Fig. 1. In the following, we will describe more detail.

3.1   Data Map Pre-processing

In the pre-process stage, we calculate the point properties from the raw image data, following and extending the method proposed by Nießner [2] and Zhang [7]. We denote a pixel position as $u = (x,y)^T \in \mathbb{R}^2$, the corresponding depth map value at frame t is $\mathcal{D}_t(u) \in \mathbb{R}$. Then transform the denoised depth map $\mathcal{D}_t$ into a corresponding vertex map $\mathcal{V}_t$ using the intrinsic camera matrix K, $V_t(u) = \mathcal{D}_t(u)K^{-1}(u_t, 1)^T \in \mathbb{R}^3$. The corresponding normal map $\mathcal{N}_t$ is extracted from central-differences of the vertex map.

3.2   Structure Primitives Segmentation

Based on the vertex map and normal map, the curvature map $\mathcal{R}_t$ can be calculate by [20]. The result can be shown in Fig. 2(d). On some edges and surfaces, the

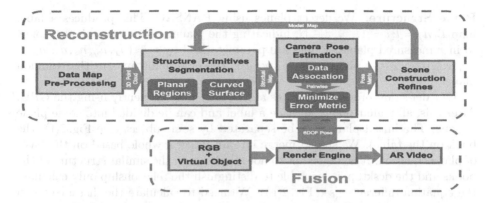

Fig. 1. Main method pipeline.

curvature values reflect the corresponding changes. The two principal curvatures at a given point measure how the surface bends by different amounts in different directions at that point, and we can estimate the local surface shape structure information. At flat point both principle curvature are approximately equal zero, we can classify the same plane property point and construct some plane areas. And vice versa, both principal curvatures are non-zero, we can fitting a local quadratic surface around the point. For clarity, we refer to a point as a planar point or non-planar point.

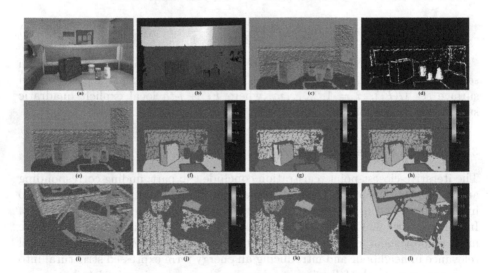

Fig. 2. Example planar segmentation result.

Plane Structure. We detect planes using RANSAC. This produces a label map $\mathcal{L}(\boldsymbol{u}) = l; l = 1, \ldots, r \in \mathbb{N}$, indicating the planar region each pixel belongs to in r measured planes. Planes are parameterized by $\pi = (n_x, n_y, n_z, d)$, $n_\pi = (n_x, n_y, n_z)^T$ is the plane normal and d is the closest distance to the common plane.

Our detection proceed computes a similarity map efficiently using the GPU. After this, all planar point can have a label and can be divided into some plane regions. The initial plane may be separated by some object (see Fig. 2(f), the book on the table). We should merge into an integral whole based on the basis of plane equation. It is important to note that due to the similar structure of the books and the desktop, it is unable to distinguish the relationship only calculate the equation difference (see Fig. 2(g)). We need to calculate the distance mean square deviation of all points in book to the table to distinguish the similar and identical(see Fig. 2(h)), thereby generating some initial parameterized plane region. With the continuous integration of data, the results of plane segmentation are more complete (see Fig. 2(i–l)).

Curved Surface Structure. After identifying all the potential planes, the unprocessed non-planar point should be fitted a local quadric surface. Lets describe the local surface geometry at point p by Darboux frame $\triangle_p = (n, e_1, e_2, \kappa_1, \kappa_2)$, where n is the surface normal vector at p, and e_1, e_2 are the principal directions corresponding to the principal curvatures $\kappa_1, \kappa_2(\kappa_1 \geq \kappa_2)$ respectively. We can following [5] define an explicit quadratic surface parameterized over the $e_1 - e_2$ tangent plane in local coordinate as:

$$F(x, y, z) = \frac{1}{2}\kappa_1 x^2 + \frac{1}{2}\kappa_2 y^2 + \frac{1}{4}(\kappa_1 + \kappa_1)z^2. \tag{1}$$

After segmentation, we can formulate some parameterized area: planar region set $\Phi(\pi_i)$, $i = 1, 2, \ldots, r$, where π_i is i-th plane equation parameter; curved surface set $\Omega(F_j)$, $j = 1, 2, \ldots, m$, where F_j is j-th local explicit quadratic surface representation.

3.3   Camera Pose Estimation

The iterative-closest-point registration conclude two part: finding corresponding and energy optimization. Currently, the high-profile issue is the sensitivity to initial values. The higher corresponding reliability, the better optimization result. In our work, we aim to cooperate the scene structure to origin framework. The main improvement can be summed into two fold: considering the structural relevance in neighbour and introducing an energy term expressed structural into the error term. In the following, we describe these extensions in detail.

The 6DOF camera pose $\boldsymbol{T} = [\text{R}|\text{t}]$ described by rotation matrix $(R \in \mathbb{SO}_3)$ and translation vector $(\text{t} \in \mathbb{R}^3)$. We should establish corresponding pairwise between the vertex map $\mathcal{V}_t$ in frame t and the model projection map $\mathcal{V}^{\mathcal{M}}_{t-1}$ accumulated in previous $t - 1$ frames. And then minimizing the point-to-plane error metric term which reflects the pairwise mismatch under the relative transformation $T^{t \to t-1}$, where $\mathcal{V}^{\mathcal{M}}_{t-1} \approx T^{t \to t-1} * \mathcal{V}_t$.

At frame $t = 0$, the initial 6DOF pose is identity matrix. For subsequent frames, each input point in the vertex map $p_t^i = \mathcal{V}_t(u_t)$ and its geometric attribute $\{n_t^i, \pi_t^i\}$, including normal n_t^i and labelled plane equation $\pi_t^i \in \Phi$, should transformed into the model space $p_{t-1}^i = T^{t \to t-1} p_t^i$, $n_{t-1}^i = R^{t \to t-1} n_t^i$, $\pi_{t-1}^i = T^{t \to t-1} \pi_t^i$. And then looking for the most similarity point q_j^* in the eight-neighbour around relative model point $\mathcal{O}(p_{t-1}^i)$ by minimizing a weighted sum of dissimilarity measures of vertex positions D_v, normal D_n, and structure D_s.

$$q_j^* = \underset{q_{t-1}^j \in \mathcal{O}(p_{t-1}^i)}{\arg\min} \quad \omega_v D_v + \omega_n D_n + \omega_s D_s, \tag{2}$$

with the normal vector direction measures $D_n = 1 - < n_{t-1}^j, n_{t-1}^i >$, the spatial location measurement $D_v = \left\| q_{t-1}^j - p_{t-1}^i \right\|_2$, the structural measurement D_s. Due to we consider the plane and the curve surface, we can discuss the two cases separately. If $label(q_{t-1}^j) \in \Phi$, $D_s = 1 - < \pi_{t-1}^j, \pi_{t-1}^i >$, the plane equation π_{t-1}^j can be calculated by parameterizing the local surface consisting by its four neighbouring triangles. If $label(q_{t-1}^j) \in \Omega$, $D_s = \left\| \kappa_1^{t-1} - \kappa_1^t \right\| + \left\| \kappa_2^{t-1} - \kappa_2^t \right\|$. All corresponding weight $\omega = \frac{1}{3}$.

Then the point-to-plane error metric can be expressed as:

$$E(T^{t \to t-1}) = \sum (\omega_1 (p_{t-1}^i - q_{t-1}^j) \cdot \pi_{t-1}^j + \omega_2 (p_{t-1}^i - q_{t-1}^j) \cdot n_{t-1}^{j,*}), \tag{3}$$

where $\omega_1 = exp(-\frac{\kappa_1^{t-1} + \kappa_2^{t-1}}{2\sigma^2})$, $\omega_2 = exp(-\frac{1}{2(\sigma \kappa_1^{t-1})^2})$, shown that the different point-to-plane distance weights measured from different geometric structures. For the plane situation, π_{t-1}^j is the plane normal. For the curve surface situation, $n_{t-1}^{j,*} = (2\kappa_1, 2\kappa_1, \kappa_1 + \kappa_2) \cdot q_{t-1}^j$ is the tangent plane in q_{t-1}^j from quadratic surface $F(x, y, z)$.

3.4   Scene Construction Refines

After estimate the current camera pose, the input vertex map and associated geometry attributes are integrated into the global model, as shown in Fig. 3(c), (d). A key ingredient toward improved reconstruction method is focus on the high-quality model surface rendering, which is vital to the structural segmentation and camera pose estimation. Traditionally, the raycasting algorithm is feasible to volume rendering. It needs cast a ray along the depth sensors lines of sight onto the model, and taking the point located at the intersection of the model surface, regardless of the underlying model representation. We transfer the surface intersection problem to calculate the intersection point between the light and the local parameterized surface structure. Please see the Algorithm 1 in our supplemental document, we describe these extensions in detail. As shown in Fig. 3(a), (b), the refine result and comparing to the KinectFusion can prove our model is smoother and higher quality.

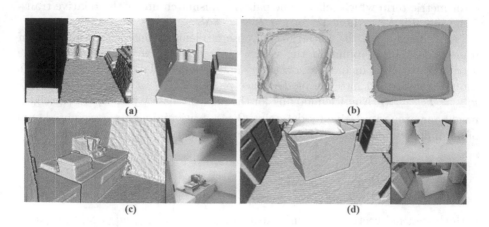

Fig. 3. Scene reconstruction and refine result.

Algorithm 1. Model maps generation using the updated index map

---

Input: V(volume).
Output: $\mathcal{V}^{\mathcal{M}}$(model vertex map).
1: **for** each pixel u in model map in parallel **do**
2:     $ray =$ generate ray for u from camera view;
3:     $\mathcal{G} =$ the passing voxels of viewing ray with volume V;
4:     $g =$ first voxel along ray;
5:     $\Gamma \leftarrow \oslash$
6:     **for** $g \in G$ **do**
7:         **if** zero crossing from g to $g^p rev$ **then**
8:             $Q =$ extract voxel grid position and eight-neighbour
9:             **for** $q \in Q$ **do**
10:                 ($(v_p, n_p) = r \cap$ local surface patch at q)
11:                 Γ.append(v_p, n_p)
12:                 $w_q = exp(-\frac{1}{2}(|q - v_p|)^2))$
13:             **end for**
14:             **for** $(v, n, w) \in \Gamma$ **do**
15:                 $w_{valid} \leftarrow w_{valid} + w$
16:                 $\mathcal{V}^{\mathcal{M}}(u) \leftarrow \mathcal{V}^{\mathcal{M}}(u) + wv$
17:                 $\mathcal{N}^{\mathcal{M}}(u) \leftarrow \mathcal{N}^{\mathcal{M}}(u) + w\hat{n}$
18:                 $\mathcal{V}^{\mathcal{M}}(u) \leftarrow \mathcal{V}^{\mathcal{M}}(u)/w_{valid}$
19:                 $\mathcal{N}^{\mathcal{M}}(u) \leftarrow normalize(\mathcal{N}^{\mathcal{M}}(u))$
20:             **end for**
21:         **end if**
22:     **end for**
23: **end for**
24: **return** $\mathcal{V}^{\mathcal{M}}(u)$;

---

4 Experimental Results

We implemented our method on a desktop PC with an Intel I7-6700 CPU and an Nvidia GeForce GTX 970 graphics card. It was tested on some indoor scenes capture by ASUS Xtion at 30 Hz with a resolution of 640 × 480, and also on the dataset provided by Dai et al. [10]. The complete reconstruction results and partial details of scenes captures using our method are shown in Fig. 4. Based on the various large-scale indoor scene datasets (offices, apartments, over 8000 frames), our method can reconstruct the overall model of the large-scale scene alignment, and embody the local details. This demonstrates that our registration algorithm incorporating with the structure context is feasible. The identification of the scene structure can use to complete the whole part of the hole generated by the occlusion, refining the finally model.

Performance. The overall performance of our method is about 25 fps, with data pre-process (10 ms) and structure primitives segmentation (6 ms), pose estimate (10 ms), scene reconstruction refine (5 ms).

Structure Refine. Figure 4 demonstrates the improvement to the reconstruction quality with our structure refine. Due to occlusion, there are holes on the ground and walls of the office scene in the raw data, since those regions might be hard to cover completely while scanning. In contrast, with our plane data, large planar regions, such as walls and the ground, are extended automatically to these missing areas as an extrapolation of the most common shape.

Drifting Relief. The existence of structure constrains can be leveraged to greatly reduce drifting in the fusion process, and also enhance the convergence efficient. Figure 5 show the error analysis during iteration between traditional ICP and ICP with structural constraint (SC). We also apply our work to the TUM RGB-D benchmarking dataset which provides RGB-D image sequences with 6DoF camera poses [22]. The absolute trajectory error (ATE) between our estimate camera pose and ground truth can be shown in Fig. 6.

AR Application. To generate a convincing augmented reality application with geometric occlusion, we utilize the scene structure to locate the virtual object and tracks the camera pose to fuse the virtual object to real world timely. Before the start of the rendering and fusion, the primary task is to obtain the initial position of virtual object. We achieve this by a previous labeled plane. We select the largest plane (generally the ground) as the initial reference frame, and the center as the initial position of the virtual object. The specific work of scene fusion is done by Open Scene Graph (OSG). Making use of the Kinect sensor, we can get RGB view of the current frame drawn as the background, virtual 3D model rendered on top.

The application can be shown in Fig. 8. Figure 8 shows dense reconstruction of real scene and the spatial relationship between virtual object and the scene before the fictitious fusion, which is the key to dealing with occlusion problem. Compared to the other typical augmented reality applications, one significant improvement in our proposed method is the ability to perform realistic occlusion

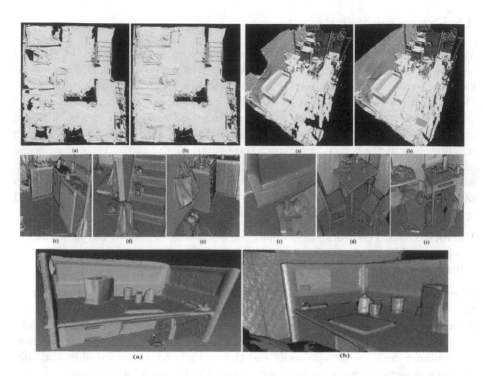

Fig. 4. Our proposed real-time structure optimization reconstruction result and the details

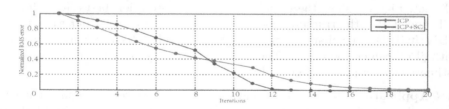

Fig. 5. Convergence speed is compared between ICP, and ICP+SC (Structural Constraint)

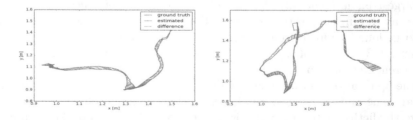

Fig. 6. The ATE between our work and TUM RGB-D dataset groundtruth

of AR contents. Meanwhile, we place a textbook on the floor and initialized the position of virtual dinosaur to be there in order to test the robustness of camera tracking in AR system. We observe the dinosaurs in different perspectives and see whether the location of the dinosaurs has changed. We conduct a 360° view of the dinosaurs, and find it is still near the textbook. Our camera tracking is robust and effective (Fig. 7).

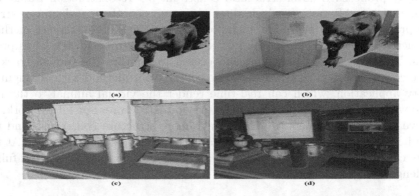

Fig. 7. Virtual object (bear, teapot) in a real scene with occluded by table.

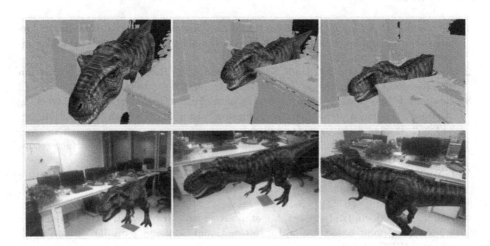

Fig. 8. Virtual dinosaur in our laboratory observed in different perspectives.

Limitations. Our method tends to refactor interior scenes that are mainly composed of a large number of planar areas, and also has some improvement effect on the relatively smooth curved surface structure. But it is still limited in not being able to successfully reconstruct scenes when the scene structure is too complex without large-scale plane structure to support or the input depth maps have a high noise level.

5   Conclusion

We have proposed a novel structural constraint 3D reconstruction framework for marker-less AR system, which allow planar or smooth curved surface structure primitive to be incorporated. In contrast to the state-of-the-art method, our method cluster the point cloud data and divide into basic structures primitives. The structure information can be used to improve the camera tracking and reconstruction quality. Moreover, we combine our method with an augment reality application, which can real-time render the virtual animal to the real scene with occlusion handling based on previous structure. For future work, we believe the pose alignment is able to have more structure constraints and the complicated surface structure can have a better representation strategy. At the same, calculating shadows for the virtual models and interactive AR that full of challenge and great value.

Acknowledgments. The authors would like to thank the anonymous reviewers for their valuable comments and helpful suggestions.

References

1. Newcombe, R.A., Izadi, S., Hilliges, O., et al.: KinectFusion: real-time dense surface mapping and tracking. In: IEEE International Symposium on Mixed and Augmented Reality, pp. 127–136. IEEE (2012)
2. Nießner, M., Zollhofer, M., Izadi, S., Stamminger, M.: Real-time 3D reconstruction at scale using voxel hashing. ACM Trans. Graph. **32**(6), 169 (2013)
3. Newcombe, R.A., Fox, D., Seitz, S.M.: DynamicFusion: reconstruction and tracking of non-rigid scenes in real-time. In: Computer Vision and Pattern Recognition, pp. 343–352. IEEE (2015)
4. Chen, J., Bautembach, D., Izadi, S.: Scalable real-time volumetric surface reconstruction. ACM Trans. Graph. **32**(4), 1–16 (2013)
5. Lefloch, D., Kluge, M., Sarbolandi, H., et al.: Comprehensive use of curvature for robust and accurate online surface reconstruction. IEEE Trans. Pattern Anal. Mach. Intell. **39**(12), 2349 (2017)
6. Keller, M., Lefloch, D., Lambers, M., et al.: Real-time 3D reconstruction in dynamic scenes using point-based fusion. In: International Conference on 3D Vision - 3DV, pp. 1–8. IEEE (2013)
7. Zhang, Y., Xu, W., Tong, Y., et al.: Online structure analysis for real-time indoor scene reconstruction. ACM Trans. Graph. **34**(5), 159 (2015)
8. Weik, S.: Registration of 3-D partial surface models using luminance and depth information. In: International Conference on Recent Advances in 3-D Digital Imaging and Modeling, p. 93. IEEE Computer Society (1997)
9. Whelan, T., Leutenegger, S., Moreno, R.S., et al.: ElasticFusion: dense SLAM without a pose graph. In: Robotics: Science and Systems (2015)

10. Dai, A., Izadi, S., Theobalt, C.: BundleFusion: real-time globally consistent 3D reconstruction using on-the-fly surface re-integration. ACM Trans. Graph. **6**(4), 76a (2017)
11. Kümmerle, R., Grisetti, G., Strasdat, H., et al.: g^2o: a general framework for graph optimization. In: IEEE International Conference on Robotics and Automation, pp. 3607–3613. IEEE (2011)
12. Triggs, B., McLauchlan, P.F., Hartley, R.I., Fitzgibbon, A.W.: Bundle adjustment — a modern synthesis. In: Triggs, B., Zisserman, A., Szeliski, R. (eds.) IWVA 1999. LNCS, vol. 1883, pp. 298–372. Springer, Heidelberg (2000). https://doi.org/10.1007/3-540-44480-7_21
13. Klein, G., Murray, D.: Parallel tracking and mapping for small AR workspaces. In: IEEE and ACM International Symposium on Mixed and Augmented Reality, pp. 1–10. IEEE (2008)
14. Newcombe, R.A., Lovegrove, S.J., Davison, A.J.: DTAM: dense tracking and mapping in real-time. In: International Conference on Computer Vision, pp. 2320–2327. IEEE Computer Society (2011)
15. Godin, G., Rioux, M., Baribeau, R.: Three-dimensional registration using range and intensity information. Proc. SPIE-Int. Soc. Opt. Eng. **2350**, 279–290 (1994)
16. Zhou, Q.Y., Koltun, V.: Depth camera tracking with contour cues. In: Computer Vision and Pattern Recognition, pp. 632–638. IEEE (2015)
17. Taguchi, Y., Jian, Y.D., Ramalingam, S., et al.: Point-plane SLAM for hand-held 3D sensors. In: IEEE International Conference on Robotics and Automation, pp. 5182–5189. IEEE (2013)
18. Pathak, K., Birk, A., Vaskevicius, N., et al.: Fast registration based on noisy planes with unknown correspondences for 3-D mapping. IEEE Trans. Robot. **26**(3), 424–441 (2010)
19. Gelfand, N., Rusinkiewicz, S., Ikemoto, L., et al.: Geometrically stable sampling for the ICP algorithm. In: Proceedings of the International Conference on 3-D Digital Imaging and Modeling, 2003, 3DIM 2003, pp. 260–267. IEEE (2003)
20. Zhang, X., Li, H., Cheng, Z.: Curvature estimation of 3D point cloud surfaces through the fitting of normal section curvatures. In: Proceedings of AsiaGraph, pp. 72–79 (2008)
21. Glocker, B., Shotton, J., Criminisi, A., et al.: Real time RGBD camera relocalization via randomized ferns for keyframe encoding. IEEE Trans. Vis. Comput. Graph. **21**(5), 571–583 (2015)
22. Sturm, J., Engelhard, N., Endres, F., Burgard, W., Cremers, D.: A benchmark for the evaluation of RGB-D SLAM systems. In: Proceedings of International Conference on Intelligent Robots and Systems (2012)

Pairwise Cross Pattern: A Color-LBP Descriptor for Content-Based Image Retrieval

Qiaohong Hao[1], Qinghe Feng[2], Ying Wei[2], Mateu Sbert[1,4(✉)], Wenhuan Lu[3], and Qing Xu[1(✉)]

[1] School of Computer Science and Technology, Tianjin University, Tianjin, China
qiaohonghao@gmail.com, mateusbert@mac.com, qingxu@tju.edu.cn
[2] School of Information Science and Engineering,
Northeastern University, Shenyang, China
fqh1368880889@126.com, weiying@ise.neu.edu.cn
[3] School of Software, Tianjin University, Tianjin, China
wenhuan@tju.edu.cn
[4] Institute of Informatics and Applications, University of Girona, Girona, Spain

Abstract. The local binary pattern (LBP) has been widely considered an excellent and extensive feature descriptor, but it is limited to gray-scale image processing. Inspired by human visual system, we develop a novel yet simple rotation-invariant color-LBP descriptor—pairwise cross pattern (PCP) to extend LBP to color image processing. In the proposed descriptor, the color information map is firstly extracted using a multi-level color quantizer which is designed based on a color distribution prior in the L*a*b* color space. Then, the color information and LBP maps are paired in parallel to construct a pairwise cross pattern, which is easily extended to the uniform pairwise cross pattern (UPCP) and the rotation-invariant pairwise cross pattern (RIPCP). Finally, compared to numerous state-of-the-art schemes and convolutional neural network (CNN)-based models, the experimental results illustrate that the proposed method is efficient, effective and robust in content-based image retrieval task.

Keywords: Local binary pattern · Multi-level color quantizer
Color distribution prior · Pairwise cross pattern
Content-based image retrieval

1 Introduction

Currently, personal electronic equipment, including smart-phones, digital cameras and camcorders, are becoming less expensive and more popular with the digital era, leading to an increasing number of images. Content-based image retrieval (CBIR) [13] which automatically extracts the inherent contents (i.e. color content, texture content and shape content) of an image has been increasingly used, due to its widespread applications in areas such as space exploration,

© Springer Nature Switzerland AG 2018
R. Hong et al. (Eds.): PCM 2018, LNCS 11164, pp. 290–300, 2018.
https://doi.org/10.1007/978-3-030-00776-8_27

medical production and industrial production. Designing effective feature representation is often acknowledged as a difficult but fundamental issue in the CBIR task, and is attracting increasing researchers attention. During the past few decades, many methods and strategies such as color GIST [15], color SIFT [15], local binary pattern (LBP) [12] and its variants [2–4,11,19,21,22], block truncation coding (BTC)-based descriptor [6], wavelet transform (WT)-based method [8] have been designed for the feature representation, and those methods have improved the performance of CBIR tasks, significantly.

Quite recently, the convolutional neural network (CNN)-based model was developed to rapidly increase image processing tasks. As reported in [10], GoogLeNet fully connection features (GL-FCF) in [18], and Dot-Diffused Block Truncation Coding (DDBTC) were joined to construct the extended deep learning two-layer codebook features (DL-TLCF). The method was applied to the CBIR task, and achieved remarkable improvements in CBIR community.

In this paper, we develop a rotation-invariant color-LBP pattern, which builds a bridge between the coding of color information and LBP. First, based on the color distribution prior in the L*a*b* color space, we propose a novel multi-level color quantizer. Second, motivated by human visual system, a novel pairwise cross pattern (PCP) is proposed for coding color information and LBP. Third, PCP is easily extended as a uniform pairwise cross pattern and a rotation-invariant pairwise cross pattern. Compared with LBP [12], color LBP-based schemes [4,11,19], other color-based descriptors [6,8,15] and CNN-based models [10,18], the proposed descriptors achieve the highest mean average precision in content-based image retrieval application.

2   Related Work

2.1   Local Binary Pattern

In [12], the local binary pattern (LBP) was firstly developed for the gray-scale texture feature representation. For a pixel (x, y), the LBP value is coded using the computational results between itself and its neighbors. Mathematically, the LBP of pixel (x, y) is given as follows:

$$LBP_{n,r} = \sum_{i=0}^{n-1} \zeta(I(x,y) - I_i(x,y)) \times 2^i \tag{1}$$

$$\zeta(x) = \begin{cases} 1, & x \geq 0 \\ 0, & x < 0 \end{cases} \tag{2}$$

where n is the number of neighbors, and r is the radius of the n neighbors. $I(x,y)$ is the gray value of pixel (x,y), $I_i(x,y)$ is the gray value of pixel (x,y)'s ith neighbor.

Further, according to the practical experience, nearly 23% of the LBP patterns usually describe over 90% image micro-structures, and these LBP patterns are named as "uniform local binary pattern. "For clarity, the uniform local pattern is expressed as follows:

$$U(LBP_{n,r}(x,y)) = \sum_{i=0}^{n} |\zeta(I(x,y) - I_i(x,y)) - \zeta(I(x,y) - I_{i-1}(x,y))| \quad (3)$$

where $U(\cdot)$ is a measure operator, $U(LBP_{n,r}(x,y)) \leq 2$, and $I_0(x,y) = I_n(x,y)$.

In addition, considering image anti-rotation, the LBP in pixel (x,y) is extended to rotation-invariant LBP $LBP_{n,r}^{ri}(x,y)$, and it is written as follows:

$$LBP_{n,r}^{ri}(x,y) = \min\{ROR(LBP_{n,r}(x,y),i)|i \in 0,1,...,n-1\} \quad (4)$$

where $ROR(LBP_{n,r}(x,y),i)$ performs a circular bit-wise right shift for i times on n-bit number $LBP_{n,r}(x,y)$. For simplicity, (n,r) is limited to $(8,1)$ in this paper.

2.2 Color Distribution Prior

The L*a*b* color space includes three pairs of color channels consisting of the white-black pair of the L* channel (ranging from 0 to 100), the yellow-blue pair of the a* channel (ranging from -128 to $+127$), and the red-green pair of the b* channel (ranging from -128 to $+127$) [7]. At the same time, it provides excellent decoupling between intensity (represented by the L* channel) and color (represented by the a* and b* channels) [16]. As presented in Fig. 1(a), (b), the frequency of pixels in the a* and b* channels focuses on the concentration of the color channels on ImageNet database [14]. In practice, to verify the consistency of this prior, we performed the statistics experiments on thousands of color image databases in our previous work [5]. The results also illustrated that the frequency of pixels mainly focuses on the concentration in the a* and b* channels.

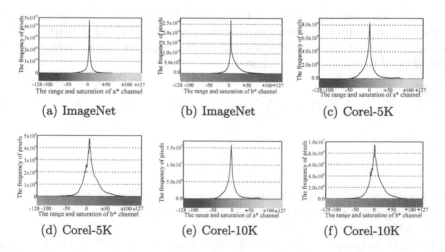

(a) ImageNet   (b) ImageNet   (c) Corel-5K

(d) Corel-5K   (e) Corel-10K   (f) Corel-10K

Fig. 1. The color probability distribution of the a* and b* channels for different databases.

Further, we test the stability of this prior when the image database is changed. For instance, Corel-5K [1] (a half of Corel-10K) is a subset of Corel-10K [1]. The color probability distribution of the a* and b* channels on Corel-5K and Corel-10K are shown in Fig. 1(c)–(f), respectively. It can be seen that there is almost no change between Corel-5K and Corel-10K, apart from the frequency of pixels. The results illustrate that the color distribution prior is stable, even if the image database is changed [5].

More deeply, we explore the reason lied in behind in this paper. Through a large number of experiments, we summarize that the color probability distribution is negative correlation with the saturation of color in the a* and b* channels. As shown in Fig. 1(a)–(f), the saturation of a* and b* gradually decline from both sides to center, yet the frequency of pixels inversely goes up from both sides to center.

3 Pairwise Cross Pattern

3.1 Multi-level Color Quantizer

Here, we project the original range $[-128, +127]$ of the a* and b* channels into 2^8. The a* and b* channels are the red-green pair and the yellow-blue pair, so we uniformly map 2^8 into four equal-intervals 2^6 in level 1, and the indexes are correspondingly flagged as 0, 1, 2 and 3. Inspired by our previous work in [5], two central intervals in level 1 are split into four equal-intervals 2^5 in level 2, and the remaining intervals in level 1 are copied into level 2. In this way, two central intervals can be effectively refined and maintained. We repeat the above operations until the central intervals are 2^1 in level 6. Combing level 1 through level 6, a multi-level color quantizer is constructed. For clarity, this process is shown in Fig. 2, in which each level contains a group of color quantized intervals and its indexes. Mathematically, the quantized level of the a* and b* channels in the multi-level color quantizer are defined as Y_{a*} and Y_{b*}, respectively, and the indexes are correspondingly flagged as $X_{a*} \in [0, 2Y_{a*}+1]$ and $X_{b*} \in [0, 2Y_{b*}+1]$, respectively, where $Y_{a*}, Y_{b*} \in [1, 2, ..., 6]$.

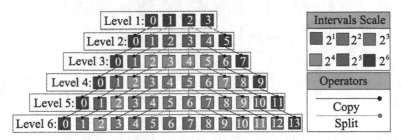

Fig. 2. The details of multi-level color quantizer.

In addition, the L* channel only maintains the intensity information (except for color information). According to the human visual intensity perceptive mechanism [20], the original range $[0, +100]$ of L* channel is mapped into $[0, +25]$,

[+26, +75] and [+76, +100], and the indexes are flagged as X_{L*}, $X_{L*} \in [0, 1, 2]$. Combining the indexes of X_{L*}, X_{a*} and X_{b*}, the index of color information map (CIM) in pixel (x, y) is denoted as follows:

$$CIM(x, y) = 2(Y_{b*} + 1) \times [(2(Y_{a*} + 1) \times X_{L*}(x, y) + X_{a*}(x, y)] + X_{b*}(x, y) \quad (5)$$

where $X_{L*}(x, y)$, $X_{a*}(x, y)$ and $X_{b*}(x, y)$ are the color indexes of the L*, a* and b* channels, respectively, in pixel (x, y); Y_{a*} and Y_{b*} are the color quantized level of the a* and b* channels in the multi-level color quantizer, where $Y_{a*}, Y_{b*} \in [1, 2, ..., 6]$.

3.2 Human Visual System

As depicted in Fig. 3, the Gray's Anatomy [17] elucidates that the human visual system is comprised of pairs of eyeballs, optic nerves, lateral geniculate nucleus and visual cortices. Specially, the optic chiasma is an important cross structure used for combining the visual information between the optic nerves and the lateral geniculate nucleus. The low-level visual cues are first extracted by the eyeballs. And then, the extracted visual information is coded into the "+" and "−" ions on the optic nerves. Thirdly, a part of transmitted information is crossed at optic chiasma. Fourthly, the crossed information is reconstructed at the lateral geniculate nucleus. Finally, the reconstructed information is projected into the human brain visual cortices to form the high-level visual perception.

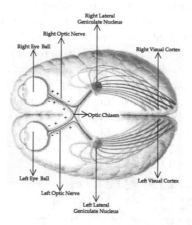

Fig. 3. Human visual system.

3.3 Pairwise Cross Pattern Definition

Inspired by human visual system, namely "pairwise" and "cross", we develop a novel strategy that can effectively code a color information map and a LBP map into a pairwise cross pattern (PCP). For a pixel (x, y), $LBP(x, y)$ and $CIM(x, y)$ are first extracted and paired. And then, the eight neighbors of the LBP map and the CIM map are simultaneously binarized to generate a LBP binary map

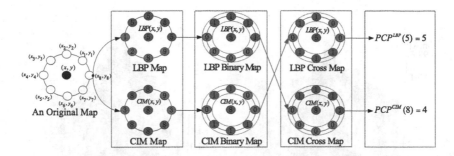

Fig. 4. An example of the pairwise cross pattern.

and a CIM binary map. Afterward, the eight neighbors of LBP binary and CIM binary maps are crossed to obtain the LBP cross map and the CIM cross map. Finally, the eight neighbors of the LBP cross map and the CIM cross map are linearly summed and projected as the feature value of PCP, and the LBP(x, y) and CIM(x, y) are defined as the indexes of PCP. For clarity, an example of the pairwise cross pattern is shown in Fig. 4, in which the PCP in pixel (x, y) is computed as $PCP(x, y) = [PCP^{LBP}(5), PCP^{CIM}(8)]$. Mathematically, PCP is defined as follows:

$$PCP^{CIM}(CIM(x, y)) = \sum_{i=1}^{N} \vartheta(LBP(x, y), LBP(x_i, y_i))|_{N=8} \quad (6)$$

$$PCP^{LBP}(LBP(x, y)) = \sum_{i=1}^{N} \vartheta(CIM(x, y), CIM(x_i, y_i))|_{N=8} \quad (7)$$

$$\vartheta(s, t) = \begin{cases} 1, & \text{if } s = t \\ 0, & \text{otherwise} \end{cases} \quad (8)$$

where (x, y) is the central pixel, (x_i, y_i) is the neighbors of (x, y), and $\vartheta(\cdot)$ is the binarized function. Experimentally, the final feature vector of the pairwise cross pattern is given by concatenating PCP^{CIM} and PCP^{LBP}. According to the different databases, the optimal color quantization levels of Y_{a*} and Y_{b*} are chosen in the proposed descriptor. The feature dimension of PCP^{CIM} and PCP^{LBP} are calculated as $\max\{CIM(x, y)\} + 1$ (Eq. 5) and 256 (Eq. 1), respectively.

Further, to accelerate computational efficiency, we extend the pairwise cross pattern to a uniform pairwise cross pattern (UPCP), in which LBP only is replaced by a uniform local binary pattern. Considering the rotation-invariant of color information, we extend the pairwise cross pattern to a rotation-invariant pairwise cross pattern (RIPCP), in which LBP is replaced by the rotation-invariant local binary pattern.

4   Experiments and Evaluations

Content-based image retrieval is selected as the application in this paper to test the performance of the proposed descriptors. In the experiments, the Extended Canberra distance [6,9] is chosen as the distance metric between a query image

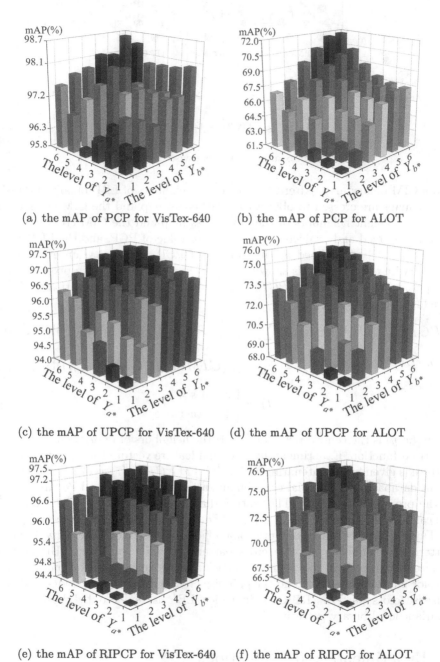

(a) the mAP of PCP for VisTex-640   (b) the mAP of PCP for ALOT

(c) the mAP of UPCP for VisTex-640   (d) the mAP of UPCP for ALOT

(e) the mAP of RIPCP for VisTex-640   (f) the mAP of RIPCP for ALOT

Fig. 5. mAP of different color quantization levels.

and the database images. Considering the limitation of page, the public bench-mark databases VisTex-640 [4,10] and AOLT (with rotation) [4,10] are used as the test image databases. Mean average precision (mAP) rate is applied to evaluate the accuracy of image retrieval. Referring to the experimental setting in [4,6,10], the number of retrieved images is set to 10. To guarantee the accu-racy of the experiments, all experiments are performed under the principle of leave-one-out cross-validation.

4.1 Evaluation of Different Color Quantization Level

Figure 5 shows the mAP(%) of PCP, UPCP and RIPCP with different color quantization levels for VisTex-640 and ALOT. As shown in Fig. 5(a), (b), it can be observed that PCP achieves the highest mAP for VisTex-640 and ALOT when ($Y_{a*} = 6$, $Y_{b*} = 6$). As shown in Fig. 5(c), (d), it can be seen that UPCP achieves the highest mAP for VisTex-640 when ($Y_{a*} = 5$, $Y_{b*} = 5$), and the highest mAP for ALOT when ($Y_{a*} = 6$, $Y_{b*} = 5$). From Fig. 5(e)–(f), it can be seen that RIPCP achieves the highest mAP for VisTex-640 when ($Y_{a*} = 5$, $Y_{b*} = 6$), and the best mAP for ALOT when ($Y_{a*} = 6$, $Y_{b*} = 5$). In addition, we also can see that the simplest color quantization scheme (e.g., $Y_{a*} = 1$, $Y_{b*} = 1$) does not lead to the lowest mAP, and the most refined color quantization scheme (e.g., $Y_{a*} = 6$, $Y_{b*} = 6$) does not guarantee the highest mAP. This phenomenon demonstrates that it is necessary to adaptively select the fitness quantization layers of Y_{a*} and Y_{b*}. According to the different databases, the optimal color quantization levels of Y_{a*} and Y_{b*} are chosen in the proposed descriptors.

4.2 Evaluation of the Proposed Descriptors

Table 1 shows the mAP(%) of the proposed descriptors for VisTex-640 and ALOT. As listed in this table, it can be clearly seen that the mAP(%) of PCP outper-forms PCP^{CIM} and PCP^{LBP} by 4.26% and 3.89% for VisTex-640, as well as by 8.08% and 21.83% for ALOT, correspondingly. Meanwhile, it also can be observed that the mAP(%) of UPCP outperforms UPCP^{CIM} and UPCP^{LBP} by 4.06% and 6.95% for VisTex-640, as well as by 7.04% and 26.55% for ALOT, respectively. Similarly, more significant values are reported among RIPCP, RIPCP^{CIM} and RIPCP^{LBP}. Based on these results, it can be demonstrated that the proposed descriptors with pairwise setting (including PCP, UPCP and RIPCP) can achieve that higher performances because they integrate the merits of color and LBP.

4.3 Comparison with Other Schemes

To illustrate the efficiency, effectiveness and robustness of the proposed descrip-tors, the experimental comparison results among the proposed descriptors and LBP, color LBP-based schemes (e.g., LEP + colorhist, LECoP, maLBP and mdLBP), other color-based descriptors (e.g., color GIST, color SIFT, DDBTC and Copula Model of GW), CNN-based models (e.g., GL-FCF, HD(DL-TLCF),

Table 1. Comparison with the proposed descriptors in terms of mean average precision (mAP%) for VisTex-640 and ALOT(rotation) databases.

Method	VisTex-640	ALOT (rotation)
PCP^{CIM}	94.33	63.74
PCP^{LBP}	94.70	49.99
PCP	**98.59**	**71.82**
UPCP^{CIM}	93.47	68.67
UPCP^{LBP}	90.58	49.16
UPCP	**97.53**	**75.71**
RIPCP^{CIM}	94.28	67.06
RIPCP^{LBP}	89.64	49.43
RIPCP	**97.38**	**76.86**

HD(DL-TLCF)(nor), DL-TLCF-hier and DL-TLCF-hier(nor)) are listed in Table 2, in which the bold values represent the mAP values using the proposed descriptors in this paper. To guarantee the fairness of the experiments, all experimental settings are identically chosen as in the compared methods.

Table 2 reports the comparisons between the proposed descriptors and the former schemes in terms of mAP and feature dimension. As documented in this table, the proposed PCP descriptor yields the highest mAP = 98.59% on VisTex-640. Although PCP is not designed for anti-rotation, the mAP = 71.82% of PCP still outperforms all other previous descriptors on ALOT. We summarize the following two reasons: (1) the color information can be effectively refined and retained. (2) "Pairwise" and "Cross" are benefit to narrow the gap between low-level visual cues and high-level visual perception. In what follow, as compared with PCP with 844 feature dimensions, UPCP with 491 feature dimensions achieves the mAP = 97.53% (−1.06%) on VisTex-640, and UPCP with 563 feature dimensions achieves the mAP = 75.71% (+3.89%) on ALOT, respectively. The mainly reason is that considering all PCP patterns are suitable for VisTex-640, yet not all PCP patterns are useful for ALOT (rotation). On the other hand, the proposed RIPCP descriptor yields the highest mAP = 76.86% on ALOT (rotation), which illustrates the robustness of the proposed descriptor for anti-rotation. Meanwhile, we note that the proposed RIPCP descriptor also achieves a competitive mAP = 97.38% on VisTex-640. Specially, compared with convolutional neural network (CNN)-based models, such as, GL-FCF, HD(DL-TLCF), HD(DL-TLCF)(nor), DL-TLCF-hier and DL-TLCF-hier(nor), even if the feature dimension is lower, PCP, UPCP and RIPCP still provide a higher mAP, which also demonstrates the efficiency and effectiveness of the proposed descriptors. Based on the above experimental results, the efficiency, effectiveness and robustness of the proposed descriptors are demonstrated by compared to other previous methods.

Table 2. Comparison with the proposed methods and other methods in terms of mean average precision (mAP%) for VisTex-640 and ALOT (rotation) databases.

Method	Feature dimension	VisTex-640	ALOT (rotation)
LBP [12] (PAMI2002)	256	82.00	35.30
LEP + colorhist [11] (VSTIA2013)	8192	94.81	48.64
LECoP [19] (Neurocomputing2015)	348	97.22	48.64
maLBP [4] (TIP2016)	1024	74.86	53.74
mdLBP [4] (TIP2016)	2048	80.08	63.04
Color GIST [15] (PAMI2010)	-	43.98	42.44
Color SIFT [15] (PAMI2010)	-	19.63	23.52
DDBTC [6] (TIP2015)	384	92.65	48.64
Copula Model of GW [8] (PR2017)	-	92.40	60.80
GL-FCF [18] (CVPR2015)	1024	90.07	66.39
HD(DL-TLCF) [10] (TIP2017)	1408/1048	94.69	67.38
HD(DL-TLCF)(nor) [10] (TIP2017)	1408/1048	97.09	70.69
DL-TLCF-hier [10] (TIP2017)	1408/1048	93.60	66.06
DL-TLCF-hier(nor) [10] (TIP2017)	1408/1048	96.97	69.45
Proposed Method(PCP)	844	**98.59**	**71.82**
Proposed Method(UPCP)	491/563	**97.53**	**75.71**
Proposed Method(RIPCP)	540	**97.38**	**76.86**

5   Conclusion

In this study, we presented a rotation invariant color-LBP descriptor—pairwise cross pattern that effectively coded color information and LBP into a unified whole. Compared with numerous former methods, the efficiency, effectiveness and robustness of the proposed schemes have been demonstrated in this paper. In the future researches, we shall extend the proposed descriptor to other potential applications, e.g., image classification, face recognition, objection detection and tracking.

Acknowledgement. This work has been funded by National Natural Science Foundation of China under grants No. 61471261 and No. 61771335, and by grants TIN2016-75866-C3-3-R from Spanish Government.

References

1. Corel Photo Collection Color Image Database (2014). http://wang.ist.psu.edu/docs/realted/
2. Banerji, S., Verma, A., Liu, C.: Novel color LBP descriptors for scene and image texture classification. In: 15th International Conference on Image Processing, Computer Vision, and Pattern Recognition, pp. 537–543 (2011)
3. Choi, J., Plataniotis, K., Ro, Y.: Using colour local binary pattern features for face recognition. In: IEEE International Conference on Image Processing, pp. 4541–4544 (2010)

4. Dubey, S.R., Singh, S.K., Singh, R.K.: Multichannel decoded local binary patterns for content-based image retrieval. IEEE Trans. Image Process. **25**(9), 4018–4032 (2016)
5. Feng, Q.H., Hao, Q.H., Chen, Y.Q., Yi, Y.G., Wei, Y., Dai, J.Y.: Hybrid histogram descriptor: a fusion feature representation for image retrieval. Sensors **18**, 1943 (2018)
6. Guo, J.M., Prasetyo, H., Wang, N.J.: Effective image retrieval system using dot-diffused block truncation coding features. IEEE Trans. Multimed. **17**(9), 1576–1590 (2015)
7. Hurvich, L.M., Jameson, D.: An opponent-process theory of color vision. Psychol. Rev. **64**, 384–404 (1957)
8. Li, C.R., Huang, Y.Y., Zhu, L.H.: Color texture image retrieval based on gaussian copula models of gabor wavelets. Pattern Recognit. **64**, 118–129 (2017)
9. Liu, G.H., Yang, J.Y.: Content-based image retrieval using color difference histogram. Pattern Recognit. **46**(1), 188–198 (2013)
10. Liu, P.Z., Guo, J.M., Wu, C.Y., Cai, D.L.: Fusion of deep learning and compressed domain features for content based image retrieval. IEEE Trans. Image Process. **26**(12), 5706–5717 (2017)
11. Murala, S., Wu, Q.M.J., Balasubramanian, R., Maheshwari, R.P.: Joint histogram between color and local extrema patterns for object tracking. In: Video Surveillance and Transportation Imaging Applications, vol. 8663 (2013)
12. Ojala, T., Pietikainen, M., Maenpaa, T.: Multiresolution gray-scale and rotation invariant texture classification with local binary patterns. IEEE Trans. Pattern Anal. Mach. Intell. **24**(7), 971–987 (2002)
13. Piras, L., Giacinto, G.: Information fusion in content based image retrieval: a comprehensive overview. Inf. Fusion **37**, 50–60 (2017)
14. Russakovsky, O., et al.: Imagenet large scale visual recognition challenge. Int. J. Comput. Vis. **115**(3), 211–252 (2015)
15. van de Sande, K.E.A., Gevers, T., Snoek, C.G.M.: Evaluating color descriptors for object and scene recognition. IEEE Trans. Pattern Anal. Mach. Intell. **32**(9), 1582–1596 (2010)
16. Sarrafzadeh, O., Dehnavi, A.M.: Nucleus and cytoplasm segmentation in microscopic images using k-means clustering and region growing. Adv. Biomed. Res. **4**, 174 (2015)
17. Standring, S.: Gray's Anatomy: The Anatomical Basis of Clinical Practice, forty-fisrt edn. Elsevier Limited, New York (2016)
18. Szegedy, C., et al.: Going deeper with convolutions. In: Proceedings of the IEEE Conference on Computer Vision and Pattern Recognition, pp. 1–9 (2015)
19. Verma, M., Raman, B., Murala, S.: Local extrema co-occurrence pattern for color and texture image retrieval. Neurocomputing **165**(1), 255–269 (2015)
20. Zhang, M., Zhang, K., Feng, Q., Wang, J., Kong, J.: A novel image retrieval method based on hybrid information descriptors. J. Vis. Commun. Image Represent. **25**(7), 1574–1587 (2014)
21. Zhu, C., Bichot, C., Chen, L.: Multi-scale color local binary patterns for visual object classes recognition. In: 20th International Conference on Pattern Recognition, pp. 3065–3068 (2010)
22. Zhu, C., Bichot, C., Chen, L.: Image region description using orthogonal combination of local binary patterns enhanced with color information. Pattern Recognit. **46**(7), 1949–1963 (2013)

Multimodal Dimensional and Continuous Emotion Recognition in Dyadic Video Interactions

Jinming Zhao, Shizhe Chen, and Qin Jin[✉]

Renmin University of China, Beijing, China
{zhaojinming,cszhe1,qjin}@ruc.edu.cn

Abstract. Automatic emotion recognition is a challenging task which can make great impact on improving natural human computer interactions. In dyadic human-human interactions, a more complex interaction scenario, a person's emotion state will be influenced by the interlocutor's behaviors, such as talking style/prosody, speech content, facial expression and body language. Mutual influence, a person's influence on the interacting partner's behaviors in a dialog, is shown to be important for predicting the person's emotion state in previous works. In this paper, we proposed several multimodal interaction strategies to imitate the interactive patterns in the real scenarios for exploring the effect of mutual influence in continuous emotion prediction tasks. Our experiments based on the Audio/Visual Emotion Challenge (AVEC) 2017 dataset used in continuous emotion prediction tasks, and the results show that our proposed multimodal interaction strategy gains 3.82% and 3.26% absolute improvement on arousal and valence respectively. Additionally, we analyse the influence of the correlation between the interactive pairs on both arousal and valence. Our experimental results show that the interactive pairs with strong correlation significantly outperform the pairs with weak correlation on both arousal and valence.

Keywords: Emotion recognition · Multimodal · Dyadic interaction

1 Introduction

In the dyadic interactions, the mutual behavior effect controls the dynamic flow of a conversation and describes the overall interaction patterns. Understanding human interaction mechanisms and computationally modeling interaction dynamics of human behavior can bring insights into automating emotion recognition as well as the design of human-machine interfaces. There are many applications in human-computer interactions, including call-center dialogue systems, conversational agents [9], depression severity prediction [17], mental health diagnoses and educational softwares [8].

In this paper, we focus on the dyadic human-human conversation scenario, and our experiments are based on a benchmark dataset AVEC2017 for continuous emotion prediction. In this scenario, the challenges in continuous emotion

© Springer Nature Switzerland AG 2018
R. Hong et al. (Eds.): PCM 2018, LNCS 11164, pp. 301–312, 2018.
https://doi.org/10.1007/978-3-030-00776-8_28

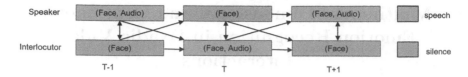

Fig. 1. Interaction framework overview. Each block denotes a turn which is defined as the portion speech belonging to a single speaker before he/she finishes speaking, and may consists of multiple original segmented utterances. The green block denote the speech turn and blue block denote the silence turn. "T" represents the turn time step. When one person in speech turn, there exist both audio and facial expression information. When one person in silence turn, there are only facial expression information. (Color figure online)

prediction systems mainly including extracting efficient multimodal features, fusing the complementary multimodal features effectively, capturing effective context information and correctly handling the interlocutor's influence.

For extracting efficient multimodal features, Kevin et al. [3], the winner of the AVEC2016, employ sparse coding method on expert-knowledge based low-level audio features and deep visual features. Chen et al. [7] utilize deep audio features from SoundNet [1] and deep facial expression features from more efficient DenseNet [12] models, and get the best performance in AVEC 2017 challenge. In this paper, we use a more efficient audio representation model, VGGish [10], which is trained on a large scale dataset and can learn more richer audio representations. For multimodal fusion methods, Chen et al. [6] analyses the early fusion, model-level fusion and late fusion strategies, and propose the conditional attention fusion method to dynamically pay attention to different modalities at each time step and gain significant performance improvement on valence. For capturing context information, Long Short Term Memory Recurrent Neural Network (LSTM-RNN) [11], one of the state-of-the-art sequence modeling techniques, is widely used to capture the temporal information and achieves state-of-the-art performance [3,5–7]. However, for correctly handling interlocutor's influence, there are fewer research works considering this influence, especially for continuous emotion prediction tasks. Chen et al. [7] propose several feature sequence construction combinations to imitate the interaction patterns and achieve significant performance improvement, and yet the authors just imitate the interaction patterns on audio modality. Motivated by this, we propose several multimodal interaction strategies for imitating the real dyadic human-human interaction pattern shown in Fig. 1. In this paper, we explore and analyse the influence of the various modalities of the interlocutor, and we find the best multimodal feature interaction strategy and get the best prediction performance.

Additionally, the interlocutor behaviors bring different influence to speaker's emotion state in different scenarios [15], such as wedding [2] and customer center. We analyse the correlation strength of interactive pairs of the AVEC 2017 dataset, and find that 54.17% pairs and 70.83% pairs have strong positive correlation on arousal and valence respectively. Our experimental results suggest that the pairs with strong correlation significantly outperform the pairs with weak correlation using the interaction strategies.

Our main contributions in this paper are from four aspects: 1. We propose a more efficient audio features extracted from the pretrained VGGish model for emotion prediction. 2. We propose several interaction strategies to make full use of interlocutor's information under multimodal dyadic human-human conversation scenarios for dimensional continuous emotion prediction. 3. We explore and analyse which modality information of the interlocutor is helpful for improving prediction performance. 4. We explore and analyse the influence of correlation strength between interactive pairs on the emotion prediction of arousal and valence.

The remainder of this paper is organized as follows: we present the related works and our proposed methods in Sect. 2 and Sect. 3, respectively. The experimental results and analyses are described in Sect. 4. Finally we conclude the paper in Sect. 5.

2 Related Works

Multimodal Features: Previous works in the emotion recognition tasks have explored a variety of multimodal features. Kevin et al. [3], the winner of the AVEC2016 challenge, derive high-level acoustic, visual and physiological features from the low-level descriptors using sparse coding and deep learning. Chen et al. [7], the winner of the AVEC2017 challenge, focus on using Deep Convolution Neural Network to learn deep acoustic and facial expression features. The acoustic and facial expression features are extracted from the SoundNet [1] and the DenseNet [12] respectively, and the proposed features bring significant performance improvement. They also find the correlations between different emotion dimensions with the Pearson Correlation Coefficient (PCC) and finetune the DenseFace features extractor based on multitask learning on the arousal and valence simultaneously. And the finetuned features gain significant improvement on both arousal and valence.

Multimodal Fusions: Emotion is naturally expressed through multi-modalities, such as audio, facial expression, body language and speech content, and these modalities capture complementary information about the emotion [16]. Early fusion uses the concatenated features from different modalities as input features. It has been widely used in the literature to successfully improve performance [5,19]. Chen et al. [6] compare the early fusion, model-level fusion, late fusion and the proposed conditional attention fusion strategies, and the proposed fusion method significantly outperforms the other three methods on valence prediction. However, the early fusion gains better performance than the other three methods on arousal.

Emotion Recognition Models: Previous works proposed various context-sensitive model to capture the context information. [16] proposed a hierarchical HMM framework to fuse the context information and multimodal information for emotion prediction. In the emotion recognition researches of the last few years, almost works utilize LSTM model for continuous emotion prediction [3,5–7].

Mutual Information in Dyadic Interaction: In dyadic human-human conversation scenarios, speaker's emotion is influenced by the interlocutor's behaviors. Dynamic Bayesian Network (DBN) is used to explicitly model the conditional dependency between two interacting partners' emotional states in a dialog [13]. Angeliki et al. [16] propose a hierarchical framework which models emotional turn-level evolution. It incorporate a variety of generative or discriminative classifiers at each level and provides flexibility and extensibility in terms of multimodal fusion. Soroosh et al. [15] present a thorough analysis of IEMOCAP dataset [4] and reveals that 72% conversational partners present similar emotions. Based on these findings, they propose novel cross-modality, cross-speaker emotion recognition methods that improve the performance. These works are based on the turn-level classification and aim to predict the emotion categories of the turns. In contrast, Chen et al. [7] propose a interaction strategy that using feature sequence construction combinations to imitate the interaction patterns on the audio modality in continuous emotion recognition task and gain significant performance improvement. Motivated by this, we extent the feature sequence construction combinations strategies in multimodal scenarios.

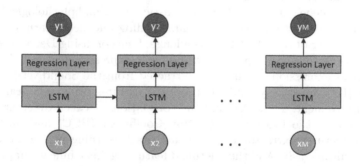

Fig. 2. System framework overview.

3   Proposed Methods

3.1   System Framework

In this paper, we adopt the LSTM [18] temporal model as our perdiction model for capturing the temporal information. As is shown in Fig. 2, given an feature sequence of M frames $x = \{x_1, x_2, ..., x_M\}$ as the system input, where M is the max time step of the LSTM model. A regeresion layer follows the output of the LSTM layer, and then we get the continues emotional prediction, $y = \{y_1, y_2, ..., y_M\}$.

We use the mean square error (MSE) as our loss function, which minimizes:

$$L = \frac{1}{2T} \sum_{t=1}^{T} (g_t - y_t)^2 \tag{1}$$

where g_t are ground truth labels, and T is the total time steps of the data.

3.2 Multimodal Features

Vggish. We utilize the pretrained VGGish [10], which is trained on a large scale dataset and can learn more richer audio representations, for extracting short-term acoustic features. The audio is first divided into non-overlapping frames with window size 0.98 s. Each frame is then transformed into log-mel spectrogram as the input to the VGGish network. We extract activations from the last fully connected layer with dimensionality of 128 as the audio embedding features and refer the features as "vggish.100ms".

Denseface. The recent proposed Densely Connected Convolutional Networks (DenseNet) [12] have achieved the state-of-the-art performance in the image recognition tasks. It connects all the preceding layers as the input for a certain layer, which can strengthen feature propagation and alleviate the gradient vanishing problem. Besides, due to the feature reuse, DenseNet only needs to learn a small set of new feature maps in each layer and thus requires fewer parameters than traditional CNNs, which is more suitable for small datasets. The details of the DenseNet structure, training process and the network finetuning are described in [7]. We extract the activations from the last mean pooling layer of finetuned model and refer the features as "denseface.tune". To match with the shift of ground-truth labels, we apply mean pooling over consecutive 5 facial CNN frame features. The frames where no face is detected are filled with zeros.

3.3 Multimodal Interaction Strategies

In the dyadic human-human conversation conversation scenarios, when one person talks, another is silent, alternately. As is shown in Fig. 1, we propose the real dyadic human-human interaction pattern, and our proposed interaction strategies imitate this interaction pattern. For speaker's continuous emotion evolution, the influences are mainly derived from three aspects: the previous speaker's context information, the previous interlocutor's context information and the current interlocutor's information.

As shown in Fig. 3, the speaker's audio information just exists in speaker's speech turn, while speaker's facial expressions exist all time. The original audio recordings contain both speaker's and interlocutor's audio signals, so directly use the original audio recordings is not accurate. Chen et al. [7] show that directly use the original audio recordings not only cannot take advantage of the interlocutor's information, but also bring a lot of noise and result in worse performance.

Turn	Only Audio	Only Face
T-1	S_A	S_F
T		S_F
T+1	S_A	S_F

Fig. 3. Unimodal feature sequence constructions. "S_A" and "S_F" denotes the audio and facial expression features of speakers. The green block denotes the speech turn, the blue block denotes the silence turn and the white block indicates that it is filled with zero vectors. (Color figure online)

Turn	Audio Interact		Face Interact1		Face Interact2	
T-1	S_A		S_F	L_F	S_F	L_F
T		L_A	S_F	L_F	S_F	
T+1	S_A		S_F	L_F	S_F	L_F

Fig. 4. Unimodal feature sequence construction combinations with interlocutor's influence. "L_A" and "L_F" denotes the audio and facial expression features of interlocutors. In each interaction structure, the left column denotes the speaker's features and right column denotes the interlocutor's features.

For exploring the interlocutor's influence under uni-modality scenarios, we propose three feature sequence construction combinations. As shown in Fig. 4, "Audio Interact" strategy imitates the interaction pattern under only audio modality, which is similar to the "Double" interaction strategy mentioned in [7]. For "Face Interact1" strategy, we make full use the available facial expression information of the speaker and the interlocutor. When speaker talks, the interlocutor's facial expression information can make up for the missing audio information. However, when speaker is slient, the speaker just has facial expression information and the targets were labeled only by speaker's facial expression. Further, the interlocutor's facial expression is not exactly accurate when he/she talks. So, we design the "Face Interact2" construction strategy which only use the speaker's facial expression information when the speaker is silent.

Emotion is naturally expressed through multi-modalities, compared to only using the uni-modality information, multimodality information can bring significant prediction performance improvement [6,7]. As shown in Fig. 5, we propose two multimodal feature sequence constructions. The "Multimodal1" construction denotes that we use all available information of speakers. For verifying the efficiency of the speaker's facial expression which is not exactly accurate when she/he talks, we design the "Multimodal2" construction which don't consider the facial expression information in speech turns.

Turn	Multimodal1		Multimodal2	
T-1	S_A	S_F	S_A	
T		S_F		S_F
T+1	S_A	S_F	S_A	

Fig. 5. Multimodal feature sequence constructions.

Turn	Interact-AFA			Interact-AFF		
T-1	S_A	S_F		S_A	S_F	L_F
T		S_F	L_A		S_F	
T+1	S_A	S_F		S_A	S_F	L_F

Fig. 6. Multimodal feature sequence construction combinations with interlocutor's uni-modality influence. In each interaction strategy, the left column denotes the speaker's audio features, the middle column denotes the speaker's facial expression features and the right column denotes the interlocutor's audio/facial expression features.

Turn	Interact-AFAF			
T-1	S_A	S_F		L_F
T		S_F	L_A	
T+1	S_A	S_F		L_F

Fig. 7. Multimodal feature sequence construction combinations with interlocutor's multimodal influence.

In order to verify the influence of the interlocutor's uni-modality information on the speaker's emotion prediction. As shown in Fig. 6, We propose two inter-action strategies, "Interact-AFA" which only use the interlocutor's audio information and "Interact-AFF" which only consider the interlocutor's facial expression information. Because the "Face Interact2" outperforms the "Face Interact1" strategy (as shown in Table 1), we don't use the interlocutor's facial expression information when the speaker is silent.

Finally, we reasonably consider both the interlocutor's audio and facial expression information and propose the 'Interact-AFAF' strategy shown in Fig. 7.

4   Experiments

4.1   Corpus Description

In this paper, we use the corpus used in the Audio/Visual Emotion Challenge (AVEC 2017) [17] which is a subset of the Sentiment Analysis in the Wild (SEWA) dataset[1]. There are 64 subjects and every two subjects in a group of video chats. All audiovisual recordings were collected "in-the-wild" using standard webcams and microphones from the computers in the subjects' offices or homes. The duration of each conversation is at most 3 min. All three emotion dimensions are annotated every 100ms and scaled into $[-1, +1]$.

4.2   Experimental Setup

We implement the LSTM [18] temporal model with the tensorflow[2] deep learning framework. All of our experiments are using this LSTM temporal model. The number of layers of the LSTM are set to be 1 and the hidden units number is optimized for different input features. We use the truncated back propagation through time (BPTT) with max time step of 100 to train our LSTM networks. Dropout is adopted to avoid overfitting with dropout rate of 0.5. Adam optimizer is applied and the learning rate is initialized from 0.01 and reduce half every 50 epochs. The predictions of our models are smoothed by simply averaging the predictions within a fixed window. We evaluate performance using 6-fold cross validation on the available dataset.

The concordance correlation coefficient (CCC) [14] works as the evaluation metric for this task, which is defined as:

$$\rho_c = \frac{2\rho\sigma_x\sigma_y}{\sigma_x^2 + \sigma_y^2 + (\mu_x - \mu_y)^2} \tag{2}$$

where μ_x and μ_y are the means of the sequence x and y, and σ_x and σ_y are the corresponding standard deviations. ρ is the Pearson Correlation Coefficient (PCC) between the x and y.

4.3   Experimental Results

As is shown in Table 1, the proposed efficient audio features significantly outperform the IS10 official audio features features and SoundNet features used in [7] both on the arousal and valence.

For the audio and facial expression uni-modality, our experimental results show that using the interlocutor's information significantly improves the performance on both arousal and valence. For audio modality, our interaction strategy gains 4.74% and 2.57% absolute improvement on the arousal and valence,

---

[1] http://sewaproject.eu.
[2] https://www.tensorflow.org.

respectively. These experimental results suggest that "Audio Interact" interaction strategy can make full use of the interlocutor's audio information. For facial expression modality, our "denseface.tune.interact2" features gains 4.38% and 1.96% absolute improvement on the arousal and valence respectively. However, the performance of the "denseface.tune.interact1" features is poor. The main reason for this poor performance is that when speaker is silent, there exist only facial expression information which is the only basis for labeling. Furthermore, the interlocutor's facial expressions are not exactly accurate when he/she talks, so this information cannot enhance the speaker's emotion expression, or even bring noises and reduce the performance.

As is shown in Table 2, compared with the uni-modality best performance, the results of the "multimodal1" features bring sginificant performance improvement on both arousal and valence. The superior performance of "multimodal1" features which include all available information than "multimodal2" features which ignore the facial expression information when she/he talks suggests that the speaker's facial expression information is useful for emotion prediction, although this facial expression information is not exactly accurate. Unlike the above mentioned conclusion that interlocutor's facial expressions when she/he talks will reduce the speaker's emotion prediction performance, the speaker's emotion expressed through the audio and facial expression information and both speaker's audio and facial experssion information are as basis for labeling.

Table 1. CCC performance of different interaction strategies of uni-modality on the validation and testing set. "A" and "V" represent the arousal dimension and the valence dimension, respectively. "vggish.100ms.interact" denotes the "vggish.100ms" features with "Audio Interact" strategy shown in Fig. 4. "denseface.tune.interact1" and "denseface.tune.interact2" denote the "denseface.tune" features with "Face Interact1" strategy and "Face Interact2" strategy shown in Fig. 4, respectively.

	Dim	Val		Test	
		A	V	A	V
IS10.100ms	120	46.74 ± 5.03	49.56 ± 6.70	35.95 ± 9.18	35.64 ± 10.07
soundnet	120	46.02 ± 6.15	49.08 ± 9.30	42.72 ± 8.66	44.68 ± 6.31
vggish.100ms	120	53.88 ± 6.36	53.99 ± 4.80	51.51 ± 4.71	52.03 ± 5.22
vggish.100ms.interact	120	**60.63 ± 4.21**	**59.90 ± 3.79**	**56.25 ± 6.32**	**54.6 ± 6.82**
denseface.tune	120	69.22 ± 6.11	69.72 ± 6.68	64.52 ± 6.65	65.62 ± 5.66
denseface.tune.interact1	240	67.74 ± 7.41	67.62 ± 7.51	63.68 ± 4.68	62.58 ± 5.35
denseface.tune.interact2	240	**72.99 ± 5.57**	**72.65 ± 4.51**	**68.9 ± 3.93**	**67.58 ± 5.18**

The "Interact-AFA" and "Interact-AFF" features, which only consider the interlocutor's audio and facial expression influence respectively, outperform the "multimodal1" features. These results show that both the interlocutor's audio and facial expression information are helpful for improving the prediction performance on arousal and valence. Finally, our proposed "Interact-AFAF" strategy, which reasonably using the interlocutor's audio information and facial expression information together, achieves the best performance on arousal and valence on the testing set. However, the "Interact-AFA" strategy outperforms the "Interact-AFAF" on both arousal and valence on the validation set. The main reason of phenomenon may be that the high dimensionality of the interaction construction features and the LSTM model can't handle this complex interaction construction. In the future, we will explore more efficient interaction models to handle this complex scenario. These results suggest that the proposed "Interact-AFAF" strategy which consider interlocutor's all availible information has strong generation capability and can bring more robust performance.

Table 2. CCC performance of different multimodal interaction strategies on the validation and testing set. "multimodal1" and "multimodal2" features are based on the "Multimodal1" strategy and the "Multimodal2" strategy shown in Fig. 5, respectively. "Interact-AFA" and "Interact-AFF" features are based on the "Interact-AFA" strategy and the "Interact-AFF" strategy shown in Fig. 6, respectively

	Dim	Val		Test	
		A	V	A	V
denseface.tune	120	69.22 ± 6.11	69.72 ± 6.68	64.52 ± 6.65	65.62 ± 5.66
multimodal1	120	**72.14 ± 3.89**	**71.56 ± 4.26**	**67.85 ± 6.89**	**68.73 ± 6.83**
multimodal2	120	68.93 ± 2.61	69.15 ± 1.86	65.79 ± 5.85	66.18 ± 4.25
Interact-AFA	240	**75.51 ± 3.85**	**74.99 ± 3.39**	70.64 ± 3.62	71.80 ± 4.04
Interact-AFF	240	73.99 ± 2.26	73.94 ± 2.42	69.47 ± 6.74	69.32 ± 4.19
Interact-AFAF	480	73.89 ± 3.23	73.53 ± 3.56	**71.67 ± 3.47**	**71.99 ± 4.36**

In this subsection, we explore and analyse the influence of correlation strength between interactive pairs for continuous emotion prediction. We analyse the correlations of the targets of each interactive pair, and the distribution of these correlations shown in Fig. 8. We find that there are 54.17% pairs and 70.83% pairs with strong positive correlation on arousal and valence respectively. For verifying the influence of pair's correlation strength on continuous emotion prediction, we select 3 pairs with the 3 largest PCC values as strong correlation pairs and 3 pairs with the 3 lowest PCC values as the weak correlation pairs. We randomly select 3 pairs from the rest pairs as the validation set and the rest 15 pairs as the training set. In order to get robust conclusion, we repeat 6 times and present the mean and standard deviation of CCC performance.

As is shown in Table 3, the pairs with strong correlation significantly outperform the pairs with weak correlation on both arousal and valence, especially on the arousal.

4.4   Influence of Correlation Strength Bewteen Interactive Pairs

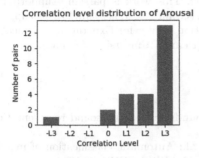
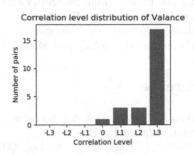

Fig. 8. Correlation distribution. The map from correlation levels to PCC is: {L3:[0.5, 1.0], L2:[0.3, 0.5], L1:[0.1, 0.3], 0:[−0.1, 0.1], −L1:[−0.3, 0.1], −L2:[−0.5, −0.3], −L3:[−1.0, −0.5]}.

Table 3. Influence of correlation strength of pairs on CCC performance.

	Dim	Val	Strong-cor	Weak-cor
Interact-AFAF.A	480	70.54 ± 5.28	**75.41 ± 1.63**	22.04 ± 3.66
Interact-AFAF.V	480	74.54 ± 2.78	**80.48 ± 1.41**	74.86 ± 5.55

5   Conclusion

In this paper, we explore several interaction strategies under uni-modality and mutil-modality scenarios for imitating the real dyadic human-human interaction patterns, and experiments show that the expressive behaviors from one dialog partner is complementary for predicting the emotional state of the interlocutor in dyadic human-human interactions. The audio and facial features are extracted from the pretrained VGGish and DenseNet models and a LSTM temporal model is used to capture the temporal information. The experimental results show that our proposed VGGish audio features outperforms the IS10 and SoundNet audio features. In the experiments of interaction strategies, our results show that the interlocutor's facial expression when speaker is silent will bring noises and reduce the speaker's emotion prediction performance. The others interlocutor's multi-modality informations are helpful for improving the speaker's continuous emotion prediction performance. And the final multimodal interaction strategy shows strong generation capability and robust performance. Additionally, we analyse the distribution of correlations between interacticve pairs of the AVEC 2017 dataset. The experimental results show that the pairs with strong correlation significantly outperform the pairs with weak correlation on both arousal and valence prediction, especially in arousal prediction. In the future, we will explore more efficient interaction strategies to make full use of the multimodal information. And we will explore more robust interaction strategies for adapting to a variety of scenarios that encourage emotional expression or restrain emotional expression.

Acknowledgment. This work is supported by National Key Research and Development Plan under Grant No. 2016YFB1001202. This work is partially supported by National Natural Science Foundation of China (Grant No. 61772535). We also appreciate the support from the National Demonstration Center for Experimental Education of Information Technology and Management (Renmin University of China).

References

1. Aytar, Y., Vondrick, C., Torralba, A.: Soundnet: learning sound representations from unlabeled video (2016)
2. Black, M., Katsamanis, A., Lee, C.C., et al.: Automatic classification of married couples' behavior using audio features. In: INTERSPEECH (2010)
3. Brady, K., Gwon, Y., Khorrami, P., et al.: Multi-modal audio, video and physiological sensor learning for continuous emotion prediction. In: AVEC (2016)
4. Busso, C., Bulut, M., Lee, C.C.: IEMOCAP: interactive emotional dyadic motion capture database. Lang. Resour. Eval. **42**(4), 335 (2008)
5. Chen, S., Jin, Q.: Multi-modal dimensional emotion recognition using recurrent neural networks. In: AVEC (2015)
6. Chen, S., Jin, Q.: Multi-modal conditional attention fusion for dimensional emotion prediction (2017)
7. Chen, S., Jin, Q., Zhao, J., et al.: Multimodal multi-task learning for dimensional and continuous emotion recognition. In: AVEC (2017)
8. Conati, C.: Probabilistic assessment of user's emotions in educational games. Appl. Artif. Intell. **16**, 555–575 (2002)
9. Fragopanagos, N., Taylor, J.G.: Emotion recognition in human-computer interaction. IEEE Sig. Process. Mag. (2002)
10. Hershey, S., Chaudhuri, S., Ellis, D.P.W., et al.: CNN architectures for large-scale audio classification. In: ICASSP (2017)
11. Hochreiter, S., Schmidhuber, J.: Long short-term memory. Neural Comput. **9**(8), 1735–1780 (1997)
12. Huang, G., Liu, Z., Weinberger, K.Q.: Densely connected convolutional networks. CoRR (2016)
13. Lee, C.C., Busso, C., Lee, S., et al.: Modeling mutual influence of interlocutor emotion states in dyadic spoken interactions. In: INTERSPEECH (2009)
14. Lin, L.I.: A concordance correlation coefficient to evaluate reproducibility. Biometrics **45**(1), 255–268 (1989)
15. Mariooryad, S., Busso, C.: Exploring cross-modality affective reactions for audiovisual emotion recognition. IEEE Trans. Affect. Comput. **4**(2), 183–196 (2013)
16. Metallinou, A., Katsamanis, A., Narayanan, S.: A hierarchical framework for modeling multimodality and emotional evolution in affective dialogs. In: ICASSP (2012)
17. Ringeval, F., Pantic, M., Schuller, B., et al.: Avec 2017: Real-life depression, and affect recognition workshop and challenge. In: AVEC Workshop (2017)
18. Sak, H., Senior, A., Beaufays, F.: Long short-term memory based recurrent neural network architectures for large vocabulary speech recognition. Comput. Sci. (2014)
19. Wöllmer, M., Metallinou, A., Eyben, F., et al.: Context-sensitive multimodal emotion recognition from speech and facial expression using bidirectional LSTM modeling. In: INTERSPEECH (2010)

Visual Object Tracking via Graph Learning and Flexible Manifold Ranking

Bo Jiang, Doudou Lin, Jin Tang$^{(\boxtimes)}$, and Bin Luo

School of Computer Science and Technology,
Anhui University, No.111 Jiulong Road, Hefei, China
jiangbo@ahu.edu.cn

Abstract. Recently, weighted patch representation has been widely studied to improve visual tracking by alleviating the undesired impact of background information in target bounding box. However, existing representation methods generally only use spatial structure information among patches which fails to exploit the unary feature information of each patch. In addition, traditional methods generally use a human fixed neighborhood graph for patch structure representation which may have no clear structure and also be sensitive to the noise. To overcome these problems, we propose a graph learning and flexible manifold ranking model for weighted patch representation. First, we propose to adopt a flexible manifold ranking for patch weight computation which explores both unary feature and structure relationship in a unified manner and thus performs more discriminatively than existing models which generally only explore structure relationship in patch representation. Second, we learn an adaptive and robust graph to better capture the intrinsic relationship among patches and thus can help to obtain a more robust patch representation. Extensive experiments on two standard benchmark datasets show the effectiveness of the proposed tracking method.

Keywords: Visual tracking · Graph learning
Flexible manifold ranking

1 Introduction

Visual tracking is an active research topic in computer vision area and has been extensively studied [8,11–13,16]. Existing visual tracking methods generally adopt tracking-by-detection framework which aims to localize the target object via a bounding box by using a classifier during the tracking process. One issue for this tracking-by-detection is that the bounding box is generally difficult to describe the target object accurately due to the irregular shape of the target object and thus usually introduces undesired background information, which may degrade the effectiveness of classifier.

To address this issue, many methods have been developed to alleviate the impact of background information in target object representation [4–6,11,15,22,23]. Among them, one kind of popular methods is to develop weighted

© Springer Nature Switzerland AG 2018
R. Hong et al. (Eds.): PCM 2018, LNCS 11164, pp. 313–323, 2018.
https://doi.org/10.1007/978-3-030-00776-8_29

patch representation to alleviate the impact of background information and thus improve the final tracking results [11–13]. These methods generally aim to first assign different foreground weights to the patches in bounding box feature representation to alleviate the impact of patches that are belonging to background and then incorporate this weighted patch representation into tracking-by-detection framework to propose a kind of robust tracking method. For example, Kim et al. [11] proposed to construct neighborhood graph to represent the structure of patches in bounding box and employed a random walk model over the neighborhood graph to obtain different weights for the patches. Li et al. [12] proposed to learn a low-rank sparse graph for target object representation to obtain more robust weight computation. Li et al. [13] recently also provided an improved graph learning model by further considering local and global relationship information

However, one main limitation of existing methods is that they only exploit structure information among patches in patch weight computation which explicitly neglect the unary features of patches although these features have been used in their graph construction. To overcome this problem, we propose a novel flexible manifold ranking model to obtain robust weighted patch representation for visual tracking. The proposed model simultaneously explores the unary feature and structure relationship in a unified manner and thus performs more robustly and discriminatively than existing methods which generally only explore structure relationship for patch ranking. Also, we propose to learn an adaptive and robust graph to better capture the intrinsic relationship among patches and thus can further help to obtain a more robust patch representation. We incorporate our weighted patch representation into the Struck [5] tracking framework to provide a robust tracking method. Extensive experiments on two standard benchmark datasets show the effectiveness and benefits of the proposed method.

2   The Proposed Representation Model

2.1   Patch Weight Computing via Flexible Manifold Ranking

Given one bounding box of the target object, we first partition it into n non-overlapping patches $p = \{p_1, p_2, \cdots p_n\}$. Then, we aim to assign each patch p_i with a foreground weight to represent its probability confidence of belonging to the target object. We conduct this task via a graph based ranking problem [19,24].

In order to do so, we first extract feature vector $\mathbf{x} \in \mathbb{R}^d$ (e.g., color, HOG) for each patch p_i. Let $\mathbf{X} = (\mathbf{x}_1, \mathbf{x}_2, \cdots \mathbf{x}_n) \in \mathbb{R}^{d \times n}$ be the collection of feature vectors. Then, a graph $G(V, E)$ is constructed, where nodes V represent the patches and edges E denote the spatial relationship $\mathbf{Z}_{ij}$ between patches. Let $\mathbf{y} = (\mathbf{y}_1, \mathbf{y}_2, \cdots \mathbf{y}_n)$ be the indication vector of queries, in which $\mathbf{y}_i = 1$ if patch p_i belongs to target object, and $\mathbf{y}_i = 0$ otherwise. Based on feature $\mathbf{X}$, graph $\mathbf{Z}$ and queries $\mathbf{y}$, we propose to compute the optimal patch weight of each patch according to their relevances to the queries by solving,

$$\min_{\mathbf{v}, \mathbf{w}, b} \frac{1}{2} \sum_{i,j} \mathbf{Z}_{ij}(\mathbf{v}_i - \mathbf{v}_j)^2 + \lambda \|\mathbf{X}^T\mathbf{w} + \mathbf{1}b - \mathbf{v}\|_2^2 + \beta \sum_i (\mathbf{v}_i - \mathbf{y}_i)^2 \quad (1)$$

where $\mathbf{w} \in \mathbb{R}^{d \times 1}$ and b is a scalar. Vector $\mathbf{1} = (1, 1, \cdots 1)^T \in \mathbb{R}^{n \times 1}$. The parameter λ and β control the balances of the linear prediction and label fitting terms. The main benefit of the above graph ranking is that it can conduct both label prediction via linear mapping and label propagation via graph regularization simultaneously.

Closed-Form Solution. The optimal $\mathbf{w}^*, b^*$ and $\mathbf{v}^*$ are obtained by setting the first derivation w.r.t variable $\mathbf{w}, b$ and $\mathbf{v}$ to zeros. After some algebra operations, we obtain the optimal solution as, $\mathbf{v}, \mathbf{w}, b$ [19], which are denotes as:

$$\mathbf{v}^* = \beta(\beta\mathbf{I} + \mathbf{L} + \lambda\mathbf{H}_c + \lambda\mathbf{N})^{-1}\mathbf{y}$$
$$\mathbf{w}^* = (\mathbf{X}_c\mathbf{X}^T)^{-1}\mathbf{X}_c\mathbf{v} \tag{2}$$
$$b^* = \frac{1}{n}(\mathbf{v}^T\mathbf{1} - \mathbf{w}^T\mathbf{X}\mathbf{1})$$

where $\mathbf{N} = \mathbf{X}_c^T(\mathbf{X}_c\mathbf{X}^T)^{-1}\mathbf{X}_c[\mathbf{X}^T(\mathbf{X}_c\mathbf{X}^T)^{-1}\mathbf{X}_c] - 2\mathbf{I}$, $\mathbf{X}_c = \mathbf{X}\mathbf{H}_c$, $\mathbf{H}_c = \mathbf{I} - \frac{1}{n}\mathbf{1}\mathbf{1}^T$. and $\mathbf{I}$ is an identity matrix.

We take this ranking $\mathbf{v}_i$ as foreground weight to represent its possibility of belonging to the target object. Comparing with previous models [11,12] which only use structure information, this model explores both feature and structure information of patches simultaneously.

2.2 Graph Construction

One important aspect of the above ranking model Eq. (1) is the construction of graph $\mathbf{Z}$. One traditional way is to construct a graph with k-nearest neighbor criteria and then assign edge weight $\mathbf{Z}_{ij}$ with Gaussian kernel or local linear reconstruction. However, such a pre-defined graph may have no clear structure and thus usually sensitive to the noise [21]. Therefore, we propose to learn an adaptive graph $\mathbf{Z}$ to better capture the intrinsic relationship among patches and performs robustly to the noise [9]. Inspired by graph learning work [21], we propose to learn a graph $\mathbf{Z}$ to better capture the intrinsic relationship among patches as,

$$\min_{\mathbf{Z}} \sum_{i,j} \|\mathbf{x}_i - \mathbf{x}_j\|\mathbf{Z}_{ij} + \gamma\|\mathbf{Z}\|_F^2 + \alpha Tr(\mathbf{F}\mathbf{L}\mathbf{F}^T) + \beta_1\|\mathbf{Z} - \mathbf{S}\|_2^2$$
$$s.t. \ \mathbf{Z} \geq 0, \mathbf{Z}^T\mathbf{1} = 1, \mathbf{F}\mathbf{F}^T = \mathbf{I}_c \tag{3}$$

where $\mathbf{S}$ denotes the traditional 8-neighborhood graph, i.e., each patch is connected to 8 neighbor patches and the edge weights are computed as Gaussian kernel [11]. The parameter α, γ, β_1 are used to balance the graph sparsity, connectivity and local structure preserving property, respectively. The Laplacian matrix $\mathbf{L}$ is computed as $\mathbf{L} = \mathbf{D} - \mathbf{W}$, where $\mathbf{D} = diag(d_1, d_2, \cdots d_n)$ and $d_i = \sum_j \mathbf{Z}_{ij}$.

Optimization. Since the problem is convex, the global optimal $\mathbf{Z}$ and $\mathbf{F}$ can be obtained by iteratively solving variable $\mathbf{Z}$ and $\mathbf{F}$ until convergence. Formally, the algorithm iteratively conducts the following **Step 1** and **Step 2** until convergence.

Step 1. Solving **Z** while fixing variable **F** And, we can obtain **W** by fixing variable **F**.Let $\mathbf{F} = (\mathbf{f}_1, \mathbf{f}_2, \cdots, \mathbf{f}_n) \in \mathbb{R}^{c \times n}$, then problem (Eq. (3)) can be rewritten more compactly as,

$$\min_{\mathbf{Z}} \sum_{i,j} \mathbf{A}_{ij} \mathbf{Z}_{ij} + \gamma \|\mathbf{Z}\|_F^2 \quad s.t. \quad \mathbf{Z} \geq 0, \quad \mathbf{Z}^T \mathbf{1} = 1, \tag{4}$$

where $\mathbf{A}_{ij} = \|\mathbf{x}_i - \mathbf{x}_j\|^2 + \frac{\alpha}{2} \|\mathbf{f}_i - \mathbf{f}_j\|^2 - 2\beta_1 \mathbf{S}_{ij}$ and $\mathbf{f}_i$ denotes the i-th column of matrix **F**. Problem Eq. (4) is equivalent to

$$\min_{\mathbf{z}_i \geq 0, \mathbf{z}_i^T \mathbf{1} = 1} \left\| \mathbf{z}_i - \left(-\frac{1}{2(\gamma + \alpha)} \right) \mathbf{a}_i \right\|^2 \tag{5}$$

where $\mathbf{z}_i, \mathbf{a}_i$ denote the i-th column of matrix **Z** and **A**, respectively. Thus, the optimal $\mathbf{z}_i$ can be obtained by using a simplex projection algorithm [10].

Step 2. Solving **F** while fixing variable **Z**, then the problem becomes,

$$\min_{\mathbf{F}} Tr(\mathbf{FLF}^T) \quad s.t. \quad \mathbf{FF}^T = \mathbf{I}_c \tag{6}$$

It is well known, the optimal **F** of problem Eq. (6) is given by the eigenvectors of Laplacian matrix **L** corresponding to the first c smallest eigenvalues.

3   Visual Object Tracking

In this section, we incorporate our optimized weights of patches into the tracking-by-detection framework, Struck [5] to provide a robust tracking algorithm. Overall, our tracking approach contains two main steps: weighted patch feature descriptor and structured output tracking.

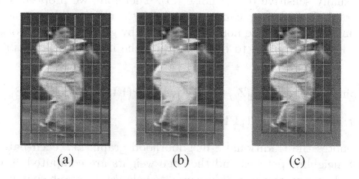

(a)   (b)   (c)

Fig. 1. (a) Original bounding box (red box) and expended bounding box (blue box); (b) Shrunk region (yellow region); (c) Expended region (blue region). (Color figure online)

Step 1: Weighted patch feature descriptor. Similar to [11], we first establish the foreground queries as follows. For each patch in the bounding box, if it belongs to the shrunk region of the bounding box then we regard it as a fore-ground query patch because it is likely to belong to target object, as shown in Fig. 1(b). In addition, we can also obtain background queries by using patches in the expended region of bounding box, as shown in Fig. 1(c). Based on these foreground and background queries, we can compute the foreground and background weights $\mathbf{v}_i, \mathbf{u}_i$ for each patch respectively by using the proposed graph ranking model. Then, we combine $\mathbf{v}_i$ and $\mathbf{u}_i$ together to provide a more accurate foreground weight as

$$w_i = \frac{1}{1 + \exp(-\epsilon(\mathbf{v}_i - \mathbf{u}_i))} \tag{7}$$

Let $\mathbf{X}^c = (\mathbf{x}_1^c, \mathbf{x}_2^c, \cdots \mathbf{x}_n^c)$ be the collection of original feature descriptor for bounding box c. By incorporating the computed weights w_i into the feature descriptor $\mathbf{X}^c$, one can obtain a kind of weighted descriptor for the bounding box c as

$$\mathbf{X}_w^c = \left(w_1\mathbf{x}_1^c, w_2\mathbf{x}_2^c \cdots w_n\mathbf{x}_n^c\right) \tag{8}$$

The weighted feature descriptor $\mathbf{X}_w^c$ alleviates the undesired effects of background information which thus provides a kind of more robust and accurate feature descriptor for target object [11].

Step 2: Structured SVM tracking. Our weighted patch descriptor $\mathbf{X}_w$ is incorporated into Struck tracking algorithm [5], which determines the optimal bounding box by maximizing the following

$$c^* = \arg\max_c \langle \mathbf{P}_{t-1}, \mathbf{X}_w^c \rangle \tag{9}$$

where $\mathbf{P}_{t-1}$ is the normal vector of a decision plane of $(t-1)$th frame. To further incorporate the information of the first initial frame and previous frame, instead of Eq. (9) we compute the optimal bounding box as

$$c^* = \arg\max_c \left(\alpha_1 \langle \mathbf{P}_{t-1}, \mathbf{X}_w^c \rangle + \alpha_2 \langle \mathbf{P}_{t-2}, \mathbf{X}_w^c \rangle + \alpha_3 \langle \mathbf{P}_1, \mathbf{X}_w^c \rangle \right), \tag{10}$$

where $\mathbf{P}_1, \mathbf{P}_{t-2}$ is learned in first frame and previous $t-2$ frame, respectively. This strategy can prevent it from learning drastic appearance changes. To prevent the effects of unreliable tracking results, the classifier is updated only when the confidence score of tracking result is larger than a threshold which is commonly set to 0.3.

4   Experiments

In this section, we evaluate the effectiveness of the proposed tracking method on two widely used benchmark datasets [14,20] and compare with some other state-of-the-art trackers. We implement our tracker using C++ language on a desktop computer with an Inter i7 3.6GHz CPU and 32GB RAM. The proposed tracker performs at about 6 FPS (frames per second).

4.1   Evaluation Settings

Parameters. Given one bounding box, we partition it into 8×8 no-overlapping patches to obtain better performance. For each patch, we extract a 32-dimensional feature descriptor including a 24-dimensional RGB color histogram and 8-dimensional oriented gradient histogram. For efficiency consideration, we scale each frame and set the minimum side length of bounding box to 32 pixels. We set the side length of a searching window as $2\sqrt{wh}$, where w and h denote the width and height of the scaled bounding box, respectively. We set parameters $\{\alpha, \gamma, \beta_1, \beta, \lambda\}$ to $\{0.01, 0.5, 11, 2.5, 1.25\}$. Empirically, the tracking performance is insensitive to these parameters.

Table 1. Comparison of attribute-based PR/SR scores on OTB benchmark dataset. The attributes includes IV (illumination variation), SV (scale variation), OCC (occlusion), DEF (deformation), MB(motion blur), FM (fast motion), IPR (in-plane-rotation), OPR (out-of-plane rotation), OV (out-of-view), BC (background clutters), and LR (low resolution). The best, second best and third best performances are indicated by red, green and blue colors.

	ACFN	Staple	MEEM	SOWP	LCT	SRDCF	HCF	DGT	Ours
FM	0.758/0.566	0.670/0.501	0.752/0.542	0.723/0.556	0.681/0.534	0.769/	0.815/0.570	0.777/0.549	/0.607
BC	0.769/0.542	0.716/0.524	0.744/0.515	0.775/0.570	0.734/0.550	0.769/0.573	/0.580	0.867/0.614	0.836/
MB	0.731/0.568	0.642/0.493	0.731/0.556	0.702/0.567	0.669/0.533	0.767/	/0.585	0.815/0.591	0.779/0.605
DEF	0.772/0.535	0.712/0.514	0.754/0.489	0.741/0.527	0.689/0.499	0.734/0.544	0.791/0.530	/	0.866/0.604
IV	0.777/0.554	0.772/0.551	0.740/0.517	0.766/0.554	0.732/0.557	0.792/0.613	0.817/0.540	0.838/0.573	0.819/
IPR	0.785/0.546	0.756/0.520	0.794/0.529	0.828/0.567	0.782/0.557	0.745/0.544	/0.559	0.856/0.573	0.819/
LR	0.818/	0.773/0.406	0.684/0.345	0.903/0.423	0.699/0.399	0.762/0.536	0.849/0.404	0.732/0.417	/0.479
OCC	0.756/0.542	0.715/0.520	0.741/0.504	0.754/0.528	0.682/0.507	0.735/0.559	0.767/0.525	/	0.825/0.577
OPR	0.777/0.543	0.734/0.523	0.794/0.525	0.787/0.547	0.746/0.538	0.742/0.550	0.807/0.534	0.855/	/0.589
OV	0.692/0.508	0.594/0.446	0.685/0.488	0.633/0.497	0.592/0.452	0.597/ 0.460	0.671/0.474	0.753/	/0.551
SV	0.764/0.551	0.732/0.506	0.736/0.470	0.746/0.475	0.681/0.488	0.745/ 0.561	0.799/0.485	/0.504	0.818/
ALL	0.802/0.575	0.755/0.537	0.781/0.530	0.803/0.560	0.762/0.562	0.789/	0.837/0.562	/0.586	0.865/0.612

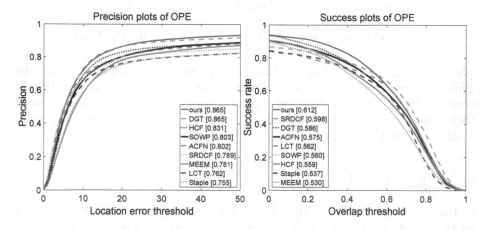

Fig. 2. Precision plots and success plots of OPE (one-pass evaluation) of the proposed tracker against other state-of-the-art trackers on OTB100 dataset.

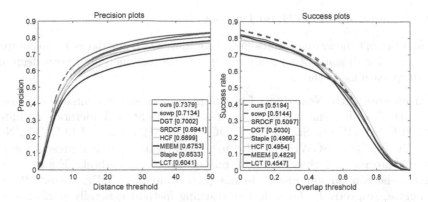

Fig. 3. Evaluation results on TColor-128 dataset, The legend contains representative PR and SR values. Note that, our method performs favorably against the state-of-the-art trackers.

OTB100 Benchmark Dataset. [20] contains 100 image sequences whose ground-truth target locations are marked manually. The sequences are associated with 11 different attributes. For performance evaluation metrics, we use precision rate (PR) and success rate (SR) [7, 20] to measure the quantitative performances of tracker methods.

Temple-Color Benchmark Dataset. [14] contains 128 challenging image sequences of human, animals and rigid objects, whose ground-truth target locations are marked manually. Each sequence in this dataset is also annotated by its challenge factors, which is the same as in [20]. The evaluation metrics including PR and SR used in this dataset are also same with [20].

Fig. 4. Tracking results of our method against 8 trackers (denoted in different colors and lines) on six challenging sequences. Intuitively, one can note that our tracker locates the visual object more accurately on these challenging sequences. (Color figure online)

4.2   Evaluation on OTB100 Dataset

We first present the evaluation results on standard OTB 100 dataset and compare our method with some other state-of-the-art methods including both deep and non-deep learning methods.

Comparison with Non-deep Learning Trackers. We compare our tracking method with some recent state-of-the-art traditional methods including DGT [12], SOWP [11], Staple [1], SRDCF [3], MEEM [23] and LCT [17]. Note that, DGT [12]and SOWP [11] also use the weighted patch representation in tracking process and thus are most related with our methods. Figure 2 summarizes the comparison results in one-pass evaluation (OPE) using PR and SR curves, respectively. Overall, our tracking method generally performs better than the other state-of-the-art methods. In particular, our method achieves 0.0%/2.6%, 6.2%/5.2% performance gains in PR/SR over related work DGT [12] and SOWP [11], which demonstrates the effectiveness of the proposed weighted patch representation model in conducting object tracking task. Note that, comparing with other related works, our model further explores the feature information via label prediction and also exploits temporal correlation in visual object representation and thus performs more discriminatively and robustly.

Comparison with Deep Learning Trackers. In Fig. 2, we also report some results of the deep learning based tracking methods including ACFN [2] and HCF [18]. Our method obtains better performance than these deep learning trackers. Note that, deep learning based tracking methods generally require large-scale annotated training samples while our method only uses the ground truth annotation in the first frame to train our model and then updates the model in subsequent frames.

Evaluation on Different Attributes. We present the representative PR/SR values on videos belonging to 11 different attributes, respectively. Table 1 reports the comparison results (PR/SR) on sequences that belong to 11 different attributes, respectively. One can note that, our method obtains the best performance on most challenging attributes in SR. Figure 4 shows some tracking examples on some challenging videos. Intuitively, one can note that our tracker locates the visual object more accurately on these challenging sequences.

4.3   Evaluation on Temple-Color Dataset

We also evaluate our method on Temple-Color dataset [14]. Figure 3 shows the success plot and precision plot over all 129 videos on this dataset. Generally, our tracker outperforms the other related trackers and obtains the best performance on PR/SR values. Especially, it achieves 2.54%/0.1% and 3.86%/1.15% performance gains in PR/SR over most related work SOWP and DGT. This further demonstrates the effectiveness of the proposed tracking method.

Table 2. Performance of three variants (Ours-noG, Ours-noF and Ours-noR) of the proposed ranking model on OTB100 dataset.

	Ours-noG	Ours-noF	Ours-noR	Ours
PR	0.838	0.839	0.842	0.865
SR	0.591	0.593	0.597	0.612

4.4 Component Analysis

To justify the importance of three main components (graph learning, linear mapping of features and graph fitting) in our representation model, we implement some special variants of our model, i.e., Ours-noG, Ours-noF and Ours-noR. (1) Ours-noG only uses the human established graph S with Gaussian Kernel function [11] and does not exploit the graph learning Z in our model. (2) Ours-noF that does not use the linear mapping of features term (the 2th term in Eq. (1)). (3) Ours-noR that does not use the graph fitting term (the 4th term in Eq. (3)) in our graph learning process. Table 2 summarizes the PR and SR scores on OTB 100 dataset. We can note that (1) utilizing the graph learning is obviously beneficial in our weighted patch representation and thus tracking performance. (2) the linear mapping of features is an important cue to obtain robust weighted patch representation. (3) the neighborhood structure of patches is a important cue to obtain robust weighted patch representation.

5 Conclusion

This paper proposes a graph learning and flexible manifold ranking model for weighted patch object representation and visual tracking problem. The proposed model integrates the cues of spatial structure and unary features together and thus performs more robustly and discriminatively in patch weight computation. We incorporate the optimized weighted patch representation into Struck tracker to carry out visual object tracking. Experiments on two standard benchmark datasets show the effectiveness of the proposed tracking method.

Acknowledgment. This work is supported by the National Natural Science Foundation of China (61602001, 61671018); Natural Science Foundation of Anhui Province (1708085QF139); Natural Science Foundation of Anhui Higher Education Institutions of China (KJ2016A020).

References

1. Bertinetto, L., Valmadre, J., Golodetz, S., Miksik, O., Torr, P.H.S.: Staple: complementary learners for real-time tracking. In: IEEE Conference on Computer Vision and Pattern Recognition, pp. 1401–1409 (2016)
2. Choi, J., Chang, H.J., Yun, S., Fischer, T., Demiris, Y., Choi, J.Y., et al.: Attentional correlation filter network for adaptive visual tracking. In: Proceedings of the IEEE Conference on Computer Vision and Pattern Recognition, vol. 2 (2017)
3. Danelljan, M., Hager, G., Khan, F.S., Felsberg, M.: Learning spatially regularized correlation filters for visual tracking. In: IEEE International Conference on Computer Vision, pp. 4310–4318 (2015)
4. Dorin, C., Visvanathan, R., Peter, M.: Kernel-based object tracking. IEEE Trans. Pattern Anal. Mach. Intell. **25**, 564–575 (2003)
5. Hare, S., Saffari, A., Torr, P.H.S.: Struck: structured output tracking with kernels. In: ICCV, pp. 263–270 (2011)
6. He, S., Yang, Q., Lau, R.W.H., Wang, J., Yang, M.H.: Visual tracking via locality sensitive histograms. In: IEEE Conference on Computer Vision and Pattern Recognition, pp. 2427–2434 (2013)
7. Henriques, J.F., Caseiro, R., Martins, P., Batista, J.: High-speed tracking with kernelized correlation filters. IEEE Trans. Pattern Anal. Mach. Intell. **37**(3), 583–596 (2015)
8. Hong, R., Hu, Z., Wang, R., Wang, M., Tao, D.: Multi-view object retrieval via multi-scale topic models. IEEE Trans. Image Process. **25**(12), 5814–5827 (2016)
9. Hong, R., Zhang, L., Tao, D.: Unified photo enhancement by discovering aesthetic communities from flickr. IEEE Trans. Image Process. **25**(3), 1124–1135 (2016)
10. Huang, J., Nie, F., Huang, H.: A new simplex sparse learning model to measure data similarity for clustering. In: IJCAI, pp. 3569–3575 (2015)
11. Kim, H.U., Lee, D.Y., Sim, J.Y., Kim, C.S.: Sowp: spatially ordered and weighted patch descriptor for visual tracking. In: IEEE International Conference on Computer Vision ICCV, pp. 3011–3019 (2015)
12. Li, C., Lin, L., Zuo, W., Tang, J.: Learning patch-based dynamic graph for visual tracking. In: AAAI (2017)
13. Li, C., Wu, X., Bao, Z., Tang, J.: ReGLe: spatially regularized graph learning for visual tracking. In: Proceedings of the 2017 ACM on Multimedia Conference, pp. 252–260. ACM (2017)
14. Liang, P., Blasch, E., Ling, H.: Encoding color information for visual tracking: algorithms and benchmark. IEEE Trans. Image Process. **24**(12), 5630–5644 (2015)
15. Liu, F., Gong, C., Zhou, T., Fu, K., He, X., Yang, J.: Visual tracking via nonnegative multiple coding. IEEE Trans. Multimed. **19**(12), 2680–2691 (2017)
16. Ma, B., Hu, H., Shen, J., Zhang, Y., Shao, L., Porikli, F.: Robust object tracking by nonlinear learning. IEEE Trans. Neural Netw. Learn. Syst. (2017)
17. Ma, C., Yang, X., Zhang, C., Yang, M.H.: Long-term correlation tracking. In: IEEE Conference on Computer Vision and Pattern Recognition (CVPR), pp. 5388–5396, June 2015
18. Ma, C., Huang, J.B., Yang, X., Yang, M.H.: Hierarchical convolutional features for visual tracking. In: IEEE International Conference on Computer Vision, pp. 3074–3082 (2015)
19. Nie, F., Xu, D., Tsang, W.H., Zhang, C.: Flexible manifold embedding: a framework for semi-supervised and unsupervised dimension reduction. IEEE Trans. Image Process. **19**(7), 1921–1932 (2010)

20. Wu, Y., Lim, J., Yang, M.H.: Object tracking benchmark. IEEE Trans. Pattern Anal. Mach. Intell. **37**, 1834–1848 (2015)
21. Xiong, K., Nie, F., Han, J.: Linear manifold regularization with adaptive graph for semi-supervised dimensionality reduction. In: Proceedings of the 26th International Joint Conference on Artificial Intelligence, pp. 3147–3153 (2017)
22. Yuan, Y., Yang, H., Fang, Y., Lin, W.: Visual object tracking by structure complexity coefficients. IEEE Trans. Multimed. **17**(8), 1125–1136 (2015)
23. Zhang, J., Ma, S., Sclaroff, S.: MEEM: robust tracking via multiple experts using entropy minimization. In: Fleet, D., Pajdla, T., Schiele, B., Tuytelaars, T. (eds.) ECCV 2014. LNCS, vol. 8694, pp. 188–203. Springer, Cham (2014). https://doi.org/10.1007/978-3-319-10599-4_13
24. Zhou, D., Bousquet, O., Lal, T.N., Weston, J.: Learning with local and global consistency. In: Neural Information Processing Systems, pp. 321–328 (2003)

Gaussian Dilated Convolution
for Semantic Image Segmentation

Falong Shen$^{(\boxtimes)}$ and Gang Zeng

Peking University, Beijing, China
shenfalong@pku.edu.cn

Abstract. In semantic image segmentation, multi scale contextual information is collected by probing the features with dilated large convolution filters or spatial pooling operations. Such enlargement of the receptive field promotes a more stable and global consistence segmentation prediction. Dilated convolution can be treated as the combination of a sampling process and a common convolution. For example, a 3×3 convolution with a large dilation rate picks 9 positions in a very large window. In this paper we propose a more rational way to sample features from a very large receptive field. Specifically Gaussian kernels are used to accumulate features in each position to produce a more stable representation. We also delve into the difference of up-sampling logits and down-sampling ground truth and provide a theoretical explanation. We demonstrate the effectiveness of Gaussian dilated convolution on the semantic image segmentation datasets of Pascal VOC 2012, Cityscapes and ADE20k. Gaussian dilated convolution performs consistently superior to dilated convolution throughout our experiments, which verifies the effectiveness of this method. Code will be released for reproduction.

Keywords: Dilated convolution · Semantic image segmentation
Gaussian window

1 Introduction

Semantic image segmentation is a long-standing and fundamental topic in computer vision. The goal of image semantic segmentation is to assign semantic labels to every pixel in a given image providing no interaction with users [5,7,24]. Recently the most of research works on semantic image segmentation are based on fully convolutional neural networks, which perform per-pixel classification in the given image [12,16]. The segmentation models of fully convolutional neural networks are fine-tuned from classification models pre-trained on large-scale datasets such as ImageNet-1k [6] and JFT-300M [18]. The Convolutional Neural Networks (CNN) for classification learn the abstract feature representation with consecutive pooling operations or strided convolutions, which increases the invariance ability to local image transformation [11]. Nevertheless, in semantic segmentation dense predictions for every pixel are required. There are two methods to overcome this difficulty. The first one is dilated convolution, which is a

R. Hong et al. (Eds.): PCM 2018, LNCS 11164, pp. 324–334, 2018.
https://doi.org/10.1007/978-3-030-00776-8_30

well-known trick in the wavelet community [9]. Dilated convolution is equivalent to inserting zeros between the filter weights to repurpose the 3×3 convolution layers in the pre-trained classification models. The second method is the consecutive deconvolution operations and the concatenations of low-level features, which is also known as U-net [12,13,15]. The dilated convolution or atrous convolution is also used to enlarge the receptive field of the segmentation classifiers to incorporate multi scale context. For example, Fisher yu *et al.* designed a cascaded module including a series of dilated convolution layers for semantic segmentation [22]. Deeplab series use several parallel dilated convolutions with various dilation rates and a global pooling layer to capture multi scale context and recolonize objects at multi scale [1–4,14].

In this paper we focus on the dilated convolution to harness the multi scale context information. In particular we investigate the sampling strategy of dilated convolution on the large window of feature maps. Previous work such as Astrous Spatial Pyramid Pooling (ASPP) employed three parallel dilated 3×3 convolutional layers, each of the dilated convolutional layers samples nine positions in different size of receptive windows. Recently, deformable dilated convolution was proposed to learn the sampling point in the windows using the *offset* learned from another convolutional layer. Instead of picking several fixed positions of the receptive window, we propose a more stable and robust way to exploit the large receptive field on the feature maps. Specifically, a large Gaussian kernel is performed on the feature maps to accumulate information and capture context preceding the dilated 3×3 convolution. The resolution of predicted score maps from Deep CNN is not compatible with the ground truth annotations. In order to compute the cross entropy loss, we need to up-sample the logits or down-sample the ground truth annotations. We theoretically discuss the difference between the two strategies. We also investigate the negative sample mining strategy for image semantic segmentation. Synchronized batch normalization is important for semantic image segmentation but it doesn't mean that larger *BatchNorm* batch size makes better performance. We have done experiments to compare different SGD batch size and BatchNorm batch size.

In summary this paper includes three folds of contributions:

- We propose a Gaussian kernel operation preceding dilated convolution to produce a more robust feature representation.
- A theoretical discussion on the difference of up-sampling logits and down-sampling ground truth is given and corresponding experiments are done.
- We investigate the hard sample mining strategy for image semantic segmentation.

2   Related Works

In recent works, dilated convolution has been an integral part for deep semantic segmentation [2–4,23]. Dilated convolution serves two functions in deep semantic segmentation: (a) it improves the resolution of predicted feature maps by removing the down-sampling layers from the last few blocks and replacing the following 3×3 convolution with dilated 3×3 convolution; and (b) it allows us to capture long range context at multi orientations and multi scales by using dilated convolutions with different dilation rates. We discuss some related works about how to exploit the long range context for deep semantic segmentation.

Dilated convolution or spatial pooling are adopted to enhance deep features in a *cascaded* manner or *parallel* manner. Yu and Koltun proposed *cascaded* dilated convolutions at progressively increasing dilation rates on top of the fully convolutional networks to aggregate multi scale contextual information [22]. Wu *et al.* employed two *cascaded* dilated 3×3 convolutions with a dilation rate of twelve on top of the final feature maps [21]. Atrous Spatial Pyramid Pooling (ASPP) is proposed in Deeplab series [1–4,14]. In the newest version of Deeplab v3+ [4], ASPP adopts three *parallel* dilated 3×3 convolutions with dilatation rates $= (6, 12, 18)$ and a global average pooling feature to encode context information. The convolution with the largest dilation rate covers almost the whole feature maps at the cost of lots of them extending out the boundary. Pyramid Scene Parsing Net [23] concatenates *parallel* spatial pooling features at 4-level scales to gather context information and achieved high performance on several segmentation benchmark datasets.

3   Methods

In this section we provide a detailed description of our proposed methods. Firstly we revisit the dilated convolution. For each location $\mathbf{p}$ on the output feature y and a filter ω, dilated convolution is applied over the feature map x

$$y[\mathbf{p}] = \sum_{\mathbf{k}} x[\mathbf{p} + r \cdot \mathbf{k}]\omega[\mathbf{k}] \tag{1}$$

where $\mathbf{k}$ is index of filter weight, r is the dilated rate and if we set $r = 1$ it equals to the standard convolution. Specifically for a 3×3 convolution filter, the dilated convolution pick 9 locations in a $(2r+1) \times (2r+1)$ window preceding the standard convolution. The dilated convolution allows us to adaptively to change the field-of-view of the classifier or filter by changing the dilated rate.

Although the theoretical receptive field of deep convolutional neural networks is very large, it is still necessary to employ cascaded or parallel modules on top of the original Deep CNN to keep global consistence and exploit long-range information. These extra module on top of deep features are often made up of pooling layers or dilated convolution with different pooling rates or dilatation rates. For example, Deeplab series adopt 5 parallel branches which consist of a 1×1 convolution, three 3×3 convolutions with dilation rates $(6, 12, 18)$ and a global average pooling, which they called it Atrous Spatial Pyramid Pooling (ASPP) [2,3]. The dilated convolutions in ASPP with different dilation rates effectively capture multi scale context by sampling a long-range location in the feature map. There are two defects in this sampling strategy:

– As the convolution window is as large as the dilation rate, the reason is not clear why the 9 locations are special and we do not choose other locations.

Fig. 1. The 3×3 convolution with a large dilation rate to capture long-range context. The dilated filter succeeds to build relationships between two locations spread across the feature map with a large distance. However, for some cases the 3×3 convolution degenerates to a 2×2 convolution or even 1×1 convolution since only the center filter weight is effective.

A more robust and steady feature representation for the large convolution window is necessary.

- A 3×3 convolution with large dilation rates is usually employed to capture wider context. As the dilation rates become larger, many 3×3 convolutions overstep the boundary of the feature maps which are padded with zero and then the 3×3 convolution degrades to a 2×2 convolution or even 1×1 convolution. This makes 3×3 convolution with large dilation rates difficult to effectively access long-range features.

Take Deeplab series as an example, the size of deep feature maps is 32×32 (input image size is 512×512, $output\_stride = 16$) while the largest dilation rate they used is 18. As shown in the Fig. 1, a large portion of positions have ran out of the feature maps. Deeplabv3+ proposes to use global pooling or image level feature to overcome this issue. However, the global pooling feature is dissimilar from the wider context. One might think stacking more dilated 3×3 convolutional layers with small dilation rates can give helps. However, stacking more dilated convolution layers equals to making the deep networks deeper which is not useful. To overcome the aforementioned two problems, we propose Gaussian dilated convolution to exploit wider context information. Instead of picking features from a special long-range location, we propose that the long-range location should collect features around its position $x[\mathbf{p}+r\cdot\mathbf{k}]$. This operation has two advantages: (a) it improves the stability and robustness of the long-range features relative to the center position. (b) it widens the field-of-view with less possibility of overstepping the boundary of the feature maps.

Formally for each 3×3 convolution location a Gaussian blur function G with standard variance σ and kernel radius d are performed on the feature map preceding the dilated 3×3 convolution,

$$y[\mathbf{p}] = \sum_{\mathbf{k}} G(x[\mathbf{p} + r \cdot \mathbf{k}], d, \sigma)\omega[\mathbf{k}] \qquad (2)$$

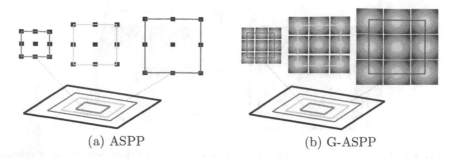

(a) ASPP   (b) G-ASPP

Fig. 2. Comparisons of Atrous Spatial Pyramid Pooling (ASPP) and Gaussian Atrous Spatial Pyramid Pooling (G-ASPP). The former picks 9 locations for a large field-of-view, while the later accumulates features among a larger region.

where x is the input feature maps and y is the output features. We set $d = \frac{r}{2}$ and $\sigma = 1$ throughout the experiments (Fig. 2).

3.1 Up-Sampling Predicted Map and Down-Sampling Ground Truth

The aforementioned deep segmentation model produces a coarse prediction with $output\_stride = 16$. The image scales of ground truth and predicted score map (or logits) are not compatible. To compute the cross entropy loss between the predicted score map and the ground truth, some previous works [1,14] suggested to down-sample the ground truth by the corresponding $output\_stride$. Formally denote the predicted score map as $\mathbf{p}$ and the ground truth as $\mathbf{t}$. Denote i as a location variable which iterates through the whole image, then the cross entropy loss is

$$\sum_i -\mathbf{t}_i^{\downarrow} \log \mathbf{p}_i, \tag{3}$$

where the $t_i^{\downarrow}$ is a one-hot vector from the down-sampled (by nearest neighbor sampling [1]) ground truth. Later some works [4,23] suggested to up-sample the predicted score map, then the cross entropy loss becomes

$$\sum_i -\mathbf{t}_i \log \mathbf{p}_i^{\uparrow}, \tag{4}$$

where $\mathbf{t}_i$ is a one-hot vector and $\mathbf{p}_i^{\uparrow}$ is the predicted score vector for the location i from the bilinear up-sampled score map $\mathbf{p}^{\uparrow}$. We want to delve into the difference between the two strategies. The resolution of the final predicted score map is $1/16$ of the ground truth. Simply bilinear up-sampling the predicted score map won't enrich the segmentation details and theoretically it should perform the same as down-sampling the ground truth. The key lies in the down-sampling strategy of the ground truth. Chen *et al.* [1] proposed to use nearest neighbor to down-sample the ground truth, which keeps the down-sampled ground truth still being one-hot. However, the nearest neighbor down-sampling operation misses all the fine annotations while the bilinear up-sampling of predicted map still keeps the information of detailed boundary by isotropic interpolation. Besides, the up-sampling of predicted map makes the training stage in the same manner as the testing stage. We have done some experiments to verify this analysis.

3.2 Hard Example Mining in Semantic Segmentation

In the stage of training a fully convolutional neural networks for semantic segmentation, each pixels i comes with a deep feature $\mathbf{x}_i$ and after `softmax` yielding to the predicted score $\mathbf{p}_i$. Then the back-propagation gradient of the loss regarding to the deep feature $\mathbf{x}_i$ is

$$\frac{\partial \ell}{\partial \mathbf{x}_i} = \frac{1}{N}(\mathbf{p}_i - \mathbf{t}_i),\tag{5}$$

which reflects the difference of two distributions and N is the number total pixels in a mini-batch. Some pixels lying at the center of a large object are easily discriminated from other pixels along the object boundary. The gradients from the loss of these "easy" pixels soon converge to zero but they still contribute to the total number of pixels N, which amounts to decrease the effective learning rate.

To keep the effective learning rate and force the networks to attach importance to the "difficult" pixels, following the previous works [20], we drop the pixels with smaller cross entropy loss. All the N pixels in a mini-batch are sorted descending the cross entropy loss

$$Sort\{-\mathbf{t}_i \log \mathbf{p}_i, i = 1...N\},\tag{6}$$

where i indexes a pixel. After sorting, only the forefront α portion of all the pixels are kept to compute the loss for back-propagation,

$$\ell = \frac{1}{N_\alpha}\sum\{-\mathbf{t}_i \log \mathbf{p}_i, i = 1...N_\alpha\}.\tag{7}$$

4   Experiments

Our implementation is based on Caffe [10]. The back-bone classification model throughout our work is ResNet-101 [8]. The *output_stride* is defined as the ratio of the image size to the deep feature size and the *output_stride* is controlled by using different dilatation rates in the blocks of ResNet-101. We set *output_stride* = 16 throughout our experiments. We do not adopt segmentation model with *output_stride* = 8 because it only brings a marginal improvement at the cost of three times of computation complexity. We evaluate the proposed methods on three semantic benchmark datasets.

Pascal VOC 2012 dataset [7] is a widely used benchmark dataset for semantic segmentation, including 20 categories (object) plus one background.

The original dataset includes 1464 (*train*), 1449 (*val*) and 1456 (*test*) pixel-wise labeled images. The *train* set is augmented with more pixel-wise labeled data, resulting into 10,582 (*trainaug*) images. The augmented annotation rule is slightly different with the original annotation rule therefore it is necessary to fine-tune the model on the original *trainval* set.

Cityscapes dataset [5] consists of 2975 (*train*), 500 (*val*) and 1525 (*test*) pixel-wise labeled images and 19998 *coarse* labeled images. There are 19 categories (object + stuff) in this datasets. Every image comes with the same size of 1024 × 2048 and all are about street scene in some European cities.

ADE20k dataset [24] is divided into 20,000 (*train*), 2000 (*val*) and 3000 (*test*) for training, validation and testing, respectively. There are 150 categories (object + stuff). ADE20k served as a competition dataset for scene parsing track in ILSVRC 2016.

Training Details: The batch size of SGD is 16 for all the three datasets, momentum is 0.9 and weight decay is 0.0001. We use the "poly" learning rate policy where the initial learning rate (0.01) is multiplied by $(1 - \frac{iter}{max\_iter})^{0.9}$. For Pascal VOC 2012, we firstly trains the model on *trainaug* set for 60k iterations, then fine-tune the model on official *train* set for another 30k iterations.

We implement the sync BatchNorm in Caffe to freely control the *batch_size* of BatchNorm in each mini-batch of SGD. Given the condition that the mini-batch size of SGD is fixed to 16, experiments on Pascal VOC 2012 show that *batch_size* = 4 of BatchNorm leads to a failure of convergence in training and that *batch_size* = 16 leads a decreased performance as shown in the Table 1.

The *crop_size* is 464 for Pascal VOC 2012 and is 688 for the other two datasets. We randomly resize the image to [0.5, 2] of its original scale with random aspect ratio $[\frac{3}{4}, \frac{4}{3}]$. We also adopt color jittering [±20] for three channels. Our baseline is ASPP from Deeplabv3+ [4] with modifications. The parallel branches of ASPP in our experiments include a *conv*1, two *conv*3s with dilation rates 6 and 12 and a global pooling operation.

Table 1. Ablative studies on the Pascal VOC 2012 *val* set. We firstly trains the model on *trainaug* set for 60k iterations, then fine-tune the model on official *train* set for another 30k iterations. The metric is the mean of intersection-over-union (mIoU).

ASPP	G-ASPP	Hard	BN8	BN16	US	BDS	NDS	mIoU (%)
✓								77.2
	✓		✓				✓	77.6
✓				✓	✓			78.5
✓			✓			✓		79.3
✓	✓		✓		✓			79.5
✓			✓		✓			80.2

Table 2. Results of *single model* and *single scale* on the *val* sets on Pascal VOC 2012, Cityscapes and ADE20k. We have submitted *single model* and *multi scale* predictions to the corresponding servers to measure the performance on the *test* sets. The backbone classification models for all these competing methods are ResNet-101. Our models set *output_stride* = 16 while other competing methods set *output_stride* = 8, which means our methods costs about one-third computation of other methods.

Method	PasVOC12[a]	CityScapes	ADE20k
Single model and *single scale* on *val* sets			
DUC-HDC [19]	-	76.2	-
SegModel [17]	77.7	-	-
PSP [23]	-	-	42.0
Deeplabv3 [3]	78.5	77.8	-
Deeplabv3+ [4]	79.4	-	-
Ours (ResNet-101, 16×)	80.2	78.7[b]	42.1
Single model and *multi scale* on *test* sets			
DUC-HDC [19]	-	77.6	-
VeryDeep [20]	79.1	74.6	-
Ours (ResNet-101, 16×)	81.3	81.4	35.3

[a] Auxiliary datasets are not used in all the competing methods.
[b] Coarse annotations are not used for experiments on the *val* set.

Ablative Studies: We evaluate the effectiveness of Gaussian dilated convolution and compare different experimental settings on Pascal VOC 2012 *val* set. The G-ASPP brings about one percent improvement comparing to the original ASPP. The nearest-neighbor down-sampling strategy (NDS) reaches a relative low performance (77.6%) while the bilinear down-sampling (BDS) performs better (79.3%) which conforms our expectation. BN8 and BN16 stand for that the *batch_size* of BatchNorm are 8 and 16, respectively. We found hard example mining and large *batch_size* of BatchNorm do not lead to superior performance on this dataset. However, the two tricks bring improvements on Cityscapes and ADE20k and we set $\alpha = 0.5$ for the two datasets. The detailed results are shown in the Table 1.

Comparisons to the State-of-the-Art: We further compare our methods with other works of semantic image segmentation on the *val* set and the *test* set of the three datasets. We compare the performance of *single model* and *multi scale* without resorting to any other auxiliary training data. Our G-ASPP improves the performance comparing to the base-line of Deeplab series [3,4] as shown in the Table 2. Some qualitative results are displayed in the Fig. 3.

Fig. 3. Visualization results on the *val* sets of Pascal VOC 2012, Cityscapes and ADE20k.

5   Conclusion

In this paper we propose Gaussian dilated convolution for semantic image segmentation, which boosts the performance of Atrous Spatial Pyramid Pooling from Deeplab. In previous works, the down-sampled ground truth and up-sampled logits lead to different performance. We explore the differences and propose a new down-sampling strategy. We adopt hard example mining for semantic image segmentation and compare the importance of different of batch size of batch normalization. Finally experiments on Pascal VOC 2012, Cityscapes and ADE20k have verified the effectiveness and superiority of our proposed models.

Acknowledgments. This work is supported by the National Key Research and Development Program of China (2017YFB1002601), and National Natural Science Foundation of China (61375022, 61403005, 61632003).

References

1. Chen, L.C., Papandreou, G., Kokkinos, I., Murphy, K., Yuille, A.L.: Semantic image segmentation with deep convolutional nets and fully connected CRFs. In: ICLR (2015)
2. Chen, L.C., Papandreou, G., Kokkinos, I., Murphy, K., Yuille, A.L.: Deeplab: semantic image segmentation with deep convolutional nets, atrous convolution, and fully connected CRFs (2016). arXiv:1606.00915
3. Chen, L.C., Papandreou, G., Schroff, F., Adam, H.: Rethinking atrous convolution for semantic image segmentation (2017). arXiv preprint: arXiv:1706.05587
4. Chen, L.C., Zhu, Y., Papandreou, G., Schroff, F., Adam, H.: Encoder-decoder with atrous separable convolution for semantic image segmentation (2018). arXiv preprint: arXiv:1802.02611

5. Cordts, M., et al.: The cityscapes dataset for semantic urban scene understanding. In: Proceedings of the IEEE Conference on Computer Vision and Pattern Recognition, pp. 3213–3223 (2016)
6. Deng, J., Dong, W., Socher, R., Li, L.J., Li, K., Fei-Fei, L.: Imagenet: a large-scale hierarchical image database. In: IEEE Conference on Computer Vision and Pattern Recognition, CVPR 2009, pp. 248–255. IEEE (2009)
7. Everingham, M., Van Gool, L., Williams, C.K., Winn, J., Zisserman, A.: The pascal visual object classes (VOC) challenge. Int. J. Comput. Vis. **88**(2), 303–338 (2010)
8. He, K., Zhang, X., Ren, S., Sun, J.: Deep residual learning for image recognition. In: Proceedings of the IEEE Conference on Computer Vision and Pattern Recognition, pp. 770–778 (2016)
9. Holschneider, M., Kronland-Martinet, R., Morlet, J., Tchamitchian, P.: A real-time algorithm for signal analysis with the help of the wavelet transform. In: Combes, J.M., Grossmann, A., Tchamitchian, P. (eds.) Wavelets. Inverse Problems and Theoretical Imaging, pp. 286–297. Springer, Heidelberg (1990). https://doi.org/10.1007/978-3-642-75988-8_28
10. Jia, Y., et al.: Caffe: convolutional architecture for fast feature embedding. In: Proceedings of the 22nd ACM International Conference on Multimedia, pp. 675–678. ACM (2014)
11. Krizhevsky, A., Sutskever, I., Hinton, G.E.: Imagenet classification with deep convolutional neural networks. In: Advances in Neural Information Processing Systems, pp. 1097–1105 (2012)
12. Long, J., Shelhamer, E., Darrell, T.: Fully convolutional networks for semantic segmentation. In: Proceedings of the IEEE Conference on Computer Vision and Pattern Recognition, pp. 3431–3440 (2015)
13. Noh, H., Hong, S., Han, B.: Learning deconvolution network for semantic segmentation. In: Proceedings of the IEEE International Conference on Computer Vision, pp. 1520–1528 (2015)
14. Papandreou, G., Chen, L.C., Murphy, K., Yuille, A.L.: Weakly- and semi-supervised learning of a DCNN for semantic image segmentation. In: ICCV (2015)
15. Ronneberger, O., Fischer, P., Brox, T.: U-Net: convolutional networks for biomedical image segmentation. In: Navab, N., Hornegger, J., Wells, W.M., Frangi, A.F. (eds.) MICCAI 2015, Part III. LNCS, vol. 9351, pp. 234–241. Springer, Cham (2015). https://doi.org/10.1007/978-3-319-24574-4_28
16. Sermanet, P., Eigen, D., Zhang, X., Mathieu, M., Fergus, R., LeCun, Y.: Overfeat: integrated recognition, localization and detection using convolutional networks (2013). arXiv preprint: arXiv:1312.6229
17. Shen, F., Gan, R., Yan, S., Zeng, G.: Semantic segmentation via structured patch prediction, context CRF and guidance CRF. In: Proceedings of the IEEE Conference on Computer Vision and Pattern Recognition, pp. 1953–1961 (2017)
18. Sun, C., Shrivastava, A., Singh, S., Gupta, A.: Revisiting unreasonable effectiveness of data in deep learning era. In: 2017 IEEE International Conference on Computer Vision (ICCV), pp. 843–852. IEEE (2017)
19. Wang, P., et al.: Understanding convolution for semantic segmentation (2017). arXiv preprint: arXiv:1702.08502
20. Wu, Z., Shen, C., van den Hengel, A.: Bridging category-level and instance-level semantic image segmentation (2016). arXiv preprint: arXiv:1605.06885
21. Wu, Z., Shen, C., van den Hengel, A.: Wider or deeper: Revisiting the resnet model for visual recognition (2016). arXiv preprint: arXiv:1611.10080
22. Yu, F., Koltun, V.: Multi-scale context aggregation by dilated convolutions (2015). arXiv preprint: arXiv:1511.07122

23. Zhao, H., Shi, J., Qi, X., Wang, X., Jia, J.: Pyramid scene parsing network. In: IEEE Conference on Computer Vision and Pattern Recognition (CVPR), pp. 2881–2890 (2017)
24. Zhou, B., Zhao, H., Puig, X., Fidler, S., Barriuso, A., Torralba, A.: Scene parsing through ADE20K dataset. In: Proceedings of CVPR (2017)

Reading Document and Answering Question via Global Attentional Inference

Jun Song[1], Siliang Tang[1], Tianchi Qian[1], Wenwu Zhu[2], and Fei Wu[1(⊠)]

[1] College of Computer Science and Technology, Zhejiang University, Hangzhou, China
{songjun54cm,siliang,qiantianchi}@zju.edu.cn, wufei@cs.zju.edu.cn
[2] Computer Science Department, Tsinghua University, Beijing, China
wwzhu@tsinghua.edu.cn

Abstract. Teaching a computer to understand and answer a natural language question pertaining to a given general document is a central, yet unsolved task in natural language processing. The answering of complex questions, which require processing multiple sentences, is much harder than answering those questions which can be correctly answered by merely understanding a single sentence. In this paper, we propose a novel global attentional inference (GAI) neural network architecture, which learns useful cues from structural knowledge via a dynamically terminated multi-hop inference mechanism, to answer cloze-style questions. Here, reinforcement learning is employed to learn an inference gate which can determine whether to keep accumulating cues or to predict an answer. By exploiting structural knowledge, our model can answer complex questions much better than other compared methods.

Keywords: Reading comprehension · Global attention
Question answering

1 Introduction

Humans do not learn language in an isolated way. The context in which words and sentences are understood, whether a conversation, book chapter or road sign, plays an important role in comprehension [1]. Thus building a machine reading system is a challenging task, which requires the understanding of natural languages and the ability to reasoning over various cues.

A recent trend to measure machines' ability toward reading and comprehension is to test a system's performance on answering cloze-style questions over documents.

- **Cloze-style questions** [15] are questions in which a particular entity word is hidden. A popular way to generate this kind of question is to remove an entity from a selected sentence. Figure 1 shows one example of cloze-style question, as well as its anonymisation counterpart (i.e., the entity words are replaced

© Springer Nature Switzerland AG 2018
R. Hong et al. (Eds.): PCM 2018, LNCS 11164, pp. 335–345, 2018.
https://doi.org/10.1007/978-3-030-00776-8_31

with anonymisation tokens), respectively. In Fig. 1 the original selected sentence comes from the original document. The anonymized document is generated by replaced entity words in the original document with anonymisation tokens (e.g., @entity0, @entity1). The question is generated by replacing the entity "Davion Navar Henry" in the original selected sentence (@entity4 in the anonymized document) with @placeholder.

Original Document: (CNN) The boy who asked a church to help him find a forever parent finally has one. Desperate for a home in 2013, Davion Navar Henry only dressed up in a suit and borrowed a Bible from the boys home where he lived. Then he headed to a St. Petersburg, Florida, church to ······

Original Selected Sentence: *Davion Navar Henry* only took to the pulpit to find a forever home.

Anonymized Document: (@entity0) the boy who asked a church to help him find a forever parent finally has one. desperate for a home in 2013, @entity4 only dressed up in a suit and borrowed a @entity7 from the boys home where he lived. then he headed to a @entity11, @entity12, church to ······

Question: *@placeholder* only took to the pulpit to find a forever home.

(**Answer**: @entity4)

Fig. 1. An example of cloze-style question. The document is a news article, the question is a cloze-style question, in which one entity in the sentence is replaced by a placeholder, and the answer is the questioned entity. Anonymisation algorithms have been used to replace tokens in original texts with user defined tokens.

To answer the cloze-style questions, we need to detect appropriate cues in given context. It is not easy for a computer to correctly answer the question in Fig. 1, since the question can not be answered by learning any individual sentence in the document. In fact, to answer this question, human readers may ask "who wants a home?", "who makes a plea for adoption?" and "who is the boy?". In general when people have trouble understanding what they are reading, they tend to go back and re-read a portion of the text several times to further accumulate more cues [18]. Thus, after re-reading the document in Fig. 1 multiple times, human readers may pay attention to the global cues like "home", "find", "family", "church" and "boy". By inferring these global cues, the answer may be found as "Davion Navar Henry". In this paper, we design a novel neural network model, which can infer global attentional cues via adaptively re-reading structural knowledge (triplets extracted from context) multiple times (i.e., hops), to answer cloze-style questions.

In the research of reading and answering questions, symbolic models, introduced by Hermann et al. [8] and Chen et al. [3] like models based on word maximum frequency, exclusive frequency, frame semantic and word distance can get some results. These models always need less training but have lower performance comparing with neural models.

In recent years, neural network based machine reading models have been developed and shown their advantages. Recurrent neural networks [2] are widely used in this scenario. Efforts like Deep LSTM Reader [7] deals with a pair of document-question as a single long sequence. Some efforts build dynamic meaning representations for context, like Memory Networks (MemNets) [9], Dynamic Entity Representation Reader (DER Reader) [11]. EpiReader [17] re-ranks the hypotheses using supporting text. Reasoning Network (ReasoNet) [13] makes use of multiple turns to read and then produce an answer. Many efforts employ an attention mechanism in the sequence learning procedure, like Attentive Reader and Impatient Reader [8], Attention Sum Reader (AS Reader) [10], Stanford Attention Reader (Stanford AR) [3], Iterative Attention Reader (IA Reader) [14], Gated Attention Reader (GA Reader) [5] and Attention-over-Attention Reader (AoA Reader) [4].

In order to accumulate adequate cues before answering a question, we devise a dynamically terminated multi-hop inference algorithm, which adaptively decides whether to keep re-reading the triplets several times or to predict an answer. The REINFORCE mechanism [20] is used to train the dynamically terminated global inference (i.e., stop reading) when information has been adequately accumulated w.r.t a question. Our experiments show that our model answers cloze-style questions in higher accuracy, comparing with other methods and our model indeed learns information from multiple sentences in a wider and deeper way.

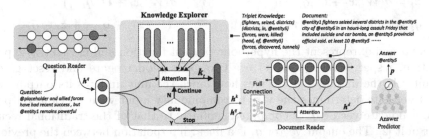

Fig. 2. The overall architecture of our model. There are four main components for global attentional inference: Question Reader, Knowledge Explorer, Document Reader and Answer Predictor. The Question Reader and the Document Reader are two bi-directional GRU network, respectively. The Knowledge Explorer adopts the dynamically terminated multi-hop inference mechanism. The Answer Predictor computes the probability of an answer.

2   Our Approach

In this section, we explain the general architecture of our model in detail. There are four major components in our model: question reader, knowledge explorer, document reader and answer predictor. The question reader and document reader are two different bi-directional GRU neural networks. The answer predictor predicts an answer by computing the probabilities of candidate answers using softmax function. The knowledge explorer adopts a dynamically terminated multi-hop inference mechanism, that will exploit the structural knowledge multiple times until the inference gate decides to stop or the maximum inference steps are reached. Figure 2 illustrates the overall architecture of our model. In the rest of this section, we describe each component of our model.

2.1 Question Reader

A word embedding matrix E is learned and shared among all the components in our model. The matrix E contains thousands of embedding vectors with a same dimension. Each word embedding is a vector representation of a word in a word dictionary. We first initialize the embedding matrix E using GloVe [12], then tuning it while training our model.

Question reader consists of a bi-directional GRU network (BiGRU), whose bi-directional architecture follows the bi-directional LSTM [6], but the LSTM unit is replaced by a GRU unit, which performs similarly but computationally more efficient than LSTM. For a GRU unit, at each step t, it takes the previous output vector q_{t-1} and current word embedding vector x_t as input, and computes a new output vector q_t, as follows:

$$z_t = \sigma(W^z x_t + U^z q_{t-1}); \qquad r_t = \sigma(W^r x_t + U^r q_{t-1})$$
$$\tilde{q}_t = \tanh(\tilde{W} x_t + \tilde{U}(r_t \odot q_{t-1})); \qquad q_t = (1 - z_t) \odot q_{t-1} + z_t \odot \tilde{q}_t \qquad (1)$$

where σ and $\odot$ represent the hyperbolic tangent function and element-wise production, respectively. z_t and r_t are two gates (update gate and reset gate). Intuitively the update gate decides how much the unit updates its output and the reset gate determines the way to combine new input with previous output. The candidate vector $\tilde{q}_t$ is computed similarly to that of the traditional recurrent unit [2]. The output vector q_t is a linear interpolation between the previous output vector q_{t-1} and the candidate vector $\tilde{q}_t$. Here $W^z, U^z, W^r, U^r, \tilde{W}, \tilde{U}$ are parameter matrices need to be learned during model training.

The BiGRU network consists of two GRUs in two directions: forward and backward. At each step t, two hidden vectors, $\overrightarrow{q}_t$ and $\overleftarrow{q}_t$, are learned by the BiGRU network, which can be viewed as a summary of the past and the future information, respectively. We compute the question vector by concatenating the last output vector of the forward GRU and the backward GRU: $h^q = [\overrightarrow{q}_{last}; \overleftarrow{q}_{last}]$.

2.2 Knowledge Explorer

We regard (subject, verb, object) triplets, that extracted from sentences in each document, as structural knowledge. Given a sentence, we extract these triplets using the Stanford Part-Of-Speech Tagger [16]. Then each triplet will be mapped into three word embedding vectors according to the shared word embedding matrix. We then concatenate these three vectors to one vector v_i to denote the ith triplet.

The knowledge explorer can infer global attentional cues w.r.t. a given question from relevant knowledge. It keeps inferring global attentional cues among all the triplets until the inference gate decides to stop (when adequate observed information is ready for an answer) or the maximum inference steps are reached.

– Dynamically terminated multi-hop inference: The knowledge explorer keeps inferring until the inference gate decides to stop. The inference gate computes a stop probability score s_t at each inference step and compares it with a threshold τ. If $s_t \geq \tau$ or the maximum inference steps are reached, the knowledge explorer will stop inferring cues from triplets and output a learned knowledge vector h^k. Otherwise, it will keep inferring. Thus the knowledge explorer may infer multiple time over the same set of triplets and the number of times it infers differs over different data samples. We call this process "dynamically terminated multi-hop inference".

At each inference step t, a new inference vector k_t is learned based on the knowledge vectors $v_{1...K}$ and the previous inference vector k_{t-1}. We first compute an attention vector α_t according to the previous inference vector, and the new inference vector is a linear interpolation among all the knowledge vectors. At the first inference step, the question vector h^q is used to compute the attention vector, as follows:

$$\alpha_{ti} = \frac{\exp(k_{t-1}^T W^\alpha v_i)}{\sum_i^K \alpha_{ti}}; \quad k_t = \sum_i^K \alpha_{ti} v_i; \quad k_0 \equiv h^q \tag{2}$$

where K is the number of triplets extracted from sentences. α_{ti} and v_i are the ith element of the attention vector and the ith knowledge vector, respectively. We compute the attention vector via a bilinear mapping, which reflects how relevant or important each knowledge vector is to the previous inference vector. W^α is a parameter matrix that will be learned during model training. For convenience, we denote the final inference vector as $h^k \equiv k_{final}$.

The knowledge explorer will keep learning new inference vector until the inference gate tells it to stop or the maximum inference steps are reached. At each inference step, the inference gate computes a stop probability score s_t. By comparing the score with a threshold, the gate tells the explorer to keep inferring or terminate. The probability score is computed via a bilinear mapping based on the current inference vector and the question vector: $s_t = \phi(k_t^T W^s h^q)$. ϕ denotes the *sigmoid* function and W^s is a parameter matrix need to be learned.

It is hard to train one optimal dynamically terminated multi-hop mechanism, since we do not exactly know when to stop inferencing. Thus, in practice, we apply the REINFORCE idea [20] to minimize a inference loss (equals to maximizing a reinforce reward). The inference loss can only be computed at the final inference step when an answer is predicted. If the answer is correct, that means the knowledge explorer stop inferencing at the right time, thus we set the expected stop probability score as $y^s = 1$. Otherwise, we set the expected probability as $y^s = 0$, since the explorer should stop earlier or not stop. For the nth training sample, the inference loss l_n^{infer} is computed as the cross entropy loss, by comparing the final termination stop probability score s_n^{final} with the expected score y_n^s, as follows:

$$l_n^{infer} = -(y_n^s \log(s_n^{final}) + (1 - y_n^s) \log(1 - s_n^{final})) \qquad (3)$$

Intuitively, every time the knowledge explorer produces a new inference vector, it combines the learned information with triplet knowledge and decides whether to stop inferring or not. By iteratively inferring though the triplet knowledge, the model can gradually approach to the correct answer.

2.3 Document Reader and Answer Predictor

The document reader is another BiGRU network which has the same architecture with the question reader but has its own parameters. The document reader reads all the words in a document and computes their hidden vectors $u_{1...T}$. We concatenate the question vector h^q and the inference vector at the final termination inference step h^k. Then the concatenated vector is mapped into a new vector ω by a fully connected neural network layer. Take ω as a cue, we first compute an attention vector β based on $u_{i...T}$ and then compute the document vector h^d using the attention vector, as follows:

$$\omega = \tanh(W^f[h^k; h^q] + b^f); \quad \beta_i \propto \exp(\omega^T W^\beta u_i); \quad h^d = \sum_i^T \beta_i u_i \qquad (4)$$

here W^f, b^f and W^β are a parameters need to be learned. The document reader computes h^d by integration of the document context, according to the attention vector.

The document vector reflects the importance of document context to the question. We use the vector to predict the answer by projecting it to a probability distribution. The probabilities are computed using a learned mapping matrix and an activation function: $p = f(W^p h^d + b^p)$. In our experiments softmax function is used.

Where p denotes the prediction vector. W^p and b^p denote the learned mapping matrix and bias vector, respectively. Each element in p denotes the probability of the corresponding candidate. Finally, we take the candidate which has the highest probability, as the predicted answer.

2.4 Model Training

To train our model, we aim to minimize the overall loss of our model on the training data. The overall loss of our model is a composition of the inference loss and the prediction loss. For the nth sample, we compute the prediction loss l_n^{pred} as the cross entropy loss between the predicted answer and the ground truth answer. The training objective is to minimize the overall loss with an L_2 regularization term as follows:

$$O(\theta) = \arg\min_{\theta} \frac{1}{N} \sum_{n}^{N} (l_n^{pred} + \lambda_1 l_n^{infer}) + \frac{\lambda_2}{2} ||\theta||_2^2 \qquad (5)$$

where θ denotes all the parameters need to be trained. N is the number of all training samples. λ_1 denotes how much we want the model to have a good inference ability. λ_2 is the regularization rate.

In order to efficiently train our model, we use the Chain Rule and Back Propagation Through Time [19] to compute those gradients. We train our model using stochastic gradient descent with learning rate 0.1 and regularization rate 0.001. While training our model, we set the inference loss rate $\lambda_1 = 0.3$ and the inference gate threshold $\tau = 0.5$.

3 Experiments

In this section, we evaluate the performance of our model on two reading and comprehension corpora, CNN and Daily Mail [8]. We compare the performance of our model with other symbolic models and neural network models on the same task. We demonstrate experimental results to validate the superiority of our model (i.e., inferring the global attentional cues from triplet knowledge). Intuitively, by learning the global attentional cues, our model can answer much more complex questions with better performance.

3.1 Accuracy

We evaluate our method on the task of answering cloze-style questions over documents. Table 1 shows the answering accuracy comparisons of different models on the task. From Table 1, we can see that neural network based models gain better performance than the symbolic models. Compared with other neural models, our model (GAI), which infers global attentional cues via dynamically terminated multi-hop inference over triplet knowledge, shows its advantages.

3.2 Global Inference

We evaluate the performance of our model on answering complex questions. Here we regard the questions which require more than one sentence to infer the correct answers as complex questions. To evaluate the performance, we select 200 complex question samples and 200 simple question samples. The document sentences in a complex question sample do not contain the answer token and any common words with the question at the same time. Otherwise, the sample is a simple one. Table 2 shows the comparison among three different models. Stanford Attention Reader [3] does not consider any structural knowledge. GAI(hop = 1) is a simplified version of our proposed approach which only exploits structural

Table 1. Accuracy of different models.

Models	CNN		Daily mail		Models	CNN		Daily mail	
	Valid	Test	Valid	Test		Valid	Test	Valid	Test
Maximum frequency	30.5	33.2	25.6	25.5	Deep LSTM Reader	55.0	57.0	63.3	62.2
Exclusive frequency	36.6	39.3	32.7	32.8	Attentive Reader	61.6	63.0	70.5	69.0
Frame-semantic	36.3	40.2	35.5	35.5	MemNets	63.4	66.8	-	-
Word distance	50.5	50.9	56.4	55.5	AS Reader	68.6	69.5	75.0	73.9
Stanford AR	72.2	72.4	76.9	75.8	GA Reader	73.0	73.8	76.7	75.7
DER Network	71.3	72.9	-	-	AoA Reader	73.1	74.4	-	-
IA Reader	72.6	73.3	-	-	ReasoNet	72.9	74.7	77.6	76.6
EpiReader	73.4	74.0	-	-	GAI	**74.2**	**75.0**	**77.8**	**77.0**

knowledge once. In practice, we set the maximum inference step as 1. Our model (GAI) infers global attentional cues by exploiting the structural knowledge multiple times, until it is able to infer the answer correctly.

Table 2. Performances of different models over simple and complex questions.

Models	Simple	Complex
Stanford AR	73.9 (↑ 0.00%)	29.9 (↑ 00.00%)
GAI(hop = 1)	74.6 (↑ 0.95%)	35.4 (↑ 18.39%)
GAI	**74.9 (↑ 1.35%)**	**36.3 (↑ 21.40%)**

From Table 2, we can see that answering those complex questions is much harder than answering those simple questions. By inferring over those triplet knowledge, models can answer both simple questions and complex questions with higher accuracy. Since these triplets deliver the key semantics of multiple sentences in a dense way. By exploiting these triplets, the model indeed learns information from multiple sentences in a wider and deeper way.

3.3 Inference Flow Path

We human beings perform inference via a inference flow path, namely, a sequence of triplets is in order employed to accumulate adequate cues before answering an answer w.r.t a question. Here we are interested in how many triplets are involved one by one during answering a question. We define the length of inference flow path as the least number of subject-verb-object triplets needed to predict the correct answer. Given the example in Fig. 1, assume we can extract many of triplets from the corresponding document such as "(family, at, home), (boy, has, family), (family, welcome, boy), (house, is beautiful), (boy, is, Davion)" respectively. Once the question "@placeholder find a home" is given, the correct answer

"Davion" can be inferred by employing at least 3 triplets: (family, at, home) →
(family, welcome, boy) → (boy, is, Davion)". Thus the length of inference flow
path w.r.t this question is 3. The length of inference flow path can be regarded
as a measure of difficulty towards answering a question. The longer the activated
inference path of answering a question is, the harder the question is.

We divide the questions in test and validation data sets into 6 groups with
the length of inference flow path from 1 to 5 and greater than 5. Then we
evaluate different models on the 6 groups of data. The accuracies are shown in
Table 3. From Table 3, we can see that the larger the length of inference flow
path is, the lower the accuracy of predicting a answer will be. Among those
evaluated models, GAI(hop = 1) shows better performance than Stanford AR.
Our model which infers global attentional cues from structural knowledge and
applies dynamically terminated multi-hop inference can always achieve better
performance than other models.

Table 3. Accuracies of different models over different lengths of inference flow paths.

Models	1	2	3	4	5	>5
Stanford AR	73.4	68.8	57.8	53.8	48.0	44.6
GAI(hop = 1)	74.2	69.0	58.0	54.7	48.6	46.1
GAI	**76.2**	**73.7**	**61.5**	**57.6**	**54.1**	**52.4**

3.4 Multi-hop Attention

Figure 3 shows a learned result of an example. In this example, the question can
not be answered easily by reading any individual sentence. In Fig. 3, triplets with
high attention weights at the first hop are denoted with red lines, and triplets
with high attention weights at the second hop are denoted with blue dash lines.
Words with high attention weights are highlighted with background color. Our
model learns to pay much attention on those triplets that will provide helpful
cues towards answering this question. At the first inference hop, triplets that have
strong relation with the question gain high attention weights, since the words
"home", "plea" are strongly related with the question. At the second inference
hop, triples that gain high attention weights have strong correlation with the
answer. As the words "boy", "adoption" have strong relation with the answer
token. While reading the document, our model learns to pay much attention on
those candidate answers, as highlighted in Fig. 3.

> **Document**: (@entity0) the boy who asked a church to help him find a forever parent finally has one. desperate for a home in 2013, @entity4 only dressed up in a suit and borrowed a @entity7 from the boys home where he lived. then he headed to a @entity11, @entity12, church to make a plea for his own adoption. now 16 years old, he had lived his entire life in foster care, bouncing from one home to another. the older he got, the less likely it was that he would be adopted. but the @entity20 documented his journey, and a video of his plea went viral. thousands of calls came into his agency, and a minster's family in @entity28 asked him to come live with them ······
>
> **Structural Knowledge**: (boy, asked, church), (@entity4, borrowed, @entity7), (plea, for, adoption), (@entity11, head, plea), (@entity20, document, journey), (@entity7, borrow, home), (video, of, plea), (guess, as, mom), (boy, spend, life), (home, pass, study), (@entity20, document, plea), (photo, help, get),···
>
> **Question**:@placeholder only took to the pulpit to find a forever home.
>
> **Answer**: @entity4   —— hop 1   — — - hop 2

Fig. 3. The illustration of learned global attentional inference. Triplets with high attention weights at the first hop are denoted with red under line. Triplets with high attention weights at the second hop are denoted with blue dash line. Words with high attention weights are highlighted. In general, our model learns to focus on those triplets that will provide helpful information (to stop reading when adequate cues are captured). For the document, those words which are the candidates answers are much more likely to be paid attention. (Color figure online)

4 Conclusion

In this paper, we propose to infer global attentional cues from structural knowledge via dynamically terminated multi-hop inference to tackle cloze-style question answering tasks. As the results shown in our experiments, the structural knowledge provides cues to produce a better prediction. The dynamically terminated multi-hop inference can adaptively accumulate adequate cues to predict correct answers.

Acknowledgements. This work is supported in part by 973 program (No. 2015CB352300) and NSFC (U1611461,61751209), Chinese Knowledge Center of Engineering Science and Technology (CKCEST).

References

1. Altmann, G., Steedman, M.: Interaction with context during human sentence processing. Cognition **30**(3), 191–238 (1988)
2. Bengio, Y., Simard, P., Frasconi, P.: Learning long-term dependencies with gradient descent is difficult. IEEE Trans. Neural Netw. **5**(2), 157–166 (1994)
3. Chen, D., Bolton, J., Manning, C.D.: A thorough examination of the CNN/daily mail reading comprehension task. In: Association for Computational Linguistics (ACL) (2016)
4. Cui, Y., Chen, Z., Wei, S., Wang, S., Liu, T., Hu, G.: Attention-over-attention neural networks for reading comprehension. In: Proceedings of the 55th Annual Meeting of the Association for Computational Linguistics, ACL 2017, Vancouver, Canada, July 30–August 4, Volume 1: Long Papers, pp. 593–602 (2017). https://doi.org/10.18653/v1/P17-1055

5. Dhingra, B., Liu, H., Yang, Z., Cohen, W.W., Salakhutdinov, R.: Gated-attention readers for text comprehension. In: Proceedings of the 55th Annual Meeting of the Association for Computational Linguistics, ACL 2017, Vancouver, Canada, July 30–August 4, Volume 1: Long Papers, pp. 1832–1846 (2017). https://doi.org/10.18653/v1/P17-1168

6. Graves, A., Fernández, S., Schmidhuber, J.: Bidirectional LSTM networks for improved phoneme classification and recognition. In: Duch, W., Kacprzyk, J., Oja, E., Zadrożny, S. (eds.) ICANN 2005. LNCS, vol. 3697, pp. 799–804. Springer, Heidelberg (2005). https://doi.org/10.1007/11550907_126

7. Graves, A., et al.: Supervised Sequence Labelling with Recurrent Neural Networks, vol. 385. Springer, Heidelberg (2012). https://doi.org/10.1007/978-3-642-24797-2

8. Hermann, K.M., et al.: Teaching machines to read and comprehend. In: Advances in Neural Information Processing Systems, pp. 1693–1701 (2015)

9. Hill, F., Bordes, A., Chopra, S., Weston, J.: The Goldilocks principle: reading children's books with explicit memory representations. CoRR abs/1511.02301 (2015). http://arxiv.org/abs/1511.02301

10. Kadlec, R., Schmid, M., Bajgar, O., Kleindienst, J.: Text understanding with the attention sum reader network. In: Proceedings of the 54th Annual Meeting of the Association for Computational Linguistics, ACL 2016, Berlin, Germany, 7–12 August 2016, Volume 1: Long Papers (2016). http://aclweb.org/anthology/P/P16/P16-1086.pdf

11. Kobayashi, S., Tian, R., Okazaki, N., Inui, K.: Dynamic entity representation with max-pooling improves machine reading. In: HLT-NAACL, pp. 850–855 (2016)

12. Pennington, J., Socher, R., Manning, C.D.: Glove: global vectors for word representation, pp. 1532–1543 (2014). http://www.aclweb.org/anthology/D14-1162

13. Shen, Y., Huang, P.S., Gao, J., Chen, W.: ReasoNet: learning to stop reading in machine comprehension. In: Proceedings of the 23rd ACM SIGKDD International Conference on Knowledge Discovery and Data Mining, pp. 1047–1055. ACM (2017)

14. Sordoni, A., Bachman, P., Bengio, Y.: Iterative alternating neural attention for machine reading. CoRR abs/1606.02245 (2016). http://arxiv.org/abs/1606.02245

15. Taylor, W.L.: "Cloze procedure": a new tool for measuring readability. J. Bull. **30**(4), 415–433 (1953)

16. Toutanova, K., Klein, D., Manning, C.D., Singer, Y.: Feature-rich part-of-speech tagging with a cyclic dependency network, pp. 173–180 (2003)

17. Trischler, A., Ye, Z., Yuan, X., Bachman, P., Sordoni, A., Suleman, K.: Natural language comprehension with the EpiReader. In: Proceedings of the 2016 Conference on Empirical Methods in Natural Language Processing, EMNLP 2016, Austin, Texas, USA, 1–4 November 2016, pp. 128–137 (2016). http://aclweb.org/anthology/D/D16/D16-1013.pdf

18. Warren, S.L.: Make it stick: the science of successful learning. Educ. Rev. Reseñas Educativas **23** (2016)

19. Werbos, P.J.: Backpropagation through time: what it does and how to do it. Proc. IEEE **78**(10), 1550–1560 (1990)

20. Williams, R.J.: Simple statistical gradient-following algorithms for connectionist reinforcement learning. Mach. Learn. **8**(3–4), 229–256 (1992)

Residual Compression Network for Faster Correlation Tracking

Chao Xie, Ning Wang, Wengang Zhou, Weiping Li$^{(\boxtimes)}$, and Houqiang Li

CAS Key Laboratory of Technology in Geo-spatial, Information Processing
and Application System, Department of Electronic Engineering and Information
Science, University of Science and Technology of China, Hefei 230027, China
{chaoxie,zhwg,wpli,lihq}@ustc.edu.cn, wn6149@mail.ustc.edu.cn

Abstract. The recent Correlation Filter (CF) based methods have
shown attractive performance in visual tracking task. In real-time CF
based trackers, they usually adopt hand-crafted features (*e.g.,* HOG and
Color Names), while these artificially designed features still have redundancy and can be further compressed and refined. In this paper, we design
a lightweight network to offline learn how to compress the hand-crafted
features for better and faster correlation tracking. To achieve this goal,
we adopt CF as one layer in the network to force the learned model to be
suitable for tracking task. Besides, we apply residual structure to avoid
the overfitting problem in the training process. Our simple yet effective
network is universal and can be applied to existing CF based trackers.
After adopting our lightweight network, several state-of-the-art CF based
trackers are improved in both tracking accuracy and efficiency.

Keywords: Correlation tracking · Feature compression
Hand-crafted features

1 Introduction

Visual tracking is a basic and important task with many applications including
video surveillance, robotics, autonomous driving and so on. Although significant progress has been made in the past decades, challenges such as occlusion,
deformation and target rotation make visual object tracking still a difficult task.

In recent years, the correlation filter (CF) [3,15] based tracking framework has gained much attention due to its attractive performance and efficiency. Specially, with carefully designed filter formulation [8,11], scale handling
[7,16], multi-channel feature fusion [9,18,21] and so on, the CF based trackers
have shown comparable performance with recent deep learning based trackers
[2,20,24]. The time cost of CF based methods mainly comes from two aspects:
(1) the feature extraction process; (2) the filter construction process, *e.g.,* for
a feature with 512 channels (from VGG [23]), the computation burden of CF
is 512 times compared to the single-channel feature. Thus, real-time CF based

© Springer Nature Switzerland AG 2018
R. Hong et al. (Eds.): PCM 2018, LNCS 11164, pp. 346–356, 2018.
https://doi.org/10.1007/978-3-030-00776-8_32

trackers [1,11,15,17] usually adopt hand-crafted features due to the efficient feature extraction as well as low feature channel. As studied by many previous arts [10,15,16], HOG [5] and Color Names (CN) [29] have been proved to be the most powerful and suitable features for CF. Although HOG and CN are efficient and lightweight features, which are only 31-dim and 11-dim, respectively, they are originally designed for different visual tasks, e.g., HOG is first used in human detection [5].

In this paper, we introduce a lightweight network (only 7 KB in memory storage) to refine existing hand-crafted features (HOG and CN) and make them be more suitable for correlation tracking, *which improves the tracking speed and performance simultaneously*. Our basic motivation is that HOG and CN features may be the suboptimal choice for CF, and still have feature redundancy as well as further improvement room. Thus, we propose to apply linear mapping to the original HOG and CN, which simultaneously refines and compresses the features. Since the feature channel is reduced greatly, e.g., in our experiment, the concatenation of HOG and CN (42-dim in total) is compressed into a 16-dim feature, the CF computation is significantly reduced (more than 60% CF computation) and the overall tracking speed is naturally enhanced. Besides, due to the reduction of feature redundancy, the tracking performance also gains slight improvement. To train such a lightweight network, we add the CF as one network layer, which makes the proposed compression network fit the correlation tracking. Further, we utilize the residual structure in this model to avoid the overfitting problem. Once the network is offline trained, it can be directly applied to the existing hand-crafted feature based CF trackers. After adopting our proposed network, several state-of-the-art tracking approaches achieve higher speed as well as better performance, and we believe this universal model is ready for more trackers.

2   Related Work

In CF based trackers, a filter is trained through minimizing a least-squares loss for all circular shifts of the training sample. The target is tracked by correlating the filter over a region of interest, and the location with the maximum response indicates the new location of the target. Since Bolme et al. propose a tracker using minimum output sum of squared error (MOSSE) filter [3], the correlation filters have been widely studied in visual tracking. Heriques et al. exploit the circulant structure of the training patches [14] and propose to use correlation filter in a kernel space with HOG features to achieve better performance [15]. Zhang et al. propose a STC algorithm [33], which incorporates context information into filter learning. The DSST tracker utilizes multi-scale correlation filters to handle scale changes of the object [7]. The SRDCF tracker alleviates the boundary effects by penalizing correlation filter coefficients depending on spatial location [8]. Luca et al. propose a Staple tracker which combines CF and color-based model to handle color changes and deformation while runs in real-time [1]. The CSR-DCF algorithm constructs CF with channel and spatial reliability [17]. The

recent C-COT has adopted a continuous-domain formulation of the CF, leading to the top performance on several tracking benchmarks [9]. The enhanced
version of C-COT is ECO [6], which improves both speed and performance by
introducing several efficient strategies.

Different from the above methods which mainly focus on how to construct
more robust CF based trackers, our work aims to refine and compress the features in CF trackers by carefully designing a lightweight network. Our method
is also related to some other previous arts. In [12,25], the correlation filter is
embedded into the existing SiamFc tracker [2] to enhance the discrimination
capability. Different from [12,25], our method focuses on compressing the feature channel for higher speed instead of training an offline tracker. Our model is
universal and can be applied to various hand-crafted feature based CF trackers.
Besides, to avoid the overfitting problem, we propose to use residual structure
in our network. Our approach is also related to [10,24], which compress the
features using PCA before constructing CF. However, the PCA method is an
unsupervised dimensionality reduction approach in a linear way which is more
easily influenced by noisy data. As discussed in [10], the PCA helps achieve
higher speed but the performance also drops. Different from these methods, we
learn the feature mapping/compression through offline data-driven supervised
training and improve speed as well as performance at the same time.

3   Method

In this section, we first preview the formulation of CF, and then introduce the
structure of the proposed network. Finally, we discuss the network design details.

3.1   Preview of Correlation Filter

A typical tracker based on CF [3,15] is trained using an image patch $\mathbf{x}$ of size
$M \times N$, which is centered around the target. All the circular shifts of the target feature map $\varphi(\mathbf{x})_{(m,n)} \in \{0,1,...M-1\} \times \{0,1,...N-1\}$ (e.g., HOG, CN
feature space) are generated as training samples with Gaussian function label
$y(m,n)$. The filter $\mathbf{w}$ can be trained by minimizing the following regression error:

$$\min_{\mathbf{w}} \|\mathbf{X}\mathbf{w} - \mathbf{y}\|_2^2 + \lambda\|\mathbf{w}\|_2^2, \tag{1}$$

where λ is a regularization parameter ($\lambda \geq 0$) and $\mathbf{X}$ is the data matrix by
concatenating all the circular shifts. The closed-form solution of Eq. (1) is defined
by $\mathbf{w} = (\mathbf{X}^T\mathbf{X} + \lambda\mathbf{I})^{-1}\mathbf{X}^T\mathbf{y}$ [15]. Since $\mathbf{X}$ is circulant, the filter solution on the
d-th ($d \in \{1, \cdots, D\}$) feature channel can be efficiently calculated as follows,

$$\hat{\mathbf{w}}_d = \frac{\hat{\mathbf{y}}^* \odot \hat{\varphi}_d(\mathbf{x})}{\sum_{i=1}^D \hat{\varphi}_i(\mathbf{x}) \odot (\hat{\varphi}_i(\mathbf{x}))^* + \lambda}, \tag{2}$$

where $\odot$ is the element-wise product, the hat symbol " $\hat{\bullet}$ " denotes the Discrete
Fourier Transform (DFT) of a vector and " $\bullet^*$ " is the complex-conjugate operation. To avoid the boundary effects during learning, we apply Hann window to
the signals [15].

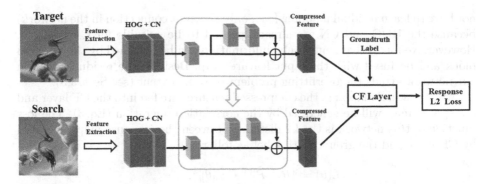

Fig. 1. A systematic flowchart of the proposed framework. First, the hand-crafted features (HOG and CN) of target and search patches are extracted. Then, we design a lightweight residual network (orange box in figure) to compress these features for better and faster correlation tracking. The loss function of this network is the L2 distance of generated response map and groundtruth gaussian label. (Color figure online)

In the next frame, a Region of Interest (RoI) patch $\mathbf{z}$ with the same size of $\mathbf{x}$ is cropped out for predicting the target position. The response map R of the search patch $\mathbf{z}$ is calculated by Eq. (3).

$$R = \mathcal{F}^{-1}\left(\sum_{d=1}^{D} \hat{\mathbf{w}}_d^* \odot \hat{\varphi}_d(\mathbf{z})\right). \tag{3}$$

3.2 Network Structure

In this subsection, we introduce the structure of the proposed network and how to embed the CF layer for model training. As shown in Fig. 1, our simple network is very lightweight with only three trainable convolutional blocks (denoted by the orange box in the figure). Between the successive convolution layers, we adopt the ReLU function. Given a pair of training patches, we first extract their HOG and CN features, which are further concatenated into one 42-dim feature. Then, the 42-channel feature maps of target and search patches are taken as the input of the compression network. The first block in compression model is 1×1 convolution kernel, which reduces the feature map channel from 42 to 16. After the first convolution layer, we obtain the 16-dim output feature map $\mathcal{M}$. Then, we learn a residual mapping which takes $\mathcal{M}$ as input and outputs the desired feature representation as follows,

$$\widetilde{\mathcal{M}} = \mathcal{M} + \mathcal{R}\left(\mathcal{M}; \theta_R\right), \tag{4}$$

where $\widetilde{\mathcal{M}}$ denotes the final compressed and refined features (*e.g.,* for target patch $\mathbf{x}$, $\widetilde{\mathcal{M}}$ denotes the $\varphi(\mathbf{x})$ in Eq. 2), $\mathcal{R}\left(\mathcal{M}; \theta_R\right)$ is the residual mapping to be learned, and θ_R denotes the model parameters of residual block. Ideally, we do not have to learn residual mapping (*i.e.*, only shortcut connection in the network) because the HOG and CN are already proved to be suitable for CF tracking. However, we found that some useful information can be obtained by this residual block and be fused with the input feature map. Besides, the residual block in our network avoids the overfitting problem to some extent (see Sect. 3.3).

After residual mapping, the compressed features are fed into the CF layer and a response map will be generated by the correlation operation (Eq. 3). The loss function of this network is the L2 distance between the response map generated by CF layer and the groundtruth label as follows,

$$L(\theta) = \|R - \widetilde{R}\|^2 + \gamma\|\theta\|^2,$$

$$s.t. \qquad R = \mathcal{F}^{-1}\left(\sum_{d=1}^{D} \hat{\mathbf{w}}_d^* \odot \hat{\varphi}_d(\mathbf{z}; \theta)\right),$$

$$\hat{\mathbf{w}}_d = \frac{\hat{\mathbf{y}}^* \odot \hat{\varphi}_d(\mathbf{x}; \theta)}{\sum_{i=1}^{D} \hat{\varphi}_i(\mathbf{x}; \theta) \odot (\hat{\varphi}_i(\mathbf{x}; \theta))^* + \lambda},$$

(5)

where the $\widetilde{R}$ is the groundtruth label with a gaussian distribution centered at the real target location, and all parameters in our model are denoted as θ. Since the CF has a closed-form solution, the backward formulas can be derived using the chain rule. The back-propagation of loss with respect to the $\varphi(\mathbf{x})$ and $\varphi(\mathbf{z})$ are formulated in Eq. 6, and readers can refer to [25, 28] for more details.

$$\frac{\partial L}{\partial \varphi_d(\mathbf{x})} = \mathcal{F}^{-1}\left(\frac{\partial L}{\partial(\hat{\varphi}_d(\mathbf{x}))^*} + \left(\frac{\partial L}{\partial(\hat{\varphi}_d(\mathbf{x}))}\right)^*\right),$$

$$\frac{\partial L}{\partial \varphi_d(\mathbf{z})} = \mathcal{F}^{-1}\left(\frac{\partial L}{\partial(\hat{\varphi}_d(\mathbf{z}))^*}\right).$$

(6)

3.3   Network Design Discussion

The role of residual structure. The residual structure has been widely investigated and utilized since the success of ResNet [13], which is designed to solve the gradient vanishment problem. However, in this work, we use the residual structure to avoid the overfitting problem. Ideally, if we do not adopt the additional blocks and directly feed the HOG and CN features into the CF layer, the loss is already very little (Fig. 2), which is reasonable because these features are suitable for CF and this process can be regard as the traditional correlation tracking. However, if we only add more convolution layers in the network without adopting the residual structure, we find that although the loss further reduces, the tracking performance is also damaged (Sect. 4.3). Thus, we use residual structure to restrain the overfitting problem, which provides the highway and learnable blocks at the same time.

The choice of network details. The proposed network only adopts one residual block. In the experiments, we observe that more convolution layers and more residual blocks do not improve the tracking performance and influence the efficiency. Thus, we just use this simple and lightweight network. Besides, we only adopt 1×1 kernels with learnable weights and bias in all convolution layers to learn the linear mapping. Since the HOG feature has the spatial invariance already to some extent, the advantage of receptive field in large kernels is not obvious.

Why not directly train a network for CF. There are some previous arts [2,25,28] that train an offline network for tracking, while the performance is just comparable with the existing real-time CF based trackers [1,17]. Due to the training data is from detection task and the tracking task is limited of training data, thus it is difficult and needs much more time to train a perfect tracking network. In contrast, it is relatively easy to refine the existing good features (HOG and CN) using a small network, and in our experiment, the training process tends to be stable after only 5 epoches (Fig. 2). Further, compared with directly training a network, our model can be widely and conveniently applied to the existing top-performance CF based trackers.

About the efficiency. As introduced in Sect. 3.1, for per-channel feature, one CF training computation is needed (Eq. 2). After applying our proposed network, the original 42-dim hand-crafted features are compressed into only 16-dim, and thus we can save more than 60% CF computation. It should be noted our small network is very efficient and needs much less time compared to the saved CF training burden.

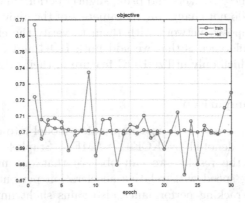

Fig. 2. Network training process on ILSVRC 2015 [22]. The blue curve denotes the training objective. Although the initial loss is already small, our residual network is able to learn extra information and further reduces the loss. The training process is stable after about 5 epoches.

4   Experiment

4.1   Experimental Setup

Implementation Details: In our experiment, we follow the parameters in standard CF method [15] to construct CF layer in the network. Our training data comes from ImageNet Large Scale Visual Recognition Challenge (ILSVRC 2015) [22], including both the training and validation set. In each frame, patch is cropped around groundtruth with 2.0 padding and resized into 125 × 125. We apply the standard stochastic gradient descent (SGD) with momentum of 0.9 and set the weight decay to 0.005 to end-to-end train the network. Our model is trained for 30 epoches with a learning rate of 10^{-5} using MatConvNet toolbox [26] on a computer with an Intel I7-4790K 4.00GHz CPU, 48GB RAM, Nvidia GTX TITAN 1080Ti GPU.

Evaluation Benchmarks and Metrics: Our method is evaluated on two benchmark datasets by a no-reset evaluation protocol: OTB-2013 [30] and OTB-2015 [31]. All the tracking methods are evaluated by the average distance precision (DP) plots and overlap success plots over these datasets using one-pass evaluation (OPE) [30,31].

Baseline Trackers: After training our model, we apply it to four state-of-the-art CF based trackers including KCF [15], SAMF [16], Staple [1] and CSR-DCF [17]. These trackers all adopt hand-crafted features for efficient correlation tracking and achieve impressive performance. However, our model can further accelerate the speed of these trackers and bring slightly performance improvement at the same time. We do not change any parameters of the above trackers and just incorporate the proposed network with them to verify the effectiveness of our model. The only difference is that we add both HOG and CN features to the KCF and Staple, which only utilize HOG feature in their original versions.

4.2   Baseline Comparison

As shown in Table 1, we embed our compression model into existing CF based trackers, which are denoted by KCF+, SAMF+, Staple+ and CSR-DCF+, respectively. From the results, we can observe that our model significantly boosts the tracking speed with relative acceleration percentage from 26% to 41%. Besides, the tracking performance also gains slight improvement, which verifies the effectiveness of the proposed model. For example, on the challenging OTB-2015 benchmark, our model boosts the performance of Staple and CSR-DCF trackers with a gain of about 1% DP and AUC.

4.3   Ablation Study

In Table 2, we study the effectiveness of residual block. The "one conv layer" denotes the network with only 1 × 1 convolution layer for feature channel reduction, the "more conv layer" denotes the network with more convolution layers but no residual block. From the results in Table 2, we can observe that only one layer can not achieve the best performance because one layer may not be able to approximate the complex feature mapping function. However, more layers tend to cause the overfitting problem and demage the performance, which is lower than the original/baseline performance. In contrast, our final residual structure exhibits the highest results.

Table 1. Effectiveness study of our lightweight compression network. The trackers combined with our model are labeled with "+". The DP (@20px) and AUC scores are reported on the OTB-2013 [30] and OTB-2015 [31] datasets (DP/AUC) corresponding to the OPE metric. The tracking speed (frame per second, FPS) is evaluated on the OTB-2015 benchmark. Our model obviously accelerates the speed of these trackers.

	When/where	OTB-2013 (DP/AUC)	OTB-2015 (DP/AUC)	Average Speed (FPS)
KCF	2015 TPAMI	75.1/53.4	69.7/49.4	172
KCF+		**75.4/53.5**	**70.7/50.3**	**235 (37%↑)**
SAMF	2014 ECCVw	75.1/57.6	74.0/55.3	25
SAMF+		**75.5/57.8**	**74.2/56.4**	**32 (28%↑)**
Staple	2016 CVPR	78.1/60.2	76.5/58.5	65
Staple+		**78.4/60.9**	**77.8/59.9**	**82 (26%↑)**
CSR-DCF	2017 CVPR	79.0/59.1	77.2/57.6	12
CSR-DCF+		**81.5/61.5**	**78.6/58.6**	**17 (41%↑)**

Table 2. The effectiveness study of residual block. Compared with single layer, the residual block learns more feature mapping while avoids the overfitting problem at the same time. In the table, the DP (@20px) and AUC scores are reported on the OTB-2015 [31] datasets (DP/AUC) corresponding to the OPE metric.

	Baseline	One conv layer	More conv layer	Residual block (Our final)
KCF	69.7 / 49.4	70.1 / 49.6	68.2 / 48.0	**70.7 / 50.3**
SAMF	74.0 / 55.3	73.5 / 55.8	72.2 / 53.4	**74.2 / 56.4**

4.4 Comparison with Other Trackers

We further compare the CF based methods using our model (*i.e.*, KCF+, SAMF+, Staple+, CSR-DCF+) with other state-of-the-art trackers on the OTB-2015 dataset, including DSST [7] , MEEM [32], LCT [19], SCT [4], SiamFc [2] and CFNet [25]. In this work, we focus on the feature compression for faster tracking (beyond real-time), and thus the recent complex deep learning based trackers (*e.g.*, MDNet [20], 1 FPS and C-COT [9], 0.3 FPS) are not put into comparison, which are far away from real-time level. From the results in Fig. 3, we can observe that compared with other trackers, our methods achieve comparable or better performance. As for the efficiency, the above trackers all achieve real-time speed. However, the SiamFc (85 FPS) and CFNet (67 FPS) rely on GPU device, while our Staple+ method operates 82 frames per second on a single CPU, which is very efficient and faster than the original Staple tracker. Besides, the KCF+, SAMF+ and CSR-DCF+ all achieve higher speed as well as slightly better performance compared to their original versions.

Finally, it should be noted that our model is quite general and we just select four standard CF based trackers to validate its effectiveness. The recent more sophisticated filter training methods including background-aware correlation tracker [11], large margin CF tracker [27] also use standard hand-crafted features and we believe they can also benefit from our model.

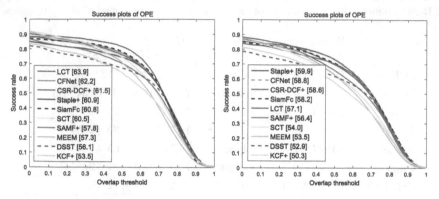

Fig. 3. Success plots on the OTB-2013 [30] (left) and OTB-2015 [31] (right) datasets. In the legend, the area-under-curve (AUC) score is reported. The trackers with our proposed model are labeled with "+", *i.e.,* Staple+, CSR-DCF+, SAMF+ and KCF+.

5   Conclusion and Future Work

In this paper, we propose a lightweight feature compression network for better and faster correlation tracking. To train this model, a correlation filter (CF) layer is embedded, which adapts the network to the CF based trackers. Besides, we use residual struture to learn proper feature mapping and alleviate the overfitting problem at the same time. Our lightweight model (only 7 KB) is very efficient and can reduce more than 60% correlation filter computation. To validate the effectiveness of the proposed method, our model is embedded into several state-of-the-art CF based trackers, which shows both performance and speed improvement.

For future work, we aim to compress the existing deep models (*e.g.,* VGG [23]) and accelerate the recent deep feature based CF trackers (*e.g.,* HCF [18], C-COT [9]) for real-time applications.

References

1. Bertinetto, L., Valmadre, J., Golodetz, S., Miksik, O., Torr, P.: Staple: complementary learners for real-time tracking. In: CVPR (2016)
2. Bertinetto, L., Valmadre, J., Henriques, J.F., Vedaldi, A., Torr, P.H.: Fully-convolutional siamese networks for object tracking. In: ECCV (2016)
3. Bolme, D.S., Beveridge, J.R., Draper, B.A., Lui, Y.M.: Visual object tracking using adaptive correlation filters. In: CVPR (2010)
4. Choi, J., Jin Chang, H., Jeong, J., Demiris, Y., Young Choi, J.: Visual tracking using attention-modulated disintegration and integration. In: CVPR (2016)
5. Dalal, N., Triggs, B.: Histograms of oriented gradients for human detection. In: CVPR (2005)
6. Danelljan, M., Bhat, G., Shahbaz Khan, F., Felsberg, M.: Eco: efficient convolution operators for tracking. In: CVPR (2017)
7. Danelljan, M., Häger, G., Khan, F., Felsberg, M.: Accurate scale estimation for robust visual tracking. In: BMVC (2014)
8. Danelljan, M., Hager, G., Shahbaz Khan, F., Felsberg, M.: Learning spatially regularized correlation filters for visual tracking. In: ICCV (2015)
9. Danelljan, M., Robinson, A., Khan, F.S., Felsberg, M.: Beyond correlation filters: learning continuous convolution operators for visual tracking. In: ECCV (2016)
10. Danelljan, M., Shahbaz Khan, F., Felsberg, M., Van de Weijer, J.: Adaptive color attributes for real-time visual tracking. In: CVPR (2014)
11. Galoogahi, H.K., Fagg, A., Lucey, S.: Learning background-aware correlation filters for visual tracking. In: ICCV (2017)
12. Guo, Q., Feng, W., Zhou, C., Huang, R., Wan, L., Wang, S.: Learning dynamic siamese network for visual object tracking. In: CVPR, pp. 1–9
13. He, K., Zhang, X., Ren, S., Sun, J.: Deep residual learning for image recognition. In: CVPR (2016)
14. Henriques, J.F., Caseiro, R., Martins, P., Batista, J.: Exploiting the circulant structure of tracking-by-detection with kernels. In: ECCV (2012)
15. Henriques, J.F., Caseiro, R., Martins, P., Batista, J.: High-speed tracking with kernelized correlation filters. TPAMI **37**(3), 583–596 (2015)
16. Li, Y., Zhu, J.: A scale adaptive kernel correlation filter tracker with feature integration. In: ECCV Workshop (2014)
17. Lukezic, A., Vojir, T., Cehovin Zajc, L., Matas, J., Kristan, M.: Discriminative correlation filter with channel and spatial reliability. In: CVPR (2017)
18. Ma, C., Huang, J.B., Yang, X., Yang, M.H.: Hierarchical convolutional features for visual tracking. In: ICCV (2015)
19. Ma, C., Yang, X., Zhang, C., Yang, M.H.: Long-term correlation tracking. In: CVPR (2015)
20. Nam, H., Han, B.: Learning multi-domain convolutional neural networks for visual tracking. In: CVPR (2016)
21. Qi, Y., Zhang, S., Qin, L., Yao, H., Huang, Q., Yang, J.L.M.H.: Hedged deep tracking. In: CVPR (2016)
22. Russakovsky, O., et al.: Imagenet large scale visual recognition challenge. IJCV **115**(3), 211–252 (2015)
23. Simonyan, K., Zisserman, A.: Very deep convolutional networks for large-scale image recognition. arXiv preprint arXiv:1409.1556 (2014)
24. Song, Y., Ma, C., Gong, L., Zhang, J., Lau, R., Yang, M.H.: Crest: convolutional residual learning for visual tracking. In: ICCV (2017)

25. Valmadre, J., Bertinetto, L., Henriques, J.F., Vedaldi, A., Torr, P.H.: End-to-end representation learning for correlation filter based tracking. In: CVPR (2017)

26. Vedaldi, A., Lenc, K.: Matconvnet: convolutional neural networks for matlab. In: ACM MM (2014)

27. Wang, M., Liu, Y., Huang, Z.: Large margin object tracking with circulant feature maps. In: CVPR (2017)

28. Wang, Q., Gao, J., Xing, J., Zhang, M., Hu, W.: Dcfnet: discriminant correlation filters network for visual tracking. arXiv preprint arXiv:1704.04057 (2017)

29. Weijer, J.V.D., Schmid, C., Verbeek, J., Larlus, D.: Learning color names for real-world applications. TIP 18(7), 1512–1523 (2009)

30. Wu, Y., Lim, J., Yang, M.H.: Online object tracking: a benchmark. In: CVPR (2013)

31. Wu, Y., Lim, J., Yang, M.H.: Object tracking benchmark. TPAMI 37(9), 1834–1848 (2015)

32. Zhang, J., Ma, S., Sclaroff, S.: Meem: robust tracking via multiple experts using entropy minimization. In: ECCV (2014)

33. Zhang, K., Zhang, L., Yang, M.H., Zhang, D.: Fast visual tracking via dense spatio-temporal context learning. In: ECCV (2013)

Robust Neighborhood Preserving Low-Rank Sparse CNN Features for Classification

Zemin Tang[1,2], Zhao Zhang[1,2(✉)], Xiaohu Ma[1,2], Jie Qin[3], and Mingbo Zhao[4]

[1] School of Computer Science and Technology, Soochow University, Suzhou 215006, China
cszzhang@gmail.com
[2] Collaborative Innovation Center of Novel Software Technology and Industrialization, Nanjing 210023, China
[3] Computer Vision Laboratory, ETH Zürich, 8092 Zurich, Switzerland
[4] Department of Electronic Engineering, City University of Hong Kong, Kowloon, Kowloon Tong, Hong Kong

Abstract. Convolutional Neural Networks (CNN) has achieved great success in the area of image recognition, but it usually needs sufficient training data. Meanwhile, similar images tend to deliver compact CNN features, so the original CNN features of different images of each subject or similar subjects should have the low-rank and sparse characteristics. Moreover, CNN features may contain redundant information and noise. To this end, we investigate how to discover the robust low-rank and sparse CNN features and show how these features behave for image classification, specifically for the case that the number of training data is relatively small. Specifically, we perform the robust neighborhood preserving low-rank and sparse recovery step over the original CNN features so that salient key information can be extracted and the included noise can also be removed. To demonstrate the effectiveness of the computed joint low-rank and sparse CNN features on image classification, three deep networks, i.e., VGG, Resnet and Alexnet, are evaluated. The simulation results on two widely-used image databases (CIFAR-10 and SVHN) show that the extracted joint low-rank and sparse CNN features can indeed obtain the enhanced results, compared with the original CNN features.

Keywords: Image classification · Convolutional Neural Networks
Feature extraction · Joint low-rank and sparse coding

1 Introduction

In recent years, Convolutional Neural Networks (CNN) [5] has aroused considerable attention due to its success application to different visual recognition tasks. CNN is a kind of computational network which imitates the human brain to process the task of data representation and classification. Specifically, CNN consists of alternating convolutional and pooling layers [15]. Convolutional layers take inner product of the linear filter and the underlying receptive field followed by a nonlinear activation function at every local portion of the input. The resulting outputs are called feature maps

© Springer Nature Switzerland AG 2018
R. Hong et al. (Eds.): PCM 2018, LNCS 11164, pp. 357–369, 2018.
https://doi.org/10.1007/978-3-030-00776-8_33

[15, 16, 30–33]. Note that for image recognition by CNN, similar images tend to deliver compact CNN features, thus the original CNN features of different images of each subject or similar subjects should have the low-rank and sparse characteristics [26–33]. Moreover, the original data may contain redundant information and noise, but the conventional CNN models may not handle the task of removing noise in the feature representation process, which may degrade the representation power of the features and subsequent classification power. Meanwhile, CNN usually deliver worse results when lack of sufficient training samples, so it would be better to explore the ways to achieve better results when the number of training data is relatively small.

To this end, in this paper we investigate how to discover the robust low-rank and sparse properties of the original CNN features and show whether low-rank sparse CNN features can improve the representation and classification abilities, especially when the amount of training data is relatively small. Specifically, we perform the robust neighborhood preserving low-rank and sparse principal feature coding process over the original CNN features so that salient and important information can be extracted and the included noise in the data can also be removed. The proposed method can explicitly produce the low-rank and sparse CNN features, which can potentially improve the performance than the original CNN features due to the noise removal process. Figure 1 illustrates the two-stage procedures of our method, which consists of three parts. The first part is CNN Feature Extraction Layer, which extracts the original CNN feature from input data; and the second part is the Low-Rank and Sparse Coding Layer, which plays a core role in building the low-rank and sparse CNN features and removing noise; the last part is the Classification Layer which is designed to achieve the label assignment results. To verify the effectiveness of those extracted low-rank and sparse CNN features on improving image classification, three deep learning networks, that is, VGG [13], Resnet [9] and Alexnet [5], are used for evaluations. The results on two widely-used image databases (i.e., CIFAR-10 [4] and SVHN [24]) demonstrate that the learnt robust neighborhood preserving low-rank and sparse CNN features can indeed achieve enhanced results, compared with those on the original CNN features. The major contributions of this paper can be summarized as twofold. First, we design a new framework that adds the low-rank and sparse coding layer after the convolutional layer for the low-rank and sparse CNN feature extraction. Second, to improve the representation and classification abilities, we propose a robust neighborhood preserving low-rank and sparse recovery method that is performed over the original CNN features to deliver the neighborhood preserving low-rank and sparse CNN features, which can improve the robustness and performance of conventional model for image recognition, especially for cases with corrupted pixels.

The paper can be outlined as follows. Section 2 proposes our framework. Section 3 shows the settings and results. Finally, the paper is concluded in Sect. 4.

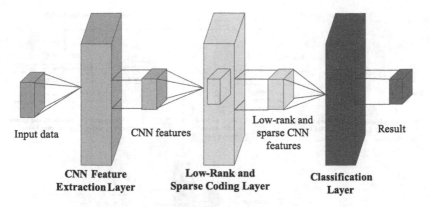

CNN Feature   Low-Rank and   Classification
Extraction Layer   Sparse Coding Layer   Layer

Fig. 1. Illustration of the two-stage procedures of our proposed framework.

2 Low-Rank and Sparse CNN Features for Image Classification

We mainly describe our method technically in this part. The whole framework can be divided into three steps. In the first step, we focus on computing the CNN features in the feature extraction layer. Then, in the second step, we mainly obtain the low-rank and sparse CNN features and remove the noise in the low-rank and sparse coding layer. Finally, label prediction results can be obtained in classification layer.

2.1 Whole Framework for Classification

Figure 2 illustrates the entire framework of our method for classification. The whole process has three parts, i.e., training, testing preprocessing and classification modules. The testing preprocessing and classification modules form the entire testing module. In training module, we firstly obtain the original CNN features of training samples in the Feature Extraction Layer, and then perform our method to get the salient low-rank and sparse CNN features in the Low-Rank and Sparse Coding Layer, and at the same time, output the underlying projection P in the Low-Rank and Sparse Coding Layer. The joint low-rank and sparse CNN features of training samples are used to construct the classifier model or are stored in the server for use in the testing module.

In the testing preprocessing module, we mainly obtain the CNN features of test data in the Feature Extraction Layer and then use the learnt projection P to extract salient and important low-rank and sparse CNN features of test data in Low-Rank and Sparse Coding Layer. Note that there is no need to compute the projection P in testing phase. After the low-rank and sparse features of test data are obtained, we send them into the Classification Layer for the label prediction. Note that classification is mainly performed based on the trained classifier model or stored salient low-rank sparse CNN features of training data. The learnt classifier model can be used to predict the labels of test data. In what follows, the Feature Extraction Layer, Low-Rank and Sparse Coding Layer and Classification Layer are described in detail as follows.

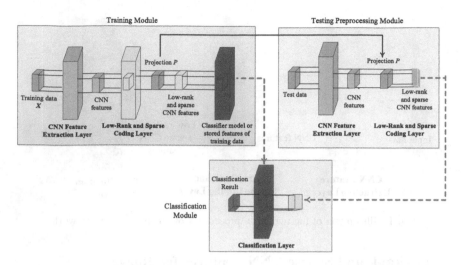

Fig. 2. The whole learning framework of our method for representation and classification.

2.2 CNN Feature Extraction Layer

The classical convolutional neural networks usually consist of alternatively stacked convolutional layers, spatial pooling layers and several fully connected layers [16]. In Fig. 3, we show the structure of a simple CNN model. As we all know, it updates the related parameters with Back Propagation algorithm (BP) [17].

In this paper, we use three different kinds of CNN models as feature extractors, i.e., VGG, Alexnet and Resnet. Since the form of samples will go changing continuously under the function of convolution and pooling layers, it's very important to choose the right output form among different layers. When we get into the step of low-rank transformation, we need a matrix with size $D \times N$ (where D is the dimensionality of each vector and N is the number of vectors). Therefore, we choose the first fully connected layer in each model, with the output form of is $1 \times 1 \times D$. In this way, the learnt CNN feature can be obtained and used for low-rank and sparse coding.

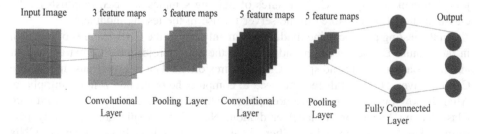

Fig. 3. The structure example of a simple CNN model.

2.3   Robust Neighborhood Preserving Low-Rank and Sparse Coding Layer

In this layer, we perform the robust neighborhood preserving low-rank sparse coding process over the learnt CNN features from feature extraction layer so that the more salient and key information in CNN features can be extracted. The proposed method can explicitly obtain the salient low-rank sparse CNN features, which can potentially improve the performance compared with the original CNN features.

The Formulation Inspired by the recent inductive low-rank and sparse principal feature coding model [6], we propose a refined truncated nuclear norm minimization based framework that can extract the neighborhood preserving low-rank and sparse features from original CNN features and at the same time preserve the neighborhood relationships between them. The proposed objective function can be formulated as

$$\langle P, E \rangle = \arg\min_{P,E} (1 - \alpha)\|PC_{NN}\|_r + \alpha\|PC_{NN}\|_1 + \beta\|PC_{NN} - PC_{NN}W\|_F^2 + \lambda\|E\|_l$$
$$s.t.\ C_{NN} = PC_{NN} + E \tag{1}$$

where $\|PC_{NN}\|_r = \sum_{i=r+1}^{\min(n,\,N)} \sigma_i(PC_{NN})$ is the truncated nuclear norm [20] that is defined as the sum of $\min(n, N) - r$ minimum singular values, $P \in \mathbb{R}^{n \times n}$ is a projection that can be used to extract the low-rank and sparse salient features, E is an error part, C_{NN} represents the outputted CNN feature matrix in the feature extraction layer, α, β and λ are tuning parameters. To preserve the neighborhood information of encoded features PC_{NN}, a reconstruct error term $\|PC_{NN} - PC_{NN}W\|_F^2$ is incorporated, where W is a weight matrix defined based on C_{NN} features by the reconstruction weights as [19]. By jointly minimizing the reconstruct error $\|PC_{NN} - PC_{NN}W\|_F^2$, more informative salient feature can be obtained. Note that the motivation of the truncated nuclear norm detailed in [20], i.e., the values of the largest r nonzero singular values will not affect the rank of the matrix, so the truncated nuclear norm mainly focuses on minimizing the sum of smallest $\min(n, N) - r$ singular values. The truncated norm $\|PC_{NN}\|_r$ is generally nonconvex, so it is not easy to optimize it directly. To address this issue, the authors of [20] have proved that the truncated nuclear norm $\|PC_{NN}\|_r$ and the standard nuclear norm $\|PC_{NN}\|_*$ have the following relation:

$$\max_{\tau\tau^T=I,\varsigma\varsigma^T=I} Tr\left(\tau PC_{NN}\varsigma^T\right) = \sum_{i=1}^{r} \sigma_i(PC_{NN}), \tag{2}$$

$$\|PC_{NN}\|_* - \max_{\tau\tau^T=I,\varsigma\varsigma^T=I} Tr\left(\tau PC_{NN}\varsigma^T\right) = \sum_{i=1}^{\min(n,\,N)} \sigma_i(PC_{NN}) - \sum_{i=1}^{r} \sigma_i(PC_{NN})$$
$$= \sum_{i=r+1}^{\min(n,\,N)} \sigma_i(PC_{NN}) = \|PC_{NN}\|_r \tag{3}$$

where $\|PC_{NN}\|_* = \sum_{i=1}^{\min(m,n)} \sigma_i(PC_{NN})$ is the nuclear norm and $\sigma_i(PC_{NN})$ is the i-th largest singular value of PC_{NN}, I is an identity matrix. Suppose that $U \sum V^T$ is the singular value decomposition (SVD) of PC_{NN}, where $U = (u_1, \ldots, u_m) \in \mathbb{R}^{m \times m}$ and $V = (v_1, \ldots, v_r) \in R^{n*n}$. In addition, $\tau = (u_1, \ldots, u_r)^T$ and $\varsigma = (v_1, \ldots, v_r)^T$. Thus, the objective function of our method can be reformulated as

$$
\begin{aligned}
\langle P, E \rangle = \arg\min_{P,E} \ & (1 - \alpha)\left(\|PC_{NN}\|_* - Tr(\tau PC_{NN}\varsigma^T)\right) + \alpha\|PC_{NN}\|_1 \\
& + \beta\|PC_{NN} - PC_{NN}W\|_F^2 + \lambda\|E\|_l, \quad s.t. \ C_{NN} = PC_{NN} + E
\end{aligned} \tag{4}
$$

Note that the above problem can also obtain the underlying projection based on the original C_{NN} feature, so the results may be decreased by the possibly included noise. To handle this issue, we borrow the idea of Robust LatLRR (rLatLRR) [21] and learn it based on the noise-removed data C_{NN}-E instead of the original C_{NN} features to make the learnt projection more notable and robust to noise. Thus, the objective function of our method can be further formulated as

$$
\langle P, E \rangle = \arg\min_{P,E} \ (1 - \alpha)\left(\|PC_{NN}\|_* - Tr(\tau PC_{NN}\varsigma^T)\right)
$$
$$
+ \alpha\|PC_{NN}\|_1 + \beta\|PC_{NN} - PC_{NN}W\|_F^2 + \lambda\|E\|_l, \quad s.t. \ C_{NN} - E = P(C_{NN} - E),
$$
$$\tag{5}$$

The Optimization Procedures. Since this problem is generally convex, it can be solved by various methods. In this study, we use the Inexact Augmented Lagrange Multiplier (Inexact-ALM) for efficiency [22, 23]. Following the common procedures [6, 20–23, 26–33], we first convert it into the following equivalent one:

$$
\begin{aligned}
\wp = \ & (1 - \alpha)\left(\|J\|_* - Tr(\tau J\varsigma^T)\right) + \alpha\|F\|_1 + \beta\|PC_{NN} - PC_{NN}W\|_F^2 + \lambda\|E\|_{2,1}, \\
& s.t. \ Q = PQ, \ PC_{NN} = J, \ PC_{NN} = F, \ Q = C_{NN} - E
\end{aligned} \tag{6}
$$

where the variables Q, J and F are involved to simplify the optimization. Note that the corresponding Lagrange function $\wp$ can be defined as

$$
\begin{aligned}
\wp = \ & (1 - \alpha)\left(\|J\|_* - Tr(\tau J\varsigma^T)\right) + \alpha\|F\|_1 + \beta\|PC_{NN} - PC_{NN}W\|_F^2 + \lambda\|E\|_{2,1} \\
& + \langle Y_1, Q - PQ \rangle + \langle Y_2, \hat{P}C_{NN} - J \rangle + \langle Y_3, \hat{P}C_{NN} - F \rangle + \langle Y_4, Q - C_{NN} + E \rangle, \\
& + \tfrac{\mu}{2}\left[\|Q - PQ\|_F^2 + \|\hat{P}C_{NN} - J\|_F^2 + \|\hat{P}C_{NN} - F\|_F^2 + \|Q - C_{NN} + E\|_F^2\right]
\end{aligned} \tag{7}
$$

where Y_1, Y_2, Y_3, Y_4 are Lagrange multipliers and the μ is the weighting parameter. Note that we update these of variables alternately by solving the Lagrange function $\wp$:

$$
\begin{aligned}
& \left(J_{k+1}, F_{k+1}, E_{k+1}, \hat{P}_{k+1}\right) = \arg\min_{J,F,E,\hat{P}} \wp_k(J, F, E, \hat{P}) \\
& Y_1^{k+1} = Y_1^k + \mu_k(Q_{k+1} - \hat{P}_{k+1}Q_{k+1}), Y_2^{k+1} = Y_2^k + \mu_k(\hat{P}_{k+1}C_{NN} - J_{k+1}) \\
& Y_3^{k+1} = Y_3^k + \mu_k(\hat{P}_{k+1}C_{NN} - F_{k+1}), Y_4^{k+1} = Y_4^k + \mu_k(Q_{k+1} - C_{NN} + E_{k+1})
\end{aligned} \tag{8}
$$

Since the involved variables depend on each other, the problem cannot be solved directly. Thus, we follow the common procedures to update them alternately:

(1) Optimization for low-rank matrices J, F and error E:

We first update J, F and E. The closed form solution of J can be obtained from

$$J_{k+1} = \arg\min_J \frac{1-\alpha}{\mu_k}\left(\|J\|_* - Tr(\tau J \varsigma^T)\right) + \frac{1}{2}\left\|J - (\hat{P}_k Q_k + Y_2^k/\mu_k)\right\|_F^2$$
$$= \Omega_{(1-\alpha)/\mu_k}\left[\hat{P}_k Q_k + Y_2^k/\mu_k\right], \tag{9}$$

where $\Omega_{(1-\alpha)/\mu_k}\left[\hat{P}_k Q_k + Y_2^k/\mu_k\right] = \tilde{U} S_{(1-\alpha)/\mu_k}[\Sigma]V$ is the singular value shrinkage Operator [14], $\tilde{U}\Sigma V$ is the SVD of $\hat{P}_k Q_k + Y_2^k/\mu_k$, and $S_\varepsilon[X] = \text{sgn}(x)\max(|x| - \varepsilon, 0)$ is the scalar shrinkage operator [3,7, 8]. The variable F can be updated as

$$F_{k+1} = \arg\min_F \frac{\alpha}{\mu_k}\|F\|_1 + \frac{1}{2}\left\|F - (\hat{P}_k Q_k + Y_3^k/\mu_k)\right\|_F^2 = S_{\alpha/\mu_k}\left[\hat{P}_k Q_k + Y_3^k/\mu_k\right]. \tag{10}$$

The solution of the sparse error matrix E can be analogously inferred as

$$E_{k+1} = \arg\min_E \frac{\lambda}{\mu_k}\|E\|_{2,1} + \frac{1}{2}\left\|E - (C_{NN} - Q_k + Y_4^k/\mu_k)\right\|_F^2. \tag{11}$$

(2) Optimization for projection $\hat{P}$ and clean data Q:

By removing terms irrelevant to $\hat{P}$ and Q, we can get the following reduced problem:

$$\wp(\hat{P}, Q) = \beta\|PC_{NN} - PC_{NN}W\|_F^2 + \langle Y_1, Q - PQ\rangle + \langle Y_2, \hat{P}C_{NN} - J\rangle$$
$$+ \langle Y_3, \hat{P}C_{NN} - F\rangle + \frac{\mu}{2}\left[\|Q - PQ\|_F^2 + \|\hat{P}C_{NN} - J\|_F^2 + \|\hat{P}C_{NN} - F\|_F^2\right], \tag{12}$$

By taking the derivative *w.r.t.* $\hat{P}$, and setting it to zero, we can update $\hat{P}$ as

$$R_{k+1} = \left((C_{NN} - C_{NN}W)(C_{NN} - C_{NN}W)^T + \mu(Q_k Q_k^T + 2C_{NN}C_{NN}^T)\right)^{-1}$$
$$\hat{P}_{k+1} = \left(Y_1^k Q_k^T - 2Y_2^k C_k^T + \mu(Q_k Q_k^T + J_{k+1}C_{NN}^T + F_k C_{NN}^T)\right)R_{k+1} \tag{13}$$

Similarly, by setting the derivative $\wp(\hat{P}, Q)/\wp Q$ to zero, we can update Q as

$$Q_{k+1} = \left[\left(I - \hat{P}_{k+1}\right)\left(I - \hat{P}_{k+1}\right)^T + I\right]^{-1}\left[\left(\hat{P}_{k+1} - I\right)^T Y_1^k/\mu_k - Y_4^k/\mu_k + (C_{NN} - E_{k+1})\right], \tag{14}$$

where I is an identity matrix. For complete presentation of the method, we summarize the procedures of our algorithm in Algorithm 1.

Algorithm 1:
Robust Neighborhood Preserving Low-Rank and Sparse Coding

Inputs: Training data C_{NN} , and parameters α, β, γ.
Initialization: $k = 0, J_k = 0, F_k = 0, E_k = 0, \hat{P}_k = 0, Q_k = 0, R_k = 0, Y_1^k = 0, Y_2^k = 0, Y_3^k = 0,$
$Y_4^k = 0, W = 0, \max_\mu = 10^{10}, \mu_k = 10^{-6}, \eta = 1.2, \varepsilon = 10^{-7}.$
1. **While** *not converged* **do**
2. Fix others and update the low-rank matrix J by Eq.9;
3. Fix others and update the sparse matrix F by Eq.10;
4. Fix others and update the sparse error E by Eq.11;
5. Fix others and update the projection $\hat{P}$ by Eq.13;
6. Fix others and update the clean data Q by Eq.14;
8. Update the multipliers Y_1, Y_2, Y_3, Y_4 by Eq.8;
9. Update the parameter μ with $\mu_{k+1} = \min\left(\eta\mu_k, \max_\mu\right)$;
10. Convergence check:
if $\max\left(\left\|Q_{k+1} - \hat{P}_{k+1}Q_{k+1}\right\|_\infty, \left\|\hat{P}_{k+1}C_{NN} - J_{k+1}\right\|_\infty, \left\|\hat{P}_{k+1}C_{NN} - F_{k+1}\right\|_\infty, \left\|Q_{k+1} - C_{NN} + E_{k+1}\right\|_\infty\right) < \varepsilon$,
stop; else $k = k + 1$.
End while
Outputs: $P^* \leftarrow \hat{P}_{k+1}.$

2.4   Classification Layer

We use two classifiers for classification, i.e., Softmax [25] and Discriminant Analysis Classifier (DAC) [18]. We compare the classification results over the low-rank and sparse CNN features with the original CNN features under the same setting.

3   Simulation Results and Analysis

3.1   Overview

We mainly evaluate our method for representation and classification and compare the result with that of original CNN model. Two popular datasets, i.e., CIFAR-10 [4] and The Street View House Numbers (SVHN) [24], are evaluated. Three deep network models, VGG [13], Alexnet [5] and Resnet [9], are employed for the evaluations.

3.2   Image Classification

Results on CIFAR-10
The CIFAR-10 dataset contains natural images of 10 classes including 50,000 training images in total and 10,000 testing images. Each image is an RGB image of size 32×32. In this study, the compared methods are divided into three groups based on the three CNN models. To evaluate the robustness performance, we add corrupt 40 percent of pixels of training images over Alexnet and VGG16, and 20 percent for

Resnet. In the first group, the compared methods are named as Alexnet + DAC, LS-Alexnet + DAC, Noise + Alexnet + DAC, Noise + LS-Alexnet + DAC, Noise + Alexnet + Softmax, Noise + LS-Alexnet + Softmax. The prefix 'LS' means the results over our robust low-rank and sparse CNN features. The prefix 'Noise' means the according results under corruptions. In each group, we select 400/800/1200/1600/2000/2400 training samples from the dataset to form the training data and select all the test data in the dataset which includes 10,000 images. We compare the results over the original CNN feature and robust low-rank & sparse CNN feature in each group. For fair comparison, the original parameters of the three CNN models and classifiers are selected. The classification accuracies over varied training numbers are shown in Fig. 4, from where we can see that (1) The overall performance of each method increases as the training number increases. (2) Our method with salient low-rank sparse CNN features can deliver higher accuracies than those on original CNN features in most cases. (3) For the noisy case, our method can still deliver better performance than those on the original CNN features in most case.

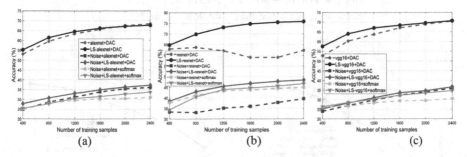

Fig. 4. Comparison of classification results of different CNN models on CIFAR-10 database, where (a) Alexnet, (b) Resnet, (c)VGG16.

Table 1 shows the statistics of the largest improvement, where the "Accuracy1" term is the accuracy of each method with the original CNN features, and "Accuracy2" term is the classification accuracy of our method with the robust low-rank and sparse CNN features. The results demonstrate that the extracted low-rank and sparse CNN features can indeed improve the data representation and classification performances compared with the original CNN features by existing models.

Results on SVHN
SVHN is a real image dataset for developing for recognition. SVHN incorporates an order of magnitude more labeled data (over 60,000digit images) and comes from a significantly harder, unsolved, real world problem (recognizing digits and numbers in natural scene images). In this study, the compared methods are divided into two groups with two CNN models, i.e., Alexnet and VGG16. For evaluating the performance on de-noising, we also corrupt 20 percent pixels for Alexnet and VGG16. We choose 100/200/300/400/500/600 images to form the training set and select 1,000 test images for testing. Detailed comparison results are shown in Fig. 5. Table 2 summarizes the statistics of the largest improvement of our method with salient low-rank sparse CNN

features over the method with the original CNN features. The results demonstrate that our proposed method can get better performance than the original CNN models when we lack sufficient training data.

Table 1. The statistics of the largest improvement of our method with robust salient low-rank sparse CNN features over the method with the original CNN features.

Method	Accuracy1	Accuracy2
Alexnet + DAC	59.53%	61.41%
VGG16 + DAC	52.77%	57.54%
Resnet + DAC	62.31%	75.81%
Noise + Alexnet + DAC	24.91%	27.8%
Noise + VGG16 + DAC	29.85%	31.02%
Noise + Resnet + DAC	35.97%	46.78%
Noise + Alexnet + Softmax	31.04%	33.65%
Noise + VGG16 + Softmax	30.51%	34.72%
Noise + Resnet + Softmax	44.96%	46.7%

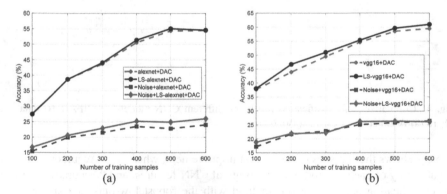

Fig. 5. Comparison of classification results of different CNN models on the SVHN database, where (a) Alexnet, (b) VGG16.

Table 2. The statistics of the largest improvement of our method with salient low-rank sparse CNN features over the method with the original CNN features.

Method	Accuracy1	Accuracy2
Alexnet + DAC	50.6%	51.4%
VGG16 + DAC	43.9%	46.7%
Noise + Alexnet + DAC	22.8%	24.9%
Noise + VGG16 + DAC	17.2%	18.8%

4 Conclusion

We have discussed the robust neighborhood preserving low-rank sparse CNN features for data representation and classification with small amount of training samples. The framework of our method includes three parts, that is, CNN feature extraction layer, sparse low-rank and sparse coding layer, and classification layer. We have examined the effectiveness of our proposed framework on two widely-used image databases. The extensive results show that the enhanced performance can be delivered by our proposed method in most case, compared with the original CNN features. That is, the computed robust neighborhood preserving salient low-rank and sparse CNN features can indeed produce the higher accuracy. In future, we will explore how to integrate the robust low-rank and sparse coding process with the CNN feature extraction layer for joint computation so that more promising results can be obtained.

Acknowledgments. This work is partially supported by the National Natural Science Foundation of China (61672365), Major Program of Natural Science Foundation of the Jiangsu Higher Education Institutions of China (15KJA520002), Natural Science Foundation of the Jiangsu Province of China (BK20141195), and the High-Level Talent of "Six Talent Peak" Project of the Jiangsu Province of China (XYDXX-055). Zhao Zhang is the corresponding author of this paper.

References

1. Liu, G., Lin, Z., Yan, S.: Robust recovery of subspace structures by low-rank representation. IEEE Trans. Pattern Anal. Mach. Intell. **35**(1), 171–184 (2013)
2. Liu, G., Lin, Z., Yu, Y.: Robust subspace segmentation by low-rank representation. In: Proceedings of the 27th International Conference on Machine Learning (ICML), pp. 663–670 (2010)
3. Candès, E.J., Li, X., Ma, Y., Wright, J.: Robust principal component analysis. J. ACM (JACM) **58**(3), 1–37 (2011)
4. Krizhevsky, A., Hinton, G.: Learning multiple layers of features from tiny images (2009)
5. Krizhevsky, A., Sutskever, I., Hinton, G.: Imagenet classification with deep convolutional neural networks. In: Advances in neural information processing systems, pp. 1097–1105 (2012)
6. Zhang, Z., Li, F., Zhao, M., Zhang, L., Yan, S.: Joint low-rank and sparse principal feature coding for enhanced robust representation and visual classification. IEEE Trans. Image Process. **25**(6), 2429–2443 (2016)
7. Lin, Z., Ganesh, A., Wright, J., Wu, L., Chen, M., Ma, Y.: Fast convex optimization algorithms for exact recovery of a corrupted low-rank matrix. In: Computational Advances in Multi-Sensor Adaptive Processing (CAMSAP), vol.61. June 2009
8. Lin, Z., Chen, M., Y, Ma.: The augmented lagrange multiplier method for exact recovery of corrupted low-rank matrices, pp. 1009–5055 (2010)
9. He, K., Zhang, X., Ren, S., Sun, J.: Deep residual learning for image recognition. In: Proceedings of the IEEE Conference on Computer Vision and Pattern Recognition, pp. 770–778 (2016)
10. Chen, J., Yi, Z.: Sparse representation for face recognition by discriminative low-rank matrix recovery. J. Vis. Commun. Image Represent. **25**(5), 763–773 (2014)

11. Zhang, Y., Jiang, Z.L., Davis, S.: Learning structured low-rank representations for image classification. In: Proceedings of the IEEE Conference on Computer Vision and Pattern Recognition, pp. 676–683 (2013)
12. Ma, L., Wang, C., Xiao, B., Zhou, W.: Sparse representation for face recognition based on discriminative low-rank dictionary learning. In: Proceedings of the International Conference on Computer Vision and Pattern Recognition (CVPR), pp. 2586–2593 (2012)
13. Simonyan, K., Zisserman, A.: Very deep convolutional networks for large-scale image recognition. (2014)
14. Cai, J.F., Candès, E.J., Shen, Z.: A singular value thresholding algorithm for matrix completion. SIAM J. Optim. **20**(4), 1956–1982 (2010)
15. LeCun, Y., Bottou, L., Bengio, Y., Haffner, P.: Gradient-based learning applied to document recognition. Proc. IEEE **86**(11), 2278–2324 (1998)
16. Lin, M., Chen, Q., Yan, S.C.: Network in network. In: Proceedings of International Conference on Learning Representations (2013)
17. Goodfellow, I., Bengio, Y., Courville, A.: Deep Learning. MIT Press, Cambridge, MA (2016)
18. Guo, Y., Hastie, T., Tibshirani, R.: Regularized linear discriminant analysis and its application in microarrays. Biostatistics **8**(1), 86–100 (2007)
19. Roweis, S.T., Saul, L.K.: Nonlinear dimensionality reduction by locally linear embedding. Science **290**, 2323–2326 (2000)
20. Hu, Y., Zhang, D., Ye, J., Li, X., He, X.: Fast and accurate matrix completion via truncated nuclear norm regularization. IEEE Trans on Pattern Analysis Machine Intelligence **35**(9), 2117–2130 (2013)
21. Zhang, H., Lin, Z., Zhan, C., Gao, J.: Robust latent low rank representation for subspace clustering. Neurocomputing **145**, 369–373 (2014)
22. Sim, T., Baker, S., Bsat, M.: The CMU pose, illumination, and expression database. IEEE Trans. Pattern Anal. Mach. Intell. **25**(12), 1615–1618 (2003)
23. Lin, Z., Chen, M., Wu, L., Ma, Y.: The augmented Lagrange multiplier method for exact recovery of corrupted low-rank matrices. University of Illinois Urbana-Champaign, Champaign, IL, USA, Tech. (2009)
24. Netzer, Y., Wang, T., Coates, A.: Reading digits in natural images with unsupervised feature learning. In: NIPS Workshop on Deep Learning and Unsupervised Feature Learning, vol. 2011, No. 2, p. 5 (2011)
25. Nasrabadi, N.M.: Pattern recognition and machine learning. J. Electron. Imaging **16**(4), 049901 (2007)
26. Zhang, Z., Li, F.Z., Zhao, M., Zhang, L., Yan, S.C.: Robust neighborhood preserving projection by Nuclear/L2,1-Norm regularization for image feature extraction. IEEE Trans. Image Process. **26**, 1607–1622 (2017)
27. Zhang, Z., Zhao, M., Li, F.Z., Zhang, L., Yan, S.C.: Robust alternating low-rank representation by joint Lp- and L2, p-norm Minimization. Neural Netw. **96**, 55–70 (2017)
28. Zhang, Z., Yan, S.C., Zhao, M., Li, F.Z.: Bilinear low-rank coding framework and extension for robust image recovery and feature representation. Knowl.-Based Syst. **86**, 143–157 (2015)
29. Zhang, Z., Yan, S.C., Zhao, M.: Similarity preserving low-rank representation for enhanced data representation and effective subspace learning. Neural Netw. **53**, 81–94 (2014)
30. Zhang, H., Patel, V.M.: Convolutional sparse and low-rank coding-based image decomposition. IEEE Trans. Image Process. **27**, 2121–2133 (2018)

31. Zhang, H., Patel, V.M.: Convolutional sparse and low-rank coding-based rain streak removal. WACV, pp. 1259–1267 (2017)
32. Ongie, G., Jacob, M.: A fast algorithm for convolutional structured low-rank matrix recovery. IEEE Trans. Comput. Imaging **3**, 535–550 (2017)
33. Jaderberg, M., Vedaldi, A., Zisserman, A.: Speeding up Convolutional Neural Networks with Low Rank Expansions. BMVC (2014)

Pedestrian Detection
with a Directly-Cascaded
Deconvolution-Convolution Structure

Zhiming Chen[1], Xintong Han[2], Weiyao Lin[1(✉)], Ming-Ming Cheng[3],
Guangcan Liu[4], and Hongkai Xiong[1]

[1] Shanghai Jiao Tong University, Shanghai, China
wylin@sjtu.edu.cn
[2] University of Maryland, College Park, USA
[3] Nankai University, Tianjin, China
[4] Nanjing University of Information Science and Technology, Nanjing, China

Abstract. Driven by recent advances in deep learning, the accuracy of object detection has been tremendously improved. However, detecting small and blurred pedestrians still remains an open challenge. In this paper, we propose a novel neural network structure, which can be flexibly combined with powerful object detection systems for boosting pedestrian detection. The proposed structure contains two key modules: (i) a cascaded deconvolution-convolution (CDC) module to expand the resolution of feature maps, meanwhile, keep the crucial information in the feature maps; and (ii) a double-helix connection (DHC) module to effectively fuse shallow-level and deep-level features in the detection network. The CDC module enables the network to reuse features of the lower layers and learn richer features given low-resolution input. In addition, the DHC module incorporates the features learned in different layers in a novel and unified fashion. Extensive experiments on KITTI and Caltech Pedestrian datasets demonstrate that the proposed modules can be easily plugged into existing object detection networks (e.g., single-stage SSD and two-stage MSCNN) and consistently achieve better performance without bells and whistles.

Keywords: Pedestrian detection
Deconvolution-convolution cascade · Double-helix connection

1 Introduction

Object detection plays an increasingly significant role in computer vision, attributing to its wide applications such as security monitoring, human-computer interaction, autonomous driving [5]. Recently, convolutional neural networks (CNNs) have significantly improved the performance of object detection.

This work is supported by NSFC(61471235), and Shanghai 'The Belt and Road' Young Scholar Exchange Grant(17510740100).
G. Liu—NSFC 61622305, NSFC 61502238, NSFJPC BK20141003

© Springer Nature Switzerland AG 2018
R. Hong et al. (Eds.): PCM 2018, LNCS 11164, pp. 370–380, 2018.
https://doi.org/10.1007/978-3-030-00776-8_34

Fig. 1. Pedestrians being far from the camera in images are difficult to recognize whether is pedestrian or background, and the predicted bounding box usually deviates. To relieve this issue, we propose a cascade of deconvolution-convolution module and double-helix connection.

Off-the-shelf approaches [8,11] had made a great performance, however, the core challenge to pedestrian detection is rooted in the rapid decline in accuracy when low-resolution pedestrians present, due to the missing of fine-detailed information. As shown in Fig. 1, pedestrians being far from the camera in the image are difficult to be detected since small objects have no enough resolution making their corresponding of feature maps so weak. In general, enlarging the resolution of input to remain the fine details, but that would cause the increase in computational cost because of the experiment [9] reporting the majority computation of CNNs focusing on the lower layer.

Although aiming at phenomenon presented in Fig. 1, SSD [7] utilized different layers to detect different scale objects, it is not desirable to small-size objects in Fig. 2(a). The core of problems is that small-size objects are short of semantic information and resolution, causing their response weak in the feature maps. As shown in Fig. 2(b) and (c), DSSD [35] added hourglass structure at the bottom of network FPN [20] added feature pyramid structure respectively to fuse high-level semantic feature maps. They all achieved state-of-the-art performance, however, feature maps still lack resolution and finer details for small objects. For that, detecting small-size pedestrians still depends on enlarging the resolution of input.

Our aim is that improving the performance of detecting small object under low-resolution input. Motivated by Fully Convolution Network (FCN)[1] demonstrating impressive performance on semantic segmentation tasks by upsampling feature maps, We assume that whether adding upsample operation to expand the resolution of feature maps in the middle layer. Only expanding the feature maps to make resolution higher, but feature maps are not strong and robust.

Also, it is worth noting that feature maps from only one layer regressing bounding boxes are not sophisticated enough. Meanwhile, as illustrated in Fig. 2(a), (b) and (c), the early downsample operation can damage the details of small objects. According to above analysis, whether enlarging the resolution of feature maps as well as keeping the finer details as much as possible. The answer is feasible. Our proposed structure with directly-cascaded

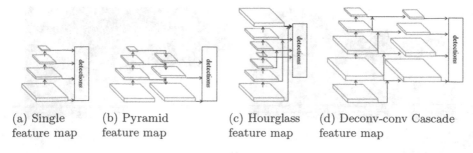

(a) Single   (b) Pyramid   (c) Hourglass   (d) Deconv-conv Cascade
feature map   feature map   feature map   feature map

Fig. 2. (a) utilizing the multi-scale feature maps from different layers, (b) and (c) fusing strong feature maps from different layers. (d) Our proposed Directly-cascaded Deconvolution-convolution Network detects object by using a feature hierarchy like (a), (b) and (c), but contribute to detecting small objects.

deconvolution-convolution can keep the details as much as possible and enlarge the responsive resolution.

In this paper, we demonstrate that it is workable that embedding a cascade of deconvolution-convolution module to reconstruct larger feature maps without loss the fine-detail information prominently in the middle layer. Moreover, we propose the double-helix connection to take full the advantage of feature maps and integrate the down-top context information. Our experiments showed that the proposed modules implanted into SSD [7] and MSCNN [13] framework achieves coherently improvement on both KITTI [5] and Caltech [4] datasets, without bells and whistles.

Overall, the key contributions of our work are three-fold:

– A cascade of deconvolution-convolution module is introduced to enlarge the resolution of small-size objects and keep theirs finer details as much as possible.
– To effectively incorporate rich features from different layers for locating pedestrians, we design a double-helix connection model.
– We show that our modules make consistent improvement implanting into SSD and MSCNN framework on the KITTI [5] and Caltech [4] datasets.

2 Related Work

Object Detection: With the fast development of deep learning, object detection has made a giant step forward. Its main task is that locating object regions and recognizing object classes. Object detection has attracted many researchers, which generally can be divided into two categories: one is the one-stage detector with fast speed like SSD [7], YOLO [12], the other is the two-stage detector with high accuracy like RCNN [10], Fast RCNN [18], Faster RCNN [23]. As for the two-stage detector, R-CNN [10] is a golden key opening the door using deep learning for object detection outperforming prominently traditional methods [14,15]. R-CNN is so slow due to needing to feed thousands of cropping image patches from Selective Search [34] to CNNs. For speeding up, Fast R-CNN [18] proposed ROI pooling layer and just forwarded for one time. For further acceleration, its variant: Faster R-CNN [23] applied Region Proposal Network(RPN) sharing feature maps with the backbone of network and replacing SS. As for the one-stage detector, they can train by end-to-end to regress bounding boxes directly. Compared with the two-stage detector, the one-stage detector is superior on speed, which is inferior on accuracy. Subsequently, Lin et al. [2] analyzed the intrinsic problem and presented the focus loss that boosts the performance further. Besides, for dealing with multi-scale object, MSCNN [13], SSD [7], SDP-CRC [17] applied that different layer detect different size of object. HyperNet [21], RRC [16], FPN [20] proposed some methods to explore rich feature maps. To improve the flexibility of network, DSOD [3] designed the network to train from scratch. To alleviate the troublesome label bounding boxes, WSCL [6] put forward weak supervision for object detection. Above approaches all made great performance.

Pedestrian Detection: Pedestrian detection can be viewed as one specific task in object detection. Basically, most of the existing object detection methods can be directly applied to detect pedestrians. However, since the appearance of pedestrian distorts easily at the mercy of small size, occlusion, pose variations and environmental change, special consideration is often needed when handling pedestrians. Traditional pedestrian detectors mostly depend on hand-crafted features. The HOG descriptor by Dalal and Triggs [24] plays a key part in constructing for pedestrian detection. The integral channel features (ICF) by Dollar et al. [25] utilizes channel feature pyramid. to efficiently extract robust features such as local sums, histograms. Recently, CNNs based approaches [11,27] have made tremendous progress on pedestrian detection compared traditional ways. Wang et al. [26] introduced into repulsion loss to solve crowed occlusion. Brazil et al. [28] and Gidaris et al. [29] both employ the semantic segmentation for assisting to reinforce pedestrian detection. Furthermore, utilizing multi-layer feature maps for detecting different scale pedestrian. As for that, two methods were employed. One is that training a single network which inputs a unified resolution such as [30,31], but the resolution of images is resized to different size during in the test phase. Using this image pyramid to construct a feature pyramid, causing that features are computed on multi-scale image respectively. The other is that to simultaneously deploy multiple object detectors to which a single image feeds respectively, such as [32,33].

In methods above, their studies majorly pay attention to detecting relatively large objects, which is compromised for detecting small objects. Our method is proposed to overcome the detection of small-size objects.

3   Our Method

Due to the aforementioned analysis, it is difficult for the existing methods to detect tiny and blurred objects. To this end, we proposed a double-helix connection with deconvolution-convolution cascade directly for pedestrian detection.

Figure 3 shows the framework of our approach. In order to simplify the description, we illustrate our framework based on the SSD detector. In practice, our approach can be easily implemented with various object detection structures. Compared to SSD, we implant our proposed two modules into SSD structure. In experiment section, we also explore to embed our modules into MSCNN structure. The whole framework consists of three key parts. the first part is the backbone of VGGNet [22], the second part is the deconvolution-convolution cascade module to reconstruct conv4, conv5, and fc7, the last part is the double-helix module to reconstruct the additional layer.

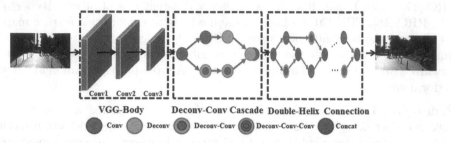

Fig. 3. Flow diagram of the whole framework consists of three key parts. the first part is the backbone of VGGNet [22], the second part is the deconvolution-convolution cascade module, the last part is the double-helix module.

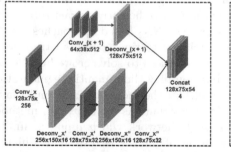 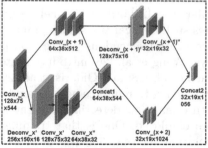

(a) Deconvolution-Convolution Cascade   (b) Double-Helix Connection

Fig. 4. The diagram of proposed module structure

3.1 Deconvolution-Convolution Cascade Module

To expand the resolution of feature maps and introduce into the fine-detail information, we design a deconvolution-convolution cascade module. As shown in Fig. 4(a), this module contains newly two deconvolution-convolution cascades and one deconvolution, compared with original. The procedure can be illustrated that assuming initiative feature maps (Conv_x) pass the trunk to get feature maps (Conv_(x + 1)), then upsampling to get feature maps (Deconv_(x + 1));

next Conv_x forward to two deconvolution-convolution cascades to get Conv_x''; finally concating Deconv_(x + 1) and Conv_x'' by channel to get rich feature maps.

3.2 Double-Helix Connection Module

To explore more contextual information, we design double-helix connection. As presented in Fig. 4(b), the procedure can be illustrated that assuming initiative feature maps (Conv_x) pass the trunk to get feature maps (Conv_(x + 1)), then upsampling to get feature maps (Deconv_(x + 1)); next Conv_x forward to deconv-conv-conv cascade to get Conv_x''; next concatting Deconv_(x + 1) and Conv_x'' by channel to get feature maps(Concat1); Concat1 pass the trunk to get feature maps(Conv_(x + 2)),Conv_(x + 1) forward to deconv-conv-conv cascade to get Conv_$(x + 1)''$, finally concatting Conv_(x + 2) and Conv_$(x + 1)''$ by channel to get feature maps (Concat2).

3.3 Model Details

As shown in Fig. 3, that is our model's structure based on SSD. We take an example of assuming that the size of the input images is 512×300 with 3 channels. The size of conv4_3 layer, conv5_3 layer, fc7 layer of original SSD are $64 \times 38 \times 512$, $32 \times 19 \times 512$ and $32 \times 19 \times 1024$ respectively. We reconstruct them to reconstruct_conv4_3 layer, reconstruct_conv5_3 layer, reconstruct_fc7 layer which are $128 \times 75 \times 544$, $64 \times 38 \times 544$, $64 \times 38 \times 1056$, applying our proposed directly-cascaded deconvolution-convolution structure. Following SSD, we add some auxiliary layers reconstruct_conv6_2, reconstruct_conv7_2, reconstruc_conv8_2, reconstruct_conv9_2 as the double-helix connection for multi-scale detection. The size of auxiliary layers is $32 \times 19 \times 544$, $16 \times 10 \times 288$, $8 \times 5 \times 288$, $4 \times 3 \times 256$ respectively. For regressing bounding boxes steadily and accurately, we further discretize the bounding boxes by distributing 4 regressors for each feature map. Following SSD, loss function contains confidence loss with softmax loss and location loss with SmoothL1 loss.

4 Experimental Evaluation

In this section, we present our experimental result of pedestrian detection. To evaluate the advantage of our method for dealing with small, blurred and occluded objects, we conduct a series of experiments on KITTI and Caltech datasets, which include more small-size and blurred objects. Beside we compare with our baseline and other methods based on the same training dataset.

4.1 KITTI Dataset Result

KITTI: The KITTI contains 8 different classes, only the classes Car and Pedestrian are evaluated in our benchmark, and three levels of evaluation: easy, moderate and hard. The moderate level is the most commonly used. In total, 7,481 images are available for training/validation, and 7,518 for testing. KITTI includes many small and occluded object. In kitti evaluation, only evaluating objects larger than 25 pixels (height) in the image. In our experiments, we focus on detecting pedestrians, however, a small quantity of images only contain pedestrians. Therefore, we train our model with images containing pedestrians and cyclists.

Our network based on SSD: We follow the setting of original SSD, but we change the smallest and largest default box to 0.1 and 0.8 scale respectively due to objects are very small. For training network, stochastic gradient descent with momentum of 0.9 was used for optimization. Weight decay is set to 0.0005. We set the initial learning rate to 0.0001. Batch_size is set to 4. The learning rate is decreased by a power of 10 at the 30,000, 50,000, 60,000 iterations. The size of the input image is 512×300 (Table 1).

Table 1. Results on the KITTI testing and comparison with SSD and MSCNN

Based on SSD		Based on MSCNN	
Benchmark	Moderate	Benchmark	Moderate
Pedestrian(SSD)	43.45%	Pedestrian(MSCNN)	70.03%
Cyclist(SSD)	33.81%	Cyclist(MSCNN)	66.64%
Pedestrian(Ours)	48.57%	Pedestrian(Ours)	71.52%
Cyclist(Ours)	37.78%	Cyclist(Ours)	67.59%

Our network based on MSCNN: We follow the setting of original MSCNN and just adjust the downsample rate to 4–64. To compared fairly, we set uniformly the size of the input image to be 1920×576.

Detection result and example: As is shown in Table 3, our method consistently achieves better performance with original SSD and MSCNN under the same resolution. To highlight the advantage of our method, we can see more subjective performance in Fig. 5.

4.2   Caltech Dataset Result

Caltech: The Caltech dataset contains about 350 K pedestrian bounding box annotations across 10 hours of urban driving. The log average miss rate sampled against a false positive per image (FPPI) range of is used for measuring performance. A minimum IoU threshold of 0.5 is required for a detected box to match with a ground truth box. According to [28], we filter bounding boxes with less than 35% occlusion based on Caltech10, which contains 42,782 images, finally remaining 12,160 images to train. We evaluate on the standard 4,024 images in

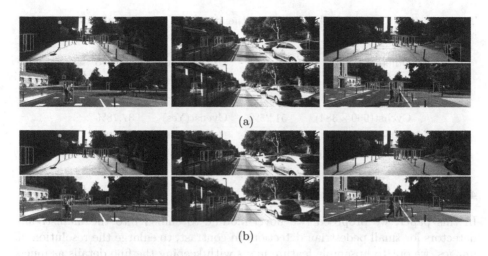

Fig. 5. (a) Detection results of ssd method in KITTI testing set, (b) Detection results of our method in KITTI testing set.

Table 2. Comparative results on the Caltech testing

Benchmark	mAP
Pedestrian(SSD)	64.34%
Pedestrian(Ours)	66.09%

the Caltech1 test set using the reasonable setting, which only considers pedestrians with at least 40 pixels in height and with less than 35% occlusion.

Detection result: To validate our method furthermore, we also evaluated our method on the Caltech pedestrian benchmark. As is shown in Table 2, our method can detect very well for small and blurred objects, being superior to original SSD.

4.3 Ablation Study

The impact of input resolution. Enlarging the resolution of input can keep the details of the small object, but it is obvious that the larger input is, the longer inference time is and the larger memory is. Table 3 shows that when the size of the input images is 960×384, the performance promotes significantly.

The impact of DHC module. For fusing richer feature maps, we design a double-helix connection (DHC) module to effectively fuse shallow-level and deep-level features. We also reconstruct additional layer with DHC module. As is shown in Table 3, we make a small improvement on accuracy. We argue that fusing some shallow-level feature maps is beneficial to detect object accurately.

Table 3. Results on the KITTI testing, the impact of input resolution and DHC module

Different Resolution		DHC Module, Yes or No	
Benchmark	Moderate	Benchmark	Moderate
Pedestrian(960 × 384)	54.63%	Pedestrian(Yes)	48.57%
Cyclist(960 × 384)	51.25%	Cyclist(Yes)	37.78%
Pedestrian(512 × 300)	48.57%	Pedestrian(No)	47.69%
Cyclist(512 × 300)	37.78%	Cyclist(No)	36.92%

4.4 Conclusion and Outlook

In this paper, we proposed two novel module to embed into the mainstream detectors for small pedestrian detection. In contrast, to enlarge the resolution of images, we opt to upsample feature maps with keeping the fine details as much as possible in the middle layer. In this way, we can boost the robustness under the relatively low resolution of images. Meanwhile, to exploit more rich feature maps, we proposed double-helix connection. Finally, we demonstrate that our method can improve effectively pedestrian detection on the KITTI dataset and Caltech dataset. In the feature, we will focus on exploring the efficient detector facing to mobile terminals for real-time object detection.

References

1. Long, J., Shelhamer, E., Darrell, T.: Fully convolutional networks for semantic segmentation. In: CVPR (2015)
2. Lin, T.-Y., et al. Focal loss for dense object detection. arXiv preprint arXiv:1708.02002 (2017)
3. Shen, Z., Liu, Z., Li, J., Jiang, Y.-G., Chen, Y., Xue, X.: DSOD: learning deeply supervised object detectors from scratch. In: The IEEE International Conference on Computer Vision (ICCV) (2017)
4. Dollar, P., Wojek, C., Schiele, B., Perona, P.: Pedestrian detection: a benchmark. In: IEEE Conference on Computer Vision and Pattern Recognition, CVPR 2009, pp. 304–311. IEEE (2009)
5. Geiger, A., Lenz, P., Urtasun, R.: Are we ready for autonomous driving? the kitti vision benchmark suite. In: IEEE Conference on Computer Vision and Pattern Recognition (CVPR), 2012. IEEE (2012)
6. Wang, J., Yao, J., Zhang, Y., et al.: Collaborative learning for weakly supervised object detection. arXiv preprint arXiv:1802.03531 (2018)
7. Liu, W., Anguelov, D., Erhan, D., Szegedy, C., Reed, S., Fu, C., Berg, A.: SSD: single shot multibox detector. In: ECCV (2016)
8. Li, J., Liang, X., Shen, S., Xu, T., Yan, S.: Scale-aware fast R-CNN for pedestrian detection. In: CVPR (2015)
9. Lu, Z., et al.: Modeling the resource requirements of convolutional neural networks on mobile devices. In: Proceedings of the 2017 ACM on Multimedia Conference. ACM (2017)

10. Girshick, R., Donahue, J., Darrell, T., Malik, J.: Rich feature hierarchies for accurate object detection and semantic segmentation. In: CVPR (2014)
11. Hosang, J., Omran, M., Benenson, R., Schiele, B.: Taking a deeper look at pedestrians. In: Proceedings of the IEEE Conference on Computer Vision and Pattern Recognition, pp. 4073–4082 (2015)
12. Redmon, J., Divvala, S., Girshick, R., Farhadi, A.: You only look once: unified, real-time object detection. In: IEEE Conference on Computer Vision and Pattern Recognition (CVPR) (2016)
13. Cai, Z., Fan, Q., Feris, R.S., Vasconcelos, N.: A unified multi-scale deep convolutional neural network for fast object detection. In: Leibe, B., Matas, J., Sebe, N., Welling, M. (eds.) ECCV 2016. LNCS, vol. 9908, pp. 354–370. Springer, Cham (2016). https://doi.org/10.1007/978-3-319-46493-0_22
14. Felzenszwalb, P.F.: Object detection with discriminatively trained part-based models. IEEE Trans. Pattern Anal. Mach. Intell. **32**(9), 1627–1645 (2010)
15. Girshick, R.B., Felzenszwalb, P.F., McAllester, D.: Discriminatively trained deformable part models, release 5 (2012)
16. Ren, J., et al.: Accurate single stage detector using recurrent rolling convolution. In: CVPR (2017)
17. Yang, F., Choi, W.: Exploit all the layers: fast and accurate CNN object detector with scale dependent pooling and cascaded rejection classifiers. In: Proceedings of the IEEE Conference on Computer Vision and Pattern Recognition (2016)
18. Girshick, R.: Fast R-CNN. In: The IEEE International Conference on Computer Vision (ICCV) (2015)
19. Girshick, R., Donahue, J., Darrell, T., Malik, J.: Rich feature hierarchies for accurate object detection and semantic segmentation. In: The IEEE Conference on Computer Vision and Pattern Recognition (CVPR), vol. 1, p. 2 (2014)
20. Lin, T.-Y., Dollar, P., Girshick, R., He, K., Hariharan, B., Belongie, S.: Feature pyramid networks for object detection. In: The IEEE Conference on Computer Vision and Pattern Recognition (CVPR) (2017)
21. Kong, T., et al.: Hypernet: towards accurate region proposal generation and joint object detection. In: Proceedings of the IEEE Conference on Computer Vision and Pattern Recognition (2016)
22. Simonyan, K., Zisserman, A.: Very deep convolutional networks for large-scale image recognition. arXiv preprint arXiv:1409.1556 (2014)
23. Ren, S., He, K., Girshick, R., Sun, J.: Faster R-CNN: towards real-time object detection with region proposal networks. In: Advances in Neural Information Processing Systems (2015)
24. Triggs, B., Dalal, N.: Histograms of oriented gradients for human detection. In: CVPR (2005)
25. Dollar, P., Tu, Z., Perona, P., Belongie, S.: Integral channel features. In: BMVC (2009)
26. Wang, X., Xiao, T., Jiang, Y., et al.: Repulsion loss: detecting pedestrians in a crowd. arXiv preprint arXiv:1711.07752 (2017)
27. Zhang, L., Lin, L., Liang, X., He, K.: Is faster R-CNN doing well for pedestrian detection? In: Leibe, B., Matas, J., Sebe, N., Welling, M. (eds.) ECCV 2016. LNCS, vol. 9906, pp. 443–457. Springer, Cham (2016). https://doi.org/10.1007/978-3-319-46475-6_28
28. Brazil, G., Yin, X., Liu, X.: Illuminating pedestrians via simultaneous detection & segmentation. arXiv preprint arXiv:1706.08564 (2017)

29. Gidaris, S., Komodakis, N.: Object detection via a multi-region and semantic segmentation-aware CNN model. In: Proceedings of the IEEE International Conference on Computer Vision, pp. 1134–1142 (2015)
30. Girshick, R., Donahue, J., Darrell, T., Malik, J.: Rich feature hierarchies for accurate object detection and semantic segmentation. In: CVPR (2014)
31. Dollar, P., Belongie, S., Perona, P.: The fastest pedestrian detector in the west. In: BMVC (2010)
32. Benenson, R., Mathias, M., Timofte, R., Van Gool, L.: Pedestrian detection at 100 frames per second. In: CVPR (2012)
33. Sermanet, P., Kavukcuoglu, K., Chintala, S., LeCun, Y.: Pedestrian detection with unsupervised multi-stage feature learning. In: CVPR (2013)
34. Uijlings, J.R., Van De Sande, K.E., Gevers, T., Smeulders, A.W.: Selective search for object recognition. In: IJCV (2013)
35. Fu, C.Y., Liu, W., Ranga, A., et al.: DSSD: deconvolutional single shot detector. arXiv preprint arXiv:1701.06659 (2017)

Adaptive Integration Skip Compensation Neural Networks for Removing Mixed Noise in Image

Kai Lin, Yiwei Zhang, Thomas H. Li, Kan Huang, and Ge Li$^{(\boxtimes)}$

Digital Media R&D Center, SECE, Shenzhen Graduate School,
Peking University, Shenzhen, China
`1701213612@sz.pku.edu.cn`, `yuriyzhang@pku.edu.cn`, `18662197462@qq.com`,
`huangkan@pkusz.edu.cn`, `geli@ece.pku.edu.cn`

Abstract. During the process of acquisition and transmission, images are often likely to be corrupted by mixed Gaussian-impulse noise. Among various image denoising methods, most traditional methods can only deal with a single type of noise due to the difficulty of modeling the distribution of the mixed noise. In this paper, we propose a novel mixed Gaussian-impulse noise removal method based on adaptive integration skip compensation Network (Ai-Sc-Net). More concretely, a couple of skip compensation networks (Sc-Net) Sc-Net-AWGN and Sc-Net-IN are trained on Gaussian and Impulse noise datasets separately to deal with the corresponding single type noise. Further, an adaptive integration network (Ai-Net) is used to integrate the two outputs of Sc-Net-AWGN and Sc-Net-IN. The Ai-Sc-Net is then be constructed based on Sc-Net and Ai-Net, which can handle mixed noise. Experimental results in synthetic noise images have shown great improvements over several state-of-the-art mixed noise removal methods.

Keywords: Mixed noise removal · Neural networks · Residual learning

1 Introduction

Image denoising is one of the most basic tasks in image processing. Owing to digital images are easily interfered by noise in the process of acquisition and transmission, they cannot be used directly for high-level tasks such as remote sensing imaging, object recognition and medical imaging, etc. Generally, the noise encountered in images can be divided into two types: additive white Gaussian noise (AWGN) and impulse noise (IN) [6]. The AWGN is caused by the thermal motion in camera [10], while the IN is often caused by bit errors or the damaged pixels in camera sensors [20].

Many methods have been proposed for removing image noise of a certain type and promising results have been achieved(e.g., BM3D [2], NCSR [3] and LRTR [5]). However, in real world circumstances, the image noise appears typically in

© Springer Nature Switzerland AG 2018
R. Hong et al. (Eds.): PCM 2018, LNCS 11164, pp. 381–391, 2018.
https://doi.org/10.1007/978-3-030-00776-8_35

Fig. 1. Restoration results of Barbara image (GN + SPN, $\sigma = 20, \rho = 0.2$). From left to right and top to bottom: original image, noise image, the reconstructed images by M-BM3D (PSNR = 23.64 dB), $l_1 - l_0$ (PSNR = 29.41 dB), WJSR (PSNR = 30.01 dB), and our proposed method (PSNR = 31.20 dB).

the mixed form instead of a particular type. Therefore, the methods that are successful to remove a certain type of noise are not applicable for the scenario of mixed noise, where the distributions of the two types of noise are totally different. In this paper, we focus on removing mixed noise in image.

The AWGN is introduced into images by adding each pixel a random value sampled from the Gaussian distribution. In general, there are two typical IN, i.e., the salt-and-pepper IN (SPIN) and random valued IN (RVIN). Both the two types of IN affect some of the image pixels with the remaining ones unchanged. There are generally two types of methods for mixed noise reduction. The first one is based on the "detecting then filtering" strategy, in which first detect the locations of the damaged pixels and then utilize the remaining ones to reconstruct the damaged pixels. Garnett [6] used the rank order absolute difference (ROAD) as a detector to describe the probability of an image pixel to be damaged by IN. And then he adopts the bilateral filter [17] to remove the mixed noise. Coupled with a detector, the nonlocal mean (NLM) filter-based model [6] and the total variation-based denoising model [11] are also extended for mixed noise removal. Xiao et al. [19] employed a detector to identify the IN and the remaining non-IN corrupted pixels were used in sparse coding for reconstruction. However, the performance of this type methods relies too much on the accuracy of the damaged pixels detection. The second method uses robust fidelity terms, which

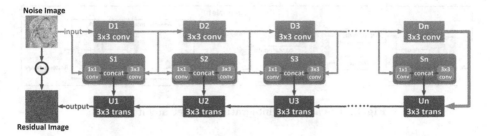

Fig. 2. The Skipping Compensation network architecture

treats the IN as outliers and needs not to detect the locations of the damaged pixels. [12] achieved good results by combing the priors of sparsity and nonlocal self-similarity. In [8], through incorporating with an appropriate regularizer (e.g., low-rank regularizer), the IN is estimated by the hard or soft thresholding. These methods have shown promising performance for removing mixed noise. However, the selection of appropriate thresholds remains a challenging task.

The contributions of this paper are as follows:

(1) A skip compensation network (Sc-Net) is proposed for single type noise removal, which combines the two-way skip to concatenate shallow and deep feature together. Thus the network can learn more details from the input.
(2) An adaptive integration Network (Ai-Net) is constructed to integrate the outputs of Sc-Net-AWGN and Sc-Net-IN adaptively. By connecting (1) and (2), the adaptive integration skip compensation network (Ai-Sc-Net) is able to not only remove mixed noise but also preserve texture information.
(3) A new loss metric called smooth loss is proposed for image denoising, which can reduce the checkerboard patterns and make the restored result more visible.

The rest of this paper is organized as follows. The related work is presented in Sect. 2, followed by the details of our method in Sect. 3. Section 4 presents the experimental results. The conclusion is drawn in Sect. 5.

2   Related Work

Deep Neural Networks for Image Denoising. Owing to the strong power of feature learning and representation, a variety of image denoising methods have been proposed based on deep neural networks. In [9], convolutional neural networks (CNNs) is successfully applied to image denoising, proving that CNNs have better representation abilities than traditional models. Chen [1] proposed a trainable nonlinear reaction diffusion (TNRD) model that can be treated as a feed-forward deep network. Different from the existing denoising models which train a specific model for AWGN at a certain noise level, Zhang [22] combined residual learning and batch normalization to handle Gaussian denoising with

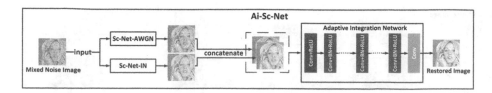

Fig. 3. The adaptive integration Sc-Net architecture

unknown noise level. However, these models are trained for single type of noise. To the best of our knowledge, it remains uninvestigated to design a specialized neural network for mixed noise removal.

Encoder-Decoder Networks. Based on the idea that the features learned from networks are enough to describe the inputs, Denoising Autoencoder(DAE) [18] has been proposed for image restoration. Further, Inspired by U-Net [15] and Skip-Net [13], shallow features and deep features are concatenated to learn more details from the input noise image. U-Net uses the contracting path to follow down-sampling steps and the expansive path to follow up-sampling steps and then concatenate the symmetry features. Deriving from U-Net, Skip-Net links convolutional and deconvolutional layers by skip-layer connections. All of above shows that the encoder-decoder architecture possesses the ability to recover inputs with learned features.

Residual Learning. In order to solve the problem of degradation when network goes deeper, He [7] present a residual learning framework to ease the training of networks that are substantially deeper than those proposed previously. They learn residual functions with reference to the input instead of with unreferenced functions. Based on the fact that the residual mapping is much easier to be optimized than the original unreferenced mapping, excellent results have been achieved for blind AWGN image denoising [22].

3   The Proposed Method

In this section, we first describe in detail the Sc-Net which is designed for removal of a certain type of noise. Then we will elaborate on the process of integrating the outputs of Sc-Net-AWGN and Sc-Net-IN using Ai-Sc-Net.

3.1   Skip Compensation Network Architecture

Skip-Net uses the symmetric encoder-decoder, where the decoder receives additional encoder features extracted by 1×1 conv. However, they add shallow and deep features directly without operating and each decoder layer receives input features only from the encoder layer with the same scale. In our opinions, a network ought to be capable of digesting priors rather than extracting features only. Inspired by this idea, we propose the Skipping Compensation Network as shown in Fig. 2. The network consists of a contracting path, a skipping path and an expansive path. The contracting path is composed of two subpaths, where one follows 2×2 average pooling and the other follows 3×3 conv. Each of the above pooling and conv layers is followed by a LeakyReLU and the stride is 2 for downsampling. The expansive path is composed of upsampling feature map followed by a 3×3 transposed layer and a ReLU. The skipping path is our innovative part. we add a 1×1 conv which skips the down-layer to the same scale up-layer and a 3×3 conv which skips the down-layer to the next scale up-layer. The advantage of the skipping path is that both shallow and deep priors can be concatenated at each scale. In our network, we explore 6 different scales, each of which is half of the previous scale. Besides, when the original mapping is more like an identity mapping, the residual mapping will be much easier to be calculated [22]. Under the fact that the noisy image is much more similar to the clean one than the residual image, the residual learning method is reasonable for image denoising. Furthermore, we allow the network to learn residual image and derive the restored image by calculating the difference between noise image(input) and residual image(output).

3.2 Adaptive Integration Sc-Net Architecture

The purpose of adaptive integration network (Ai-Net) is to integrate two paths' results to get the final restored image. Given Ai-Net with depth D, the layers are divided into three types as shown in Fig. 3 with three different colors. For the first layer, 32 filters of size $3 \times 3 \times 2c$ Conv are applied to generate 32 feature maps and follows a ReLU. Here c indicates the number of image channels. Since two images are concatenated in the $3rd$ channel, we denote $2c$ here. From the $2nd$ layer to the $(D-1)th$ layer, 32 filters of size $3 \times 3 \times 32$ Conv are applied, which is followed by batch normalization(BN) and ReLU. For the last layer, c filters of size $3 \times 3 \times 32$ are used to reconstruct the output.

3.3 Loss Function

For Sc-Net, the input is a noisy observation $y = x + v$, where x is the clean image, v is the additive noise and y is the noisy image. First, by training a residual mapping $f(y) \approx v$, and then we have $x = y - v$. Here the $L1$ loss is adopted for loss metric because it is identical with MSE loss in value space and has the ability to contract the noise space. Formally, the averaged $L1$ error between the desired residual images and estimated ones from noise input is as follows

$$L_{l1}(\Theta_1) = \frac{1}{2N_1} \sum_{i=1}^{N_1} \|f(y_i; \Theta_1) - (y_i - x_i)\|_1^1 \tag{1}$$

where Θ_1 is trainable parameters in Sc-Net, and $\{(y_i, x_i)\}_{i=1}^{N_1}$ represents N_1 noise-clean training image (patch) pairs. Besides, in order to reduce checkerboard pattern [21] in restored image, we introduce a new loss term called smooth loss. The intuition is to prevent major difference between the neighboring pixels which would result in checkerboard pattern in the image. To calculate the smooth loss we slide a copy of the generated image one unit to the left and one unit down and then take an Euclidean distance between the shifted images. Thus, the total loss for Sc-Net is redefined as follows

$$L_{Sc} = L_{l1} + \lambda \cdot L_{smooth} \tag{2}$$

where λ is a pre-defined parameter. Here we set λ to be 0.5.

For Ai-Net, we use $L2$ loss as bellow:

$$L_{Ai}(\Theta_2) = \frac{1}{2N_2} \sum_{i=1}^{N_2} \|\Re(p_i; \Theta_2) - q_i\|_2^2 \tag{3}$$

where $\Re(p)$ is a mapping to integrate paired images to a restored image, Θ_2 is trainable parameters in Ai-Net, and $\{(p_i, q_i)\}_{i=1}^{N_2}$ represents N_2 intermediate-clean training image pairs.

3.4   Training

The training process of Ai-Sc-Net is conducted in two stages. In the first step, Single type of noisy images including AWGN and IN images are fed into Sc-Net respectively to train both the Sc-Net-AWGN and the Sc-Net-IN. In the second step, mixed noisy images are fed into the Sc-Net-AWGN and the Sc-Net-IN. For the same mixed noisy image, the output of Sc-Net-AWGN and Sc-Net-IN are concatenated in the $3rd$ channel to generate paired training datasets for Ai-Net. Then, the paired training datasets are used to train Ai-Net, which is to merge two images into one restored image.

4   Experimental Results

4.1   Training and Testing Data

Training Data. For Sc-Net, We use BSD400 [1] for training. We set the patch size as 128×128, and crop 384000 patches to train the model. For AWGN, we set the range of the noise levels as $\sigma \in [0, 100]$. For IN, we set the percentage of SPN and RVIN to be 1:1 and set the proportion of image pixels that be replaced with noise as $\rho \in [0, 0.5]$. For Ai-Net, we use 17000 images of size 512×512 from VOC2012 [4]. Firstly, clean images in training datasets are added with both AWGN and IN to construct mixed noisy images, the noise level σ of AWGN belongs to [0, 100] and the corrupted proportion of IN $\rho \in [0, 0.5]$. Secondly, each mixed noisy image is fed into the Sc-Net-AWGN and the Sc-Net-IN respectively, and each paired outputs are concatenated in the $3rd$ channel to build up the training datasets for Ai-Net.

Fig. 4. Restoration results of Lena image (GN + SPN, $\sigma = 20, \rho = 0.2$). From left to right and top to bottom: original image, noise image, reconstructed images by M-BM3D (PSNR = 27.61 dB), $l_1 - l_0$ (PSNR = 31.31 dB), WJSR (PSNR = 31.69 dB), and our proposed method (PSNR = 32.61 dB).

Testing Data. We apply BSD68 and 12 commonly used images for testing. The noise level σ of AWGN and the corrupted proportion ρ of IN are under a wide range. All those test images are not included in the training dataset and they are widely used for the performance evaluation of image denoising task.

4.2   Parameter Setting and Network Training

For Sc-Net, SGD is used with weight decay of 0.0001 and a 0.9 momentum. The mini-batch size is 128. While for Ai-Sc-Net, for optimization we use SGD with weight decay of 0.0001, the momentum is 0.9 and a mini-batch size of 256. Sc-Net and Ai-Sc-Net are both trained with 150 epochs. The learning rate was decayed exponentially from 10^{-1} to 10^{-4} during the 150 epochs. All our experiments are carried out in the Pytorch [14] environment running on a server with Intel(R) Core(TM) i7-8700K CPU 3.70 GHz and a NVIDIA Tesla P4 GPU. It takes about 12 h and one day to train Sc-Net and Ai-Sc-Net on GPU, respectively.

4.3   Comparison of Mixed Noise Removal

We evaluate the performance of our Ai-Sc-Net against several state-of-the-art mixed noise removal methods. To generate mixed noise, two main types of mixed

Table 1. Comparison of restoration results in PSNR for images corrupted by **AWGN + SPN**

Image	ρ	$\sigma = 10$				$\sigma = 30$			
		BM3D	$l_1 - l_0$	WJSR	ours	M-BM3D	$l_1 - l_0$	WJSR	ours
lena	20%	32.97	35.44	36.12	**36.45**	28.93	30.27	30.97	**31.61**
	40%	29.42	34.58	35.08	**35.44**	26.74	28.98	29.57	**29.78**
barbara	20%	24.07	31.37	32.28	**33.42**	23.28	26.66	27.13	**27.68**
	40%	23.28	29.41	31.19	**31.93**	22.29	25.65	25.89	**26.60**
boat	20%	28.00	31.23	31.94	**33.19**	25.70	27.04	27.54	**28.34**
	40%	25.19	30.20	30.81	**31.97**	23.98	26.19	26.56	**27.54**
bridge	20%	24.49	27.17	27.68	**29.50**	23.04	23.82	24.00	**24.81**
	40%	22.61	26.22	26.43	**27.70**	21.89	23.08	23.29	**24.17**
couple	20%	27.72	31.05	31.77	**33.16**	25.48	26.91	27.35	**28.06**
	40%	24.96	29.99	30.58	**31.85**	23.78	26.13	26.29	**27.36**
house	20%	35.91	37.38	37.95	**38.37**	31.17	32.28	33.42	**33.86**
	40%	32.52	36.83	37.26	**37.57**	28.35	30.54	31.65	**32.68**
F16	20%	30.23	32.94	33.94	**35.30**	27.25	28.52	29.15	**29.80**
	40%	26.70	32.15	32.91	**33.93**	25.12	27.14	27.85	**28.60**
peppers	20%	31.18	32.69	33.03	**34.56**	28.27	29.18	29.62	**30.64**
	40%	28.82	31.96	32.36	**33.28**	26.37	28.14	28.77	**29.73**
pentagon	20%	27.32	29.88	30.33	**30.40**	25.40	26.09	26.12	**26.20**
	40%	25.61	28.77	29.20	**29.31**	24.12	25.55	**25.59**	25.25
hill	20%	28.94	31.63	32.04	**32.95**	26.67	27.70	27.97	**28.55**
	40%	27.00	30.72	30.99	**31.90**	25.19	26.87	27.04	**27.76**
baboon	20%	25.74	28.92	31.00	**31.13**	23.54	25.20	25.43	**25.51**
	40%	22.89	27.38	29.24	**29.30**	22.13	24.26	24.46	**24.51**
man	20%	28.73	31.20	31.75	**33.33**	26.36	27.15	27.47	**28.31**
	40%	26.47	30.41	30.63	**32.04**	24.76	26.34	26.54	**27.63**
Average	20%	28.78	31.74	32.49	**33.48**	30.90	27.57	28.01	**28.61**
	40%	26.29	30.72	31.39	**32.19**	24.56	26.57	26.96	**27.63**

noise were considered: (1) AWGN + SPN and (2) AWGN + RVIN. We choose four state-of-the-art methods: M-BM3D (BM3D coupled with median filter) [2], the $l_1 - l_0$ [19], and the WJSR [12] for comparison. The source codes of all compared methods are downloaded from the authors' websites. To evaluate the quality of reconstructed images, Peak Signal to Noise Ratio (PSNR) and visual information fidelity (VIF) [16] are calculated to evaluate the visual quality. We apply 12 commonly used images in our experiments. For AWGN + SPN, the AWGN with noise level $\sigma = 10, 30$ mixed with SPN with noise density $\rho = 20\%, 40\%$ were considered. For AWGN + RVIN, the AWGN with noise level

Table 2. Comparison of restoration results in PSNR for images corrupted by **AWGN + RVIN**

Image	ρ	$\sigma = 15$				$\sigma = 25$			
		BM3D	$l_1 - l_0$	WJSR	Ours	M-BM3D	$l_1 - l_0$	WJSR	Ours
lena	25%	29.79	29.95	31.91	**32.31**	26.59	26.08	28.46	**29.05**
	45%	22.62	26.25	28.47	**29.52**	20.15	23.26	25.61	**27.71**
barbara	25%	23.79	24.02	24.37	**24.89**	22.78	22.46	23.24	**24.13**
	45%	21.45	22.44	23.07	**23.97**	18.26	21.11	22.39	**24.00**
boat	25%	27.08	26.86	27.60	**28.87**	24.06	24.45	25.36	**26.12**
	45%	22.88	24.58	25.58	**25.87**	19.53	22.57	23.91	**25.19**
bridge	25%	23.83	24.19	24.21	**24.38**	21.80	22.64	22.65	**22.91**
	45%	20.72	23.37	22.60	**23.86**	18.45	20.90	21.40	**22.48**
couple	25%	26.72	26.78	27.43	**28.96**	23.91	24.37	25.18	**25.56**
	45%	23.00	24.46	25.38	**25.95**	19.45	22.50	23.65	**24.12**
house	25%	31.54	31.06	34.45	**34.98**	28.68	26.72	31.13	**32.53**
	45%	23.31	27.33	30.74	**31.77**	20.99	24.13	27.76	**29.07**
F16	25%	28.12	28.53	29.92	**30.45**	24.87	25.29	27.12	**28.36**
	45%	21.63	25.21	26.82	**29.37**	19.29	22.49	24.49	**26.10**
peppers	25%	28.95	28.90	30.30	**31.03**	26.60	25.71	28.04	**28.97**
	45%	23.07	25.98	27.71	**29.13**	20.64	23.22	25.53	**27.05**
pentagon	25%	26.14	26.36	**26.55**	26.41	24.65	24.21	24.82	**24.89**
	45%	24.69	25.04	25.45	**25.56**	21.97	23.05	23.90	**24.03**
hill	25%	27.75	27.74	28.52	**28.97**	25.59	25.08	26.30	**27.32**
	45%	23.24	25.53	26.71	**26.97**	20.73	22.98	24.84	**25.52**
baboon	25%	25.44	25.83	25.86	**26.37**	22.30	23.09	23.23	**23.98**
	45%	22.93	23.75	24.01	**24.71**	20.11	21.98	22.15	**22.74**
man	25%	27.51	27.44	28.22	**28.95**	25.02	24.83	25.95	**27.03**
	45%	23.22	25.26	26.27	**27.71**	20.62	22.96	24.42	**25.03**
Average	25%	27.22	26.47	28.28	**28.88**	24.74	24.58	25.96	**26.74**
	45%	22.73	24.93	26.07	**27.03**	20.02	22.60	24.17	**25.25**

$\sigma = 15, 25$ mixed with RVIN with noise density $\rho = 25\%, 45\%$ were considered. Tables 1 and 2 present the PSNR values of all these methods for all the tested images. These images are corrupted by the AWGN + SPN or AWGN + RVIN respectively. The best metrics calculated are marked in bold. From Tables 1 and 2, one can see that our proposed method achieves much higher PSNR values than other methods in most cases. The average values of PSNR of our method is also the highest. And for VIF(%), the average values of our method higher than the WJSR about 0.9(due to the limit of space, specific values are not presented).

In Figs. 1 and 4, we show the denoising performance comparison on different images. Obviously, our proposed method is able to not only remove the noise drastically but also preserve most of the image details as well. In addition, our restored image is more visually pleasant than the others.

5   Conclusion

In this paper, an Ai-Sc-Net is proposed for mixed noise removal. The process of denoising includes two steps. In the first step, the Sc-Net is constructed for single type noise removal, where the bidirectional skipping is adopted to preserve shallow and deep features. AWGN and IN images are fed into Sc-Net respectively to train the Sc-Net-AWGN and the Sc-Net-IN. In the second step, each mixed noisy image is fed into the Sc-Net-AWGN and the Sc-Net-IN simultaneously to generate the paired images, then these paired images are used to train the adaptive integration network (Ai-Net). Finally, the Ai-Sc-Net is constructed by combining the Sc-Net and the Ai-Net, which possesses the ability to achieve mixed noise removal. Extensive experimental results demonstrate that our proposed method not only obtain favorable performance quantitatively but also produce more visually pleasant restored results than several other state-of-the-art methods.

Acknowledgment. This work was supported in part by the Project of National Engineering Laboratory for Video Technology - Shenzhen Division, in part by Shenzhen Key Laboratory for Intelligent Multimedia and Virtual Reality under Grant ZDSYS 201703031405467, and in part by the Shenzhen Municipal Development and Reform Commission (Disciplinary Development Program for Data Science and Intelligent Computing) under Grant 1230233753.

References

1. Chen, Y., Pock, T.: Trainable nonlinear reaction diffusion: a flexible framework for fast and effective image restoration. IEEE Trans. Pattern Anal. Mach. Intell. **39**(6), 1256–1272 (2016)
2. Dabov, K., Foi, A., Katkovnik, V., Egiazarian, K.: Image denoising by sparse 3-d transform-domain collaborative filtering. IEEE Trans. Image Process. **16**(8), 2080–2095 (2007)
3. Dong, W., Zhang, L., Shi, G., Li, X.: Nonlocally centralized sparse representation for image restoration. IEEE Trans. Image Process. **22**(4), 1620 (2013). A Publication of the IEEE Signal Processing Society
4. Everingham, M., Van Gool, L., Williams, C.K.I., Winn, J., Zisserman, A.: The PASCAL Visual Object Classes Challenge (VOC2012) Results (2012). http://www.pascal-network.org/challenges/VOC/voc2012/workshop/index.html
5. Fan, H., Chen, Y., Guo, Y., Zhang, H., Kuang, G.: Hyperspectral image restoration using low-rank tensor recovery. IEEE J. Sel. Top. Appl. Earth Obs. Remote Sens. **10**(10), 4589–4604 (2017)
6. Garnett, R., Huegerich, T., Chui, C., He, W.: A universal noise removal algorithm with an impulse detector. IEEE Trans. Image Process. **14**(11), 1747–1754 (2005)

7. He, K., Zhang, X., Ren, S., Sun, J.: Deep residual learning for image recognition. In: Computer Vision and Pattern Recognition, pp. 770–778 (2016)
8. Huang, T., Dong, W., Xie, X., Shi, G., Bai, X.: Mixed noise removal via laplacian scale mixture modeling and nonlocal low-rank approximation. IEEE Trans. Image Process. **26**(7), 3171–3186 (2017)
9. Jain, V., Seung, H.S.: Natural image denoising with convolutional networks. In: International Conference on Neural Information Processing Systems, pp. 769–776 (2008)
10. Jiang, J., Zhang, L., Yang, J.: Mixed noise removal by weighted encoding with sparse nonlocal regularization. IEEE Trans. Image Process. **23**(6), 2651–2662 (2014)
11. Li, G., Huang, X., Li, S.G.: Adaptive bregmanized total variation model for mixed noise removal. AEU - Int. J. Electron. Commun. **80**, 29–35 (2017)
12. Liu, L., Chen, L., Chen, C.L., Tang, Y.Y., Pun, C.M.: Weighted joint sparse representation for removing mixed noise in image. IEEE Trans. Cybern. **47**(3), 600–611 (2017)
13. Mao, X.J., Shen, C., Yang, Y.B.: Image restoration using very deep convolutional encoder-decoder networks with symmetric skip connections (2016)
14. Paszke, A., et al.: Automatic differentiation in pytorch. In: NIPS-W (2017)
15. Ronneberger, O., Fischer, P., Brox, T.: U-Net: convolutional networks for biomedical image segmentation. In: Navab, N., Hornegger, J., Wells, W.M., Frangi, A.F. (eds.) MICCAI 2015. LNCS, vol. 9351, pp. 234–241. Springer, Cham (2015). https://doi.org/10.1007/978-3-319-24574-4_28
16. Sheikh, H.R., Bovik, A.C.: Image information and visual quality. IEEE Trans. Image Process. **15**(2), 430–444 (2006)
17. Tomasi, C., Manduchi, R.: Bilateral filtering for gray and color images ICCV. In: Proceedings of Sixth International Conference on ICCV, p. 839 (1998)
18. Vincent, P., Larochelle, H., Lajoie, I., Bengio, Y., Manzagol, P.A.: Stacked denoising autoencoders: learning useful representations in a deep network with a local denoising criterion. J. Mach. Learn. Res. **11**(12), 3371–3408 (2010)
19. Xiao, Y., Zeng, T., Yu, J., Ng, M.K.: Restoration of images corrupted by mixed gaussian-impulse noise via minimization. Pattern Recognit. **44**(8), 1708–1720 (2011)
20. Yan, M.: Restoration of images corrupted by impulse noise and mixed gaussian impulse noise using blind inpainting. SIAM J. Imaging Sci. **6**(3), 1227–1245 (2013)
21. Yu, A.C., Peng, Q.: Robust recognition of checkerboard pattern for camera calibration. Opt. Eng. **45**(9), 1173–1183 (2006)
22. Zhang, K., Zuo, W., Chen, Y., Meng, D., Zhang, L.: Beyond a gaussian denoiser: residual learning of deep cnn for image denoising. IEEE Trans. Image Process. **26**(7), 3142–3155 (2017). A Publication of the IEEE Signal Processing Society

Retrieval Across Optical and SAR Images
with Deep Neural Network

Yifan Zhang, Wengang Zhou$^{(\boxtimes)}$, and Houqiang Li

CAS Key Laboratory of Technology in Geo-spatial, Information Processing
and Application System, Department of Electronic Engineering and Information
Science, University of Science and Technology of China, Hefei 230027, China
zyf1@mail.ustc.edu.cn, {zhwg,lihq}@ustc.edu.cn

Abstract. In this paper, we are dedicated to the cross-modal image retrieval between optical images and synthetic aperture radar (SAR) images. This cross-modal retrieval is a challenging task due to the different imaging mechanisms and huge heterogeneity gap. Here, we design a two-stream fully convolutional network to tackle this issue. The network maps the optical and SAR images to a common feature space for comparison. For different modal images, the comparable features are obtained by feeding them into the corresponding branch. Each branch fuses two types of features in a weighted manner. These two kinds of features root in the pooling features of VGG16 at different depths, but are refined by the well-designed channels-aggregated convolution (CAC) operation as well as semi-average pooling (SAP) operation. In order to get a better model, an extensible training approach is proposed. The training of the model is from the local to the whole. Besides, we collect an optical/SAR image retrieval (OSR) dataset. Comprehensive experiments on this dataset demonstrate the effectiveness of our proposed method.

Keywords: Retrieval · Cross-modal
Convolutional Neural Network (CNN) · Feature fusion

1 Introduction

With the development of sensor technology, there is an increasing number of remote sensing data with diversified modalities. Through integrating multimodal data which represent the same content, many tasks [12,23,26] have been reinvestigated. Therefore, how to find relevant data between different modalities is an active topic. Here, we focus on the retrieval across optical and SAR images. These two modal images cover the same ground scene but are vastly different. Hence, this is also a cross-modal problem. If we successfully search the matching images, it will be useful for subsequent tasks, such as height estimation [27].

Electronic supplementary material The online version of this chapter (https://doi.org/10.1007/978-3-030-00776-8_36) contains supplementary material, which is available to authorized users.

R. Hong et al. (Eds.): PCM 2018, LNCS 11164, pp. 392–402, 2018.
https://doi.org/10.1007/978-3-030-00776-8_36

In recent years, many cross-modal retrieval methods have been developed. Fukui *et al.* [9] employ matching correlation analysis for retrieval between image and tag. In [21], Sharma *et al.* use partial least squares to handle the cross-modal face recognition with huge variation in resolution. However, most of these methods not only separate feature extraction and subspace generation [24], but also need external data preprocessing. Enlightened by the deep learning techniques, Li *et al.* [15] employ the deep features to improve the retrieval performance across the street and shop domains. Qi *et al.* [19] use siamese network to better complete the sketch-based cross-modal image retrieval. Moreover, the deep models are also effective in remote sensing field. Zhong *et al.* [29] use deep belief networks to classify hyperspectral images. In [6], CNN is used for object detection in optical images. Nevertheless, in remote sensing community, the cross-modal retrieval using deep networks has been rarely explored.

Motivated by the above observations and the fact that the deep features are usually more discriminative [25], a cross-modal remote sensing image retrieval method is proposed. The retrieval method is based on a deep two-stream network, where each branch is the same fully convolutional network to process a specific domain, yet their weights are different. For each branch, we aim to learn proper weights to map the different modal images to a common feature space [11], in which relevant data are closer to each other. In our task, we want to learn a correct deep model so that using an optical image can find the matching SAR images, which cover the same ground scene with optical image.

In this paper, the proposed deep model simultaneously utilizes the visual appearance and powerful semantic representation by fusing features from different layers, so we name it as Double-Feature Convolutional Neural Network (DFCNN). The framework is shown in Fig. 1. Especially, the channels-aggregated convolution and semi-average pooling operations are applied in DFCNN to make full use of the semantic information and reduce the feature dimension. Inspired by the coarse-to-fine training strategy, an extensible training method is proposed. We first train each branch of the model by other tasks (*e.g.*, semantic segmentation task and autoencoder model), and then fine-tune on the pre-trained model. Finally, we collect an optical/SAR image retrieval (OSR) dataset to study this task, and the dataset will be released to the public. To the best of our knowledge, the dataset is proposed for the first time. The experiments on this dataset demonstrate the effectiveness of our retrieval method.

2   Our Approach

In this paper, we focus on the cross-modal retrieval across optical and SAR images. Inspired by the deep learning methods, a two-stream deep network is designed. The objective is to use this network to learn effective feature representations, which make the distances between matched pairs smaller than the mismatched pairs. By training with contrastive loss [7], the network can learn proper weights to map optical and SAR images into a common feature space. Each branch handles a modal image here. For the two branches, the learned features which are in the common space, can be used for our cross-modal retrieval.

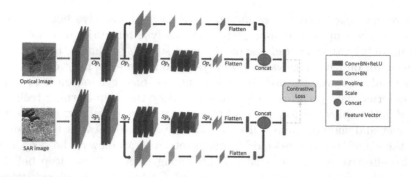

Fig. 1. The proposed two-steam fully convolutional network. Each branch contains ten convolutional layers from VGG16, and the feature fusion part. The fused two types of features are refined by the channels-aggregated convolution as well as semi-average pooling operation, which is explained well in Sect. 2.1. We train the network with contrastive loss which can force the features from two branches to a common space.

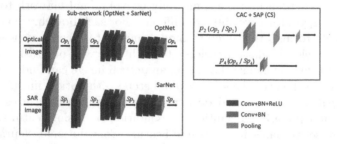

Fig. 2. Some important modules in our proposed network, *i.e.,* DFCNN. In the Sub-network, there are two branches, named as OptNet in optical branch and SarNet in SAR branch, respectively (left). The well-designed channels-aggregated convolution (CAC) and semi-average pooling (SAP) operations on p_2 and p_4 (right).

In this section, we first introduce the network architecture of DFCNN and the effective training method in detail. Then, the retrieval method based on the trained deep model is proposed.

2.1   Network Structure

The proposed network architecture, *i.e.,* DFCNN is depicted in Fig. 1, and some important modules are also illustrated in Fig. 2 for clarity. The designed deep model mainly contains two parts. One is the Sub-network shown in Fig. 2, and the other fuses the features on two levels. In the Sub-network which is the backbone of DFCNN, there are two branches to handle optical and SAR images separately, named as OptNet and SarNet, respectively. Each branch only uses the first four blocks of convolutions of VGG16 [22], *i.e.,* ten convolutional layers. The reasons are as follows. (1) The fully convolutional network can better learn contextual information. (2) The lighter network already has the adequate ability to solve our problem and reduces the difficulty in training.

Inspired by the skip connections [4], it is feasible to improve the search performance by fusing the low-level and high-level features [13]. In general, the deeper the network is, the higher level of abstraction the features provide. In our work, we choose to fuse the features from the second and fourth pooling layers in Subnetwork, *i.e.*, p_2 and p_4 in Fig. 2. Specifically, the Op_2 and Op_4 (Sp_2 and Sp_4) are fused in optical (SAR) branch. The p_4 is a necessary choice for bridging the semantic gap because it is more discriminative. Besides, p_2 is more helpful than p_1 because it carries more abstract knowledge. Moreover, compared to the p_4 and p_3, p_4 and p_2 has fewer semantic conflicts because of the farther distance. However, the original p_4 and p_2 are rough with very high dimensions. Enlightened by the fact that human understands the world by holistic perceptions from various views and a reasonable spatio-temporal range can bring right judgment [8], we design the corresponding channels-aggregated convolution (CAC) and semi-average pooling (SAP) schemes to refine features.

As illustrated in Fig. 2, the CAC and SAP operations are performed on p_2 and p_4. The CAC operation aggregates the scattered features on different feature maps into global features. Assume that the CAC operation is performed on the feature maps with C channels, one CAC feature f (*i.e.*, one global feature) is denoted as follows:

$$f = \sum_{i=1}^{C} W_i \bullet Conv_i(x_i), \tag{1}$$

where $Conv_i(x_i)$ indicates using the i-th kernel convolves the i-th feature map. W_i is the corresponding weight to balance the contribution of each feature map. In this paper, the CAC operation is implemented by applying a convolutional layer, whose number of output channels corresponds to the number of generated CAC features.

Since the downsampling can eliminate redundant features and make better use of the contextual information, we choose pooling operation to enhance semantic representation. Babenko *et al.* [2] point out a simple global descriptor based on sum pooling aggregation performs well on standard retrieval datasets. However, such descriptor to aggregate the raw features ignores the influence of spatial resolution. Based on the sensing mechanism of human, we advocate the semi-average pooling (SAP) operation which can retain more beneficial information. The SAP operation downsamples the feature maps to a suitable resolution rather than one value. Empirically, the appropriate resolution is considered as half of the size of p_4. In our experiments, the CAC and SAP operations have been demonstrated based on p_4, respectively.

The detailed configuration of DFCNN can be found in the supplementary material. For the semi-average pooling operation, the avg-pooling is selected instead of the max-pooling to reduce information loss [2]. Meanwhile, the ReLU non-linearity is not used after Batch Normalization (BN) in the channels-aggregated convolution (CAC), because the negative values also contain significant contents. Besides, two types of features are fused in a weighted manner to form the last feature. The weighting factor is learned in the network by adding a scale layer.

2.2   Training Approach

Since a proper initialization allows the network to converge quickly and correctly, a bottom-up training approach is devised. First, we adopt the encoder-decoder architecture about optical image (OED) to get the pre-trained OptNet in Fig. 2. The OED comes from the SegNet [3], but there are some differences. Only the first ten convolutional layers of SegNet are used in encoder part, and only four upsampling layers are used in decoder part. Besides, the softmax function is replaced by the sigmoid function. Because extracting buildings is a simplified semantic segmentation task, which just has two categories. The weights of OED model are initialized with the weights of the model[1]. Then, the OED model is trained on the Massachusetts Buildings Dataset [17] because of the similarity of the optical data. We optimize the OED model by minimizing the cross-entropy loss.

The encoder-decoder architecture about SAR image (SED) is similar to OED but discards the sigmoid layer. The input of SED is the SAR images in the training set, and the training target is to reconstruct the SAR images, like the autoencoder (AE) model [20]. Hence, the Euclidean distance loss is used to optimize the SED model. The weights of SED model are initialized by MSRAFiller [10]. When the training of OED and SED model is completed, their encoder parts can be used as the pre-trained OptNet and SarNet, respectively.

In the next training steps, all the networks are optimized with contrastive loss. And the input of networks is the image pairs on the OSR training set. First, we refine the Sub-network based on the pre-trained OptNet and SarNet. Next, for each branch, the channels-aggregated convolution and semi-average pooling operations are executed on features of two levels, and then we fuse the two types of features. Lastly, the complete DFCNN is fine-tuned with the fixed weights of pre-trained Sub-network.

2.3   End-to-end Retrieval

When the DFCNN is trained to convergence, we can use this DFCNN model to map optical and SAR images into the common feature space, in which relevant images are closer to each other. Here, each modal images are analyzed on the corresponding branch to get the compared features. To complete the cross-modal search, we only need to compare the similarity of output features between the query and database. The similarity is measured with Euclidean distance here. Hence, the retrieval across optical and SAR images is now performed in an end-to-end manner. In our experiments, both the optical image query and SAR image query are studied. The results are quantitatively evaluated with mAP and *Recall-K*. *Recall-K* denotes the *Recall* when the sorted index ranks top-K.

---

[1] http://mi.eng.cam.ac.uk/~agk34/resources/SegNet/segnet_pascal.caffemodel.

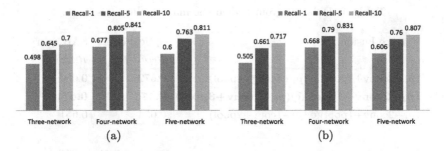

Fig. 3. The results of three different Sub-network. (a) Optical image as the query. (b) SAR image as the query.

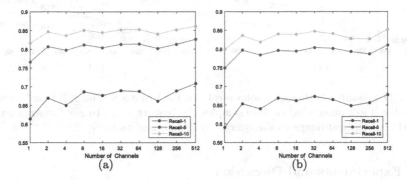

Fig. 4. The results when we produce different numbers of CAC features. Here are 10 different cases in all. (a) Optical image as the query. (b) SAR image as the query.

3 Experiments

3.1 Dataset Description

According to the Homography matrix, we first use the given 29 X-SAR grayscale images and the rough matched optical images from BIGEMAP[2] to get the ripe matched image pairs with the size of 8192×10240. Then, the slider operation is performed on each image pair with the stride of 90 pixels to generate our OSR dataset. Some screening conditions (*e.g.,* the number of black dots) are used to remove the meaningless data here. Finally, the dataset includes 46474 optical image patches with the size of 227×227. Besides, there are 5 positive pairs and 10 negative pairs for each optical image. In our setup, the dataset is divided into training set, validation set and test set in a proportion of $7 : 2 : 1$. The detailed construction of the dataset can be found in the supplementary material.

---

[2] http://www.bigemap.com/.

Table 1. Different groups when the map resolution is 8×8.

Feature Fusion	optical query	SAR query
	Recall-1	Recall-1
$(p_4+2\text{conv}+1\text{pool})©(p_1+1\text{conv}+4\text{pool})$	0.672	0.650
$(p_4+2\text{conv}+1\text{pool})©(p_2+1\text{conv}+3\text{pool})$	**0.675**	**0.671**
$(p_4+2\text{conv}+1\text{pool})©(p_3+1\text{conv}+2\text{pool})$	0.654	0.651

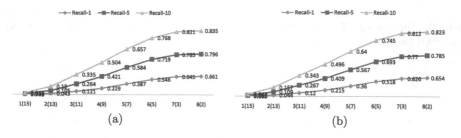

Fig. 5. The results of different resolutions from 1×1 to 8×8. The 8(2) in axis denotes that using 2×2 pixel window downsamples the feature maps to 8×8 resolution and so forth. (a) Optical image as the query. (b) SAR image as the query.

3.2 Experiments and Discussion

The networks are implemented in Caffe [14] and optimized with SGD algorithm [5] with momentum. We set the initial learning rate to 10^{-3}, momentum to 0.9 and batch size to 5. Following the proposed training method, the OED and SED model are first trained about 180k and 150k iterations to get the pre-trained OptNet and SarNet, respectively. And then, the Sub-network and DFCNN are trained about 800k and 160k iterations, respectively. Besides, we conduct the ablation studies and the comparative experiments with baseline methods.

Exploration of Sub-network, CAC and SAP. Here, we compare the Sub-network with different numbers of convolution blocks (from 3 to 5) of VGG16. Each branch of Four-network uses ten convolutional layers, *i.e.*, the first four convolution blocks of VGG16. As shown in Fig. 3, the Four-network which is used by our DFCNN, works best with different metrics. The reasons are as follows. (1) Each branch of Three-network only has seven convolutional layers and can not learn more semantic representation. (2) Each branch of Five-network has 13 convolutional layers, which makes the training more difficult.

From the Four-network, we can obtain the output feature p_4 (including Op_4 and Sp_4), each of which has 512 feature maps with 15×15 resolution. Here, the same channels-aggregated convolution (CAC) operation is performed on Op_4 and Sp_4, and then we compare the output CAC features. The number of channels of generated CAC features is determined by one convolutional layer, *i.e.*, how many filters are used. The results are shown in Fig. 4. In most cases, there is no significant performance degradation. However, the result becomes worse when only one CAC feature is used. It is mainly because the sole feature contains incomplete information. By adding an avg-pooling layer with different kernel size on p_4 (Op_4 and Sp_4), we can get the features of different resolutions. For the avg-pooling layers, the stride is always 2 with padding 0. As shown in Fig. 5, the resolution is crucial. The result only drops by about 1 percent when the resolution changes from original 15×15 on p_4 to 8×8. However, the performance degrades a lot with other sizes. The motivation behind the semi-average pooling (SAP) is that the suitable spatial information is important. Hence, this result provides the basis for our choice of SAP rather than global pooling.

Evaluation on Feature Fusion, Weighting and Training Method. In this paper, if two channels-aggregated convolution (CAC) features are generated, *i.e.*, two filters are used to convolve input features, it is denoted as 2conv and so forth. Therefore, p_4+2conv+2pool denotes that we perform 2conv, and followed by twice avg-pooling on p_4 (Op_4 and Sp_4). The parameters of convolutional layer in CAC operation and pooling layer refer to the configuration of DFCNN. Besides, A©B denotes the concatenation between feature A and B. For each branch, we carry out the corresponding feature fusion and then compare the output features of two branches. As shown in Table 1, the group of p_4 and p_2 works best. Compared to the single feature p_4+2conv+1pool, this fused feature brings in an average 1.65% performance improvement in *Recall-1*.

Table 2. The impact of weighting. FD denotes feature dimension. *Recall-1* is abbreviated as R1 and so forth. ∗ denotes the architecture of our DFCNN in Fig. 1.

Scale	FD	optical query			SAR query	
		R1	R5	mAP	R1	R5
$(p_4$+2conv$)$©$(p_2$+1conv+2pool$)$	675	0.690	0.805	0.647	0.681	0.798
$(p_4$+2conv$)$©$(p_2$+1conv+2pool+scale$)$	675	**0.694**	**0.815**	0.668	**0.685**	**0.809**
$(p_4$+2conv+1pool$)$©$(p_2$+1conv+3pool$)$	192	0.675	0.797	0.645	0.671	0.791
∗ $(p_4$+2conv+1pool$)$© $(p_2$+1conv+3pool+scale$)$	**192**	0.689	0.813	**0.671**	0.682	0.803

In our experiments, we add a scale layer to control the contribution of two types of features. As shown in Table 2, the weighting has a positive effect on all metrics. The lower-dimensional feature of 8×8 resolution (used by us) reaches comparable capacity with the feature of 15×15 resolution. The mAP rises by 2.6% and the *Recall* increases by an average of about 1.3%. Besides, the training methods are also important. When we change the training method of OED model from the dependence on semantic segmentation task to autoencoder model, the different pre-trained OptNet (*i.e.*, the encoder part of OED model) can be obtained. As shown in Table 3, compared to the use of pre-trained OptNet by autoencoder model, the result of fine-tuned Sub-network is better with

Table 3. The impact of different training methods. We always get the pre-trained Sar-Net by the autoencoder (AE) model. The pre-trained OptNet is obtained by semantic segmentation (SS) task or AE model.

Training Method	Feature Dimension	optical query				SAR query		
		R1	R5	R10	mAP	R1	R5	R10
Sub-network (SS + AE)	115200	**0.677**	**0.805**	**0.841**	**0.675**	**0.668**	**0.790**	**0.831**
Sub-network (AE + AE)	115200	0.382	0.580	0.651	0.379	0.392	0.594	0.662

the weights of OptNet trained by semantic segmentation task. It is because performing semantic segmentation task extracts more discriminative features.

Baseline and Proposed Method. In this paper, the effectiveness of SCCM algorithm [16] is demonstrated in two types of features. One is the 20000-D encoded SIFT features by bag-of-words model, the other is the CNN features learned by the DFCNN. And the result of CMCP algorithm [28] is based on the CNN features. Besides, the NetVLAD network [1] which works well for standard retrieval benchmarks, is taken as a branch of similar two-stream network. This network is initialized with the weights of Five-network in Fig. 3 and optimized with contrastive loss. Especially, we set 64 cluster centroids and using a trainable matrix reduces the feature dimensions into 1024 dimensions. According to [18], we implement the MISO method, which works on the identification of corresponding patches across optical and SAR images. The output of probability value is used to measure the similarity of two images here.

As shown in Table 4, the methods with deep features have obvious advantages over the methods with SIFT features. Besides, the result of SCCM method heavily relies on the basic features. Among three deep models, our method has the best performance. In addition, the output of DFCNN is 192-D features, whose dimension is lower than the output of NetVLAD. Moreover, the optical and SAR images can be processed separately in the DFCNN, but the MISO fails to do this. Hence, our method has the shortest inference time. The above analysis shows our retrieval method based on the DFCNN model is superior for handling this cross-modal task.

Table 4. The mAP of six methods when optical image as the query. The latter three methods are based on deep neural networks.

Method	SIFT	SCCM on		CMCP	**Ours**	MISO	NetVLAD
		SIFT	DFCNN				
mAP	0.004	0.029	0.669	0.393	**0.671**	0.600	0.262

4   Conclusion

In this paper, the cross-modal retrieval between optical and SAR remote sensing images is first proposed to our best knowledge. Considering the huge difference between optical and SAR images, we design an end-to-end deep model (*i.e.*, Double-Feature Convolutional Neural Network (DFCNN)) to solve this challenging issue. By adopting the tailored training strategy, the DFCNN can efficiently fuse two kinds of features after the channels-aggregated convolution and semi-average pooling (CS) operations. The CS scheme is a novel way to refine features. Quantitative experimental results demonstrate the effectiveness of the proposed cross-modal retrieval method. Furthermore, an optical/SAR image retrieval (OSR) dataset is collected for further researches. In the future, we intend to adjust the network model (*e.g.*, considering the multi-scale information) and extend our work to other research fields, such as image registration.

Acknowledgement. This work was supported in part by 973 Program under Contract 2015CB351803, by Natural Science Foundation of China (NSFC) under Contract 61390514 and 61331017, and by the Fundamental Research Funds for the Central Universities.

References

1. Arandjelovic, R., Gronat, P., Torii, A., Pajdla, T., Sivic, J.: Netvlad: CNN architecture for weakly supervised place recognition. In: CVPR, pp. 5297–5307 (2016)
2. Babenko, A., Lempitsky, V.: Aggregating local deep features for image retrieval. In: ICCV, pp. 1269–1277 (2015)
3. Badrinarayanan, V., Kendall, A., Cipolla, R.: Segnet: a deep convolutional encoder-decoder architecture for image segmentation. TPAMI **39**(12), 2481–2495 (2017)
4. Bell, S., Lawrence Zitnick, C., Bala, K., Girshick, R.: Inside-outside net: detecting objects in context with skip pooling and recurrent neural networks. In: CVPR, pp. 2874–2883 (2016)
5. Bottou, L.: Stochastic gradient descent tricks. In: Montavon, G., Orr, G.B., Müller, K.-R. (eds.) Neural Networks: Tricks of the Trade. LNCS, vol. 7700, pp. 421–436. Springer, Heidelberg (2012). https://doi.org/10.1007/978-3-642-35289-8_25
6. Cheng, G., Zhou, P., Han, J.: Learning rotation-invariant convolutional neural networks for object detection in VHR optical remote sensing images. TGARS **54**(12), 7405–7415 (2016)
7. Chopra, S., Hadsell, R., LeCun, Y.: Learning a similarity metric discriminatively, with application to face verification. In: CVPR. vol. 1, pp. 539–546. IEEE (2005)
8. Eysenck, M.W., Keane, M.T.: Cognitive psychology: A student's handbook. Psychology press, New York (2013)
9. Fukui, K., Okuno, A., Shimodaira, H.: Image and tag retrieval by leveraging image-group links with multi-domain graph embedding. In: ICIP, pp. 221–225. IEEE (2016)
10. He, K., Zhang, X., Ren, S., Sun, J.: Delving deep into rectifiers: Surpassing human-level performance on imagenet classification. In: ICCV, pp. 1026–1034 (2015)
11. Hong, R., Hu, Z., Wang, R., Wang, M., Tao, D.: Multi-view object retrieval via multi-scale topic models. TIP **25**(12), 5814–5827 (2016)

12. Hong, R., Zhang, L., Tao, D.: Unified photo enhancement by discovering aesthetic communities from flickr. TIP **25**(3), 1124–1135 (2016)
13. Hong, R., Zhang, L., Zhang, C., Zimmermann, R.: Flickr circles: aesthetic tendency discovery by multi-view regularized topic modeling. TMM **18**(8), 1555–1567 (2016)
14. Jia, Y., et al.: Caffe: convolutional architecture for fast feature embedding. In: ACM MM, pp. 675–678. ACM (2014)
15. Li, Z., Li, Y., Gao, Y., Liu, Y.: Fast cross-scenario clothing retrieval based on indexing deep features. In: Chen, E., Gong, Y., Tie, Y. (eds.) PCM 2016. LNCS, vol. 9916, pp. 107–118. Springer, Cham (2016). https://doi.org/10.1007/978-3-319-48890-5_11
16. Luo, M., Chang, X., Li, Z., Nie, L., Hauptmann, A.G., Zheng, Q.: Simple to complex cross-modal learning to rank. CVIU **163**, 67–77 (2017)
17. Mnih, V.: Machine learning for aerial image labeling. Ph.D. thesis, University of Toronto (Canada) (2013)
18. Mou, L., Schmitt, M., Wang, Y., Zhu, X.X.: A CNN for the identification of corresponding patches in SAR and optical imagery of urban scenes. In: JURSE, pp. 1–4. IEEE (2017)
19. Qi, Y., Song, Y.Z., Zhang, H., Liu, J.: Sketch-based image retrieval via siamese convolutional neural network. In: ICIP, pp. 2460–2464. IEEE (2016)
20. Rumelhart, D.E., Hinton, G.E., Williams, R.J.: Learning representations by back-propagating errors. Nature **323**(6088), 533 (1986)
21. Sharma, A., Jacobs, D.W.: Bypassing synthesis: PLS for face recognition with pose, low-resolution and sketch. In: CVPR, pp. 593–600. IEEE (2011)
22. Simonyan, K., Zisserman, A.: Very deep convolutional networks for large-scale image recognition (2014). arXiv preprint arXiv:1409.1556
23. Tambo, A.L., Bhanu, B.: Dynamic bi-modal fusion of images for the segmentation of pollen tubes in video. In: ICIP, pp. 148–152. IEEE (2015)
24. Wang, K., He, R., Wang, L., Wang, W., Tan, T.: Joint feature selection and subspace learning for cross-modal retrieval. TPAMI **38**(10), 2010–2023 (2016)
25. Wang, K., Yin, Q., Wang, W., Wu, S., Wang, L.: A comprehensive survey on cross-modal retrieval (2016). arXiv preprint arXiv:1607.06215
26. Wang, Y., Zhu, X.X., Zeisl, B., Pollefeys, M.: Fusing meter-resolution 4-d insar point clouds and optical images for semantic urban infrastructure monitoring. TGARS **55**(1), 14–26 (2017)
27. Wegner, J.D., Ziehn, J.R., Soergel, U.: Combining high-resolution optical and insar features for height estimation of buildings with flat roofs. TGARS **52**(9), 5840–5854 (2014)
28. Zhai, X., Peng, Y., Xiao, J.: Cross-modality correlation propagation for cross-media retrieval. In: ICASSP, pp. 2337–2340. IEEE (2012)
29. Zhong, P., Gong, Z., Li, S., Schönlieb, C.B.: Learning to diversify deep belief networks for hyperspectral image classification. TGARS **55**(6), 3516–3530 (2017)

Tiny Surface Defects on Small Ring Parts Using Normal Maps

Yang Zhang[1], Jia Song[1], Huiming Zhang[1], Jingwu He[1],
and Yanwen Guo[1,2(✉)]

[1] National Key Laboratory for Novel Software Technology,
Nanjing University, Nanjing 210023, China
ywguo@nju.edu.cn
[2] Science and Technology on Information Systems Engineering Laboratory,
Nanjing 210007, China

Abstract. Detection of tiny surface defects on small ring parts remains challenging due to the unnoticeable visual features of such defects and the interference of small surface scratches. This paper proposes a novel method for detecting tiny surface defects based on normal maps of metal parts. To better characterize features of tiny defects and differentiate them from small scratches, we recover the normal map of the metal part through analyzing its directional reflections obtained with our specifically designed directional light units. Based on the normal map, a cascaded detector trained by the AdaBoost approach combined with the joint features and fast feature pyramid is used to localize the defects, achieving fast and accurate detection of tiny surface defects. The proposed method can achieve high detection accuracy with extremely fast speed, only 23 ms per metal part, and comparisons against other methods show our superiority.

Keywords: Defect detection · Tiny surface defect · Normal maps
Combined light units

1 Introduction

In recent years, with the development of machine vision and the extensive application of precision machining of metal parts, defect detection of small metal parts has gained increasing attention. Detection of tiny surface defects is serious to ensure the quality of mechanical parts. The defects of parts may affect the safety and performance of the entire product. In particular, some products require parts to be with high precision such as automotive parts for which defected parts may lead to car safety problems and large-scale car recalls. Therefore, in electronics, machinery and other related industries, detection of small metal parts faults (scratches, defects, etc.) is particularly important.

For a long time, defect detection is a complex and time-consuming job of engineers and inspectors, with low detection probability and a waste of manpower. Due to the continuing and rapid advances in both hardware and software technology of sensors such as the laser and camera, automatic detection system of surface defects has received considerable attention. In [1], the metal surface defects are detected by using

© Springer Nature Switzerland AG 2018
R. Hong et al. (Eds.): PCM 2018, LNCS 11164, pp. 403–413, 2018.
https://doi.org/10.1007/978-3-030-00776-8_37

the iterative thresholding technique. However, the algorithm only works well for those obvious defects of metal surface. The length, width and depth of defects are predicted by the magnetic-leakage method [2]. Besides, a finite element method [3] is used to establish a model of laser ultrasonic transmission wave to detect the depth of surface defects and the relationship between characteristics of the transmitted wave and defect depth. The relationship between the acoustic surface waves and the defects of micro-cracks with different depths is used for defect detection [4]. As a rapidly developing defect detection technique, infrared thermography is used to estimate the depth of defects to detect defects [5]. However, those methods used to estimate defect depth are based on surface heating, which may damage the parts during inspection. Other methods based on the estimated depth are in [6] and [7]. Most of the existing depth estimation methods are introduced for subsurface defect detection of unknown depth in metal parts, and the feature information captured by various sensors is difficult to apply directly.

More and more defect detection algorithms based on machine learning are applied in the field of industrial testing [8, 9]. Compared with the methods relying on specific sensors such as laser and ultrasonic, using machine learning based visual detection to realize the detection of metal parts is not only accurate and efficient, but also with high reliability. In this paper, we apply this attractive technique to inspect the tiny surface defects on small ring parts. Given a metal part shown in Fig. 1, the part with a deep scratch is deemed as a defected one, while a slight surface scratch is not a defect. For this part with the 3.5 cm diameter, the depth of scratch exceeding 0.5 mm is viewed as "deep" from the view of engineers.

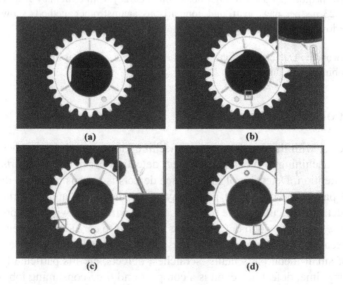

Fig. 1. Mechanical parts. (a) A normal part. (b–d) Parts with tiny surface defects. From the view of engineers, the scratch with over 0.5 mm depth is considered as a defect.

It is challenging to perform such defect detection because the tiny defects can hardly be described with only simple visual features. Moreover, it is difficult to differentiate such defects from small surface scratches due to manufacturing technology, considering only the visual appearance. To deal with these problems, our key insight is to explore the geometry feature of the metal part using the recovered normal map, and then process them by machine learning to identify whether there is any visual defect on metal surface. The proposed visual detection method contains two core components: normal map recovery based on directional reflections and a defect classifier using the recovered normal maps. Experiments demonstrate that the proposed method can be effectively applied to the detection of tiny surface defects with high detection accuracy and fast speed.

The rest of the paper is organized as follows. Section 2 is an overview of our detection framework. Section 3 introduces the reconstruction of the normal map of a metal part. Section 4 explains the diagram of normal information extraction and detection of tiny surface defects. The experimental results and analysis are summarized in Sect. 5. Section 6 concludes this paper and describes the future work.

2 Overview

The proposed defect detection framework contains an image acquisition system and an image analysis module. Small metal parts images are first acquired by our image acquisition system, and then processed by image analysis to identify whether there are any visual defects on the metal surface.

2.1 Image Acquisition System

We have designed an image acquisition system which uses a special device consisting of a shading box, a transparent conveyor belt, a CCD camera, and three sets of LED light units (top, middle, bottom). The output of our classifier mostly depends on the quality of the recovered normal map. Therefore, for better quality of the normal maps we need a way to estimate the normal maps accurately. As illustrated in Fig. 2(a), a shading box is designed to ensure a lighting-free, dark environment with black flannelette on the inner surface which can help relieve the influence of reflection of box surface. Specially, the LED lamps are installed around the platform which are above 20 cm in order to make the lights shot the parts with an angle of 45°.

In Fig. 2(a), the top and bottom light units comprise four LED light belts parallel to the moving direction of the conveyor belt respectively. The middle light units are arranged along with the four main directions (north, south, east, west) of the center of the shading box. When the original metal part is sent to the specified position, the CCD camera is triggered to capture the images while controlling different light units to turn on in a certain order. When the images are transmitted to the computer, the image analysis module is used to recover normal map and detect the image of metal parts. The proposed method for normal map reconstruction is described in detail in Sect. 3.

2.2   Image Analysis for Defect Detection

The pipeline of the proposed defect detection framework is shown in Fig. 2(b). We have collected a lot of samples of positive and negative samples of the defects by sampling the images of normal and defected parts captured using our image acquisition system. It should be noted that since most of the small surface scratches and noises are filtered out by the previous normal reconstruction module, the difficulty of defect detection can be reduced significantly. These samples are used to train a classifier by incorporating joint image features. With the trained classifier, we are able to process the image of each given part and find whether there are any defects on its surface. We will describe the method in Sect. 4.

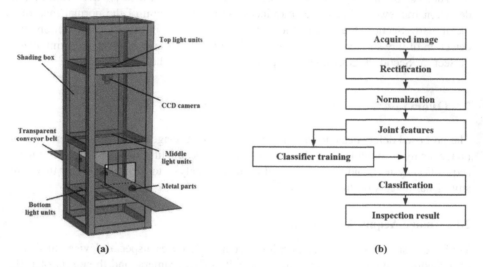

(a)   (b)

Fig. 2. (a) Image acquisition system (The shading box is rendered with semi-transparency to explain the interior structure of the system). (b) Pipeline of the defect detection framework.

3   Normal Map Reconstruction

As discussed previously, the acquired images are shown in Fig. 3. Intuitively, when the light unit in a certain direction opens, the imaged pixel is brighter if the local region around this pixel faces the lighting direction, and vice versa. The brightness of the image is only related to the slope angle of the material surface [10]. In view of this regularity, the normal maps can be recovered by the directional reflections of the parts.

A traditional method called direct calculation is calculating the directional images (north, south, east, and west) without light attenuation which is captured by the camera fixed on a slide. However, this method needs four slides, and the precise control of the light movement on the slides. So the equipment is too large and inconvenient to be manipulated, and overexposure is easy to appear because of the long exposure. To solve these problems, a novel method is proposed to calculate the light compensation of the directional images which is captured by a shading frame with combined light units.

Compared with direct calculation, the proposed method is quite robust and effective to recover the normal map.

Firstly, a filter coating is installed on the camera, and then turn on the top, middle (four directions) and bottom light units in proper order, while taking pictures. Next, we remove the filter coating and repeat the above steps. Based on the light compensation algorithms [10–12] and mixed superposition method in [13], the normal map of a part is reconstructed as shown in Fig. 4. The main calculation steps are as follows:

Let I_N1, I_S1, I_W1, and I_E1 indicate the filtered images in four directions, separately, according to a light compensation algorithm mentioned in [10], and store them separately as a single-channel floating point brightness map I_NL, I_SL, I_WL, I_EL. Next, creating two new three-channel images NW and SE, taking I_WL as R channel of NW, I_NL as G channel of NW, and adjust the color gradation of NW to 0–0.5. Taking I_EL as R channel of SE, I_SL as G channel of SE, and adjust the color gradation of SE to 0.5–1.0. Finally, NW and SE are blended to obtain image $N\_T1 = 2 \times NW \times SE$. Because the normal information is a normalized vector, in this paper, it can further calculate the information of normal B channel based on the value of normal R and G channel. We assume the value of R channel is r, and the value of G channel is g, the value b of channel B is defined as $b = 2 \times \sqrt{1 - (r - 0.5)^2 - (g - 0.5)^2} - 1$.

Let I_N2, I_S2, I_W2, I_E2 represent the unfiltered images in four directions, separately, and then calculating a normal map named N_T2 by the same processing way. By the further processing including removing frilling, brightness adjustment, and contrast adjustment, the image N_T1 and N_T2 are transformed as the *normal1* and *normal2* respectively. The final normal map is blended by the *normal1* and *normal2* named as *Normal*, where *Normal* $= 0.5 \times$ (*normal1* + *normal2*). The above normal map blending method is adopted to obtain a normal map which can retain the details of metal parts without color jump [13].

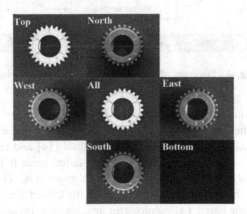

Fig. 3. Acquired images under the combined light units. (Top is meant to use only top light units. East is meant to use the eastern light units. All is meant to use all the light units)

4 Defect Detection

For the convenience of detection, we firstly determine location and orientation of the acquired images, and then translate and rotate the normal maps of metal parts according to the positional relationship between the parts and a standard orientation. Besides, the standard orientation contains the centroid and the angle between a straight line connecting the centroid with the anchor point and the horizontal axis. Finally, we extract the section of a part that needs to be detected. Specifically, about 90% of the background regions are filtered out by the image normalization which can speeds up the computation. The diagram of the normal information extraction is shown in Fig. 4.

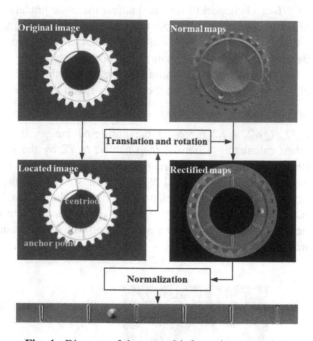

Fig. 4. Diagram of the normal information extraction.

After the processing of rectification and normalization, we use a cascaded classifier to identify whether there are any visual defects. The training of the cascaded classifiers includes joint features extraction, fast feature pyramid [14] and learning via the Ada-Boost algorithm [15]. The classifier allows the non-defect areas to be quickly discarded while more computations are spent on the promising areas. The joint features are effective to capture the salient characteristics of the object. In this paper, the joint feature consists of four parts: LUV color features, rotation invariant local binary patterns (LBP), normalized gradient magnitude, and histogram of oriented gradients (HOG).

Fast feature pyramid can greatly accelerate the feature extraction speed without losing important information. Compared with the traditional method of computing image features step by step, fast feature pyramid only calculates a scale in each octave, and then use this feature to determine the other dimensions of image features within the octave. And the structure of the cascaded inspection process could essentially be considered as a degenerate decision tree. The initial stage classifier eliminates many non-defect images with little processing. After several stages of processing, the number of sub-windows is reduced rapidly. Finally, the sub-window that passes all stage classifiers is classified as a target (defect). Each stage classifier is composed of a set of weak linear SVM classifiers and trained by the AdaBoost learning algorithm on labeled data.

5   Experimental Results

To verify the performance of the proposed defect detection method, we present the experimental results followed the previous approaches on defect detection in this section. The experiment is developed under the PC condition 3.60 GHz of Intel Core i7-7700 processor, 8G RAM and Nvidia GTX 1070ti.

5.1   Databases and Evaluation Metrics

In order to verify the effectiveness of the algorithm, we conduct an empirical study of evaluating the proposed detection method on a tiny surface defects database. The database contains 564 defected images and 3164 normal images, 70% of which are used for training. And all images are the size of 1200×900 pixels. Each training set image is labeled according to the format of the PASCAL VOC dataset [16]. To evaluate the effectiveness of the proposed method, there are four indexes [17]: correct detection rate (CDR), missing detection rate (MDR), false detection rate (FDR), and detection speed. For example, if a testing set contains m defect images and n non-defect images, by detection of the method, a images are inspected as defect, among them b images are inspected by error, meanwhile, c images are inspected as non-defect, among them d images are inspected by error. So, the above indexes can be defined as: $CDR = 1 - b/(m + n)$, $MDR = b/(m + n)$, $FDR = d/(m + n)$. In the task of the defect detection, the index of FDR is less crucial than MDR, because the impact is not serious if a non-defect region is detected by error.

5.2   Comparison of Different Images from Combined Light Units

To observe the influences of images under different light units, we compare our method in different images from combined light units. Figure 5 shows typical detection results. The results show that our method can effectively detect and recognize the defect based on the normal map. The Top image usually contains highlight information. So our method can be easily affected by the serious interference of surface reflection on metal parts. In addition, the directional images present too many small surface scratches in details. The detection method can be easily affected by the noise and small surface scratches of metal parts. In conclusion, the normal map can ensure high detection accuracy of the tiny surface defects of small ring parts.

5.3   Performance Comparison with the Related Methods

To illustrate the superiority of proposed method for defect detection, we also carry out experiments using hand-crafted feature descriptors and conventional machine learning tools. With some widely used object detection methods such as cascade detector [18] based on Haar-like and HOG, the gradient coded co-occurrence matrix (GCCM) [19] and the deep learning method based on convolutional neural network (CNN) [20, 21], the same images are trained for defect detection. The other parameters are all default values. The detection results are detailed in Table 1. Our method has 99.15% CDR, 0.85% MDR, and 4.00% FDR. The proposed detection method outperforms other related methods based on normal maps. Some typical results are shown in Figs. 6, 7 and 8. Our method can be effectively applied to the detection of tiny surface defects.

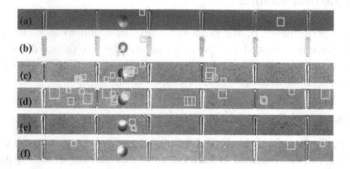

Fig. 5. Detection results of different images from combined light units. (a) Normal. (b) Top. (c) East. (d) West. (e) South. (f) North

Table 1 also lists a comparison of the detection speed. the computation speed of the cascade of Haar-like features method is the fastest, but its accuracy is the lowest. The accuracy of the CNN-based method is satisfactory, but its speed is too low because of the computational complexity of CNN. In addition, CNN-based method cannot effectively deal with detection of tiny surface defects. The results confirm that our method is preferable for the visual detection framework.

Table 1. Detection results of different methods.

Methods	CDR/%	MDR/%	FDR/%	Speed/ms
Cascade(Haar-like)	81.20	9.80	23.93	**11**
Cascade(HOG)	92.31	7.69	12.82	17
GCCM	89.74	10.26	19.66	586
CNN-based	96.43	3.56	17.86	168
Our method	**99.15**	**0.85**	**4.00**	23

There are three main reasons that make the visual detection framework have high inspection accuracy and speed. First, the cascaded detection approach is important to make the framework fast, which allows background regions of the image to be quickly discarded while spending more computation on promising regions. Second, image normalization technology significantly speeds up the computation. Specifically, about 90% of the background regions are filtered out by image normalization, and only 10% of the image regions need to be verified in the following module. Third, the joint features are effective to capture the salient characteristics of the defects.

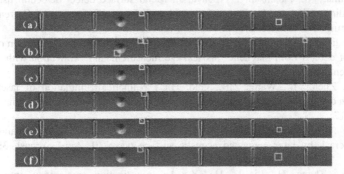

Fig. 6. Detection results of different methods. (a) Ground truth. (b) Cascaded detector with Haar-like. (c) Cascaded detector with HOG. (d) GCCM. (e) CNN-based. (f) Our method.

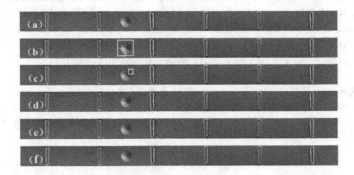

Fig. 7. Detection results of different methods. (a) Ground truth. (b) Cascaded detector with Haar-like. (c) Cascaded detector with HOG. (d) GCCM. (e) CNN-based. (f) Our method.

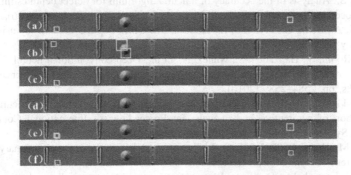

Fig. 8. Detection results of different methods. (a) Ground truth. (b) Cascaded detector with Haar-like. (c) Cascaded detector with HOG. (d) GCCM. (e) CNN-based. (f) Our method.

6   Conclusion

In this paper, a novel visual method is proposed to detect tiny surface defects of the small metal parts under interference of small surface scratches. Unlike the conventional methods, the proposed method can achieve a high detection rate based on the normal maps of metal parts. And the joint features are used to train an accurate and effective detector by the Adaboost approach and fast feature pyramid. The results show that the approach outperforms other methods with a significantly fast speed only 23 ms per mechanical part. For the future work, an extended version of our method could be used to detect a greater variety of metal parts with a higher accuracy.

Acknowledgments. This work is supported in part by the Natural Science Foundation of Jiangsu Province under Grants BK20150016, the National Natural Science Foundation of China under Grants 61772257, 61672279, and the Fundamental Research Funds for the Central Universities 020214380042.

References

1. Senthikumar, M., Palanisamy, V., Jaya, J.: Metal surface defect detection using iterative thresholding technique. In: Proceedings of International Conference on Current Trends in Engineering and Technology, pp. 561–564 (2014)
2. Kandroodi, M.R., Araabi, B.N., Bassiri, M.M., Ahmadabadi, M.N.: Estimation of depth and length of defects from magnetic flux leakage measurements: verification with simulations, experiments, and pigging data. IEEE Trans. Magn. **53**(3), 1–10 (2017)
3. Hui, L., Zheng, B., Wang, Z.B., Guo, H.L.: Numerical simulation of laser ultrasonic transmitted wave: applied to detect surface defects depth. J. North Univ. China **38**(2), 119–123 (2017)
4. Wang, Y., Yao, W., Liu, H., Guo, H.: An experimental study on depth evaluation of micro-surface crack by laser generated acoustic surface waves. J. Appl. Acoust. **35**(1), 36–40 (2016)
5. Yang, R., Zhang, H., Li, T., He, Y.: An investigation and review into microwave thermography for NDT and SHM. In: Proceedings of NDT New Technology and Application Forum, pp. 133–137 (2016)
6. Dudzik, S.: Analysis of the accuracy of a neural algorithm for defect depth estimation using PCA processing from active thermography data. Infrared Phys. Technol. **56**(36), 1–7 (2013)
7. Bernieri, A., Betta, G., Ferrigno, L., Laracca, M.: Crack depth estimation by using a multi-frequency ECT method. IEEE Trans. Instrum. Meas. **62**(3), 544–552 (2013)
8. Faghih-Roohi, S., Hajizadeh, S., Núñez, A., et al.: Deep convolutional neural networks for detection of rail surface defects. In: Proceedings of International Joint Conference on Neural Networks, pp. 2584–2589 (2016)
9. Edris, M.Z.B., Jawad, M.S., Zakaria, Z.: Surface defect detection and neural network recognition of automotive body panels. In: Proceedings of International Conference on Control System, Computing and Engineering, pp. 117–122 (2016)
10. Aittala, M., Weyrich, T., Lehtinen, J.: Two-shot SVBRDF capture for stationary materials. ACM Trans. Graph. **34**(4), 1–13 (2015)
11. Jian, M., Lam, K.M., Dong, J.: Illumination-insensitive texture discrimination based on illumination compensation and enhancement. Inform. Sci. **269**(11), 60–72 (2014)

12. Jian, M., Yin, Y., Dong, J., et al.: Comprehensive assessment of non-uniform illumination for 3D heightmap reconstruction in outdoor environments. Comput. Ind. **99**, 110–118 (2018)
13. Lv, G.J.: Material Surface Visualization and Its Application. M.S. thesis, Department of Computer Science, Nanjing University, China, 2017
14. Dollar, P., Appel, R., Belongie, S., Perona, P.: Fast feature pyramids for object detection. IEEE Trans. Pattern Anal. Mach. Intell. **36**(8), 1532–1545 (2014)
15. Nam, W., Dollar, P., Han, J. H.: Local decorrelation for improved pedestrian detection. In: Proceedings of Advances in Neural Information Processing Systems, pp. 424–432 (2014)
16. Everingham, M., Eslami, S.M.A., Gool, L.V., et al.: The pascal visual object classes challenge: a retrospective. Int. J. Comput. Vis. **111**(1), 98–136 (2015)
17. Zhang, Y., Lin, K., Zhang, H., et al.: A unified framework for fault detection of freight train images under complex environment. In: Proceedings of IEEE International Conference on Image Processing (2018)
18. Viola, P., Jones, M.J.: Rapid object detection using a boosted cascade of simple features. In: Proceedings of Computer Vision and Pattern Recognition, pp. 511–518 (2001)
19. Liu, L., Zhou, F., He, Y.: Automated visual inspection system for bogie block key under complex freight train environment. IEEE Trans. Instrum. Meas. **65**(1), 1–13 (2015)
20. Krizhevsky, A., Sutskever, I., Hinton, G.E.: ImageNet classification with deep convolutional neural networks. In: Proceedings of International Conference on Neural Information Processing Systems, pp. 1097–1105 (2012)
21. Simonyan, K., Zisserman, A.: Very deep convolutional networks for large-scale image recognition. In: Proceedings of International Conference on Learning Representations, (2015)

Extracting Features of Interest from Small Deep Networks for Efficient Visual Tracking

Zhao Luo[1,2], Shiming Ge[1(✉)], Yingying Hua[1,2], Haolin Liu[1,2], and Xin Jin[3,4]

[1] Institute of Information Engineering, Chinese Academy of Sciences, Beijing, China
geshiming@iie.ac.cn
[2] School of Cyber Security, University of Chinese Academy of Sciences,
Beijing, China
[3] Department of Cyber Security, Beijing Electronic Science
and Technology Institute, Beijing, China
[4] CETC Big Data Research Institute Co., Ltd., Guiyang 550018, Guizhou, China

Abstract. Recent deep trackers have achieved impressive performance in visual tracking. Typically, these trackers apply complex deep networks with massive parameters for representing objects, which makes their deployment on resource-limited devices very challenging due to two reasons: (1) high computation complexity, and (2) high storage footprint. To address these two issues, this paper proposes a lightweight deep tracker to facilitate efficient tracking by using a small deep convolutional neural network. This tracker adaptively extracts features of interest from different deep layers for representing different objects, and then integrates into discriminative correlation filter formulation. Due to the usage of small deep networks and selection of deep features, the drop on tracking accuracy could be effectively alleviated, while the costs in computation and storage could be greatly reduced. This tracker can run at a very fast speed of 55 fps when only taking 4.8M parameters. Experimental results on the public OTB2013 and OTB100 benchmarks demonstrate the effectiveness and efficiency of the proposed tracker.

Keywords: Visual tracking · Convolutional Neural Networks
Deep learning

1 Introduction

Visual tracking is an important component in many computer vision tasks, such as video surveillance [26], robotics [21] and virtual reality [11]. These tasks require not only high tracking accuracy but also fast speed and low storage respective to their deployment environments. Lots of methods have been involved in this field. In recent years, discriminative correlation filter (DCF) has been successfully employed to visual tracking. It aims at solving a ridge regression

© Springer Nature Switzerland AG 2018
R. Hong et al. (Eds.): PCM 2018, LNCS 11164, pp. 414–425, 2018.
https://doi.org/10.1007/978-3-030-00776-8_38

problem to discriminate the objects from the backgrounds [3,9]. The DCF formulation provides a fast solution in Fourier domain, which can be calculated by fast Fourier translation (FFT) with element-wise operations. Based on this fast solution, the resulting DCF trackers are capable to re-train tracking models in every frame in a high speed, when giving high performance in speed. With the promising speed, some early DCF trackers give poor accuracy in tracking challenging objects when representing objects with hand-crafted features. It is necessary to apply more discriminative features, such as the features learned by deep learning, for object representation.

Deep convolutional neural networks (CNNs) have demonstrated effectiveness in many visual tasks, such as object detection [23] and recognition [16]. Inspired by the success of CNNs, more and more DCF trackers try to exploit the advantages of CNNs to address the challenges in visual tracking (e.g. deformation, occlusion, illumination variation, etc.) to improve tracking accuracy and robustness, such as MDNet [20] and CF2 [18]. Different to conventional DCF trackers which use hand-crafted features like HOG variant [7] to represent objects, recent DCF trackers aim to representing objects with pre-trained deep CNNs learned for object classification tasks, such as CF2 [18], HDT [22], DeepSRDCF [4] and C-COT [6]. In CF2 [18], Ma *et al.* proposed extracting features from different deep CNN layers to represent appearance of objects by finding that features from shallow layers and high layers encode more spatial information and more semantic information respectively. In C-COT [6], Danelljan *et al.* proposed integrating multi-resolution deep feature maps by training convolutional filters in a continuous spatial domain. These DCF trackers with CNNs outperform most of conventional DCF trackers with hand-crafted features in tracking accuracy and robustness, which demonstrates the effectiveness of deep features in robust visual tracking. However, these trackers often extract deep features from a large and complex network with massive parameters. For example, CF2 [18] uses VGG19 [25] network with more than 500 M parameters. This will cause two problems in practical deployment: (1) large models need large space to store, while many devices in visual tracking applications are embedded devices equipped with limited storage space, and (2) massive parameters will take expensive cost in extracting features for training correlation filters, which lowers the tracking speed. However, a tracker should react quickly in a high frame in many tracking applications, such as auto pilotings.

In this paper, we propose a lightweight DCF tracker with deep network. The tracker uses a small network to extract features in a standard DCF framework. Firstly, we find that the tracking accuracy of a DCF tracker with features extracted from some fixed deep layers (e.g., conv1 layer of VGG-19 for all objects in OTB100 dataset [29]) does not been improved significantly compared with conventional DCF trackers with hand-crafted features. The reason arises from that the extracting features from fixed layers for all objects are not optimal because different objects have different characteristics. Inspired by that and some previous works on saliency detection [13–15], different object should be represented by adaptively selected layers, making more suitable representation for each object according to its own characteristic to improve final tracking accuracy. Toward this end, we then propose an adaptive feature extraction method to extract features of interest from a deep network for different objects. The selection of features of interest is adaptively implemented by minimizing the energy in the standard DCF formulation. Therefore, for a specific object, the features extracted from some layers make the energy minimum and the selected layers are used to extract informative features for this object. We evaluate the effectiveness of our proposed method on public OTB2013 and OTB100 benchmarks. The results show that the DCF tracker with selected features from small deep networks gives comparable performance to state-of-the-art DCF trackers with large deep networks in accuracy and robustness, when significantly outperforms in speed and storage. The main contributions of this paper are three folds: (1) We propose an efficient DCF tracker with deep features by extracting features from a small deep network, which significantly reduces the storage and computation complexity, (2) We propose an adaptive feature selection method to extract features of interest for different objects, which maintains accuracy compared with DCF trackers with large deep networks, and (3) We conduct a comprehensive evaluation to demonstrate the effectiveness of our proposed method, which is helpful to understand the advantages of small deep networks in efficient visual tracking.

2   Related Work

In this section, we will give a brief review about three types of trackers with deep features, including deep trackers in DCF with pre-trained networks, deep trackers with an end-to-end manner and deep trackers with learned DCF.

2.1   Deep Trackers in DCF with Pre-trained Networks

Visual tracking can broadly divided into two categories: generative model based methods and discriminative model based methods. Trackers based on discriminative models are getting more and more attention in recent years, especially the DCF trackers that shows remarkable results in most of popular benchmarks. The DCF trackers with deep features usually extract the features from pre-trained deep CNNs on objects recognition tasks to represent the appearance of objects. CF2 [18] extracted the hierarchical features from VGG19 [25] and got final results by combining the output response of each feature map. HDT [22] extracted features from different deep layers from VGG19 [25] and used an adaptive hedge method to hedge several CNN trackers into a stronger one to improve tracking performance. C-COT [6] applied an implicit interpolation model to learn continuous convolutional filters in the continuous spatial domain with multi-resolution deep features, which gets impressive results. These trackers improve tracking accuracy and robustness in most public benchmark significantly due to the powerful deep features. However, it is time-consuming to extract these features from very deep CNN models (e.g. VGG19). Besides, large CNN models have a massive number of parameters to store. These two issues will certainly affect the speed and deployment on embedded devices with limited storage space.

2.2 Deep Trackers with an End-to-End Manner

Most of the pre-trained CNN models focused on object classification tasks. In order to identify many categories, they are always very large to learn a massive number of parameters. However, it doesn't need such large CNN models which will greatly increase the computation complexity and storage consuming because there are only two categories (objects and background) in object tracking. Thus, many end-to-end algorithms are proposed to train small networks in large-scale datasets. MDNet [20] trained a multi domain network to classify objects and background in a large-scale object tracking dataset collected from public object tracking datasets. GOTURN [8] trained a regression network off-line in large-scale images which got an impressive speed. By using ILSVRC2015 [24] as a training set, SiameseFC [1] trained a fully-convolutional siamese networks to track objects. These end-to-end trackers get a better balance between tracking accuracy and speed, but most of these trackers need to collect large-scale training images, which is not easy in object tracking. What's more, some end-to-end trackers are trained in public object tracking datasets, which leads to the risks of over-fitting.

2.3 Deep Trackers with Learned DCF

Another kind of tracking methods are deep trackers with learned DCF. These trackers combine the powerful representation of CNN with the good locating capability of DCF to improve tracking performance in an end-to-end manner. CFNet [28] added a CF-layer in SiameseFC to improves tracking performance significantly. CREST [27] reformulated DCF as a one-layer convolutional neural network and applied residual learning to take appearance changes to reduce model degradation during online update, which got superior performance. However, it needs to extract features from a large VGG16 [25] network, costing expensive computation and storage.

3 Methods

In this section, we firstly present the general DCF formulation to describe the details about how to learn linear correlation filters. Secondly, we propose a novel features selection method to extract features of interest for different object by formulating it as an energy minimization problem in a standard DCF framework. We will review the small deep convolutional neural networks used in this paper before the second part.

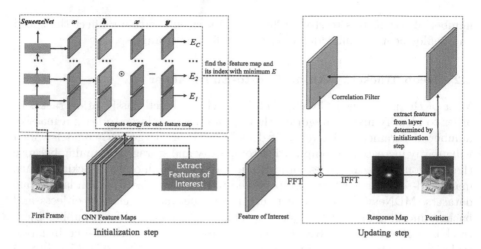

Fig. 1. The framework of our proposed tracker. In initialization step, we mainly find the index of layer used to extract features. x denotes feature map, h denotes correlation filter and y denotes Gaussian function.

3.1   DCF Formulation

The task of typical DCF trackers [3,9,10] is to learn a discriminative correlation filter h from a set of object training samples $\{(x_i, y_i)\}_{i=1}^{n}$, where h is a single or multi channels correlation filter, n is the number of training samples. In this work, x_i denotes the CNN features extracted from single CNN layer, y_i is the desired output which is a Gaussian function label for each x_i. Then the correlation filter h can be learned by solving an optimized problem:

$$h^* = \arg\min_{h} \sum \|x_i \otimes h - y_i\|^2 + \lambda \|h\|_2^2 \tag{1}$$

where $*$ denotes complex conjugation, $\otimes$ denotes circular convolution, $\lambda(\lambda \geq 0)$ is a regularization term. According to Parsevals formula, Eq. 1 can be transformed into Fourier domain:

$$\hat{h}^* = \arg\min_{\hat{h}} \sum \|\hat{x}_i \odot \hat{h} - \hat{y}_i\|^2 + \lambda \|\hat{h}\|_2^2 \tag{2}$$

where $\odot$ denotes element-wise, $\hat{f} = \mathcal{F}(f)$ denotes the Discrete Fourier transform (DFT) operator.

For multi-channel features, Boddeti $et\ al.$ [2] proposed vector correlation filter (VCF) to solve Eq. 2 in each individual feature channel:

$$\hat{h}^d = \frac{\hat{y}_i \odot \hat{x}_i^d}{\sum_{l=1}^{D} \hat{x}_i^l \odot \hat{x}_i^{l\star} + \lambda} \tag{3}$$

where $d(d \in \{1, 2, ..., D\})$ denotes the d-th channel. For a test image patch, we extract the CNN features z with the same size as x_i from a specific layer.

Then the output response between z and learned correlation filter h can be calculated by:

$$r = \mathcal{F}^{-1}(\sum_{d=1}^{D} \hat{h}^d \odot \hat{z}^{d*}) \tag{4}$$

where $\mathcal{F}^{-1}$ denotes the inverse Fast Fourier Transform. The position of target in the test image patch can be predicted by finding the maximum value in the correlation response map r:

$$p = \arg\max_{p}\{r(p)\} \tag{5}$$

where $p = (dx, dy)$ denotes the offset of coordinate position.

3.2 SqueezeNet

SqueezeNet [12] is a small deep network trained on an object recognition task, which can achieve AlexNet-level accuracy with 1/50 parameters (4.8 M). Low computation complexity, storage consuming and high accuracy all make it an attractive deep network for resource-limited applications. It consists of one convolutional layer, eight fire modules and one final convolution layer. A fire module begins with a squeeze convolution layer with 1×1 filters, ends with an expand layer including 1×1 and 3×3 convolution filters. In this paper, we extract the features from layers before softmax layer.

3.3 Extract Features of Interest

As mentioned in Sect. 1, the direct inspiration to extract features of interest is: the features of interest will minimize the energy in standard DCF formulation. For a specific object, let x_l denotes the features map extracted from the l-th layer with D channels, then the learned filter h_l of the l-th layer can be calculated by Eq. 3:

$$\hat{h}_l^d = \frac{\hat{y} \odot \hat{x}_l^d}{\sum_{i=1}^{D} \hat{x}_l^i \odot \hat{x}_l^{i*} + \lambda} \tag{6}$$

After learning correlation filter h_l, the energy of the l-th layer can be calculated by:

$$E_l(\hat{h}_l^*) = \sum \|\hat{x}_l \odot \hat{h}_l - \hat{y}\|^2 + \lambda\|\hat{h}_l\|_2^2 \tag{7}$$

We calculate E_l for each l-th layer, and finally get a set of energy:

$$E = \{E_1, E_2, ..., E_C\} \tag{8}$$

where C denotes the number of layers, $l \in \{1, 2, ..., C\}$. We find the index of minimum in E:

$$c = \arg\min_{c}\{E(c)\} \tag{9}$$

and use features extracted from the c-th layer to represent appearance of the object.

4   Tracking Framework

We apply our proposed method to the standard DCF framework and the main steps are showed in Fig. 1. In initialization step, we firstly input an image patch and extract a set of CNN feature maps from SqueezeNet [12] in each layer. After that, we extract the features of interest map by our proposed method introduced in Sect. 3.3. When finishing initialization step, we can get the features of interest and the index of layer to extract the features for the input object. In order to obtain a real-time speed, we fix the index of layer after the first frame. In updating step, we train and update correlation filters with the training method introduced in Sect. 3.1. Finally, the offset of coordinate between two contiguous frames can be estimated by Eq. 5.

5   Experiments

5.1   Benchmark and Criteria

We evaluate our proposed method in two popular tracking benchmarks, OTB2013 [30] and OTB100 [29]. OTB2013 contains 50 videos and OTB100 is an extension of OTB2013 which contains 100 videos. Most of popular tracking algorithms evaluate their performance in these benchmarks to demonstrate their effectiveness. Following these popular algorithms, we also evaluate our algorithm in these benchmarks to show effectiveness of our approach. The popular evaluation criteria used in most of popular trackers are mean distance precision (DP) and mean overlap precision (OP), where DP calculates the center location error between predicted positions and ground truth, and OP calculates the intersection-over-union (IoU) between predicted positions and ground truth. Most of popular trackers use the DP=20pixels as final DP scores and the area under the curve (AUC) of OP plots as final success scores.

5.2   Comparison with Baseline and State-of-the-Art

In order to demonstrate the effectiveness of our method, we compare our method with one baseline method. The baseline is built in the same framework with proposed method but extract feature from Fire3 layer in Squeezenet. In our experiments, we test all independent layer and find that Fire3 layer get the best tracking accuracy, so that we choose this layer to build the baseline. In addition, eight popular state-of-the-art trackers: DCF [9], KCF [9], SAMF [17], CACF [19], DSST [5] CF2 [18], HDT [22] and CFNet [28] are compared with our method. DCF is a standard DCF tracker which has a fast speed (333 fps). KCF is a kernel based DCF tracker. SAMF is the improving version of DCF with adaptive scale estimation. CACF is the variant of DCF, which exploits context information to improve precision while maintain a real-time speed (82 fps). DSST is a robust tracker with accurate scale estimation. CF2 and HDT are deep DCF trackers which replace hand-crafted features by CNN features extracted from a larger pre-trained CNN model in DCF framework to get an impressive improvement in precision but at the cost of speed (10 fps). CFNet is an end-to-end tracker that combines correlation filter and CNN in one unified framework. The summaries of these trackers are listed in Table 1.

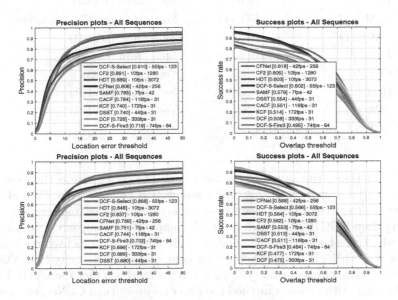

Fig. 2. Comparison with eight state-of-the-art trackers and the baseline. The first two are results on OTB2013, the last two are results on OTB100.

Figure 2 presents the precision plots and success (overlap) plots under one-pass evaluation (OPE). The results show that our proposed method (named DCF-S-Select) has better performance in both location precision and success rate than baseline method, which demonstrates that our method is effective in improving tracking performance. For state-of-the-art comparison, our method outperforms all the state-of-the-art methods in location error precision. Although the success rate of our method is lower than three deep trackers on OTB2013 and CFNet on OTB100, but our method is faster than these deep trackers. Compared with HDT and CF2, our method has better precision while runs at a real-time speed than these two deep trackers (55 fps vs 10 fps). Compare with two methods with hand-crafted feature, our method is slower than them but has higher accuracy.

5.3 Analysis on Storage and Computation Complexity

We further analyze the storage consumption and computation complexity for our proposed approach and other five state-of-the-art methods. We list the size of parameters which used to extract features, speed, the success rate on OTB100 and their implementations in Table 2. We can see that DCF, KCF,

Table 1. The trackers are evaluated in our comparison.

Tracker	Features	Scale	Published
DCF [9]	HOG [7]	No	2015 (PAMI)
KCF [9]	HOG [7]	No	2015 (PAMI)
SAMF [17]	HOG+CN [7]	Yes	2014 (ECCV Workshop)
CACF [19]	HOG [7]	No	2017 (CVPR)
DSST [5]	HOG [7]	No	2014 (BMVC)
CF2 [18]	CNN	Yes	2015 (ICCV)
HDT [22]	CNN	No	2016 (CVPR)
CFNet [28]	CNN	Yes	2017 (CVPR)
DCF-S-Fire3 (baseline)	CNN	No	-

CACF and DSST run much faster than deep trackers because they use simple HOG [7] features which has low computation complexity. CF2 and HDT run much slower than other deep trackers because they extract features from the very deep VGG19 [25] which has high computation complexity. In contrast, our method runs faster than CF2 and HDT because it extracts features from the shallow SqueezeNet [12] which has a lower computation complexity than VGG19. Besides, the size of parameters of SqueezeNet is much smaller than VGG19 (4.8 M vs 71.2 M). Although our method has lower accuracy than CFNet, our method has faster speed and smaller parameters than CFNet.

Table 2. The parameters size and speed of these trackers.

Tracker	Feature	Parameters	Speed	Success Rate	Implemention
DCF [9]	HOG	-	333 fps	0.475	Matlab (CPU)
KCF [9]	HOG	-	172 fps	0.477	Matlab (CPU)
SAMF [17]	HOG + CN	-	7 fps	0.553	Matlab (CPU)
CACF [19]	HOG	-	82 fps	0.511	Matlab (CPU)
DCF [5]	HOG	-	44 fps	0.513	Matlab (CPU)
CF2 [18]	VGG19 [25]	71.2 M	10 fps	0.562	Matlab (GPU)
HDT [22]	VGG19 [25]	71.2 M	10 fps	0.564	Matlab (GPU)
CFNet [28]	Self Training	13.8M	42 fps	0.588	Matlab (GPU)
DCF-S-Select	SqueezeNet [12]	4.8 M	55 fps	0.566	Matlab (GPU)

6   Conclusion

Recent researches have demonstrated that CNN features are very helpful to improve tracking performance in DCF trackers. Most of these trackers extract

CNN features from large pre-trained models, which have high computation complexity and huge storage usages. It degrades the tracking speed and arrangement in practice. This paper proposed a lightweight deep DCF tracker, which adaptively extracts features of interest from a small deep network. Extensive experimental results shows the effectiveness of our approach in accuracy and robustness. This work makes small deep networks having capacity to get better accuracy, lower storage and real-time speed simultaneously in visual object tracking, which is helpful to design lightweight and efficient object trackers in the future.

Acknowledgment. This work is supported in part by the National Key Research and Development Plan (2016YFC0801005), the National Natural Science Foundation of China (61772513) and the International Cooperation Project of Institute of Information Engineering, Chinese Academy of Sciences (Y7Z0511101). Shiming Ge is also supported by Youth Innovation Promotion Association, CAS.

References

1. Bertinetto, L., Valmadre, J., Henriques, J.F., Vedaldi, A., Torr, P.H.S.: Fully-convolutional siamese networks for object tracking. In: Hua, G., Jégou, H. (eds.) ECCV 2016. LNCS, vol. 9914, pp. 850–865. Springer, Cham (2016). https://doi.org/10.1007/978-3-319-48881-3_56
2. Boddeti, V.N., Kanade, T., Kumar, B.V.K.V.: Correlation filters for object alignment. In: IEEE Conference on Computer Vision and Pattern Recognition, pp. 2291–2298 (2013)
3. Bolme, D.S., Beveridge, J.R., Draper, B.A., Lui, Y.M.: Visual object tracking using adaptive correlation filters. In: IEEE Computer Vision and Pattern Recognition, pp. 2544–2550 (2010)
4. Danelljan, M., Hager, G., Khan, F.S., Felsberg, M.: Convolutional features for correlation filter based visual tracking. In: IEEE International Conference on Computer Vision Workshop, pp. 621–629 (2015)
5. Danelljan, M., Häger, G., Khan, F.S., Felsberg, M.: Accurate scale estimation for robust visual tracking. In: British Machine Vision Conference, pp. 65.1–65.11 (2014)
6. Danelljan, M., Robinson, A., Shahbaz Khan, F., Felsberg, M.: Beyond correlation filters: learning continuous convolution operators for visual tracking. In: Leibe, B., Matas, J., Sebe, N., Welling, M. (eds.) ECCV 2016. LNCS, vol. 9909, pp. 472–488. Springer, Cham (2016). https://doi.org/10.1007/978-3-319-46454-1_29
7. Felzenszwalb, P.F., Girshick, R., Mcallester, D., Ramanan, D.: Object detection with discriminatively trained part-based models. IEEE Trans. Pattern Anal. Mach. Intell. **32**(9), 1627–1645 (2010)
8. Held, D., Thrun, S., Savarese, S.: Learning to track at 100 FPS with deep regression networks. In: Leibe, B., Matas, J., Sebe, N., Welling, M. (eds.) ECCV 2016. LNCS, vol. 9905, pp. 749–765. Springer, Cham (2016). https://doi.org/10.1007/978-3-319-46448-0_45
9. Henriques, J.F., Caseiro, R., Martins, P., Batista, J.: High-speed tracking with kernelized correlation filters. IEEE Trans. Pattern Anal. Mach. Intell. **37**(3), 583–596 (2015)

10. Henriques, J.F., Caseiro, R., Martins, P., Batista, J.: Exploiting the circulant struc-
    ture of tracking-by-detection with kernels. In: Fitzgibbon, A., Lazebnik, S., Perona,
    P., Sato, Y., Schmid, C. (eds.) ECCV 2012. LNCS, vol. 7575, pp. 702–715. Springer,
    Heidelberg (2012). https://doi.org/10.1007/978-3-642-33765-9_50
11. Huang, C.M., Wang, S.C., Chang, C.F., Huang, C.I.: An air combat simulator in
    the virtual reality with the visual tracking system and force-feedback components.
    In: IEEE International Conference on Control Applications, vol. 1, pp. 515–520
    (2004)
12. Iandola, F.N., Han, S., Moskewicz, M.W., Ashraf, K., Dally, W.J., Keutzer, K.:
    Squeezenet: Alexnet-level accuracy with 50x fewer parameters and <0.5MB model
    size (2016). arXiv: Computer Vision and Pattern Recognition
13. Jian, M., Lam, K.M., Dong, J., Shen, L.: Visual-patch-attention-aware saliency
    detection. IEEE Trans. Cybern. 45(8), 1575–1586 (2015)
14. Jian, M., Qi, Q., Dong, J., Sun, X., Sun, Y., Lam, K.: Saliency detection using
    quaternionic distance based weber local descriptor and level priors. Multimedia
    Tools and Applications, pp. 1–18 (2017)
15. Jian, M., Qi, Q., Dong, J., Yin, Y., Lam, K.M.: Integrating qdwd with pattern
    distinctness and local contrast for underwater saliency detection. J. Vis. Commun.
    Image Represent. 53, 31–41 (2018)
16. Krizhevsky, A., Sutskever, I., Hinton, G.E.: Imagenet classification with deep con-
    volutional neural networks. In: International Conference on Neural Information
    Processing Systems, pp. 1097–1105 (2012)
17. Li, Y., Zhu, J.: A scale adaptive kernel correlation filter tracker with feature inte-
    gration. In: Agapito, L., Bronstein, M.M., Rother, C. (eds.) ECCV 2014. LNCS,
    vol. 8926, pp. 254–265. Springer, Cham (2015). https://doi.org/10.1007/978-3-319-
    16181-5_18
18. Ma, C., Huang, J.B., Yang, X., Yang, M.H.: Hierarchical convolutional features
    for visual tracking. In: IEEE International Conference on Computer Vision, pp.
    3074–3082 (2015)
19. Matthias, M., Neil, S., Ghanem, B.: Context-aware correlation filter tracking. In:
    IEEE Conference on Computer Vision and Pattern Recognition (2017)
20. Nam, H., Han, B.: Learning multi-domain convolutional neural networks for visual
    tracking. In: Computer Vision and Pattern Recognition, pp. 4293–4302 (2016)
21. Papanikolopoulos, N.P., Khosla, P.K., Kanade, T.: Visual tracking of a moving
    target by a camera mounted on a robot: a combination of control and vision. IEEE
    Trans. Robot. Autom. 9(1), 14–35 (1993)
22. Qi, Y., et al.: Hedged deep tracking. In: Computer Vision and Pattern Recognition,
    pp. 4303–4311 (2016)
23. Ren, S., He, K., Girshick, R., Sun, J.: Faster R-CNN: towards real-time object
    detection with region proposal networks. In: International Conference on Neural
    Information Processing Systems, pp. 91–99 (2015)
24. Russakovsky, O., Deng, J., Su, H., Krause, J., Satheesh, S., Ma, S., Huang, Z.,
    Karpathy, A., Khosla, A., Bernstein, M.S.: Imagenet large scale visual recognition
    challenge. Int. J. Comput. Vis. 115(3), 211–252 (2015)
25. Simonyan, K., Zisserman, A.: Very deep convolutional networks for large-scale
    image recognition. Computer Science (2014)
26. Smeulders, A.W.M., Chu, D.M., Cucchiara, R., Calderara, S., Dehghan, A., Shah,
    M.: Visual tracking: an experimental survey. IEEE Trans. Pattern Anal. Mach.
    Intell. 36(7), 1442–1468 (2014)

27. Song, Y., Ma, C., Gong, L., Zhang, J., Lau, R.W.H., Yang, M.H.: Crest: convolutional residual learning for visual tracking. In: IEEE International Conference on Computer Vision, pp. 2574–2583 (2017)
28. Valmadre, J., Bertinetto, L., Henriques, J.F., Vedaldi, A., Torr, P.H.S.: End-to-end representation learning for correlation filter based tracking. In: IEEE Conference on Computer Vision and Pattern Recognition (2017)
29. Wu, Y., Lim, J., Yang, M.H.: Object tracking benchmark. IEEE Trans. Pattern Anal. Mach. Intell. **37**(9), 1834–1848 (2015)
30. Wu, Y., Lim, J., Yang, M.H.: Online object tracking: A benchmark. In: IEEE Computer Vision and Pattern Recognition, pp. 2411–2418 (2013)

A Novel Feature Fusion with Self-adaptive Weight Method Based on Deep Learning for Image Classification

Qijun Tian[1], Shouhong Wan[1,2(✉)], Peiquan Jin[1,2], Jian Xu[1],
Chang Zou[1], and Xingyue Li[1]

[1] School of Computer Science and Technology, University of Science
and Technology of China, Hefei 230027, China
zkdtqj@mail.usct.edu.cn, {wansh,jpq}@ustc.edu.cn,
{jianxxxu,kkpanda,votelxy}@mail.ustc.edu.cn
[2] Key Laboratory of Electromagnetic Space Information,
Chinese Academy of Science, Hefei 230027, China

Abstract. In recent years, excellent convolutional neural networks (CNN) have been used to solve a variety of visual tasks, and the image classification is an important role for the most visual task. To improve the performance of classification networks, several recent methods have shown the benefit of extracting deeper features by increasing depth of networks and fusing features by linear connection. But deep features only describe the high-level semantic features and loses the shallow features such as edge contour. And features fusion don't describe the impact factor of features for results. In this paper, we study shallow and deep features fusion and propose a new architectural unit, which we call the "Self-adaptive Weight Fusion" (SFW) method. SFW adaptively recalibrates the features relation by determining impact factors of shallow and deep features, which is used for classify objects. Experimental results demonstrate that SFW can be transplanted to different networks. In particular, we find that this mechanism can produce significant performance improvements for existing state-of-the-art deep architectures with minimal additional computational cost. In addition, results show that shallow and deep features fusion is beneficial to learn more general and robust features for a network.

Keywords: Convolution neural network · Image classification
Fusion · Weight

1 Introduction

In the past decades, lots of methods have been proposed for image classification tasks. These methods have improved performance by extracting different shallow features and deep features, which have achieved excellent classification performance [6]. However, these methods use only shallow or deep features for image classification.

The most common method was to extract low-level features such as the shape feature, context feature and local image feature e.g. scale-invariant feature transform (SIFT) and histogram of oriented gradients (HOG), then construct a bag-of-visual-words representation or dictionary and run a statistical classifier [20] in last.

© Springer Nature Switzerland AG 2018
R. Hong et al. (Eds.): PCM 2018, LNCS 11164, pp. 426–436, 2018.
https://doi.org/10.1007/978-3-030-00776-8_39

But low-level features are not more general and robust. So this method did not achieve excellent performance. In 2012, AlexNet [1], the convolutional neural network (CNN), changed this situation. Afterwards, various CNNs were proposed, such as VGG [11], GoogleNet [10] and ResNet [9]. These networks extracted deep features to classify objects and improved accuracy. In recent years, related researches have demonstrated the benefits of increasing depth of CNN [9–11]. Deeper networks improve performance by learning more general and robust features. However the features used to classify objects have only deep features.

Meanwhile, feature fusion is employed in various domains for excellent performance [2, 3, 13, 19]. Feature fusion mainly includes feature fusion based on Bayesian theory, sparse representation theory and deep learning theory [20]. The feature fusion based on CNNs is mainly to deepen the depth of a CNN instead of using the fused features directly to classify objects. And these methods haven't considered the effect of different features for classification.

In this paper, we study a different aspect of architectural design, shallow and deep features fusion, by designing a new architectural unit, which we call the "Self-adaptive Weight Fusion" (SFW). Our goal is to improve the representation ability of a network by fusing shallow and deep features and explicit the impact factors of shallow and deep features.

The main contributions of the paper are as follows. (1) Unlike the current existing fusion methods, fusing shallow and deep features with self-adaptive weight (SFW) can learn more general features without changing the main structure of a network. (2) SFW improves performance than the base architecture with an extremely small increase in number of parameters. (3) SFW can be applied to different CNNs.

The remainder of this paper is organized as follows: we present feature fusion in Sect. 2. The proposed SFW is described in Sect. 3. Section 4 shows our experimental results and their corresponding analysis. At last, we give the conclusion of this work in Sect. 5.

2 Related Work

Feature fusion based on deep learning theory consists of two feature fusion methods. CNNs are one of the important models in deep learning theory. Therefore feature fusion is also used in CNN. Hence Simonyan et al. [22] first proposed a deep convolutional neural network model using a dual-stream architecture. Then Feichtenhofer et al. [21] proposed a spatial feature fusion method and a temporal feature fusion method on the basis of [21]. Feature could not only be fused in the softmax layer, but also in the ReLU layer behind the convolutional layer to achieve feature level fusion.

Spatial feature fusion merges two feature maps by fusion functions f as shown formula (1) [20]. Here H, W, and D denote the length, width, and number of channels of the feature map, respectively. And x^a and x^b denote two feature maps.

$$f : x^a + x^b \rightarrow y, \ x^a, x^b, y \in \mathbb{R}^{H \times W \times D} \tag{1}$$

The fusion function includes an additive fusion function, a maximum fusion function, a cascade fusion function, convolution fusion function and bi-linear fusion function, etc. The specific fusion functions are as formula (2)–(6).

The additive fusion function adds the elements of the corresponding positions of the two feature maps. As the formula (2) shows.

$$y^{sum} = f^{sum}(x^a, x^b), y_{i,j,d}^{sum} = x_{i,j,d}^a + x_{i,j,d}^b \tag{2}$$

The maximum fusion function is to take the larger value of the corresponding position element of the two feature maps as the fusion result, as shown in the formula (3).

$$y^{max} = f^{max}(x^a, x^b), y_{i,j,d}^{max} = \max\left\{x_{i,j,d}^a, x_{i,j,d}^b\right\} \tag{3}$$

The cascade fusion function preserves the results of the two feature maps, and changes the number of channels in the merged feature map to the sum of the two original feature maps as shown in the Eq. (4).

$$y^{cat} = f^{cat}(x^a, x^b), y_{i,j,2d}^{cat} = x_{i,j,d}^a, y_{i,j,2d-1}^{cat} = x_{i,j,d}^b \tag{4}$$

The convolutional fusion function achieves the dimensionality reduction after feature fusion on the basis of the cascade fusion as shown in the formula (5).

$$y^{conv} = f^{conv}(x^a, x^b), y^{conv} = y^{cat}f + b, f \in \Re^{1 \times 1 \times 2D \times D}, b \in \Re^D \tag{5}$$

The bilinear fusion function adds the two elements of the corresponding position. The number of feature channels becomes the square of the original channel number. As shown in the formula (6). This fusion function is often used behind the ReLU layer to fuse the corresponding channels of the two feature maps.

$$y^{bil} = f^{bil}(x^a, x^b), y^{bil} = \sum_i^H \sum_j^W x_{i,j}^{aT} \otimes x_{i,j}^b, y^{bil} \in \Re^{D^2} \tag{6}$$

Here i, j and d take a value from [1, H], [1, W] and [1, D], respectively, x^a, x^b denotes two feature maps. We focus on the study of spatial feature fusion in this paper. From the fusion function, it can be seen that the cascaded fusion is simply splicing the two feature maps together. The impact factors of different features is one. However the effect of different features on classification is different, therefore we propose self-adaptive weight fusion to improve performance of a network.

3   Feature Fusion with Self-adaptive Weight Method

Our goal is to ensure that a network is able to learn more expressive features to improve performance at similar consumption. To improve performance, we propose a Fusion-and-Weight by determining weights of different features in three steps, *extract, weight and fuse*. The SFW shown in Fig. 1 can be used directly with existing state-of-the-art architectures whose modules can be strengthened by directly adding their SFW.

For any given networks, we can construct a corresponding SFW network (SFWNet) to achieve shallow and deep features fusion with self-adaptive weight. The features $F1$ are first passed through a pooling operation, which extracts the features $F1_p$. This is followed by a reshape operation which makes the features $F1_p$ change to the features F_p, and a multiply operation which changes the features F_p to the features wF_p. The next operation is reshape operation again, which changes spatial dimensions $H'' \times W'' \times 1$ to spatial dimensions $1 \times 1 \times B$ (B equals that C times H times W). The last operation is concatenating features. The weight will be optimized by an end-to-end training. The shallow features F_{pl} extracted from the features $F1$ are the output of the SFW, which is a part of input of the last fully connected layer.

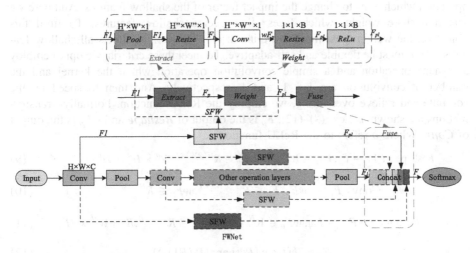

Fig. 1. A SFWNet and a SFW schematic diagram.

3.1   Extract: Obtain Shallow Features

To fuse shallow and deep features [2, 3, 6, 19], we first extract shallow features from the early layers of a network and deep features from the last layer of a network. For simplicity, we just discuss the shallow features learned by the first convolution layer in here and take $F1 \in \mathbb{R}^{H \times W \times C}$ (C, H and W express the number of channels, high and width of outputs, respectively.) to be the outputs of $\textbf{Conv}_1$ (the first convolution operation). The deep features are learned by the last pool or convolution layer of a network, so we do not to specially extract deep features by a new architecture. And it is enormous computation that all shallow features are used to classify. Therefore, to mitigate or slight increase computation and parameters, we propose to filter shallow features information by using maximum or average pooling operation. Supposed, the kernel of pooling operation equals $K_h \times K_w$, and the stride equals S. We take $\textbf{Pool}_i$ to be pooling operation and take $F1_p$ to be output of $\textbf{Pool}_i$ (i equals a denotes averge). Formally, a transformation is shown as Eqs. (7) and (8).

$$Pool_i : F1 \rightarrow F1_p, F1 \in R^{H \times W \times C}, F1_p \in R^{H' \times W' \times C}, i \in \{a, m\}. \tag{7}$$

$$H' = (H - K_h)/S, W' = (W - K_W)/S. \tag{8}$$

Here a and m indicate average and maximum operation, respectively. The transformation output $F1_p$ can be interpreted as a collection of the local descriptions whose statistics are expressive for the whole image or local of a image. Exploiting such information is prevalent in feature engineering work [15].

3.2   Weight: The Impact Factor of Features

For explicitly the impact factor of the shallow and deep features used to classify and adaptively recalibrating the feature relation [16], we follow '*extract*' with a second operation which aims to change the impact factor of the shallow features. And here we search a share weight, which faces all extracted shallow features. To fulfil this objective, the weight must meet two criteria: (1): The weight effects all shallow features. (2): It must be flexible and self-adaptive. To meet these criteria, we opt to employ a reshape operation and a simple convolution operation which the kernel and the number of convolution are 1×1, and 1, respectively. And then to speed up the compute and relieve over-fitting, we employ the ReLU function. Formally, transformations are shown as Eqs. (9)–(12). F_p is the output of **Reshape** and wF_p is the output of **Conv**. Where δ refers to the ReLU function.

$$\text{Reshape} : F1_p \rightarrow F_p, F1_p \in R^{H' \times W' \times C}, F_p \in R^{H'' \times W'' \times 1}.W'' \times H'' = H' \times W' \times C. \tag{9}$$

$$\text{Conv} : F_p \rightarrow wF_p, F_p \in R^{H'' \times W'' \times 1}, wF_p \in R^{H'' \times W'' \times 1}. \tag{10}$$

$$\text{Reshape} : wF_p \rightarrow F_{pl}, wF_p \in R^{H'' \times W'' \times 1}, F_p \in R^{1 \times 1 \times B}.B = W'' \times H'' \tag{11}$$

$$F_{pl} = \delta(Conv(\text{Reshape}(P_i(F1)))). \tag{12}$$

These operations just increases one parameter, which can be neglected for a network. Why employ weight? Weight is optimized by end-to-end training. And when weight is not useful, it is infinitely close to one.

3.3   Fuse: Concatenate Shallow and Deep Features

For integrating useful features to classify in the last operation, a *fuse* operation is employed. A network learn deep feature F_l from the last pooling operation, which the size is $1 \times 1 \times c$ (c refers to number of classes.), so we need to resize size of wF_p to size of F_l. Then we fuse F_l and F_{pl} to F. And the size of F is that the size of F_l adds the size of F_{pl}, so it increases that parameters and the computational complexity (more discussion can be found in Sect. 4). Formally, as shown in a Eq. (12).

$$F = f^{cat}(F_l, F_{pl}), F_d^{cat} = F_l, F_{d-1}^{cat} = F_{pl} \tag{13}$$

The flexibility of the SFW means that it can be directly applied to transformations beyond standard convolution. To illustrate this point, we develop SFWNet by applying SFW into modern architectures with simple designs.

Non-residual networks are different from residual networks in construction, so we implement in two classes networks. For non-residual networks, such as Inception network [10], SFWNets are constructed for the network by taking correlative operations to be the Inception network. By making this change for network in the architecture, we construct a SFW-GoogLeNet network. Moreover, SFW is also sufficiently flexible to be used in residual networks. More variants that integrate with ResNeXt [8] and Inception v2 [17] can be constructed by following the similar schemes. We describe the architecture of SFW-GoogLeNet in Table 1.

Table 1. (**Left**) GoogLeNet [10]. (**Right**) SFW-GoogLeNet. The shapes and operations with specific parameters setting of a non-resdual builing are listed in the table. And specific parameters setting Inception modules are not listed.

Output size	GoogLeNet [10]	SFW-GoogLeNet			
$112 \times 112 \times 64$	conv, 7×7, 64, stride 2	conv, 7×7, stride 2	max pool, 40×40, stride 20, $(6 \times 6 \times 64)$	avg pool, 40×40, stride 20, $(6 \times 6 \times 64)$	
$56 \times 56 \times 64$	max pool, 3×3, stride 2	max pool, 3×3, stride 2			
$56 \times 56 \times 192$	conv, 3×3, 192, stride 1	conv, 3×3, 192, stride 1			
$28 \times 28 \times 192$	max pool, 3×3, stride 2	max pool, 3×3, stride 2			
$28 \times 28 \times 480$	inception3 $\times$ 2	inception3 $\times$ 2	reshape	reshape	
$14 \times 14 \times 480$	max pool, 3×3, stride 2	max pool, 3×3, stride 2	conv, 1×1, 1, stride 1, $(48 \times 48 \times 1)$	conv, 1×1, 1, stride 1, $(48 \times 48 \times 1)$	
$14 \times 14 \times 832$	inception4 $\times$ 5	inception4 $\times$ 5			
$7 \times 7 \times 832$	max pool, 3×3, stride 2	max pool, 3×3, stride 2			
$7 \times 7 \times 1024$	inception5 $\times$ 2	inception5 $\times$ 2			
$1 \times 1 \times 1024$	avg pool, 7×7, stride 1	avg pool, 7×7, stride 1	reshape	reshape	
$1 \times 1 \times 1024$	dropout(40%), 1024	concatenate, 5632			
		dropout(40%), 5632			
$1 \times 1 \times 38$	linear, softmax	linear, softmax			

4 Experiments and Analysis

4.1 Implementation

Each plain network and its corresponding SFWNet counterpart are trained with identical optimization schemes. During pre-trained on ImageNet and training on PatternNet, we follow standard practice and perform data augmentation with random-size cropping [12] to 224 × 224 pixels and random horizontal flipping. Input are normalized through mean channel subtraction. In addition, we adopt the data balancing strategy described in [18] for mini-batch sampling. Optimization is performed using synchronous SGD with momentum 0.9. and a mini-batch size of 38. The initial learning rate is set to 0.001 and decreased by a factor of 10 every 25 epochs. All models are pre-trained in ImageNet dataset [13] and then trained by transfer learning in PatternNet dataset [4]. When testing, we apply a centre crop evaluation on the validation set, where 224 × 224 pixels are cropped from each image whose shorter edge is first resized to 256. [4] is comprised of 30 thousand and 4 hundred remote sensing images and 38 classes, including 6080 training images and 24320 validation images, and each class has 800 pictures. We train networks on the training set and report the top-1 accuracy about validation set. All experiments are run on a TiTan X: 12G, win 10, 32G memory computer.

4.2 Parameters

As is shown in Sect. 3, SFW is computationally lightweight and impose only a slight increase in model complexity and computational burden. For the proposed SFW to be viable in practice, it must provide an effective trade-of between model complexity and performance which is important for scalability. To illustrate the parameters of module, we take the comparison between [10] and SFW-GoogLeNets as an example, where the accuracy of SFW-GoogLeNets is superior to [10] (shown in Table 2). The majority of parameters come from the last stage of the SFW network. The parameters only add a little that can be neglected, and the accuracy increased by 0.64% to 1.58% in the PatternNet dataset [4]. It is an effective trade-off. In practice, with a training mini-batch of 38 images, the parameters of the pooling operations follow the first convolution layer is that the kernel (K_w, K_h) is (40, 40) and the stride (S) is 24, and the parameters of the pooling operations follow the second convolution layer is that the kernel (K_w, K_h) is (36, 36) and the stride (S) is 18. And all models are pre-trained on ImageNet dataset [13].

Googlenet_first64 × 4 × 4 indicates the shallow features are extracted by average and maximum pooling operations from the first convolution layer and the size of features is 64 × 4 × 4. Googlenet_first_avg64 × 4 × 4 indicates that just using average operation extracts shallow features from the first convolution layer. Google-net_all indicates that shallow features are extracted from the first and second convolution layers and deep features which are extracted by maximum pooling operation also join in making decision. Googlenet_second192 × 3 × 3 denotes that the shallow features are extracted by average and maximum pooling operations from the second convolution layer. Googlenet_second_max192 × 3 × 3 denotes the size of features is 192 × 3 × 3 and just using maximum operation extracts shallow features. Along the way, we know all networks. mAP is mean average precision.

However, we found that the scale of shallow features has influenced for results (as shown in Table 3), so we also take the comparison between [10] and SFW-GoogLeNets which just extracts shallow features from the first convolution layer, but the scale of shallow features changed. We found in Table 3 that the scale exceeds a certain value, the performance of the network changes in a small range and the parameters increase more. The scale 64 × 6 × 6 is the best effective trade-off for the first convolution layer.

Table 2. The parameters and performance of GoogLeNet [10] and SFW-GoogLeNets are shown as follow. PatternNet dateset is used, and train dataset/test dateset is 1:4.

Network\Percision	mAP	Parameters
Googlenet [10]	92.54	5.84 M
Googlenet_first64 × 4 × 4	**93.70**	5.84 M + 77.8 K
Googlenet_first_avg64 × 4 × 4	93.54	5.84 M + 38.9 K
Googlenet_first_max64 × 4 × 4	93.18	5.84 M + 38.9 K
Googlenet_second192 × 3 × 3	93.36	5.84 M + 131.3 K
Googlenet_second_avg192 × 3 × 3	**93.55**	5.84 M + 65.7 K
Googlenet_second_max192 × 3 × 3	93.00	5.84 M + 65.7 K
Googlenet_first_second	**93.77**	5.84 M + 209.1 K
Googlenet_first_second_avg	93.29	5.84 M + 104.6 K
Googlenet_first_second_max	93.15	5.84 M + 104.6 K
Googlenet_all	**94.12**	5.84 M + 248.1 K
Googlenet_all_avg	93.61	5.84 M + 143.5 K
Googlenet_all_max	93.55	5.84 M + 143.5 K

4.3 Classification

The Effect of Different Shallow Features on Classification. We first compare the fusing different shallow features SFWNets against GoogLeNet architectures. The results in Table 2 show that SFWNets consistently improve performance across different shallow features with an extremely small increase in computational complexity. And it can obtain the better performance with using multiple different shallow features from different convolution layers and different extracting operations, and SFWNet extracting the shallow features with only average pooling operation can also obtain the better performance than only maximum pooling operation.

Remarkably, SFWNets achieve a single-crop top-1 validation accuracy in a range of 93.00% ∼ 94.12%, exceeding GoogLeNet [10] (92.54%) by 0.46% ∼ 1.58%. While it should be noted that the SFWs themselves add width and dimension of features, they do so in an extremely computationally efficient manner and yield good returns even at point at which extending the width of the base architecture achieves diminishing returns. Moreover, we see that the performance improvements are consistent through training across a range of different shallow features, suggesting that the improvements induced by SFWs can be used in combination with increasing the dimension of features.

The Effect Different Scales of Shallow Features on Classification. The first experiment shows that fusion shallow and deep features can improve performance of network. Then we research that extracting shallow features of the different scale from the same convolution layer impacts on results. The results in Table 3 illustrate that performance of network varies with the scale and is better than the base architecture. Performance increases with the increase of the scale which is in range of 1 ∼ 6, then

Table 3. The scale of extracting shallow features is different. PatternNet dateset is used, and train dataset/test dateset is 1:4. mAP is mean average precision.

Net\Percision	mAP	Parameters
GoogLeNet [10]	92.54	5.84 M
Googlenet_first64 × 1 × 1	92.94	5.84 M + 4.9 K
Googlenet_first64 × 2 × 2	93.10	5.84 M + 19.5 K
Googlenet_first64 × 3 × 3	93.04	5.84 M + 43.8 K
Googlenet_first64 × 4 × 4	93.59	5.84 M + 77.8 K
Googlenet_first64 × 5 × 5	93.58	5.84 M + 112.6 K
Googlenet_first64 × 6 × 6	**94.19**	5.84 M + 175.1 K
Googlenet_first64 × 7 × 7	93.36	5.84 M + 238.3 K
Googlenet_first64 × 8 × 8	93.68	5.84 M + 311.3 K
Googlenet_first64 × 16 × 16	93.38	5.84 M + 1.25 M
Googlenet_first64 × 32 × 32	93.48	5.84 M + 4.98 M
Googlenet_first64 × 56 × 56	93.25	5.84 M + 15.3 M
Googlenet_first64 × 112 × 112	93.42	5.84 M + 61.01 M

decreases in 7 and fluctuates on a small range as the scale increases to the maximum value. Clearly, the best result is 94.19% when the scale is $64 \times 6 \times 6$ with an extremely small increase in the parameters and computational complexity.

As shown in Table 3, SFWNets fusing the shallow features of different scale achieve a top-1 validation accuracy in a range of 92.94% to 94.19%, exceeding the base architecture by 0.4% to 1.65%, and SFW-Googlenet_first64 $\times$ 6 $\times$ 6 (94.19% top-1 accuracy) approaching the performance achieved by the SFW-Googlenet_all network (94.12% top-1 accuracy). Increasing the scale that is deemed to add dimension of the last features, is an extremely computationally efficient manner when the parameters just increases priming and can acquire good performance.

Deep Network. In this part of the experiments, we comprise the SFW-ResNet against ResNet network with different depths to assesss the effect of SFWs when operating on residual networks. The results of the comparison are shown in Table 4, exhibiting the same phenomena that emerged in the GoogLeNet [10] architecture.

Integration with Modern Architectures. We next investigate the effect of combining SFWs with another a state-of-the-art architecture, ResNeXt (using the setting of $32 \times 4d$) [11], which both introduce prior structures in modules. We construct SFWNet equivalents of these networks, SFW-BN-Incption and SFW-ResNeXt. The results in Table 4. In here we extract shallow features of the same dimensionality with Googlenet_first64 $\times$ 6 $\times$ 6.

Table 4. Single-crop accuracy rates on the PatternNet.

Network	mAP	Network	mAP
BN-Inception [18]	93.23	GoogLeNet [10]	92.54
FW-BN-Inception	93.88	FW-GoogLeNet	94.19
ResNet-50 [9]	94.11	ResNeXt-50 [8]	92.74
FW-ResNet-50	94.63	FW-ResNeXt-50	93.70
ResNet-101 [9]	94.81	ResNeXt-101 [8]	94.23
FW-ResNet-101	95.37	FW-ResNeXt-101	94.86

5 Conclusion

In this paper we proposed the SFW, a novel architectural unit designed to improve the representational capacity of a network by enabling it to perform shallow and deep features fusion with self-adaptive weight. Extensive experiments demonstrate the effectiveness of SFWNets which achieves excellent performance. In addition, SFW can be applied to different CNNs. Finally, we hope SFWNet will achieve excellent performance for visual tasks except image classification.

References

1. Krizhevsky, A., Sutskever, I., Hinton, G.E.: Imagenet classification with deep convolutional neural networks. In: Advances in Neural Information Processing Systems, pp. 1097–1105 (2012)
2. Ma, L., Lu, J., Feng, J., Zhou, J.: Multiple feature fusion via weighted entropy for visual tracking. In: Proceedings of the IEEE International Conference on Computer Vision, pp. 3128–3136 (2015)
3. Al-Wassai, F.A., Kalyankar, N.V., Al-Zaky, A.A.: Multisensor images fusion based on feature-level. arXiv preprint arXiv:1108.4098 (2011)
4. Zhou, W., Newsam, S., Li, C., Shao, Z.: PatternNet: a benchmark dataset for performance evaluation of remote sensing image retrieval. arXiv preprint arXiv:1706.03424 (2017)
5. Kang, L., Hu, B., Wu, X., Chen, Q., He, Y.: A short texts matching method using shallow features and deep features. In: Zong, C., Nie, J.Y., Zhao, D., Feng, Y. (eds.) Natural Language Processing and Chinese Computing. CCIS, vol. 496, pp. 150–159. Springer, Heidelberg (2014). https://doi.org/10.1007/978-3-662-45924-9_14
6. Hu, J., Shen, L., Sun, G.: Squeeze-and-excitation networks. arXiv preprint arXiv:1709.01507 (2017)
7. Huang, G., Liu, Z., Weinberger, K.Q., van der Maaten, L.: Densely connected convolutional networks. In: Proceedings of the IEEE Conference on Computer Vision and Pattern Recognition, vol. 1(2), p. 3, July 2017
8. Xie, S., Girshick, R., Dollár, P., Tu, Z., He, K.: Aggregated residual transformations for deep neural networks. In: 2017 IEEE Conference on Computer Vision and Pattern Recognition (CVPR), pp. 5987–5995. IEEE, July 2017
9. He, K., Zhang, X., Ren, S., Sun, J.: Deep residual learning for image recognition. In: Proceedings of the IEEE Conference on Computer Vision and Pattern Recognition, pp. 770–778 (2016)

10. Szegedy, C., Liu, W., Jia, Y., Sermanet, P., Reed, S., Anguelov, D., Rabinovich, A.: Going deeper with convolutions. In: CVPR, June 2015
11. Simonyan, K., Zisserman, A.: Very deep convolutional networks for large-scale image recognition. arXiv preprint arXiv:1409.1556 (2014)
12. Bodla, N., Zheng, J., Xu, H., Chen, J.C., Castillo, C., Chellappa, R.: Deep heterogeneous feature fusion for template-based face recognition. In: 2017 IEEE Winter Conference on Applications of Computer Vision (WACV), pp. 586–595. IEEE, March 2017
13. Russakovsky, O., et al.: ImageNet Large Scale Visual Recognition Challenge. arXiv:1409.0575 (2014)
14. LeCun, Y., Bottou, L., Bengio, Y., Haffner, P.: Gradient-based learning applied to document recognition. Proceedings of the IEEE 86(11), 2278–2324 (1998)
15. Shen, L., Sun, G., Huang, Q., Wang, S., Lin, Z., Wu, E.: Multi-level discriminative dictionary learning with application to large scale image classification. IEEE Trans. Image Process. 24(10), 3109–3123 (2015)
16. Levine, S., Finn, C., Darrell, T., Abbeel, P.: End-to-end training of deep visuomotor policies. J. Mach. Learn. Res. 17(1), 1334–1373 (2016)
17. Ioffe, S., Szegedy, C.: Batch normalization: Accelerating deep network training by reducing internal covariate shift. arXiv preprint arXiv:1502.03167 (2015)
18. Shen, L., Lin, Z., Huang, Q.: Relay backpropagation for effective learning of deep convolutional neural networks. In: Leibe, B., Matas, J., Sebe, N., Welling, M. (eds.) ECCV 2016. LNCS, vol. 9911, pp. 467–482. Springer, Cham (2016). https://doi.org/10.1007/978-3-319-46478-7_29
19. Feichtenhofer, C., Pinz, A., Zisserman, A.: Convolutional two-stream network fusion for video action recognition. In: CVPR 2016, pp. 1933–1941. IEEE Computer Society, Las Vegas (2016)
20. Wei-bin, L., Zzhi-yuan, Z., Wei-wei, X.: Feature fusion methods in pattern classification. J. Beijing Univ. Posts Telecommun. 40(4), 1–8 (2017)
21. Feichtenhofer, C., Pinz, A., Zisserman, A.P.: Convolutional two-stream network fusion for video action recognition (2016)
22. Simonyan, K., Zisserman, A.: Two-stream convolutional networks for action recognition in videos. In: Advances in Neural Information Processing Systems, pp. 568–576 (2014)

Text Component Reconstruction
for Tracking in Video

Minglei Yuan[1], Palaiahnakote Shivakumara[2], Hao Kong[1],
Tong Lu[1(✉)], and Umapada Pal[3]

[1] National Key Lab for Novel Software Technology,
Nanjing University, Nanjing, China
{mlyuan, hkong}@smail.nju.edu.cn, lutong@nju.edu.cn
[2] Faculty of Computer Science and Information Technology,
University of Malaya, Kuala Lumpur, Malaysia
shiva@um.edu.my
[3] Computer Vision and Pattern Recognition Unit,
Indian Statistical Institute, Kolkata, India
umapada@isical.ac.in

Abstract. Text tracking is challenging due to unpredictable variations in orientation, shape, size, color and loss of information. This paper presents a new method for reconstructing text components especially from multi-views for tracking. Our first step is to find Text Candidates (TCs) from multi-views by exploring deep learning. Text candidates are then verified with the degree of similarity and dissimilarity estimated by SIFT feature to eliminate false text candidates, which results in Potential Text Candidates (PTCs). Potential text candidates are further aligned in standard format with the help of affine transform. Next, the proposed method uses mosaicing concept for stitching PTC from multi-views based on overlapping regions between PTC, which results in reconstructed images. Experimental results on a large dataset with multi-view images show that the proposed method is effective and useful. The recognition experiments of several recognition methods show that the performances of the recognition methods improve significantly for the reconstructed images compared to prior reconstruction results.

Keywords: Text tracking · SIFT · Affine transform · Fusion · Mosaicing
Reconstruction

1   Introduction

Text tracking is important for several real time applications such as navigation, tracing persons, object or player tracing in sports video and surveillance applications. At the same time, tracking is used for enhancing text detection and recognition performances by making use of temporal information [1]. However, it is noted that actual tracking in real time environment involves many challenges, such as shape deformation, perspective distortion, size of text or object variations, font or contrast variations, and occlusion. In literature, most methods explore temporal information for improving text detection and recognition results without addressing the above-mentioned challenges

© Springer Nature Switzerland AG 2018
R. Hong et al. (Eds.): PCM 2018, LNCS 11164, pp. 437–447, 2018.
https://doi.org/10.1007/978-3-030-00776-8_40

[2, 3]. It is evident from the literature [2, 3] on text tracking that the methods perform poorly when there are blur in frames, texts with perspective distortions, occlusion, loss of shapes, etc. As a result, we believe that before tracking texts, one should address the above-mentioned challenges. Therefore, in this work, we aim at reconstructing text components of different forms caused by multiple views. Sample images with text of different forms are shown in Fig. 1 where one can see the bounding box of the text affected by different causes. It is noted from literature [4, 5] on text recognition in natural scene images and videos that the methods do not report satisfactory results for the images affected by multiple adverse factors mentioned-above despite the methods explore deep learning. Besides, the methods focus is to detect text in individual images but not from multi-view of the same scene as the proposed problem. This motivates us to propose the work for reconstruction from the text of multi-views. To the best of our knowledge, this is the first work attempting to find a solution to the above challenges rather than tracking texts for detection and recognition.

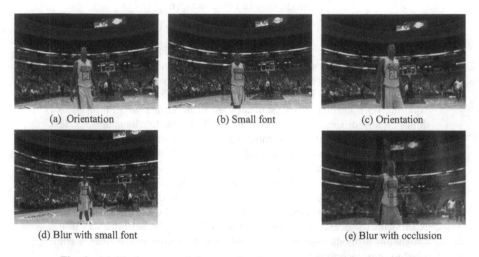

(a) Orientation   (b) Small font   (c) Orientation

(d) Blur with small font   (e) Blur with occlusion

Fig. 1. Multi-views containing text for the same scene from sports video.

There are methods [6] that reconstruct character shapes or contours for impaired characters caused due to disconnections, perspective distortions, low contrast and occlusion. The methods usually work based on stroke width thickness or distance. However, during tracking, we cannot ensure that stroke width remains the same for all the situations especially skewed or distorted characters. In addition, the methods are sensitive to blur and noises. Recently, there are methods that explored deep learning to recognize characters, which are claimed robust to noises, blur, orientation, etc. Lee and Osindero [7] proposed recursive recurrent nets with attention modeling for OCR in the wild. The method considers different oriented texts, distorted texts and blur texts for recognition. The method uses lexicons and semantics between characters to recognize words or texts. However, during tracking, the same text appears in different forms. Therefore, it is hard to define lexicons to derive semantics for the words extracted during tracking. Bai et al. [8] proposed a multi-scale and mid-level representation for scene text recognition. Though the method works well for texts of different scripts, it still gives poor results for blur, oriented and low contrast texts. Yang et al. [9] proposed recurrent highway networks with an attention mechanism for scene text recognition. This method also works based on a large number of lexicons and training samples. Endo et al. [10] proposed a scene text detection method which is robust against orientation and discontiguous components of characters. Though the method works well for different orientations and languages, it fails when texts have non-uniform alignment. Unfortunately, non-uniform alignment is quite common during text tracking. Shi et al. [5] proposed an end-to-end trainable neural network for image based sequence recognition and its application to scene text recognition. The method does not segment characters for recognition. However, it depends largely on dictionary and a number of training samples.

It is found from the above discussions that some methods are proposed for recognizing texts extracted from different images but not from the multi-views of the same scene. Recently, to reconstruct document images from the multi-views as the proposed work, Kramer et al. [11] proposed robust stereo correspondence for documents by matching connected components of text lines with dynamic programming. This method is good for document images but not video or natural scene images, where one cannot expect plain background as in document images. Overall, it is noticed from the above discussions that none of the methods addressed the challenges of text tracking explicitly. In addition, the methods use tracking for improving text detection and recognition without tackling the actual challenges. Therefore, in this work, we propose a new framework.

2   Proposed Method

This work considers five frames containing the same text in different forms, namely, different orientations, font sizes, blur and occlusion as the input for the reconstruction. In the recent advances, deep learning methods are popular and powerful for detecting texts in images of different situations [4, 10]. In addition, since text candidate or text detection is a pre-processing step for recognition, if text detection results lose little information, it does not affect much for text component reconstruction. This is because text component reconstruction has the ability to restore the missing information from multiple images. Therefore, in order to take the advantage of this situation, we propose to explore the method [4], which uses synthetic data for training in natural images. Due to complex background and undesirable nature of text, the method detects non-text as text. As a result, one can expect many true text candidates and false ones for the same text. To overcome this issue, motivated by the ability of SIFT features [12] which are invariant to scaling and views, we propose to use SIFT features for finding the correspondence among text candidates of five frames. Using SIFT features, the proposed method then estimates the degrees of similarity between every two text candidates, which are then used for detecting true text candidates by eliminating false ones. For deciding similarity and dissimilarity, the proposed method fixes certain thresholds empirically based on the degree of matching areas of text candidates, which allows tolerance of missing one or two characters. This step results in text candidates of the same text with lower false positive rates. In order to perform registration for text candidates of different forms, the proposed method estimates the quality of text candidates based on spatial frequency and sharpness [13]. With this quality information, the proposed method chooses high quality text candidate as reference. Once the method finds reference text candidate, we calculate the number of similar points between reference text candidate and similar text candidates. If the number of similar points is greater than one empirical threshold, we use Random Sample Consensus (RANSAC) algorithm [14] to get the homography matrix for similar text candidates. Then homography matrix is used to transform the images affected by multi-views into the standard format. Further, the method proposes to explore mosaicing concept [15] for fusing all the text candidates based on finding overlapping regions, which is called text component reconstruction.

2.1 Text Candidates Detection

Inspired by the method [4], which has ability to detect text of different orientations, low contrast, low resolution, some extent to blur, we use the same method for text candidates in this work. Since the scope of the proposed work is to reconstruct the text components from the components of multi-views, we prefer to use state-of-the-art for text candidate detection. The method introduces several layers to extract text specific features and it uses a dense regression network built on these features. For the images shown in Fig. 1, the method detects text as shown in Fig. 2 and the output is considered as Text Candidates (TC) detection, where we can see the method detects text well for the different situations.

Fig. 2. Text candidate detection for five views of the scene images in Fig. 1.

2.2 Potential Text Candidates Detection

For each Text Candidate (TC), the proposed method explores SIFT features to estimate the degree of similarity and dissimilarity based on the matched points between two text candidates. Since the input of this work includes the images containing the same text with different forms, the SIFT based similarity is high for the same text candidate and low for false text candidates. Motivated by the SIFT, which is invariant to scaling, the proposed method use SIFT for detecting true text candidates by eliminating false ones, which we called Potential Text Candidates (PTC). The matching process which based SIFT points are shown in Fig. 3, where we can see key points for text candidates in Fig. 3(a) and the correspondence between two PTC in Fig. 3(b).

For the PTC, the method further introduces spatial frequency as defined in Eqs. (1), (2) and (3) for estimating the quality. Based on the quality information, the proposed method chooses the potential text candidates, which have high quality in terms of

(a) Key points for two text candidates   (b) Matching between two text candidates

Fig. 3. Finding correspondence to estimate the degree of similarity.

content as reference PTC. If the SIFT matching points between a PTC and a reference PTC is lower than an empirical threshold of 8, we discard the PTC and name it as Invalided Potential Text Candidate (IPTC), if it is greater than the empirical threshold, the proposed method uses Random Sample Consensus (RANSAC) algorithm [14] to find homography matrix H. H is a 3 × 3 matrix, which can map a pixel on PTC to the corresponding pixel that match the reference PTC as given in Eq. (4). The number of key points is determined based on experiments using predefined samples. The experimental analysis is given in Experimental section. Next, in order to align the PTC with the reference PTC, we propose to use H to perform an affine transformation on the PTC as defined in Eq. (5). The effect of affine transform can be seen in Fig. 4, where the characters are aligned properly.

Fig. 4. Alignment using affine transform

$$RF = \sqrt{\frac{1}{M*N} \sum_{i=1}^{M} \sum_{j=2}^{N} \left[Z(x_i, y_j) - Z(x_i, y_{j-1}) \right]^2} \qquad (1)$$

$$CF = \sqrt{\frac{1}{M*N} \sum_{i=2}^{M} \sum_{j=1}^{N} \left[Z(x_{i-1}, y_j) - Z(x_i, y_j) \right]^2} \qquad (2)$$

$$SF = \sqrt{RF^2 + CF^2} \qquad (3)$$

where M and N are dimensions of an image, RF and CF are the vertical spatial frequency and the horizontal spatial frequency, respectively.

$$H = \begin{bmatrix} a_{11} & a_{12} & a_{13} \\ a_{21} & a_{22} & a_{23} \\ a_{31} & a_{32} & 1 \end{bmatrix} \qquad (4)$$

Suppose $p(x_1, y_1)$ is a point and its correspondence to the homogeneous coordinates is $p_1 = (x_1, y_1, 1)$. Therefore, H can be used to define affine transform as

$$p_2 = Hp_1 \qquad (5)$$

where $p_2 = (x_2, y_2, 1)$ is the homogeneous coordinate of point p.

2.3   Text Components Reconstruction

The above steps give aligned PTC as output. Since the proposed work considers occlusion as one of the challenges, one can expect loss of characters. To restore the missing information using other aligned PTC, we propose to use mosaicing concept [15]. The method extracts features for the aligned PTC and then finds overlapping regions. Based on overlapping regions, the proposed method finds boundary points of two PTC. The boundary points are then used for stitching two PTC, which results in a fused image, which is called text component reconstruction. When we have many PTC, the mosaicing method combines two PTC at every iteration. In other words, in the first iteration, the method fuses PTC-1 and PTC-2, which is considered as Fusion-1 result. Then Fusion-1 result is combined with PTC-3. In this way, the mosaicing method fuses all the PTC, which gives a reconstructed image as shown in Fig. 5, where it can be seen that the process of obtaining the final reconstructed images for multiple PTC. In Fig. 5, IPTC denotes, text candidate which does not satisfy the above condition. Therefore, the text candidate is ignored for the mosaicing process in all subsequent steps.

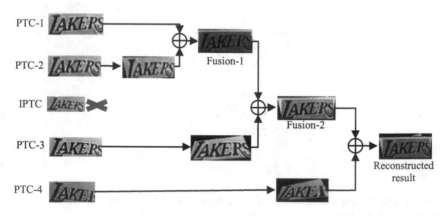

Fig. 5. Illustrations of the fusion process of the proposed method.

3   Experimental Results

To the best of our knowledge, the proposed work is new and the first attempt for reconstructing text components from multi-views. In this work, we consider tracking as an application. The input requires five frames containing the same text in different forms, namely, different orientations, blur, font sizes and occlusion, which is named as one set. We collect 41 sets of data from our own resources, 21 sets of data from YVT dataset, and 127 sets from YouTube. In total, 189 sets, which give 945 frames, which includes sports video, automobile race video, production release conference, etc., Since there is no assumption and constraints for collecting dataset, set contains complex background, low contrast, low resolution and different scripts.

It is expected that reconstruction results given by the proposed method should have good quality compared to text candidates. Therefore, to validate the quality of the reconstructed results, we estimate well-known quality measures [16–19], namely, BRISQUE, NRIQUE, GPC and SI. BRISQUE metric measures the quality of the image in terms of naturalness and smoothness, NIRIQUE metric measures the image quality in terms of degradations, GPC and SI use different criteria to measure the smoothness of images. The low value of BRISQUE and NRIQUE and the high value of GPC and SI indicates the image has good quality. To show that the reconstruction results are effective, we consider edit distance and OCR accuracy as defined in [20] for evaluating the proposed method. Edit distance is calculated for the output of recognition method and its ground truth. Note: we create the ground truths for text candidates and reconstructed results manually. The OCR accuracy is calculated using edit distances as defined in Eq. (6).

$$OCRAccuracy = \frac{\sum_{i=1}^{n} GT_i - \sum_{i=1}^{n} E_i}{\sum_{i=1}^{n} GT_i} \tag{6}$$

where GT_i is the length of the i-th ground truth string. E_i is the edit distance between OCR result of the i-th text candidate area and the ground truth string.

To show usefulness of the proposed method, we use different recognition methods or OCR engines. Shi et al. [5] explore a convolutional neural network and deep learning for scene character recognition (CRNN). Google Cloud Vision API (GCV-API), which is an online system [21]. The website does not provide any technical details, but gives recognition results for the input images. There is one more popular OCR which is called GOOGLE Tesseract [22]. We also use MATLAB 2017 version which provides an OCR function for recognition. The motivation for choosing the above OCR is that they are popular, available online and robust as they involve powerful deep learning tools and a large number of features for recognition.

The proposed method uses SIFT for estimating degree of similarity to detect Potential Text Candidates (PTC). It is fact that the performance of the SIFT depends on the number of key points. In this work, we prefer the images which give more than 8 key points based experimental analysis. To determine the number of key points, which give good results, we calculate recognition rate using different recognition methods for the different number of key points as reported in Table 1. Table 1 show that the state-of-art recognition method listed in Table 1 scores good recognition rate for the 8 key points and hence we use 8 key points for the proposed work.

The quantitative results of quality measures a prior to reconstruction and after reconstruction are reported in Table 2. Table 2 shows that BISQUE and NIRQUE gives high values for the text candidates before fusion while vice versa for after fusion. GPC and SI give low values for text candidates before fusion, while vice versa for after fusion. Therefore, we can conclude that the quality of the reconstruction results is

Table 1. Determining the number of key points for detecting PTC using SIFT.

OCR method	3	4	5	6	7	8	9	10	11	12
CRNN [5]	79.08	78.99	79.25	79.34	78.95	**79.55**	79.12	79.08	78.81	78.87
GCV- API [21]	61.23	62.55	62.93	62.04	64.00	**64.68**	65.23	63.53	65.21	63.66
Tesseract [22]	23.71	26.12	25.61	26.12	25.99	**24.03**	23.35	24.16	23.93	23.65
Matlab	15.62	17.98	18.24	18.19	18.28	**16.32**	16.06	16.92	16.69	16.23

Table 2. The quality measures before and after reconstruction (fusion)

Metric	Before fusion	After fusion
BRISQUE [16]	43.86	43.33
NRIQUE [17]	16.95	16.79
GPC [18]	913.54	1875.21
SI [19]	10.76	26.00

improved compared to text candidates given by the text detection method. Note: the proposed method requires approximately 0.98 s (average processing time) for reconstructing each image with the following system configuration: Intel(R) Core(TM) i5-7400 CPU@3.00 GHZ, RAM is 8.0 GB.

Qualitative results of the proposed reconstruction and recognition results are shown in Fig. 6, where we can see for the text candidates, the OCR in [5] gives inconsistent results, while for the reconstructed results given by the proposed method, the same OCR gives correct recognition results. It is true for both sample-1 and sample-2 as shown in Fig. 6. It is also observed from Fig. 6 that the reconstruction results have better quality with no loss of information compared to text candidates. This shows that the proposed reconstruction is useful and effective for tracking applications, where it is necessary to handle the situations like occlusion, blur, missing information, and different orientations for the same text.

To assess the contribution of each steps involved in the proposed methodology, we calculate edit distance and recognition rate for each step using all the recognition methods. Edit distance and recognition rate for the Text candidates given by text detection method, Potential text candidates given by SIFT with degree of similarity, Text components given by fusion operation without affine transformation and finally Text candidates with affine transformation. The results of different recognition methods for all the above steps and the proposed method are reported in Table 3. To compare the influence of affine transformation, the fusion results of similar text candidates without affine transformation were also reported. Table 3 shows that the result of edit distance and recognition rates of each key steps improves gradually when we compared to text candidates given by text detection methods for all recognition methods. As a result, we can assert that each intermediate steps contributes to achieve the better results for the proposed method. When we compare the results of CRNN and GCV-API with the results of Tesseract and MATLAB OCR, CRNN and GCV-API report better results than these two conventional systems (Tesseract and MATLAB OCR). This is valid because the conventional OCRs do not have the ability to handle images of complex

'sream'   'scream'   'scream'   'scream'   'scream'   result:'scream'

'ikeps'   'kers'   'kers'   'nk'   'ak'   result:'akers'

Fig. 6. Sample qualitative results of the proposed reconstruction. The text in '' is the recognition results by the method [5].

'microsof'   'ics'   'miac'   'micross'   'microsos'   result: 'microsta'

Fig. 7. Limitation of the proposed method. Recognition results are given by the method [5]

Table 3. Performance of the different recognition methods at different stages. ED denote Edit Distance and OCR denotes recognition rate.

OCR method	TC		PTC		Fusion without affine transformation		With affine transformation		Proposed reconstruction	
	ED	OCR	ED	OCR	ED	OCR	ED	OCR	ED	**OCR**
CRNN [5]	3917	68.80	2773	74.25	663	71.75	2470	70.79	480	**79.55**
GCV- API [21]	5978	52.39	4837	55.09	999	57.44	4281	49.68	829	**64.68**
Tesseract [22]	10199	18.77	8429	21.74	1851	21.13	6708	21.15	1783	**24.03**
Matlab	11094	11.64	9292	13.71	2112	10.01	7256	14.71	1964	**16.32**

backgrounds and different font sizes, while CRNN and GCV-API have the ability to handle such situations due to the involvement of powerful deep learning tools. In summary, from the above discussions, one can conclude that the performances of the recognition methods improve significantly for reconstructed results. Sometimes, if the images are affected by severe blur, the proposed method does not perform well as shown in Fig. 7 where it can be noticed that recognition method fails to recognize the reconstructed results. This shows that there is a scope for improvement in future.

4   Conclusion and Future Work

In this work, we have proposed a new framework for text component reconstruction from text components in multi-views of the same scene. We have explored a deep learning method for text candidate detection from different views. Text candidates are then verified by estimating degree of similarity and dissimilarity to remove the false text candidates, which results in potential text candidates. Reference potential text candidates are chosen from potential text candidates according to spatial frequency. The reference potential text candidates are used for alignment correction based on affine transform. Further, the proposed method explores mosaicing concept for stitching aligned potential text candidates as a single fused image based on finding overlapping regions between two potential text candidates. This results in reconstructed images from the text components in multi-views. Experimental results on different recognition methods show that the recognition performance improves significantly for the reconstructed images compared to the result of prior to reconstruction in terms of edit distance and OCR accuracy. As mentioned in experimental section, there is scope for improvement in future.

Acknowledgment. The work described in this paper was supported by the Natural Science Foundation of China under Grant No. 61672273 and No. 61272218, the Science Foundation for Distinguished Young Scholars of Jiangsu under Grant No. BK20160021, and Scientific Foundation of State Grid Corporation of China (Research on Ice-wind Disaster Feature Recognition and Prediction by Few-shot Machine Learning in Transmission Lines).

References

1. Yin, X.C., Zuo, Z.Y., Tian, S., Liu, C.L.: Text detection, tracking and recognition in video: A comprehensive survey. IEEE Trans. Image Process. **25**(6), 2752–2773 (2016)
2. Jain, M., Mathew, M., Jawahar, C.V.: Unconstrained scene text and video text recognition for Arabic script. In: Proceedings of ASAR, pp. 26–30 (2017)
3. Tian, S., Yin, X.C., Su, Y., Hao, H.W.: A unified framework for tracking based text detection and recognition from web videos. IEEE Trans. Pattern Anal. Mach. Intell. **40**(3), 542–554 (2018)
4. Gupta, A., Vedaldi, A., Zisserman, A.: Synthetic data for text localisation in natural images. In: Proceedings of CVPR, pp. 2315–2324 (2016)
5. Shi, B., Bai, X., Yao, C.: An end-to-end trainable neural network for image-based sequence recognition and its application to scene text recognition. IEEE Trans. Pattern Anal. Mach. Intell. **39**(11), 2298–2304 (2017)
6. Wu, Y., Shivakumara, P., Lu, T., Tan, C.L., Blumenstein, M., Kumar, G.H.: Contour restoration of text components for recognition in video/scene images. IEEE Trans. Image Process. **25**(12), 5622–5634 (2016)
7. Lee, C.Y., Osindero, S.: Recursive recurrent nets with attention modeling for ocr in the wild. In: Proceedings of the IEEE Conference on Computer Vision and Pattern Recognition CVPR, pp. 2231–2239 (2016)
8. Bai, X., Yao, C., Liu, W.: Strokelets: A learned multi-scale mid-level representation for scene text recognition. IEEE Trans. Image Process. **25**(6), 2789–2802 (2016)
9. Yang, H., Li, S., Yin, X., Han, A., Zhang, J.: Recurrent highway networks with attention mechanism for scene text recognition. In: Proceeding of the DICTA, pp. 1–8 (2017)
10. Endo, R., Kawai, Y., Sumiyoshi, H., Sano, M.: Scene-text-detection method robust against orientation and discontiguous components of characters. In: Proceedings of the CVPR, pp. 1–9 (2017)
11. Krämer, M., Afzal, M.Z., Bukhari, S.S., Shafait, F., Breuel, T.M.: Robust stereo correspondence for documents by matching connected components of text-lines with dynamic programming. In: Proceedings of the ICPR, pp. 734–737 (2012)

12. Lowe, D.G.: Object recognition from local scale-invariant features. In: Proceedings of ICCV, pp. 1150–1157 (1999)
13. Beck, J., Sutter, A., Ivry, R.: Spatial frequency channels and perceptual grouping in texture segregation. In: Proceedings of ICVGIP, pp. 299–325 (1987)
14. Lee, Jj, Kim, G.: Robust estimation of camera homography using fuzzy RANSAC. In: Gervasi, O., Gavrilova, Marina L. (eds.) ICCSA 2007. LNCS, vol. 4705, pp. 992–1002. Springer, Heidelberg (2007). https://doi.org/10.1007/978-3-540-74472-6_81
15. Michahial, S.: Automatic Image Mosaicing Using Sift Ransac and Homography. IJENT 3(10), 247–251 (2014)
16. Mittal A., Moorthy A.K., Bovik A.C.: Blind/Referenceless image spatial quality evaluator. In: Proceedings of ACSSC, pp. 723–727(2011)
17. Mittal A., Soundararajan R., Bovik A.C.: Making a 'Completely Blind' image quality analyzer. In: Proceedings of ISPL, pp. 209–212(2013)
18. Blanchet G., Moisan L., Roug´e, B.: Measuring the global phase coherence of an image. In: Proceedings of ICIP, pp. 1176–1179(2008)
19. Blanchet, G., Moisan, L.: An explicit sharpness index related to global phase coherence. In: Proceedings of ICASSP, pp. 1065–1068(2012)
20. Robust Reading Competition. http://rrc.cvc.uab.es/?ch=5&com=evaluation&task=2. accessed 17 May 2018
21. Google Cloud Vision API. https://cloud.google.com/vision/. accessed 17 May 2018
22. Tesseract OCR. https://github.com/tesseract-ocr/tesseract. accessed 17 May 2018

Robust Deep Gaussian Descriptor
for Texture Recognition

Jiahua Wang[1], Jianxin Zhang[1(✉)], Qiule Sun[1,2], Bin Liu[3,4(✉)],
and Qiang Zhang[1,2]

[1] Key Lab of Advanced Design and Intelligent Computing (Ministry of Education),
Dalian University, Dalian, China
jxzhang0411@163.com
[2] Faculty of Electronic Information and Electrical Engineering,
Dalian University of Technology, Dalian, China
[3] International School of Information Science and Engineering (DUT-RUISE),
Dalian University of Technology, Dalian, China
liubin@dlut.edu.cn
[4] Key Laboratory of Ubiquitous Network and Service Software of Liaoning Province,
Dalian University of Technology, Dalian, China

Abstract. Recently, second-order statistical modeling methods with
convolutional features have shown impressive potential as image representation for vision tasks. Among them, bilinear convolutional neural
network (B-CNN) has attracted a lot of attentions due to its simplicity
and effectiveness. It captures the second-order local feature statistics via
outer product, which approximately explores the covariance between convolutional features and achieves promising performance for texture recognition. In order to inherit the merits of B-CNN while further improving
its performance, we introduce a Gaussian descriptor into B-CNN and
propose a novel robust deep Gaussian descriptor (RDGD) method for
texture recognition. We first compute Gaussian by using the output of
outer product of B-CNN, and then embed it into the space of symmetric positive definite (SPD) matrices. Finally, matrix power normalization
operation is employed to obtain more robust Gaussian descriptor. Experimental results on three texture databases demonstrate that RDGD is
superior to its baseline B-CNN and the state-of-the-arts.

Keywords: Robust Gaussian descriptor · Second-order statistics
Convolutional neural network · Texture recognition

1 Introduction

Texture, as a basic visual attribute of image, plays an important role on a variety of vision tasks, such as material recognition [1], object recognition [2] and
semantic segmentation [3]. Therefore, texture recognition has been attracting
wide attentions in computer vision fields. However, due to the large intraclass
variation and small interclass variation [4] existing in texture images, achieving
accurate accuracy remains a challenge for texture recognition.

© Springer Nature Switzerland AG 2018
R. Hong et al. (Eds.): PCM 2018, LNCS 11164, pp. 448–457, 2018.
https://doi.org/10.1007/978-3-030-00776-8_41

Recently, deep convolutional neural networks (CNNs) have been rapidly expanding and achieving brilliant performance in a wide range of vision applications due to its powerful ability of features learning. Although CNN models pretrained on ImageNet dataset [5] have been generalized well to many vision tasks, they mainly focus on the first-order feature statistics. To provide a more discriminant performance, some recent studies [6–10] further explore the potential of CNNs through second-order pooling (e.g., covariance pooling and global Gaussian) to exploit the second-order feature statistics, illuminating the effectiveness of combining CNN features and second-order statistic for visual tasks. Ionescu et al. [6] present, namely DeepO$_2$P, an second-order feature statistics method in deep CNNs by singular value decomposition (SVD), which embeds a trainable O$_2$P layer into deep CNNs architecture to perform second-order pooling of convolutional features for region classification. Wang et al. [8] construct approximate infinite dimensional Gaussian descriptors as image representations for material recognition, through a regularized Maximum Likelihood Estimation (MLE) called vN-MLE, and the descriptors explore the robust covariance estimation with high dimensional convolutional features. MPN-COV [9] is proposed to systematically evaluate the second-order information helpful for large-scale visual recognition. It employs matrix power normalized covariance as more discriminative image representations and achieves consistent improvements over its first-order counterparts under a variety of CNN models. Among them, B-CNN [7] has received considerable attentions due to its significant improvements on fine-grained categorization and texture recognition. It exploits second-order statistics through outer product of features of the last convolutional layers of two CNN models, which approximately explores the covariance between convolutional features.

In this paper, in order to inherit the cutting edges of B-CNN while introducing much richer information, we propose a novel robust deep Gaussian descriptor (RDGD) method motivated by [8]. The differences between B-CNN and RDGD are shown in Fig. 1. To begin with, we compute Gaussian by the output of outer product based on B-CNN framework for improving its performance. Furthermore, considering Gaussian models lying on a Riemannian manifold, we embed it into the space of SPD matrices. In addition, due to the high dimensionality of the convolution features and their small number, the classical MLE falls into an ill-condition to estimate covariance. Hence, we apply matrix power normalization (MPN) [9] to SPD matrix for obtaining more robust Gaussian descriptor.

The main contributions of this paper are summarized in three folds: (1) We propose a novel texture representation method by combining B-CNN and Gaussian descriptor, named robust deep Gaussian descriptor (RDGD), where the outer product of B-CNN, viewed as the coarse estimation of covariance, is embedded into Gaussian representation. (2) To overcome the coarse covariance estimation obtained from B-CNN with high dimension and small sample size, MPN operation is adopted to acquire robust Gaussian descriptor, to a certain extent, which solves the influence brought by the coarse estimation of covariance. (3) We extensively evaluate the proposed RDGD on three widely used

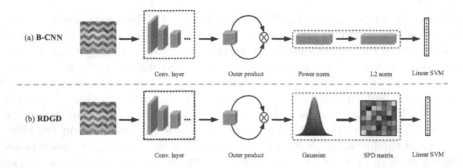

Fig. 1. Comparison of (a) B-CNN and (b) the proposed RDGD. B-CNN employs the outer product of convolutional features followed by element-wise power normalization and ℓ_2 normalization. Instead of two-step normalization operation after outer product, we compute the Gaussian descriptor based on outer product of B-CNN framework and embed it into the space of SPD matrices with matrix power normalization to obtain robust Gaussian descriptor (detailed in Sect. 3).

texture databases. Experimental results illuminate that RDGD gains significant accuracy improvement over its counterpart B-CNN and achieves competitive performance compared to the state-of-the-art methods.

2   B-CNN

In this section, we briefly review the B-CNN model which is closely related to our RDGD. The B-CNN model focuses on the cross-correlation information between the local features extracted from two CNN models. In details, given two feature matrices of $\mathbf{X} \in \mathbb{R}^{N \times d}$ and $\mathbf{Y} \in \mathbb{R}^{N \times d}$ to represent the last convolutional layers of two CNN models, where N and d are the number and dimension of features, respectively. Then, the bilinear pooling of the input image can be computed through the outer product, i.e.,

$$\mathbf{Z} = \mathbf{X}^T \mathbf{Y}, \tag{1}$$

which is subject to the element-wise power normalization (i.e., signed square root) and ℓ_2 normalization. In this way, $\mathbf{Z}$ captures the second-order local feature statistics in which the cross-correlation is collected. However, it has high cost of computing and storage caused by two CNN models.

In the other case, as illustrated in Fig. 1(a), when the two employed CNN models share the same parameters and settings, the bilinear pooling can be similarly produced by the outer product of the last convolutional layer of a single CNN model, i.e.,

$$\mathbf{Z} = \mathbf{X}^T \mathbf{X}. \tag{2}$$

One can see that $\mathbf{Z}$ exploits correlation of convolutional features and approximately produces covariance matrix followed by a two-step normalization operation. It has the translation invariance, which is a very important factor for texture recognition [11], relative to the original features. For image representation, obviously, the more useful information it contains, the more accurate image classification results will be achieved. This motivates us to consider much richer information for further improvement.

3   The Proposed RDGD

In this section, we introduce the proposed RDGD with the merits of B-CNN. As mentioned above, B-CNN explores the second-order pooling (i.e., coarse covariance) of the convolutional features. Instead of performing element-wise power normalization and ℓ_2 normalization after bilinear pooling, as shown in Fig. 1(b), we employ the global Gaussian as image representation, combining the output of the outer product of B-CNN with additional mean vector of convolutional features, to enhance its performance. Considering that the space of Gaussian forms a Riemannian manifold, we embed it into the space of SPD matrices through the methods given in [12,13]. Besides, as suggested in [9], covariance with matrix power normalization approximately exploits Riemannian geometry structure and amounts to robust covariance estimation used in [8]. Therefore, the matrix power normalization is adopted to obtain more robust Gaussian descriptor. Here, we briefly review the estimation of Gaussian descriptor based on MLE, and describe our more robust Gaussian descriptor for image representation in details.

3.1   Gaussian Descriptor

Assumed that the sample distribution obeys a Gaussian distribution, according the central limit theorem, we can use the Gaussian probability density function to establish a Gaussian model for the sample set. Given a set of N d-dimensional features $\mathbf{X} = \{\mathbf{x}_1, \mathbf{x}_2, \ldots, \mathbf{x}_N | \mathbf{x}_i \in \mathbb{R}^d\}$, we can model the distribution

$$\varphi(\mathbf{x}; \boldsymbol{\mu}, \boldsymbol{\Sigma}) = |2\pi\boldsymbol{\Sigma}|^{-\frac{1}{2}} \exp(-\frac{1}{2}(\mathbf{x} - \boldsymbol{\mu})^T \boldsymbol{\Sigma}^{-1}(\mathbf{x} - \boldsymbol{\mu})) \tag{3}$$

with mean vector $\boldsymbol{\mu}$ and covariance matrix $\boldsymbol{\Sigma}$, where $|\cdot|$ and T indicate the determinant and transpose of matrix, respectively. Then, the likelihood function of the sample set can be written as $L(\mathbf{X}; \boldsymbol{\mu}, \boldsymbol{\Sigma}) = \prod_{i=1}^{N} \varphi(\mathbf{x}_i; \boldsymbol{\mu}, \boldsymbol{\Sigma})$. By MLE, the estimation of mean and covariance of the sample set can be reduced to the following optimization problem:

$$\min_{\boldsymbol{\mu}, \boldsymbol{\Sigma}} \frac{N}{2} \log |\boldsymbol{\Sigma}| + \frac{1}{2}\boldsymbol{\Sigma}^{-1} \sum_{i=1}^{N}(\mathbf{x}_i - \boldsymbol{\mu})^T(\mathbf{x}_i - \boldsymbol{\mu}) \tag{4}$$

By minimizing above objective function, one can get

$$\boldsymbol{\mu} = \frac{1}{N} \sum_{i=1}^{N} \mathbf{x}_i, \tag{5}$$

$$\Sigma = \frac{1}{N} \sum_{i=1}^{N} (\mathbf{x}_i - \boldsymbol{\mu})^T (\mathbf{x}_i - \boldsymbol{\mu}), \tag{6}$$

which are equal to the mean and covariance of sampling set. From Eqs. (2) and (6), we can know that B-CNN approximately explores the covariance between convolutional features. To make full use of B-CNN framework, we further compute Gaussian model by using Eq. (2) rather than Eq. (6) to get additional effective information (e.g., mean vector) and boost the performance of B-CNN. As convolutional features usually have high dimensionality while their sample numbers may be small, there is an unfriendly fact that estimating covariance is not robust in this case. In order to tackle the challenge, we perform the MPN on SPD matrix formed by Eqs. (2) and (5) inspired by [9].

3.2   Robust Gaussian Descriptor

To improve the performance of B-CNN, we construct Gaussian descriptor according to Eqs. (2) and (5) based on B-CNN framework. We first embed Gaussian manifold into the space of SPD matrices. Let $\mathcal{N}(\boldsymbol{\mu}, \boldsymbol{\Sigma})$ be a Gaussian model with mean vector $\boldsymbol{\mu}$ (Eq. (5)) and approximate covariance $\boldsymbol{\Sigma}$ (Eq. (2)). Following the embedded methods [12,13], $\mathcal{N}(\boldsymbol{\mu}, \boldsymbol{\Sigma})$ can be uniquely mapped to a $(d+1) \times (d+1)$ SPD matrix $\mathbf{S}$

$$\mathcal{N}(\boldsymbol{\mu}, \boldsymbol{\Sigma}) \sim \mathbf{S} = \begin{bmatrix} \boldsymbol{\Sigma} + \boldsymbol{\mu}\boldsymbol{\mu}^T & \boldsymbol{\mu} \\ \boldsymbol{\mu}^T & 1 \end{bmatrix}. \tag{7}$$

However, the dimension and order of magnitude of each dimension of $\boldsymbol{\mu}$ and $\boldsymbol{\Sigma}$ may vary, and their effects may be different for various tasks. Following [12], a well-motivated parameter β ($\beta > 0$) is introduced to make a trade-off between $\boldsymbol{\mu}$ and $\boldsymbol{\Sigma}$ in the embedding matrix (7)

$$\mathcal{N}(\boldsymbol{\mu}, \boldsymbol{\Sigma}) \sim \mathbf{S}(\beta) = \begin{bmatrix} \boldsymbol{\Sigma} + \beta^2 \boldsymbol{\mu}\boldsymbol{\mu}^T & \beta\boldsymbol{\mu} \\ \beta\boldsymbol{\mu}^T & 1 \end{bmatrix}. \tag{8}$$

It is easy to see that $\mathbf{S}(\beta)$ is a more general form and reduces to the approximate covariance $\boldsymbol{\Sigma}$ (Eq. (2)) when $\beta = 0$, and becomes the original form (Eq. (7)) when $\beta = 1$. We control the interaction between $\boldsymbol{\mu}$ and $\boldsymbol{\Sigma}$ by adjusting the value of β. The favorable parameter β is a relatively beneficial factor for our task (detailed in Sect. 4).

In our case, the dimension of the features of the last convolutional layer is 512, which is larger than the number of features. As suggested in [8], the classical MLE is not robust to estimate covariances in case of high dimension and small sample size. Eigenvalues shrinkage [14,15] is a well-known kind of way to handle the situation mentioned above. Besides, MPN [9] is proved to be closely related to the shrinkage principle, which approximately estimates robust covariance [8] and exploits Riemann geometry structure. Thus, we adopt MPN to address the problem of robust Gaussian estimation. Specifically, we perform SVD of $\mathbf{S}(\beta)$

$$\mathbf{S}(\beta) \mapsto (\mathbf{U}, \boldsymbol{\Lambda}), \quad \mathbf{S}(\beta) = \mathbf{U}\boldsymbol{\Lambda}\mathbf{U}^T, \tag{9}$$

where $\boldsymbol{\Lambda} = diag(\lambda_1, ..., \lambda_{d+1})$ is a diagonal matrix and λ_i, $i = 1, ..., d+1$ are eigenvalues arranged by descend order. $\mathbf{U} = [\mathbf{u}_1, ..., \mathbf{u}_{d+1}]$ is an orthogonal matrix whose column $\mathbf{u}_i$ is the eigenvector corresponding to λ_i. After that, the matrix power of $\mathbf{S}(\beta)$ can be transformed to the power of its eigenvalues

$$(\mathbf{U}, \boldsymbol{\Lambda}) \mapsto \mathbf{Q}, \quad \mathbf{Q} \triangleq \mathbf{S}(\beta)^\rho = \mathbf{U}\boldsymbol{\Lambda}^\rho\mathbf{U}^T, \tag{10}$$

where $\boldsymbol{\Lambda}^\rho = diag(\lambda_1^\rho, ..., \lambda_{d+1}^\rho)$ and ρ is a positive real number less than 1. With MPN, we obtain more robust Gaussian descriptor, and our final embedding matrix is

$$\mathcal{N}(\boldsymbol{\mu}, \boldsymbol{\Sigma}) \sim \mathbf{S}(\beta, \rho) \triangleq \begin{bmatrix} \boldsymbol{\Sigma} + \beta^2\boldsymbol{\mu}\boldsymbol{\mu}^T & \beta\boldsymbol{\mu} \\ \beta\boldsymbol{\mu}^T & 1 \end{bmatrix}^\rho. \tag{11}$$

Note that our embedding matrix $\mathbf{S}(\beta, \rho)$ with $\rho = 0.5$ (i.e., $\mathbf{S}(\beta, 0.5)$ almost explores the robust estimation of approximate infinite dimensional Gaussian (RAID-G) [8], i.e.,

$$\mathbf{S}(\beta, \rho) \sim \mathcal{N}(\hat{\boldsymbol{\mu}}, \hat{\boldsymbol{\Sigma}}),$$
$$\hat{\boldsymbol{\Sigma}} = \mathbf{U}diag(\delta_k)\mathbf{U}^T,$$
$$\delta_k = \sqrt{\left(\frac{1-\alpha}{2\alpha}\right)^2 + \frac{\lambda_k}{\alpha}} - \frac{1-\alpha}{2\alpha}, \tag{12}$$

where $\hat{\boldsymbol{\mu}}$ and $\mathbf{U}$ are the same as in Eqs. (11) and (9) respectively. The λ_k is the eigenvalue of $\mathbf{S}$ mentioned in Eq. (9), and $0 < \alpha < 1$ is a regularizing parameter. Through the robust vN-MLE, RAID-G has shown that it outperforms corresponding counterparts and achieves state-of-the-art results. Please refer to [8] for more details.

4   Experiments

4.1   Experimental Databases and Settings

Experimental Databases. We conduct experiments on three benchmark texture databases, i.e., Flickr Material Database (FMD) [16], Describable Textures Database (DTD) [17] and KTH-TIPS 2b Database (KTH-2b) [18], to evaluate the effectiveness of RDGD. Some typical images are illuminated in Fig. 2. The brief descriptions of the three databases are as follows:

Flickr Material Database [16] is a relatively small database. It includes 10 categories of texture images, e.g., watermarks, foliage and glass, and each category has 100 images.

Describable Textures Database [17] is a challenging texture database, which consists of 5640 images from 47 categories (120 images per category) including freckled, knitted, blotchy, meshed, sprinkled and porous, etc.

KTH-TIPS 2b Database [18] is a widely-used texture database with varying illumination, pose and scale. It has in total of 4752 images from 11 categories, and each category has 4 sub-categories of 108 images, such as brown bread, cotton, linen and wool, etc.

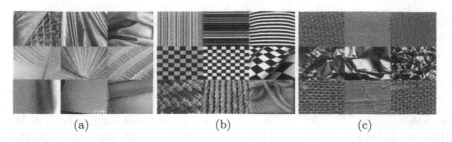

(a)   (b)   (c)

Fig. 2. Sample images from three benchmarks. From left to right are (a) FMD, (b) DTD and (c) KTH-2b, respectively.

Experimental Settings. Following the common protocol, we employ half of images in each category as training data and the remaining images as testing data for FMD and DTD databases. For KTH-2b database, one sample for each class is used for training and the other three are used for testing as [18]. Besides, similar to B-CNN [7], experiments are carried out at different scales of the input image, i.e., $s = 1$, $s = 2$ and $s = ms$, which respectively refer to 224×224 pixels, 448×448 pixels, and $2^s, s \in \{1.5 : -.5 : -3\}$, relative to 224×224 pixels. For multiple scales (i.e., $s = ms$), the final image representation is a average of representations at each scale. The VGG-16 model [19] without finetuning is adopted to extract features of the last convolutional layer to product robust Gaussian $\mathbf{S}(\beta, \rho)$. After the vectorization operation on it, we employ a linear SVM based on LIBSVM package [20] for classification. All programs are developed with Matlab R2015b, and run on a PC equipped with Intel (R) Core (TM) i7-7700k CPU @4.20 GHz, 64 GB RAM and a single NVIDIA GTX 1080 GPU.

4.2   Experimental Evaluation

Effect of β and ρ. We evaluate the effect of two parameters, i.e., β and ρ, existing in the robust Gaussian embedding. The parameter $\beta > 0$ is introduced to balance the influence of mean vector and covariance, while exponent ρ is used to control the power of eigenvalues in Eq. (11). To test the effect of both β and ρ, the results achieved on FMD at $s = 2$ are illustrated in Fig. 3.

We first evaluate the parameter β in Eq. (8). As shown in Fig. 3(a), with $\beta = 0$, the Eq. (8) reduces to the output of outer product of B-CNN obtaining the accuracy of 80.7%. Further embedding it into a SPD matrix $\mathbf{S}$ (Eq. (7)) with $\beta = 1$, the accuracy increases from 80.7% to 82.0%. Appropriate balancing at $\beta = 0.3$ outperforms $\beta = 0$ and $\beta = 1$ over 1.7% and 0.4% on FMD, respectively. It means that the usage of mean is important and can balance the effect between it and covariance for further improvement. Consistently, we test the robust Gaussian against the exponent ρ in Eq. (11) with fixed $\beta = 0.3$. The exponent ρ shrinks eigenvalues larger one, otherwise stretches them. As illustrated in Fig. 3(b) there is obvious performance influenced by using various values of ρ. On the whole, as ρ increases, the accuracy is getting better, and when $\rho = 0.8$ it works best. In

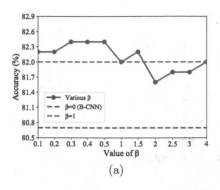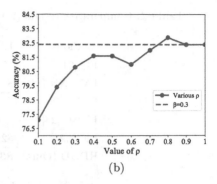

(a)   (b)

Fig. 3. Results of (a) various β and (b) various ρ on FMD at $s = 2$.

Table 1. Comparative results (%) of RDGD and its baseline B-CNN.

Methods	FMD			DTD			KTH-2b		
	s = 1	s = 2	s = ms	s = 1	s = 2	s = ms	s = 1	s = 2	s = ms
B-CNN	77.8	80.7	81.6	69.6	71.5	72.9	75.1	76.4	77.9
RDGD (Ours)	**79.6**	**82.9**	**83.3**	**70.1**	**72.7**	**73.4**	**78.0**	**81.8**	**82.8**

this paper, the values of β and ρ are set to 0.3, 0.8 for FMD, 0.6, 0.5 and 0.6, 1.0 for DTD and KTH-2b, respectively.

Comparison with B-CNN. We compare RDGD with B-CNN on three texture databases with different scales. The comparative results are listed in Table 1. From it, we can see that our RDGD outperforms the baseline B-CNN on all databases. Especially for KTH-2b, RDGD achieves significant improvements and is superior to B-CNN over 2.9%, 5.4% and 4.9%, respectively. Moreover, the multiple scales operator can consistently improve the performance and obtain the best results on all databases. This suggests that the features from different scales are complementary, and the combination of them leads to further improvement in accuracy. The promising improvements compared with B-CNN demonstrate the effectiveness of our RDGD.

Comparison with the State-of-the-arts. Finally, we further compare the proposed RDGD with the state-of-the-art methods, including FV-CNN [11], CBP [22], and LFV [24], etc. The comparative results are shown in Table 2. From the table, we can see that RDGD obtains the best performance on FMD and KTH-2b databases. It is slightly weaker than FV-CNN [11] and LFV [24] on DTD database, however, their results are very comparable. Moreover, note that LFV adopts the deeper and more powerful VGG-19 model [19]. In addition, CBP [22] and LRBP [23] are alternative methods for improving B-CNN. They focus on reducing the high dimension caused by bilinear pooling and achieve

Table 2. Comparative results (%) of RDGD and the state-of-the-arts.

Methods	FMD	DTD	KTH-2b
FC [21]	77.4	62.9	75.4
FV-CNN [11]	80.8	73.6	77.9
CBP [22]	-	67.7	-
LRBP [23]	-	65.8	-
LFV [24]	82.1	**73.8**	82.6
RDGD (Ours)	**83.3**	73.4	**82.8**

impressive performance on fine-grained categorization, but they have inferior accuracies (67.7% and 65.8%) for texture recognition. The desirable performance of RDGD confirms its effectiveness for texture recognition task.

5   Conclusion

The paper proposed a more robust Gaussian descriptor based on the B-CNN framework. It captures the first- and second-order feature statistics and approximately employs regularized vN-MLE based robust estimation of covariance with very high-dimensional features. The promising performance on texture recognition indicates the effectiveness of the proposed method. This motivates our future work towards mining much higher-order statistics of CNN features and utilizing information of multiple layers of CNN model. We will also apply the proposed method to other visual applications, such as scene categorization and image retrieval.

Acknowledgements. This work is supported by the National Natural Science Foundation of China (Nos. 61202251 and 91546123), Program for Changjiang Scholars and Innovative Research Team in University (No. IRT_15R07), the Liaoning Provincial Natural Science Foundation (No. 201602035) and the High-level Talent Innovation Support Program of Dalian City (No. 2016RQ078).

References

1. Sharan, L., Liu, C., Rosenholtz, R., et al.: Recognizing materials using perceptually inspired features. Int. J. Comput. Vision. **103**(3), 348–371 (2013)
2. Oyallon, E., Mallat, S.: Deep roto-translation scattering for object classification. In: IEEE Conference on Computer Vision and Pattern Recognition, pp. 2865–2873 (2015)
3. Girshick, R., Donahue, J., Darrell, T., et al.: Rich feature hierarchies for accurate object detection and semantic segmentation. In: IEEE Conference on Computer Vision and Pattern Recognition, pp. 580–587 (2014)
4. Varma, M., Garg, R.: Locally invariant Fractal features for statistical texture classification. In: International Conference on Computer Vision, pp. 1–8 (2007)

5. Russakovsky, O., Deng, J., Su, H., et al.: ImageNet large scale visual recognition challenge. Int. J. Comput. Vision. **115**(3), 211–252 (2014)
6. Ionescu, C., Vantzos, O., Sminchisescu, C.: Matrix backpropagation for deep networks with structured layers. In: International Conference on Computer Vision, pp. 2965–2973 (2015)
7. Lin, T.Y., Roychowdhury, A., Maji, S.: Bilinear CNN models for fine-grained visual recognition. In: International Conference on Computer Vision, pp. 1449–1457 (2016)
8. Wang, Q., Li, P., Zuo, W., et al.: RAID-G: robust estimation of approximate infinite dimensional Gaussian with application to material recognition. In: IEEE Conference on Computer Vision and Pattern Recognition, pp. 4433–4441 (2016)
9. Li, P.H., Xie, J.T., Wang, Q.L., et al.: Is second-order information helpful for large-scale visual recognition? In: International Conference on Computer Vision, pp. 2089–2097 (2017)
10. Sun, Q.L., Wang, Q.L., Zhang, J.X., et al.: Hyperlayer bilinear pooling with application to fine-grained categorization and image retrieval. Neurocomputing **282**, 174–183 (2018)
11. Lin, T.Y., Maji, S.: Visualizing and understanding deep texture representations. In: IEEE Conference on Computer Vision and Pattern Recognition, pp. 2791–2799 (2016)
12. Wang, Q.L., Li, P.H., Zhang, L., et al.: Towards effective codebookless model for image classification. Pattern Recog. **59**(C), 63–71 (2016)
13. Lovric, M., Min-Oo, M., Ruh, E.A.: Multivariate normal distributions parametrized as a Riemannian symmetric space. J. Multivariate Anal. **74**(1), 36–48 (2000)
14. Ledoit, O., Wolf, M.: A well-conditioned estimator for large-dimensional covariance matrices. J. Multivariate Anal. **88**(2), 365–411 (2004)
15. Chen, Y., Wiesel, A., Eldar, Y.C., et al.: Shrinkage algorithms for MMSE covariance estimation. IEEE Trans. Signal Process. **58**(10), 5016–5029 (2010)
16. Haran, L., Rosenholtz, R., Adelson, E.H.: Material perception: what can you see in a brief glance? J. Vision **9**(8), 784–784 (2009)
17. Cimpoi, M., Maji, S., Kokkinos, I., et al.: Describing textures in the wild. In: IEEE Conference on Computer Vision and Pattern Recognition, pp. 3603–3613 (2014)
18. Caputo, B., Hayman, E., Mallikarjuna, P.: Class-specific material categorisation. In: International Conference on Computer Vision, pp. 1597–1604 (2015)
19. Simonyan, K., Zisserman, A.: Very deep convolutional networks for large-scale image recognition. In: International Conference on Learning Representations, pp. 1–9 (2015)
20. Chang, C.C., Lin, C.J.: LIBSVM: a library for support vector machines. ACM Trans. Interact. Intell. **2**(3), 1–27 (2011)
21. Cimpoi, M., Maji, S., Vedaldi, A.: Deep filter banks for texture recognition and segmentation. In: IEEE Conference on Computer Vision and Pattern Recognition, pp. 3828–3836 (2015)
22. Gao, Y., Beijbom, O., Zhang, N., et al.: Compact bilinear pooling. In: IEEE Conference on Computer Vision and Pattern Recognition, pp. 317–326 (2016)
23. Kong, S., Fowlkes, C.: Low-rank bilinear pooling for fine-grained classification. In: IEEE Conference on Computer Vision and Pattern Recognition, pp. 7025–7034 (2017)
24. Song, Y., Zhang, F., Li, Q., et al.: Locally-transferred Fisher vectors for texture classification. In: International Conference on Computer Vision, pp. 4922–4930 (2017)

JND-Pano: Database for Just Noticeable Difference of JPEG Compressed Panoramic Images

Xiaohua Liu[1], Zihao Chen[1], Xu Wang[1(✉)], Jianmin Jiang[1],
and Sam Kowng[2]

[1] College of Computer Science and Software Engineering, Shenzhen University,
Shenzhen 518060, China
liuxiaohua1993@gmail, zihaocheniml@gmail.com,
{wangxu,jianmin.jiang}@szu.edu.cn
[2] Department of Computer Science, City University of Hong Kong,
Kowloon, Hong Kong
cssamk@cityu.edu.hk

Abstract. Just noticeable difference (JND) characterizes the minimum visibility threshold, which is important for the compression optimization and quality assessment of visual content. According to our best knowledge, there are no public JND model or related database specific to panoramic content, which has become a hot topic in recent years. To facilitate future researches of JND modeling of panoramic content, we explored the JND characteristics of JPEG compressed panoramic images in this paper. Considering the actual application scenario and the scale of experiment, we first establish a database consisting of 40 reference panoramic images and 4000 distorted panoramic images generated by JPEG encoder. Subsequently, a subjective experiment was conducted based on the subjective JND evaluation method. With the proposed database, the existing state-of-the-art JND models are further evaluated and analyzed. Finally, the performance comparison experiment indicates that it is necessary to focus on the particularity of panoramic content and explore new JND models, which can also provide new ideas for related research.

Keywords: Just noticeable difference · Virtual reality · Panoramic image

1 Introduction

With the emerging development of virtual reality (VR) technology, panoramic video (a. k.a. omnidirectional or 360° content) has become very popular for its immersive and interactive experience. When wearing head-mounted display (HMD), users can freely control the field of view (FoV) and focus on the attracted content by moving their heads. Due to the hardware limitation of HMD, the resolution of content displayed in the FoV (90°) for a 4 K panoramic image is approximately only 960x540. Since distance between the screen and human eyes are very close, the artifacts on the content maybe more visible and annoying due to the low spatial resolution. To guarantee the quality of experience (QoE) and make user feeling totally immersed, the requirements

© Springer Nature Switzerland AG 2018
R. Hong et al. (Eds.): PCM 2018, LNCS 11164, pp. 458–468, 2018.
https://doi.org/10.1007/978-3-030-00776-8_42

on the panoramic content such as resolution and frame rate are significantly increased, which occupy huge bandwidth [1]. Thus, high efficiency compression algorithm for panoramic content is highly demanded to tackle the bottleneck of storage and streaming.

Currently, there are many attempts try to exploit the structure or representation redundancy of panoramic video, such as designing compact project formats or unequal bit allocation according to the latitude location. Perceptual based compression scheme is another potential way to further reduce the perceptual redundancy by exploiting the masking effect of human visual system (HVS). Since just noticeable difference (JND) characterizes the minimum visibility threshold below which the pixel level variations cannot be perceived by the HVS [2], it is widely used as a compromise between visual quality and image/video compression techniques [3–6].

Since immersive media is still new to customers due to its high requirement on the hardware, existing research works on JND models are mainly focus on the traditional 2D visual content. According to our knowledge, there are no public JND model or related database specific to panoramic image. Since the content or order displayed on the screen/HMD is user dependent, immersive visual experience provided by panoramic content is significantly different. There is a need to build a database for understanding the JND characteristics for compressed panoramic images.

In this paper, the JND characteristics of JPEG compressed panoramic images are investigated. First, a database with 40 reference panoramic images is established to explore the JND characteristics in the field of VR. Second, the performance comparison of the state-of-the-art JND models are implemented and evaluated on our database. Experimental results show that the performance of existing JND models are limited, since they are not targeted for the panoramic content. There is a lot of room of further improving the prediction accuracy of JND model. The research community of image compression or image quality assessment may benefit from this new database.

The rest of this paper is organized as follows. Section 2 describes the database building and data processing in details. Section 3 presents the performance comparison results of existing JND models. Finally, conclusions are given in Sect. 4.

2   Description of Database Building

To investigate the JND characteristics of panoramic images, we firstly build a database with 40 reference panoramic images (denoted as JND-Pano). Then the psychophysics experiments are conducted on the JND-Pano database to obtain visible threshold of each panoramic images. The details are provided as follows.

2.1   Generation of Image Database

Since we need to find out the visibility threshold for compressed panoramic image, thus the reference panoramic image must be with high quality and resolution. First, we crawled and collected more than 300 high quality panoramic images from the internet. To guarantee the immersive experience, all the images are with the same resolution of

5000×2500. Based on the guidelines described in [7], 40 reference images were selected from the collection by considering the following two principles:

- A wide variety of scene should be included to provide a consistent experience with the real world. Figure 1 shows the thumbnail of reference panoramic images. The whole set of 40 reference panoramic images can be classified into seven class of scenes as described in Table 1. These scenes cover the day and night, land and sea, indoor and outdoor, natural scene images and computer-generated contents.
- The space constructed by the spatial information and colorfulness of all reference images should cover as large a range as possible. Figure 2 illustrates the spatial information (SI) and colorfulness index (CF) of these reference images. SI and CF are computed according to [7] and [8] respectively. It is observed that the coverage of JND-Pano is large and evenly distributed.

After obtaining the reference images, the JPEG compressed panoramic images are generated for subjective experiments. The compression process is implemented via MATLAB imwrite function by varying the quality factor (QF) from 1 to 100. As a result, each reference image corresponds to 100 JPEG compressed images. Final, the initial database consists of 4040 images in total.

Fig. 1. Thumbnail of reference panoramic images in JND-Pano database. These images, from left to right, from top to bottom, correspond to index from 1 to 40.

Table 1. Scenes descriptions of reference images in database

Scene Description	Number
Snow-covered Landscape	2
Dark Scene	8
Buildings on Land	5
Lake/Sea	6
Indoor	2
Seashore Buildings	9
Computer Generated Indoor Scenes	8

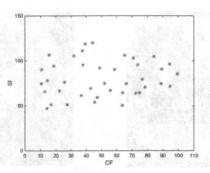

Fig. 2. Spatial information (SI) versus colorfulness (CF) for JND-Pano database.

2.2 Subjective Test Set-up and Procedure

Different from the subjective test of traditional display environment, viewers need to wear the HMD devices. The human-interaction interface of data collection and flow control need to be redesigned to make viewer feeling comfortable and immersive. Based on existing works [9–11] and our previous experience on conducting experiments for panoramic content, the following factors are taken into consideration in subjective test set-up and procedure designing:

- To simulate the real experience environment as much as possible, viewer need to sit on a rotatable seat in her/his most comfortable posture and does not fix the HMD direction and angle.
- Considering that this test is the first time to experience VR with HMD for most of the viewers, it is necessary to get viewers to be familiar with and adapt to the immersive environment before the formal test. Thus, the train session is included before the formal test, to make sure the subjects understand the purpose of experiment.
- In addition, due to the influence of the helmet weight, the screen refresh rate, the resolution of the helmet lens and the influence of individual factors, it is easy to experience the discomfort of visual fatigue and vertigo through the VR experience, so the appropriate interval between successive trials is necessary.

To clearly explain the subjective test set-up, a brief explanation including the test environment, methodology and procedure are provided as follows:

Test Environment and Platform. During the test, all the panoramic images are rendered in the GPU workstation and displayed in the HTC Vive device. Users wear the HMD and sit in a swivel chair, to freely switch FoV by changing their body or head position. A subjective test platform is implemented based on Unreal Engine (UE4), to collect data and control the test flow. Figure 3 provides the examples of graphical user interface (GUI) of the developed test platform. Viewers can use the HTC Vive controller to make decisions for rating.

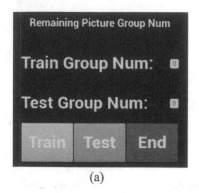
(a)

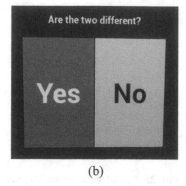
(b)

Fig. 3. GUI of test platform: (a) Start interface of each trial; (b) Rating interface.

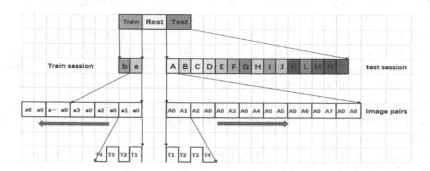

Fig. 4. Procedure of each session

Methodology and Procedure. During the test, aggressive binary search method [12] is employed to guide the JND point search process of each reference image. Based on the considerations mentioned-above, each session contains a train and test sub-session. As shown in Fig. 4, the train session contains two trials to make viewers familiar with the system and understand the task purpose. Then the test session will be started after taking a rest. To avoid subjects feeling fatigue and uncomfortable, only ten trials are contained in the test session. Thus, the total running time for each session is about 25–30 min. For each subject, the reference images of training session are randomly selected, and the data will be not recorded.

For each trial in both train and test sessions, the reference image and compressed image with anchor QF value are paired displayed with random order. During JND point searching, viewer is asked whether there is difference between the quality of the two images, which drives aggressive binary search algorithm until the JND point of the current reference image was found.

2.3 Data Processing and Result Analysis

During the test, the reference images are divided into four test sessions, then each test sessions contain ten reference images. A total of 42 viewers participate in the experiment and each reference image was viewed by at least 25 subjects.

After obtaining the rating results, these data cannot be directly used for JND analysis. There are many factors that can cause some experimental data to be different from most data, called outlier. For example, the observer maybe not really understand the concept of the quality difference in the experiment, or they maybe not familiar with the device and make wrong options. These outliers will affect the final JND processing. Therefore, we need to remove the outliers form the raw data firstly and then calculate the final JND value of each reference image.

Outlier Selection and Removal. The raw JND values of each session are separately processed and analyzed when selecting outlier. Refer to the method in [13], range (R) and standard deviation (SD) are calculated. Firstly, the z-score consistency check is adopted, which shows the consistency of an individual subject with respect to the majority. The z-score is defined as

$$z^m = (z_1^m, z_2^m, \ldots, z_N^m), z_n^m = \frac{J_n^m - \mu_n}{\sigma_n} \tag{1}$$

where J_n^m means JND value of subject m on image n. μ_n and σ_n separately mean average and standard deviation vectors against all subjects can be written as

$$\mu = (\mu_1, \mu_2, \ldots, \mu_N), \mu_n = \frac{1}{M} \sum_{m=1}^{M} J_n^m \tag{2}$$

$$\sigma = (\sigma_1, \sigma_2, \ldots, \sigma_N), \sigma_n = \sqrt{\frac{1}{M} \sum_{m=1}^{M} (J_n^m - \mu_n)^2} \tag{3}$$

where M and N mean the number of subjects and the number of reference images. Both R and SD of the z-score vector are used as the dispersion metrics, which are defined as

$$R = \max(z^m) - \min(z^m) \tag{4}$$

$$SD = std(z^m) \tag{5}$$

respectively. A larger dispersion indicates that the corresponding subject gives inconsistent evaluation results in the test. A subject is identified as an outlier if the associated range and SD values are both large. Total 13 outliers and partial or all of their data are removed.

Determination of JND Point of Reference Image. After removing all the outliers, the JND distribution of 40 reference images obtained. As shown in Fig. 5. the JND point of each image is the upper quarter point (*i.e.* only 25% subjects can tell the difference) of corresponding bar. Furthermore, we have the following observations.

- All JND points are within the range from 30 to 80. The JND value between 40 and 50 accounts for 62.5%; JND value between 50 and 60 accounts for 15%; JND value between 70 and 80 accounts for 12.5%; the other JND value accounts for 10%.
- These JND value is content dependent. The scenes with large smooth area (*e.g.* scene #18 in Fig. 1) where the color and texture changes are gentle or unchanged have large JND value. The scenes with complex spatial information (*e.g.* scene #12 in Fig. 1) have small JND values. This is reasonable due to the spatial masking effect.

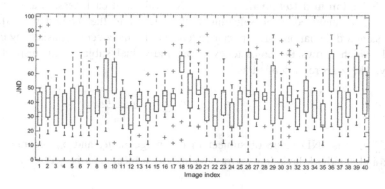

Fig. 5. Boxplot of all 40 reference images

3 Performance Evaluation on JND Models on JND-Pano Database

To investigate the performance of existing JND models on JPEG compressed panoramic images, five state-of-the-art pixel based JND models including NAMM [14], DMST (Decomposition Model for Separating edge and Textured regions) [15], FEP (Free-Energy Principle) [16], PMSU (Pattern Masking estimation in Image with Structural Uncertainty) [17] and PaCo (Pattern Complexity) [18] are evaluated on the

JND-Pano database. All the source codes are collected from the project website of references. The description of criterion for performance evaluation and the final performance comparison results are provided as follows.

3.1 Description of Criterion for Performance Measurement

To make fair performance comparison, the method described in [15] is employed to measure the prediction accuracy in terms of Ratio over Threshold (RoT), which indicates the number of pixels beyond the threshold. Given a reference image I and the corresponding compressed image J with QF = JND_I, a contaminated image I_c is firstly generated as

$$I_c(x) = I(x) + \beta \times rand(x) \times JND(x),$$ (6)

where rand(x) is a function randomly take +1 or −1. The value β regulates the energy of distortion and is depended on J [15]. Subsequently, the RoT is defined as

$$RoT = 1 - \frac{1}{N}\sum_{k=1}^{N} TM(k)$$ (7)

where N means the number of pixels within the image. The threshold map $TM(x)$ indicates whether the value of current pixel x in the contaminated image beyond the threshold, which is defined as

$$TM(x) = \begin{cases} 1, & I_c(x) \in [\min(I(x), J(x)), \max(I(x), J(x))] \\ 0, & otherwise \end{cases}$$ (8)

Based on the definition above, the JND model with lower RoT means that it can predict the threshold more accuracy.

3.2 Experimental Results

Given the predicted JND values of each reference image of the proposed database calculated by existing JND models, the RoT values can be obtained according to Eqs. (6)–(8), which measure the performance of existing JND models. The mean and standard deviation of RoTs of each JND model on all images are summarized in Table 2. The performance comparison results on the JND-Pano database are summarized in Table 2. It is observed that 80% of the pixels predicted by existing JND models are not in the threshold range. In other words, most of the subjects will notice the difference between the contaminated panoramic images generated by JND models and the reference panoramic image. For better understanding the limitation of existing JND models, panoramic images and the corresponding TM maps generated by Model NAMM are provided in Fig. 6. It is observed that the JND model are failed in predicting the safety bound for compressed panoramic image, especially in the smooth areas. This may be caused by following reasons.

First, the existing JND models take more consideration on texture regions and measurement of spatial masking effects in a variety of ways, but JPEG compress operation may cause serious distortions on smooth areas. Second, the way of experience panoramic images is totally different from that of traditional 2D images. When wearing HMD devices, viewer can freely change and choose the FoVs with attracted content. Then viewers will be more sensitivity to the distortion.

Based on the results of performance comparison, we can make a conclusion that existing pixel based JND models cannot be directly applied for predicting the safety bound of JPEG compressed panoramic images. In the future study of JND mode designing, the impact of immersive experience to the viewing behavior of user should be considered. Besides, more attention should be paid to low-frequency areas, such as sky in the outdoor scene. For texture regions, further study of the tolerance of HVS on the distortion of regular and irregular texture regions, is a potential way to further improve the performance of JND model.

Table 2. Summarized results of performance comparison in terms of RoT.

Method	NAMM [14]	DMST [15]	FEP [16]	PMSU [17]	PaCo [18]
Average	0.8540	0.8460	0.8610	0.8420	0.8210
Standard Deviation	0.0435	0.0433	0.0367	0.0433	0.0440

Fig. 6. Examples of reference panoramic images and the corresponding TM maps.

4   Conclusion

In this paper, we conducted a subjective experiment and build a database for investigating the JND characteristics of JPEG compressed panoramic images. The final database consists of 40 reference panoramic images and the corresponding JND value.

Based on the performance evaluation and comparison of existing pixel-based JND models, it can be concluded that there is still a lot of room to further investigated on the JND modeling for JPEG compressed panoramic images. In the future, we will try to design JND model specific to panoramic images based on the JND-Pano database.

Acknowledgements. This work was supported in part by the National Natural Science Foundation of China under Grant 31670553, 61501299, 61672443 and 61620106008, in part by the Guangdong Nature Science Foundation under Grant 2016A030310058, in part by the Shenzhen Emerging Industries of the Strategic Basic Research Project under Grants JCYJ20160226191842793, in part by the Natural Science Foundation of SZU (grant no. 827000144), and in part by the Tencent "Rhinoceros Birds"-Scientific Research Foundation for Young Teachers of Shenzhen University.

References

1. Zare, A., Aminlou, A., Miska, M.H.: Virtual reality content streaming: Viewport-dependent projection and tile-based techniques. In: 2017 IEEE International Conference on Image Processing (ICIP), Beijing, China, pp. 1432–1436. IEEE (2017)
2. Wang, S., et al.: Just noticeable difference estimation for screen content images. IEEE Trans. Image Process. **25**(8), 3838–3851 (2016)
3. Yang, X., et al.: Rate control for videophone using local perceptual cues. IEEE Trans. Circuits Syst. Video Technol. **15**(4), 496–507 (2005)
4. Zhang, X., et al.: Just-noticeable difference-based perceptual optimization for JPEG compression. IEEE Signal Process. Lett. **24**(1), 96–100 (2017)
5. Chou, C., Li, Y.: A perceptually tuned subband image coder based on the measure of just-noticeable-distortion profile. IEEE Trans. Circuits Syst. Video Technol. **5**(6), 467–476 (1995)
6. Chen, Z., Guillemot, C.: Perceptually-friendly H.264/AVC video coding based on foveated just-noticeable-distortion model. IEEE Trans. Circuits Syst. Video Technol. **20**(6), 806–819 (2010)
7. Winkler, S.: Analysis of public image and video databases for quality assessment. IEEE J. Sel. Top. Signal Process. **6**(6), 616–625 (2012)
8. Hasler, D., Süsstrunk, S.: Measuring colourfulness in natural images. In: Electronic Imaging 2003, Santa Clara, CA, United States, vol. 5007, pp. 9. SPIE (2003)
9. Xu, M., Li, C., Liu, Y., Deng, X., Lu, J.: A subjective visual quality assessment method of panoramic videos. In: 2017 IEEE International Conference on Multimedia and Expo (ICME), Hong Kong, China, pp. 517–522. IEEE (2017)
10. Perrin, A.N.M., Bist, C., Cozot, R., Ebrahimi, T.: Measuring quality of omnidirectional high dynamic range content. In: SPIE Optical Engineering + Applications, vol. 10396, pp. 18. SPIE, San Diego, California, United States (2017)
11. Upenik, E., Řeřábek, M., Ebrahimi, T.: Testbed for subjective evaluation of omnidirectional visual content. In: Picture Coding Symposium (PCS), Nuremberg, Germany, pp. 1–5. IEEE (2016)
12. Wang, H., et al.: MCL-JCV: a JND-based H.264/AVC video quality assessment dataset. In: 2016 IEEE International Conference on Image Processing (ICIP), Phoenix, AZ, USA, pp. 1509–1513. IEEE (2016)
13. Wang, H., et al.: VideoSet: a large-scale compressed video quality dataset based on JND measurement. J. Vis. Commun. Image Represent. **46**, 292–302 (2017)

14. Yang, X., Lin, W., Ong, E.P., Yao, S.: Just-noticeable-distortion profile with nonlinear additivity model for perceptual masking in color images. In: Acoustics, Speech, and Signal Processing, Hong Kong, China, pp. 609–612. IEEE (2003)
15. Liu, A., Lin, W., Paul, M., Deng, C., Zhang, F.: Just noticeable difference for images with decomposition model for separating edge and textured regions. IEEE Trans. Circuits Syst. Video Technol. 20(11), 1648–1652 (2010)
16. Wu, J., Shi, G., Lin, W., Liu, A., Qi, F.: Just noticeable difference estimation for images with free-energy principle. IEEE Trans. Multimed. 15(7), 1705–1710 (2013)
17. Wu, J., Lin, W., Shi, G., Wang, X., Li, F.: Pattern masking estimation in image with structural uncertainty. IEEE Trans. Image Process. 22(12), 4892–4904 (2013)
18. Wu, W., et al.: Enhanced just noticeable difference model for images with pattern complexity. IEEE Trans. Image Process. 26(6), 2682–2693 (2017)

HDP-Net: Haze Density Prediction Network for Nighttime Dehazing

Yinghong Liao, Zhuo Su$^{(\boxtimes)}$, Xiangguo Liang, and Bin Qiu

School of Data and Computer Science, National Engineering Research Center of Digital Life, Sun Yat-sen University, Guangzhou, China
suzhuo3@mail.sysu.edu.cn

Abstract. Nighttime dehazing is a challenging ill-posed problem. Affected by unpredictable factors at night, daytime methods may be incompatible with night haze removal. In this paper, we propose an end-to-end learning-based solution to remove haze from night images. Different from the most-used atmospheric scattering model, we use a novel model to represent a night hazy image. We first present an estimator to predict the haze density in patches of the image. Based on this, a CNN, called Haze Density Prediction Network (HDP-Net), is adopted to obtain a haze density map so that it can be subtracted by the original hazy input to generate the desired haze-free output. The range of hue in night images may be altered by artificial light sources. To improve the dehazing capability in the certain range of hue, we devise four datasets under white light and yellow light conditions for network training. Finally, our method is compared with the state-of-the-art nighttime dehazing methods and demonstrated to have a superior performance. The project is available at https://github.com/suzhuoi/HDP-Net.

Keywords: Nighttime image dehazing · Image enhancement
Density prediction · Convolutional neural network

1 Introduction

Haze is a common atmospheric phenomenon caused by dust, smoke and other particles in the air. Light scattering and light attenuation caused by haze can result in the severe degradation of the visibility of images and videos. In particular, for a nighttime photograph which suffers from low ambient illumination and poor visibility, its visual quality can be further seriously affected by the presence of haze. Therefore, how to effectively remove haze in the night image is a challenging issue in image enhancement and is of great significance to the applications such as self-driving and traffic surveillance.

This research is supported by the National Natural Science Foundation of China (61502541, 61772140, 61402546), the Natural Science Foundation of Guangdong Province (2016A030310202), and the Fundamental Research Funds for the Central Universities (Sun Yat-sen University, 16lgpy39).

© Springer Nature Switzerland AG 2018
R. Hong et al. (Eds.): PCM 2018, LNCS 11164, pp. 469–480, 2018.
https://doi.org/10.1007/978-3-030-00776-8_43

In recent years, some daytime dehazing methods [2, 4, 12] mainly remove haze by the atmospheric scattering model [8] which describes a hazy image as the linear combination of the direct attenuation term and scattering term. These methods could recover a haze-free image via estimating the unknown parameters in the model (the transmission map and the atmospheric light). The atmospheric light is estimated from the brightest region of the image and considered as consistent globally. However, the estimation of the atmospheric light during night time faces great difficulty in two aspects: the low illuminance intensity of the natural atmospheric light and the existence of artificial light sources. Without the accurate value, the error from the estimation thus becomes the main cause of color distortion in dehazed images. On the other hand, daytime images have wide range of hue in terms of that in HSV color model, because white light generated by the sun is the composite of various monochromatic lights. But in the dark environment, the presence of artificial light sources will cause the range of hue to shrink, e.g., a yellow street light may turn the picture yellowish. Several researchers have proposed some nighttime dehazing methods [6, 10, 14, 15] that are mainly based on the atmospheric scattering model, but color distortion is still existed due to their excessive dehazing. Therefore, our goal is to design a nighttime dehazing model that can address color distortion issue and restore a haze-free image close to the reality without estimating the atmospheric light. And our model could be effective for haze removal in night images with different range of hue.

In this paper, we propose an end-to-end Haze Density Prediction Network (HDP-Net) to perform nighttime dehazing. Instead of estimating the transmission map and the atmospheric light in the atmospheric scattering model [8], we create a novel prediction function to obtain the haze density map of a night image. The haze-free map could be recovered after the original hazy image subtracts the haze density map. Distinguished from DehazeNet [1], AOD-Net [5] and DCPDN [13], we adopt a network architecture similar to a fully convolutional network (FCN) to estimate haze density. In HDP-Net, two shortcuts are added to connect same-size feature maps to reduce training error, and three structures resembling the bottleneck in ResNet [3] are applied to make the network more trainable. To make data-driven HDP-Net work effectively on an image with varying range of hue, we design four datasets of night synthetic hazy images under white light and yellow light environment for training respectively. Based on large-scale data, HDP-Net could accurately estimate the haze density and remove the night haze.

In summary, the contributions of our work are as follows:

- Propose a haze density prediction function and a novel model describing a night hazy image that is different with the atmospheric scattering model.
- Design a network architecture based on the model above to obtain a haze density map.
- Devise datasets NightHaze under white light condition and YellowHaze under yellow light condition to make the network dehaze effectively for different range of hue of night hazy images.

2   Related Work

In general, dehazing methods are divided into two categories: prior-based methods and data-driven methods. We will introduce the application of these two types of methods in haze removal in day time and night time, respectively.

Daytime Dehazing. Early approaches are based on various image priors. They use the atmospheric scattering model to represent a hazy image and explore the cues between the parameters in the model and the features from the image, such as the Dark Channel Prior (DCP) [4], the Maximize Contrast [12], the Color Lines [2]. In recent years, with the extensive application of deep learning in the field of computer vision, researchers began to apply the convolutional neural network (CNN) to image dehazing. The CNN-based methods are trained with designed hazy inputs and given haze-free outputs so that the optimal value of the parameters or their derived ones could be obtained after limited times of training iteration. For instance, Cai et al. proposed DehazeNet in [1], Li et al. introduced AOD-Net in [5], Zhang and Patel presented DCPDN in [13]. Some approaches remove haze without the atmospheric scattering model, e.g. Ren et al. [11] proposed Gated Fusion Network (GFN).

Nighttime Dehazing. The majority of daytime methods might not be suitable for nighttime dehazing, though many methods have presented effectiveness under daytime environment. Affected by complex and unpredictable factors of night, e.g., insufficient intensity and imbalanced distribution of atmospheric light, the removal of night haze is more challenging. There are a few approaches in eliminating haze from night images and they are almost prior-based. Pei and Lee [10] first turned the original hazy image into the gray one and dehazed it with the DCP [4]. Li et al. [6] proposed the Glow and Multiple Light Colors (GMLC) by adding a glow term to the atmospheric scattering model. The glow term is estimated by APSF function in [9] and used to reduce the effects from the glow generated by artificial light sources. The DCP is used as the final step to finish dehazing. Zhang et al. [15] presented a New Imaging Model (NIM), which initially adopted a light compensation on hazy inputs to obtain an illumination balanced one and then dealt it with a color correction step. The method is ended with the same DCP operation. Furthermore, Zhang et al. [14] proposed another approach that used Maximum Reflectance Prior (MRP) to remove night haze. This method estimates the ambient illumination of image by maximizing the reflectance. With the calculated ambient illumination, color effects are removed and the transmission is then estimated. The haze-free image is obtained by using the DCP finally. The DCP that these methods use is sensitive to the extreme value in local patch and cannot produce ideal dehazing effects in the region such as sky. Based on the atmospheric scattering model, these methods usually obtain the image with severe color distortion.

3 Our Method

In this section, we illustrate our nighttime dehazing method. We first analyze limitations of the atmospheric scattering model in the removal of night haze. Then we introduce the designed haze density prediction function. With the function, we propose a novel model to describe a night hazy image and verify its rationality. Finally, we give the details of our proposed nighttime dehazing network, HDP-Net.

3.1 Limitations of the Atmospheric Scattering Model

Most daytime dehazing methods adopt the atmospheric scattering model [8] to represent a hazy image, its form is as follows

$$I(x) = J(x)t(x) + A(1 - t(x)), \tag{1}$$

where $I(x)$ denotes the observed hazy map, $J(x)$ denotes the desired haze-free map, A is the intensity of the atmospheric light and $t(x)$ is the medium transmission. Some daytime haze removal approaches, including prior-based methods and data-driven methods, restore a haze-free image by the estimation of unknown variables, the transmission $t(x)$ and the atmospheric light A in Eq. (1).

However, this model has two limitations of in nighttime dehazing.

Difficult Estimation of the Atmospheric Light A. Natural daytime atmospheric light is parallel light with high intensity, so the value of A could be obtained from the brightest pixels of the image and set as consistent globally. However, natural atmospheric light is not strong at night and there are also artificial light sources with strong glow. The bright patches formed by artificial light sources and the dark areas generated by insufficient night atmospheric light make the uneven distribution of image illumination. Therefore, the atmospheric light A cannot be estimated from the brightest regions and is hard to be predicted.

Color Distortion. Atmospheric light A in Eq. (1) could be represented by RGB and HSV color model. Hue is the parameter in HSV color model that describes the colorfulness of an image and is defined as follows:

$$tan(h) = \frac{\sqrt{3} \cdot (G - B)}{2R - G - B}, \tag{2}$$

where h is hue whose range is $[0, \pi]$. R, G and B are the red, green and blue channel of an image, respectively. An image is considered colorful by the wide range of hue. Natural daytime atmospheric light is white light, which is composed of multiple monochromatic lights, thus the range of hue is wide in daytime hazy images. In night time, there is no strong atmospheric light but artificial light sources. Artificial light sources may narrow down the range of hue, e.g., street light with low color temperature (2700 K–3200 K) produces yellow light that makes the picture to become yellow. If the estimation of A is not accurate, the hue of A and the dehazing result are affected and color distortion occurs consequently. After the analysis, a new model is needed to perform nighttime dehazing.

3.2 Haze Density Prediction Function

To address the issue of night haze, we design a haze density prediction (HDP) function as follows:

$$I_h = P(I(x)), \tag{3}$$

where $I_h(x)$ denotes the haze density, $P(\cdot)$ denotes the proposed prediction function that estimates haze density. The range of haze density is $[-1, 1]$, which will be proved in Sect. 3.4.

The function is hard to be hand-designed because the density and the distribution of night haze vary in images. Thus we use CNN to build a network architecture that constructs HDP function. In one thing, CNN can perform complex non-linear transformations and use shared parameters to estimate varying haze density. In other thing, CNN is data-driven, which obtains optimal parameters for various conditions.

3.3 Nighttime Hazy Image Model with HDP Function

According to the composition of hazy image, our model is re-defined as:

$$I(x) = J_s(x) + J_t(x) + I_h(x), \tag{4}$$

where J_s and J_t are the structure and texture of haze-free map respectively, I_h is the haze density. With Eq. (4), the desired dehazing result $J(x)$ can be represented as:

$$J(x) = I(x) - I_h(x). \tag{5}$$

Therefore, we first obtain a haze density map $I_h(x)$ by Eq. (5). Then we recover a night haze-free image $J(x)$ from the hazy input $I(x)$ by subtracting $I_h(x)$ with the preserved structure $J_s(x)$ and texture $J_t(x)$. Compared with the atmospheric scattering model in Eq. (1), our model predicts only the haze density $I_h(x)$ of the hazy map without computing the transmission $t(x)$ and the atmospheric light A. Thus the cost of computing two parameters is avoided and color distortion caused by estimating A could be prevented.

3.4 The Rationality of Nighttime Dehazing with HDP Function

We can obtain another form of haze density $I_h(x)$ by Eq. (5):

$$I_h(x) = I(x) - J(x). \tag{6}$$

The rationality of night haze removal with HDP function can be demonstrated by proving the range of the haze density is $[-1, 1]$.

Proof. Here the atmospheric scattering model in Eq. (1) is introduced to prove the rationality of the HDP function. The HDP function could be derived from the atmospheric scattering model but it advoids the computation of the atmospheric light A and the transmission $t(x)$.

From the atmospheric scattering model in Eq. (1), we obtain

$$I_h(x) = J(x)t(x) + A(1 - t(x)) - J(x). \tag{7}$$

The normalization is usually performed on an image for network training (divided by 255). Thus the range of $I(x)$, $J(x)$ and A is narrowed down to $[0, 1]$. From Eq. (7) we have

$$I_h(x) = (A - J(x))(1 - t(x)). \tag{8}$$

The range of $t(x)$ is $[0, 1]$, so $(1 - t(x))$ is in $[0, 1]$. In daytime, A is obtained from the brightest pixels in hazy image $I(x)$, that is $A \geq I(x)$. And we get

$$A \geq J(x)t(x) + A(1 - t(x)). \tag{9}$$

From Eq. (9) we have

$$(J(x) - A)t(x) \leq 0. \tag{10}$$

Since $t(x) \geq 0$, so $(J(x) - A) \leq 0$, then $A \geq J(x)$ is obtained. In the daytime, the atmospheric light A is in the range of $[0.6, 1]$ and $(A - J(x))$ is thus $[0, 1]$. But A cannot be estimated in the night image by the brightest patch and $A \geq J(x)$ is not true. The range of A and $J(x)$ are both $[0, 1]$, so the term $(A - J(x))$ is $[-1, 1]$. And $(1 - t(x))$ is $[0, 1]$, thus the range of Eq. (8) is $[-1, 1]$. Therefore, haze density range is proved as $[-1, 1]$. The haze density map could be obtained by extracting the feature value in $[-1, 1]$ by CNN without the estimation of A and $t(x)$. And the model could be derived from the atmospheric scattering model, the rationality of our model is demonstrated.

3.5   Network Architecture

The purpose of HDP function is to estimate haze density map from a night hazy image and preserve image details. After a series of experiments, we finally devise a network for nighttime dehazing with HDP function.

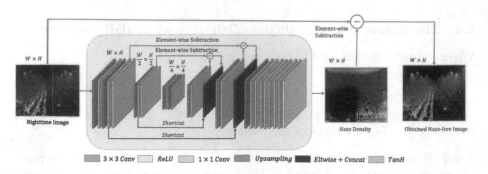

Fig. 1. The architecture of HDP-Net.

The network architecture we designed is illustrated in Fig. 1. The network structure includes 11 convolutional layers, each of which except the last layer is followed by a ReLU as non-linear activation function. Different kernels, strides and pads are applied in convolutional layers to realize three functions. Feature extraction layers are used with kernel 3, stride 1 and pad 1. Pooling layers with kernel 1, stride 2 and pad 0 have the same down-sampling operation to generate half-size feature maps. Mapping layers with kernel 1, stride 1 and pad 0 perform multistage mapping to obtain a color haze density map that matches the size of the input. In addition, we import two deconvolutional layers to do up-sampling so that the feature map size is recovered. Details and colors of the image are distributed in low-level features, so we bring two shortcuts to concatenate same-size feature maps for retaining low-level features. And three subtraction structures are appended to make the network more trainable, which are similar to ResNet bottleneck [3]. TanH with range $[-1, 1]$ is adopted as the activation function to control haze density range in the final layer.

Overall, our network architecture resembles the fully convolutional network (FCN) [7]. Inspired by FCN, we design this network to predict haze density map. The layer number of HDP-Net is based on training images with the size of 128×128. Satisfying dehazing results could be produced after the multi-scale operations of feature extraction, fusion and mapping by 11 layers. The details of HDP-Net architecture are presented in Table 1.

Table 1. The details of HDP-Net architecture.

Formulation	Type	Input Size	Num	Filter	Stride	Pad
Feature Extraction	Conv	$3 \times 128 \times 128$	8	3×3	1	1
		$8 \times 128 \times 128$	16	3×3	1	1
		$16 \times 128 \times 128$	32	1×1	2	0
		$32 \times 64 \times 64$	32	3×3	1	1
		$32 \times 64 \times 64$	64	1×1	2	0
		$64 \times 32 \times 32$	64	3×3	1	1
		$64 \times 32 \times 32$	64	3×3	1	1
Fusion	Deconv	$64 \times 32 \times 32$	32	2×2	2	0
		$32 \times 64 \times 64$	16	2×2	2	0
Feature Extraction	Conv	$16 \times 128 \times 128$	16	3×3	1	1
Mapping	Conv	$16 \times 128 \times 128$	16	1×1	1	0
		$16 \times 128 \times 128$	8	1×1	1	0
		$8 \times 128 \times 128$	3	1×1	1	0

4   Experiment

This section presents the experiments on our proposed HDP-Net. We introduce four synthetic datasets of night hazy images for training. Then we make quantitative and qualitative experiments to compare our methods with state-of-the-art nighttime dehazing methods on synthetic images and real images, respectively.

4.1   Implementation Details

Caffe stands out with its excellent support for CNN and outstanding capability of training lots of classical models. Thus, we use the Caffe implementation to train the designed network. The cost function in training is defined as follows:

$$Loss = \sum \|J - (I - I_{conv})\|_2, \tag{11}$$

where I_{conv} represents the output of our network, standing for the predicted haze density. Mean squared error (MSE) is adopted to measure the difference between night hazy image and real haze-free image to supervise the training. The calculation of $I - I_{conv}$ could be implemented by element-wise layer in Caffe.

To train the network efficiently, we use Xavier Filler to initialize the parameters of our network. The training error is decreased by optimization algorithm of Stochastic Gradient Descent (SGD) during back propagation. We found that the constant learning rate smaller than 0.001 could have ideal convergence effects, but the speed of the convergence is relatively slow. To achieve better convergence effects, the learning rate is initially set as 0.001 to accelerate the convergence and gradually decreases for better optimal parameters when the training error is stable in a certain range. Thus the learning policy is set as *inv* with initial learning rate as 0.001, gamma as 0.001 and power as 0.75. The momentum and weight decay are fixed as 0.9 and 0.005, respectively, so that the optimization algorithm of SGD could produce more stable and fast convergence.

4.2   Training Datasets

The selected training dataset is important for the data-driven nighttime dehazing network. In daytime dehazing, DehazeNet [1], AOD-Net [5] and DCPDN [13] designed hazy and haze-free datasets for experiments. But few hazy and haze-free datasets in night time have been designed so far. Our method does not predict the value of transmission and atmospheric light, so the problem of violating the laws of physics caused by range is not taken into account. Predicting haze density by our model has no limitations to use the synthetic datasets obtained by Eq. (1). Moreover, we also focus on dehazing effects on the hazy images under the nighttime yellow light condition to improve the performance of HDP-Net on dark images in narrow range of hue.

Our datasets are based on bright and clean nighttime photographs, including night view of cities, night scenes of streets and other common nighttime images.

NightHaze-1   NightHaze-2   YellowHaze-1   YellowHaze-2

Fig. 2. Samples of NightHaze-1, NightHaze-2, YellowHaze-1 and YellowHaze-2. (Color figure online)

We select 10,000 of them as a basic training dataset and then use the atmospheric scattering model to generate the corresponding synthetic hazy images. We roughly estimate the value of the transmission $t(x)$ and the atmospheric light A in natural night environment by using the DCP [4]. After the statistical analysis and the experimental observation, we find out the main range of atmospheric light A and transmission $t(x)$ are [0.6, 0.9] and [0.7, 0.4], respectively. In order to explore night haze removal under yellow light environment, the hue of images is also modified by HSV color model, and the range is defined in [0.05, 0.10], which is within the range of yellow light.

For the convenience of training, we set the image size to 128 × 128. Based on different hazing effects and considering the impacts of yellow light, we generate four hazy image datasets. NightHaze-1, the collected night images are globally added by haze with the same density. NightHaze-2, the collected images are locally hazed by various density. A picture is divided into 32 × 32 or smaller patches where the haze is added at a certain probability, and some patches have no haze. The purpose is to improve the capability of the network to detect haze density. YellowHaze-1 and YellowHaze-2 are processed in the same manner as NightHaze-1 and NightHaze-2, except the hue is modified as yellow. The corresponding samples in each dataset are shown in Fig. 2.

4.3   Quantitative Evaluation on Synthetic Dataset

To quantify the dehazing effects of the network, we pick out the remaining 944 pictures from our collections to synthesize hazy images for the test, with the size of 480 × 740. These pictures are in yellow light and white light environment, respectively. The dehazing results are then in the comparison with those generated by other night haze removal models. Two indicators are used to measure the dehazing effects: Peak Signal-to-Noise Ratio (PSNR) and Structural Similarity (SSIM).

At present, mainstream nighttime dehazing methods are prior-based and we select representative three of them for comparison. They are Glow and Multiple Light Colors (GMLC) [6], New Imaging Model (NIM) [15] and Maximum Reflectance Prior (MRP) [14].

Dehazing results of some night hazy images by different methods are illustrated in Fig. 3. Three images with night haze are shown in Fig. 3(a) and the

(a) Hazy Inputs   (b) GMLC [6]   (c) NIM [15]   (d) MRP [14]   (e) Ours

Fig. 3. The dehazing results on synthetic dataset. (Color figure online)

obtained results by GMLC [6], NIM [15] and MRP [14] are displayed in Fig. 3(b)–
(d). Our dehazing images are given in Fig. 3(e), the first of which is the output
from trained YellowHaze and the last two of which are the effects by trained
NightHaze. The first in Fig. 3(a) is a hazy map under yellow light. The dehazing
results by NIM in Fig. 3(c) and MRP in Fig. 3(d) have obvious color distortion.
GMLC in Fig. 3(b) can effectively extract details, such as the floor textures and
the reduced glow of light. By comparison, our result is closer to the original hue
and has clear object outlines, such as the lines of the bridge. The second picture
in Fig. 3(a) is a hazy image under night white light. GMLC and NIM are shown
to be sensitive to white light and have excessive dehazing effects, resulting in the
severe color distortion like whitened sky. In contrast, our method could remove
the haze that covers the picture while retaining the original color composition.
The last picture in Fig. 3(a), a challenging synthetic night hazy image is used to
test the dehazing effects of these three methods and our method, by adding haze
with different density in patches. The results show that GMLC, NIM and MRP
could not well remove the haze in varying density. Our method could basically
remove the low-density haze and reduce the high-density haze while retaining
the hue of the original image.

The quantitative evaluation of all methods is displayed in Table 2. It is obvious that our method has the best results in two indicators. Prior-based methods
produce good effects in extracting details but have the problem color distortion
due to the excessive haze removal, which results in low value of PSNR and SSIM.

Table 2. The average PSNR and SSIM in dehazing results on synthetic dataset.

	GMLC [6]	NIM [15]	MRP [14]	Ours
PSNR	6.759	11.090	11.853	**15.984**
SSIM	0.121	0.290	0.159	**0.589**

4.4   Qualitative Evaluation on Real Images

To further evaluate the effectiveness of HDP-Net, in Fig. 4, we select images under yellow light and white light in [6,14,15] to give a qualitative comparison on real images. Figure 4(a) gives night hazy images. Figure 4(b)–(d) show results obtained by GMLC [6], NIM [15] and MRP [14], Fig. 4(e) displays the effects of our method. In Fig. 4(b)–(d), most of the haze is removed, and the details of the objects and scenes are well restored, but there are phenomena of over-enhancement (e.g., the sky areas of the fourth pictures in Fig. 4(c) and (d)). In addition, NIM and MRP could not deal well with the effects of artificial light sources, and there is a noticeable color distortion in the first picture. Light source glow is processed better by GMLC, but the color distortion is more severe in global, especially in pictures under yellow light. Prior-based methods in Fig. 4(c) and (d) inevitably overestimate features of images, such as transmission and reflectance, resulting in the change of colors. In contrast, our method could estimate features accurately, and colors and details of pictures are well retained.

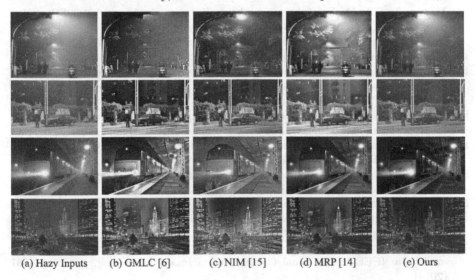

(a) Hazy Inputs   (b) GMLC [6]   (c) NIM [15]   (d) MRP [14]   (e) Ours

Fig. 4. Dehazing effects on real images. (Color figure online)

5   Conclusion

In this paper, we present a novel data-driven method to address nighttime dehazing and design a corresponding network model. We also design some datasets to train the network and test its dehazing effects at night. For further evaluation of night images with different hue range, we devise datasets under night yellow light and white light respectively. Our method are compared with other nighttime dehazing methods. The results in data measurement and visual quality prove that HDP-Net has a better performance than the current nighttime dehazing methods.

References

1. Cai, B., Xu, X., Jia, K., Qing, C., Tao, D.: Dehazenet: an end-to-end system for single image haze removal. IEEE Trans. Image Process. **25**(11), 5187–5198 (2016)
2. Fattal, R.: Dehazing using color-lines. ACM Trans. Graph. **34**(1), 13:1–13:14 (2014)
3. He, K., Zhang, X., Ren, S., Sun, J.: Deep residual learning for image recognition. In: IEEE Computer Vision and Pattern Recognition, pp. 770–778 (2016)
4. He, K., Sun, J., Tang, X.: Single image haze removal using dark channel prior. In: IEEE Computer Vision and Pattern Recognition, pp. 820–824 (2009)
5. Li, B., Peng, X., Wang, Z., Xu, J., Feng, D.: Aod-net: All-in-one dehazing network. In: IEEE International Conference on Computer Vision, pp. 4780–4788 (2017)
6. Li, Y., Tan, R.T., Brown, M.S.: Nighttime haze removal with glow and multiple light colors. In: IEEE International Conference on Computer Vision, pp. 226–234 (2015)
7. Long, J., Shelhamer, E., Darrell, T.: Fully convolutional networks for semantic segmentation. In: IEEE Computer Vision and Pattern Recognition, pp. 3431–3440 (2015)
8. Mccartney, E.J.: Optics of the Atmosphere: Scattering by Molecules and Particles, p. 421. Wiley, New York (1976)
9. Narasimhan, S.G., Nayar, S.K.: Shedding light on the weather. In: IEEE Computer Vision and Pattern Recognition, pp. 665–672 (2003)
10. Pei, S.C., Lee, T.Y.: Nighttime haze removal using color transfer pre-processing and dark channel prior. In: IEEE International Conference on Image Processing, pp. 957–960 (2012)
11. Ren, W., Ma, L., Zhang, J., Pan: Gated fusion network for single image dehazing. In: IEEE Computer Vision and Pattern Recognition (2018)
12. Tan, R.T.: Visibility in bad weather from a single image. In: IEEE Computer Vision and Pattern Recognition, pp. 1–8 (2008)
13. Zhang, H., Patel, V.M.: Densely connected pyramid dehazing network. In: IEEE Computer Vision and Pattern Recognition (2018)
14. Zhang, J., Cao, Y., Fang, S., Kang, Y., Chen, C.W.: Fast haze removal for nighttime image using maximum reflectance prior. In: IEEE Computer Vision and Pattern Recognition, pp. 7016–7024 (2017)
15. Zhang, J., Cao, Y., Wang, Z.: Nighttime haze removal based on a new imaging model. In: IEEE International Conference on Image Processing, pp. 4557–4561 (2014)

Reflectance Reference for Intra-Frame Coding of Surveillance Video

Dong Liu[1]([✉]) [iD], Zhenxin Zhang[1], Fangdong Chen[2], Houqiang Li[1], and Feng Wu[1]

[1] CAS Key Laboratory of Technology in Geo-Spatial Information Processing and Application System, University of Science and Technology of China, Hefei 230027, China
dongeliu@ustc.edu.cn
[2] Hikvision, Hangzhou 310052, China

Abstract. Surveillance videos have the feature of static background, which has been exploited in surveillance video coding by the background reference techniques. However, in long surveillance videos that last for several hours to several days, the appearance of background is varying due to the change of illumination, which incurs inefficiency of the background reference techniques. Moreover, intra frames in video coding cannot make use of the background reference and thus incur bit-rate burst. To solve this problem, we propose to separate illuminance out of appearance, and introduce a new kind of references known as *reflectance reference* (RefRef). RefRef consists of the reflectance component of the background in surveillance videos. The RefRef is invariant to illumination change, and thus can be put always in memory when encoding/decoding the surveillance videos of a specific camera. Especially, RefRef can be used for intra frames, together with the encoding/decoding of the illuminance component. Since the illuminance is usually more smooth than the appearance, RefRef provides higher compression efficiency for intra frames. Experimental results show that the RefRef method leads to more than 50% BD-rate reduction, compared to the High Efficiency Video Coding (HEVC) anchor, on typical surveillance videos under all-intra configuration.

Keywords: Illuminance · Intra coding · Reflectance reference Surveillance video · Video coding

1 Introduction

In recent years, security concerns have driven large-scale installations of high-definition surveillance cameras, which generate a huge amount of surveillance video data continuously [3]. Currently, such surveillance videos are usually compressed by the existing video coding schemes, such as H.264/MPEG-4 AVC [10] or High Efficiency Video Coding (HEVC) [11], but these schemes are designed

© Springer Nature Switzerland AG 2018
R. Hong et al. (Eds.): PCM 2018, LNCS 11164, pp. 481–491, 2018.
https://doi.org/10.1007/978-3-030-00776-8_44

for generic videos and may be not efficient enough for surveillance videos. Thus, it has been a topic of increasing interests to customize video coding methods for surveillance videos to address the challenge of storage and transmission of them.

Surveillance videos are captured by fixed cameras that usually have little motion, and then the background in surveillance videos is static. This feature can be exploited to improve the compression efficiency by adopting the so-called background reference techniques. Specifically, background reference is a special kind of long-term references proposed for surveillance video coding [2,9,12]. This reference is intended to represent the static background, and thus can be used for a longer while than the usual short-term references. Many research efforts have been put on the generation of background references. For example, Zhang et al. [12] proposed generating a background picture by simply averaging many pictures pixel by pixel. Paul et al. [9] proposed modeling pixels of many pictures at a same position using a Gaussian mixture distribution and generating a background picture using the most probable pixel values. In these methods, the generated background picture is required to be compressed at high quality, which incurs bit-rate burst. To solve this problem, Chen et al. [2] proposed a block-composed background reference, which generates the background reference progressively by selecting background blocks from each encoded frame and composing them. These background reference techniques have been verified to be efficient in surveillance video coding.

However, when compressing long surveillance videos that last for several hours to several days, the existing background reference techniques still have troubles. First, in a long period like several hours, the appearance of background in surveillance videos is varying though the background is static, due to the change of illumination. As in the existing techniques, the background reference represents the appearance of background, such reference needs to be updated accordingly. Second, when compressing long videos, intra (I) frames need to be inserted once in a while, together with instantaneous decoding refresh (IDR), to facilitate random access and to provide error robustness. As an I frame is inserted, in the existing techniques, the background reference as well as short-term references are all removed from the buffer, and I frames cannot make use of the background reference. Due to the above twofold reasons, the efficiency of background reference is greatly restricted.

In order to further enhance the compression efficiency of surveillance videos, in this paper, we introduce a new kind of references known as *reflectance reference* (RefRef). Like the background reference, RefRef also represents the static background, but differently, RefRef is a picture consisting of the reflectance component rather than the appearance of the background. It is generally acknowledged that the appearance can be decomposed into reflectance and illuminance; while illuminance often changes, the reflectance of static background can be assumed to be invariant. Thus, for one fixed camera, the RefRef does not need to be updated in a long while. As it is a single picture and does not change with time, the RefRef can be cached *always in memory* to serve as reference. In this sense, it is indeed a long-term reference, even not affected by IDR, and

I frames can utilize the RefRef without hindering random access. In addition, the usage of RefRef is also different from that of existing background reference. The background reference is used like the short-term references in video coding, i.e. for motion estimation and compensation. The RefRef only records the reflectance component, thus using the RefRef requires the compression of illuminance component. Since the illuminance component is usually more smooth than the appearance, it is more compressible by intra coding. The adoption of RefRef then provides higher compression efficiency for I frames in surveillance video coding, as demonstrated by our experimental results.

The remainder of this paper is organized as follows. In Sect. 2, the proposed scheme using RefRef is first illustrated, and details of the scheme are discussed. To verify the effectiveness of the proposed scheme, comparative experimental results are presented in Sect. 3. Finally, conclusions are drawn in Sect. 4.

2 The Proposed Reflectance Reference-Based Coding Scheme

Our proposed scheme using RefRef is depicted in Fig. 1 and discussed in detail in this section. After an overview in Sect. 2.1, the details of generating RefRef, extracting the illuminance component, and deciding coding parameters for illuminance component are presented in the following subsections, respectively.

2.1 Overview of the Scheme

As shown in Fig. 1, the scheme consists of offline and online parts. The reflectance reference is generated offline using historical videos of the same surveillance

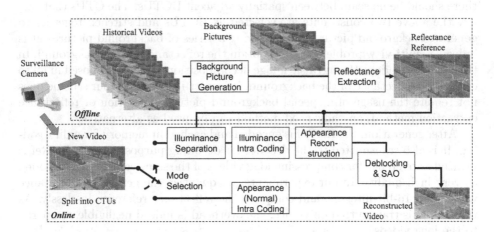

Fig. 1. The flowchart of our proposed reflectance reference-based coding scheme for intra frames in surveillance videos. The reflectance reference is extracted offline. During online coding, one block (CTU) of the current picture may choose one mode from the two: intra coding of its appearance, or intra coding of its illuminance.

camera. The online coding process utilizes the RefRef as an always-in-memory reference. Since the RefRef is always in memory, even I frames can utilize the RefRef without hindering the random access in video coding. Surely, predictive (P) and bidirectional predictive (B) frames can also use the RefRef, which will be studied in our future work.

In Fig. 1 we consider an I frame using the RefRef. For each block (also known as coding tree unit–CTU–in HEVC) of the current I frame, either its illuminance is extracted based on RefRef and then encoded, or its original pixel values (appearance) are encoded. To decide which mode is better, we adopt the joint rate-distortion cost $J = D + \lambda R$ as the metric to evaluate the two modes, where D is the distortion calculated as sum-of-squared-difference (SSD) in the pixel value (appearance) domain, R is the coding rate, and λ is the Lagrangian multiplier. One binary flag for each block is signaled to the decoder on the chosen mode (illuminance or appearance).

2.2 Generation of Reflectance Reference

To generate the RefRef, we need to collect some videos of the given camera and build the background pictures to extract reflectance. Our empirical studies show that using videos of several hours, provided enough change of illumination, is enough to build the RefRef. The videos are divided into many segments, each of which lasts for several seconds so that each segment generates one background picture. In this paper, the background picture is generated using the block composing approach proposed in [2]. The approach detects background blocks at the level of CTU, and stitches background CTUs (BCTUs) together until achieving the entire background picture. On the one hand, the BCTUs should be almost identical to the collocated CTUs in temporally adjacent frames. On the other, there should be no seam between spatially adjacent BCTUs. The CTUs that satisfy the above two constraints are selected as BCTUs and stitched together to generate background pictures. Once having a series of background pictures, it is indeed a multi-view problem [5–7] to obtain the reflectance of the background. In this paper, we adopt the intrinsic image decomposition method proposed in [4] to obtain the reflectance of the background. It is worth noting that our scheme does not require the usage of a special background picture generation or reflectance extraction method, any advanced method can be used as alternative.

After generation, the RefRef is encoded and cached in memory for online coding. It is also transmitted to the decoder for decoding purpose. As the RefRef is utilized as reference for compressing long videos of the same camera, it is encoded at very high quality. In our experiments, we quantize the generated reflectance values into 16-bit integers, and compress the reflectance reference losslessly. As the RefRef is transmitted only once, the overhead is indeed negligible compared to the long videos.

2.3   Extraction of Illuminance Component

An image formation model commonly used in the literature of intrinsic image decomposition is $I = \rho l$, where I is the appearance (pixel value), ρ is the reflectance, and l is the illuminance. Thus, we divide the current picture by the RefRef, pixel by pixel, to calculate the illuminance component. To pursue a high numerical precision, we adopt an adaptive scaling and clipping procedure. For each I picture, after dividing by the RefRef, we find two integers v_0 and v_1 so that the range $[v_0, v_1]$ covers at least 99% of all the illuminance coefficients of the entire picture. The illuminance coefficients inside the range are quantized uniformly to 8-bit unsigned integers, while the coefficients out of the range are clipped. Specifically,

$$l_{enc} = \begin{cases} 0 & \text{if } l_{org} < v_0 \\ \lfloor \dfrac{l_{org} - v_0}{\theta} \rfloor & \text{if } l_{org} \in [v_0, v_1] \\ 255 & \text{if } l_{org} > v_1 \end{cases} \qquad (1)$$

where θ equals to $(v_1 - v_0)/255$, and $\lfloor \cdot \rfloor$ stands for rounding down. The two integers v_0 and v_1 need to be coded into the bitstream so that decoder can performs the inverse scaling.

2.4   Decision of Coding Parameters

When encoding the illuminance component rather than the appearance (pixel values) of one block, we reuse the intra coding tools in HEVC, but have adjusted the coding parameters, namely λ and quantization parameter (QP). We give the detailed analyses in the following. Recall that in the final joint rate-distortion cost $D + \lambda R$, D is measured in the appearance domain. The distortion corresponding to a certain coding mode i is

$$D_i = \sum_{j=1}^{M} (I_i'(j) - I(j))^2 \qquad (2)$$

where $I_i'(j)$ and $I(j)$ are the reconstructed and the original pixel values of the j-th pixel, respectively. M stands for the number of pixels in the current block. However, during encoding of the illuminance component, we calculate distortion in the illuminance domain, i.e.,

$$D_i^l = \sum_{j=1}^{M} (l_i'(j) - l(j))^2 \qquad (3)$$

Note that $I_i'(j) = \rho_j(\theta l_i'(j) + v_0)$. With up to rounding and clipping error, we have $I(j) = \rho_j(\theta l(j) + v_0)$. Therefore,

$$D_i = \sum_{j=1}^{M} \rho_j^2 \theta^2 (l_i'(j) - l(j))^2 \qquad (4)$$

For ease of calculation, we assume the reflectance coefficients within a small block are constant, then we have

$$D_i = \bar{\rho}^2 \theta^2 D_i^l \tag{5}$$

Finally, the joint rate-distortion cost of mode i is

$$J_i = D_i + \lambda R_i = \bar{\rho}^2 \theta^2 D_i^l + \lambda R_i \tag{6}$$

Therefore, during encoding the illuminance component, we set $\lambda_l = \lambda/(\bar{\rho}^2\theta^2)$, where $\bar{\rho}^2$ is calculated as the mean square reflectance in the current block. Accordingly, the QP is adjusted based on new λ value, according to the empirical relation in [8].

3   Experimental Results

3.1   Settings

We have implemented the proposed coding scheme based on the HEVC reference software HM version 16.0. For evaluation purpose, we compare our scheme with the HM under the HEVC common test conditions. Specifically, in this paper we only use the All-Intra (AI) configuration as specified in [1]. Four QPs, 22, 27, 32, 37, are used in experiments. BD-rate is adopted to measure the overall rate-distortion (R-D) performance, where the distortion is measured by PSNR.

As our proposed method requires long surveillance videos of the same camera to build the RefRef, there is no such video publicly available, to the best of our knowledge. Thus, we captured surveillance videos by ourselves. We chose two scenes, one indoor and one outdoor, and produced videos of about 12 h for each scene, using a SONY HDR-CX900E video camera. Then, we trimmed 8 sequences for test, each of which lasts for 10 s, from the long videos. Snapshots of the test sequences are shown in Fig. 2 and characteristics of them are summarized in Table 1. It can be observed from Fig. 2 that the test sequences of the same scene have significant variety of illuminance, especially in the indoor case. In our experiments, we have generated only one reflectance reference for each scene, and used that reference for all the test sequences of the same scene.

3.2   Results

Overall Performance. Table 2 presents the overall BD-rate results as well as computational time results by comparing the proposed method with HEVC anchor. Under AI configuration, the average BD-rate on Luma (Y) of the 8 test sequences is 21.9% when measured by PSNR. The BD-rate exceeds 60% for one sequence (Indoor_2). It can be observed from Fig. 2 that the BD-rate is dependent on video content: the sequences with larger variance of illuminance, and/or with more foreground content, have lower BD-rate.

Some typical R-D curves are shown in Fig. 3. From this figure, it can be observed the proposed scheme achieves higher improvement at lower bit rates,

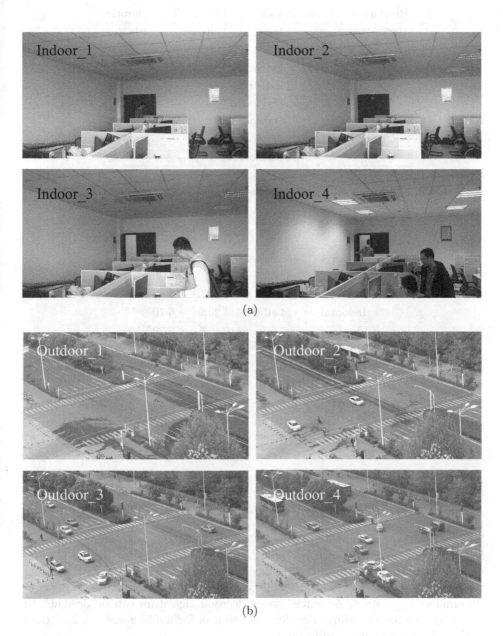

Fig. 2. Snapshots of the test sequences corresponding to two scenes: (a) Indoor, (b) Outdoor.

Table 1. Characteristics of the test sequences

Resolution	Scene	Video	Frame Rate	Duration
1920 × 1080	Indoor	Indoor_1	30 fps	10 s
		Indoor_2		
		Indoor_3		
		Indoor_4		
	Outdoor	Outdoor_1		
		Outdoor_2		
		Outdoor_3		
		Outdoor_4		

Table 2. BD-Rate results of the proposed RefRef method anchored on HEVC

Sequence	All-Intra		
	Y	U	V
Indoor_1	−36.90%	−32.70%	−31.40%
Indoor_2	−62.50%	−53.80%	−56.50%
Indoor_3	−17.40%	−14.40%	−16.60%
Indoor_4	− 4.80%	− 4.20%	− 5.10%
Outdoor_1	− 7.60%	−10.10%	− 8.20%
Outdoor_2	−17.40%	−19.90%	−21.10%
Outdoor_3	− 6.80%	−12.30%	− 5.60%
Outdoor_4	−15.30%	−16.50%	−14.90%
Overall	−21.90%	−20.49%	−19.93%
Enc. Time	246%		
Dec. Time	121%		

since the RefRef is encoded at fixed, very high quality, its advantage is more significant when the normal intra coding becomes more inaccurate at lower bit rates.

Regarding the computational complexity, RefRef incurs an increase of encoding/decoding time, as shown in Table 1. The encoding time of RefRef is more than twice of that of HM. This is because we perform two rounds of intra coding (illuminance and normal) in RefRef encoder for mode decision, and also perform illuminance separation. Advanced mode decision algorithm can be designed to avoid the two-round coding. The decoding time of RefRef increases by 21% than HM, which is not quite significant.

Analyses of Mode Selection. In the proposed scheme, one intra-coded block can choose the normal (appearance) intra coding mode or the illuminance intra coding mode. Figure 4 shows typical results of the mode selection. The blocks

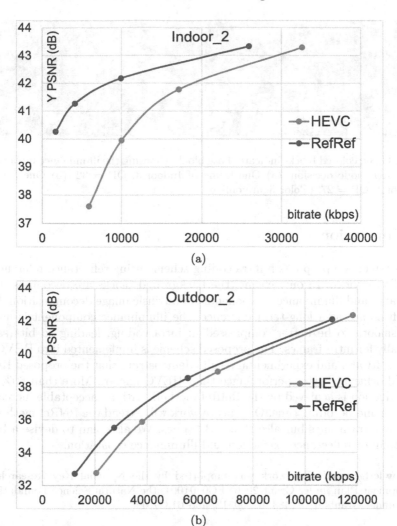

Fig. 3. Rate-distortion (RD) curves of two test sequences.

that chose the illuminance mode are almost all inside the background area. However, the background blocks that are very smooth (like the wall and the road) still chose the normal mode as it works well enough. The blocks that have very complex textures (like the trees) also chose the normal mode, because the reflectance extraction algorithm works not well at these blocks (partially due to the algorithm itself, partially due to the slight motion of these trees). Advanced illuminance separation algorithm can be designed in the future to cope with these complex textures better.

Fig. 4. Green colored blocks indicate those blocks choosing the illuminance intra coding mode after mode decision. (a) One frame of Indoor_3, QP = 32. (b) One frame of Outdoor_1, QP = 27. (Color figure online)

4 Conclusion

In this paper, we propose an intra coding scheme using reflectance reference for surveillance video. In our scheme, the background blocks are decomposed into reflectance and illuminance components via intrinsic image decomposition. With the reflectance as a long-term reference, the illuminance component is usually more smooth to be better compressed in intra coding, leading to bits saving especially for intra frames. The proposed scheme is implemented into HEVC reference software, and experimental results demonstrate that the proposed RefRef method achieves better performance than HEVC anchor. More than 60% BD-rate reduction is achieved by the RefRef method, with an acceptable increase of encoding and decoding time. Our future work will extend the RefRef method for not only intra frames but also P and B frames. We also plan to devise a better algorithm for reflectance extraction and illuminance separation.

Acknowledgment. This work was supported by the National Key Research and Development Plan under Grant 2016YFC0801001, the Natural Science Foundation of China under Grants 61772483, 61390512, and 61331017.

References

1. Bossen, F.: Common test conditions and software reference configurations. JCTVC-H1100, presented at the 8th meeting of Joint Collaborative Team on Video Coding (JCT-VC), San Jose, CA, USA, February 2012
2. Chen, F., Li, H., Li, L., Liu, D., Wu, F.: Block-composed background reference for high efficiency video coding. IEEE Trans. Circuits Syst. Video Technol. **27**(12), 2639–2651 (2017)
3. Gorur, P., Amrutur, B.: Skip decision and reference frame selection for low-complexity H.264/AVC surveillance video coding. IEEE Trans. Circuits Syst. Video Technol. **24**(7), 1156–1169 (2014)
4. Hauagge, D., Wehrwein, S., Bala, K., Snavely, N.: Photometric ambient occlusion for intrinsic image decomposition. IEEE Trans. Pattern Anal. Mach. Intell. **38**(4), 639–651 (2016)

5. Hong, R., Hu, Z., Wang, R., Wang, M., Tao, D.: Multi-view object retrieval via multi-scale topic models. IEEE Trans. Image Process. **25**(12), 5814–5827 (2016)
6. Hong, R., Zhang, L., Tao, D.: Unified photo enhancement by discovering aesthetic communities from Flickr. IEEE Trans. Image Process. **25**(3), 1124–1135 (2016)
7. Hong, R., Zhang, L., Zhang, C., Zimmermann, R.: Flickr circles: aesthetic tendency discovery by multi-view regularized topic modeling. IEEE Trans. Multimed. **18**(8), 1555–1567 (2016)
8. Li, B., Xu, J., Zhang, D., Li, H.: QP refinement according to Lagrange multiplier for high efficiency video coding. In: IEEE International Symposium on Circuits and Systems (ISCAS), pp. 477–480. IEEE (2013)
9. Paul, M., Lin, W., Lau, C.T., Lee, B.S.: A long-term reference frame for hierarchical B-picture-based video coding. IEEE Trans. Circuits Syst. Video Technol. **24**(10), 1729–1742 (2014)
10. Richardson, I.E.: H.264 and MPEG-4 Video Compression: Video Coding for Next-Generation Multimedia. Wiley, New York (2004)
11. Sullivan, G.J., Ohm, J., Han, W.J., Wiegand, T.: Overview of the high efficiency video coding (HEVC) standard. IEEE Trans. Circuits Syst. Video Technol. **22**(12), 1649–1668 (2012)
12. Zhang, X., Tian, Y., Huang, T., Dong, S., Gao, W.: Optimizing the hierarchical prediction and coding in HEVC for surveillance and conference videos with background modeling. IEEE Trans. Image Process. **23**(10), 4511–4526 (2014)

Deeper Spatial Pyramid Network with Refined Up-Sampling for Optical Flow Estimation

Zefeng Sun[1,2] and Hanli Wang[1,2(✉)]

[1] Department of Computer Science and Technology, Tongji University,
Shanghai 201804, People's Republic of China
hanliwang@tongji.edu.cn
[2] Key Laboratory of Embedded System and Service Computing,
Ministry of Education, Tongji University,
Shanghai 200092, People's Republic of China

Abstract. Convolutional neural networks (CNNs) have been successfully applied to optical flow estimation and outperformed a number of variational approaches. The spatial pyramid network (SPyNet) is one of these CNN based approaches which is efficient to estimate optical flow. In this paper, a deeper spatial pyramid network (DSPyNet) is proposed based on SPyNet. In DSPyNet, the network architecture of SPyNet is reused and further refined at each pyramid level by convolutional factorization, and an addition of inception module and 1×1 convolutional operation is further used to enhance visual representation. Moreover, since bilinear interpolation reduces the quality of up-sampled flow field due to its low-pass filtering property, it is replaced with small kernel convolutional operations like image super-resolution using CNNs. The proposed DSPyNet is evaluated on several optical flow estimation benchmark datasets and the experimental results verify its effectiveness.

Keywords: Optical flow estimation · Convolutional neural network
Spatial pyramid · Image super-resolution

1 Introduction

Convolutional neural networks (CNNs) play an important role in many computer vision tasks like classification [9], human pose estimation [17] and semantic segmentation [13] in recent years. Meanwhile, a number of CNN based optical flow estimation methods have been designed, such as FlowNet [3], spatial pyramid

This work was supported in part by National Natural Science Foundation of China under Grants 61622115 and 61472281, Program for Professor of Special Appointment (Eastern Scholar) at Shanghai Institutions of Higher Learning (No. GZ2015005), Shanghai Engineering Research Center of Industrial Vision Perception & Intelligent Computing (17DZ2251600), and IBM Shared University Research Awards Program.

© Springer Nature Switzerland AG 2018
R. Hong et al. (Eds.): PCM 2018, LNCS 11164, pp. 492–501, 2018.
https://doi.org/10.1007/978-3-030-00776-8_45

network (SPyNet) [12] and FlowNet2.0 [8]. These methods outperform many variational methods in terms of running time and accuracy.

Traditional optical flow estimation methods require a variety of assumptions about the image in theory, such as brightness constancy and spatial smoothness. But in real circumstances, it's hard to match these assumptions so that the performance is weakened. The CNN based methods intend to get rid of these assumptions, embody the variation in image brightness and spatial smoothness, and thus perform better with a trade-off between the number of model parameters, memory requirements and running time. However, less parameters generally lead to networks with weaker representation power and hence poorer prediction accuracy. Moreover, the intermediate feature maps tend to be processed by some operations, like warping in FlowNet2.0 [8] and up-sampling in SPyNet [12], which are critical for flow estimation. But the bilinear interpolation method with low-pass filtering property fades the high-frequency component and fuzzs the image contour [6], which causes quality loss of up-sampled flow fields and affects the final result negatively.

In order to achieve a better trade-off between model complexity and performance, a deeper spatial pyramid network (DSPyNet) is proposed in this work with refined up-sampling for flow fields based on SPyNet [12]. Two network architectures are designed. One is a more efficient network architecture for residual optical flow estimation, and the other is a relatively small network architecture for up-sampling the flow field with less quality loss compared to the bilinear interpolation method. Through these refinements, the network at each level of DSPyNet gets higher-quality input flow information and generates intermediate small-size flow fields, which makes DSPyNet as a whole outperform SPyNet [12] and FlowNet [3] on most optical flow datasets. In the current work, only relatively small models are considered for comparison, so FlowNet2.0 [8] with a large number of parameters is not included.

To sum up, the main contributions of this work are summarized as two folds. First, the network architecture for residual flow estimation at each pyramid level is refined into a deeper architecture, which is achieved by employing the inception module and the 1×1 convolutional layer with factorizing convolutions. Second, the bilinear interpolation method is replaced with a small neural network to up-sample the horizontal and vertical components of the flow field with less quality loss. The rest of this paper is organized as follows. Section 2 presents the related works. The details of the proposed DSPyNet is introduced in Sect. 3. The experimental results are shown in Sect. 4. Finally, Sect. 5 concludes this work.

2   Related Work

Since the breakthrough from Horn and Schunck [7], variational approaches have flourished and become the main-stream methods that substantially improve the accuracy of optical flow estimation. But when applied into real circumstances, many variational approaches don't have ideal performance due to several assumptions like brightness constancy and spatial smoothness in theory. So it's a hot research topic on how to reduce the constraints of these assumptions in practice. The review [15] is recommended for further knowledge about this field.

Recently, a number of novel methods using deep leaning techniques have emerged and outperformed variational approaches. In [3], optical flow estimation is considered as a supervised learning task and FlowNet is trained to accomplish this task. The successive work of FlowNet is to stack several FlowNets on top of each other and fuse a small displacement network to construct a deeper and wider network called FlowNet2.0 [8]. FlowNet2.0 performs very well but has a large number of model parameters. SPyNet [12], a small-scale network for optical flow estimation, combines a classical spatial pyramid formulation with deep learning, and the network at every pyramid level is trained to learn residual flow which is then added to the up-sampled flow field from the higher level to refine the flow field. SPyNet copes with large motions through the spatial pyramid structure and gets similar results as FlowNet, even though the number of parameters of SPyNet is only 4% of that of FlowNet. However, the small number of parameters leads to a simpler network with weaker representation power and limits the performance of SPyNet in part.

3 Proposed Deeper Spatial Pyramid Network

3.1 DSPyNet Overview

An overview of the structure of the highest two levels in DSPyNet is shown in Fig. 1. Let $d(I)$ represent an operator that an $m \times n$ image I is decimated to an $\frac{m}{2} \times \frac{n}{2}$ image. $\{I_k^1, I_k^2\}$ stands for a pair of images at the k-th pyramid level. G_k, F_k, f_k denote the network for residual optical flow estimation, the refined flow field and the residual flow estimated by G_k respectively at the k-th pyramid level. U_k is the network for up-sampling flow fields at the k-th pyramid

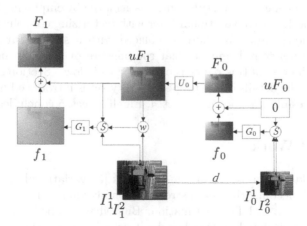

Fig. 1. Overview of the highest two spatial pyramid levels in DSPyNet.

level, and the flow field up-sampled using the network U_k is denoted as uF_{k+1}. $w(I, uF)$ is a warping operator to warp the image I with the flow field uF. $S(I_k^1, w(I_k^2, uF_k), uF_k)$ is a stacking operator to stack images and the flow field.

Similar to SPyNet [12], the spatial pyramid structure is employed to deal with large displacements and CNNs are applied to estimate sub-pixel flow. At the beginning, the input image pair is down-sampled for several times. Then the smallest image pair $\{I_0^1, I_0^2\}$ is stacked with a flow field uF_0 (all the values are 0) and input into the network G_0 at the highest level to generate the residual flow f_0. uF_0 is added to f_0 pixel by pixel, and the summed flow field F_0 is up-sampled by U_0 into uF_1 which is then propagated to the next level. In the next level, the network G_1 computes the residual flow f_1 using $\{I_1^1, w(I_1^2, uF_1), uF_1\}$ stacked together. At each level, similar processes are performed and finally a refined full-resolution flow field is generated. The following Algorithm 1 recaps the algorithm of optical flow estimation.

Algorithm 1. Optical flow estimation in DSPyNet.

Input: A pair of images, $\{I_{L-1}^1, I_{L-1}^2\}$
Output: Corresponding optical flow, F_{L-1}
1: L = the number of pyramid levels in DSPyNet
2: **for** $i = L - 1; i > 0; i - -$ **do**
3:      $I_{i-1}^1 = d(I_i^1)$
4:      $I_{i-1}^2 = d(I_i^2)$
5: **end for**
6: $f_0 = G_0(S(I_0^1, I_0^2, uF_0))$
7: $F_0 = f_0 + uF_0$
8: **for** $i = 1$ to $L - 1$ **do**
9:      $uF_i = U_{i-1}(F_{i-1})$
10:      $f_i = G_i(S(I_i^1, w(I_i^2, uF_i), uF_i))$
11:      $F_i = f_i + uF_i$
12: **end for**

3.2 Network Architecture

There are two network architectures in DSPyNet. One is applied for residual flow estimation and the other for up-sampling flow fields.

Network for Residual Flow Estimation. A specific network G is constructed for residual flow estimation. The image, the image warped with flow field and the flow field are stacked and then input into G. Then, G processes the input information in an end-to-end manner based on what it has learned and outputs the corresponding residual flow. Specifically, G is designed based on SPyNet [12]. First, the third 7×7 convolutional layer is replaced with the inception module, resulting in faster convergence and more powerful representation. Second, a 1×1 convolutional layer [10] is inserted between the inception module and the forth

7×7 convolutional layer to reduce dimension and compress feature, which contributes to less performance penalty. Third, the rest of 7×7 convolutional layers, except the last one, are factorized into three 3×3 convolutional layers to reduce the number of parameters and deepen the network according to [16]. A ReLU layer [5] is inserted between every two 3×3 convolutional layers to enhance non-linear representation ability. Figure 2 shows the comparison of convergence in terms of training end-point error (EPE) loss and validation EPE loss between the network G_0 of SPyNet and that of the proposed DSPyNet, where it can be seen that the proposed DSPyNet converges much faster than SPyNet.

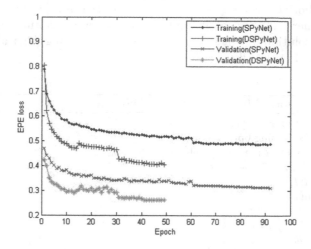

Fig. 2. Comparison of convergence in the network G_0 between SPyNet and DSPyNet.

Network for Up-Sampling Flow Field. To pursue high-quality up-sampled flow information, a network U is constructed to up-sample small-size flow fields. The network U includes one input layer, two 3×3 convolutional layers, one de-convolutional layer and two 7×7 convolutional layers. Here are two points about this network. First, the bilinear interpolation method is replaced with the de-convolutional operation to up-sample feature maps. Second, motivated by [14] where feature maps are only extracted from the low-resolution space to reduce computational complexity, the de-convolutional layer is set as the third hidden layer to reduce the computations of the first two hidden layers.

4   Experimental Results

The proposed DSPyNet is evaluated on a number of benchmark optical flow datasets including Flying Chairs [3], Sintel [2], KITTI2012 [4], Middlebury [1] and Monkaa & Driving Scenes [11]. For comparison, SPyNet [12], FlowNet [3] and Classic+NLP [15] are employed with the metrics of EPE loss and running time.

The proposed DSPyNet is implemented based on the Torch7 framework. All the networks including G_0, U_0, G_1, U_1, G_2, U_2, G_3, U_3 and G_4 are trained on four Titan-X GPUs. The Adam optimization method is used with $\beta_1 = 0.9$, $\beta_2 = 0.999$ and batch size of 16. Every epoch contains 4000 iterations. All the networks are trained in the order of $\{G_0, U_0, G_1, U_1, G_2, U_2, G_3, U_3, G_4\}$, and at the beginning of every training, the model $G_k(U_k)$ is initialized with the pre-trained model $G_{k-1}(U_{k-1})$. While training U, we set the learning rate of 5e-4 for the first 35 epochs, 1e-4 for the next 15 epochs and 1e-5 until the network converges. The same data augmentation method used by SPyNet [12] is applied for training DSPyNet, which includes random image scaling by a factor of $[1, 2]$, random image rotation within $[-17°, 17°]$, random image cropping, additive white Gaussian noise sampled uniformly from $N(0, 0.1)$, color jitter and image normalization. DSPyNet is trained on the Flying Chairs dataset [3] and fine-tuned on the Sintel dataset [2] and Driving & Monkaa Scenes dataset [11].

First, the proposed network U for up-sampling flow field is evaluated by comparing with the traditional bilinear method, with the comparative results shown in Table 1, where the Flying Chairs dataset [3] and Sintel clean dataset [2] are tested.

Table 1. Comparison of EPE loss between the proposed network U and the bilinear method for up-sampling flow field.

Method	Flying chairs	Sintel
Bilinear	2.12	3.61
Network U	2.04	3.51

From the results shown in Table 1, it can be observed that the proposed network U is superior to the bilinear method to improve the accuracy of optical flow estimation. Moreover, two visualization examples from the Flying Chairs dataset are illustrated in Fig. 3 to further demonstrate the effectiveness of the proposed network U.

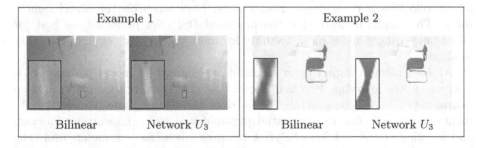

Fig. 3. Visualization of up-sampled flow fields of two exmaples from the Flying Chairs dataset.

Then, the proposed DSPyNet is compared with Classic+NLP [15], FlowNet [3] and SPyNet [12]. The comparative results are summarized in Table 2, where the best results with the smallest EPE value on the same dataset have been highlighted in bold for comparison. And the running time in the last column is tested on the Sintel dataset and excludes the image loading and flow copying time from device memory to host memory. Moreover, "+ft" indicates that the model has been fine-tuned on the corresponding datasets. As observed from the EPE results, the proposed DSPyNet outperforms the other comparative methods on most of the datasets. When fine-tuning is employed, the EPE performance of all the competing methods are improved. The visualization of several typical optical flow images estimated by the competing methods on the Sintel dataset is shown in Fig. 4, where it can be observed that the proposed DSPyNet achieves higher-quality flow than the other methods. On the other hand, as shown in the last column in Table 2, the running time of the methods of FlowNet, SPyNet and DSPyNet are in a similar level which is much less than the traditional Classic+NLP method.

Table 2. Comparison of EPE loss and running time. "+ft" means the model has been fine-tuned on the corresponding datasets.

Method	Sintel Clean		Sintel Final		KITTI		Middlebury		Flying Chairs	Time (s)
	Train	Test	Train	Test	Train	Test	Train	Test	Test	
Classic+NLP [15]	4.13	6.73	5.90	8.29	-	-	0.22	0.32	3.93	102
FlowNetS [3]	4.50	7.42	5.45	8.43	**8.26**	-	1.09	-	2.71	0.027
FlowNetC [3]	4.31	7.28	5.87	8.81	9.35	-	1.15	-	2.19	0.040
SPyNet [12]	4.12	6.69	5.57	8.43	9.12	-	**0.33**	**0.58**	2.63	0.010
DSPyNet	**3.51**	**6.55**	**5.22**	**8.33**	9.20	-	0.45	0.67	**2.05**	0.022
FlowNetS+ft [3]	3.66	6.96	4.44	7.76	7.52	9.1	0.98	-	3.04	0.027
FlowNetC+ft [3]	3.78	6.85	5.28	8.51	8.79	-	0.93	-	**2.27**	0.040
SPyNet+ft [12]	3.17	6.64	4.32	8.36	**4.13**	**4.7**	0.33	0.58	3.07	0.010
DSPyNet+ft	**1.82**	**6.22**	**2.89**	**7.45**	4.56	7.8	**0.31**	**0.43**	2.29	0.022

Moreover, Table 3 shows the performance of the competing methods at different velocities and distances from motion boundaries on the Sintel benchmark. The results indicate that the proposed DSPyNet performs the best for all distance ranges and is more accurate for most velocity ranges than the other methods.

At last, model complexity in terms of the amount of model parameters is analyzed in the following. It is well known that the representation power of deep neural networks will be generally boosted at the cost of more parameters. As mentioned in [12], due to the spatial pyramid structure, the warping function and learning of residual flow, SPyNet requires much less parameters and thus less model complexity as compared to FlowNet [3]. Therefore, as an improved extension of SPyNet, the proposed DSPyNet also needs less model parameters

Ground Truth   FlowNetS   FlowNetC   SPyNet   DSPyNet

Fig. 4. Visualization of optical flow estimated by the competing methods on the Sintel dataset.

Table 3. EPE comparison on the Sintel benchmark at different velocities (s) and distances (d) from motion boundaries.

Method	Sintel Final						Sintel Clean					
	d_{0-10}	d_{10-60}	d_{60-140}	s_{0-10}	s_{10-40}	s_{40+}	d_{0-10}	d_{10-60}	d_{60-140}	s_{0-10}	s_{10-40}	s_{40+}
FlowNetS+ft	7.52	4.61	2.99	1.87	5.83	43.24	5.99	3.56	2.19	1.42	3.81	40.10
FlowNetC+ft	7.19	4.62	3.30	2.30	6.17	**40.78**	5.57	3.18	1.99	1.62	3.97	**33.37**
SPyNet+ft	6.69	4.37	3.29	1.39	5.53	49.71	5.50	3.12	1.71	**0.83**	3.34	43.44
DSPyNet+ft	**5.55**	**3.68**	**2.72**	**1.12**	**4.83**	44.93	**5.10**	**2.94**	**1.56**	0.88	**2.84**	40.95

Number of Parameters

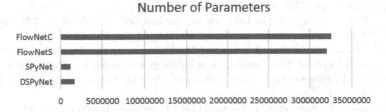

Fig. 5. Comparison of model size.

than FlowNet. Specifically speaking, the total number of model parameters of DSPyNet is about 137.8% of that of SPyNet and is only about 5.5% of that of FlowNet. Figure 5 displays the comparison of model size of FlowNet, SPyNet and DSPyNet.

5   Conclusion

In this paper, a deeper spatial pyramid network (DSPyNet) is designed based on SPyNet for optical flow estimation. At each pyramid level of the spatial pyramid structure, the up-sampling operation is refined by substituting the bilinear interpolation method with a network to get higher-quality flow fields. Furthermore, the model architecture is refined by convolutional factorization and the addition of the inception module, which efficiently enhances the model representation power. The proposed DSPyNet is compared with a number of optical flow estimation approaches including FlowNet and SPyNet, and the experimental results demonstrate its effectiveness.

References

1. Baker, S., Scharstein, D., Lewis, J.P., Roth, S., Black, M.J., Szeliski, R.: A database and evaluation methodology for optical flow. Int. J. Comput. Vis. **92**(1), 1–31 (2011)
2. Butler, D.J., Wulff, J., Stanley, G.B., Black, M.J.: A naturalistic open source movie for optical flow evaluation. In: Proceedings of the European Conference on Computer Vision, pp. 611–625, October 2012
3. Dosovitskiy, A., Fischery, P., Ilg, E., Hausser, P.: Flownet: learning optical flow with convolutional networks. In: Proceedings of the IEEE International Conference on Computer Vision, pp. 2758–2766, December 2015
4. Geiger, A., Lenz, P., Urtasun, R.: Are we ready for autonomous driving? The KITTI vision benchmark suite. In: Proceedings of the IEEE Conference on Computer Vision and Pattern Recognition, pp. 3354–3361, June 2012
5. Glorot, X., Bordes, A., Bengio, Y.: Deep sparse rectifier neural networks. In: Proceedings of the International Conference on Artificial Intelligence and Statistics, pp. 315–323, April 2011
6. Han, D.: Comparison of commonly used image interpolation methods. In: Proceedings of the International Conference on Computer Science and Electronics Engineerings, pp. 1556–1559, March 2013
7. Horn, B.K., Schunck, B.G.: Determining optical flow. Artif. Intell. **17**(1–3), 185–203 (1981)
8. Ilg, E., Mayer, N., Saikia, T., Keuper, M., Dosovitskiy, A., Brox, T.: Flownet 2.0: evolution of optical flow estimation with deep networks. In: Proceedings of the IEEE Conference on Computer Vision and Pattern Recognition, pp. 2462–2470, December 2017
9. Krizhevsky, A., Sutskever, I., Hinton, G.E.: Imagenet classification with deep convolutional neural networks. In: Proceedings of the International Conference on Neural Information Processing Systems, pp. 1097–1105, December 2012
10. Lin, M., Chen, Q., Yan, S.: Network in network. arXiv preprint arXiv:1312.4400 (2013)
11. Mayer, N., et al.: A large dataset to train convolutional networks for disparity, optical flow, and scene flow estimation. In: Proceedings of the IEEE Conference on Computer Vision and Pattern Recognition, pp. 4040–4048, June 2016
12. Ranjan, A., Black, M.J.: Optical flow estimation using a spatial pyramid network. In: Proceedings of the IEEE Conference on Computer Vision and Pattern Recognition, pp. 2720–2729, June 2017
13. Shelhamer, E., Long, J., Darrell, T.: Fully convolutional networks for semantic segmentation. IEEE Trans. Pattern Anal. Mach. Intell. **39**(4), 640–651 (2016)

14. Shi, W., et al.: Real-time single image and video super-resolution using an efficient sub-pixel convolutional neural network. In: Proceedings of the IEEE Conference on Computer Vision and Pattern Recognition, pp. 1874–1883, June 2016

15. Sun, D., Roth, S., Black, M.J.: A quantitative analysis of current practices in optical flow estimation and the principles behind them. Int. J. Comput. Vision **106**(2), 115–137 (2014)

16. Szegedy, C., Vanhoucke, V., Ioffe, S., Shlens, J., Wojna, Z.: Rethinking the inception architecture for computer vision. In: Proceedings of the IEEE Conference on Computer Vision and Pattern Recognition, pp. 2818–2826, June 2016

17. Toshev, A., Szegedy, C.: Deeppose: human pose estimation via deep neural networks. In: Proceedings of the IEEE Conference on Computer Vision and Pattern Recognition, pp. 1653–1660, June 2014

An Improved Algorithm for Saliency Object Detection Based on Manifold Ranking

Huiling Wang[✉], Hao Wang, Jing Wang, and Zhengmei Xu

School of Computer and Information, Fuyang Normal College,
Fuyang, Anhui, China
Wanghuilingfy@foxmail.com, 393950571@qq.com,
710813573@qq.com, 78805748@qq.com

Abstract. The goal of saliency detection is to locate important pixels or regions in an image. To overcome the shortage that the spatial connectivity of every region is modeled only via k-regular graph, and do not consider the deficiency between multi-layer super pixel segmentations based on manifold ranking, an improved method is proposed. First, we tackle an image from a scale point of view and use a multi-layer approach to analyze saliency cues. Second, through building a graph model which is on the basis of k- regular graph, we connect the nodes belonging to the same cluster and located in the same spatial connected area with edges, to highlight the whole goal more uniformly and evenly, then used manifold ranking to generate multi-layer saliency map. Finally, the final saliency map is got through weighted linear fusion. By the experimental comparison of the 3 quantitative evaluation indexes of 8 start-of-the-art methods on three datasets, our method proves the effectiveness and superiority.

Keywords: Saliency detection · Improved graph model · Manifold ranking
Weighted linear fusion

1 Introduction

According to the visual attention mechanism of human eyes, human beings can quickly locate the most interesting areas from the scene. Therefore, using the computer to simulate the human eye and identifying the most prominent pixels from the digital image has become an important task of the computer vision. In addition, the results of saliency detection are also applied in many fields, such as image recognition [1], target segmentation [2], image retrieval [3, 4] and image scaling [5, 6] etc. Because of its importance, experts in many fields have conducted deep study on the saliency and put forward many related detection algorithms.

Itti [7] proposed the earliest salient detection algorithm, since then, as for most saliency detection methods, the color, gradient, edge, texture and other low-level features of an image region are simply used to calculate the local or global uniqueness. Achanta et al. [8] proposed Frequency-Tuned (FT) algorithm, after smoothing the image with Gauss method, we get the saliency by calculating the distance between the color of a certain pixel and the average color of the original image. Then some methods consider the color and spatial information, Goferman et al. [9] proposed Context-

© Springer Nature Switzerland AG 2018
R. Hong et al. (Eds.): PCM 2018, LNCS 11164, pp. 502–512, 2018.
https://doi.org/10.1007/978-3-030-00776-8_46

Aware (CA) method, but it is likely to produce a high saliency value in the object boundary. Chen et al. [10] proposed a histogram acceleration method which takes account of spatial relations and feature contrast simultaneously.

Some significant progress has been made by means of a priori foreground or background hypothesis. Xie et al. [11] extracts the convex hull of the key point structure from images, the area encircled by it is taken as a rough foreground target range, the salient value is calculated by the Bayesian framework. According to Wei et al. [12], most of the regions located at the boundary of the image belong to the background. The difference between each image block and the prior background is measured by geodesic distance, and the difference value is taken as the saliency. In order to solve this problem, Zhu et al. [13] designs a more robust background estimation method, called boundary connectivity, by using the ratio of the circumference and area of each region. Jiang et al. [14] makes a more accurate judgment of the authenticity of the background by using the saliency method of selective background priority.

In recent years, the saliency method based on graph models has a high detection accuracy. In [16], the super pixel located at the boundary of the image is used as an absorption node. The absorbing Markov chain is used to detect the saliency. Yang et al. [17] (MR) based on graph uses the super pixel as the node in the graph model to construct the k-regular graph, and then the idea of the manifold ranking is strengthened to get the final saliency map. But the method uses k-regular graph to characterize the spatial connectivity among nodes. Sometimes, it cannot describe the spatial correlation among multiple nodes within the same salient target. In addition, as shown in Fig. 1, due to the diversity and complexity of natural pictures, the size of different image salient targets is unknown. The single-dimension super pixel segmentation size directly affects the accuracy of detection results.

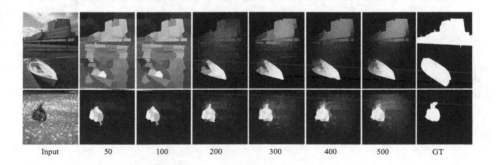

| Input | 50 | 100 | 200 | 300 | 400 | 500 | GT |

Fig. 1. Multi-scale saliency maps of MR method

In view of the above problems, this paper proposes the improved method: When building a graph model, on the basis of k- regular graph, we connect all nodes belonging to the same color class and located in the same connected area, so as to highlight the whole saliency object more uniformly and evenly. In addition, multi-scale fusion can reduce the inaccuracy of the detection results caused by the single scale segmentation error, keep and make up for the advantages and disadvantages of the different scale segmentations.

2   Graph-Based Manifold Ranking

In this section, we provide a brief review of the manifold ranking model as preliminary knowledge.

In [17], given a set of point (such as image) $X = \{x_1, x_2 \ldots x_n\} \in R^{m \times n}$, some data points are set to the queries and the rest is sorted according to its relevance to the query points. Let $\mathbf{y} = [y_1, y_2 \ldots y_n]^T$ is an indicator vector, if x_i is a query, then $y_i = +1$, and $y_i = 0$ otherwise. Denote a ranking function $f: X \to R^n$ to estimate a ranking score f_i for each point x_i. Corresponding graph model $G = (V, E)$ is defined on the dataset X, V is the nodes and E is edge set. Ranking score f is obtained by solving optimization problems by Eq. 1.

$$f^* = \operatorname*{argmin} \frac{1}{2_f} \left[\sum_{i,j=1}^{n} w_{ij} \left\| \frac{f_i}{\sqrt{d_{ii}}} - \frac{f_j}{\sqrt{d_{jj}}} \right\|^2 + u \sum_{i=1}^{n} \|f_i - y_i\|^2 \right] \tag{1}$$

Where $\mathbf{W} = [w_{ij}]_{n \times n}$ denotes a affinity matrix which weight the edges E, in which $w_{ij} = \exp\left(-\frac{dist^2(x_i, x_j)}{\sigma^2}\right)$, note that $w_{ii} = 0$ to avoid self-reinforcement. And $\mathbf{D} = \operatorname{diag}(d_{11}, d_{22} \ldots d_{nn})$ is the degree matrix. u controls the balance of the smoothness constraint and the fitting constraint.

Computing derivative of Eq. 2, and makes result equal to zero to get the optimal solution. Sort function optimization results in the form of a matrix can be expressed as:

$$f^* = (\mathbf{I} - \alpha \mathbf{S})^{-1} \mathbf{y} \tag{2}$$

See [15] for the rigorous proof. I is an identity matrix, $\mathbf{S} = \mathbf{D}^{-1/2} \mathbf{W} \mathbf{D}^{-1/2}$. Saliency detection can be seen as an one-class classification problem, so we can use the unormalized Laplacian matrix in Eq. 3:

$$f^* = (\mathbf{D} - \alpha \mathbf{W})^{-1} \mathbf{y} \tag{3}$$

3   The Method

The main contents of our method include: multi-layer extraction, hierarchical saliency calculation and multi scale saliency map fusion. The framework is illustrated in Fig. 2 and each step is detailed described in Sects. 3.1–3.3 respectively. The algorithm steps are as follows:

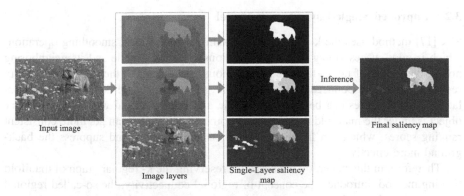

Fig. 2. The flow chart of our method

(1) Three image layers of different scales are extracted from input. We over-segment the original image, and based on that, use the region merging algorithm to generate three region-merge result maps.

(2) According to the improved graph model, the manifold ranking algorithm is used to obtain the initial saliency image of the image at different scales.

(3) Fusing the multi-scale saliency maps by different weigh to obtained the final saliency map.

3.1 Image Layer Extraction

The pixel-based saliency detection algorithm, with pixels as units, does not consider the spatial organization relationship between pixels, so that the efficiency of the algorithm is too low. To produce the three layers, we first generate an initial over-segmentation as illustrated in Fig. 2 by the hierarchical model method [22]. For each segmented region, we compute a region scale (the number of pixels).

In order to better handle and understand the general natural images, we use an encompassment scale measure based on shape uniformities and use it to obtain region sizes in the merging process. Our the scale of region R is written as

$$S(R) = \arg\max_{t}\{R_{t\times t}|R_{t\times t}\subseteq R\} \qquad (4)$$

Where $R_{t\times t}$ is a $t \times t$ square region, R is the original region, t is the segment scale that we set. We set thresholds for the three layers for the typical images, If a (the number of pixels) below 3, we merge it to tis nearest region in terms of average CIELAB color distance, and update its scale and the color of the region as their average color to get the resulting region map as the bottom layer Layer1.

The middle layer Layer2 and the top layer Layer3 are generated similarly from Layer1 and Layer2 with larger scale thresholds that we set, a region in the middle or top layers embraces corresponding ones in the lower levels.

3.2   Improved Single-Layer Graph Model

MR [17] method uses the k- regular graphs (similar to the local smoothing operation) which is good at describing the spatial relationship of each node and its neighboring nodes, but in fact if nodes that are homogenous and belong to the same connectivity region in space are connected together in the graph model, the description of correlation between nodes can be more reasonable and accurate. And when a given query object is under the manifold ranking, these homogenous nodes can get more consistent ranking scores, which can highlight the targets more evenly and suppress the background more effectively.

Therefore, in this paper, on the basis of preserving the k- regular graph of manifold ranking method, introduce the concept of regional connectivity, the so-called regional connectivity refers to belong to one kind of color and space gathered in the collection of the super pixel of a certain region, and among the super pixel by pixel adjacency over reach each other, will also belong to a connected region of the super pixel connection. Then the edge weights between the two nodes defined are adjusted to:

$$w_{ij} = exp - \frac{\|c_i - c_j\| - \sum_{l=1}^{L} \|c_i - c_{j+1}\|}{\sigma^2} \tag{5}$$

c_i and c_j are the values of the connected area of node i and j in CIELAB color space, that is the average color of all the super pixels contained in it, λ is the equilibrium coefficient. According to the Eq. 5, the adjustment of the edge weight will make the nodes of the same connected region stronger and more weighted. The edge weight between the nodes of different connected regions under different color classes is weakened. It is more distinguishable and more conforms to the visual perception of human vision. According to the improved single layer graph model, we use MR to get three scale saliency maps, and the results are shown in Fig. 3.

Input image   Layer1   Layer2   Layer3

Fig. 3. Multi-scale method saliency maps

3.3   Weighted Linear Fusion

There are differences in the detection results at different levels, and it is hardly possible to determine which layer can be used to achieve the best results. Therefore, in order to make more reasonable calculation of saliency results, we build between different levels based on the single layer manifold ranking, and use hierarchical inference method to fuse the multi-scale saliency map to get the final saliency map.

In order to fully consider the problem of the detection results produced by the single super pixel segmentation, the final fusion results are obtained by the weighted linear fusion. The formula is as follows:

$$S(i) = aS_{l1}(i) + bS_{l2}(i) + cS_{l3}(i) \tag{6}$$

Where the a, b and c are the statistical results of three sizes, by the normalization operation, $a + b + c = 1$. We get the final saliency map of our method by Eq. 6. The multi-layer fusion algorithm makes full use of the structure information and the structure and characteristics between layers, and the calculation results are more consistent with the human visual mechanism. Therefore, the algorithm in this paper uses the weighted linear fusion to merge the multi-layer saliency maps. According to the results obtained at the last stage, the final saliency value is obtained by updating each variable from top to bottom to the lowest level. The effect is shown in Fig. 4, the result of our method is more complete and closer to the ground truth.

Input image   Ours   MR   GT

Fig. 4. The Compare of MR and Our method

4   Experiments

In this work, we extensively evaluated our method on three different types of datasets. The first one is the ASD dataset which is the most popular dataset in the literature, which constructed with 1000 natural images from the MSRA-B containing 5000 highly unambiguous images. The second one is CSSD dataset, which includes 200 images, each image of the dataset contains complex scene. The last one is ECSSD dataset, which is the expanding dataset of the CSSD, which containing 1000 semantically meaningful but structurally complex images. We compared our method against 8 start of-the-art saliency algorithms- FT [8], CA [9], RC [10], SF [18], SR [19], BS [20], IM [21], MR [17].

4.1   Quantitative Analysis

The most commonly used PR curve in the field of significant detection is selected for the experiment to carry out quantitative analysis. P is the correct proportion of the detection results; R is the proportion of the correct detection results in the Ground Truth.

The saliency maps obtained by various algorithms are adjusted to [0–255], and 0 to 255 are selected as threshold values in turn, and the saliency map is processed by two values. After comparing the two valued images with Ground Truth, the accuracy and recall rate are calculated, and the P-R curve is drawn.

Figure 5(a) shows the P-R curve effect map of a variety of methods on the ASD dataset. From the P-R curve, we can see that the algorithm in this paper is better than the classic FT, CA, RC, SF and SR global saliency detection algorithm, and also is superior to the MR algorithm based on the boundary prior.

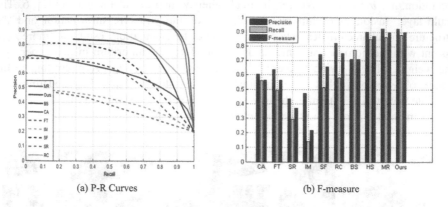

(a) P-R Curves   (b) F-measure

Fig. 5. Comparison of Precision-Recall curves on ASD dataset

In order to verify the effectiveness of the algorithm for the detection of saliency targets in complex scenes, a variety of algorithm contrast experiments are also carried out. As shown in Figs. 6(a) and 7(a), this algorithm has a better effect on the CSSD, and ECSSD datasets.

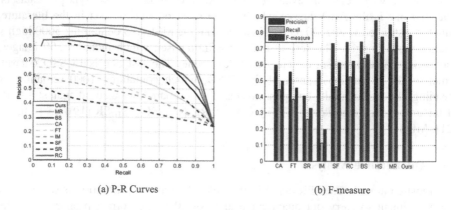

(a) P-R Curves   (b) F-measure

Fig. 6. Comparison of Precision-Recall curves on CSSD dataset

When the saliency effect is evaluated, the accuracy and recall rate often affect each other, and the recall rate may be reduced when the accuracy of some algorithms is

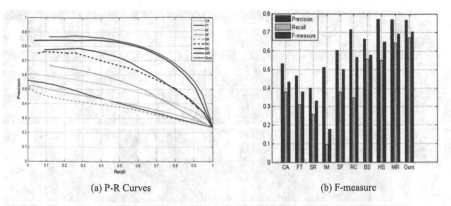

(a) P-R Curves (b) F-measure

Fig. 7. Comparison of Precision-Recall curves on ECSSD dataset

raised. Therefore, F-measure is often used to measure the overall performance of the algorithm. The F-measure formula is as follows:

$$F_\beta = \frac{(1+\beta^2)Precision \times Recall}{\beta^2 \; Precision + Recall} \qquad (7)$$

From the evaluation index histogram of Figs. 5(b), 6(b) and 7(b), we can see that in the dataset ASD, the accuracy rate of the algorithm in this paper is the same as that of MR method, but on the three datasets, the values of the algorithm F-measure in the algorithm of this paper are all higher than those in CA, FT, SR, IM, SF, RC, BS and MR algorithm.

4.2 Visual Contrast

From Fig. 8, we find that the IM algorithm is the worst in many methods. That method can detect significant targets roughly, but there are significant regions blurred and uneven. The FT method can find out the area with special pixels in the image, but there is a situation where the prominence area is not complete. The HC algorithm can detect the area with high color contrast in the image, but it will lead to the error identification of the background area with high color contrast as the foreground. The CA method highlights the edge of the saliency region, but the internal detection of the saliency objects is incomplete. BS, SR and MR fully highlight the whole significant region, but the resistance to the background is poor. On the basis of the highlight of the entire image saliency region, the algorithm of this paper also better resists the non-saliency region in the image. From Fig. 8, we can see that the algorithm has better detection accuracy, and can highlight the whole salient target uniformly and evenly. It is closer to Ground Truth, and the final saliency map has higher accuracy and completeness.

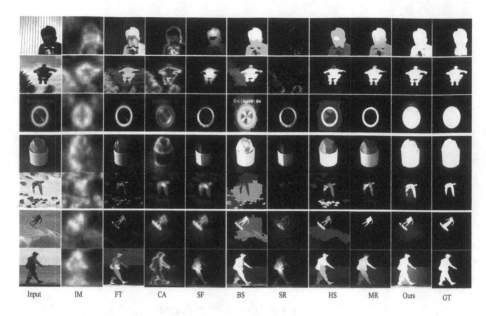

Fig. 8. Saliency maps computed by different state-of-art methods

4.3 Comparison with Single Layer Algorithm

This algorithm makes full use of the feature information between multi-layer images, and it can be seen from Fig. 9 that there are different effect images for different layers, the Layler1 contains small structure information, so the effect is the worst. The larger size of the two layers of Layer2 and Layer3 can achieve better results, but there is still a problem related to the scale.

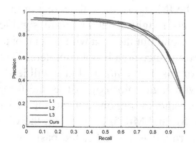

Fig. 9. Comparison of multi-layers result on CSSD dataset

5   Conclusion

In this paper, to overcome the shortage that the spatial connectivity of every node is modeled only via the K-regular graph and neglects the level information in existing salient object detection method based on graph-based manifold ranking, an improved method is proposed to increase the precision while preserving the high recall. It is connected with a super pixel which belongs to a color class and is located in the same connected region. It can be more reasonable and accurate to describe the correlation between the nodes in the graph model. By using multi-scale analysis to generate different results map of saliency detection, and multiple saliency graphs are fused by using graph model. Comparison experiments of 8 saliency methods on 3 types of saliency test datasets, and the experimental results show the effectiveness and superiority of the proposed algorithm in this paper.

The next work is to construct a multi-graph model with hierarchical information, and to improve the accuracy of seed selection; secondly, it is to consider to use high-level information such as image contour to improve the accuracy of the complex scene saliency object detection. At the same time, we should consider the application of this algorithm to specific detection tasks such as target tracking and object recognition to improve the effectiveness and accuracy of the existing algorithms.

Funding. This study was funded by Natural Science Foundation of Fuyang Normal University (2018FSKJ04ZD, 2018FSKJ01ZD, 2017FSKJ11), Natural Science Foundation of Anhui Province (1808085QF209), National natural science foundation project (61673117).

References

1. Hong, R., et al.: Beyond search: Event-driven summarization for web videos. TOMCCAP **7**(4), 35:1–35:18 (2011)
2. Borji, A., Cheng, M.M., Jiang, H., et al.: Saliency object detection: a survey. arXiv preprint arZXiv, 2014: pp. 1411–5878 (2014)
3. Hong, R., et al.: Coherent semantic-visual indexing for large-scale image retrieval in the cloud. IEEE Trans. Image Process. **26**(9), 4128–4138 (2017)
4. Hong, R., Hu, Z., Wang, R., Wang, M., Tao, D.: Multi-View object retrieval via multi-scale topic models. IEEE Trans. Image Process. **25**(12), 5814–5827 (2016)
5. Hong, R., Zhang, L., Tao, D.: Unified photo enhancement by discovering aesthetic communities from flickr. IEEE Trans. Image Process. **25**(3), 1124–1135 (2016)
6. Hong, R., Zhang, L., Zhang, C., Zimmermann, R.: Flickr circles: aesthetic tendency discovery by multi-view regularized topic modeling. IEEE Trans. Multimed. **18**(8), 1555–1567 (2016)
7. Itti, L., Koch, C., Niebur, E.: A model of saliency-based visual attention for rapid scene analysis. Comput. Soc. **20**(11), 1254–1259 (1998)
8. Achanta, R., Hemami, S., Estrada, F., Susstrunk, S.: Frequency-tuned salient region detection. In: Proceedings of the IEEE International Conference on Computer Vision and Pattern Recognition. Miami, USE, vol. 22(9–10), pp. 1597–1604. IEEE Press (2009)
9. Goferman, S., Zelnik-Manor, L., Tal, A.: Context-aware saliency detection. IEEE Trans. Pattern Anal. Mach. Intell. **34**(10), 1915–1926 (2012)
10. Cheng, M.M., Mitra, N.J., Huang, X., Torr, P.H., Hu, S.M.: Global contrast based salient region detection. IEEE Trans. Pattern Anal. Mach. Intell. **37**(3), 569–582 (2015)
11. Xie, Y., Lu, H., Yang, M.H.: Bayesian saliency via low and mid level cues. IEEE Trans. Image Process. **22**(5), 1689–1698 (2013)

12. Wei, Y., Wen, F., Zhu, W., Sun, J.: Geodesic saliency using background priors. In: Fitzgibbon, A., Lazebnik, S., Perona, P., Sato, Y., Schmid, C. (eds.) ECCV 2012. LNCS, vol. 7574, pp. 29–42. Springer, Heidelberg (2012). https://doi.org/10.1007/978-3-642-33712-3_3

13. Zhu, W.J., Liang, S., Wei, Y.C.: Saliency optimization from robust background detection. In: Proceedings of the IEEE International Computer Vision & Pattern Recognition, vol. 1049-1050(8), pp. 2814-2821 (2014)

14. Jiang, Y., Tan, L., Wang, S.: Saliency detected model based on selective edges prior. J. Electron. Inf. Technol. 37(1), 130–136 (2015)

15. Cheng, X., Du, P., Guo, J.: Ranking on data manifold with sink points. IEEE Trans. Knowl. Data Eng. 25(1), 177–191 (2012)

16. Jiang, B.W., Zhang, L.H., Lu, H.C.: Saliency detection via absorbing markov chain. In: Proceedings of the IEEE International Conference on Computer Vision, Sydney, pp. 1665–1672 (2014)

17. Yang, C., Zhang, L.H., Lu, H.C.: Saliency detection via graph-based manifold ranking. In: Proceedings of the IEEE International Conference on Computer Vision and Pattern Recognition, Portland, vol. 9(4), pp. 3166–3173 (2013)

18. Perazzi, F., Krahenbuhl, P., Pritch, Y., Hornung, A.: Saliency filters: contrast based filteringfor salient region detection. In: IEEE Conference on Computer Vision & Pattern Recognition, vol. 157(10), pp. 733–740 (2012)

19. Hou, X., Zhang, L.: Saliency detection: a spectral residual approach. In: IEEE Conference on Computer Vision & Pattern Recognition, pp. 1–8 (2007)

20. Xie, Y.L., Lu, H.C., Yang, M.H.: Bayesian saliency via low and mid-level cues. IEEE Trans. Image Process. 22(5), 1689–1698 (2013)

21. Murray, N., Vanrell, M., Otazu, X., Parraga, C.: A Saliency estimation using anonparametric low-level vision model. In: IEEE Conference on Computer Vision & Pattern Recognition, vol. 42(7), pp. 433–440 (2011)

22. Yan, Q., Li, X., Shi, J., Jia, J.: Hierarchical saliency detection. In: IEEE Conference on Computer Vision & Pattern Recognition, vol. 38(4), pp. 1155–1162 (2013)

Discrete Manifold-Regularized Collaborative Filtering for Large-Scale Recommender Systems

Deming Zhai[1](✉), Ao Li[1], Yang Li[1], Yang Liu[1], Guojun Liu[1], Xianming Liu[1], and Maozu Guo[2]

[1] School of Computer Science and Technology,
Harbin Institute of Technology, Harbin, China
zhaideming@hit.edu.cn
[2] School of Electrical and Information Engineering,
Beijing University of Civil Engineering and Architecture, Beijing, China

Abstract. In many online web services, precisely recommending relevant items from massive candidates is a crucial yet computationally expensive task. To confront with the scalability issue, discrete latent factors learning is advocated since it permits exact top-K item recommendation with sub-linear time complexity. However, the performance of existing discrete methods is limited due to they only consider the cross-view user-item relations. In this paper, we propose a new method called *Discrete Manifold-Regularized Collaborative Filtering* (DMRCF), which jointly exploits cross-view user-item relations and intra-view user-user/item-item affinities in the hamming space. On one hand, inspired by the observation that similar users are more likely to prefer similar items, manifold regularization terms are introduced to enforce similar users/items have similar binary codes in the hamming space. Accordingly, our method is able to learn more about the preference of user and then recommend items based on user's attributes. On the other hand, for cross-view relations, we cast the reconstruction errors of user-item rating matrix and ranking loss for relative preferences of users into a joint learning framework. Due to latent factors are restricted to be binary values, the optimization is generally a challenging NP-hard problem. To reduce the quantization error, we develop an efficient algorithm to solve the overall discrete optimization problem. Experiments on two real-world datasets demonstrate that DMRCF outperforms the state-of-the-art methods significantly.

1 Introduction

With the rapid development of e-commerce in the Internet, recommender systems have become more and more popular in recent years. Recommender systems aim to help users find their desirable items, and have been applied in

This work was supported by Natural Science Foundation of China under Grant No. 61502122, 61672193, 61671188, and 61571164, and China Postdoctoral Science Foundation funded project 2018M630360.

© Springer Nature Switzerland AG 2018
R. Hong et al. (Eds.): PCM 2018, LNCS 11164, pp. 513–523, 2018.
https://doi.org/10.1007/978-3-030-00776-8_47

various web services such as Amazon, eBay, Taobao, Netflix and iTunes. However, within most of these web services, the number of users and items is dramatically growing, making recommendation more challenging than ever before. For instance, Amazon.com has a total of 573 million products on sale in November 2017. Therefore, it is really challenging to give prompt response to find out user-prefered items in a million-scale collection by analyzing extremely sparse user history. Moreover, there are more than 300 millions of such active Amazon users for recommendation.

In the literature, a variety of approaches have been proposed for recommender systems [1–3], which can be roughly classified into collaborative filtering algorithms, content-based methods, and the combination of both kinds [3]. Among all existing work, latent factor based Collaborative Filtering (CF) has been shown to achieve great success in balancing accuracy and efficiency [4,5]. Specifically, CF factorizes the observed user-item rating matrix into two low-dimensional real-valued latent factors, such that their inner product possibly close to the original user-item rating matrix. In this way, the intrinsic relationship between users and items are established in the joint latent space. Finally, recommendation by CF naturally falls into a similarity search problem, $i.e.$, top-K item recommendation for a user can be casted into finding the top-K similar items queried by the user [6]. In general, to find top-K preferred items for each user, recommender systems need to compute the similarity between user and all items, and ranking them in descending order. Since all items in the database are scanned, the time complexity for one user is $\mathcal{O}(n)$, where n is the number of items. However, when item size n is very large, it becomes infeasible to compute the preferences for all items due to the prohibitive computation costs.

To confront with the scalability issue for recommendation, some researchers advocated the use of discrete encodings ($a.k.a.$, hash techniques), from real-valued vectors into compact binary codes. In this case, the inner product could be efficiently achieved by bit-wise XOR operations, $i.e.$ Hamming distance, with just a few machine instructions per comparison. It has been shown that in hamming space one can perform exact similarity search remarkably faster than linear search, with constant $\mathcal{O}(1)$ or sub-linear $\mathcal{O}(log(n))$ time complexity [7] by exploiting lookup tables. Moreover, Compact binary codes can greatly reduce the storage cost such that we can load massive data into memory to facilitate the subsequent processing. Up to now, a few attempts have been made for large-scale recommender systems. For instance, the work in [8] and [9] proposed to learn the latent factors via two independent stages: real-valued optimization and binary quantization. Later, Discrete Collaborative Filtering (DCF) method [6] was put forward to directly optimize the discrete latent factors learning problem without discarding the binary constraints, which could significantly reduce the quantization errors compared to previous two-stage approaches. More recently, Discrete Personalized Ranking (DPR) [10] method further considered the ranking order of items for a user in the objective function.

While promising progress has been made, the performance of state-of-the-art works is still far from satisfactory for the task of large-scale recommendation.

One limitation is that existing discrete methods of CF only consider the cross-view user-item relations, but ignore the auxiliary content information, which has shown to be helpful in traditional recommender systems. In this work, we propose a new method called *Discrete Manifold-Regularized Collaborative Filtering* (DMRCF), which jointly considers the *cross-view* user-item relations and *intra-view* user-user/item-item affinities for discrete latent factors learning. Our method is inspired by the following observations: (1) users with similar attributes (*e.g.* age, education, and social relationships) are more likely to favor similar items; (2) users who are interested in an item may also favor other items of the same kind (*e.g.* books/songs of the same writer/singer). Accordingly, in the aspect of intra-view relations, we introduce additional manifold regularization terms [11] to preserve the local structures of user/item modalities, which enforces similar users/items have similar binary codes in the hamming space. As a result, our method is able to learn more about the preference of user to recommend items based on user's attributes, and transfer the knowledge from old users/items to handle new ones, which is the well-known cold-start problem. In addition, in the aspect of cross-view relations, we cast the reconstruction errors of user-item rating matrix and ranking loss for relative preferences of users into a unified learning framework. An optimization algorithm is then developed to efficiently solve the discrete constrained objective function. Experimental evaluations on two large-scale recommendation datasets show that DMRCF outperforms the state-of-the-art method significantly.

The main contributions of this paper are highlighted in the following:

- To the best of our knowledge, this is the first work that jointly exploits cross-view user-item relations and intra-view user-user/item-item affinities in the hamming space for large-scale recommendation.
- We propose a unified learning framework to combine the reconstruction errors of user-item rating matrix and ranking loss for relative preferences.
- Efficient optimization algorithm is developed to directly solve the overall objective with discrete binary constraints.

The rest of this paper is organized as follows. In Sect. 2, we describe the proposed DMRCF method in detail. Experiments and analysis are presented in Sect. 3. Finally, Sect. 4 gives some concluding remarks.

2   Discrete Manifold-Regularized Collaborative Filtering

2.1   Notations

Suppose that we have an observed user-item rating matrix $\mathbf{S}^{xy} \in \mathbb{R}^{m \times n}$, which consists of m users and n items, and each entry $\mathbf{S}^{xy}_{ij}$ represents the rating of user i on item j. The r-bit discrete binary latent factors for users and items are denoted as $\mathbf{B} = [\mathbf{b}_1, ..., \mathbf{b}_m] \in \{\pm 1\}^{r \times m}$ and $\mathbf{D} = [\mathbf{d}_1, ..., \mathbf{d}_n] \in \{\pm 1\}^{r \times n}$, respectively.[1] Then the user-item similarity search can be efficiently conducted

---

[1] Here we generate binary latent factors as $\{-1, 1\}$, which are straightforward to convert to $\{0, 1\}$ valued latent factors.

in the Hamming space via fast XOR operations between $\mathbf{b}_i$ and $\mathbf{d}_j$. Moreover, we construct user-user/item-item similarity matrix $\mathbf{S}^x \in \mathbb{R}^{m \times m}$ and $\mathbf{S}^y \in \mathbb{R}^{n \times n}$ according to users/items' attributes, categories or relationships. Besides, let $\mathbf{I}_n$ be a $n \times n$ identity matrix; $\mathbf{1}$ and $\mathbf{0}$ be the vector of all ones and zeros.

2.2 Objective Function

To pursue good discrete latent factors, three important criteria are taken into account in our method: (1) The observed user-item rating matrix should be preserved as much as possible, such that the learned binary codes can well reflect the relationship between users and items; (2) In order to obtain a personalized total ranking of all items, the relative item-ranking orders for users should be preserved in the hamming space. Specifically, we denote $\mathbf{T}^{xy}_{ij_1j_2} = \mathbf{S}^{xy}_{ij_1} - \mathbf{S}^{xy}_{ij_2}$ as the comparison of $\mathbf{S}^{xy}_{ij_1}$ and $\mathbf{S}^{xy}_{ij_2}$. If $\mathbf{T}^{xy}_{ij_1j_2} > 0$, we conclude that user i prefers item j_1 over j_2; otherwise, user i prefers item j_2 over j_1. Accordingly, the personalized preference set could be presented as $\mathcal{T} = \{(i, j_1, j_2)|$ user i prefers item j_1 over $j_2\}$. We expect that, for a particular user i, positive items will get higher scores than other items could be satisfied as much as possible; (3) The intra-view user-user/item-item relations should be exploited to enforce users/items with high similarity have similar binary latent factors in the hamming space. That is to say, the local topological structures for user/item modality are preserved in the hamming space. In this way, our method is able to learn more about the preference of user to recommend items based on user's attributes, and transfer the knowledge from old users/items to handle new ones, which is the well-known cold-start problem.

We propose to achieve the above design objectives by minimizing the following energy function $w.r.t.$ discrete latent factors $\mathbf{B}$ and $\mathbf{D}$:

$$\underset{\mathbf{B},\mathbf{D}}{\arg\min}(1-\rho)\sum_i \sum_{j \in \mathcal{S}^{xy}_{i*}} (\mathbf{S}^{xy}_{ij} - \mathbf{b}_i^\top \mathbf{d}_j)^2 + \rho \sum_{(i,j_1,j_2) \in \mathcal{T}} (\mathbf{T}^{xy}_{ij_1j_2} - \mathbf{b}_i^\top(\mathbf{d}_{j_1} - \mathbf{d}_{j_2}))^2$$

$$+ \mu_x/2 \sum_i \sum_{j \in \mathcal{S}^x_i} \mathbf{S}^x_{ij}||\mathbf{b}_i - \mathbf{b}_j||^2 + \mu_y/2 \sum_i \sum_{j \in \mathcal{S}^y_i} \mathbf{S}^y_{ij}||\mathbf{d}_i - \mathbf{d}_j||^2 \qquad (1)$$

$$s.t. \mathbf{B} \in \{\pm 1\}^{r \times m}, \mathbf{D} \in \{\pm 1\}^{r \times n},$$

$$\mathbf{B}\mathbf{1}_m = \mathbf{0}, \mathbf{B}\mathbf{B}^\top = m\mathbf{I}_r, \mathbf{D}\mathbf{1}_n = \mathbf{0}, \mathbf{D}\mathbf{D}^\top = n\mathbf{I}_r,$$

where $\mathcal{S}^{xy}_{i*}$ indicates the set of items that user i rated; $\mathcal{S}^x_i$ and $\mathcal{S}^y_i$ are the user/item i's nearest neighbors sets, respectively. $\rho, \mu_x,$ and μ_y are the regularization parameters to balance these three kinds of loss terms. The first term in Eq. (1) measures the reconstruction errors of the observed user-item ratings, and the second term is the ranking loss which penalizes the relative preferences for users. Inspired by manifold regularization [11], the last two terms serve to preserve the intra-view local structures in user/item modalities, which will enforce similar users/items have similar binary codes in the hamming space. Moreover, we impose additional constraints of scaling invariance and bits decorrelation [6] to obtain compact binary latent factors.

Generally speaking, optimization of Eq. (1) is a constrained NP-hard problem, since the latent factors are restricted to be binary values. To solve this discrete optimization problem tractably [12], we introduce two auxiliary continuous variables $\mathbf{X}$ and $\mathbf{Y}$ satisfying the constraints $\mathbf{X}\mathbf{1}_m = \mathbf{0}, \mathbf{X}\mathbf{X}^\top = m\mathbf{I}_r, \mathbf{Y}\mathbf{1}_n = \mathbf{0}, \mathbf{Y}\mathbf{Y}^\top = n\mathbf{I}_r$, and minimize the deviation between the binary latent factors and their continuous real-value variables via $\|\mathbf{B} - \mathbf{X}\|_F^2$ and $\|\mathbf{D} - \mathbf{Y}\|_F^2$. Finally, the optimization problem in Eq. (1) becomes:

$$\underset{\mathbf{B},\mathbf{D},\mathbf{X},\mathbf{Y}}{\arg\min} (1-\rho) \sum_i \sum_{j \in \mathcal{S}_{i*}^{xy}} (\mathbf{S}_{ij}^{xy} - \mathbf{b}_i^\top \mathbf{d}_j)^2 + \rho \sum_{(i,j_1,j_2) \in \mathcal{T}} (\mathbf{T}_{ij_1j_2}^{xy} - \mathbf{b}_i^\top (\mathbf{d}_{j_1} - \mathbf{d}_{j_2}))^2$$

$$+ \mu_x/2 \sum_i \sum_{j \in \mathcal{S}_i^x} \mathbf{S}_{ij}^x \|\mathbf{b}_i - \mathbf{b}_j\|^2 + \mu_y/2 \sum_i \sum_{j \in \mathcal{S}_i^y} \mathbf{S}_{ij}^y \|\mathbf{d}_i - \mathbf{d}_j\|^2$$

$$+ \alpha\|\mathbf{B} - \mathbf{X}\|_F^2 + \beta\|\mathbf{D} - \mathbf{Y}\|_F^2$$

$$s.t. \mathbf{B} \in \{\pm 1\}^{r \times m}, \mathbf{D} \in \{\pm 1\}^{r \times n},$$

$$\mathbf{X}\mathbf{1}_m = \mathbf{0}, \mathbf{X}\mathbf{X}^\top = m\mathbf{I}_r, \mathbf{Y}\mathbf{1}_n = \mathbf{0}, \mathbf{Y}\mathbf{Y}^\top = n\mathbf{I}_r, \qquad (2)$$

where the last two term are called discrete fitting error losses; α and β are the penalty parameters. If α and β are imposed a very large value, the solutions of (2) will be close to those of (1) in theory. In practice, α and β are set appropriate values to limit the discrepancy of the additional hard constraint, such that optimization of (2) is more flexible. Therein, we still enforce explicit binary constraint in order to achieve discrete latent factors with good quality.

2.3   Optimization

We solve the problem in Eq. (2) via alternating procedure [13]. Specifically, we iteratively optimize *w.r.t.* one variable while keeping all other variables fixed until convergence or arriving at the maximum iteration number T. It is worth highlighting that the $\mathbf{B}/\mathbf{D}$-subproblems seek binary latent factors that preserve the intrinsic cross-view user-item relations/preferences and intra-view user-user/item-item local affinities; while $\mathbf{X}/\mathbf{Y}$-subproblems attempt to regularize the learned binary factors should be as balanced and uncorrelated as possible.

B-subproblem: *Optimizing w.r.t.* $\mathbf{B}$ *when* $\mathbf{D}$, $\mathbf{X}$, $\mathbf{Y}$ *are fixed*, the objective function in Eq. (2) becomes:

$$\underset{\mathbf{b}_i \in \{\pm 1\}^r}{\arg\min} (1-\rho) \sum_i \sum_{j \in \mathcal{S}_{i*}^{xy}} (\mathbf{S}_{ij}^{xy} - \mathbf{b}_i^\top \mathbf{d}_j)^2$$

$$+ \rho \sum_{(i,j_1,j_2) \in \mathcal{T}} (\mathbf{T}_{ij_1j_2}^{xy} - \mathbf{b}_i^\top (\mathbf{d}_{j_1} - \mathbf{d}_{j_2}))^2$$

$$+ \mu_x/2 \sum_{j \in \mathcal{S}_i^x} \mathbf{S}_{ij}^x \|\mathbf{b}_i - \mathbf{b}_j\|^2 + \alpha\|\mathbf{b}_i - \mathbf{x}_i\|^2. \qquad (3)$$

Although the above minimizing problem is in quadratic form, it is still a nontrivial problem due to the binary constraints. It can be observed that if we only learn one-bit binary code at a time, then the optimization problem can be solved with a closed-form solution. Accordingly, we will learn $\mathbf{b}_i$ bit by bit via discrete cyclic coordinate descent (DCCD) [14,15] method. Let $\mathbf{b}_{ik}$ be the k-bit of $\mathbf{b}_i$ and $\mathbf{b}_{i\bar{k}}$ be remaining subvector by deleting element $\mathbf{b}_{ik}$. In analogous, $\mathbf{d}_{jk}$ and $\mathbf{d}_{j\bar{k}}$ have similar meanings as that for $\mathbf{b}_{ik}$ and $\mathbf{b}_{i\bar{k}}$. After a simple algebra derivation and neglecting the constant terms, optimization $w.r.t.$ $\mathbf{b}_{ik}$ in Eq. (3) becomes:

$$\underset{\mathbf{b}_{ik}\in\{\pm1\}}{\arg\min} \quad -\hat{\mathbf{b}}_{ik}\mathbf{b}_{ik}, \tag{4}$$

where

$$\hat{\mathbf{b}}_{ik} = (1-\rho)\sum_{j\in\mathcal{S}_{i*}^{xy}}(\mathbf{S}_{ij}^{xy} - \mathbf{d}_{j\bar{k}}^{\top}\mathbf{b}_{i\bar{k}})\mathbf{d}_{jk}$$

$$+\rho\sum_{(i,j_1,j_2)\in\mathcal{T}}(\mathbf{T}_{ij_1j_2}^{xy} - (\mathbf{d}_{j_1\bar{k}} - \mathbf{d}_{j_2\bar{k}})^{\top}\mathbf{b}_{i\bar{k}})(\mathbf{d}_{j_1k} - \mathbf{d}_{j_2k})$$

$$+\alpha\mathbf{x}_{ik} + \mu_x/2\sum_{j\in\mathcal{S}_i^x}\mathbf{S}_{ij}^x\mathbf{b}_{jk}.$$

Due to the space limitation, the derivation is omitted here.
As a consequence, the optimal one-bit binary latent factor $\mathbf{b}_{ik}$ can be obtained with a closed form solution as:

$$\mathbf{b}_{ik} = \begin{cases} sgn(\hat{\mathbf{b}}_{ik}), \hat{\mathbf{b}}_{ik} \neq 0; \\ \mathbf{b}_{ik}, \qquad \hat{\mathbf{b}}_{ik} = 0, \end{cases} \tag{5}$$

where sgn($\cdot$) denotes the sign function. Finally, the solution of $\mathbf{B}$ could be derived by iteratively updating each $\mathbf{b}_{ik}$ until convergence.

D-subproblem: *Optimizing w.r.t. $\mathbf{D}$ when $\mathbf{B}$, $\mathbf{X}$, $\mathbf{Y}$ are fixed,* the objective function in Eq. (2) becomes:

$$\underset{\mathbf{d}_j\in\{\pm1\}^r}{\arg\min} (1-\rho)\sum_i\sum_{j\in\mathcal{S}_{i*}^{xy}}(\mathbf{S}_{ij}^{xy} - \mathbf{b}_i^{\top}\mathbf{d}_j)^2$$

$$+\rho\sum_{(i,j,j_2)\in\mathcal{T}}(\mathbf{T}_{ijj_2}^{xy} - \mathbf{b}_i^{\top}(\mathbf{d}_j-\mathbf{d}_{j_2}))^2$$

$$+\rho\sum_{(i,j_1,j)\in\mathcal{T}}(\mathbf{T}_{ij_1j}^{xy} - \mathbf{b}_i^{\top}(\mathbf{d}_{j_1}-\mathbf{d}_j))^2$$

$$+\mu_y/2\sum_{i\in\mathcal{S}_j^y}\mathbf{S}_{ij}^y\|\mathbf{d}_i - \mathbf{d}_j\|^2 + \beta\|\mathbf{d}_j - \mathbf{y}_j\|^2. \tag{6}$$

It should be noted that $\mathbf{d}_j$ may appear in different positions of the personalized preference set $\mathcal{T} = \{(i, j_1, j_2)\}$ for ranking loss term. Based on a similar derivation as **B**-*subproblem*, the optimal one-bit binary latent factor $\mathbf{d}_{jk}$ can be derived as:

$$\mathbf{d}_{jk} = \begin{cases} sgn(\hat{\mathbf{d}}_{jk}), \hat{\mathbf{d}}_{jk} \neq 0; \\ \mathbf{d}_{jk}, \qquad \hat{\mathbf{d}}_{jk} = 0, \end{cases} \tag{7}$$

where

$$\hat{\mathbf{d}}_{jk} = (1-\rho) \sum_{i \in \mathcal{S}_{*j}^{xy}} (\mathbf{S}_{ij}^{xy} - \mathbf{b}_{ik}^{\top}\mathbf{d}_{jk})\mathbf{b}_{ik}$$

$$+ \rho \sum_{(i,j,j_2) \in \mathcal{T}} \mathbf{d}_{j_2 k} + (\mathbf{T}_{ijj_2}^{xy} - \mathbf{b}_{ik}^{\top}(\mathbf{d}_{j\bar{k}} - \mathbf{d}_{j_2\bar{k}}))\mathbf{b}_{ik}$$

$$+ \rho \sum_{(i,j_1,j) \in \mathcal{T}} \left(\mathbf{d}_{j_1 k} - (\mathbf{T}_{ij_1 j}^{xy} - \mathbf{b}_{ik}^{\top}(\mathbf{d}_{j_1\bar{k}} - \mathbf{d}_{j\bar{k}}))\mathbf{b}_{ik} \right)$$

$$+ \beta \mathbf{y}_{jk} + \mu_y/2 \sum_{i \in \mathcal{S}_j^y} \mathbf{S}_{ij}^y \mathbf{d}_{ik}.$$

X/Y-subproblem: *Optimizing w.r.t.* $\mathbf{X}$ *when* $\mathbf{B}, \mathbf{D}, \mathbf{Y}$ *are fixed,* the objective function in Eq. (2) becomes:

$$\arg\min_{\mathbf{X} \in \mathbb{R}^{r \times m}} \|\mathbf{B} - \mathbf{X}\|_F^2, \quad s.t. \quad \mathbf{X}\mathbf{1}_m = 0, \mathbf{X}\mathbf{X}^{\top} = m\mathbf{I}_r. \tag{8}$$

This is a classic orthogonal procrustes problem, with the solution [16]: $\mathbf{X} = \sqrt{m}[\mathbf{P}_b \ \hat{\mathbf{P}}_b][\mathbf{Q}_b \ \hat{\mathbf{Q}}_b]^{\top}$, where $\mathbf{P}_b$ and $\mathbf{Q}_b$ are the left and right singular vectors of the row-centered matrix $\mathbf{B}(\mathbf{I}_m - \frac{1}{m}\mathbf{1}_m\mathbf{1}_m^{\top})$, $\hat{\mathbf{P}}_b$ is stacked by the eigenvectors of the zero eigenvalues and $\hat{\mathbf{Q}}_b$ can be calculated by Gram-Schmidt orthogonalization based on $[\mathbf{Q}_b \ \mathbf{1}_m]$.

Similar to $\mathbf{X}$-*subproblem*, the optimal solution of $\mathbf{Y}$ can be solved with a closed solution as $\mathbf{Y} = \sqrt{n}[\mathbf{P}_d \ \hat{\mathbf{P}}_d][\mathbf{Q}_d \ \hat{\mathbf{Q}}_d]^{\top}$.

3 Experiments

3.1 Experimental Settings

We conduct experiments on two publicly available datasets: MovieLens-1M [17], and Amazon [18]. **MovieLens-1M** consists of 1,000,209 ratings from 6,040 users to 3952 movies. All users in this dataset have the record including gender, age, occupation and Zip-code; while all movies are associated with genre labels. For **Amazon** dataset, we use a subset which contains 2,705,597 ratings from 155,517 users to 40,272 products. All products in Amazon can access the metadatdata such as descriptions, category, price, brand etc., but there is no demographic information for users. Besides, all ratings of both datasets are within $[0,5]$. The user-user/item-item similarity are defined as the number of identical attributes (without the Zip-code)/genres labels. We use 80% of the ratings as the training set and the rest to form the testing set for prediction.

We compare our method against several state-of-the-art large-scale recommendation algorithms. Specifically, five approaches are included in our comparative study: (1) Collaborative Hashing (CH) [9]; (2) Discrete Collaborative Filtering (DCF) [6]; (3) Discrete Personalized Ranking (DPR) [10]; (4) DMRCF($\rho = 0$) (our approach with reconstruction loss); (5) DMRCF($\rho = 1$) (our approach with ranking loss). In experiments, we empirically set the regularization parameters of our approach $\alpha = \beta = 10^{-3}$, while μ_x and μ_y are determined by cross-validation. And we also tune the parameters of all compared methods to achieve their best performances. The recommendation performance is measured in terms of NDCG (Normalized Discounted Cumulative Gain)@K metric, where K is the number of recommended items. NDCG has taken into account both ranking precision and the position of ratings. The larger value of NDCG indicates higher accuracy for recommendation.

Table 1. Results of NDCG@10 on MovieLens-1M dataset

Method	Code Length			
	$r = 8$	$r = 16$	$r = 32$	$r = 64$
CH	0.2028	0.2621	0.3371	0.4105
DCF	0.3500	0.3553	0.3789	0.4033
DPR	0.5439	0.4928	0.4936	0.5096
DMRCF($\rho = 0$)	0.3979	0.3793	0.3977	0.4187
DMRCF($\rho = 1$)	**0.5510**	**0.5100**	**0.5158**	**0.5171**

Table 2. Results of NDCG@10 on Amazon dataset

Method	Code Length			
	$r = 8$	$r = 16$	$r = 32$	$r = 64$
CH	0.1397	0.2913	0.4659	0.6306
DCF	0.2307	0.2718	0.3669	0.4643
DPR	0.1973	0.3008	0.4710	0.7671
DMRCF($\rho = 0$)	**0.2411**	0.3122	0.4149	0.5306
DMRCF($\rho = 1$)	0.2203	**0.3185**	**0.4810**	**0.7803**

3.2 Recommendation Results and Analysis

The NDCG@10 results with various code lengths on both databases are summarized in Tables 1 and 2. As shown in these two tables, the Collaborative Hashing (CH) method, which learn the latent factors via two independent stages: real-valued optimization and binary quantization, yields poor performance due to large quantization errors. The DPR and DMRCF($\rho = 1$) methods, which consider the ranking order of items in the objective function, achieves higher accuracy than DCF and DMRCF($\rho = 0$) in most cases. Among all compared methods, the proposed DMRCF($\rho = 1$) achieves the best overall results. Benefiting from the intra-view user-user/item-item similarity preserving, our DMRCF method leads to significant performance benefits.

We further illustrate the NDCG@K curves of both datasets in Fig. 1. We can find that, DMRCF($\rho = 0$) outperforms DCF, and DMRCF($\rho = 1$) get higher

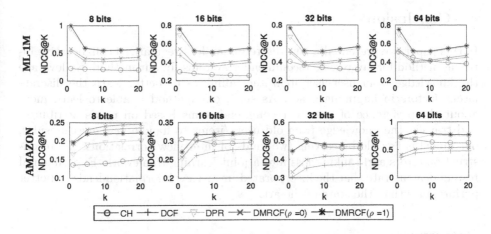

Fig. 1. The NDCG@K curves on MovieLens-1M and Amazon datasets

accuracy than DPR under all settings. In almost all cases, DMRCF($\rho = 1$) achieves the best performance. It further verifies that the proposed DMRCF method is more robust and can better model the user-item relations by leveraging the knowledge from user-user/item-item affinities.

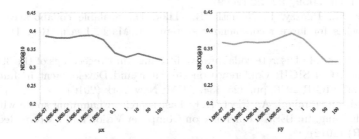

Fig. 2. Parameter selection for μ_x and μ_y on MovieLens-1M dataset

In our method, there are two important parameters: the intra-view manifold regularization factors μ_x and μ_y for user/item modalities. To illustrate their influence, we take MovieLens-1M dataset as the testbed by varying the two parameters from 10^{-7} to 10^2 with 8-bit binary latent factors length. As depicted in Fig. 2, the overall trend increases firstly and decreases with the increasing value of μ_x and μ_y, and the best choice is $\mu_x = 10^{-3}$ and $\mu_y = 0.1$.

4   Conclusion

In this paper, we presented a new discrete manifold-regularized collaborative filtering method for large-scale recommender systems. In our method, we leverage the knowledge from user-user/item-item affinities to better learn the discrete latent factors in hamming space. As such, our method is able to learn more about the preference of user to recommend items based on user's attributes, and transfer the knowledge from old users/items to handle new ones. Besides, our model is flexible by casting two-kinds of inter-view correlations: the reconstruction errors and ranking loss into a joint learning framework. Experimental results demonstrate that the incorporated intra-view information leads to better performance than the state-of-the-art.

References

1. Bobadilla, J., Ortega, F., Hernando, A., GutiéRrez, A.: Recommender systems survey. Know. Based Syst. **46**, 109–132 (2013)
2. Melville, P., Sindhwani, V.: Recommender systems. In: Encyclopedia of Machine Learning and Data Mining (2010)
3. Ekstrand, M.D., Riedl, J.T., Konstan, J.A.: Collaborative filtering recommender systems. Found. Trends Hum. Comput. Interact. **4**(2), 81–173 (2011)
4. Su, X., Khoshgoftaar, T.M.: A survey of collaborative filtering techniques. Adv. in Artif. Intell. **2009**, 4:2–4:2 (2009)
5. Takács, G., Pilászy, I., Németh, B., Tikk, D.: Scalable collaborative filtering approaches for large recommender systems. J. Mach. Learn. Res. **10**, 623–656 (2009)
6. Zhang, H., et al.: Discrete collaborative filtering. In: Proceedings of the 39th International ACM SIGIR Conference on Research and Development in Information Retrieval, SIGIR 2016, pp. 325–334. ACM, New York (2016)
7. Norouzi, M., Punjani, A., Fleet, D.J.: Fast search in hamming space with multi-index hashing. In: IEEE Conference on Computer Vision and Pattern Recognition (CVPR) (2012)
8. Zhou, K., Zha, H.: Learning binary codes for collaborative filtering. In: Proceedings of the 18th ACM SIGKDD International Conference on Knowledge Discovery and Data Mining, KDD 2012, pp. 498–506 (2012)
9. Liu, X., He, J., Deng, C., Lang, B.: Collaborative hashing. In: 2014 IEEE Conference on Computer Vision and Pattern Recognition, pp. 2147–2154, June 2014
10. Zhang, Y., Lian, D., Yang, G.: Discrete personalized ranking for fast collaborative filtering from implicit feedback. In: AAAI (2017)
11. Belkin, M., Niyogi, P.: Laplacian eigenmaps for dimensionality reduction and data representation. Neural Comput. **15**, 1373–1396 (2003)
12. Hanke, M., Hansen, P.C.: Regularization methods for large-scale problems. Surv. Math. Ind. **3**(4), 253–315 (1993)
13. Bertsekas, D.P.: Nonlinear programming. Athena Scientific (1999)
14. Yun, S.: On the iteration complexity of cyclic coordinate gradient descent methods. SIAM J. Optim. **24**(3), 1567–1580 (2014)
15. Shen, F., Shen, C., Liu, W., Shen, H.T.: Supervised discrete hashing. In: The IEEE Conference on Computer Vision and Pattern Recognition (CVPR), June 2015

16. Liu, W., Mu, C., Kumar, S., Chang, S.-F.: Discrete graph hashing. In: Ghahramani, Z., Welling, M., Cortes, C., Lawrence, N.D., Weinberger, K.Q. (eds.) Advances in Neural Information Processing Systems 27, pp. 3419–3427. Curran Associates Inc. (2014)

17. Harper, F.M., Konstan, J.A.: The movielens datasets: history and context. ACM Trans. Interact. Intell. Syst. **5**(4), 19:1–19:19 (2015)

18. He, R., McAuley, J.: Ups and downs: modeling the visual evolution of fashion trends with one-class collaborative filtering. In: Proceedings of the 25th International Conference on World Wide Web, WWW 2016. International World Wide Web Conferences Steering Committee, pp. 507–517 (2016)

Multiscale Cascaded Scene-Specific Convolutional Neural Networks for Background Subtraction

Jian Liao, Guanjun Guo, Yan Yan, and Hanzi Wang[✉]

Fujian Key laboratory of Sensing and Computing for Smart City,
School of Information Science and Engineering, Xiamen University,
Xiamen 361005, China
hanzi_wang@163.com

Abstract. Recent years have witnessed the widespread success of convolutional neural networks (CNNs) in computer vision and multimedia. The CNNs based background subtraction methods, which are effective for addressing the challenges (such as shadows, dynamic backgrounds, illumination changes) existing in real-world applications, have attracted much attention. However, these methods usually require a large amount of densely labeled video training data, which are hardly collected in the real-world. To address this problem, in this paper, we propose a multiscale cascaded scene-specific CNNs based background subtraction method equipped with a novel training strategy, which takes advantage of the balance of positive and negative training samples. The proposed method can rely on a small number of training samples to effectively train the robust neural network models. Experimental results on the CDnet-2014 dataset show that the proposed method obtains better performance with much less training samples compared with the state-of-the-art methods.

Keywords: Background subtraction
Multiscale cascaded scene-specific CNNs
Convolutional neural networks

1 Introduction

Background subtraction aims to construct a scene background model, which is used to detect the foreground moving objects. Background subtraction has been widely used in a variety of computer vision applications, such as video surveillance [1], human interaction [2], video coding [3] and object tracking [4]. Therefore, background subtraction becomes a popular research topic in the last few decades.

However, existing background subtraction methods can not effectively handle with some complex and challenging environments (such as night, rainy or snowy environments). For example, several recently-proposed methods (such as

© Springer Nature Switzerland AG 2018
R. Hong et al. (Eds.): PCM 2018, LNCS 11164, pp. 524–533, 2018.
https://doi.org/10.1007/978-3-030-00776-8_48

IUTIS-5 [5], WeSamBE [6] and ViBe-based method [7]) can not obtain satisfactory performance in the category of night video in the CDnet-2014 dataset [8], since strong illumination changes exist in this video category. In recent years, the CNNs based background subtraction methods have been proposed, which have shown significant advantages in handling with the above mentioned challenges. More specifically, Braham et al. [9] adopt the scene-specific CNNs for background subtraction, where the neural network models are independently trained for each specific scene. Note that the inputs of the scene-specific CNNs are the background images and current video frames. In other words, the scene-specific CNNs can not only learn the characteristics of the foreground objects, but also focus more on the differences between the background images and current video frames. Therefore, the scene-specific CNNs based method can robustly detect the foreground objects with less training samples. However, the drawback of this method is that the scene-specific CNNs are difficult to deal with shadows and strong illumination changes, since only single-scale neural network models are considered. Subsequently, CascadeCNN [10] adopts multiscale cascaded neural network models to obtain promising background subtraction performance. However, the CascadeCNN method only considers the characteristics of the foreground objects, and thus it needs a large number of training samples.

To overcome the performance degeneration of using a small amount of training samples in traditional deep learning based methods, in this paper, we propose a novel background subtraction method based on multiscale cascaded scene-specific CNNs. With the help of the multiscale cascaded neural network models, the performance of the proposed method can be effectively improved. In addition, a novel training strategy based on the balance of positive and negative training samples is also proposed, and thus the neural network models can be effectively trained with a small number of training samples. Compared with the Cascaded-CNN method, the proposed method can obtain better background subtraction performance with the less number of training samples.

2 The Proposed Method

In this section, we will introduce the structure of scene-specific CNNs in Sect. 2.1. Then, we will describe the multiscale cascaded scene-specific CNNs in Sect. 2.2. Finally, we will introduce the proposed training strategy and the implementation details of the proposed method in Sect. 2.3.

2.1 Scene-Specific Convolutional Neural Networks

The structure of the scene-specific CNNs [9] is given in Fig. 1, which have six layers (including two convolutional layers, two max-pooling layers, and two fully connected layers). The kernel size of the two convolutional layers is 5×5, and the stride of the two convolutional layers is 1. The number of channels in the first convolutional layer is 6, while that in the second convolutional layer is 16. Before the convolutional operation, the image patches or feature maps are surrounded

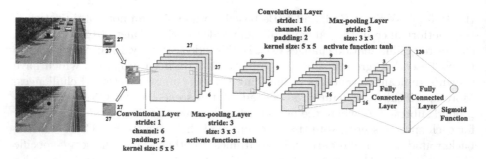

Fig. 1. The structure of the scene-specific CNNs. The size and number of channels of feature map are shown with red numbers in each layer.

with zero-padding, and the padding size is 2. The size of receptive fields of the two non-overlapping max-pooling layers is 3×3. The first fully connected layer has 120 hidden units. Based on the sigmoid function $f(x) = \frac{1}{1+e^{(-x)}}$, the second fully connected layer calculates the probability of each pixel belonging to the foreground in the current frame. Note that, compared with the activate function (i.e., $Relu(x) = max(0; x)$) used in the original scene-specific CNNs, we adopt the new activate function $tanh(x) = 2f(2x) - 1$, which shows better discriminative capability when fewer training samples are used. The loss function adopted in the scene-specific CNNs is defined as follows:

$$L = -\frac{1}{n} \sum_{i=1}^{n} (L_i ln(y_i) + (1 - L_i)ln(1 - y_i)) \tag{1}$$

where n is the number of pixels in the training video frames, L_i is the label of the pixel i, and y_i is the probability of the pixel i belonging to the foreground pixel.

As can be seen in Fig. 1, for each pixel, we can extract the pairs of image patches (the sizes are both 27×27) centered by the pixel from the background images and current video frames, respectively, which are used as the inputs of the first layer. Then the scene-specific CNNs calculate the probability of each center pixel belonging to the foreground. In other words, if the probability of the center pixel is larger than the threshold T_p, this pixel will be determined as the foreground pixel.

2.2 Multiscale Cascaded Scene-Specific CNNs

The structure of multiscale cascaded scene-specific CNNs is shown in Fig. 2, which consists of two kinds of scene-specific CNNs: multiscale scene-specific CNNs (three different scales are used) and cascaded scene-specific CNNs, respectively. We adopt the same way as introduced in [9] to obtain the background images. In the meanwhile, the three scene-specific CNNs make use of parameter sharing, which can not only reduce the number of parameters, but also can make the training of the multiscale scene-specific CNNs quickly converge. Since the outputs of multiscale scene-specific CNNs have different sizes, we resize the three different outputs into the same size and obtain the intermediate results. Based on the intermediate results, one cascaded scene-specific CNNs are used to further improve the performance. Note that the inputs of the cascaded scene-specific CNNs are the current video frames and the corresponding intermediate results. As we can see in Fig. 2, the intermediate results exist many foreground noises between the cars. However, by taking advantage of cascaded scene-specific CNNs, the foreground noises between the cars are perfectly suppressed in the final results.

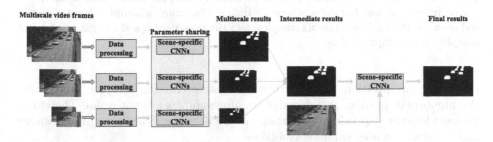

Fig. 2. The structure of multiscale cascaded scene-specific CNNs.

2.3   A Novel Training Strategy and Implementation Details

As introduced in Sect. 2.1, we can extract the pairs of image patches in the background images and current video frame for training the scene-specific CNNs. The training image patches are usually centered by background pixels or foreground pixels. However, the ratio between background pixels and foreground pixels is generally 100 : 1 or even 1000 : 1 in the training video frames, since the number of background pixels is much more than that of foreground pixels. Therefore, the distributions of positive and negative training samples (corresponding to foreground pixels and background pixels) are extremely imbalanced, which causes the longer training time and local optimum.

To address the above mentioned problems, we develop a novel training strategy based on the balance of positive and negative training samples. Specifically, as depicted in Fig. 3, we firstly construct a matrix consisting of positive and negative training samples. Here, each batch of training sample contains 100 pairs of image patches, and the center pixels of the 100 image patches are continuous. And the training samples correspond to the image patches centered by foreground pixels are considered as positive training samples. Otherwise, the training samples are considered as negative training samples. Obviously, the initial distributions of positive and negative training samples in the matrix are unbalanced. For example, as shown in Fig. 3, there are 646 negative training samples and 122 positive training samples existing in the matrix. Secondly, we

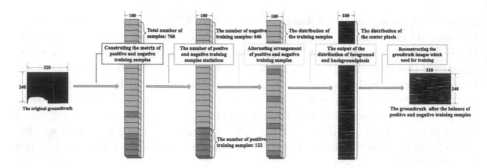

Fig. 3. The balance of positive and negative training samples. Each batch of the training sample consists of 100 pairs of image patches. The training samples denoted as the red ones are the positive training samples, and the gray ones are the negative training samples. (Color figure online)

re-arrange the positions of positive and negative training samples. Note that, the numbers of positive and negative training samples are not equal. Therefore, we calculate the gap value N_{gap} which can be used for re-arranging the positive and negative training samples as follows:

$$N_{gap} = \begin{cases} \lceil \frac{N_{neg}}{N_{pos}} \rceil & \text{if } N_{neg} > N_{pos} \\ \lceil \frac{N_{pos}}{N_{neg}} \rceil & \text{otherwise} \end{cases} \tag{2}$$

where N_{pos} is the number of positive training samples, N_{neg} is the number of negative training samples, and the $\lceil * \rceil$ is the ceil operation. Then the matrix of positive and negative training samples will be reorganized according to the value of N_{gap}. That is, if $N_{neg} > N_{pos}$, all the negative training samples constitute the initial matrix. Next, we insert a positive training sample in every N_{gap} negative training samples. Finally, the rest of positive training samples are randomly inserted into the matrix. An example is given in Fig. 3.

The implementation details of the proposed method contain two aspects. The first aspect is about the parameter settings. Firstly, the training frames of each video are resized to the size of 240×320. Then in the stage of multiscale scene-specific CNNs, we adopt three scales (i.e, $[1.00, 0.75, 0.50]$) to resize the original training images into three different sizes. And in the training process of multiscale scene-specific CNNs, the batch sizes of three different scales images are $[100, 57, 25]$, which means the number of training samples in the different scaled images are $[768, 758, 768]$, respectively. The number of training iteration and learning rate are set to 500 and 0.001, respectively, and we use the optimization strategy of RMSProp [11]. Moreover, the threshold T_p introduced in Sect. 2.1 is set to 0.60. The second aspect is about the training process of the proposed neural network models. We initially train the multiscale scene-specific CNNs, and then train the cascaded scene-specific CNNs based on the intermediate results obtained from the multiscale scene-specific CNNs.

3   Experiments and Analysis

In this section, the datasets of the experiments are introduced in Sect. 3.1. Then the experimental results and comparative analysis are given in Sect. 3.2.

3.1   Datasets

The adopted dataset in our experiments is the CDnet-2014 dataset [8]. There are eleven video categories in this dataset, which are Baseline (BL), Intermittent Object Motion (IOM), Dynamic Background (DB), Shadow (SD), Camera Jitter (CJ), Thermal (TM), Bad Weather (BW), Low Frame Rate (LFR), Night Videos (NV), Turbulence (TB) and PTZ. Each video category has four to six videos. In addition, each of the different video categories in this dataset corresponds to one or more specific challenging problems in background subtraction, such as dynamic backgrounds, hard shadows and illumination changes. Therefore, this dataset can well evaluate the comprehensive properties of the background subtraction methods. We use all the video categories in the CDnet-2014 dataset for evaluation, except for the PTZ video category which is captured by a moving camera (note that our background subtraction method is proposed for the fixed-camera scene).

In our experiments, the adopted evaluation metric is the F-Measure (FM), which can well reflect the performance of the background subtraction methods. More specifically, the value of F-Measure is calculated by $FM = (2 \times Recall \times Precision)/(Recall + Precision)$. The proposed method is implemented based on the deep learning framework of Theano and the language of python.

3.2   Experimental Results and Analysis

The aim of our experiments is three-fold. Firstly, in order to validate the effectiveness of the proposed method in the design of the neural network structure and the training strategy, we evaluate the performance of the proposed method under four different modes. For each mode, the neural network models of each video are trained with 5 video frames. Specifically, the four different modes are the proposed method without adopting the proposed training strategy based on the balance of positive and negative training samples (denoted as the non-strategy mode), the proposed method without adopting the cascaded CNNs (denoted as the non-cascaded mode), the proposed method without adopting the cascaded CNNs and the proposed training strategy (denoted as the non-strategy and cascaded mode), and the proposed method (denoted as the original mode), respectively. The experimental results of the proposed method under four different modes are shown in Table 1.

As we can see in Table 1, compared with the other modes, the original mode obtains the best experimental results, where it achieves the best in the eight video categories. This is mainly due to the multiscale cascaded neural network structures and the training strategy based on the balance of positive and negative training samples. We can also notice that, compared with the original mode, the

Table 1. The average F-Measure metric comparison obtained by the proposed method under four different modes on the CDnet-2014 dataset, where the best results are bold.

Video category	Non-strategy mode	Non-cascaded mode	Non-strategy and cascaded mode	Original mode
BL	0.938	0.924	0.931	**0.940**
IOM	**0.771**	0.763	0.758	0.770
DB	0.810	0.861	0.729	**0.881**
SD	0.889	0.914	0.892	**0.915**
CJ	0.681	0.732	0.623	**0.794**
TM	0.877	0.857	0.867	**0.883**
BW	**0.877**	0.860	0.854	0.861
LFR	0.648	0.698	0.639	**0.725**
NV	0.766	0.759	0.678	**0.788**
TB	0.798	0.872	0.743	**0.884**
Average	0.806	0.824	0.771	**0.844**

average F-Measure metric obtained by the non-strategy and cascaded mode is decreased by 7.30%. In addition, the experimental results obtained by the non-cascaded mode are better than the non-strategy mode, which means the training strategy based on the balance of positive and negative training samples is more effective than the strategy of adopting the structure of cascaded neural networks in terms of performance improvements. Moreover, in the video categories of IOM and BW, the experimental results obtained by the original mode are slightly worse than the non-strategy mode. This is because the original distributions of the positive and negative training samples in those two video categories are well enough to obtain the robust neural network models.

Secondly, in order to validate that our method can obtain the promising results based on the small number of training samples, we evaluate the performance of the proposed method and CascadeCNN method [10] with the same number of training samples. The experimental results are shown in Table 2. As we can see in Table 2, when the neural network models of the CascadeCNN method and the proposed method are trained by using 5 frames, 10 frames and 20 frames as the training samples, respectively, the experimental results obtained by the proposed method are higher than those obtained by CascadeCNN method by 5.00%, 2.83%, 1.30%, respectively. The experimental results well show that our method is more robust than the CascadeCNN method when using the same number of training samples. And we can also notice that the less the amount of training samples is used, the performance difference between the proposed method and CascadeCNN method is more significant. This is because that, the neural network models of the proposed method can not only focus on the characteristics of the foreground objects, but also can concentrate on the differences between the background images and current video frames, since the inputs of

Table 2. The average F-Measure metric comparison obtained by the proposed method and cascadeCNN method [10] on the CDnet-2014 dataset, where the best results are bold.

Video category	5 training frames		10 training frames		20 training frames	
	CascadeCNN	Ours	CascadeCNN	Ours	CascadeCNN	Ours
BL	0.910	**0.940**	0.943	**0.953**	**0.956**	0.955
IOM	0.723	**0.770**	0.750	**0.786**	0.784	**0.805**
DB	0.799	**0.881**	0.857	**0.892**	**0.930**	**0.930**
SD	0.881	**0.915**	0.919	**0.934**	0.936	**0.950**
CJ	**0.903**	0.794	**0.938**	0.862	**0.956**	0.907
TM	0.816	**0.883**	0.879	**0.918**	0.922	**0.924**
BW	0.857	**0.861**	0.925	**0.939**	0.936	**0.943**
LFR	0.643	**0.725**	0.726	**0.817**	0.798	**0.854**
NV	0.644	**0.788**	0.767	**0.833**	0.806	**0.854**
TB	0.767	**0.884**	0.856	**0.909**	0.882	**0.914**
Average	0.794	**0.844**	0.856	**0.884**	0.891	**0.904**

the proposed method are background images and current video frames. However, the neural network models of the CascadeCNN method only rely on the characteristics of the foreground objects, because the inputs of the CascadeCNN method are current video frames. On the other hand, in the video category of CJ, the experimental results obtained by CascadeCNN method are better than our method. The reason is that, the differences between background images and current video frames not only exist in the foreground objects area but also in the other areas.

Thirdly, to validate the robustness and efficiency of the proposed method, we select six state-of-the-art methods for performance comparison. We select three background subtraction methods based on background modeling, IUTIS-5 [5], WeSamBE [6] and Semantic BGS [12]. In addition, three background subtraction methods based on CNNs (Deep BS [13], MCFCN [14] and PSLCRF [15]) are selected for comparison.

As shown in Table 3, except for PSLCRF, the experimental results of the proposed method trained by using 5 frames as the training samples are better than the other methods. This illustrates that the proposed method can obtain the more robust performance based on the small amount of training samples. In addition, compared with PSLCRF, the average F-Measure metric of the proposed method trained by using 5 frames as the training samples are decreased by 4.30%. However, the average F-Measure metric of the proposed method trained by using 20 frames as the training samples is 1.70% higher than PSLCRF. We can also notice that, in the video categories of DB and TB, the experimental results of the proposed method trained by using 20 frames as the training samples are much better than PSLCRF. In the nine video categories, all the experimental results obtained by the proposed method trained by using 20 frames as the training

Table 3. The average F-Measure metric comparison obtained by the proposed method and six competing methods on the CDnet-2014 dataset, where the best, the second best and the third best results are marked with '|', '||', '|||' in the upper right corner of the numbers, respectively. The Ours-5, Ours-10 and Ours-20 represent the proposed method trained by using the 5 frames, 10 frames and 20 frames as the training samples, respectively. And the mark '*' represents the methods are trained by using half of the video frames.

Method	BL	IOM	DB	SD	CJ	TM	BW	LFR	NV	TB	Average
IUTIS-5 [5]	0.957	0.730	0.890	0.908	0.833	0.830	0.829	0.791	0.513	0.851	0.813
WeSamBE [6]	0.941	0.739	0.744	0.900	0.798	0.796	0.853	0.688	0.534	0.828	0.782
Semantic BGS [12]	0.960$^{||}$	0.788	0.949$^{|}$	0.948$^{|||}$	0.839	0.822	0.826	0.789	0.501	0.692	0.811
Deep BS* [13]	0.958$^{|||}$	0.610	0.876	0.930	0.899$^{|||}$	0.758	0.865	0.590	0.636	0.899$^{|||}$	0.802
MCFCN* [14]	0.968$^{|}$	0.907$^{|}$	0.772	0.929	0.899$^{|||}$	0.859	0.855	0.749	0.770	0.714	0.842
PSLCRF* [15]	0.929	0.845$^{||}$	0.822	0.965$^{|}$	0.952$^{|}$	0.944$^{|}$	0.946$^{|}$	0.808$^{|||}$	0.859$^{|}$	0.801	0.887$^{||}$
Ours-5	0.940	0.770	0.881	0.915	0.794	0.883	0.861	0.725	0.788	0.884	0.844
Ours-10	0.953	0.786	0.892$^{|||}$	0.934	0.862	0.918$^{|||}$	0.939$^{|||}$	0.817$^{||}$	0.833$^{|||}$	0.909$^{||}$	0.884$^{|||}$
Ours-20	0.955	0.805$^{|||}$	0.930$^{||}$	0.950$^{||}$	0.907$^{||}$	0.924$^{||}$	0.943$^{||}$	0.854$^{|}$	0.854$^{||}$	0.914$^{|}$	0.904$^{|}$

samples are top three. This indicates the performance of the proposed method can achieve the good tradeoff between accuracy and time in each video category. On the other hand, the neural network models of the Deep BS, MCFCN and PSLCRF methods are trained by using half of the video frames. Therefore, the proposed method can obtain the better experimental results by using much less training samples. In summary, compared with the state-of-the-art methods, the proposed method can achieve the more robust background subtraction results.

4   Conclusion

In this paper, we have proposed a novel background subtraction method which is based on multiscale cascaded scene-specific CNNs. Meanwhile, the training strategy based on the balance of positive and negative samples is also proposed in this paper, which helps to obtain the promising performance of the proposed method by using much less training samples. Experiments on the CDnet-2014 dataset have shown that, compared with the state-of-the-art methods, the proposed method can obtain the superior experimental results with much less number of training samples.

Acknowledgements. This work was supported by the National Natural Science Foundation of China under Grants U1605252, 61472334, 61571379, and by the Natural Science Foundation of Fujian Province of China under Grant 2017J01127.

References

1. Maddalena, L., Petrosino, A.: A self-organizing approach to background subtraction for visual surveillance applications. IEEE Trans. Image Process. **17**(7), 1168–1177 (2008)
2. Biswas, K.K., Basu, S.K.: Gesture recognition using microsoft kinect®. In: IEEE International Conference on Automation, Robotics and Applications, pp. 100–103 (2011)
3. Paul, M., Lin, W., Lau, C.T., Lee, B.S.: Pattern-based video coding with dynamic background modeling. EURASIP J. Adv. Signal Process. **2013**(1), 138–153 (2013)
4. Stauffer, C., Grimson, W.E.L.: Adaptive background mixture models for real-time tracking. In: IEEE Computer Society Conference on Computer Vision and Pattern Recognition, pp. 246–252 (1999)
5. Bianco, S., Ciocca, G., Schettini, R.: Combination of video change detection algorithms by genetic programming. IEEE Trans. Evol. Comput. **21**(6), 914–928 (2017)
6. Jiang, S., Lu, X.: WeSamBE: a weight-sample-based method for background subtraction. IEEE Trans. Circuits Syst. Video Technol. (2017). https://doi.org/10.1109/TCSVT.2017.2711659
7. Liao, J., Wang, H., Yan, Y., Zheng, J.: A novel background subtraction method based on ViBe. In: Zeng, B., Huang, Q., El Saddik, A., Li, H., Jiang, S., Fan, X. (eds.) PCM 2017, Part II. LNCS, vol. 10736, pp. 428–437. Springer, Cham (2018). https://doi.org/10.1007/978-3-319-77383-4_42
8. Wang, Y., Jodoin, P.M., Porikli, F., Konrad, J., Ishwar, P.: CDnet 2014: an expanded change detection benchmark dataset. In: IEEE Conference on Computer Vision and Pattern Recognition Workshops, pp. 387–394 (2014)
9. Braham, M., Droogenbroeck, M.V.: Deep background subtraction with scene-specific convolutional neural networks. In: International Conference on Systems, Signals and Image Processing, pp. 1–4 (2016)
10. Wang, Y., Luo, Z., Jodoin, P.M.: Interactive deep learning method for segmenting moving objects. Pattern Recognit. Lett. **96**, 66–75 (2017)
11. Tieleman, T., Hinton, G.: Lecture 6.5-RmsProp: divide the gradient by a running average of its recent magnitude. In: COURSERA: Neural Networks for Machine Learning (2012)
12. Braham, M., Pierard, S., Droogenbroeck, M.V.: Semantic background subtraction. In: IEEE International Conference on Image Processing, pp. 4552–4556 (2017)
13. Babaee, M., Dinh, D.T., Rigoll, G.: A deep convolutional neural network for video sequence background subtraction. Pattern Recognit. **76**, 635–649 (2018)
14. Zhao, X., Chen, Y., Tang, M., Wang, J.: Joint background reconstruction and foreground segmentation via a two-stage convolutional neural network. In: IEEE International Conference on Multimedia and Expo., pp. 343–348 (2017)
15. Chen, Y., Wang, J., Zhu, B., Tang, M., Lu, H.: Pixel-wise deep sequence learning for moving object detection. IEEE Trans. Circuits Syst. Video Technol. (2017). https://doi.org/10.1109/TCSVT.2017.2770319

A Reference Resource Based End-to-End Image Compression Scheme

Wenbin Yin[1(✉)], Xiaopeng Fan[1], Yunhui Shi[2], and Wangmeng Zuo[1]

[1] Harbin Institute of Technology, Harbin, Heilongjiang, China
{ywb,fxp}@hit.edu.cn, cswmzuo@gmail.com
[2] Beijing University of Technology, Beijing, China
syhzm@bjut.edu.cn

Abstract. Deep learning and convolutional neural networks have achieved a great success in computer vision and image processing, especially in low-level vision problems such as image compression. Recently, some end-to-end image compression methods have been proposed leading to a new direction of image compression. In this paper, we propose an end-to-end reference resource based image compression scheme to exploit the strong correlations with external similar images. In the proposed scheme, the side information is generated from highly correlated images in the reference resource. The features of side information can conceptually guide the compression process and assist the reconstruction process. The important map is employed to guide the allocation of local bit rate of the residual features. The proposed compression scheme is formulated as a rate distortion optimization problem in an end-to-end manner which is solved by ADAM algorithm. Experimental results prove that the proposed compression framework greatly outperforms several image compression frameworks.

Keywords: Convolutional neural networks · Reference resource
Image compression · Rate distortion optimization

1 Introduction

In recent years, image compression attracts increasing interest in image processing and computer vision area due to its potential applications in many vision systems. The typical image encoding standards [1] (such as JPEG and JPEG2000) generally rely on handcrafted image transformation and separate optimization on codecs, and thus are suboptimal for image compression. Moreover, while the compression ratio increases, the bits per pixel (BPP) decreases as a result of the use of bigger quantization steps, which will cause the decoded image to have some annoying visual artifacts.

This work is supported in part by the National Science Foundation (NSFC) of China under Grants 61672066 and 61472018, and Beijing municipal science and technology commission (Z171100 004417023).

© Springer Nature Switzerland AG 2018
R. Hong et al. (Eds.): PCM 2018, LNCS 11164, pp. 534–544, 2018.
https://doi.org/10.1007/978-3-030-00776-8_49

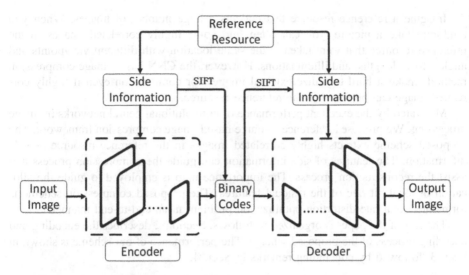

Fig. 1. Framework of the proposed method

For the existing image standards, the codecs actually are separately optimized. In the encoding stage, they first perform a linear transform to an image. Quantization and lossless entropy coding are then utilized to minimize the compression rate. For example, JPEG applies discrete cosine transform (DCT) on 8×8 image patches, quantizes the frequency components and compresses the quantized codes with a variant of Huffman encoding. JPEG 2000 uses a multi-scale orthogonal wavelet decomposition to transform an image, and encodes the quantized codes with the Embedded Block Coding with Optimal Truncation. In the decoding stage, the decoding algorithm and inverse transform are designed to minimize the distortion. In contrast, CNN based methods treat image compression as a joint rate distortion optimization problem, where both nonlinear encoder and decoder are jointly trained in an end-to-end manner.

Recently, deep convolutional networks have achieved great success in computer vision task [2–6]. As to image compression, convolutional networks are also expected to be more powerful than JPEG and JPEG 2000 by considering the following reasons. First, for image encoding and decoding, flexible nonlinear analysis and synthesis transformations can be easily achieved by stacking several convolutional layers. Second, it allows jointly optimizing the nonlinear encoder and decoder in an end-to-end manner. For the lossy image compression, Toderici et al. [7] propose a general framework for variable-rate image compression and a novel architecture based on convolutional and deconvolutional LSTM recurrent networks. Li et al. [8] propose a content weighted compression method with the importance map of the input image. Theis et al. [9] propose compressive autoencoders, which uses a smooth approximation of the discrete of the rounding function and upper-bound the discrete entropy rate loss for continuous relaxation. Balle et al. [10] make use of a generalized divisive normalization (GDN) for joint nonlinearity and replace rounding quantization with additive uniform noise for continuous relaxation.

Imagine a reference resource that collects a huge number of images. When you randomly take a picture, you can often find some highly correlated images in the reference resource that were taken at the same location with different viewpoints and angles, focal lengths, and illuminations. However, the CNN based image compression methods make it hard to utilize external images for compression even if highly correlated image can be found in the reference resource.

Motivated by the excellent performance of convolutional neural networks in image processing. We propose a reference resource based image compression framework. The proposed scheme extracts highly correlated images in the reference resource as side information. The features of side information can guide the compression process and assist the reconstruction process. The importance map is employed to guide the allocation of local bit rate of the residual features. The proposed compression scheme is formulated as a rate distortion optimization problem in an end-to-end manner.

The rest of the paper is organized as follows. Section 2 describes the encoding and decoding process of the proposed scheme. The performance of our scheme is shown in Sect. 3, followed by concluding remarks in Sect. 4.

2   Proposed Method

Our framework contains four components: reference resource (RS) based convolutional encoder, importance map network, binarizer and RR based convolutional decoder. Figure 1 shows the architecture of the proposed framework. Given an input image, the encoder first describes the input image by SIFT descriptors. SIFT descriptors are extracted from the original images and are used to retrieve near and partial duplicate images in the reference resource and identify corresponding patches. The SIFT descriptors are compressed and transmitted to decoder while it could find similar images in the reference resource by utilizing the SIFT descriptors. Then the encoder defines a nonlinear analysis transform by stacking convolutional layers which utilizes side information generated from similar images in the reference resource. The importance map network takes the intermediate feature maps as the input, and yields a content-weighted importance map. Then the binary code is trimmed based on the mask generated from the importance map. The decoder defines a nonlinear synthesis transform to produce decoding result.

2.1   Generation of Side Information

Given an input image, we need to find out its near and partial-duplicate images in the reference resource. The SIFT feature (key-point and descriptor) is one of the most robust and distinctive point features. The SIFT keypoint gains invariance to scale and rotation by exploiting scale-space extrema and the local dominant orientation. The SIFT descriptor assemble a 4×4 array of 8 gradient orientation histograms around the keypoint, making it robust to image variations induced by both photometric and geometric changes.

In the proposed scheme, we extract the side information of current input image at both encoder side and decoder side. After we obtain the SIFT descriptors of current encoding image, the encoder first transmits them to the decoder with conventional SIFT compression method [11]. In an image, the region of a SIFT descriptor with a large-scale index often partially or completely covers the regions of some SIFT descriptors with small-scale indices. As mentioned in [12], we bundle them as a group. One image often has many groups. Every group is represented by a set of visual words and their geometric relationship in the image. Decoded SIFT feature vectors are quantized to visual words and organized into groups too. Every group is matched with all groups in the reference resource. The number of matched visual words and their geometric relationship score the matching result. This score is assigned to the image that contains the group. After all groups are matched, the image with highest sum scores will be selected as to generate side information of encoding image. Then we will find the corresponding patch from the side information for each encoding patch (Fig. 2).

Fig. 2. The input image and its corresponding similar image in the reference resource

2.2 Convolutional Encoder and Decoder

Both the encoder and decoder in our framework are fully convolution networks and can be trained by back-propagation. The encoder consists of nine convolutional layers. Each three of them are treated as a group. We further remove the batch normalization operations from the last two convolutional layers in one group. The encoder utilizes side information to guide the compression process. The most different point between conventional CNN based image-coding methods and the proposed framework is that we adopt two inputs (input image and corresponding side information) for convolutional networks. There are two networks in the proposed method: feature exaction network and encoding network. The feature exaction network has the similar structure with [8]. The feature exaction networks at both encoder and decoder side are jointly trained to produce feature maps of side information in each level. The feature exaction network can be expressed as

$$F_y^l(y) = w_y^l(y) + B_y^l \tag{1}$$

Where y is the side information, l stands for layer number, B and w represents the biases and the mapping to be learn. The encoding network of proposed method can be formulated as

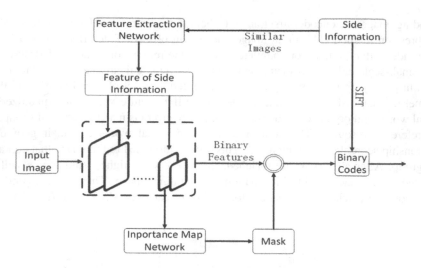

Fig. 3. Encode process of the proposed scheme

$$F_x^l(x) = w_x^l(A \circ F_y^l + F_x^{l-1}(x)) + B_x^l \qquad (2)$$

where x is the input image, $\circ$ denotes the element wise multiplication operation and A represents the weights between input image features and side information features (Fig. 3).

In the encoding network, the input image is first convolved with 128 filters with size 8×8 and stride four and followed by one CNN group. The feature maps are then convolved with 256 filters with 4×4 and stride two and followed by two CNN groups to output the intermediate feature maps. Finally, the intermediate feature maps are convolved with n filters with size 1×1 to yield the encoder output. It should be noted that we set n = 64 for low compression rate models with less than 0.5 bpp and n = 128 otherwise. The network architecture of decoder is symmetric to that of the encoder without the importance map network (Fig. 4).

2.3 Binarizer and Importance Map

Since sigmoid nonlinearity is adopted in the last convolutional layer, the encoder output should be in the range of $[0, 1]$. e_{ijk} denotes an element in output of encoder. The binarizer is defined as

$$B(e_{ijk}) = \begin{cases} 1, & \text{if } e_{ijk} > 0.5 \\ 0, & \text{if } e_{ijk} \leq 0.5 \end{cases} \qquad (3)$$

However, the gradient of the binarizer function is zero almost everywhere except that it is infinite when $e_{ijk} = 0.5$. In the back-propagation process, the gradient is computed layer by layer by utilizing the chain rule in a backward manner. This will make any layer before the binarizer never be updated during training.

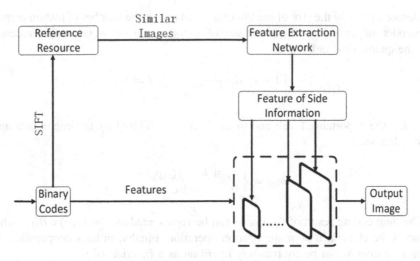

Fig. 4. Decode process of the proposed method

To solve this problem, we adopt a proxy function $\widetilde{B}(e_{ijk})$ to approximate $B(e_{ijk})$. While $B(e_{ijk})$ is still used in forward propagation and $\widetilde{B}(e_{ijk})$ is used in backpropagation. The proxy function is based on the straight though estimator on gradient [2] and is defined as

$$\widetilde{B}(e_{ijk}) = \begin{cases} 1, & \text{if } e_{ijk} > 1 \\ e_{ijk}, & \text{if } 0 \le e_{ijk} \le 1 \\ 0, & \text{if } e_{ijk} \le 0 \end{cases} \tag{4}$$

And the gradient of $\widetilde{B}(e_{ijk})$ can be easily calculated by

$$\widetilde{B}'(e_{ijk}) = \begin{cases} 1, & \text{if } 0 \le e_{ijk} \le 1 \\ 0, & \text{otherwise} \end{cases} \tag{5}$$

In general, the attentions in compressing different parts of an image should be different. The smooth regions in an image should be easier to be compressed than those with objects or rich textures. Thus, fewer bits should be allocated to the smooth region while more bits should be allocated to the region with more information. Moreover, when the whole code length for an image is limited, such allocation can also be used for rate control.

The importance map is a feature map with only one channel, and its size should be same with the encoder output. The value of importance map is in the range of $(0, 1)$. An importance map network is deployed to learn the importance map from an input image. It takes the intermediate feature maps from the last residual block of the encoder as input and use a network to produce the importance map.

Denote by h × w the size of the importance map p, n the number of feature maps of the encoder output and $n \times h \times w$ the size of importance mask m. Given an element p_{ij} in p, the quantizer is defined as

$$Q(p_{ij}) = \begin{cases} l-1, & \text{if } \frac{l-1}{L} \le p_{ij} < \frac{l}{L}, \, l = 1, \ldots, L \\ L, & \text{if } p_{ij} = 1 \end{cases} \tag{6}$$

where L is the importance levels and $(n \bmod L) = 0$. With $Q(p)$, the importance mask can be obtained by

$$m_{kij} = \begin{cases} 1, & \text{if } k \le \frac{n}{L} Q(p_{ij}) \\ 0, & \text{else} \end{cases} \tag{7}$$

The final coding result of the image can be represented as $c = M(p) \circ B(e)$, where $\circ$ denotes the element wise multiplication operation. Finally, in back-propagation, the importance map m can be equivalently rewritten as a function of p

$$m_{kij} = \begin{cases} 1, & \text{if } \lceil \frac{kL}{n} \rceil < Lp_{ij} + 1 \\ 0, & \text{else} \end{cases} \tag{8}$$

where $\lceil . \rceil$ is the ceiling function. Analogous to binarizer, we also adopt a straight-though estimator of the gradient.

2.4   Entropy Encoder

Due to the fact that no entropy constraint is included, the code generated by the encoder is non-optimal in terms of entropy rate. This provides some leeway to further compress the code with lossless entropy coding. Generally, there are two kind of entropy coding methods includes Huffman tree and arithmetic coding. Between them, arithmetic coding can exhibit better compression rate with a well-defined context, and is adopted in this work.

The binary arithmetic coding is applied according to the CABAC [13] framework. Note that CABAC is originally proposed for video compression. To encode binary code, we modify the coding schedule, redefine the context which leads to the importance mask, and use convolutional neural network for probability prediction. As to coding schedule, we simply code each binary bit map from left to right and row by row, and skip those bits with the corresponding important mask value of zero.

We also extend the convolutional entropy encoder to the quantized importance map. To utilize binary arithmetic coding, a number of binary code maps are adopted to represent the quantized importance map. The convolutional entropy encoder is then trained to compress the binary code maps.

2.5   Model Formulation

In general, the proposed image compression system can be formulated as a rate-distortion optimization problem. Our objective is to minimize the combination of the distortion loss and rate loss. A tradeoff parameter γ is introduced for balancing compression rate and distortion. Let X be a set of train data, and $x \in X$ be an image from the set. Therefore, the objective function is defined as

$$L = \sum_{x \in X} \{L_D(c, w, A, B, x) + \gamma L_R(x, y)\} \quad \text{s.t. } A \circ F_y^l + F_x^{l-1}(x) \approx w_x^{l-1}(x) + B^{l-1} \quad (9)$$

where c is the code of the input image x and w is the weights between features of input image and features of corresponding side information y. $L_D(c, w, A, B, x)$ denotes the distortion loss and $L_R(x, y)$ denotes the rate loss.

Benefited from the relaxed rated loss and the straight-though estimator of the gradient, the whole compression system can be trained in an end to end manner with ADAM solver. We initialize the model with the parameters pre-trained on training set without the side information. The model is further trained with the learning rate of le^{-4}, le^{-5} and le^{-6}. In each learning rate, the model is trained until the objective function does not decrease. And a smaller learning rate is adopted to fine-tune the model.

3 Experimental Result

Our reference resource-based image compression model is trained on a subset of INRIA Holiday dataset [14] with about 1000 high quality images. We divide these images into 128×128 patches and take use of these patches to train the network. The side information generated from reference resource is the same at encoder and decoder. After training, we test the proposed model on another subset of INRIA Holiday dataset. The compression rate of our model is evaluated by the metric bits per pixel (bpp), which is calculated as the total amount of bits used to code the image divided by the number of pixels. The image distortion is evaluated with Peak Signal to Noise Ratio (PSNR).

3.1 Parameter Setting

In our experiments, we set the number of binary feature maps according to the compression rate. For instance, it will be set as 64 when the compression rate is less than 0.5 bpp and 128 otherwise. Then, the number of importance level is chosen based on importance mask. For n = 64 and n = 128, we set the number of importance level to be 16 and 32, respectively. Moreover, different values of the tradeoff parameter γ in the range [0.0001,0.2] are chosen to get different compression rates. For the choice of the threshold value r, we set it as $r_0 hw$ for n = 64 and $0.5 r_0 hw$ for n = 128. Here, r_0 is the wanted compression rate represent with bit per pixel (bpp).

3.2 Quantitative Evaluation

We compare our model with JPEG, JPEG 2000 and Li et al. [8]. Among different variants of JPEG, the optimized JPEG with 4:2:0 Chroma subsampling is adopted.

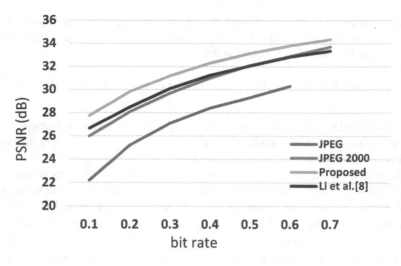

Fig. 5. Comparison of the ratio-distortion curves

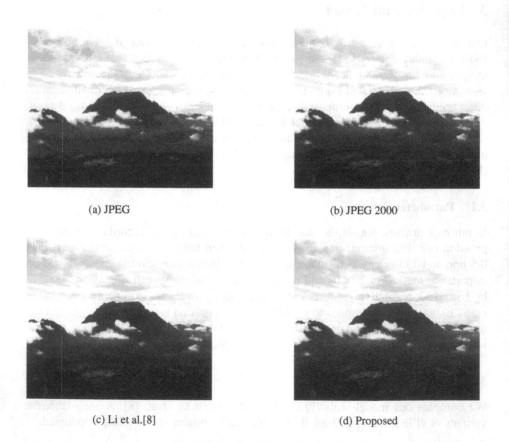

Fig. 6. Visual quality evaluation

Using PSNR as performance metric, Fig. 5 gives the ratio-distortion curves of these four methods. In terms of PSNR, the results by Li et al. [8], JPEG 2000 and ours are much higher than that by JPEG. The proposed framework achieves 1.5 dB and 1.2 dB gains in PSNR compared against JPEG 2000 and Li et al. [8].

3.3 Visual Quality Evaluation

In Fig. 6, one can see that the proposed compression framework achieves much better subjective performance than JPEG and JPEG 2000, especially at a very low bit rate. Our framework preserves more high-frequency information and recovers sharp edges and pure textures in the reconstructed image.

4 Conclusion

A convolutional neural network based system is developed for reference resource-based image compression. It well solves the problem of current image compression schemes that is hard to utilize external images for compression even if highly correlated image can be found in the reference resource. With the side information generated from the reference resource, we introduce a neural network architecture, in which the input image and corresponding image are taken as the inputs. With the importance map, we suggest a non-entropy based loss for rate control. Experiments clearly show the superiority of our method in retaining structures and removing artifacts, leading to remarkable visual quality.

References

1. Ghanbari, M.: Standard codecs: Image compression to advanced coding. Iet, no. 49 (2003)
2. Courbariaux, M., Hubara, I., Soudry, D., EI-Yaniv, R., Bengio, Y.: Binarized neural networks: Training deep neural networks with weights and activations constrained to +1 or −1. arXiv:1602.02830 (2016)
3. Dong, C., Loy, C.C., He, K., Tang, X.: Learning a deep convolutional network for image super-resolution. In: Fleet, D., Pajdla, T., Schiele, B., Tuytelaars, T. (eds.) ECCV 2014. LNCS, vol. 8692, pp. 184–199. Springer, Cham (2014). https://doi.org/10.1007/978-3-319-10593-2_13
4. Girshick, R., Donahue, J., Darrel, T., Malik, J.: Rich feature hierarchies for accurate object detection and semantic segmentation. In: Conference on Computer Vision and Pattern Recognition, pp. 580–587 (2014)
5. Parkhi, O.M., Vedaldi, A., Zisserman, A.: Deep face recognition. In: British Machine Vision Conference (2015)
6. Rastegari, M., Ordonez, V., Redmon, J., Farhadi, A.: XNOR-Net: ImageNet classification using binary convolutional neural networks. In: Leibe, B., Matas, J., Sebe, N., Welling, M. (eds.) ECCV 2016. LNCS, vol. 9908, pp. 525–542. Springer, Cham (2016). https://doi.org/10.1007/978-3-319-46493-0_32
7. Toderici, G., et al.: Variable rate image compression with recurrent neural networks. arXiv:1511.06085 (2015)

8. Li, M., Zuo, W., Gu, S., Zhao, D., Zhang, D.: Learning convolutional networks for content weighted image compression. arXiv:1703.10553 (2017)
9. Theis, L., Shi, W., Cunningham, A., Husazár, F.: Lossy image compression with compressive autoencoders. arXiv:1703.00395 (2017)
10. Balle, J., Laparra, V., Simoncelli, E.P.: End-to-End optimized image compression. axXiv: 1611.01704 (2016)
11. Chandrasekhar, V., et al.: Tandsform coding of image feature descriptors. In: Proceeding SPIE Conference on Visual Communication and Image Processing, vol. 7257 (2009)
12. Zhou, W.G., Lu, Y.J., Li, H.Q., Song, Y.B., Tian, Q.: Spatial coding for large scale partial-duplicate web image search. In: Proceeding ACM Multimedia, pp. 511–520 (2010)
13. Marpe, D., Schwarz, H., Wiegand, T.: Context-based adaptive binary arithmetic coding in the H.264/avc video compression standard. IEEE Trans. Circuits Syst. Video Technol. **13**(7), 620–636 (2003)
14. Jegou, H., Douze, M.: INRIA Holiday Dataset (2008). http://lear.inrialpes.fr/people/jegou/data.php

Context and Temporal Aware Attention Model for Flood Prediction

Zhaoyang Liu[1], Yirui Wu[1,2], Yukai Ding[2], Jun Feng[2], and Tong Lu[1(✉)]

[1] National Key Lab for Novel Software Technology,
Nanjing University, Nanjing, China
zyliumy@gmail.com, lutong@nju.edu.cn
[2] College of Computer and Information, Hohai University, Nanjing, China
{wuyirui,fengjun}@hhu.edu.cn, 291809390@qq.com

Abstract. To minimize damages brought by floods, researchers pay special attentions to solve the problem of flood prediction. Multiple factors, including rainfall, soil category, the structure of riverway and so on, affect the prediction of sequential flow rate values, but factors are not always informative for flood prediction. Extracting discriminative and informative features thus plays a key role in predicting flow rates. In this paper, we propose a context and temporal aware attention model for flood prediction based on a quantity of collected flood factors. We build our model on top of Long Short-Term Memory (LSTM) networks, which selectively focuses on informative factors and pays different levels of attentions to the outputs of different cells. The proposed CT-LSTM network assigns time-varying weights to input factors at all the cells of LSTM network, and allocates temporal-dependent weights to the outputs of each LSTM cell for boosting prediction performance. Experimental results on a benchmark flood dataset with several comparative methods demonstrate the effectiveness of the proposed CT-LSTM network for flood prediction.

Keywords: Attention model · Context and temporal aware
Flood prediction

1 Introduction

Flood, as one of the most common and largely distributed natural diasters, happens occasionally and brings large damages to life and property. In the past decades, researchers have proposed a quantity of models for accurate and robust flood prediction. We generally category them into two types, namely, physical models [7,9,10] and data-driven models [4,14,16]. Physical models generally describe the formation of flood by using functions to represent complex hydrology processes from clues to results. However, such models are extremely sensitive to parameters [17], which require large research efforts of experts to adjust. On the contrary, data-driven models directly explore relations between river flow

© Springer Nature Switzerland AG 2018
R. Hong et al. (Eds.): PCM 2018, LNCS 11164, pp. 545–555, 2018.
https://doi.org/10.1007/978-3-030-00776-8_50

and flood factors from historical observations, without considering physical processes. Since the complex mechanism of flood results in large computations of physical models, data-driven models are more efficient and costless for flood prediction.

Inspired by the significant performance [8,13] of deep LSTMs, we intend to utilize such an architecture to discover the inherent relations between flood factors and flow rates. Due to the development of internet of things, researchers can gather a large set of relevant flood factors for prediction. However, not all the collected factors are representative and informative for flood prediction. For example, the water retained in soil, named as soil tension water, has great effects on formation of floods in humid areas, while it is irrelevant with flood in places with sandy soil [9]. This is because soil in humid areas contain a great amount of water, meanwhile sandy soil is quite low in capacity to contain water. The informativeness degrees of each flood factor may vary at different time points during the same flood. Take soil tension water as an example, its value is highly relevant with flow rate values in humid areas at the beginning of a flood. Once its value exceeds the maximum water containing capacity of soil in the middle of the flood, the value of soil tension water no more changes and contributes little to the variations of flow rate values. Therefore, we propose a context-aware attention module, which automatically focuses on discriminative factors for flood prediction. The learned attention to factors are content-dependent and allowed to vary over time. This selectively focusing mechanism has been demonstrated to be very effective in various applications, such as speech recognition [3] and action recognition [11].

Furthermore, we often get predictions on flow rate values under a reasonable assumption that there exists a trend in historical flow data. We thus propose a temporal-aware attention module for simulations of the trends embedded in historical flow data. For a sequence of floods, the proposed temporal-aware attention module explicitly learns and allocates content-dependent weights to predicting flow rate values at each time point. In fact, the idea of the proposed temporal-aware attention module is similar with Holt-Winters double exponential smoothing [15], which assigns higher weights to the nearby observations for more convinced predictions. Flow rate predictions at different time points thus have different degrees of importance and robustness to variations. Moreover, some flow rate predictions can be unreliable induced by noises of input factors. Learning weight distribution for flow rate predictions under a trend assumption can help exclude such unreliable predictions.

In summary, we aim to construct a context and temporal aware attention LSTM (CT-LSTM) network for accurate and robust flood prediction. The context-aware attention module learns weight schemes for input factors based on hidden output of each LSTM cell (representing contextual information [5] between two nearby cells) in a local sense. Meanwhile, the temporal-aware attention module learns weight structures for flow rate predictions of each LSTM cell in a global sense. We have made the following three main contributions in this work.

- To the best of our knowledge, this is the first context and temporal aware attention model designed based on the LSTM architectures for flood prediction.
- A temporal-aware attention module is designed to allocate content-dependent attention to different predictions under a reasonable trend assumption.
- The proposed method is powerful to discover the inherent patterns between input factors and flow rates, especially for regions whose flood formation mechanism is too complex to construct a convinced physical model.

2   Related Work

Considering the relevance to the proposed CT-LSTM network, we introduce the data-driven model for flood prediction and attention model in this section.

Data-driven Model. Early, Yu et al. [20] utilize the support vector machine to establish a real-time flood forecasting model, which applies a two-step grid search method to find the optimal parameters for SVM. Later, Cheng et al. [2] perform daily runoff forecasting by training artificial neural network with quantum-behaved particle swarm optimization, which achieves much better forecast accuracy than the basic ANN model. Recently, Wu et al. [16] construct a hierarchical Bayesian network for flood predictions of small rivers. They establish entities and connections of Bayesian network to represent variables and physical processes of the Xinanjiang model, i.e., a famous physical model, which appropriately embeds hydrology expert knowledge for high rationality and robustness.

Due to high potentials of discovering distinctive features from data, researchers try to utilize deep learning architectures for flood prediction. For example, Zhuang et al. [21] design a novel Spatio-Temporal Convolutional Neural Network (ST-CNN) to fully utilize spatial and temporal information and automatically learn underlying patterns from data for extreme flood cluster prediction. Liu et al. [4] propose a deep learning approach by integrating stacked auto-encoders (SAE) and back propagation neural networks (BPNN) for the predictions of stream flow, which simultaneously takes advantages of the powerful feature representation capability of SAE and superior predicting capacity of BPNN. Most recently, Wu et al. [14] propose context-aware attention LSTM network to accurately predict sequential flow rate values based on a set of collected flood factors. The proposed method is built on it and involves the combination of context and temporal aware attention over all the steps of LSTM network for higher predicting accuracy.

Attention Model. When observing the real-world, human perception focuses selectively on parts of a scene to acquire information at specific places and times. The exploitation of an attention model has attracted increasing interests in various fields, such as machine translation, image recognition and action recognition. Their proposed attention models are generally constructed as a dimension of interpretability into internal representations by selectively focusing on specific

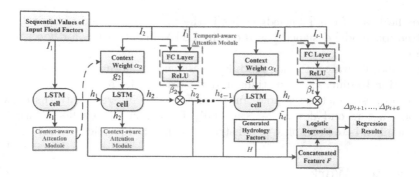

Fig. 1. Illustration of the proposed context and temporal aware attention LSTM network for flood prediction.

information. We categorize attention models into two classes, *i.e.* hard attention [19] and soft attention [1]. Hard attention mechanically chooses parts of the input data as focuses. For example, Mnit *et al.* [6] propose a hard attention model for image recognition, which adaptively selects a sequence of regions and processes the selected regions as inputs for RNN network.

On the contrary, soft attention takes the entire input into account by weighting each part or step dynamically. The fusion of neighboring frames within a sliding window with learned attention weights is proposed by Yeung *et al.* [18] to enhance the performance of dense labeling of actions in RGB videos. Liu *et al.* [5] propose a global context-aware attention LSTM for RGB-D action recognition, which recurrently optimize the global contextual information and further utilizes it as an informative function to assist accurate action recognition. Song *et al.* [12] achieve the goal of action recognition from skeleton data by selectively focusing on discriminative joints of skeleton within each frame of the inputs and assigning different levels of attention to the outputs of different frames. By designing context and temporal aware attention model as a soft attention scheme, the proposed method is reasonable to solve the regression problem of flow rate prediction.

3   LSTM Network with Context and Temporal Aware Attention Model

Take a typical river, *i.e.*, Changhua, as an example, we show its general information in Fig. 2, where we can notice 7 rainfall stations, 1 evaporation station and 1 river gauging station. In our work, we aim to predict the flow rate values at the river gauging station CH for the next 6 h with the proposed CT-LSTM network. The input set of flood factors consists of real rainfalls observed at rainfall stations, predicted rainfalls, evaporation observed at evaporation station SS and former river runoff observed at CH. We also utilize several intermediate variables such as total surface runoff, total interflow runoff and total groundwater runoff computed by a famous physical model, namely, the Xinanjiang Model. In

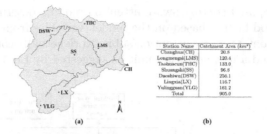

Station Name	Catchment Area (km^2)
Changhua(CH)	20.8
Longmengni(LMS)	120.4
Taohuacun(THC)	133.0
Shuangshi(SS)	96.8
Daoshiwu(DSW)	256.1
Lingxia(LX)	116.7
Yulingguan(YLG)	161.2
Total	905.0

Fig. 2. Illustration of the Changhua watershed, where (a) is the map for various kinds of stations and (b) represents catchment areas corresponding to the listed rainfall stations. Note that we need predict the flow rate values of river gauging station CH and station SS functions as an evaporation station.

Xinanjiang Model, the outflow of a watershed can be subdivided into three components, including surface runoff, interflow runoff and groundwater runoff. Using these three components for prediction will provide more information about the watershed, which will be informative about the flood formation that cannot be precisely measured by sensors. In total, we prefer 7 features for prediction.

We propose an LSTM network with context and temporal aware attention mechanisms for flood prediction as shown in Fig. 1. We only feed the features mentioned above and last hidden state to the LSTM cell of our proposed network. The designed local context-aware attention and global temporal-aware attention module help automatically select relevant and informative features from the views of factors and the trend embedded, respectively. After paying different levels of attention on inputs and outputs of LSTM cells, we concatenate sets of the hidden outputs of cells $\{h_1, \tilde{h}_2, ..., \tilde{h}_t\}$ and generated hydrology factors H as a novel feature F for prediction. The reason to predict with sets of hidden outputs lies in the restriction of LSTM in perceiving the global contextual information with forgetting mechanism. However, the forgotten contextual information is important for the global regression problem.

3.1 Context-Aware Attention Module

Inspired by [6] which considers the attention problem as the sequential decision process of how an agent interact with a visual environment, the "interaction level" for the proposed context-aware attention module is essentially described by weights assigned to each feature. The context-aware attention module thus recurrently defines the corresponding weight vector α_t for input factors I_t as

$$\alpha_t = Nor(sig(W_{c,t-1}h_{t-1} + b_{c,t-1})) \tag{1}$$

where $W_{c,t-1}$ is the learnable parameter matrix, $b_{c,t-1}$ is the bias vector, h_{t-1} is the hidden output for each cell representing context information, function $sig()$ and $Nor()$ represent sigmoid function and normalization function, respectively.

Note that the proposed context-aware attention module determines the importance of input flood factors based on the hidden variables from an LSTM layer. In our work, the context-aware attention subnetwork actually composes of a fully connected layer and a normalization unit as suggested by Eq. 1.

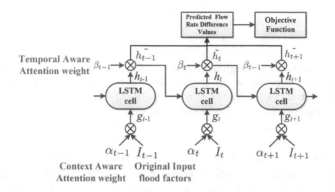

Fig. 3. Illustration of how context-aware attention weight α, temporal-aware attention weight β and objective function influence the CT-LSTM network.

The resulting weight vector α_t leads to the attention on informative factors, where Fig. 3 explains how the context-aware attention module works by a local way. We can find that the sequential input features $\{I_1, ..., I_t\}$ are separatively fed to all cells of CT-LSTM as the original time-varying description of flood factors. With the feature-wise weight vector α_t, the input of the informative flood factors g_t for each cell can be represented as:

$$g_t = I_t \bigotimes \alpha_t \qquad (2)$$

where $\bigotimes$ represents the element-wise multiplication.

3.2 Temporal-Aware Attention Module

Holt-Winters double exponential smoothing filter considers there exists a trend behind a time-varying variable and utilizes a updating weight scheme to describe how prediction interacts with former observations. It has been successfully applied on smoothing of skeleton action data [15]. Follow the idea of Holt-Winters double exponential smoothing filter, we propose to use temporal-aware attention module to simulate the trend by globally assigning different levels of weights β_t to output of all the cells h_t as shown in Fig. 3. In fact, the hidden variable h_t contains information of past time points, benefiting from the merit of LSTM which is capable of exploring temporal long range dynamics. The weight vector β_t computed by temporal-aware attention module thus can adjust the input for the next cell $\tilde{h}_t$ based on information from a temporal long range:

$$\tilde{h}_t = h_t \bigotimes \beta_t. \qquad (3)$$

As shown in Fig. 1, the temporal-aware attention module is composed of a fully connected layer and a ReLU nonlinear unit. The temporal weight vector β_t thus can be computed as

$$\beta_t = sig(W_{m,t-1}I_{t-1} + W_{m,t}I_t + b_{m,t}) \tag{4}$$

which depends on the former and current input flood factors I_{t-1} and I_t, respectively. We use the non-linear function of sigmoid due to its good convergence performance. The temporal weight vector control the amount of information of former predictions to be used for making the final prediction.

3.3 Design of Objective Function

How the context-aware attention module acts on the input flood factors and how the temporal-aware attention module acts on the hidden output of LSTM cells are given in Fig. 3. Constrained by the objective function, the main LSTM network, the context and temporal aware attention subnetwork can be jointly trained to implicitly learn the model. We thus formulate the final objective function of the context and temporal aware attention network with a regularized cross-entropy loss for a sequence of flood factors as

$$L_t = -\sum_{i=1}^{C}\sum_{t=1}^{6} loss(y_{i,t}, \tilde{p}_i + \Delta p_{i,t}) + \lambda\|W_N\|_2 \tag{5}$$

where C is the total number of training samples, $y_{i,t} = \{y_{i,t+1}, ..., y_{i,t+6}\}$ denotes the groundtruth flow rate values for the next 6 h corresponding to the ith training sample, $\tilde{p}_i = \sum_{j=0}^{4} y_{i,t-j}$ implies the mean of observed flow rate values for former 5 h and current time, $\Delta p_{i,t} = \{\Delta p_{i,t+1}, ..., \Delta p_{i,t+6}\}$ refer to the predicted difference flow rate values computed by the CT-LSTM, and function $loss()$ is defined as the smooth L1 loss function. The regularization item with L2 norm is to reduce overfitting of the networks. W_N denotes the connection matrix (merged to one matrix here) in the networks, including $W_{c,t}$ in Eq. 1 and $W_{m,t}$ in Eq. 4. Note that we use the back-propagation through time (BPTT) algorithm to minimize the loss function and adopt smooth L1 loss function. This is because the smooth L1 loss function makes the loss value convergent in a faster and more stable way comparing with adopting Root Mean Square Error.

4 Experimental Results

4.1 Dataset and Settings

Changhua Dataset. We collect hourly data of 40 floods happened from 1998 to 2010 in Changhua river as our original dataset. We use samples of flow rates every 11 h as to increase dataset size. After augmentation, the number of flood samples is increased to 8555. We utilize 8-fold cross validation and Root Mean Square Error (RMSE) to evaluate predictions:

$$RMSE = \sqrt{\frac{1}{n} \sum_{k=1}^{n} \sum_{t=1}^{6} (y_{k,t} - \tilde{p_k} - p_{k,t})^2} \qquad (6)$$

where n refers to the number of testing samples. Note that smaller values of RMSE imply better performance the predicting achieves.

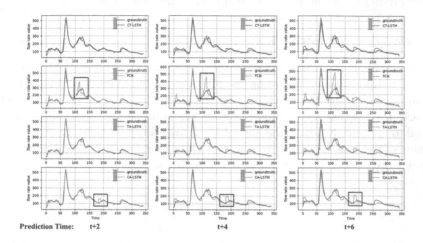

Fig. 4. Comparison with the ground-truth flow rate values and predicted flow rate values during a flood, where each row represents prediction results of the proposed CT-LSTM, FCN, TA-LSTM and CA-LSTM, respectively. Note that the rectangles indicate several obvious wrong predictions.

Implementation Details. For constructing the CT-LSTM network, we select t as 11, the dimension of hidden output as 128 and λ as 0.00005, respectively. We train the CA-LSTM network by setting learning rate, weight decay, epoch iterations and batch size as 0.00225, 10^{-6}, 500 and 100, respectively. The proposed CT-LSTM network runs on a workstation (2.4 GHz 6-core Xeon CPU, 60 G RAM and Nvidia GeForce GTX 1080Ti) for all the experiments.

4.2   Performance Analysis

We implement Fully-connected Network, CT-LSTM network without attention module (LSTM), CT-LSTM with only context-aware attention module (CA-LSTM), and CT-LSTM with only temporal-aware attention module (TA-LSTM) for comparisons. The main structures and training parameters of LSTM, CA-LSTM and TA-LSTM are exactly the same as CT-LSTM, meanwhile FCN is designed with 3 fully-connected layers. We compare the flow rate values prediction results of CT-LSTM, FCN, TA-LSTM and CA-LSTM in Fig. 4. We can see the CT-LSTM and TA-LSTM achieve nearly the same flow rates as the observed results. For CA-LSTM, we find it get obvious wrong predictions labeled by rectangles. We also view wrong predictions are enlarged when predicting with FCN.

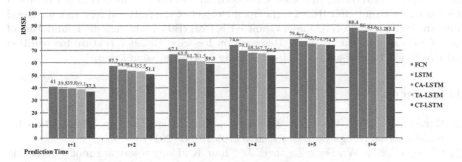

Fig. 5. Comparison of RMSE on Changhua Dataset computed by the proposed CT-LSTM, CA-LSTM, TA-LSTM, LSTM and FCN.

Figure 5 gives the detailed statistics of the proposed CT-LSTM network and several comparative methods on the Changhua dataset. As shown in Fig. 5, CT-LSTM network achieves the lowest RMSE values except for prediction at t+6. In fact, LSTM is designed to solve the problem of local dependencies with the forgetting structure, which implies LSTM network can not handle prediction with a rather long interval or delay. We thus observe that CT-LSTM performs nearly the same with TA-LSTM at t+6 due to the limitation of LSTM structure. FCN is not suitable for the time-varying prediction problem proved by much higher RMSE values comparing with other four LSTM-based methods. With the context-aware or the temporal-aware attention module, we find CA-LSTM and TA-LSTM achieve smaller RMSE than the conventional version of LSTM, which proves the effectiveness of the proposed context-aware and temporal-aware attention model. By jointly designing attentional module, the proposed CT-LSTM achieves the lowest RMSE, which proves the advantages of the structure of paying different levels of attentions on the input and output of LSTM cells for prediction.

5   Conclusions

In this paper, we extend the original LSTM network to achieve a context and temporal aware attention LSTM network for flood prediction, which is capable to selectively focus on informative flood factors and nearby predicted flow rate values. Experiment results on the Changhua dataset show the proposed method outperforms several comparative methods. Our future work includes the exploration on other hydrology purposes with the proposed method, such as mid-term flood predicting and flood frequency analysis.

Acknowledgment. This work was supported by the National Key R&D Program of China under Grant No. 2018YFC04000401, the Natural Science Foundation of China under Grant 61702160, 61370091, 61672273, the Fundamental Research Funds for the Central Universities under Grant 2016B14114, the Science Foundation of JiangSu under Grant BK20170892, the Science Foundation for Distinguished Young Scholars of Jiangsu under Grant BK20160021, the open Project of the National Key Lab for Novel Software Technology in NJU under Grant KFKT2017B05, and Scientific Foundation of State Grid Corporation of China (Research on Ice-wind Disaster Feature Recognition and Prediction by Few-shot Machine Learning in Transmission Lines).

References

1. Baradel, F., Wolf, C., Mille, J.: Pose-conditioned spatio-temporal attention for human action recognition. CoRR abs/1703.10106 (2017)
2. Cheng, C., Niu, W., Feng, Z., Shen, J., Chau, K.: Daily reservoir runoff forecasting method using artificial neural network based on quantum-behaved particle swarm optimization. Water **7**(8), 4232–4246 (2015)
3. Chorowski, J., Bahdanau, D., Serdyuk, D., Cho, K., Bengio, Y.: Attention-based models for speech recognition. In: Proceedings of NIPS, pp. 577–585 (2015)
4. Liu, F., Xu, F., Yang, S.: A flood forecasting model based on deep learning algorithm via integrating stacked autoencoders with BP neural network. In: Proceedings of IEEE International Conference on Multimedia Big Data, pp. 58–61 (2017)
5. Liu, J., Wang, G., Hu, P., Duan, L., Kot, A.C.: Global context-aware attention LSTM networks for 3d action recognition. In: Proceedings of IEEE CVPR, pp. 3671–3680 (2017)
6. Mnih, V., Heess, N., Graves, A., Kavukcuoglu, K.: Recurrent models of visual attention. In: Proceedings of NIPS, pp. 2204–2212 (2014)
7. Paquet, E., Garavaglia, F., Garçon, R., Gailhard, J.: The schadex method: a semi-continuous rainfall-runoff simulation for extreme flood estimation. J. Hydrol. **495**, 23–37 (2013)
8. Redmon, J., Divvala, S.K., Girshick, R.B., Farhadi, A.: You only look once: unified, real-time object detection. In: Proceedings of IEEE CVPR, pp. 779–788 (2016)
9. Ren-Jun, Z.: The xinanjiang model applied in china. J. Hydrol. **135**(1–4), 371–381 (1992)
10. Rogger, M., Viglione, A., Derx, J., Blöschl, G.: Quantifying effects of catchments storage thresholds on step changes in the flood frequency curve. Water Resour. Res. **49**(10), 6946–6958 (2013)
11. Sharma, S., Kiros, R., Salakhutdinov, R.: Action recognition using visual attention. CoRR abs/1511.04119 (2015)
12. Song, S., Lan, C., Xing, J., Zeng, W., Liu, J.: An end-to-end spatio-temporal attention model for human action recognition from skeleton data. In: Proceedings of AAAI, pp. 4263–4270 (2017)
13. Wei, L., Wu, Y., Wang, W., Lu, T.: A novel 3D human action recognition framework for video content analysis. In: Schoeffmann, K., et al. (eds.) MMM 2018. LNCS, vol. 10704, pp. 42–53. Springer, Cham (2018). https://doi.org/10.1007/978-3-319-73603-7_4
14. Wu, Y., et al.: Context-aware attention LSTM network for flood prediction. In: Proceedings of ICPR (2018)
15. Wu, Y., Lu, T., Yuan, Z., Wang, H.: Freescup: a novel platform for assisting sculpture pose design. IEEE Trans. Multimed. **19**(1), 183–195 (2017)

16. Wu, Y., Xu, W., Feng, J., Shivakumara, P., Lu, T.: Local and global bayesian network based model for flood prediction. In: Proceedings of ICPR (2018)
17. Yao, C., Zhang, K., Yu, Z., Li, Z., Li, Q.: Improving the flood prediction capability of the xinanjiang model in ungauged nested catchments by coupling it with the geomorphologic instantaneous unit hydrograph. J. Hydrol. **517**, 1035–1048 (2014)
18. Yeung, S., Russakovsky, O., Jin, N., Andriluka, M., Mori, G., Fei-Fei, L.: Every moment counts: Dense detailed labeling of actions in complex videos. IJCV **126**(2–4), 375–389 (2018)
19. Yeung, S., Russakovsky, O., Mori, G., Fei-Fei, L.: End-to-end learning of action detection from frame glimpses in videos. In: Proceedings of IEEE CVPR, pp. 2678–2687 (2016)
20. Yu, P., Chen, S., Chang, I.: Support vector regression for real-time flood stage forecasting. J. Hydrol. **328**(3), 704–716 (2006)
21. Zhuang, W.Y., Ding, W.: Long-lead prediction of extreme precipitation cluster via a spatiotemporal convolutional neural network. In: Proceedings of the 6th International Workshop on Climate Informatics: CI (2016)

Hand Pose Estimation
with Attention-and-Sequence Network

Tianping Hu, Wenhai Wang, and Tong Lu$^{(\boxtimes)}$

National Key Lab for Novel Software Technology,
Nanjing University, Nanjing, China
ziyiliunian@163.com, wangwenhai362@163.com, lutong@nju.edu.cn

Abstract. Hand pose estimation from depth images is an essential topic in computer vision. Despite the recent advancements in this area promoted by Convolutional Neural Network, accurate hand pose estimation is still a challenging problem. In this paper, we analyse the spatial relationship among hand joints, and discover that: (1) there exists independence of joints from different fingers, and (2) there also exists strong correlation among adjacent joints in the same finger. Based on this, we present a novel Attention-and-Sequence Network (ASNet) embedded with finger attention and joint sequence mechanisms. Here the finger attention mechanism is proposed to ensure the independence of joints from different fingers, while the joint sequence mechanism is employed to make use of strong correlation among adjacent joints in the same finger. The proposed ASNet achieves an average 3D error of 5.6 mm on ICVL, 10.3 mm on NYU, 7.3 mm on MSRA, and these competitive results further confirm the great effectiveness of ASNet.

Keywords: Hand pose estimation · Depth images
Convolutional Neural Network · Recurrent neural network
Attention model

1 Introduction

Recently, hand pose estimation from depth images has attracted extensive attention for its numerous applications, such as Human Computer Interface and Augmented Reality. However, due to large view variance, high joint flexibility, poor depth quality, severe self occlusion and similar part confusion, accurate hand pose estimation is still a challenging problem.

Previous works [1,4,5,7,13–15,21,23–25] have achieved great progress on hand pose estimation in the recent years owing to the success of Convolutional Neural Networks (CNN) [8,9,12,17]. However, most CNN-based methods simply treat a hand pose as a number of independent joints but do not consider relationships in hand joints, which is significant in hand pose estimation [1,25].

T. Hu and W. Wang—Authors contributed equally

© Springer Nature Switzerland AG 2018
R. Hong et al. (Eds.): PCM 2018, LNCS 11164, pp. 556–566, 2018.
https://doi.org/10.1007/978-3-030-00776-8_51

To address this problem, we design a network based on the spatial relationships among hand joints. Due to kinematics and physical constraints, the relationships among the hand joints can be summed up in two aspects:

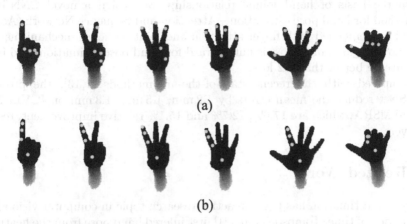

(a)

(b)

Fig. 1. The hand joints in different gestures. (a) shows the fingertips of a hand with various gestures. (b) shows the joints in index finger in different hand poses.

The independence of hand joints from different fingers. It is common sense that the fingers in human hands are flexible, which decides that joints from different fingers are largely independent. As shown in Fig. 1(a), the fingertip's positions of five fingers are irrelevant to each other in different gestures. Thus, the local features of different fingers may be redundant and even disturbing to each other. However, in the original feature given by CNN, the features of five fingers are mixed together. To reduce the negative effects from other fingers, it is necessary to extract the local feature for each finger from the original feature given by CNN.

The correlation between hand joints in the same finger. Because of the physical constraints of human hand, there is strong correlation between the two adjacent joints in the same finger. For example, as shown in Fig. 1(b), the relative position and distance between the adjacent joints in the index finger change little in various hand poses. Nevertheless, the hand joints are deemed to be independent in most CNN-based methods, which could be an obstacle for these methods. To alleviate such problem, a feasible method is to embed the correlation into the pipeline of hand pose estimation.

According to the above analyses, in this paper, we propose a novel attention-and-sequence network (ASNet) equipped with finger attention and joint sequence mechanisms. Finger attention mechanism is employed to extract the local feature of each independent finger. Joint sequence mechanism is applied to utilize the correlation between the adjacent joints in the same finger.

We conduct extensive experiments on three challenging datasets (i.e. ICVL [20], NYU [21] and MSRA [18]) to show the effectiveness of our proposed ASNet.

Our method outperforms many state-of-the-art methods. Specifically, it achieves an average 3D error of 5.6 mm on ICVL, 10.3 mm on NYU, 7.3 mm on MSRA.

The main contributions of this paper are as follows:

- On the basis of hand joints' relationship, we design a novel CNN-based method for hand pose estimation—Attention-and-Sequence Network (ASNet) which is embeded with finger attention and joint sequence mechanisms.
- A novel logarithm L2 loss is put forward for hand pose estimation, and it can converge better than L2 loss.
- Compared with the recent state-of-the-art methods [1,13], the proposed ASNet reduce the mean errors by 1.2 mm, 1.5 mm, 1.3 mm on ICVL, NYU and MSRA, which are 17.6%, 12.7% and 15.1% relative improvement, respectively.

2   Related Work

Hand pose estimation has been an active research topic in computer vision for a long period of time. Tompson et al. [21] first infered hand pose from the heatmaps produced by CNN. Oberweger et al. [14] used CNN to regress the positions of hand joints directly. In [15], a feedback loop was employed to iteratively refine the estimation results. Ge et al. [4] used multiple CNNs to predict heatmaps from different views of input depth image and estimated hand pose based on the multi-view heatmaps. In [25], physical joint constraints are incorporated into a forward kinematics based layer in CNN. Zhang et al. [24] embeds skeletal manifold into CNN and made prediction of joint locations sequentially. Ye et al. [23] used spatial attention mechanism to integrate the cascaded and hierarchical hand pose estimation into one framework. Ge et al. [5] introduced 3D Convolutional Neural Networks [11] to hand pose estimation. Guo et al. [7] trained a region ensemble network with different spatial regions of features. Chen et al. [1] adopted a pose guided structure to refine the final hand pose from a coarse hand pose predicted previously. Recently, Oberweger et al. [13] improve their previous work [14] by adding ResNet layers, data augmentation, and better initial hand localization.

3   The Proposed Method

In this section, we first introduce the overall pipeline of the proposed Attention-and-Sequence Network (ASNet). We then present the details of finger attention and joint sequence mechanisms. At last, we show the design of loss function and the implementation details of ASNet.

3.1   Overall Pipeline

The overall pipeline of the proposed ASNet is illustrated in Fig. 2. We use ResNet [8] as the backbone (i.e. CNN module in Fig. 2) of ASNet. Let us consider a 128×128 depth image I. We firstly feed depth image I to CNN module and get feature map F. Subsequently, we apply finger attention mechanism (FAM) to feature map F and obtain n attention feature maps $A^0, A^1, ..., A^{n-1}$ for n parts of a hand, respectively. After that, each attention feature map A^i passes through an average pooling layer (AP) and a fully-connected layer (FC), and then is fed to joint sequence mechanism (JSM) to get refined feature B^i. Finally, we use the refined features $B^0, B^1, ..., B^{n-1}$ to estimate all parts of a hand $P^0, P^1, ..., P^{n-1}$, and join them together to get the final hand pose R.

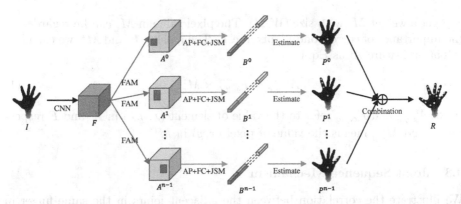

Fig. 2. The overall pipeline of ASNet. FAM, AP, FC and JSM refer to finger attention mechanism, average pooling layer, fully-connected layer and joint sequence mechanism, respectively.

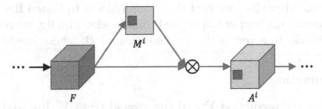

Fig. 3. The procedure of finger attention mechanism.

3.2 Finger Attention Mechanism

In Sect. 1, we discuss the independence of joints from different fingers. Inspired by the spatial attention mechanism in [23], we put forward a novel attention model for hand pose estimation called finger attention mechanism (FAM), which can extract the local features of different fingers from the original feature produced by CNN. As the procedure shown in Fig. 3, we feed the mixed feature F with size (W, H, C) to a 1×1 convolutional layer whose activation function is sigmoid,

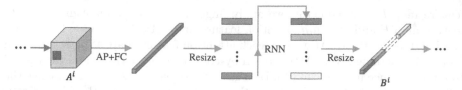

Fig. 4. The procedure of joint sequence mechanism.

and get a weight M^i with size (W, H). The pixel value in M^i can be regarded as the importance of corresponding pixel in F. Then, using F and M^i, we calculate attention feature A^i as Eq. 1.

$$A^i_{x,y,c} = F_{x,y,c} \times M^i_{x,y}. \tag{1}$$

Here, $A^i_{x,y,c}$ and $F_{x,y,c}$ refer to the value of element (x, y, c) in A^i and F respectively, and $M^i_{x,y}$ means the value of pixel (x, y) in M^i.

3.3 Joint Sequence Mechanism

We illustrate the correlation between the adjacent joints in the same finger in Sect. 1. To utilize this feature, we propose a joint sequence mechanism (JSM) to sequence the features of joints in the same finger using RNN. The procedure of JSM is shown in Fig. 4. Let us consider a part of hand containing m joints. At first, the attention feature A^i passes through a 16×16 average pooling (AP) and a fully-connected layer (FC) with $64 \times m$ neurons. Then, we resize the output of FC to $(m, 64)$, and thus the resized output can be viewed as m mini blocks whose size is 64. After that, we feed the m mini blocks to Gated Recurrent Unit (GRU) [2] sequentially and get m refined mini blocks. Finally, we restore the m refined mini blocks to a vector (i.e. refined feature) B^i whose length is $64 \times m$.

3.4 Loss Function

Let us consider the prediction $\hat{Y}$ and the ground truth Y. Inspired by [13, 14], we use L2 loss and the loss can be formulated as Eq. 2.

$$L_2 = \left\| Y - \hat{Y} \right\|^2. \tag{2}$$

However, in hand pose estimation problem, L2 loss is unable to converge well when the prediction $\hat{Y}$ is close to the ground truth Y. Therefore, we adopt the natural logarithm to L_2 and get logarithm L2 loss which can converge better when the gap between $\hat{Y}$ and Y is tiny. The revised loss can be formulated as Eq. 3.

$$L = ln(L_2) = ln(\left\| Y - \hat{Y} \right\|^2), \tag{3}$$

where $ln(\cdot)$ means natural logarithm function.

3.5   Implementation Details

The detail of the CNN module in ASNet is shown in Table 1, in which BN, ReLU, conv and Dropout refer to batch normalization [10], rectified linear units [6], convolution [12] and dropout layer, respectively. The dropout rate of dropout layer is set to 0.3.

Table 1. The detail of the CNN module in ASNet. BN, ReLU, conv and Dropout refer to batch normalization, rectified linear unit, convolution and dropout layer, respectively

Layers	Output size/channel	Operation	
Convolution	64 × 64/64	7 × 7, 64 conv, stride 2, padding 3; BN; ReLU	
Pooling	32 × 32/64	3 × 3 max pooling, stride 2, padding 1	
Residual block (1)	32 × 32/64	3 × 3, 64 conv, stride 1, padding 1; BN; ReLU 3 × 3, 64 conv, stride 1, padding 1; BN; ReLU 3 × 3, 64 conv, stride 1, padding 1; Dropout	× 2
Residual block (2)	16 × 16/128	3 × 3, 128 conv, stride 2, padding 1; BN; ReLU 3 × 3, 128 conv, stride 1, padding 1; BN; ReLU 3 × 3, 128 conv, stride 1, padding 1; Dropout	× 2

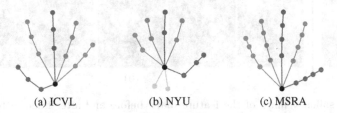

(a) ICVL   (b) NYU   (c) MSRA

Fig. 5. The split schedules for different datasets. In ICVL, NYU and MSRA datasets, a hand is divided into 5, 5, and 6 parts, respectively.

As shown in Fig. 2, ASNet has n attention-and-sequence branches which corresponds to n parts of hand. The split schedules for different datasets are shown in Fig. 5. The joints with the same color are assigned to the same part and the black joint is palm which belongs to all parts.

4   Experiments

In this section, we first conduct ablation studies for ASNet. Then, we evaluate the proposed ASNet on three challenging public datasets (i.e. ICVL, NYU and MSRA) and compare ASNet with many state-of-the-art methods.

4.1 Datasets

We evaluate ASNet on three publicly datasets: ICVL [20], NYU [21] and MSRA [18]. The ICVL dataset contains $330k$ images for training and 1596 images for testing with 16 joints. The NYU dataset includes 72757 training samples and 8252 testing samples with 14 joints. The MSRA dataset has 76500 samples from 9 different subjects, whose hand pose contains 36 joints. We follow the data augmentation schemes and evaluation metrics used in [13]. On ICVL and NYU dataset, We use all training samples and report the average 3D error of testing set at the end of training. On MSRA dataset, following previous works [1,13], we perform leave-one-out cross-validation strategy: training on 8 different subjects and evaluating on the remaining subject. We repeat this procedure for each subject and report the average errors over the different runs.

4.2 Training

All the networks are trained by using stochastic gradient descent (SGD). In the experiments, we train ASNet using batch size 128 for 300 epochs. The initial learning rate is set to 0.002, and is divided by 10 at 50% and 75% of the total number of training epochs. We use a weight decay of 5×10^{-4} and a Nesterov momentum [19] of 0.9 without dampening.

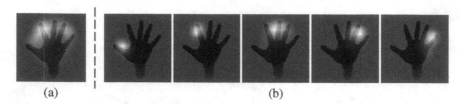

(a)   (b)

Fig. 6. The saliency maps of the feature maps before and after finger attention. (a) shows the feature map before finger attention. (b) shows the feature maps after finger attention.

4.3 Ablation Study

Effect of Finger Attention Mechanism. To study the effect of FAM, we make comparison between the models with and without FAM. In details, we evaluate the model without FAM ("Base Network") and the model with FAM ("Base Network + FAM") on NYU dataset. The experiment results are shown in Table 2, from which we can find that the model with FAM surpasses the one without FAM by 0.67 mm. The result suggests the effectiveness of finger attention mechanism.

We use class saliency extraction [16] to visualize the feature maps before and after finger attention. In Fig. 6, we can find that the feature map before finger attention (see Fig. 6(a)) covers the whole hand, while the feature maps after finger attention (see Fig. 6(b)) focus on different fingers.

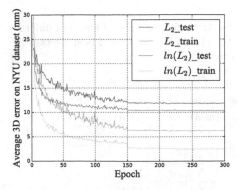

Table 2. The average 3D error (Error) of ASNets with different configurations on NYU dataset.

Configurations	Error (mm)
Base Network	12.22
Base Network + FAM	11.55
Base Network + FAM + JSM (ASNet)	10.30

Fig. 7. The average 3D error of NYU training and testing set in the process of training with L2 loss L_2 and logarithm L2 loss L.

Effect of Joint Sequence Mechanism. Next we study the effect of JSM. We evaluate the model without JSM ("Base Network + FAM") and the model with JSM ("Base Network + FAM + JSM") on NYU dataset. It is observed in Table 2 that the model with JSM further reduces the average 3D error to 10.30 mm, 1.25 mm less than the one without JSM, which shows the power of joint sequence mechanism.

Effect of Logarithm L2 Loss. To explore the effect of logarithm L2 loss L, we train models on NYU dataset with different loss functions (i.e. L_2 (see Eq. 2), L (see Eq. 3)) and report the average 3D error of training and testing set in the process of training. In Fig. 7, it can be seen that with growing of epoch, the models with logarithm L2 loss L convergence are better than the one with L2 loss L_2, which indicates the superiority of logarithm L2 loss L.

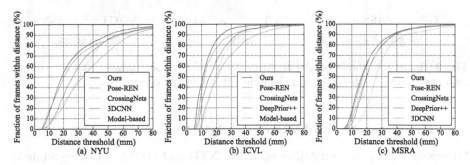

Fig. 8. Comparison with the state-of-the-art methods on different datasets. We plot the fraction of frames where all joints are within a maximum distance from the ground truth, following the protocol of [1,13].

Table 3. The average 3D error (Error) and the GPU forward time (Time) of different methods on ICVL, NYU and MSRA datasets. The best and second-best average 3D errors are highlighted in red and blue, respectively.

Methods	ICVL		NYU		MSRA	
	Error (mm)	Time (ms)	Error (mm)	Time (ms)	Error (mm)	Time (ms)
HandsDeep [14]	10.4	–	19.7	–	–	–
DeepModel [25]	11.3	–	16.9	–	–	–
Crossing Nets [22]	10.2	–	15.5	–	12.2	–
HPR [18]	9.9	–	–	–	15.2	–
JTSC [3]	9.2	–	16.8	–	–	–
DeepPrior++ [13]	8.1	–	12.3	–	9.5	–
REN [7]	7.5	0.31	13.4	–	–	–
Pose-REN [1]	6.8	–	11.8	–	8.6	–
ASNet (ours)	5.6	0.27	10.3	0.21	7.3	0.17

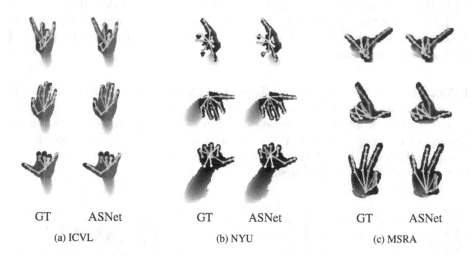

GT   ASNet   GT   ASNet   GT   ASNet

(a) ICVL   (b) NYU   (c) MSRA

Fig. 9. Some qualitative results on three datasets. For each dataset, the left column refers to ground truth (GT), and the right column is our result.

4.4   Comparisons with State-of-the-Art Results

We evaluate the proposed ASNet on ICVL, NYU and MSRA datasets and make comparisons with the representative state-of-the-art methods. The experiment results are shown in Table 3, in which our method significantly outperforms almost all the state-of-the-art methods. On ICVL dataset, ASNet achieves the best average 3D error (5.6 mm), and outperforms the second best method Pose-REN [1] by 1.2 mm. On NYU dataset, our method surpasses the second best record by 1.5 mm. On MSRA dataset, the average 3D error of our method is 7.3 mm, 1.3 mm less than Pose-REN [1]. In Table 3, we also test the forward time on single GPU (GeForce GTX 1080 Ti) to confirm the satisfactory efficiency of ASNet. Fig. 8 shows the proportion of frames over different error thresholds, in which the curves of our proposed ASNet perform better than ones of other competitors. Some qualitative results on three datasets can be seen in Fig. 9.

5 Conclusion

In this paper, we firstly analyse the spatial relationship among the hand joints. Then, on the basis of analyses, we propose a novel Attention-and-Sequence Network (ASNet) equipped with finger attention and joint sequence mechanisms. By finger attention mechanism, we can extract the local features of different fingers and thus reduce their negative effects to each other. By joint sequence mechanism, we can refine the features using the strong correlation between the adjacent joints in the same finger. The experiments on challenging public datasets demonstrate the superior performance of the proposed method.

Acknowledgements. The authors would like to thank the editor and the anonymous reviewers for their critical and constructive comments and suggestions. The work was supported by the Natural Science Foundation of China under Grant No. 61672273, No. 61272218 and No. 61321491, the Science Foundation for Distinguished Young Scholars of Jiangsu under Grant No. BK20160021, Scientific Foundation of State Grid Corporation of China (Research on Ice-wind Disaster Feature Recognition and Prediction by Few-shot Machine Learning in Transmission Lines).

References

1. Chen, X., Wang, G., Guo, H., Zhang, C.: Pose guided structured region ensemble network for cascaded hand pose estimation. arXiv preprint arXiv:1708.03416 (2017)
2. Cho, K., Van Merriënboer, B., Gulcehre, C., Bahdanau, D., Bougares, F., Schwenk, H., Bengio, Y.: Learning phrase representations using RNN encoder-decoder for statistical machine translation. arXiv preprint arXiv:1406.1078 (2014)
3. Fourure, D., Emonet, R., Fromont, E., Muselet, D., Neverova, N., Trémeau, A., Wolf, C.: Multi-task, multi-domain learning: application to semantic segmentation and pose regression. Neurocomputing **251**, 68–80 (2017)
4. Ge, L., Liang, H., Yuan, J., Thalmann, D.: Robust 3D hand pose estimation in single depth images: from single-view CNN to multi-view CNNs. In: CVPR (2016)
5. Ge, L., Liang, H., Yuan, J., Thalmann, D.: 3D convolutional neural networks for efficient and robust hand pose estimation from single depth images. In: CVPR (2017)
6. Glorot, X., Bordes, A., Bengio, Y.: Deep sparse rectifier neural networks. In: ICAIS (2011)
7. Guo, H., Wang, G., Chen, X., Zhang, C., Qiao, F., Yang, H.: Region ensemble network: improving convolutional network for hand pose estimation. In: ICIP (2017)
8. He, K., Zhang, X., Ren, S., Sun, J.: Deep residual learning for image recognition. In: CVPR (2016)

9. Huang, G., Liu, Z., Weinberger, K.Q., van der Maaten, L.: Densely connected convolutional networks. In: CVPR (2017)
10. Ioffe, S., Szegedy, C.: Batch normalization: accelerating deep network training by reducing internal covariate shift. In: ICML (2015)
11. Ji, S., Wei, X., Yang, M., Kai, Y.: 3D convolutional neural networks for human action recognition. IEEE Trans. Pattern Anal. Mach. Intell. **35**(1), 221–231 (2013)
12. LeCun, Y., Bottou, L., Bengio, Y., Haffner, P.: Gradient-based learning applied to document recognition. Proc. IEEE **86**(11), 2278–2324 (1998)
13. Oberweger, M., Lepetit, V.: Deepprior++: improving fast and accurate 3D hand pose estimation. In: ICCV Workshop (2017)
14. Oberweger, M., Wohlhart, P., Lepetit, V.: Hands deep in deep learning for hand pose estimation. arXiv preprint arXiv:1502.06807 (2015)
15. Oberweger, M., Wohlhart, P., Lepetit, V.: Training a feedback loop for hand pose estimation. In: ICCV (2015)
16. Simonyan, K., Vedaldi, A., Zisserman, A.: Deep inside convolutional networks: visualising image classification models and saliency maps. arXiv preprint arXiv:1312.6034 (2013)
17. Simonyan, K., Zisserman, A.: Very deep convolutional networks for large-scale image recognition. In: ICLR (2015)
18. Sun, X., Wei, Y., Liang, S., Tang, X., Sun, J.: Cascaded hand pose regression. In: CVPR (2015)
19. Sutskever, I., Martens, J., Dahl, G., Hinton, G.: On the importance of initialization and momentum in deep learning. In: ICML (2013)
20. Tang, D., Jin Chang, H., Tejani, A., Kim, T.-K.: Latent regression forest: structured estimation of 3D articulated hand posture. In: CVPR (2014)
21. Tompson, J., Stein, M., Lecun, Y., Perlin, K.: Real-time continuous pose recovery of human hands using convolutional networks. ACM Trans. Graph. (ToG) **33**(5), 169 (2014)
22. Wan, C., Probst, T., Van Gool, L., Yao, A.: Crossing Nets: combining GANs and VAEs with a shared latent space for hand pose estimation. In: CVPR (2017)
23. Ye, Q., Yuan, S., Kim, T.-K.: Spatial attention deep net with partial PSO for hierarchical hybrid hand pose estimation. In: ECCV (2016)
24. Zhang, Y., Xu, C., Cheng, L.: Learning to search on manifolds for 3D pose estimation of articulated objects. arXiv preprint arXiv:1612.00596 (2016)
25. Zhou, X., Wan, Q., Zhang, W., Xue, X., Wei, Y.: Model-based deep hand pose estimation. In: IJCAI (2016)

None Ghosting Artifacts Stitching Based on Depth Map for Light Field Image

Wenyuan Zhang, Shengyang Zhao, Wei Zhou, and Zhibo Chen[✉]

CAS Key Laboratory of Technology in Geo-spatial Information Processing
and Application System, University of Science and Technology of China,
Hefei 230027, China
chenzhibo@ustc.edu.cn

Abstract. Due to hardware limitations, existing light field (LF) capturing devices cannot offer sufficient field of view for building 6 degrees of freedom (6 DOF) VR applications. LF image stitching methods can be used to address this problem. The state-of-the-art LF stitching methods highly depend on the stitching accuracy of center view which is essentially a 2D image stitching task. However, conventional 2D image stitching methods usually suffer from the ghosting artifacts. In this paper, a None Ghosting Artifacts (NGA) stitching method is proposed to tackle this problem. We theoretically reveal the intrinsic cause of the ghosting artifacts and then further verify that different depth scenes require different homography matrices for warping. Therefore, the clustered depth map is employed to segment the scene into several layers, and the layer-specific homography matrix is computed for warping. An interpolation mechanism is also proposed to ensure that each layer has its own transformation. Compared with state-of-the-art stitching methods aiming to alleviate ghosting artifacts, experimental results show that the proposed method not only stitches images without ghosting artifacts, but also achieves realistic perspective transformation.

Keywords: Ghosting artifacts · Depth map · Image stitching
Light field stitching

1 Introduction

A light filed (LF) consists of a large collection of rays that store radiance information in both spatial and angular dimensions [6]. With extra angular light information, the LF can provide 6 degrees of freedom (6 DOF) experience. It is considered as a promising technique for future immersive multimedia applications, such as 3D TV and virtual reality (VR) applications.

LF can be captured by camera arrays or lenslet LF cameras, e.g. Lytro [9]. Both the camera array and lenslet camera cannot provide sufficient filed-of-view (FOV) for building the 6 DOF VR applications. In order to introduce the LF techniques into the market successfully, it is urgent to stitch the LF captured

© Springer Nature Switzerland AG 2018
R. Hong et al. (Eds.): PCM 2018, LNCS 11164, pp. 567–578, 2018.
https://doi.org/10.1007/978-3-030-00776-8_52

by multiple perspectives. A 6 DOF panoramic VR scene can be constructed by stitching the sub-views of multiple LF images.

Many research works have been carried out for LF stitching. [1, 6, 7] represent the LF as a 4D function and extend 2D image stitching algorithms to stitch LF in 4D dimension. However, they suffer from the high complexity caused by high data dimensions. [2] essentially stitches the LF in the frequency domain, which converts 4D plenoptic data to 3D focal stack before stitching and thus is limited to Lambertian scenes. The state-of-the-art method [3] represents the LF images as views array and stitch the center view with 2D stitching method. Then the stitching operation is propagated to other views. The method is based on propagation and robust for different scenes and devices. However, they suffer from the ghosting artifacts, which is an intrinsic drawback of conventional 2D image stitching. Solving the ghosting artifact is our motivation for this paper.

When the captured scene contains a relatively larger depth range, there will exist a large parallax between the adjacent images. In this case, traditional 2D image stitching will result in severe ghosting artifacts. Despite blending and feathering methods are exploited for to de-ghost, a good initial stitching can not only extremely improve the results, but also impose a much lower requirement on subsequent de-ghosting and post-processing, especially for the large parallax condition. [5, 8, 10, 11] all aimed to optimize the initial stitching, but none of them attempted to investigate the essence of the ghosting artifacts and thoroughly solved this problem.

In this paper, we first prove that the essential cause of the ghosting artifacts is that pixels at different depth planes need their own transformation matrix. Fortunately the depth information can be estimated from the LF image. Therefore, we propose a none ghosting artifacts (NGA) stitching method based on the depth map. We first use the well-known k-means algorithm to cluster the depth map of central view. Then we segment the scene into several layers with the depth map. Finally, different scene layers are transformed with their own homographies, which are computed by the matching feature points belonging to each depth plane with RANSAC algorithm. Considering that some depth planes may not have enough matching points to obtain an accurate enough transformation, we use a camera parameter based interpolation to ensure each layer has its own homography matric.

Our contributions is in the following three aspects:

(1) We theoretically point out the reason of ghosting artifact;
(2) We propose the NGA stitching algorithm for the center view of LF images;
(3) We propose an interpolation mechanism to make NGA more robust.

The rest of the paper is organized as follows: Sect. 2 surveys related work. Section 3 introduces our proposed method NGA. Experimental results are presented in Sect. 4, and we conclude our work in Sect. 5.

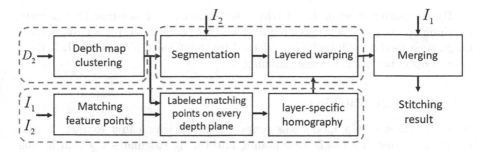

Fig. 1. Framework of the NGA stitching model. I_1 and I_2 denote the central view of two LFs respectively. D_2 denotes the depth map of I_2. The red box in the figure represents depth map clustering, the green box represents homography calculation, and the blue box represents layer-specific transformation.

2 Related Work

Our work is related to LF stitching and 2D image stitching. In this section, we review both them in the follows.

2.1 LF Stitching

As mentioned in Sect. 1, we divide the existing LF stitching algorithms into three main categories as follows:

Stitching in Ray Space. This category [1,6,7] considers the transformation of LF images in the ray space. [6,7] first find the matching rays, and then use a 5×5 transformation matrix to stitch LF images. These methods have a large amount of calculation due to the high dimensionality of the ray transformation matrix. [1] proposes to adopt the re-parameterization with double cylinders to handle different complex scenes. However, it requires highly accurate camera calibration and camera control, which is hard to satisfy in practice.

Stitching with Focal Stack. This category [2] generates panorama LF by first converting input LF images into focal stacks. Then they stitch these focal stacks and convert the resulting panorama focal stack into a LF using linear view synthesis. These method will fail in non-Lambertian scenes which loses the advantages of light-field imaging in general.

Stitching Based on Multi-view. The methods based on multi-view represent the LF images as view arrays. A general framework is established in [3], which propagates arbitrary spatial deformations operated in the center view to all other perspective views consistently. Once this method is used to stitch LF image, which may first stitch the center view with 2D stitching method, and then propagate the operation to other views without destroying the disparity consistency.

By comparing existing LF stitching algorithms, it shows that [3] has faster stitching speed and has a lower requirement for camera control accuracy. Obviously, such method will be highly affected by the 2D stitching. Thus, It is desirable to optimize the 2D image stitching.

2.2 2D Image Stitching

A classic 2D stitching model was proposed in [4], which first extracts feature points, then matches the feature points, and finally calculates a global homography to fit the matched feature points. However, this model have two limited conditions: (1) the scene of image is planar or (2) the views differ purely by rotation [10]. But in practice, the two assumptions are hard to satisfy, and hence there will generate obvious ghosting artifacts and misalignment due to parallax. By assuming that the scene contains a ground plane and a distant plane, [5] proposed dual homography warps for image stitching. Essentially [5] is a special case of a piece-wise projective warping, which is more flexible than using a single homography. But it is not robust for complex scenes. In [8], multiple affine transformations are used to make more precise local alignment. [10] replaces local affine transformations in [8] with local homography transformations, which makes local adaptation much better. A simple moving Direct Linear Transformation (DLT) method in [10] is used to estimate the local parameters, by providing higher weights to closer feature points and lower weights to the farther ones. However, once the local mesh is at the boundary of the objects, it is impossible to ensure those closer feature points belong to the same depth plane. For better stitching, Content-preserving warps (CPW) in [11] combine the global transformation with the block based warping. Therefore, those methods can only alleviate but not absolutely avoid the ghosting artifacts. In the next section, we point out the essential reason for ghosting artifacts and shows that the proposed NGA can solve this problem.

3 None Ghosting Artifacts Stitching

In this section, we first point out that each depth plane corresponds to a specific homography matrix. That is why the conventional algorithms always produce ghosting artifacts. Further, we derive the relationship between the homographies of different depth planes. Based on these analysis, we proposed our NGA stitching method.

3.1 3D Warping Function

The relationship between a 3D point $\mathbf{M}$ and its image projection $\mathbf{m}$ is given by [12].

$$s\tilde{\mathbf{m}} = \mathbf{ARM} + \mathbf{At} \,, with \; \mathbf{A} = \begin{bmatrix} \alpha & \gamma & u_0 \\ 0 & \beta & v_0 \\ 0 & 0 & 1 \end{bmatrix}, \tag{1}$$

where s is the depth of a 3D point, and $\mathbf{A}$ is the intrinsic matrix of the camera. $\mathbf{R}$ and $\mathbf{t}$ denote the camera rotation and translation, respectively. We use $\sim$ to denote the homogeneous coordinates of $\mathbf{m}$.

We can consider two cameras conditions. Then the projection of point $\mathbf{M}$ are denoted by $\mathbf{m}_1$ and $\mathbf{m}_2$. The relationship between them can be derived as [12]

$$s_2\tilde{\mathbf{m}}_2 = s_1\mathbf{A}\mathbf{R}\mathbf{A}^{-1}\tilde{\mathbf{m}}_1 + \mathbf{A}\mathbf{t}. \tag{2}$$

When capturing the LF panorama, the camera may have two kinds of movement, i.e. the translation and rotation. To simplify the derivation, we separate the two conditions. If we consider translating only, there should be $s_2 = s_1 = s$, $\mathbf{R} = \mathbf{E}$, where $\mathbf{E}$ is an indetity matrix. The Eq. (2) can be simplified as

$$\tilde{\mathbf{m}}_2 = \tilde{\mathbf{m}}_1 + \frac{\mathbf{A}\mathbf{t}}{s}, \tag{3}$$

where $\mathbf{t} = (t_x, t_y, 0)^T$. The Eq. (3) can also be written as

$$\tilde{\mathbf{m}}_2 = \mathbf{H}\tilde{\mathbf{m}}_1 \text{ , with } \mathbf{H} = \mathbf{E} + \frac{\mathbf{A}\mathbf{t}}{s}, \tag{4}$$

It shows that $\mathbf{H}$ in Eq. (4) is only decided by the depth. Thus, it is reasonable to use different homography for different depth layers. Further, the relationship between the homographies of different depth planes can be expressed as

$$\mathbf{H}_2 = \frac{s_1}{s_2}(\mathbf{H}_1 - \mathbf{E}) + \mathbf{E}, \tag{5}$$

where s_1 and s_2 are different depth values, and $\mathbf{H}_1$ and $\mathbf{H}_2$ is the homgraphy correspond to s_1 and s_2, respectively.

If only the rotation is considered, we have $\mathbf{t} = (0, 0, 0)^T$, $s_2 \approx s_1$, the Eq. (2) can be simplified as

$$\tilde{\mathbf{m}}_2 = \mathbf{H}\tilde{\mathbf{m}}_1 \text{ , with } \mathbf{H} = \mathbf{A}\mathbf{R}\mathbf{A}^{-1}. \tag{6}$$

We can find that $\mathbf{H}$ is not related to depth in this case. However, actually a pure rotation is nearly impossible because the main lens cannot be treated a point. This means the second term in Eq. (3) will affect the translation.

Now we conclude that if we want to align two projection pixel, we need to calculate the transformation based on the depth value, and different depth plane should have its own matrix. This is the reason that conventional work always produce misalignment and introduce ghosting artifacts.

3.2   Component of NGA

Based on the above analysis, we propose the None Ghosting Artifacts (NGA) stitching model. As shown in Fig. 1, the proposed model consists of three key parts: depth plane clustering, homography calculation, and layer-specific transformation.

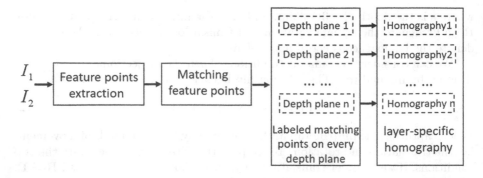

Fig. 2. The homography calculation.

Depth Plane Clustering. According to the theoretical derivation, each different value in the depth map corresponds to a different homography. However, calculating homography for all depth values is time-consuming. Also for some depth values there may be not enough feature points for calculation. Moreover the estimated depth maps are usually noisy. Therefore, we cluster the depth map by K-means algorithm. The K value is determined by Calinski-Harabasz criterion.

Homography Calculation. This part is shown in Fig. 2. For input images I_1 and I_2, the SIFT feature points are extracted and matched. For each depth layer, we count the feature points locating within it. If the num of feature points is more than a threshold N_{fp}, we calculate its own homography transformation by RANSAC algorithm; otherwise we obtain the transformation matric by our interpolation method, which is detailedly described in Sect. 3.3.

Layer-Specific Transformation. Due to I_2 is transformed to I_1, we first segment I_2 into several layers by the clustered depth map. Then we perform the corresponding transformation on each layer. The layers are merged by depth order, i.e. the farthest layer is put first and so on. It is noticeable that the post-processing methods, like blending or feathering can be used for any other algorithms. Thus, we do not adopt them to help fairly comparing.

The process of Layer-specific transformation is shown in Fig. 3. Due to the segmentation and layering operations, each depth plane monopolizes a different transformation. Thus the stitching results can achieve realistic perspective conversion.

3.3 Interpolation Module

The significance of this module is to make up for the unstability of the homography calculation in the NGA framework under certain circumstances. There are two special cases that make it hard to obtain accurate homographies of some depth layers. On the one hand, the matched feature points exist only in the overlapping areas of the I_1 and I_2, which can result in the fact that some depth

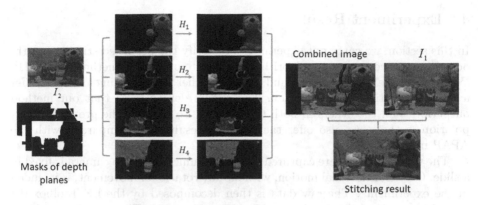

Fig. 3. The layered transformation diagram.

planes may not have matched feature points, such as the non-overlapping area in I_2 exists a depth plane that does not exist in the overlapping area, so that homography of this depth plane cannot be calculated. On the other hand, some depth planes contain fewer matching feature points so that robust homography cannot be obtained. To solve this problem, we have adopted an interpolation mechanism. Formula (5) in Sect. 3.1 shows that we can only use a known homography to compute an unknown homography. In order to get a more robust result, we use all known homgraphy to compute the unknown homography of the same depth plane. Thus we can get multiple candidate homographies that are calculated as

$$\mathbf{H}_*^i = \frac{s_k^i}{s_*}(\mathbf{H}_k^i - \mathbf{E}) + \mathbf{E}, \tag{7}$$

where subscript k present the meaning of known, so s_k^i and s_* are the depth value of the i-th known homography and the depth value of the unknown homography respectively. $\mathbf{H}_k^i$ is the i-th known homography, and $\mathbf{H}_*^i$ is the homography computed by the i-th known homography. The unknown homography is estimated from the weighted homography as

$$\mathbf{H}_* = \mathbf{E} + \sum_{i=1}^{n} w_*^i(\mathbf{H}_*^i - \mathbf{E}). \tag{8}$$

The weights $\{w_*^i\}_{i=1}^{n}$ change according to the distance of between the i-th depth layer of known homography and the depth layer of unknown homography, which are calculated as

$$w_*^i = \exp(-\|s_k^i - s_*\|^2/\sigma^2). \tag{9}$$

Here, σ is a scale parameter. Therefore, we can use known and stable homographies to calculate those unstable homography by Eq. (8).

4   Experiment Results

In this section, we carry out experiments to verify the proposed stitching methods. First we compare our algorithm with state-of-the-art methods [4,10,11]. We analyze the ghost artifact and check whether the stitching methods offer a true perspective transformation. The results demonstrate that our method outperforms all the others in both two aspects. To validate the proposed interpolation method, we also offer the stitching results and compare it with the APAP method.

The test LF images are captured by Lytro illum. The Lytro camera is fixed in a slide. Only translational motion, without any rotational movement, is included in the experiments. The raw data is then decomposed by the LF Toolbox 0.4 into view arrays of the dimensions $15 \times 15 \times 434 \times 625$. The center view depth map is estimated by the Lytro desktop software.

4.1   Comparisons with Other Methods

An adequate stitching algorithm should transform the images captured from different perspectives into one coordinate system and merge them together. Thus the algorithm should offer a true perspective transformation and introduce as few ghosting artifacts as possible.

We here compare our method against other state-of-the-art stitching methods: as-projective-as-possible (APAP) [10], parallax-tolerant image stitching (PATOIS) [11] and the global homography (GH) [4] as baseline. To fairly compare the methods, we do not consider sophisticated post-processing like seam cutting and straightening, and simply blend the aligned images by intensity averaging so that any misalignments remain conspicuous.

As shown in Fig. 4, the GH method produces severe ghosting artifacts. It is easy to be understood because a global homography cannot satisfy all the pixels lying on different depth planes, as what we emphasize in Sect. 3. The APAP divide the image into blocks and calculate transformation for each block. However if the block lies cross the edge of the objects, APAP will fail because it perform the same transformation on pixels belong to different depth planes. That's why it works well at the center of the object (like the pen container in Fig. 4) but fails at the object boundary (e.g. the teddy's hand and hello kitty head in Fig. 4). The PATOIS employs a block-based warping. Therefore, it suffers from the object boundary for the reason we give above. Our method obviously outperforms all the others. Almost no ghost artifacts or blur could be found. It verifies that our analysis for the essential cause of ghosting artifacts is reasonable and our proposed method really works.

Secondly we check if these methods provide the true perspective transformation. After transforming the right image(I_2) to the left one(I_1), the aligned image should show the perspective from the left camera. Objects at different depth have different disparity. Thus their relative position should be changed by a true perspective transformation. But we may see, for the methods shown in Fig. 5, only our method can achieve this.

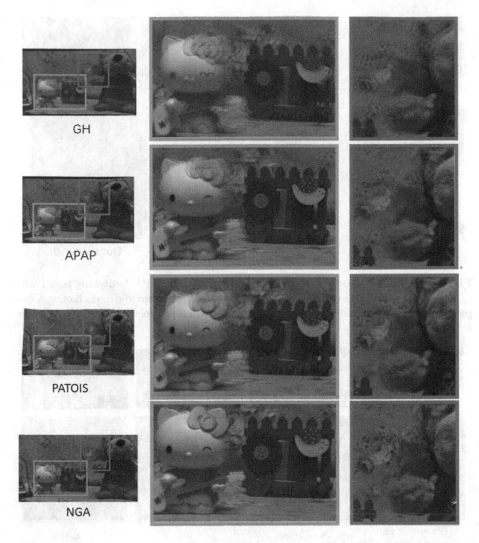

Fig. 4. The qualitative comparisons of alleviating ghosting artifacts. List of acronyms and initialisms: global homography, as-projective-as-possible, parallax-tolerant image stitching.

4.2   Interpolation Module Validation

To validate our interpolation module, we use our interpolation mechanism to interpolate the homography matrixs. As shown in Fig. 6, the stitching result with our method suffer from obvious ghosting artifacts at the depth plane of Rubik's Cube. The reason for this phenomenon is that the depth plane of Rubik's Cube contains fewer matching feature points, which results in the inaccuracy of Homography calculated using those points. At the same time, compared with

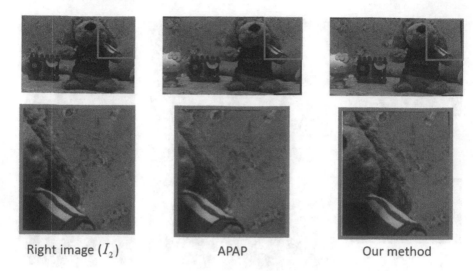

Right image (I_2)   APAP   Our method

Fig. 5. Comparison of perspective conversion. Comparing APAP stitching result with I_2, the relative position of the background layer and the bear did not change. Comparing our stitching result with I_2, the background moved relative to the bear to the left.

Our method with   Our method   GH   APAP
interpolation

Fig. 6. Interpolation result. The homography of second depth plane (Rubik's Cube) is interpolated by weighted homographies of first depth plane (MashiMaro) and third depth plane (background).

GH and APAP, the stitching result with interpolation produces a surprising result. This shows that our interpolation mechanism can achieve better results when the homographies of some depth planes cannot be calculated or calculated accurately by using the feature points directly.

4.3   Run Time

We implemented all methods in MATLAB2016 and on a DESKTOP-40REPQJ computer running at 3.40 GHz. The obvious difference between these methods (GH, APAP and NAG) is the number of times the homographies are computed, but it doesn't take much time to finish this step. In fact, it takes most of the time to extract feature points and use RANSAC algorithm in 2D image stitching, so the total running time of these methods is about the same. As shown in Table 1, the experimental result is consistent with our conjecture.

Table 1. Comparison of the running time of each method.

	GH	APAP	NGA
Feature points extraction	0.69 s	0.69 s	0.69 s
RANSAC algorithm	9.89 s	9.89 s	9.89 s
Compute homographies	0.0101 s	1.0924 s	0.0136 s
Total	23.0 s	24.4 s	23.2 s

5   Conclusion and Future Work

We have proposed a simple but effective stitching method NGA to stitch images without ghosting artifacts. We theoretically revealed the intrinsic cause of the ghosting problem and demonstrated that different depth scenes require different homographies for warping. An interpolation mechanism is also proposed to ensure that each depth plane has its own homography. The experiments results show that our proposed method not only stitch images without ghosting artifacts, but also achieve realistic perspective transformation compared with previous methods, and the interpolation mechanism we proposed is also reasonable.

We plan to implement LF stitching based on our method. We can use the parallax relationship between the off-center perspectives and the central view in LF to find areas where the off-center perspectives are the same depth planes as the central view, and to propagate homographies at various depth levels of the central perspective to the off-center perspectives.

Acknowledgements. This work was supported in part by the National Key Research and Development Program of China under Grant No. 2016YFC0801001, the National Program on Key Basic Research Projects (973 Program) under Grant 2015CB351803, NSFC under Grant 61571413, 61632001,61390514, and Intel ICRI MNC.

References

1. Birklbauer, C., Bimber, O.: Panorama light-field imaging. In: Computer Graphics Forum, vol. 33, pp. 43–52. Wiley Online Library (2014)
2. Birklbauer, C., Opelt, S., Bimber, O.: Rendering gigaray light fields. In: Computer Graphics Forum, vol. 32, pp. 469–478. Wiley Online Library (2013)
3. Birklbauer, C., Schedl, D.C., Bimber, O.: Nonuniform spatial deformation of light fields by locally linear transformations. ACM Trans. Graph. (TOG) 35(5), 156 (2016)
4. Brown, M., Lowe, D.G.: Automatic panoramic image stitching using invariant features. Int. J. Comput. Vis. 74(1), 59–73 (2007)
5. Gao, J., Kim, S.J., Brown, M.S.: Constructing image panoramas using dual-homography warping. In: 2011 IEEE Conference on Computer Vision and Pattern Recognition (CVPR), pp. 49–56. IEEE (2011)
6. Guo, X., Yu, Z., Kang, S.B., Lin, H., Yu, J.: Enhancing light fields through ray-space stitching. IEEE Trans. Vis. Comput. Graph. 22(7), 1852–1861 (2016)
7. Johannsen, O., Sulc, A., Goldluecke, B.: On linear structure from motion for light field cameras. In: Proceedings of the IEEE International Conference on Computer Vision, pp. 720–728 (2015)
8. Lin, W.Y., Liu, S., Matsushita, Y., Ng, T.T., Cheong, L.F.: Smoothly varying affine stitching. In: 2011 IEEE Conference on Computer Vision and Pattern Recognition (CVPR), pp. 345–352. IEEE (2011)
9. Ng, R., Levoy, M., Brédif, M., Duval, G., Horowitz, M., Hanrahan, P.: Light field photography with a hand-held plenoptic camera. Comput. Sci. Tech. Rep. CSTR 2(11), 1–11 (2005)
10. Zaragoza, J., Chin, T.J., Brown, M.S., Suter, D.: As-projective-as-possible image stitching with moving DLT. In: 2013 IEEE Conference on Computer Vision and Pattern Recognition (CVPR), pp. 2339–2346. IEEE (2013)
11. Zhang, F., Liu, F.: Parallax-tolerant image stitching. In: Proceedings of the IEEE Conference on Computer Vision and Pattern Recognition, pp. 3262–3269 (2014)
12. Zhang, Z.: A flexible new technique for camera calibration. IEEE Trans. Pattern Anal. Mach. Intell. 22(11), 1330–1334 (2000)

Adaptive Hierarchical Motion-Focused Model for Video Prediction

Min Tang, Wenmin Wang[✉], Xiongtao Chen, and Yifeng He

School of Electronic and Computer Engineering, Shenzhen Graduate School, Peking University, Shenzhen, China
{tangm,cxt}@pku.edu.cn, wangwm@ece.pku.edu.cn, yifenghe2008@gmail.com

Abstract. Video prediction is a promising task in computer vision for many real-world applications and worth exploring. Most existing methods generate new frames based on appearance features with few constrain, which results in blurry predictions. Recently, some motion-focused methods are proposed to alleviate the problem. However, it's difficult to capture the object motions from a video sequence and apply the learned motions to appearance, due to variety and complexity of real-world motions. In this paper, an adaptive hierarchical motion-focused model is introduced to predict realistic future frames. This model takes advantage of hierarchical motion modeling and adaptive transformation strategy, which can achieve better motion understanding and applying. We train our model end to end and employ the popular adversarial training to improve the quality of generations. Experiments on two challenging datasets: Penn Action and UCF101, demonstrate that the proposed model is effective and competitive with outstanding approaches.

Keywords: Video prediction · Motion-focused model
Hierarchical motion modeling · Adaptive transformation

1 Introduction

Video generation has recently become an important research subfield in computer vision. In this paper, we address the problem of video prediction, one of the most challenging tasks in this subfield. Given a video sequence, video prediction aims to generate realistic subsequent frames in an unsupervised way. Future frames prediction has become crucial for various kinds of real-world applications, such as robotics and automatic driving system. However, complex appearance and ambiguous dynamics in video pose a great challenge to this task.

Many methods for video prediction have been proposed to tackle the problems in recent years. To improve the sharpness of generations, Mathieu et al. [1] take advantage of a multi-scale network and a combined loss function with adversarial training. Vondrick et al. [2] introduced a two-stream network which generates foreground and background of future frames separately and combined them using a mask. Villegas et al. [3] independently modeled content and motion,

© Springer Nature Switzerland AG 2018
R. Hong et al. (Eds.): PCM 2018, LNCS 11164, pp. 579–588, 2018.
https://doi.org/10.1007/978-3-030-00776-8_53

merged their features and then reconstructed the new frames via a decoder network. All these approaches [1–4] synthesized the frames mainly through kinds of the encoder-decoder structure with few constraints. Thus predicted frames tend to lose content details of input frames and become burry after multiple filtering. Furthermore, parameters of these generative models remain constant, which is inflexible for various instances and bring in artifacts.

Actually, future frames probably have the same appearance as previous ones. In view of this condition, some motion-focused models [5–13] were introduced to get more realistic future. These methods focus on motion capture and modeling without storing extra low-level details since appearance information can be available in previous frames. In addition, learned motion features vary with motion patterns of observed scenes and are used to transform previous frames into new ones, which enable sharper predictions. Nonetheless, to independently capture and predict motions from a video and properly execute the motions are difficult since real-world motions are complex and stochastic. In other words, there are two major problems for the motion-focused models that need to be solved. Firstly, it is vital to detach motions from contents for independent motion modeling. Secondly, the way to apply learned motions for generating future frames should meet the internal mechanisms of object motion.

In order to address the problems above, an adaptive hierarchical motion-focused model with adversarial training for video prediction is proposed in this paper. Our method first estimates high-level motion representation of observed frames and then predicts detailed future motions conditioned on the high-level motion information, and finally applies the predicted motions to previous frames to generate the future frames. To be more specific, the high-level motion representation called motion map is estimated by a fully convolutional network, which indicates dynamic of observed frames. It shows motion states of different objects, which provides explicit motion information for further motion understanding and predicting. Then, the motion predicting network receives input frames concatenated with the motion map and learns specific motion kernels for observed frames. Considering that different objects in videos may have different scale, appearance and deformation, an adaptive method using deformable convolution [14] is proposed to transform previous frames into future ones. We expand the receptive field of convolution process from traditional fixed-size regular regions to adaptive object-covering regions by learning extra offsets to regular grid sampling locations. To achieve object-centric transformation, we generate multiple predictions and merge them exploiting a mask scheme as [1]. The model can be trained end to end and the adversarial training strategy is used to improve quality of predicted frames.

There are three main contributions in this paper: (1) A hierarchical motion-focused model for video prediction is proposed to generate sharp and plausible future. (2) An adaptive transformation scheme is introduced to handle complex deformation of diverse objects. (3) The proposed model is evaluated on Penn Action, UCF101 datasets and achieves competitive results to the prior works.

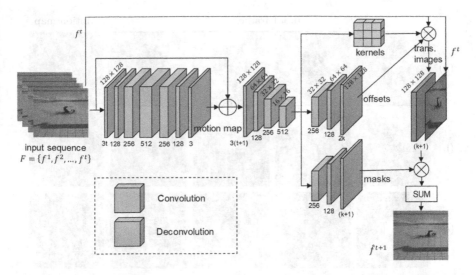

Fig. 1. Overall architecture of the adaptive hierarchical motion-focused model. Our model firstly predicts a high-level motion representation of input sequence called motion map via several convolution layers. Input sequence aided by motion map is then fed into a motion encoder to generate transformation kernels. Afterwards offsets and masks are produced separately. Offset kernels are used to transform the last input flexibly. Finally, multiple transformed images and last input are combined into the final prediction by masking.

The rest of the paper is organized as follows. Section 2 describes the proposed model in detail, including hierarchical motion modeling and adaptive future frames generation strategy. Section 3 presents the experiments on two datasets. Finally, main points of this paper are summarized in Sect. 4.

2 Our Approach

In this section, we introduce the proposed adaptive hierarchical motion-focused model for video prediction, consisting of two main compositions: hierarchical motion modeling of a video sequence and future frames generation via adaptive transformation.

2.1 Hierarchical Motion Modeling

Given a video sequence $F = \{f^1, f^2, ..., f^t\}$, most existing motion-focused methods for video prediction firstly predicts the dynamic filters in one step, which are used to transform the objects in previous frames. Unlike these methods, the proposed approach models the dynamic of observed scenes hierarchically to better understand the motions. We estimate a high-level representation of object motions called motion map by a convolutional structure. Such initial perception of motions in image segments roughly makes out objects with different

Input frames   Motion map

Fig. 2. Three examples of generated motion maps each given a four-frame video sequence. (Best viewed in color)

motion states. In motion map, regions of objects with consistent movement are represented by the same color and regions with large motions are highlighted. Examples of generated motion map can be viewed in Fig. 2.

Conditioned on the motion map and input frames, detailed motion states of objects are learned via a convolutional encoder structure and fully connected layers, and finally represented by some transformation kernels. Each kernel contains a set of parameters used for transforming previous frames to generate the future frames. Parameters in kernels are dynamically updated based on observed motions rather than remain constant. The shape of each kernel is mainly determined by the way of transformation. In this work, we adopt convolutional transformation to apply the learned motions to object appearance. Each transformation kernel is meant to deal with separate object motions while most videos show more than one motion. Thus the motion modeling network finally outputs convolutional kernels of shape $s \times s \times k$, where s is the kernel size and k is the number of kernels to handle k kinds of motions. After convolutions with predicted kernels, transformed images $P = \{p_1, p_2, ..., p_k\}$ are generated.

To combine these transformed images into the final prediction, we use a masking scheme as [5]. We learn masks $M = \{m_0, m_1, m_2, ..., m_k\}$ and their size is $w \times h$, where w and h are weight and height of each input frame. Actually, a mask is a set of weights and the combination is weighted sums of pixels defined as:

$$\hat{f}^{t+1} = \sum_{i=1}^{k} p_i \odot m_i + f^t m_0, \tag{1}$$

where $\odot$ is element-wise multiplication. In this way, objects with different motions will come from their corresponding transformed images and the static background will be the replica of the previous frame.

2.2   Adaptive Transformation Scheme

Convolution is one of the most popular methods for geometric transformation and is often employed to generate the future frames. Traditional convolution operation weights the samplings in a fixed regular grid, which is called receptive field, of the input to produce new pixel intensity. Because objects in different locations may have different shape and deformation, the clumsy way of sampling limits the capacity to perform a complex and large transformation. Therefore, we introduce an adaptive transformation scheme using deformable convolution [14] to tackle this problem. We modify the regular receptive field of traditional convolution to be amorphous by applying learnable offsets to original sampling locations. As a result, the extent and shape of receptive fields are all adaptive to various objects of input frames. Different from [14], we learn offsets conditioned on motion features and convolution kernels are adapted to observed motions instead of remaining constant.

Moreover, in order to boost the performance for multi-frames prediction, we predict frames iteratively instead of generating all of them simultaneously. That is to say, we only produce one frame at a time, and then concatenate current output and last $(t-1)$ input frames as a new t-frame input sequence to predict the next single frame.

2.3   Training

Our model is end-to-end trainable and improves the quality of generated frames with the help of adversarial training first proposed in generative adversarial networks (GAN) [15]. The model described in the preceding section is trained as the generative model G. And a discriminative model D of convolutional architecture is trained to distinguish between the generated image from G and real image. D finally estimates the probability that the input is generated by G. As a result, the loss function of discriminative model D is:

$$L_D(F,y) = L_{bce}(D(y),1) + L_{bce}(D(G(F)),0), \tag{2}$$

where y is the target frame for input F and L_{bce} is the binary cross-entropy loss. The loss function of generative model G is defined as:

$$L_G(F,y) = L_{bce}(D(G(F)),1). \tag{3}$$

In order to further constrain generations and improve the stability of adversarial training, we employed a combined loss function for G defined as:

$$L = \lambda_1 L_G + \lambda_2 L_2 + \lambda_3 L_{gdl}, \tag{4}$$

584 M. Tang et al.

Table 1. Performance (PSNR and SSIM) of single frame prediction on Penn Action dataset.

Model		PSNR	SSIM
High-level motion	Adaptive trans.		
×	√	32.5	0.97
√	×	29.7	0.96
√	√	**33.3**	**0.97**

where λs are weights for loss functions and L_{gdl} is the image gradient difference loss introduced in [1]. Generative model G and discriminative model D are alternately trained using SGD. When we train the generative model with combined loss, parameters of D should be fixed.

3 Experiments and Evaluations

In this section, two sets of experiments are designed to evaluate the proposed model for future frames prediction. The experiments are performed on two challenging datasets: Penn Action [16] and UCF101 [17]. We first separately demonstrate the effectiveness of two modules of our model: high-level motion modeling network and adaptive transformation method, on Penn Action dataset. Then the proposed model is compared with several recent representative works on UCF101 dataset.

3.1 Implementation Details

The architecture of our model is shown in Fig. 1. In our experiments, the kernel size s and the number of masks k is set to 13 and 3, respectively. Weights in the combined loss function λ_1, λ_2 and λ_3 are all 1. Size of mini-batch used to perform optimization is set to 4.

Images in Penn Action dataset and UCF101 dataset are all resized to 128 × 128 before input to the network. During training, we only input four consecutive frames and predict a single frame.

Quantitative evaluations of resultant performance are presented with Peak Signal to Noise Ratio (PSNR) and Structural Similarity Index Measure (SSIM) [18]. They are commonly used in recent works for video prediction to evaluate the quality of predictions. PSNR is defined as:

$$\text{PSNR} = 10 \log_{10}(\frac{max^2}{\text{MSE}}), \tag{5}$$

where max is the maximum generated image intensities and MSE is mean square error between the true frame and the prediction. Larger scores of PSNR and SSIM mean that the predicted image is more similar to the true image.

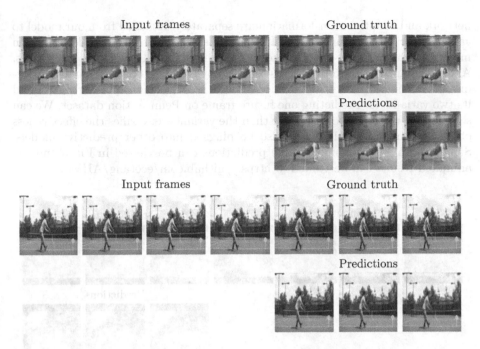

Fig. 3. Several examples of our predictions on Penn Action dataset. Input sequence contains 4 frames and 3 predicted frames are computed recursively.

3.2 Penn Action Dataset

Penn Action dataset contains 2326 video sequences of 15 different actions with a large amount of scale and appearance variations. The resolution of the frames is within the size of 640×480. We train and evaluate all the models in the raw videos from the training and test set respectively. To demonstrate the effectiveness of the two design described in the section above, the high-level motion modeling

Table 2. Performance (PSNR and SSIM) of single frame prediction on UCF101 dataset. PSNR and SSIM are measured both on moving areas and an entire image.

Model	PSNR(M)	SSIM(M)	PSNR	SSIM
CopyLastInput	28.6	0.89	30.0	0.90
Adv+GDL [1]	31.5	0.91	27.2	0.83
Adv+GDL fine-tuned [1]	32.0	0.92	29.6	0.90
DualMotionGAN [19]	–	–	30.5	0.94
DVF [6]	33.4	0.94	–	–
FullyContextAware [20]	**34.8**	0.92	–	–
Ours	32.2	**0.94**	**31.4**	**0.94**

network and adaptive transformation are separately detached from our model to generate variants for comparison. In the first setting, without high-level motion modeling, the network directly encodes the input into transformation kernels. And the adaptive transformation is replaced by standard convolutional transformation in the second setting. Table 1 shows the performance of our model and its two variants for predicting one future frame on Penn Action dataset. We can see that our model performs better than the variants. It verifies the effectiveness of our two modules, which can also be plugged into other predictive models. Some examples of our multi-frame predictions can be viewed in Fig. 3 and the animated results can be viewed in https://github.com/ecetang/AHVP.

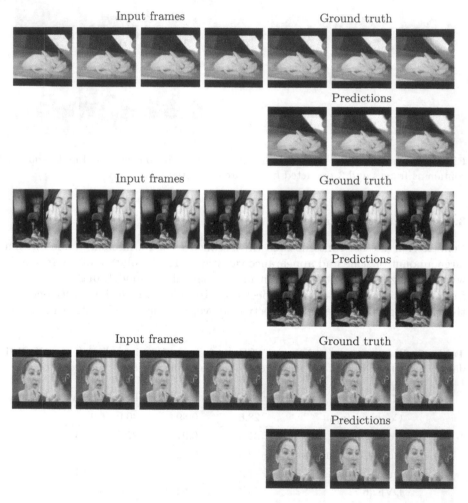

Fig. 4. Several examples of our predictions on UCF101 dataset. Input sequence contains 4 frames and 3 predicted frames are computed recursively. Please see https://github.com/ecetang/AHVP for the animated results.

3.3 UCF101 Dataset

UCF101 dataset contains over 130k videos of 101 human actions categories, of which frame rate is 25 fps and resolution is 320 × 240. In this experiment, we sample 110k clips from the training videos as the training set and evaluate the model on the test set of 379 clips provided by [1].

The proposed model is compared with several representative approaches on UCF101, including BeyondMSE [1], Dual Motion GAN [19], DVF [6] and Fully Context-Aware [20]. Quantitative measures are given in Table 2 for predicting one future frame. We evaluate the results both on moving areas and an entire image. Models of BeyondMSE (Adv+GDL and Adv+GDL fine-tuned) are trained on Sports1m and Adv+GDL fine-tuned is fined-tuned on UCF101 after pre-training. It's clear that our model achieves the best PSNR and SSIM on the entire images and the best SSIM on moving areas. For PSNR on moving areas, our model outperforms CopyLastInput and models of BeyondMSE without pre-training but is inferior to DVF [6] and Fully Context-Aware [20]. Nevertheless, our model is competitive with the state-of-the-art methods. Some examples of our multi-frame predictions on UCF101 datasets are presented in Fig. 4.

4 Conclusion and Future Work

In this paper, an adaptive hierarchical motion-focused model has been introduced to address the problem of video prediction. To better capture the motions from a video sequence, we predict the motions from coarse to fine. And finally, the objects of the previous frame are transformed by adaptive convolution with predicted motion kernels to generate the results. The experiment results demonstrate the effectiveness of both proposed modules, which may be used as components for other prediction models. Moreover, our entire model is competitive with the state-of-the-art video prediction methods in the presence of appearance variation and complex motions. In the future, the research will focus on applying the ideas to other challenging tasks such as video generation from captions.

Acknowledgement. This work is supported by Shenzhen Peacock Plan (20130408-183003656), Shenzhen Key Laboratory for Intelligent Multimedia and Virtual Reality (ZDSYS201703031405467), and National Natural Science Foundation of China (NSFC, No.U1613209).

References

1. Mathieu, M., Couprie, C., Lecun, Y.: Deep multi-scale video prediction beyond mean square error. arXiv preprint arXiv:1511.05440 (2015)
2. Vondrick, C., Pirsiavash, H., Torralba, A.: Generating videos with scene dynamics. In: Advances in Neural Information Processing Systems (NIPS), Barcelona, pp. 613–621 (2016)
3. Villegas, R., Yang, J., Hong, S., et al.: Decomposing motion and content for natural video sequence prediction. arXiv preprint arXiv:1706.08033 (2017)

4. Lu, C., Hirsch, M., Scholkopf, B.: Flexible spatio-temporal networks for video prediction. In: Conference on Vision and Pattern Recognition (CVPR) (2017)
5. Finn, C., Goodfellow, I., Levine, S.: Unsupervised learning for physical interaction through video prediction. In: Advances in Neural Information Processing Systems (NIPS), Barcelona (2016)
6. Liu, Z., Yeh, R.A., Tang, X., et al.: Video frame synthesis using deep voxel flow. In: International Conference on Computer Vision (ICCV) (2017)
7. Chen, X., Wang, W., Wang, J., et al.: Learning object-centric transformation for video prediction. In: Proceedings of the 2017 ACM on Multimedia Conference, pp. 1503–1512 (2017)
8. Villegas, R., Yang, J., Zou, Y., et al.: Learning to generate long-term future via hierarchical prediction. arXiv preprint arXiv:1704.05831 (2017)
9. Jia, X., De Brabandere, B., Tuytelaars, T., et al.: Dynamic filter networks. In: Advances in Neural Information Processing Systems (NIPS), Barcelon (2016)
10. Van Amersfoort, J., Kannan, A., Ranzato, M.A., et al.: Transformation-based models of video sequences. arXiv preprint arXiv:1701.08435 (2017)
11. Vondrick, C., Torralba, A.: Generating the future with adversarial transformers. In: Conference on Vision and Pattern Recognition (CVPR) (2017)
12. Xue, T., Wu, J., Bouman, K., et al.: Visual dynamics: probabilistic future frame synthesis via cross convolutional networks. In: Advances in Neural Information Processing Systems (NIPS), Barcelona, pp. 91–99 (2016)
13. Song, Y., Viventi, J., Wang, Y.: Multi resolution LSTM for long term prediction in neural activity video. arXiv preprint arXiv:1705.02893 (2017)
14. Dai, J., Qi, H., Xiong, Y., et al.: Deformable convolutional networks. In: Conference on Vision and Pattern Recognition (CVPR) (2017)
15. Goodfellow, I., Pouget-Abadie, J., Mirza, M., et al.: Generative adversarial nets. In: Advances in Neural Information Processing Systems (NIPS), pp. 2672–2680 (2014)
16. Zhang, W., Zhu, M., Derpanis, K.G.: From actemes to action: a strongly-supervised representation for detailed action understanding. In: International Conference on Computer Vision (ICCV) (2013)
17. Soomro, K., Zamir, A.R., Shah, M.: UCF101: a dataset of 101 human actions classes from videos in the wild. arXiv preprint arXiv:1212.0402 (2012)
18. Wang, Z., Bovik, A.C., Sheikh, H.R., et al.: Image quality assessment: from error visibility to structural similarity. IEEE Trans. Image Process. 13(4), 600–612 (2004)
19. Liang, X., Lee, L., Dai, W., et al.: Dual motion GAN for future-flow embedded video prediction. In: International Conference on Computer Vision (ICCV) (2017)
20. Byeon, W., Wang, Q., Srivastava, R.K., et al.: Fully context-aware video prediction. arXiv preprint arXiv:1710.08518 (2017)

Subjective Quality Assessment
of Stereoscopic Omnidirectional Image

Jiahua Xu, Chaoyi Lin, Wei Zhou, and Zhibo Chen[✉]

CAS Key Laboratory of Technology in Geo-spatial Information Processing
and Application System, University of Science and Technology of China,
Hefei 230027, China
chenzhibo@ustc.edu.cn

Abstract. Stereoscopic omnidirectional images are eye-catching because they can provide realistic and immersive experience. Due to the extra depth perception provided by stereoscopic omnidirectional images, it is desirable and urgent to evaluate the overall quality of experience (QoE) of these images, including image quality, depth perception, and so on. However, most existing studies are based on 2D omnidirectional images and only image quality is taken into account. In this paper, we establish the very first Stereoscopic OmnidirectionaL Image quality assessment Database (SOLID). Three subjective evaluating factors are considered in our database, namely image quality, depth perception, and overall QoE. Additionally, the relationship among these three factors is investigated. Finally, several well-known image quality assessment (IQA) metrics are tested on our SOLID database. Experimental results demonstrate that the objective overall QoE assessment is more challenging compared to IQA in terms of stereoscopic omnidirectional images. We believe that our database and findings will provide useful insights in the development of the QoE assessment for stereoscopic omnidirectional images.

Keywords: Stereoscopic omnidirectional images
Quality of experience · Subjective quality evaluation

1 Introduction

Virtual reality (VR) has become an eye-catching topic in recent years due to the rapid development of industry ecosystem and technology. The omnidirectional images and videos in VR can provide viewers with immersive experience where viewers can explore every viewing direction freely by wearing the head mounted display (HMD). When Facebook Surround 360, a high-quality 3D-360 video capture system, becomes available in 2016, it is possible to capture and render stereoscopic omnidirectional images. However, owing to the limitation of photographic apparatus, transmission bandwidth, display devices, etc., the overall

---

The first two authors made equal contributions to this work.

© Springer Nature Switzerland AG 2018
R. Hong et al. (Eds.): PCM 2018, LNCS 11164, pp. 589–599, 2018.
https://doi.org/10.1007/978-3-030-00776-8_54

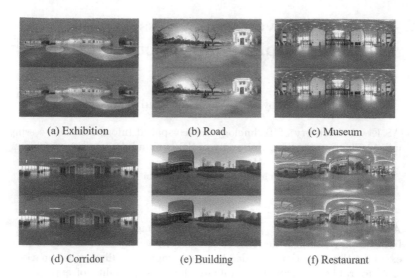

(a) Exhibition   (b) Road   (c) Museum

(d) Corridor   (e) Building   (f) Restaurant

Fig. 1. Test image contents used in our database.

quality of experience (QoE) of stereoscopic omnidirectional contents is far from satisfactory. It's worth noting that the overall QoE considers both the factors of image quality and depth perception in this paper. Thus, in order to generate high-quality and realistic stereoscopic omnidirectional images, it is desirable and urgent to evaluate the image quality, depth perception, and overall QoE of these images.

Most existing subjective quality assessment databases are based on 2D omnidirectional contents and only image quality is taken into account. An omnidirectional image quality assessment (OIQA) database [3] has been built for subjective quality evaluation study. Xu et al. [11] propose a subjective visual quality assessment method for omnidirectional videos. Zhou et al. [15] explore the impact of spatial resolution on the perceptual quality of immersive 360-degree images. IVQAD 2017 [4] is a database built for immersive video quality assessment.

Different from 2D omnidirectional images, the overall QoE of stereoscopic omnidirectional images concerns multiple aspects due to the extra dimensionality of immersive contents. Specially, the additional dimension of depth may implicitly affect the experience of viewing. Thus the image quality and depth perception are two important factors which should be taken into consideration when evaluating the subjective perception of QoE. However, to the best of our knowledge, there is no available stereoscopic omnidirectional image quality assessment (IQA) database for investigating the property of these QoE factors.

In this paper, we establish the very first **S**tereoscopic **O**mnidirectiona**L** **I**mage quality assessment **D**atabase (SOLID) by evaluating the image quality, depth perception, and overall QoE of stereoscopic omnidirectional images. The relationships among these three factors are analyzed in our database. We find that the subjective rating scores of overall QoE are highly correlated with image

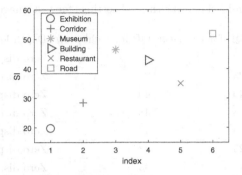

Fig. 2. Spatial information (SI) distribution for the test images used in our database.

quality and it is moderately correlated with depth perception. Besides, the subjective ratings of image quality are also correlated with depth perception scores. Finally, several well-known objective IQA metrics are tested on our database. Although these classic metrics can achieve a promising performance on predicting image quality, it is still a challenge to predict the overall QoE for stereoscopic omnidirectional images.

The rest of the paper is organized as follows. Section 2 describes the details of our SOLID database. Section 3 analyzes the subjective rating scores to investigate the relationship of multidimensional rating scores and finally several objective IQA metrics are evaluated. Section 4 concludes the paper.

2 Stereoscopic Omnidirectional Image Quality Assessment Database

This section introduces the experiment of subjective quality evaluation for stereoscopic omnidirectional images, including the dataset and subjective test methods.

2.1 Image Database

The test images used in our experiment are captured by Facebook Surround 360 which is an open source hardware and software for generating stereoscopic omnidirectional images and videos. It can render the stereoscopic omnidirectional images (equirectangular projection) with the resolution of 8192 × 8192 and the file format of PNG. Besides, the disparity can be adjusted in the software of Surround 360 before rendering stereoscopic omnidirectional images, which is used to render images with different depth perception levels in our database.

There are 6 high-quality reference stereoscopic omnidirectional images with the resolution of 8192 × 8192 in our database, which are shown in Fig. 1. The stereoscopic omnidirectional images are in the format of top-and-bottom that left and right view images are packed vertically and there exists disparity between left

Table 1. Distortion level and depth level settings for test images in our database.

HRCID	QP for left view	QP for right view	Depth level
HRC01	-	-	Zero disparity
HRC02	25	25	Zero disparity
HRC03	40	40	Zero disparity
HRC04	48	48	Zero disparity
HRC05	25	45	Zero disparity
HRC06	25	48	Zero disparity
HRC07	45	35	Zero disparity
HRC08	51	40	Zero disparity
HRC09	-	-	Medium disparity
HRC10	25	25	Medium disparity
HRC11	40	40	Medium disparity
HRC12	48	48	Medium disparity
HRC13	25	45	Medium disparity
HRC14	51	25	Medium disparity
HRC15	45	35	Medium disparity
HRC16	35	51	Medium disparity
HRC17	45	51	Medium disparity
HRC18	-	-	Large disparity
HRC19	25	25	Large disparity
HRC20	40	40	Large disparity
HRC21	48	48	Large disparity
HRC22	25	45	Large disparity
HRC23	51	25	Large disparity
HRC24	45	35	Large disparity
HRC25	35	51	Large disparity
HRC26	45	51	Large disparity

view and right view. It can be seen that there are both indoor scenes and outdoor scenes in our database. Besides, the spatial information (SI) which represents the spatial complexity of an image is taken into account when selecting the reference images. Figure 2 shows SI of the six reference images used in our experiment.

To investigate the relationship between image quality, depth perception, and overall QoE, the stimuli are generated to cover a wide range of image quality and depth perception. The distortion levels for each reference stereoscopic omnidirectional image are presented in Table 1. The hypothetic reference circuits (HRCs) are used to represent the test stereoscopic omnidirectional image with certain distortion and disparity. As shown in Table 1, there are 26 test images

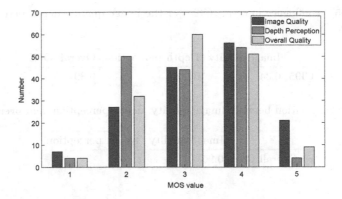

Fig. 3. MOS histogram for image quality, depth perception, and overall QoE in our database.

for each reference stereoscopic omnidirectional image. Thus, we have 156 test stereoscopic omnidirectional images in our database generated from 6 reference stereoscopic omnidirectional images.

As inspired by [16], there are 3 depth levels for each reference stereoscopic omnidirectional image in our experiment: (1) zero disparity images where there is no disparity between left and right view images, (2) medium disparity images, and (3) large disparity images.

To simulate the quality degradation, each reference image is compressed into Better Portable Graphics (BPG) format [1] with different quantization parameters (QPs). BPG is a new image format which aims to replace the JPEG image format when quality or file size is an issue. Considering the stereoscopic omnidirectional images are usually large in file size, we believe BPG format may be popular in this kind of contents and thus we choose BPG compression distortion in our experiment. The reference stereoscopic omnidirectional images are distorted either symmetrically or asymmetrically in our database. The symmetrical and asymmetrical distortion is determined according to whether the left and right view images are distorted with the same distortion level.

2.2 Subjective Test Methods

In our experiment, image quality, depth perception, and overall QoE are evaluated by the subjects. The experiment is performed according to Absolute Category Rating with Hidden Reference (ACR-HR) which is described in [5]. ACH-HR is a single stimulus evaluation and voting is performed after each viewing. Images are assessed using the five-grade scale with following levels: "5 - Excellent", "4 - Good", "3 - Fair", "2 - Poor", and "1 - Bad".

The following equipment is used in our experiment. The Samsung Gear VR, which is a kind of head-mounted display, is used as the virtual reality display system. The field of view provided by Gear VR is 101 degrees. Samsung Galaxy $S9+$ is used to display the images with a resolution of 2560×1440.

Table 2. 95% confidence interval (CI95) of the image quality, depth perception, and overall QoE.

	Image quality	Depth perception	Overall QoE
CI95	0.34	0.45	0.39

Table 3. Correlation between image quality, depth perception, and overall QoE.

	Image quality	Depth perception
Overall QoE	0.90	0.67

18 non-expert student subjects aged from 21 to 28 years old take part in our subjective test. All participants pass the visual acuity, color vision (Ishihara charts) and stereo acuity tests (RANDOT test). Before the formal test, there is a training session that each subject is explained the purpose of the evaluation. Also, they are shown the examples of different levels of compression artifacts and depth perception. In the formal test, there are 156 images including reference and distorted images to be rated randomly. There is no time limitation for subjects when they are watching stereoscopic omnidirectional images. During the experiment, subjects can take a break as long as they feel tired or uncomfortable to avoid eye fatigue.

3   Analysis of Subjective Database

This section provides a detailed analysis of the subjective evaluating scores. First, the suitability of evaluation methods is analyzed to ensure that data collected from subjects is effective. Second, we explore the relationship between image quality, depth perception, and overall QoE. Finally, performance evaluation of some objective metrics is performed on our database.

3.1   Data Analysis

The subjects whose correlation coefficient with average image quality is lower than 0.8 or with average depth perception is lower than 0.65 are considered as outliers and their subjective evaluating scores are removed from our database. There remain 16 valid subjects (7 males and 9 females) after outlier removal.

Mean Opinion Score (MOS) values are computed for each test image in the database by averaging scores of valid subjects. According to ITU-T P.910 [5], the 95% confidence interval (CI) of the subjective rating scores for image quality, depth perception, and overall QoE are given in Table 2. Results show that all the subjects reach a reasonable agreement on the perceived image quality, depth perception, and overall QoE. Figure 3 shows the MOS histogram for image quality, depth perception and overall QoE in our database, respectively. In Fig. 3, the MOS values mainly centralize among score 2, 3, and 4.

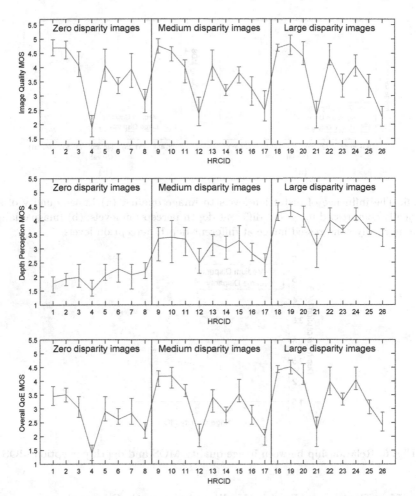

Fig. 4. MOS of image quality, depth perception, and overall QoE for each HRC.

The MOS values of 6 scenes for 26 HRCs are averaged and shown in Fig. 4. We can draw some conclusions from Fig. 4. First, the BPG compression will affect image quality and overall QoE greatly. Second, depth perception is dominated by the disparity and the BPG compression has a moderate influence on depth perception. Third, overall QoE is affected by image quality and depth perception jointly. The overall QoE scores are consistent with the image quality scores and the strong depth perception tends to enhance the overall QoE while the poor depth perception will reduce the overall QoE. This can be observed from the MOS values of HRCs in Fig. 4, such as HRC 1 and 26. The linear correlation coefficients between image quality, depth perception, and overall QoE are presented in Table 3, which demonstrates that image quality is the dominant factor for overall QoE compared with depth perception.

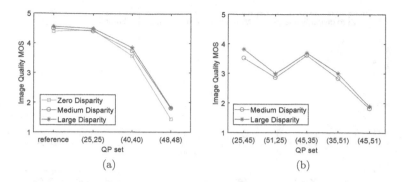

Fig. 5. The influence of distortion levels to image quality. (a) Image quality of symmetrically compressed images at different depth perception levels; (b) Image quality of asymmetrically compressed image at different depth perception levels.

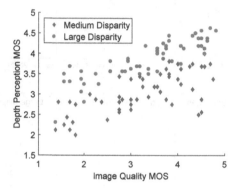

Fig. 6. Relationship between image quality MOS and depth perception MOS.

3.2 Key Factors to Image Quality and Depth Perception

The relationship between image quality and distortion levels is shown in Fig. 5. We find that under the same distortion level, images with large disparity tend to have higher image quality scores and this is especially obvious in asymmetrical distortion QP set (Fig. 5 (b)). According to the questionnaire from the subjects, stereoscopic omnidirectional images with strong depth perception can provide a more realistic environment compared to those with weak depth perception. As a result, subjects tend to rate a high score for image quality of these images.

Figure 6 is the scatter plot of subjective rating scores of image quality vs. that of depth perception. It can be observed that subjective image quality scores are highly correlated with depth perception scores. This is an interesting finding because image quality and depth perception are usually considered as two different perceptual dimensions. Two possible explanations are found from the data analysis and questionnaire. First, image quality degradation will result in blurry objects, which will weaken the depth perception. Second, subjects tend to give

Table 4. LCC and SROCC performance of objective image quality metrics on our database.

Type	Metrics	Image Quality		Overall QoE	
		LCC	SROCC	LCC	SROCC
2D IQA	PSNR	0.7397	0.6529	0.6295	0.5762
	SSIM [10]	0.8621	0.8742	0.7388	0.7508
	MS-SSIM [9]	0.7839	0.7165	0.6565	0.6147
	FSIM [14]	**0.8922**	**0.8959**	**0.7750**	**0.7742**
	VSI [13]	0.8852	0.8757	0.7623	0.7664
	BRISQUE [6]	0.7496	0.7288	0.6581	0.6550
2D OIQA	S-PSNR [12]	0.7366	0.6423	0.6196	0.5656
	WS-PSNR [7]	0.7238	0.6388	0.6160	0.5639
3D IQA	CHEN [2]	0.7430	0.7037	0.5527	0.5342

high images quality scores when depth perception of that image is strong, and vice versa [8]. Thus image quality is highly correlated with depth perception.

3.3 Performance Evaluation of Objective Metrics

We tested nine well-known objective image quality assessment metrics on our database including six 2D IQA metrics PSNR, SSIM [10], MSSSIM [9], FSIM [14], VSI [13] and BRISQUE [6], two 2D omnidirectional IQA metrics S-PSNR [12] and WS-PSNR [7], and a 3D IQA metric [2]. For these 2D metrics, the predicted image quality of left and right view images are averaged as the final quality of stereoscopic omnidirectional images. Linear correlation coefficient (LCC) and Spearman's rank correlation coefficient (SROCC) are used to measure the performance of objective metrics. The higher correlation coefficient means better correlation with human subjective quality judgement.

The performance of the above metrics is shown in Table 4 and we can find that FSIM has the best performance while 2D OIQA metrics do not show thier advantages on omnidirectional images compared with traditional 2D IQA metrics. Existing metrics can achieve promising performance for image quality, however the performance drops significantly when it refers to overall QoE. Results in Table 4 demonstrate that predicting overall QoE is still a challenge since it considers not only the image quality but also other factors, such as depth perception, visual comfort, etc.

4 Conclusion

In this paper, we build a **S**tereoscopic **O**mnidirectiona**L** **I**mage quality assessment **D**atabase named as SOLID. There are 156 test images with different levels of depth and BPG compression artifacts in SOLID. To the best of our knowledge, it is the very first work on stereoscopic omnidirectional image quality assessment database, which considers the factors of image quality, depth perception, and overall QoE. By analyzing the subjective rating scores, we find that image quality is the dominant factor for overall QoE compared to depth perception. Besides, image quality is also highly correlated with depth perception that images with large disparity tend to achieve higher image quality scores. Finally, some existing well-known objective IQA metrics are tested on our SOLID database. Experimental results show that overall QoE assessment is more challenging compared to IQA in terms of stereoscopic omnidirectional images.

Acknowledgements. This work was supported in part by the National Key Research and Development Program of China under Grant No. 2016YFC0801001, the National Program on Key Basic Research Projects (973 Program) under Grant 2015CB351803, NSFC under Grant 61571413, 61632001, 61390514, and Intel ICRI MNC.

References

1. Bellard, F.: BPG image format (2017). http://bellard.org/bpg/
2. Chen, M.J., Su, C.C., Kwon, D.K., Cormack, L.K., Bovik, A.C.: Full-reference quality assessment of stereopairs accounting for rivalry. Signal Process. Image Commun. **28**(9), 1143–1155 (2013)
3. Duan, H., Zhai, G., Min, X., Zhu, Y., Fang, Y., Yang, X.: Perceptual quality assessment of omnidirectional images. In: 2018 IEEE International Symposium on Circuits and Systems (ISCAS), pp. 1–5, May 2018
4. Duan, H., Zhai, G., Yang, X., Li, D., Zhu, W.: IVQAD 2017: An immersive video quality assessment database. In: 2017 International Conference on Systems, Signals and Image Processing (IWSSIP), pp. 1–5, May 2017
5. ITU-T Recommendation P.910: Subjective video quality assessment methods for multimedia applications (1999)
6. Mittal, A., Moorthy, A.K., Bovik, A.C.: No-reference image quality assessment in the spatial domain. IEEE Trans. Image Process. **21**(12), 4695–4708 (2012)
7. Sun, Y., Lu, A., Yu, L.: Weighted-to-spherically-uniform quality evaluation for omnidirectional video. IEEE Signal Process. Lett. **24**(9), 1408–1412 (2017)
8. Wang, J., Wang, S., Ma, K., Wang, Z.: Perceptual depth quality in distorted stereoscopic images. IEEE Trans. Image Process. **26**(3), 1202–1215 (2017)
9. Wang, Z., Simoncelli, E.P., Bovik, A.C.: Multiscale structural similarity for image quality assessment. In: The Thrity-Seventh Asilomar Conference on Signals, Systems Computers, 2003, vol. 2, pp. 1398–1402, November 2003
10. Wang, Z., Bovik, A.C., Sheikh, H.R., Simoncelli, E.P.: Image quality assessment: from error visibility to structural similarity. IEEE Trans. Image Process. **13**(4), 600–612 (2004)
11. Xu, M., Li, C., Liu, Y., Deng, X., Lu, J.: A subjective visual quality assessment method of panoramic videos. In: 2017 IEEE International Conference on Multimedia and Expo (ICME), pp. 517–522, July 2017
12. Yu, M., Lakshman, H., Girod, B.: A framework to evaluate omnidirectional video coding schemes. In: 2015 IEEE International Symposium on Mixed and Augmented Reality, pp. 31–36, September 2015
13. Zhang, L., Shen, Y., Li, H.: VSI: a visual saliency-induced index for perceptual image quality assessment. IEEE Trans. Image Process. **23**(10), 4270–4281 (2014)

14. Zhang, L., Zhang, L., Mou, X., Zhang, D.: FSIM: a feature similarity index for image quality assessment. IEEE Trans. Image Process. **20**(8), 2378–2386 (2011)
15. Zhou, R., et al.: Modeling the impact of spatial resolutions on perceptual quality of immersive image/video. In: 2016 International Conference on 3D Imaging (IC3D), pp. 1–6. IEEE (2016)
16. Zhou, W., Liao, N., Chen, Z., Li, W.: 3D-HEVC visual quality assessment: database and bitstream model. In: 2016 Eighth International Conference on Quality of Multimedia Experience (QoMEX), pp. 1–6. IEEE (2016)

A Dual-Network Based Super-Resolution for Compressed High Definition Video

Longtao Feng[1,3(✉)], Xinfeng Zhang[2], Xiang Zhang[3], Shanshe Wang[3], Ronggang Wang[1], and Siwei Ma[3]

[1] Peking University Shenzhen Graduate School, Shenzhen, China
ltfeng@pku.edu.cn, rgwang@pkusz.edu.cn
[2] University of Southern California, Los Angeles, CA, USA
xinfengz@usc.edu
[3] Institute of Digital Media, Peking University, Beijing, China
{x_zhang,sswang,swma}@pku.edu.cn

Abstract. Convolutional neural network (CNN) based super-resolution (SR) has achieved superior performance compared with traditional methods for uncompressed images/videos, but its performance degenerates dramatically for compressed content especially at low bit-rate scenario due to the mixture distortions during sampling and compressing. This is critical because images/videos are always compressed with degraded quality in practical scenarios. In this paper, we propose a novel dual-network structure to improve the CNN-based SR performance for compressed high definition video especially at low bit-rate. To alleviate the impact of compression, an enhancement network is proposed to remove the compression artifacts which is located ahead of the SR network. The two networks, enhancement network and SR network, are optimized stepwise for different tasks of compression artifact reduction and SR respectively. Moreover, an improved geometric self-ensemble strategy is proposed to further improve the SR performance. Extensive experimental results demonstrate that the dual-network scheme can significantly improve the quality of super-resolved images/videos compared with those reconstructed from single SR network for compressed content. It achieves around 31.5% bit-rate saving for 4 K video compression compared with HEVC when applying the proposed method in a SR-based video coding framework, which proves the potential of our method in practical scenarios, e.g., video coding and SR.

Keywords: Super-resolution · Enhancement network
Compression artifact reduction · Video coding · HEVC
Convolutional neural network

1 Introduction

Due to fast development of the image/video capture and display technologies, the ultra high-definition (e.g., 4 K) content has becoming more and more popular. Increasing the image resolution to 4 K or higher will dramatically improve

© Springer Nature Switzerland AG 2018
R. Hong et al. (Eds.): PCM 2018, LNCS 11164, pp. 600–610, 2018.
https://doi.org/10.1007/978-3-030-00776-8_55

the user experience by leading to a more immersive view environment. However, the data size increases significantly at the same time, which makes new compression strategy for 4 K content important and indispensable. An efficient way for 4 K content compression is based super-resolution (SR), where the original video is downsampled before compression and the decoded video is upsampled to the original resolution using SR technologies. Besides compression, SR technologies are also demanded to display low resolution video onto high definition devices. However, traditional image interpolation methods cannot get visually satisfied results especially for compressed low resolution video and may incur blurring artifacts. Therefore, the learning based SR approaches have been widely investigated recently.

A+ [16] is one representative SR method in recent years using regression to learn the correlation between low-resolution (LR) and high-resolution (HR) patches, which combines the best qualities of Anchored Neighborhood Regression (ANR) [15] and Simple Functions (SF) [18] adaptively. Recently, the convolution neural network (CNN) based SR methods [1,2,4,6–8,11,17] have achieved significant improvement compared with the traditional methods. In [1], Dong et al. proposed the shallow convolution network, SRCNN, which achieves significant quality improvement against its previous methods. To optimize the SRCNN, Dong et al. further proposed a compact hourglass-shape CNN structure, FSR-CNN [2], which is faster than SRCNN and achieves better performance. To further improve the SR performance, Kim et al. proposed a very deep convolution network by cascading many small filters, named VDSR [4]. In [8], Lim et al. developed an enhanced deep super-resolution network (EDSR) by removing unnecessary modules of Ledig et al.'s conventional residual networks [6] and won the NTIRE2017 Super-Resolution Challenge [14].

However, the above SR methods are based on uncompressed images/videos without considering compression influence. In practice, the available images and videos are all compressed versions, and the compression artifacts, e.g., blocking and ringing artifacts which have been studied for reduction in in-loop filters [3,20], can dramatically degenerate the performance of SR methods especially at low bit-rate scenario. This is essentially because of the mixture of two different degenerations, i.e., sampling and compressing. In addition, most of the SR approaches are investigated and verified on LR images and videos, e.g., 256×256, and there is little work for high-definition (HD) videos e.g., 1080P, which are assumed enough for display in the past years. Along with the wide deployment of ultra high-definition display devices, high efficiency SR algorithms for HD video is also urgently demanded for both display and compression applications.

In this paper, we focus on the compressed video SR from HD to 4 K resolution, and propose an end-to-end CNN method to optimize the quality of the super-resolved video by removing the compression artifacts, enhancing video resolution respectively and utilizing the improved geometric self-ensemble. Specifically, We divide the compressed video SR problem into two subtasks, i.e., video enhancement and video SR, which are solved by neural network methods. An enhancement network without pooling layers are proposed and located ahead

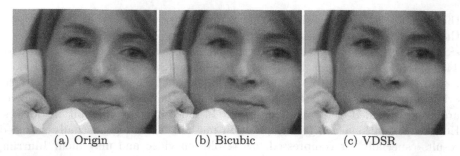

(a) Origin   (b) Bicubic   (c) VDSR

Fig. 1. Subjective results when applying SR methods to a compressed video frame.

of the SR network to reduce the compression artifacts firstly, and the SR network is then applied to obtain reconstructed 4 K video. Moreover, we also apply the proposed method to a SR based video compression framework, where the 4 K video sequences are first downsampled into 1080P and compressed at the encoder side, then the proposed SR method is applied to the decoded 1080P videos. Extensive experimental results show that the proposed SR method in SR based compression can achieve about 31.5% bit-rate saving compared with the latest video coding standard, High Efficiency Video Coding (HEVC) [13].

The rest of the paper is organized as follows: in Sect. 2, we introduce the proposed video SR method and the SR based video coding framework using the proposed method. The experimental results are shown in Sect. 3, and Sect. 4 concludes the paper.

2 Proposed Method

2.1 Motivation

At low bit-rate scenario, video compression introduces obvious compression artifacts, e.g., twisted lines, blurred edges and fuzzy textures, which are mainly caused by coarse quantization. In video coding, as the quantization parameter (QP) increases, more quantization noise will be introduced reducing the quality of the reconstructed video. Traditional SR methods cannot handle these compression distortions well and their performance degenerates seriously due to the severe compression distortions. Considering the compression process, the degeneration process of a compressed low resolution image y_d can be modeled as follows,

$$y_d = y \otimes s \otimes c, \tag{1}$$

where y represents the origin image, s and c denote sampling and compressing degeneration, respectively. The mixture of distortions cannot be resolved easily by single neural network due to the essentially different degeneration kernels. To show our motivation, we apply SR methods to compressed images directly. From Table 1, we can see that the performance of the bicubic interpolation and VDSR is poor on the HEVC compressed images, especially at low bit-rate scenario

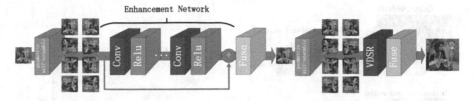

Fig. 2. The pipeline of the proposed SR method, where a dual-network structure is introduced. An enhancement network is applied before VDSR for compensating the compression distortions.

corresponding to high QPs. In Fig. 1, we further show a subjective result for a video frame compressed by HEVC at QP $= 44$. The corresponding network is trained based on the compressed images/videos. However, we can find that the compression distortions cannot be eliminated and are even enlarged after SR, e.g. the blocking artifacts around the face.

This motivates us to improve the SR performance by introducing an enhancement network before SR to reduce the compression artifacts and to solve them separately. Such dual-network structure is superior to cascading into one single network. Compared with tuning a highly deep network, a stepwise training is more feasible and efficient especially when the training set is limited.

2.2 The Proposed Video SR Method

According to the above analysis, we proposed a novel video SR method by adding an enhancement network before SR to reduce compression artifacts as shown in Fig. 2. The method can be modeled as follows,

$$\hat{y}_r = f_s(f_e(y_d)), \tag{2}$$

where $\hat{y}_r$ represents reconstructed high resolution image, f_e denotes enhancement operation which resolves compression degeneration and f_s denotes SR operation. Both enhancement and SR can be regarded as regression problems which aim to restore the high quality video from its distorted or LR version, but their degradation models are completely different. To solve the compression degeneration problem, we design a neural network with 20 convolutional layers taking the rectified linear unit (RELU) [9] as the activation function. It is inspired by the work of Kim et al. [4] which shows that the CNNs are efficient in dealing with regression problems. For SR problem, we adopt the network structure of VDSR [4] due to its high efficiency and good performance.

Since the two networks aim for different degradation models, we train them separately. The enhancement network is first trained. Given a training dataset $\left\{x_c^i, y_{uc}^i\right\}_{i=1}^N$, x_c^i represents a compressed image/video and y_{uc}^i is the corresponding uncompressed version. Our goal is to learn a model f_e which restores a image/video from its compressed version: $\hat{y}_e = f_e(x_c)$, where $\hat{y}_e$ is the restored

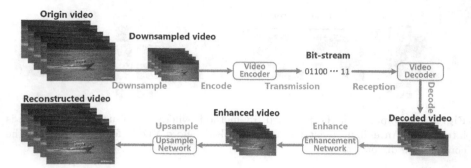

Fig. 3. The SR based video compression framework.

version of a compressed image/video. The mean squared error (MSE) is utilized as the loss function of the enhancement network:

$$L_e = \frac{1}{2}||y_{uc} - \hat{y}_e||^2. \tag{3}$$

Then, we use the existing VDSR method [4] as the SR network, of which the input is the output of the enhancement network. The data for training can be denoted as $\{x_e^i, y_h^i\}_{i=1}^N$, where x_e^i represents a LR compressed image/video which is enhanced by the enhancement network and y_h^i represents the corresponding uncompressed HR image/video. The proposed method in Fig. 2 is a flexible framework that leaves the choice for a specific network architecture open. Our choice of the network architectures provides a solution for the tradeoff between performance and complexity. More recent methods for each subtask, especially more complex SR methods [6,8] can be easily incorporated and will lead to even better results.

Moreover, to further improve the performance, the geometric self-ensemble strategy [8] is adopted and modified for our problem. Specifically, each input video I^{input} is flipped vertically and rotated to generate seven augmented inputs $I_{n,i}^{input} = T_i(I_{n,1}^{input})$, where T_i represents the ith geometric transformations including identity, i.e., $i = 1, \cdots, 8$. With those augmented input videos, we can generate the corresponding processed videos $\{I_{n,1}^{output}, ..., I_{n,8}^{output}\}$. Finally, we can generate the output video by inversely transforming the 8 processed video frames to their original structures and fusing them by an introduced weighting factor as follows,

$$I_{n,i}^{output} = \alpha \tilde{I}_{n,1}^{output} + \frac{(1-\alpha)}{7}\sum_{i=2}^{8}\tilde{I}_{n,i}^{output} \tag{4}$$

where α is set to 0.3, which is based on the assumption that the output from identity video is closer to the original geometric structure than those from transformed inputs.

Table 1. PSNR of different SRs as a function of QP on a image testset.

Set14	Bicubic	VDSR	Proposed
Uncompressed	29.63	32.45	–
QP34	27.68	28.60	28.79
QP38	26.56	27.19	27.37
QP44	24.48	24.90	25.02

2.3   The Application in SR Based Video Compression

Limited by the bandwidth, the 4 K video is usually compressed at low bit-rate, which has obvious compression artifacts. One efficient solution is the SR based video compression strategy, where the videos are first downsampled before compression and get upsampled at the decoder. Therefore, the proposed method can be naturally applied to this application because the dual-network can help in reducing the compression artifacts. The introduced compression framework is shown in Fig. 3.

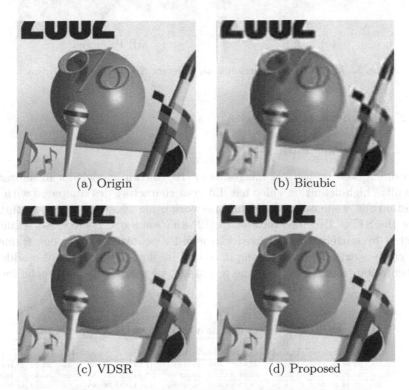

(a) Origin   (b) Bicubic

(c) VDSR   (d) Proposed

Fig. 4. Subjective comparison with different SR methods for images.

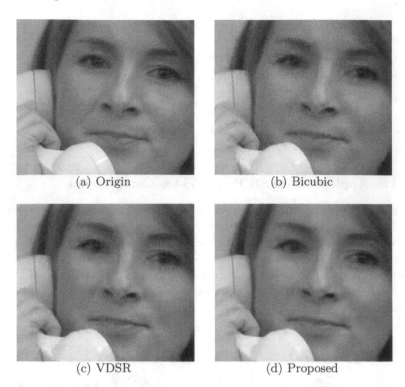

(a) Origin

(b) Bicubic

(c) VDSR

(d) Proposed

Fig. 5. Subjective comparison with different SR methods for videos.

3   Experimental Results

3.1   Datasets and Parameters

For training, we first use 291 images as in [10] to train models for initialization. Since ultra high-definition video has different characteristics compared with low-resolution one, we fine-tuned the dual-network using 4 K video sequences further, where the SJTU 4K video dataset in [12] and some other 4 K video sequences collected by ourselves are utilized. To avoid repeatability, only one frame for each video is extracted as training data and its flipped version is also added as augmentation training data. At the test stage, we use two dataset, including the

Table 2. PSNR of different SRs as a function of QP on video test set.

AWS	QP44	QP38				QP34				QP30			
	Bicu-bic	VDSR	Pro-I	Pro-II	Bicu-bic	VDSR	Pro-I	Pro-II	Bicu-bic	VDSR	Pro-I	Pro-II	Bicu-bic	VDSR	Pro-I	Pro-II
Cactus	35.33	35.84	36.11	36.37	39.13	39.85	39.94	40.10	41.07	42.20	41.65	42.53	42.46	43.92	43.10	44.13
Coastguard	30.29	30.45	30.62	30.63	33.02	33.27	33.39	33.43	35.02	35.34	35.41	35.54	37.08	37.49	37.48	37.65
Foreman	35.12	35.48	35.74	35.80	38.24	38.75	38.83	38.94	40.04	40.64	40.57	40.89	41.59	42.30	42.08	42.46
News	35.24	35.87	35.88	36.28	38.52	39.68	39.20	39.88	40.46	42.09	41.06	42.49	42.04	44.15	42.57	44.53
Suzie	35.20	35.62	35.98	36.11	38.02	38.59	38.75	38.72	39.73	40.37	40.32	40.55	41.30	41.90	41.76	42.04
Average	34.23	34.65	34.86	35.04	37.39	38.03	38.02	38.21	39.26	40.13	39.80	40.40	40.89	41.95	41.40	42.16

image test set "set14" in [19] which are widely utilized as benchmark for SR [1,15,16] and the video test set with five 4 K video sequences released by AWS elemental[1]. It is worth noting that for videos the SR methods are applied to each video frame. These images and videos are downscaled by bicubic interpolation and compressed by HEVC reference software, HM 16.17, under low-delay-P configuration in both training and testing stages.

We train our model using ADAM optimizer [5]. The training is regularized by weight decay (ℓ_2 norm penalty multiplied by 0.0001). The minibatch size is set as 64. The learning rate and training epoch vary with QPs and different types of network. More details can be found in our Github repository[2].

3.2   Experimental Results

For evaluation, we compare our model with the baseline method: the VDSR SR method trained by compressed images, the equivalent of our propose method without enhancing. In this work, we only process the luminance component for all methods, because human vision is more sensitive to details in intensity than that in chroma components.

For image results, Table 1 denotes the PSNR of the upsampled HR images from compressed images using different methods. It shows that performance degenerates with quantization when using baseline due to the mixture of two degeneration and our proposed method can compensate the losses caused by compression. Figure 4 shows one example of subjective results for different methods. We can find that the image quality is improved and some compression distortions, e.g., the ringing and blocking artifacts, are suppressed by our method.

For video results, Table 2 shows the PSNR results of upsampled HR video sequences from compressed videos using different methods. To further evaluate how the proposed enhancement network perform on different SR schemes, we compare with two methods, the Pro-I and Pro-II, respectively. The Pro-I is essentially the enhancement plus bicubic interpolation, and Pro-II is essentially the enhancement plus VDSR. From the results, we can achieve several conclusions. First one can be observed that both Pro-I and Pro-II can increase the PSNR indicating the effectiveness of the enhancement network. Next, it is clear that the SR method itself will also impact on the ultimate results, where an advanced SR will bring more gains. Third, We can see that the proposed method achieves superior results compared with either Bicubic or VDSR only. More importantly, the proposed method can obtain more PSNR gains for lower bit-rate, where about 0.4 dB gain can be achieved when QP = 44. This further implies that the proposed enhancement network is indispensable in improving the SR performance for compressed videos. Figure 5 compares the subjective results of different methods. Similar observations can be achieved that the video quality is improved and the compression distortions can be reduced by our method, e.g. the blocking artifacts around the face.

---

[1] http://www.elementaltechnologies.com/resources/4K-testsequences.

[2] https://github.com/FLT19940317/supplementary-material-of-PCM2018-Paper.

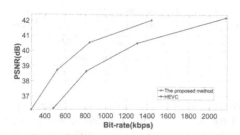

Fig. 6. RD curve comparison between the proposed and HEVC for sequence *Suzie*.

Table 3. The BD-rate of SR based compression using the proposed method compared with HEVC at low bit-rate.

AWS Sequence	BD-rate Y(%)
Cactus	−33.91
Coastguard	−24.31
Foreman	−33.86
News	−28.27
Suzie	−37.07
Average	−31.48

For the video compression application, Table 3 shows bit-rate saving compared with HEVC according to the QPs in Table 2 when applying the proposed SR method to the SR based video coding framework as shown in Fig. 3. The test sequences in Table 3 are all 4 K video sequences. Obviously the proposed SR method benefits the SR based video compression, which achieves over 30% bit-rate saving on average compared with HEVC.

Furthermore, in Fig. 6, we show the rate-distortion (RD) curves of the proposed method and HEVC according to the QPs in Table 2 for video sequence *Suzie*. We can see that the proposed SR based video compression outperforms HEVC significantly in a relative large bit-rate range. Therefore, the proposed dual-network based SR is efficient in improving the SR performance for compressed content.

4   Conclusion

In this paper, we present a dual-network based super-resolution (SR) method for compressed content, where an enhancement network is introduced before a SR network. The proposed method resolves two different degenerations stepwise and achieves better performance on SR task compared with existing methods, indicating the efficiency of the proposed scheme. In addition, the proposed method also shows its advantage in SR based video coding application, and significant bit-rate saving is obtained compared with HEVC. In future work, we will investigate the inter-frame correlations in both the enhancement and SR networks.

Acknowledgements. This work was supported in part by National Natural Science Foundation of China (61571017), National Postdoctoral Program for Innovative Talents (BX201600006)Top-Notch Young Talents Program of China, High-performance Computing Platform of Peking University, which are gratefully acknowledged.

References

1. Dong, C., Loy, C.C., He, K., Tang, X.: Learning a deep convolutional network for image super-resolution. In: Fleet, D., Pajdla, T., Schiele, B., Tuytelaars, T. (eds.) ECCV 2014. LNCS, vol. 8692, pp. 184–199. Springer, Cham (2014). https://doi.org/10.1007/978-3-319-10593-2_13
2. Dong, C., Loy, C.C., Tang, X.: Accelerating the super-resolution convolutional neural network. In: Leibe, B., Matas, J., Sebe, N., Welling, M. (eds.) ECCV 2016. LNCS, vol. 9906, pp. 391–407. Springer, Cham (2016). https://doi.org/10.1007/978-3-319-46475-6_25
3. Jia, C., Wang, S., Zhang, X., Wang, S., Ma, S.: Spatial-temporal residue network based in-loop filter for video coding. In: 2017 IEEE Visual Communications and Image Processing (VCIP), pp. 1–4. IEEE (2017)
4. Kim, J., Kwon Lee, J., Mu Lee, K.: Accurate image super-resolution using very deep convolutional networks. In: Proceedings of the IEEE Conference on Computer Vision and Pattern Recognition, pp. 1646–1654 (2016)
5. Kingma, D.P., Ba, J.: Adam: a method for stochastic optimization. arXiv preprint arXiv:1412.6980 (2014)
6. Ledig, C., et al.: Photo-realistic single image super-resolution using a generative adversarial network. arXiv preprint (2016)
7. Liang, Y., Timofte, R., Wang, J., Gong, Y., Zheng, N.: Single image super resolution-when model adaptation matters. arXiv preprint arXiv:1703.10889 (2017)
8. Lim, B., Son, S., Kim, H., Nah, S., Lee, K.M.: Enhanced deep residual networks for single image super-resolution. In: The IEEE Conference on Computer Vision and Pattern Recognition (CVPR) Workshops, vol. 1, p. 3 (2017)
9. Maas, A.L., Hannun, A.Y., Ng, A.Y.: Rectifier nonlinearities improve neural network acoustic models. In: Proceedings of the ICML, vol. 30, p. 3 (2013)
10. Schulter, S., Leistner, C., Bischof, H.: Fast and accurate image upscaling with super-resolution forests. In: Proceedings of the IEEE Conference on Computer Vision and Pattern Recognition, pp. 3791–3799 (2015)
11. Shi, W., et al.: Real-time single image and video super-resolution using an efficient sub-pixel convolutional neural network. In: Proceedings of the IEEE Conference on Computer Vision and Pattern Recognition, pp. 1874–1883 (2016)
12. Song, L., Tang, X., Zhang, W., Yang, X., Xia, P.: The SJTU 4K video sequence dataset. In: 2013 Fifth International Workshop on Quality of Multimedia Experience (QoMEX), pp. 34–35. IEEE (2013)
13. Sullivan, G.J., Ohm, J., Han, W.J., Wiegand, T.: Overview of the high efficiency video coding (hevc) standard. IEEE Trans. Circuits Syst. Video Technol. 22(12), 1649–1668 (2012)
14. Timofte, R., et al.: Ntire 2017 challenge on single image super-resolution: methods and results. In: 2017 IEEE Conference on Computer Vision and Pattern Recognition Workshops (CVPRW), pp. 1110–1121. IEEE (2017)
15. Timofte, R., De, V., Van Gool, L.: Anchored neighborhood regression for fast example-based super-resolution. In: 2013 IEEE International Conference on Computer Vision (ICCV), pp. 1920–1927. IEEE (2013)
16. Timofte, R., De Smet, V., Van Gool, L.: A+: adjusted anchored neighborhood regression for fast super-resolution. In: Cremers, D., Reid, I., Saito, H., Yang, M.-H. (eds.) ACCV 2014. LNCS, vol. 9006, pp. 111–126. Springer, Cham (2015). https://doi.org/10.1007/978-3-319-16817-3_8

17. Wang, Y., Wang, L., Wang, H., Li, P.: End-to-end image super-resolution via deep and shallow convolutional networks. arXiv preprint arXiv:1607.07680 (2016)
18. Yang, C.Y., Yang, M.H.: Fast direct super-resolution by simple functions. In: 2013 IEEE International Conference on Computer Vision (ICCV), pp. 561–568. IEEE (2013)
19. Zeyde, R., Elad, M., Protter, M.: On single image scale-up using sparse-representations. In: Boissonnat, J.-D., et al. (eds.) Curves and Surfaces 2010. LNCS, vol. 6920, pp. 711–730. Springer, Heidelberg (2012). https://doi.org/10.1007/978-3-642-27413-8_47
20. Zhang, X., Wang, S., Zhang, Y., Lin, W., Ma, S., Gao, W.: High-efficiency image coding via near-optimal filtering. IEEE Signal Process. Lett. **24**(9), 1403–1407 (2017)

Visual Dialog with Multi-turn Attentional Memory Network

Dejiang Kong[✉] and Fei Wu

College of Computer Science and Technology, Zhejiang University, Hangzhou, China
kdjysss@gmail.com, wufei@cs.zju.edu.cn

Abstract. Visual dialog is a task of answering a question given an input image, a historical dialog about the image and often requires to retrieve visual and textual facts about the question. This problem is different from visual question answering (VQA), which only relies on visual grounding estimated from an image and question pair, while visual dialog task requires interactions among a question, an input image and a historical dialog. Most methods rely on one-turn attention network to obtain facts w.r.t. a question. However, the information transition phenomenon which exists in these facts restricts these methods to retrieve all relevant information. In this paper, we propose a multi-turn attentional memory network for visual dialog. Firstly, we propose a attentional memory network that maintains image regions and historical dialog in two memory banks and attends the question to be answered to both the visual and textual banks to obtain multi-model facts. Further, considering the information transition phenomenon, we design a multi-turn attention architecture which attend to memory banks multiple turns to retrieve more facts in order to produce a better answer. We evaluate the proposed model in on VisDial v0.9 dataset and the experimental results prove the effectiveness of the proposed model.

Keywords: Visual dialog · Memory network · Multi-turn attention

1 Introduction

With the fast development of Artificial Intelligence (AI) techniques especially the design and optimization of deep neural network architectures, many areas such as computer vision (CV) and natural language processing (NLP) have made tremendous progress. Many multi-modal researches, including image caption [2, 3,20,21], image generation from captions [14,16], image question answering [7, 13,15,22] and video question answering [19,24], are beneficial from this progress and achieved great improvements.

Recently, visual question answering (VQA) has received broad attention as it requires an universal understanding of image or video content and learns the alignments between texts and visual areas. Most state-of-the-art methods adopt the attention mechanism [21] to learn these alignments. They identify the visual

© Springer Nature Switzerland AG 2018
R. Hong et al. (Eds.): PCM 2018, LNCS 11164, pp. 611–621, 2018.
https://doi.org/10.1007/978-3-030-00776-8_56

Fig. 1. Dog snoozing on the edge of a cobblestone street.

Table 1. A 10 round dialog corresponding to the left side image.

#	Question	Answer
1	Is it a color photo ?	Yes
2	What color is the dog ?	White
3	What size is the dog ?	It's a puppy
4	Are you able to tell what breed ?	It might be a lab
5	Does the puppy have a human with him ?	No
6	Does he have a leash ?	No
7	Is the puppy facing the camera ?	Yes
8	Are you able to see the color of his eyes ?	They are closed
9	Is there anything else on the street?	A bike
10	Is there anything around the bike ?	A dog

region that referred by the question via attention mechanism and produce the answer based on the visual features in that region.

Most recently, Visual Dialog [4] has been introduced as a more general cross-modal visual question answering task and it contains an image and a sequence of inter-dependent question-answer pairs. Unlike VQA, visual dialog aims to answer a question based on visual content and historical textual dialog. The textual information brings both challenges and opportunities to the generalized VQA problem. Figure 1 and Table 1 show an example of visual dialog. To answer question #10, if we use conventional one-turn attention mechanism, we may retrieve #9 QA pair as a supporting fact to answer #10 with the connection word "bike". However, just using #9 is not enough, we can observe that the image caption (dog snoozing on the edge of a cobblestone street) is an important fact for question #10 and it has straight connection with #9 via the common information "street". We define the fact clue sequence like #10-#9-caption as a information transition, and conventional one-turn attention can not handle this situation properly. We proposed a multi-turn attention architecture that can deal with this information transition and find enough facts w.r.t. question #10 in order to produce a better answer.

In this paper, we address the visual dialog problem and the contributions can be concluded as follows:

– We propose two distinct memory networks to extract and store visual and textual facts respectively.
– Based on proposed memory networks, we propose a multi-turn attentional architecture which applies multi-turn attention on both visual and textual facts to obtain enough information in order to generate or classify the best answer w.r.t. a question.
– We show empirically that the proposed model outperforms several popular visual dialog methods on VisDial dataset.

2   Related Work

Vision and Language. Many tasks such image caption [20,21], visual question answering [7,13,15,19,22,24] at the intersection of vision and language have recently gained prominence. However, they all address the correlation of a singe sentence and a single image. There are no dialogs among them.

Conversation Model and Chatbots. Visual Dialog extends text-based dialog and conversation modeling by adding extra visual information. Many works [1,10,11,17,23] have been done on text-based dialog systems in recent years. Some of them [1,23] focus on goal-oriented dialog systems that can be applied to specific applications such as smart customer service. Others [10,11,17] pay attention to neural dialog generation problem and aim to create a chat robot that can pass the Turing test. All these works are achieving exciting progress with the help of new techniques such as attention mechanism, memory network and deep reinforcement learning. The most important difference between free-form textual dialog and VisDial is that in VisDial, dialogs are subject on images and it contains the interaction between different modal data.

Visual Dialog. Visual dialogs were proposed in [4,5] recently, and their dialogs focus on different targets. In [4], answers of the questions are free form and dialogs are generated by two person talking about a target image. while in [5] the dialogs contain a series of yes/no questions which aim at discovering objects in images. As far as we know, almost no work addresses the importance of information transition phenomenon in the visual dialog task. In this paper, we study visual dialog task as introduced in [4] and pay attention to the so called information transition and mine the correlation between visual facts from a image and textual facts from a corresponding dialog.

3   The Proposed Model

3.1   Problem Definition

In this paper, we deal with visual dialog task as [4] proposed. Given a picture I with caption c and its corresponding dialog D, which consists of t round QA pairs. Each QA pair contains a question and a set of N candidate answers, and only one ground truth answer among these N candidates. All questions and answers are word sequences. We treat the visual dialog as a retrieval based problem. Given Image I, caption c, the first $t-1$ QA pairs $\{(q_1, a_1), (q_2, t_2), ..., (q_{t-1}, a_{t-1})\}$ and question q_t, we aim to learn the rank of the N candidate answers w.r.t. q_t and rank the ground truth answer a_t as top as possible.

We utilize the encoder-decoder architecture as introduced in [4]. Following subsections will describe the proposed multi-turn attentional memory network encoder and two decoders with different architectures. The overall framework is shown in Fig. 2.

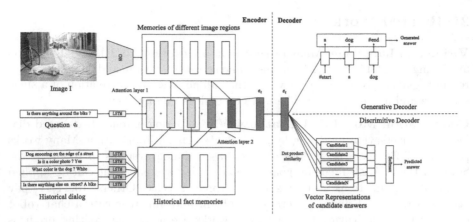

Fig. 2. An overview of multi-turn attentional memory network for visual dialog. The left part of this figure is the proposed encoder and two decoders are shown on the right part. Given the image I and the historical dialog, we first embedding them using deep models (VGG16 and LSTM respectively) and store the results in two memory banks. Then we encoder the question to be answered via another LSTM model and attend the question on two memory banks in turns in order to fetch as many as possible facts and produce the final output of the encoder. The detail algorithm is illustrated in Sect. 3.

3.2   Multi-turn Attentional Memory Network Encoder

For question q_t, we feed it into an LSTM [8] model to get its vector representation $V_{q_t} \in \mathbb{R}^d$:

$$V_{q_t} = LSTM^q(q_t) \tag{1}$$

$LSTM^q$ is the encoder for all questions.

For image I, we extract the image feature map f_I from a raw image I using CNN model VGG-16 [18]:

$$f_I = VGG16(I) \tag{2}$$

The VGG-16 is trained from ImageNet [6]. we choose the features $f_I \in \mathbb{R}^{512 \times 14 \times 14}$ from the last pooling layer, which retains spatial information of the original image. 14×14 is the number of regions and 512 is the feature vector size of each region. We reshape f_I to $\hat{f}_I$ with dimension 512×196 and then we use a single layer perception to transform each feature vector to a new vector that has the same dimension as the question V_{q_t}:

$$M^I = tanh(W_I \hat{f}_I + b_I) \tag{3}$$

where $W_I \in \mathbb{R}^{d \times 512}$ is the transform weight matrix, $M^I \in \mathbb{R}^{d \times 196}$ is a visual memory bank and its ith column m_i^I is the visual memory for the region indexed by i.

For caption c and QA pairs, we take them as useful contexts which can contribute to help find or generate the ground truth answer for question q_t since it reveal several facts about image I. We first concatenate each QA pair (q_i, a_i)

to form the QA fact QA_i, then we feed caption fact and QA facts into the same LSTM to calculate the vector representations of them:

$$m_i^H = LSTM^f(QA_i), \quad i = 1, ..., t-1 \tag{4}$$

$$QA_0 = c \tag{5}$$

where $LSTM^f$ is the model for all textual facts and share its weights among them. QA_i is the input word sequence and $m_i^H \in \mathbb{R}^d$ is the last layer output of LSTM w.r.t. the input QA_i. We take m_i^H as the vector representation of a fact. After these encoding, we obtain all the textual fact memories $M^H = \{m_0^H, m_1^H, ..., m_{t-1}^H\}$ w.r.t. question q_t.

How to utilize fact memories to boost the performance of tasks such as QA has been hot topic in recent years. Many popular methods employ attention mechanism and achieve good results. However, most of them only take one-turn attention and we argue that this is not enough for some occasions as we illustrate in former section. We proposed a multi-turn attention mechanism in order to retrieve enough facts w.r.t. question q_t and eventually help to produce the best answer.

Given question representation V_{q_t}, the corresponding two memory banks M^I and M^H, We first attends V_{q_t} to historical textual memory bank M^H as follows:

$$u_0 \equiv V_{q_t} \tag{6}$$

$$s_i = u_{j-1} \cdot m_i^H, \quad j = 1, ..., r. \tag{7}$$

$$u_{j-1} = u_{j-1} + \sum_{i=0}^{t-1} \alpha_i m_i, \quad \alpha_i = \frac{exp(s_i)}{\sum_{i=0}^{t-1} exp(s_i)} \tag{8}$$

where r represents the total number of attention turns. s_i represents the similarity between u_{j-1} and m_i^H. Each time we retrieve the historical textual memory bank and get u_{j-1}, we take it as the representation of the combination of question q_t and textual facts w.r.t. q_t. Then we attend it to M^I using a single neural network as follows:

$$h = tanh(W_{I,h} M^I \oplus (W_u u_{j-1} + b_h)) \tag{9}$$

$$p^I = softmax(W_p h + b_p) \tag{10}$$

$$u_j = u_{j-1} + \sum_{i=0}^{196} p_i^I m_i^I \tag{11}$$

where $W_{I,h}, W_u \in \mathbb{R}^{k \times d}$, $b_h \in \mathbb{R}^k$, and $\oplus$ denotes the addition between a matrix and a vector that adds the vector $W_u u_{j-1} + b_h$ to each collum of the matrix $W_{I,h} M^I$. h is the output of the neural network. $W_p \in \mathbb{R}^{1 \times k}$, $p^I \in \mathbb{R}^{196}$ represents the attention probability of each image region given u_{j-1}. After the attention to visual memory bank, we get u_j with both textual and visual facts about question q_t, then we do a next turn attention process (formula $7 - 11$) with the new generated input attention vector u_j in order to deal with the information transition as we state in Sect. 1.

After r-turn attention from both visual and textual memory banks, we obtain u_r with enough supporting facts and the final output of the encoder is produced as follows:

$$e_t = tanh(W_e u_r + b_e) \tag{12}$$

where W_e and b_e are the weight and bias of a full-connected layer and e_t is the final encoder output w.r.t. q_t.

3.3   Generative Decoder and the Discriminative Decoder

After encoding the question q_t and its corresponding contexts (i.e. target image I, historical textual facts H), we get their combined representation e_t and we decode it using following two kind of decoders.

Generative Decoder. Given q_t, we generate its answer with following functions:

$$h_0 \equiv e_t \tag{13}$$

$$h_i = LSTM^g(h_{i-1}, x_{i-1}), i = 1, ...|a_t| \tag{14}$$

$$p_i = softmax(W^g h_i + b^g) \tag{15}$$

where $LSTM^g$ is the generative decoder network and h_i is the ith output of $LSTM^g$. x_i is the vector representation of ith word of answer a_t and p_i is a distribution over words. $|a_t|$ means the length of answer a_t. During training step, we maximize the log-likelihood of the ground truth answer a_t. We rank candidate answers using the model's log-likelihood scores of them during evaluation.

Discriminative Decoder. Given q_t and its candidate answers $\{\hat{a_1}, \hat{a_2}, ..., \hat{a_N}\}$, firstly we encoder each candidate answer into vector using LSTM model as follows:

$$h_{\hat{a_i}} = LSTM^d(\hat{a_i}), i = 1, ..., N. \tag{16}$$

where $LSTM^d$ is the LSTM model for candidate answers and share weights among all answers. We calculate the dot similarity s_i between e_t and $h_{\hat{a_i}}$ as follows:

$$s_i = e_t \cdot h_{\hat{a_i}} \tag{17}$$

Then we concatenate the similarities and feed them into a softmax layer to compute the posterior probability over the candidate answers:

$$p_a = softmax(s_1 \ominus s_2 \ominus ... \ominus s_N) \tag{18}$$

where $\ominus$ denotes the concatenate operation, p_a is a distribution over the candidate answers. During training period, we maximize the log-likelihood of the ground truth answer. During evaluation, candidate answers are simply ranked based on their posterior probabilities.

4 Experiments

4.1 Data Set

We use the VisDial v0.9 dataset[1] for our experiment. The images in VisDial are all from MS-COCO [12] dataset and VisDial v0.9 has 82,783 image for training and 40,504 images for evaluation (10,000 for validation and 30,504 for testing in this paper). Each image has one dialog and each dialog contains 10 QA pairs which are generated by two person talk in turn. For each question in a dialog, there are $N(=100)$ candidate answers and only one of them is the ground truth.

4.2 Comparison Methods

All of the comparison methods share the same two decoders as we illustrate in former section. And we use following three encoders [4] for comparison:

Later Fusion (LF) Encoder. This encoder firstly encodes image I via VGG-16, and encodes question q_t and QA facts via different LSTM models. Then it fuses the output of these deep models by concatenation and feed the result into a one-layer full-connected neural network to generate the final output.

Hierarchical Recurrent Encoder (HRE). HRE shares the same architecture with LF except that it using hierarchical LSTM network to model the sequential QA facts. Attention over history QA-pairs can help to boost the model performance and we denote this hierarchical recurrent encoder with attention as HREA.

Memory Network (MN) Encoder. The MN encoder maintains each previous QA pairs as facts in a memory bank and learns to retrieve the most relevant facts w.r.t. current question q_t via an attention mechanism. Then it add these weighted memories to original question representation to produce the output of the encoder.

4.3 Evaluation Criteria

We evaluate the performance of the model and the baselines in terms of three kinds of criteria:

- **Acc@k** Acc@k is defined as: $Acc@k = \frac{\#hit@k}{\#tests}$, where $\#hit@k$ means the number of predicted answers ranking at top k and $\#test$ stands for the number of total testing samples. In this paper, we choose $k = 1, 3, 5, 10$.
- **Mean Rank (MeanR)** Mean rank means the average rank of all predicted results.

---

[1] https://visualdialog.org/data.

– **Mean Reciprocal Rank (MRR)** MRR is defined as: $MRR = \frac{1}{N} \cdot \sum_{i=1}^{N} \frac{1}{Rank_i}$, where N is the total number of test instances and $Rank_i$ is the predicted rank of instance i. MRR is common for tasks whose ground truth is only one instance.

4.4 Parameters and Settings

The parameters to learn are initialized with normal distribution. Hyper parameters such as word embedding size, question representation vector size, memory size of two memory banks and LSTM state size etc. are tuned using grid search. Parameter optimization is done using mini-batch Adam [9] and the early stop strategy is employed, the initial learning rate is 0.001 and weight decaying factor is set to 0.0001.

Table 2. Performance in terms of $Acc@k$ for all methods with discriminative and generative decoders.

Method	Discriminative				Generative			
	Acc@1	Acc@3	Acc@5	Acc@10	Acc@1	Acc@3	Acc@5	Acc@10
LF-Q	0.4082	0.6231	0.6940	0.7965	0.3896	0.5631	0.6045	0.6670
LF-QH	0.4310	0.6556	0.7296	0.8230	0.3529	0.5369	0.5921	0.6666
LF-QI	0.4317	0.6612	0.7402	0.8348	0.4211	0.5764	0.6151	0.6728
LF-QIH	0.4500	0.6815	0.7571	0.8494	0.4039	0.5713	0.6190	0.6832
HRE-QH	0.3490	0.6032	0.6634	0.7808	0.3967	0.5751	0.6154	0.6747
HRE-QIH	0.3787	0.6174	0.6781	0.7938	0.4070	0.5711	0.6181	0.6834
HREA-QIH	0.4295	0.6499	0.7315	0.8320	0.4192	0.5798	0.6231	0.6842
MN-QH	0.4407	0.6747	0.7522	0.8450	0.4029	0.5729	0.6102	0.6652
MN-QIH	0.4538	0.6828	0.7593	0.8515	0.4175	0.5750	0.6124	0.6680
MULTI-QIH-1	0.4672	0.6941	0.7713	0.8625	0.4214	0.5833	0.6318	0.6921
MULTI-QIH-2	**0.4807**	**0.7063**	**0.7824**	**0.8718**	**0.4323**	**0.5934**	**0.6508**	**0.7076**

4.5 Experimental Results and Analysis

We report the comparison results between the proposed multi-turn attentional memory network and other methods, the performance in terms of $Acc@k$, MRR and MeanR are shown in Table 2, Fig. 3(a) and (b) respectively. Note that higher is better for MRR and $Acc@k$ and lower is better for MeanR. All the comparison methods are denoted as {encoder}−{inputs} forms. For example, LF-QIH means the method contains a late fusion (LF) encoder and the input consists of question (Q), the input image (I) and related historical dialog (H). The number after MULTI-QIH means the number of attention turns. In this paper, we only perform max to two turn attentions as the dialog in VisDial v0.9 is short (only ten rounds) and there is almost none information transition with more than two hops. From the statistics we can conclude that:

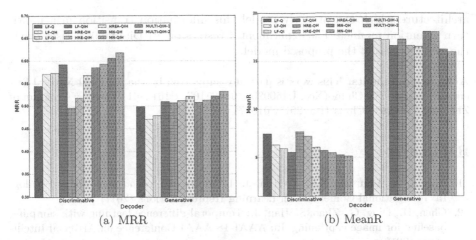

(a) MRR   (b) MeanR

Fig. 3. The mean reciprocal rank (MRR) and mean rank (meanR) for all methods with discriminative and generative decoders.

- With the mixture attention on both the visual and textual facts, MULTI-QIH-1 outperforms other methods much in both $ACC@k$, MRR and MeanR evaluation metrics with all two decoders, this proves the effectiveness of the proposed cross-modal attention mechanism.
- Comparing to MULTI-QIH-1, MULTI-QIH-2 with two-turn attention architecture performs better in all evaluation metrics, this proves the importance of information transition in dialog systems and the effectiveness of the proposed multi-turn attention architecture.
- Models with both visual and textual inputs are better performed than those with single modal input. This reveals that the combination of visual and textual features is essential for visual dialog task.
- Methods with discriminative decoder perform much better than those with generative models. As generative decoder does not score all the answer options during training process, it can not tune to the biases in these answer options. However, generative model can generate answers not included in answer options and it is more practical in realistic applications.

5 Conclusion

In this paper, we propose a multi-turn attentional memory network for visual dialog. We propose a attentional memory network that attends question on both the visual object and textual context to obtain the most relevant information about the question in order to produce a better answer, and both the visual and textual context are stored as fact memories. Further, considering the information transition among the contextual information, we design a multi-turn attention architecture which can properly model this information transition in a multi-turn attention manner. The experimental results on VisDial v0.9 dataset prove the effectiveness of the proposed model.

Acknowledgements. This work is partially supported by the National Natural Science Foundation of China (Nos. U1509206, 61625107, U1611461), the Key Program of Zhejiang Province, China (No. 2015C01027)

References

1. Bordes, A., Boureau, Y.L., Weston, J.: Learning end-to-end goal-oriented dialog. In: International Conference on Learning Representations (2017)
2. Chen, H., Ding, G., Zhao, S., Han, J.: Temporal-difference learning with sampling baseline for image captioning. In: AAAI-18 AAAI Conference on Artificial Intelligence (2018)
3. Chen, M., Ding, G., Zhao, S., Chen, H., Liu, Q., Han, J.: Reference based LSTM for image captioning. In: AAAI, pp. 3981–3987 (2017)
4. Das, A., et al.: Visual dialog. In: Proceedings of the IEEE Conference on Computer Vision and Pattern Recognition, vol. 2 (2017)
5. De Vries, H., Strub, F., Chandar, S., Pietquin, O., Larochelle, H., Courville, A.: Guesswhat?! visual object discovery through multi-modal dialogue. In: Proceedings of CVPR (2017)
6. Deng, J., Dong, W., Socher, R., Li, L.J., Li, K., Fei-Fei, L.: ImageNet: a large-scale hierarchical image database. In: IEEE Conference on Computer Vision and Pattern Recognition, CVPR 2009, pp. 248–255. IEEE (2009)
7. Goyal, Y., Khot, T., Summers-Stay, D., Batra, D., Parikh, D.: Making the V in VQA matter: elevating the role of image understanding in visual question answering. In: 2017 IEEE Conference on Computer Vision and Pattern Recognition (CVPR), pp. 6325–6334 (2017)
8. Hochreiter, S., Schmidhuber, J.: Long short-term memory. Neural Comput. **9**(8), 1735–1780 (1997)
9. Kingma, D.P., Ba, J.L.: Adam: a method for stochastic optimization. In: International Conference on Learning Representations (2015)
10. Li, J., Monroe, W., Ritter, A., Jurafsky, D., Galley, M., Gao, J.: Deep reinforcement learning for dialogue generation. In: Proceedings of the 2016 Conference on Empirical Methods in Natural Language Processing, pp. 1192–1202 (2016)
11. Li, J., Monroe, W., Shi, T., Jean, S., Ritter, A., Jurafsky, D.: Adversarial learning for neural dialogue generation. In: Proceedings of the 2017 Conference on Empirical Methods in Natural Language Processing, pp. 2157–2169 (2017)
12. Lin, T.-Y., et al.: Microsoft COCO: common objects in context. In: Fleet, D., Pajdla, T., Schiele, B., Tuytelaars, T. (eds.) ECCV 2014. LNCS, vol. 8693, pp. 740–755. Springer, Cham (2014). https://doi.org/10.1007/978-3-319-10602-1_48
13. Lu, J., Yang, J., Batra, D., Parikh, D.: Hierarchical question-image co-attention for visual question answering. In: Neural Information Processing Systems, pp. 289–297 (2016)
14. Mansimov, E., Parisotto, E., Ba, J.L., Salakhutdinov, R.: Generating images from captions with attention. In: International Conference on Learning Representations (2016)

15. Noh, H., Seo, P.H., Han, B.: Image question answering using convolutional neural network with dynamic parameter prediction. In: 2016 IEEE Conference on Computer Vision and Pattern Recognition (CVPR), pp. 30–38 (2016)
16. Reed, S.E., Akata, Z., Yan, X., Logeswaran, L., Schiele, B., Lee, H.: Generative adversarial text to image synthesis. In: International Conference on Machine Learning, pp. 1060–1069 (2016)
17. Serban, I.V., Sordoni, A., Bengio, Y., Courville, A.C., Pineau, J.: Building end-to-end dialogue systems using generative hierarchical neural network models. In: AAAI, vol. 16, pp. 3776–3784 (2016)
18. Simonyan, K., Zisserman, A.: Very deep convolutional networks for large-scale image recognition. In: International Conference on Learning Representations (2015)
19. Tapaswi, M., Zhu, Y., Stiefelhagen, R., Torralba, A., Urtasun, R., Fidler, S.: MovieQA: understanding stories in movies through question-answering. In: 2016 IEEE Conference on Computer Vision and Pattern Recognition (CVPR), pp. 4631–4640 (2016)
20. Vinyals, O., Toshev, A., Bengio, S., Erhan, D.: Show and tell: a neural image caption generator. In: 2015 IEEE Conference onComputer Vision and Pattern Recognition (CVPR), pp. 3156–3164. IEEE (2015)
21. Xu, K., et al.: Show, attend and tell: neural image caption generation with visual attention. In: International Conference on Machine Learning, pp. 2048–2057 (2015)
22. Yang, Z., He, X., Gao, J., Deng, L., Smola, A.J.: Stacked attention networks for image question answering. In: 2016 IEEE Conference on Computer Vision and Pattern Recognition (CVPR), pp. 21–29 (2016)
23. Zhou, L., Small, K., Rokhlenko, O., Elkan, C.: End-to-end offline goal-oriented dialog policy learning via policy gradient (2017)
24. Zhu, L., Xu, Z., Yang, Y., Hauptmann, A.G.: Uncovering temporal context for video question and answering. arXiv preprint arXiv:1511.04670 (2015)

DT-3DResNet-LSTM: An Architecture for Temporal Activity Recognition in Videos

Li Yao[1,2(✉)] and Ying Qian[1]

[1] School of Computer Science and Engineering, Southeast University,
Nanjing 211189, People's Republic of China
Yao.li@seu.edu.cn, 252558242@qq.com
[2] Key Laboratory of Computer Network and Information Integration
(Southeast University), Ministry of Education,
Nanjing 211189, People's Republic of China

Abstract. Human activity recognition is a very important problem in computer vision that is still largely unsolved. While recent advances such as deep learning have given us great results on image related tasks, it is still difficult to recognize behavior in videos due to a great deal of disturbance in videos. We propose an architecture DT-3DResNet-LSTM to classify and temporally localize activities in videos. We detect objects in video frames and use these detected results as input to object tracking model, achieving data association information among adjacent frames of multiple objects. Then the clipped video frames of different objects are put into 3D Convolutional Neural Network (CNN) to achieve features, and a Recurrent Neural Network (RNN), specifically Long Short-Term Memory (LSTM), is trained to classify video clips. What's more, we process the output of RNN (LSTM) model to get the final classification of input video and determine the temporal localization of input video.

Keywords: Activity recognition · 3D CNN · LSTM

1 Introduction

Activity recognition of videos has important applications in many scenarios, such as video surveillance, content-based video retrieval, and automotive autopilot technology. TREC Video Retrieval Evaluation (TRECVID) [1] is held since 2003 to promote progress in content-based analysis of and retrieval from digital video via open, metrics-based evaluation.

Traditionally, video activity recognition [2] is completed by extracting the features from the video frames and building a mathematical model according the correspondence of the features. Along with the rapid growth of deep learning, CNN have been generally used in computer vision and activity recognition [3, 4].

Video classification [5, 6] is supervised learning based on given labels. The current video datasets have been preprocessed to clear temporal information. However, a complete video recognition system should identify the activity in the unprocessed video. The most difficult part in temporal activity recognition is to accurately locate the start frame and end frame.

© Springer Nature Switzerland AG 2018
R. Hong et al. (Eds.): PCM 2018, LNCS 11164, pp. 622–632, 2018.
https://doi.org/10.1007/978-3-030-00776-8_57

In this paper, we proposed an architecture to identify the activity and find the temporal localization of the activity for an unprocessed video by three steps: object detection, object tracking and activity classification.

Our contributions are as follows:

(1) 3DResNet LSTM network: we use 3D ResNet CNN [7] model, pre-trained in Kinetics [8] dataset, to get the features of input video. Then the features are fed into LSTM network to find the actual temporal localization. We show that combination of CNN and RNN will get more accurate result of activity classification and temporal localization.

(2) Object tracking in video: our proposed tracking model ignore those objects that are predicted to be the same object but has a far distance between frames. Meanwhile, we compute the Intersection over Union (IoU) between predicted bounding box and detected bounding box to get object type of tracking objects.

As far as we know, our work is the first time to combine deeper 3D CNN with RNN for activity classification task. Previous research shows deeper 2D CNN has a good performance on the ImageNet dataset [9]. However, it is not taken for granted that deeper 3D CNN will also perform well in video activity recognition because the number of video datasets is less than the number of image datasets. The results of this study, which indicate deeper 3D CNN is effective on activity classification, can be expected to promote the development of video recognition. In addition, we can find the temporal localization of activities with LSTM more accurately.

2 Related Work

2.1 Object Detection

With the remarkable development of computer performance, deep learning based on CCN has brought new ideas to target recognition [10]. Compared with the selection of dependent features, deep learning obtains high-dimensional convolution features of images through multi-layer neural networks, which can well preserve target features and improve recognition accuracy. Fast R-CNN [11] and Faster R-CNN [12] are more mature target recognition network architectures at present. Fast R-CNN selects candidate target areas by a selective search (SS) [13] strategy, and classifies and scores candidate areas through a classification network. Faster R-CNN improves the area selection process based on Fast R-CNN, adding a separate neural network RPN. In contrast, RPN generates candidate target areas through a sliding window and an anchor mechanism on a conv feature map, and classifies candidate areas through a classification network to obtain a final result. Faster R-CNN can better locate the target area in the image and improve the recognition efficiency of the entire frame.

2.2 Object Tracking

MOT (Multiple Object Tracking) methods are based on the principle of tracking-by-detection. The most challenging problem of MOT is the data connection of detected objects in adjacent frames. Majority of the batch methods [14] views MOT as a global optimization problem in a graph-based representation, while online methods solve this problem either probabilistically [15] or determinatively (e.g., Hungarian algorithm [16], greedy association [17]). A major element in data association algorithm is the similarity function among detected objects. Both batch methods and online methods are based on learning strategy, trying to learn a similarity function for data association of training data.

2.3 Action Recogition Approaches

Recently, a series of algorithms based on two-stream CNN achieve excellent performance on activity recognition. Simonyan et al. [4] propose an architecture that combines RGB frames with stacked optical flow frames. Their results indicate that this combination can improve action recognition accuracy. After that, many methods based on two-stream CNN have been proposed to enhance action recognition performance [3, 5, 18].

Different from the above-mentioned methods, we are more concerned about 3D CNN, which achieves better performance than 2D CNN because of the usage of large-scale video datasets. The direct reason is that 3D CNN can extract spatio-temporal features from original videos. For example, Ji et al. [19] utilize 3D convolution to obtain spatio-temporal features from videos. Tran et al. [20] train 3D CNN, which they call as C3D, using the Sports-1 M dataset [21]. Since that study, C3D has become a standard model of 3D CNN. Their experimental results also show that the $3 \times 3 \times 3$ convolution kernel can achieve the best results. In addition, Varol et al. [22] show that the temporal length of inputs for C3D influences recognition effect. Meanwhile, Kay et al. show that using Kinetics dataset with 3D CNN can achieve better results than that with 2D convolutional kernels, which is pre-trained on ImageNet dataset, even though the results of 3D CNN trained on the UCF101 and HMDB51 datasets perform weaker than the 2D CNN's results. For temporal localization task, recent researches have focused on LSTM [23], a type of RNN that is able to exploit long and short temporal correlations in sequences. LSTM has been used alongside CNN for video classification [6] and activity localization in videos.

3 Proposed Architecture

As Fig. 1 shows, our DT-3DResNet-LSTM architecture contains three sections: (1) Faster-RCNN [12] is used to detect the target objects in video frames, which is the D partition of **DT-3DResNet-LSTM**. (2) Kalman filter [24] is applied to track detected different objects, generating numerous continuously clipped frames containing tracked objects, which is the T partition of DT-3DResNet-LSTM. (3) 3DResNet [7] and LSTM is combined to classify activities and temporally localize activities.

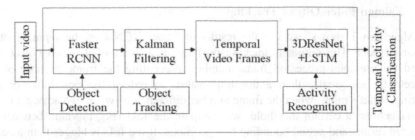

Fig. 1. The proposed activity recognition DT-3DResNet-LSTM architecture contains three sub-processes. Object detection model is applied to generate the correct bounding box of objects. Object tracking model uses the results of object detection model and then track different objects among frames. Activity recognition model works as the final process of activity recognition architecture to output the final prediction of temporal localization and classification of input videos.

3.1   Faster RCNN Object Detection

We use the Faster RCNN [12] with VGG16 [25] as the bottom feature of the video frame for object detection. The Faster RCNN network framework is shown in Fig. 2. For an arbitrary input image, the VGG16 model is used to obtain image features. The last layer of feature map is conv5-3. The RPN (Region Proposal Network) network performs a 3 × 3 convolution on the conv5-3 layer, followed by a 512-dimensional full-connection layer. The full-connection layer is followed by two sub-connection layers, used for the classification and regression of anchors. Then the proposals are got through calculation screening. The anchors are a set of fixed-size reference windows with 3 scales and 3 aspect ratios. The ROI Pooling layer uses the generated proposal to extract the feature from the feature maps for pooling. Fast RCNN [11] identifies and classifies the proposals extracted from the RPN network, and adjusts the regression parameters to obtain the precise location of the target.

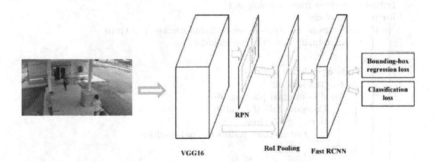

Fig. 2. The architecture of Faster RCNN. The features extracted by VGG16 were fed into RPN to generate proposals and also used with the generated proposals to ROI pooling. ROI pooling layer make the input proposals resize to the same output size and then feed into FAST RCNN to achieve classification and bounding box.

3.2 Kalman Filter Object Tracking

As Algorithm 1 shows, we use the results of Faster RCNN as the input of object tracking model. The output is tracking positions and types of detected objects. We record the previous processed frame number, preventing the frame number of two adjacent frames greater than a threshold value, to reduce the possibility of false tracking. What's more, when the frame number difference between two adjacent frames is smaller than a certain threshold, we compute the IoU (Eq. 1) value between the predicted results and actual bounding box positions. If the IoU is bigger than a certain value, the tracking result will be added to final results. The final results include valid frame number, object type and object position of detected objects in each video frame.

$$IoU = \frac{DetectionResult \cap GroundTruth}{DetectionResult \cup GroundTruth} \tag{1}$$

We estimate the inter-frame movements of each object via a linear constant velocity model which is irrelevant of other objects and camera motion. The status of each object is modeled as:

$$x = [u, v, s, r, u', v', s']^T \tag{2}$$

where u and v indicate the horizontal and vertical pixel location of the center of the target object, the scale s and r indicate the scale (area) and the aspect ratio of the target object's bounding box. When a detection is related to a target, the detected bounding box is used to regenerate the target status where the velocity elements are computed optimally through a Kalman filter [23]. If no detection is related to the target, its status is forecasted without rectification using liner velocity model.

Algorithm 1: Detected objects tracking

Input: N video frames and information about detected objects which
  containing positions and object types.
Output: Tracking positions and types of detected objects.
Initialize previous frame number = 1;
for $n \leftarrow 1$ *to* N **do**
  if current frame number - previous frame number > 5 **then**
    $\lfloor$ Clean kalman filter tracking records;

  else
    Update predicted trackers;
    for $t \leftarrow$ trackers **do**
      for $d \leftarrow$ original positions **do**
        Compute IoU of t and d;
        if IoU ≥ 0.3 **then**
          $\lfloor$ Add tracking results to final results.;

Return final tracking results.

3.3   3DResNet+LSTM Activity Classification

3D CNN does not perform well on UCF-101, HMDB-51, and ActivityNet datasets, whereas 3D CNN trained on Kinetics performs well [8, 26]. Deeper 3D CNN may have good performance compared to shallow 3D CNN. However, deep 3D CNN have more parameters to learn through training. As a result, huge datasets are required to prevent overfitting while training. Kinetics is a dataset big enough to pre-train our ResNet model.

A basic ResNets block includes two convolutional layers, followed by batch normalization and a ReLU. A shortcut connection is between the input of the block and the layer before last ReLU model. we apply identity connections and zero padding for the shortcuts in basic blocks to avoid learning many parameters of superficial networks.

ResNeXt adds a different component in terms of depth and breadth, called cardinality. Different from the original bottleneck block, the ResNeXt block partitions feature maps into small groups, called group convolutions. Cardinality represents the number of middle convolutional layer groups in the bottleneck block. Xie et al. [27] show that using more cardinality in 2D architectures improves the effectiveness compared with using wider or deeper ones. In this paper, we use the cardinality of 32 to assess the result of ResNeXt-101 on activity recognition, as shown in Fig. 3.

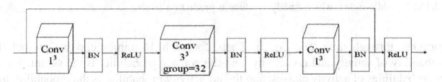

Fig. 3. Block of ResNeXt architecture. We represent conv, x^3, F as the kernel size, and the number of feature maps of the convolutional filter are $x \times x \times x$ and F, respectively, and group as the number of groups of group convolutions, which partition the feature maps into small groups. BN represents batch normalization.

Figure 4 shows our proposed architecture. LSTM has been used alongside CNN for video classification [6] and activity localization [28]. We add LSTM behind the 3DResNet output layers to classify a sequence of video frames. We design a network that extracts a sequence of C3D-f6 features of input video, and outputs a sequence of class probabilities for each 16-frames clip. We use LSTM layers, trained with dropout by probability $p = 0.5$, and a fully connected layer with a softmax as activation function.

Given a processed clipped video, the prediction of our model is a series of class probabilities for each 16-frame video clip. We process the output to predict the activity class and get temporal localization. First, to obtain the activity prediction of the whole video, we average the class probabilities over all video clips generated by object tracking model. Second, we choose the class which has the maximum probability among all candidate classes.

In order to achieve the temporal localization of predicted activity clipped video, we first apply a mean filter of k samples to the predicted series to make the values become

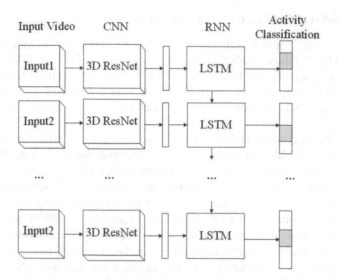

Fig. 4. Overview of proposed activity classification framework. Temporal features of inputted clipped video frames were extracted from a temporal CNN using 3D ResNet (middle-left) pre-trained on Kinetics. The features were then fed into a stack of recurrent sequence models (LSTMs, middle-right), which finally produce a prediction (right).

smoothly through time (see Eq. 3). Then, for each 16-frames clip we predict the probability of activity and no activity, and the activity probability is the summation of all probabilities of activity classes, and the no activity probability is the probability that this video clip belongs to background class. Finally, only clips with a probability value bigger than a certain threshold γ can be saved and marked as previously predicted class. Notice that, for each video clip, all predicted temporal results are activity class type.

$$\tilde{p}_i(x) = \frac{1}{2k} \sum_{j=i-k}^{i+k} p_i(x) \tag{3}$$

4   Experiments

4.1   Datasets

VIRAT Video Dataset. The dataset is designed to be realistic, natural and challenging for video surveillance domains in terms of its resolution, background clutter, diversity in scenes, and human activity/event categories than existing action recognition datasets. VIRAT Video Dataset [29] contains two broad categories of activities (single-object and two-objects) which involve both human and vehicles. Details of included activities, and annotation formats may differ per release. Relevant information can be found from each release information.

ActivityNet. ActivityNet (v1.3) [30] offers examples from 200 human action classes with an average of 137 unprocessed videos per class and 1.41 activity instances per video. Different from the other datasets, ActivityNet consists of unprocessed videos, which include frames that has no activity. The total video length is 849 h, and the total number of action instances is 28,108. This dataset is randomly split into three different subsets: training, validation, and testing. More specifically, 50% is used for training, 25% is used for validation, and 25% is used for testing.

4.2   Results

We train a Faster RCNN model to detect specified objects such as person and vehicle in VIRAT Video Dataset and detected result is shown in Fig. 5. We use detection bounding box results which has confidence more than 0.5.

Fig. 5. Result of object detection model. We use the given bounding box information of person and vehicle to train a Faster RCNN model and use this model to detect objects in video frames. Number one in image represents person and number two represents vehicle.

We feed the results of object detection model into object tracking model, and track the motion of different objects among adjacent frames to achieve multiple sequences of video frames which clipped according to detected bounding box information as shown in Fig. 6.

For video classification, we use mean average precision (mAP). For temporal localization, a prediction is marked correct only when it has the correct category and has IoU with ground truth instance larger than 0.5, and mAP is used to evaluate the performance over the entire dataset.

Table 1 shows the performance of different network architectures. We tested configurations with different number of LSTM layers and different number of cells. We use our proposed architecture with DT(detection and tracking partition) and without DT separately to compare results on ActivityNet dataset with others. As shown in Table 1, 3DResNet+LSTM performs better than 3DCNN+LSTM. 3DResNet+LSTM

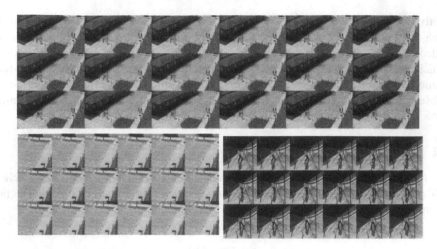

Fig. 6. Tracking results in VIRAT video dataset. The top row is original video frames. The left bottom is the tracking result of a moving vehicle, and the right bottom is the tracking result of a moving pedestrian.

with DT achieves best results. Our architecture increases 1.89% compared to 3DCNN +LSTM with the same parameter of 1×512 LSTM layer.

Table 1. Results for classification task comparing different architectures. The last four architectures are our proposed architectures tested in different parameters. We conduct experiment with and without detection and tracking separately to analyze experiments. mAP compares the accuracy of different architectures, while network convergence epoch compares the speed of convergence of different architectures.

Architecture	mAP	Network convergence epoch
3DCNN+3×1024-LSTM	0.5635	200
3DCNN+2×512-LSTM	0.5492	150
3DCNN+1×512-LSTM	0.5938	120
3DResNet+2×512-LSTM(without DT)	**0.5614**	**110**
3DResNet+1×512-LSTM(without DT)	**0.6018**	**100**
3DResNet+2×512-LSTM(with DT)	**0.5723**	**80**
3DResNet+1×512-LSTM(with DT)	**0.6127**	**60**

5 Conclusion and Future Work

We propose an architecture for classification and temporal localization of activities in videos. We firstly combine 3DResNet and LSTM to temporally recognize activity. What's more, we conduct some improvement in tracking partition to increase accuracy. In the future, we plan to make our architecture become end-to-end, which will make training process become convenient.

References

1. TRECVID Homepage. https://www-nlpir.nist.gov/projects/tv2018/
2. Wang H., Schmid, C.: Action recognition with improved trajectories. In: Proceedings of the IEEE International Conference on Computer Vision, pp. 3551–3558. IEEE (2013)
3. Ng, J.Y.H., Hausknecht, M., Vijayanarasimhan, S., et al.: Beyond short snippets: deep networks for video classification. In: 2015 IEEE Conference on Computer Vision and Pattern Recognition, CVPR, pp. 4694–4702. IEEE (2015)
4. Feichtenhofer, C., Pinz A, Zisserman A P.: Convolutional two-stream network fusion for video action recognition. In: The IEEE Conference on Computer Vision and Pattern Recognition 2016, CVPR, pp. 1933–1941. IEEE (2016)
5. Wang, L., Xiong, Y., Wang, Z., et al.: Towards good practices for very deep two-stream convnets. arXiv preprint arXiv:1507.02159 (2015)
6. Yao, L., Torabi, A., Cho, K., et al.: Describing videos by exploiting temporal structure. In: Proceedings of the IEEE International Conference on Computer Vision, pp. 4507–4515. IEEE (2015)
7. Hara, K., Kataoka, H., Satoh, Y.: Can spatiotemporal 3D CNNs retrace the history of 2D CNNs and ImageNet. In: Proceedings of the IEEE Conference on Computer Vision and Pattern Recognition 2018, pp. 18–22. IEEE, Salt Lake City (2018)
8. Kay, W., Carreira, J., Simonyan, K., et al.: The kinetics human action video dataset. arXiv preprint arXiv:1705.06950 (2017)
9. He, K., Zhang, X., Ren, S., et al.: Deep residual learning for image recognition. In: Proceedings of the IEEE International Conference on Computer Vision, pp. 770–778. IEEE (2016)
10. Dai, J., Li, Y., He, K., et al.: R-FCN: object detection via region-based fully convolutional networks. In: Advances in Neural Information Processing Systems, pp. 379–387 (2016)
11. Girshick, R.: Fast R-CNN. In: Proceedings of the IEEE International Conference on Computer Vision 2015, pp. 1440–1448 (2015)
12. Ren, S., He, K., Girshick, R., et al.: Faster R-CNN: towards real-time object detection with region proposal networks. In: Advances in Neural Information Processing Systems, pp. 91–99 (2015)
13. Uijlings, J.R.R., Van De Sande, K.E.A., Gevers, T., et al.: Selective search for object recognition. Int. J. Comput. Vis. **104**(2), 154–171 (2013)
14. Niebles, J.C., Han, B., Fei-Fei, L.: Efficient extraction of human motion volumes by tracking. In: 2010 IEEE Conference on Computer Vision and Pattern Recognition, CVPR, pp. 655–662. IEEE (2010)
15. Oh, S., Russell, S., Sastry, S.: Markov chain Monte Carlo data association for multi-target tracking. IEEE Trans. Autom. Control **54**(3), 481–491 (2009)
16. Kim, Suna, Kwak, Suha, Feyereisl, Jan, Han, Bohyung: Online multi-target tracking by large margin structured learning. In: Lee, Kyoung Mu, Matsushita, Yasuyuki, Rehg, James M., Hu, Zhanyi (eds.) ACCV 2012. LNCS, vol. 7726, pp. 98–111. Springer, Heidelberg (2013). https://doi.org/10.1007/978-3-642-37431-9_8
17. Breitenstein, M.D., Reichlin, F., Leibe, B., et al.: Online multiperson tracking-by-detection from a single, uncalibrated camera. IEEE Trans. Pattern Anal. Mach. Intell. **33**(9), 1820–1833 (2011)
18. Feichtenhofer, C., Pinz, A., Wildes, R.P.: Spatiotemporal multiplier networks for video action recognition. In: 2017 IEEE Conference on Computer Vision and Pattern Recognition 2017, CVPR, pp. 7445–7454. IEEE (2017)

19. Ji, S., Xu, W., Yang, M., et al.: 3D convolutional neural networks for human action recognition. In: IEEE Transactions on Pattern Analysis and Machine Intelligence, pp. 221–231. IEEE (2013)
20. Tran, D., Bourdev, L., Fergus, R., et al.: Learning spatiotemporal features with 3d convolutional networks. In: Proceedings of the IEEE International Conference on Computer Vision, pp. 4489–4497. IEEE (2015)
21. Karpathy, A., Toderici, G., Shetty S, et al.: Large-scale video classification with convolutional neural networks. In: Proceedings of the IEEE International Conference on Computer Vision and Pattern Recognition, pp. 1725–1732. IEEE (2014)
22. Varol, G., Laptev, I., Schmid, C.: Long-term temporal convolutions for action recognition. IEEE Trans. Pattern Anal. Mach. Intell. **40**(6), 1510–1517 (2017)
23. Hochreiter, S., Schmidhuber, J.: Long short-term memory. Neural Comput. **9**(8), 1735–1780 (1997)
24. Yoon, J.H., Yang, M.H., Lim, J., et al.: Bayesian multi-object tracking using motion context from multiple objects. In: 2015 IEEE Winter Conference on Applications of Computer Vision, WACV, pp. 33–40. IEEE (2015)
25. Simonyan, K., Zisserman, A.: Very deep convolutional networks for large-scale image recognition. arXiv preprint arXiv:1409.1556 (2014)
26. Hara, K., Kataoka, H., Satoh, Y.: Learning spatio-temporal features with 3D residual networks for action recognition. In: Proceedings of the ICCV Workshop on Action, Gesture, and Emotion Recognition. vol. 2, No. 3, p. 4 (2017)
27. Xie, S., Girshick, R., Dollár, P., et al.: Aggregated residual transformations for deep neural networks. In: IEEE Conference on Computer Vision and Pattern Recognition, CVPR, pp. 5987–5995. IEEE (2017)
28. Yeung, S., Russakovsky, O., Jin, N., et al.: Every moment counts: dense detailed labeling of actions in complex videos. Int. J. Comput. Vis. **126**(2–4), 375–389 (2018)
29. Caba Heilbron, F., Escorcia, V., Ghanem, B., Carlos Niebles, J.: Activitynet: a large-scale video benchmark for human activity understanding. In: Proceedings of the IEEE Conference on Computer Vision and Pattern Recognition 2015, pp. 961–970. IEEE (2015)
30. Oh, S., Hoogs, A., Perera, A., et al.: A large-scale benchmark dataset for event recognition in surveillance video. In: 2011 IEEE Conference on Computer Vision and Pattern Recognition, CVPR, pp. 3153–3160. IEEE (2011)

Mutiple Transfer Net with Region Ensemble for Deep Hand Pose Estimation

Haoqian Wang[1,2], Da Li[1(✉)], and Xingzheng Wang[1,2]

[1] Key Laboratory of Broadband Network and Multimedia,
Graduate School at Shenzhen, Tsinghua University, Shenzhen 518055, China
d-li16@mails.tsinghua.edu.cn
[2] Shenzhen Institute of Future Media Technology, Shenzhen 518071, China

Abstract. Deep hand pose estimation from single depth image plays a significant role in human-computer interaction. This paper proposes a novel method based on multiple transfer net to estimate hand pose utilizing only single-channel depth photos. A channel extending process for original single channel depth image is implemented to extend hand and hand palm regions and match the input format of a pre-trained network and fully utilize the parameters. A multiple transfer network refinement for the previous convolutional neural network is made to obtain various different feature maps. Also, a region ensemble is used to merge all output feature maps and integrate the results. The experimental results demonstrate that the proposed method outperforms state-of-art results with considerable accuracy on the NYU [1] and ICVL [2] datasets.

Keywords: Hand pose estimation · Convolutional Neural Network
Human computer interaction · Depth images

1 Introduction

Hand pose estimation is ubiquitously required in many critical applications and it is one of the most important techniques in human-computer interactions including virtual/augmented reality applications. It aims to predict the 3D locations of hand joints [3] from single depth images, which is critical for gesture recognition [4]. Though it has attracted broad research interests in recent years [5,6], due to the severe occlusions caused by articulate hand poses and noisy input from affordable depth sensors, high accuracy hand pose estimation is still very challenging.

Deep convolutional networks have proved themselves as highly effective for their modeling capacity and end-to-end feature learning. They are widely used in several computer vision tasks such as object detection [7], image segmentation [8], object classification [9], and specifically, hand pose estimation. Carreira et al. [10] and Haque et al. [11] used CNN-based methods to predict probability heat maps of each joint, and infer hand pose from heat maps. He et al. [12] proposed a sophisticated design with a feedback loop and Chen et al. [8] presented

© Springer Nature Switzerland AG 2018
R. Hong et al. (Eds.): PCM 2018, LNCS 11164, pp. 633–642, 2018.
https://doi.org/10.1007/978-3-030-00776-8_58

a spatial attention mechanism. Likewise, many ensemble methods for deep hand pose estimation have recently been proposed. A tree-structured Region Ensemble Network (REN) for direct 3D coordinate regression was proposed by Guo et al. [13]. Chen et al. [14] presented a pose-guided structured Region Ensemble Network to boost the performance of hand pose estimation.

Recently, ImageNet pre-trained CNNs have been used for chest pathology identification and detection in X-ray and CT modalities [15]. Maxime te al. [16] showed how image representations learned using CNNs on large-scale annotated datasets, and that they can be efficiently transferred to other visual recognition tasks with a limited amount of training data. Andrej et al. [17] studied multiple approaches for extending the connectivity of a CNN within time domain to take advantage of local spatio-temporal information and suggested a multi-resolution method for large-scale video classification. Alexandre [18] investigated the possibility of transferring knowledge between CNNs when processing RGB-D data to recognize 3D objects. However, the fine-tuning of an ImageNet pre-trained CNN model on datasets of hand pose estimation has not yet been exploited.

In this paper, we achieved further improvements by proposing a multiple transfer CNN based method to estimate hand pose with only depth information. We first used a new channel extending method to extend hand and hand palm regions and match pre-trained networks' input format to make full use of the pre-trained net, and then we refined the previous convolutional neural network with a multiple transfer network. The multiple transfer network can obtain various different feature maps by using different pre-trained convolutional networks. Finally, we optimized our network by using a region ensemble method [13]. All of the work greatly improved the accuracy of the joints' prediction and achieved remarkable performance on the benchmark.

2 Proposed Method

2.1 Channel Extending

As illustrated in Fig. 1, We transform the original single depth images into three image channels. We begin by making a center crop to the depth image in two different scales and then up-sample the cropped images to the size of original image. Then we merge them with the original depth image in order to obtain a three-channel image. Since the hand palm and the whole hand areas usually are situated in the central region of the preprocessed images, and center crop for a single hand is in process of the preprocess and data augmentation we used [19], this channel extending step actually extends hand and hand palm regions in the original images. Furthermore, the input of many commonly used pre-trained networks like VGG-19 relies on a three-channel RGB image, and channel extending helps match the single depth data with the input format of the pre-trained network, and enable full use of the pre-trained parameters.

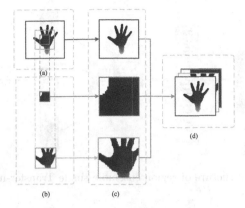

Fig. 1. Extend an original single channel depth image to a three-channel image. (a) is the original single channel depth image. (b) has two images cropped by two different scales. (c) has three channels images resized from (a) and (b). (d) is a three-channel image merged from (c).

2.2   Transfer-Net Ensemble

We use parameters, which have already been learned from a pre-trained network to optimize our network according to the transfer learning method. Compared with training from the beginning, it can decrease the training difficulty and reduce both training time and the over-fitting risks. Given that pre-trained networks with more layers have more parameters to learn, which will require far more time on training and whilst yielding poorer real-time performance while predicting, we firstly chose the VGG-11 pre-trained network, which has shallow layers and few parameters. The network structure is shown in Fig. 2. We use the first eight convolutional layers, whose parameters were pre-trained on ImageNet to initialize our convolutional layer parameters, and then fine-tune the fully connected layers defined by ourselves. To obtain further improvement, we add a region ensemble [13] layer before fully connected layers. Our experiments show that the convergence speed of this transfer-net is accelerated and the accuracy of prediction achieves that of state-of-arts.

2.3   Multiple Transfer Nets Based Region Ensemble

For in the image domain, a receptive field contributes every activation in the convolutional feature maps, we can project the multi-view inputs onto the regions of the feature maps. So multi-view voting is equal to utilizing each region to separately predict the whole hand pose and combining the results. Therefore we use a region ensemble method [13] to optimize the layers which need to be fine-tuned and achieve a higher level of accuracy.

Using one pre-trained network only generates a few series of feature maps, so we try to use multiple pre-trained models to gain a more diverse set of feature

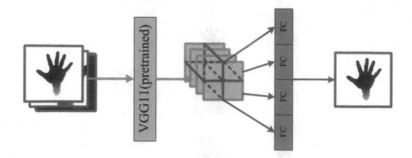

Fig. 2. Network structure of region ensemble single Transfer-net we proposed.

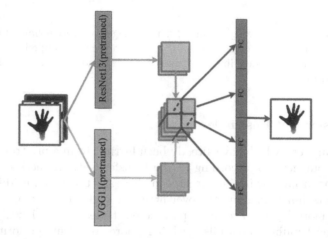

Fig. 3. Network structure of region ensemble multiple transfer net we proposed.

maps and merge all feature maps together for region ensembling. As is shown in Fig. 3, we use VGG-11 and ResNet-13 pre-trained networks to gain feature maps. Finally we upsample all feature maps to 16×16, merge them all together and make a region ensemble to obtain the final outputs.

3　Experimental Results

We evaluate our multiple transfer net on two public benchmark datasets for hand pose estimation: the NYU dataset [1] and the ICVL dataset [2]. There are different evaluation metrics for hand pose estimation following the literatures, and we report the numbers stated in the papers or measured from the graphs if provided, and plot the relevant graphs for comparison.

We use two different metrics to evaluate the accuracy. First, we evaluate the accuracy of the 3D hand pose estimation as an average 3D joint error. This is established as the most commonly used metric in literature, and allows comparison with other work due to the simplicity of evaluation. Second, we plot the fraction of frames where all predicted joints are below a given average Euclidean distance from the ground truth [20].

We train and test the network using pytorch [21]. We first segment the foreground and extract a cube from the depth image centered in the centroid of the hand region [22], and then resize the cube into a 128 × 128 patch of depth values normalized to [?1, 1] as input for ConvNet. We use Adaptive Moment Estimation (Adam) with a mini-batch size of 16. The learning rate starts from 0.00005 and the model is trained for up to 50 epochs.

3.1 Comparison with State-of-arts

ICVL Dataset. The ICVL dataset [2] contains a training set of over 180,000 depth frames of single-channel depth hand pose images. The test set includes two sequences, each of which has approximately 700 frames. The dataset contains 16 annotated joints and is recorded by a time-of-flight camera. The dataset has a high enough frame rate and resolution that the number of missing depth values is negligible while producing sharp outlines with little noise. Although the authors provide different artificially rotated training samples, we only used the genuine 22000 frames and applied the data augmentation proposed in [22]. However, compared to other datasets [1,23], this dataset's pose variability is limited. As is discussed in [3,24], the annotations of the hand joints are very inaccurate.

Table 1. Average 3D error on ICVL dataset [2].

Method	Average 3D error
Deng et al. [25] (Hand3D)	10.9 mm
Tang et al. [2] (LRF)	12.6 mm
Wan et al. [26]	8.2 mm
Zhou et al. [27] (DeepModel)	11.3 mm
Sun et al. [23] (HPR)	9.9 mm
Wan et al. [28] (Crossing Nets)	10.2 mm
Fourure et al. [29] (JTSC)	9.2 mm
Krejov et al. [30] (CDO)	10.5 mm
Oberweger et al. [22] (DeepPrior++)	8.1 mm
This work (Multiple transfer net)	**7.3 mm**

We list a comparison to different state-of-the-art methods on ICVL dataset [2] in Table 1. Our method shows state-of-the-art accuracy. However, the performance gap to other methods is much smaller. This may be due to the fact that their dataset is less complex, with smaller pose variations [3], and attributed to deviation in the annotations for evaluations [3,24].

In Fig. 4 we compared multiple transfer net to other methods on the ICVL dataset [2]. Our approach performs similar to the works of Oberweger et al. [22],

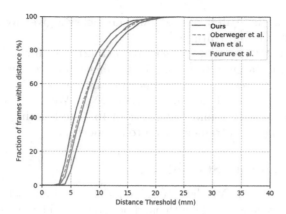

Fig. 4. Comparison with state-of-the-arts on ICVL [2] datasets: percentage of success frames.

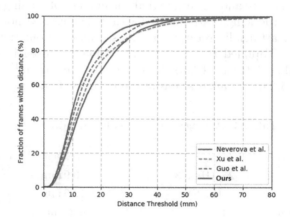

Fig. 5. Comparison with state-of-the-arts on NYU [1] datasets: percentage of success frames.

Wan et al. [26], and Fourure et al. [29], all achieving state-of-the-art accuracy on this dataset. This might be an indication that the performance on the dataset is saturating, and the remaining error is due to the annotation uncertainty. This empirical finding is similar to the discussion in [3].

NYU Dataset. The NYU dataset [1] is comprised of around 72000 training and 8000 test frames of multi-view RGB-D images. The dataset was built with a structured light-based camera, and the depth images have a series of missing values and noisy outlines, which makes the dataset very difficult to predict. We relied solely on the single channel depth images captured from a single camera for our experiments. The training set has samples from a single person and the test set samples from two different people. We followed the proposed evaluation metrics [24, 34] and chose 14 joints for evaluation. As is shown in Table 2 that our method outperforms several state-of-the-art methods.

Table 2. Average 3D error on NYU dataset [1].

Method	Average 3D error
Oberweger et al. [31] (Feedback)	16.2 mm
Deng et al. [25] (Hand3D)	17.6 mm
Guo et al. [13] (REN)	13.4 mm
Zhou et al. [27] (DeepModel)	16.9 mm
Xu et al. [32] (Lie-X)	14.5 mm
Neverova et al. [33]	14.9 mm
Wan et al. [28] (Crossing Nets)	15.5 mm
Fourure et al. [29] (JTSC)	16.8 mm
Madadi et al. [21]	15.6 mm
This work (Multiple transfer net)	**13.1 mm**

In Fig. 5 we compare our method with other discriminative approaches on the NYU dataset [1]. Our predictions performed significantly better for the majority of the frames.

3.2 Self-comparison

We implemented four baselines for comparison: (a) VGG-transfer has the same convolution structure with a VGG-11 pre-trained network and a fully connected layer created by ourselves. (b) Res-transfer has the same convolution structure

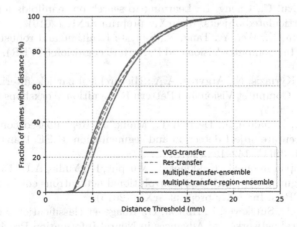

Fig. 6. Self-comparison on ICVL [2] dataset: percentage of success frames.

with a ResNet-13 pre-trained network and a fully connected layer, which we created. (c) Multiple-transfer-ensemble net has the same convolution structure with Fig. 6 without region ensemble step. (d) Multiple-transfer-region-ensemble net has the same convolution structure with Fig. 6. As is shown in Fig. 6, the results of VGG-transfer and Res-transfer are close and multiple-transfer-ensemble outperforms both, with the Multiple-transfer-region-ensemble net achieving the best result.

4   Conclusion

In this paper, an accurate multiple transfer region ensemble CNN based method for hand pose estimation is proposed. It provides a new channel extending method to extend hand and hand palm regions and match pre-trained networks' input format to fully utilize the pre-trained networks, and multiple transfer net to obtain a higher level of accuracy. Furthermore, a region ensemble for all transfer-net's output feature maps achieves further improvement in the prediction result. The experimental results demonstrate that the proposed method outperforms state-of-art results by considerable accuracy on NYU [1] and ICVL [2] datasets. Improving the real-time performance of the system is our future objective.

References

1. Tompson, J., Stein, M., Lecun, Y., Perlin, K.: Real-time continuous pose recovery of human hands using convolutional networks. ACM Trans. Graph. (TOG) **33**, 1935–1946 (2014)
2. Tang, D., Chang, H.J., Tejani, A., Kim, T.K.: Latent regression forest: structured estimation of 3D articulated hand posture. In: Conference on Computer Vision and Pattern Recognition (CVPR), pp. 3786–3793. IEEE (2014)
3. Supancic, J., Rogez, G., Yang, Y., Shotton, J., Ramanan, D.: Depth-based hand pose estimation: data, methods, and challenges. In: International Conference on Computer Vision (ICCV). IEEE (2015)
4. Zhang, Y., Xu, C., Cheng, L.: Learning to search on manifolds for 3D pose estimation of articulated objects. In: arXiv preprint arXiv (2016)
5. Qian, C., Sun, X., Wei, Y., Tang, X., Sun, J.: Realtime and robust hand tracking from depth. In: Computer Vision and Pattern Recognition (CVPR), pp. 1106–1113. IEEE (2014)
6. Makris, A., Kyriazis, N., Argyros, A.A.: Hierarchical particle filtering for 3D hand tracking. In: Computer Vision and Pattern Recognition Workshops (CVPRW), pp. 8–17. IEEE (2015)
7. Girshick, R., Donahue, J., Darrell, T., Malik, J.: Regionbased convolutional networks for accurate object detection and segmentation. IEEE Trans. Pattern Anal. Mach. Intell. **38**(1), 142–158 (2016)
8. Chen, L., Papandreou, G., Kokkinos, I., Murphy, K., Yuille, A.L.: DeepLab: semantic image segmentation with deep convolutional nets, atrous convolution, and fully connected CRFs. In: arXiv preprint arXiv (2016)
9. Krizhevsky, A., Sutskever, I., Hinton, G.: Imagenet classification with deep convolutional neural networks. In: Advances in Neural Information Processing Systems, pp. 1097–1105 (2012)

10. Carreira, J., Agrawal, P., Fragkiadaki, K., Malik, J.: Human pose estimation with iterative error feedback. In: Computer Vision and Pattern Recognition (CVPR), pp. 4733–4742. IEEE (2016)
11. Haque, A., Peng, B., Luo, Z., Alahi, A., Yeung, S., Fei-Fei, L.: Towards viewpoint invariant 3D human pose estimation. In: Leibe, B., Matas, J., Sebe, N., Welling, M. (eds.) ECCV 2016. LNCS, vol. 9905, pp. 160–177. Springer, Cham (2016). https://doi.org/10.1007/978-3-319-46448-0_10
12. He, K., Zhang, X., Ren, S., Sun, J.: Spatial pyramid pooling in deep convolutional networks for visual recognition. In: Fleet, D., Pajdla, T., Schiele, B., Tuytelaars, T. (eds.) ECCV 2014. LNCS, vol. 8691, pp. 346–361. Springer, Cham (2014). https://doi.org/10.1007/978-3-319-10578-9_23
13. Hengkai, G., Guijin, W., Xinghao, C., Cairong, Z., Fei, Q., Huazhong, Y.: Region ensemble network: improving convolutional network for hand pose estimation. In: International Conference on Image Processing (ICIP). IEEE (2017)
14. Xinghao, C., Guijin, W., Hengkai, G., Cairong, Z.: Pose guided structured region ensemble network for cascaded hand pose estimation. In: arXiv preprint arXiv (2017)
15. Bar, Y., Diamant, I., Greenspan, H., Wolf, L.: Chest pathology detection using deep learning with non-medical training. In: Biomedical Imaging (ISBI), vol. 13. IEEE (2015)
16. Maxime, O., Leon, B., Ivan, L., Josef, S.: Learning and transferring mid-level image representations using convolutional neural networks. In: Conference on Computer Vision and Pattern Recognition (CVPR), pp. 1717–1724 (2014)
17. Andrej, K., George, T., Sanketh, S., Thomas, L., Rahul, S., Fei-Fei, L.: Large-scale video classification with convolutional neural networks. In: Conference on Computer Vision and Pattern Recognition (CVPR), pp. 1725–1732 (2014)
18. Alexandre, L.A.: 3D object recognition using convolutional neural networks with transfer learning between input channels. In: Menegatti, E., Michael, N., Berns, K., Yamaguchi, H. (eds.) Intelligent Autonomous Systems 13. AISC, vol. 302, pp. 889–898. Springer, Cham (2016). https://doi.org/10.1007/978-3-319-08338-4_64
19. Li, P., Ling, H., Li, X., Liao, C.: 3D hand pose estimation using randomized decision forest with segmentation index points. In: International Conference on Computer Vision (ICCV), pp. 819–827. IEEE (2015)
20. Taylor, J., Shotton, J., Sharp, T., Fitzgibbon, A.: The vitruvian manifold: inferring dense correspondences for one-shot human pose estimation. In: Computer Vision and Pattern Recognition (CVPR). IEEE (2012)
21. Madadi, M., Escalera, S., Baro, X., Gonzalez, J.: End-to-end global to local CNN learning for hand pose recovery in depth data. In: arXiv Preprint (2017)
22. Markus, O., Vincent, L.: DeepPrior++: improving fast and accurate 3D hand pose estimation. In: International Conference on Computer Vision (ICCV) Workshops. IEEE (2017)
23. Sun, X., Wei, Y., Liang, S., Tang, X., Sun, J.: Cascaded hand pose regression. In: Conference on Computer Vision and Pattern Recognition (CVPR), pp. 824–832. IEEE (2015)
24. Oberweger, M., Wohlhart, P., Lepetit, V.: Hands deep in deep learning for hand pose estimation. In: Computer Vision Winter Workshop (CVWW), pp. 21–30 (2015)
25. Deng, X., Yang, S., Zhang, Y., Tan, P., Chang, L., Wang, H.: Hand3D: hand pose estimation using 3D neural network. In: arXiv Preprint (2017)

26. Wan, C., Yao, A., Van Gool, L.: Hand pose estimation from local surface normals. In: Leibe, B., Matas, J., Sebe, N., Welling, M. (eds.) ECCV 2016. LNCS, vol. 9907, pp. 554–569. Springer, Cham (2016). https://doi.org/10.1007/978-3-319-46487-9_34

27. Zhou, X., Wan, Q., Zhang, W., Xue, X., Wei, Y.: Model-based deep hand pose estimation. In: International Joint Conference on Artificial Intelligence (IJCAI) (2016)

28. Wan, C., Probst, T., Van Gool, L., Yao, A.: Crossing nets: dual generative models with a shared latent space for hand pose estimation. In: Conference on Computer Vision and Pattern Recognition (CVPR). IEEE (2017)

29. Fourure, D., Emonet, R., Fromont, E., Muselet, D., Neverova, N., Tremeau, A., Wolf, C.: Multi-task, multi-domain learning: application to semantic segmentation and pose regression. Neurocomputing 1(251), 68–80 (2017)

30. Krejov, P., Gilbert, A., Bowden, R.: Guided optimisation through classification and regression for hand pose estimation. Comput. Vis. Image Underst. 155(2), 124–138 (2016)

31. Oberweger, M., Wohlhart, P., Lepetit, V.: Training a feedback loop for hand pose estimation. In: International Conference on Computer Vision (ICCV). IEEE (2015)

32. Xu, C., Govindarajan, L., Zhang, Y., Cheng, L.: Lie-X: depth image based articulated object pose estimation, tracking, and action recognition on lie groups. In: International Journal of Computer Vision (IJCV) (2016)

33. Neverova, N., Wolf, C., Nebout, F., Taylor, G.: Hand pose estimation through semi-supervised and weakly-supervised learning. In: arXiv Preprint (2015)

34. Tompson, J., Stein, M., Lecun, Y., Perlin, K.: Real-time continuous pose recovery of human hands using convolutional networks. ACM Trans. Graph. 33, 169 (2014)

Parallelized Contour Based Depth Map Coding in DIBR

Wenxin Yu[1,6](✉), Yibo Fan[2], Minghui Wang[3], Gang He[4],
Gang He[1](✉), Zhuo Yang[5], and Zhiqiang Zhang[1]

[1] Southwest University of Science and Technology, Mianyang, China
yuwenxin@swust.edu.cn, star_yuwenxin27@163.com,
cosfrist@live.cn
[2] State Key Laboratory of ASIC and System, Shanghai, China
[3] Waseda University, Tokyo, Japan
[4] Xidian University, Xi'an, China
[5] Guangdong University of Technology, Guangzhou, China
[6] Sichuan Civil-Military Integration Institute, Mianyang, China

Abstract. Depth map is a critical factor in Depth-Image-Based-Rendering system. Conventionally the depth map is encoded with the block-based method, such as H.264/AVC or some MVC methods. The traditional coding strategy cannot guarantee the quality of the boundary of the objects in the depth map. The distortion on boundary will cause unrecoverable distortions in the synthesized view. To guarantee the precision of the object boundary, we proposed a Contour Based Depth map Coding (CBDC) method. In the proposal, depth maps are divided into several layers in the depth dimension and further segmented into regions. Each region is regarded to be made up of contours and interiors. We adopted a lossless vectorized method to represent the contour and a lossy dynamical modeling method to represent the interior. This paper is based on our previous work, which is a preliminary framework of the CBDC. In this paper, two further contributions are illustrated. One is a data structure improvement of contour bypassing. Redundant contour segments are bypassed under our proposed rules. Experimental result shows that the bypassing strategy reduces 40% to 60% bit cost for the CBDC. Compared with other contour based coding strategy, the improved CBDC achieves higher quality of the synthesized view and better coding efficiency. Compared with block-based method, CBDC achieves much higher image quality (3–15 dB) in high bitrate scenarios. The other contribution is about the parallel process design of CBDC. The layer-region structure is with relatively low data dependency. Taking that advantage, we implemented and evaluated the parallel structure of CBDC. Experimental results show that the parallelized process speeds up by as much as 1.8 to 2.5 times when 2 to 4 process units are employed. Up to 90% parallel efficiency is achieved when 2 process units is employed.

Keywords: DIBR · Depth map · Contour coding · Parallel

© Springer Nature Switzerland AG 2018
R. Hong et al. (Eds.): PCM 2018, LNCS 11164, pp. 643–653, 2018.
https://doi.org/10.1007/978-3-030-00776-8_59

1 Introduction

1.1 DIBR System

In order to make a tradeoff between the huge data burden and the three dimensional visual experience, a Depth Image Based Rendering (DIBR) system [1] is proposed. Compared to the simulcast-based solution, the DIBR system can provide multiview video output in the receiver side by view synthesis, while only base views and depth views are transmitted or stored [2]. The depth map (the frame of the depth view video) is introduced into this system to make the view synthesis work. Although the data burden issue is solved by DIBR, a visual quality issue comes. Depth map and texture map take different and unequivalent effects during the view synthesis process. As the only component which represents the relationship between adjacent views, depth map is an essential factor to influence the visual quality of the synthesized virtual views. Each pixel on the depth map represents a distance between a point in the scene and the camera array. The depth map illustrates the objects' surface in the scene, and appears a piecewise map. In most cases, depth map is smoother than the texture map in objects' interior, and sharper on objects' boundary. According to those significant differences between depth maps and texture maps, we intent to propose another code method to take place of the block-based strategy, which is widely adopted in the texture map coding.

1.2 From Object Base to Contour Based

Object based coding strategy was once a hot topic in the MPEG-4's era [3–5]. It is believed that object based coding strategy achieves better coding efficiency if objects in a frame are segmented precisely. For the texture maps, object segmentation is a tough task. But it is much easier to retrieve objects' boundary in the depth map. In the depth estimation process (a practical way to get depth map), in which depth map is generated semi-automatically [6], hand-made edge maps are used to aid the segmentation in the depth map estimation. Sharp edges are already calculated in the depth estimation process and represented in depth maps. Even if some objects are not segmented correctly, from the point of coding and rebuilding an image, retrieving objects is not necessary. Geometry regions, instead of physical objects, are the real targets when we encode the depth map. When the regions are retrieved, the coding target can be divided into two part, contour and interior.

Regions are defined as a set of pixels with same or similar depth values in our previous work [7]. Therefore the contour coding part is critical in depth map coding [8]. Contour coding is a classical problem. As a widely used standard, TureType font [9] in computer system shows high coding efficiency and scalability.

In the framework of our previous work [7], contour and interior of a region are encoded in different ways. Contour coding is lossless, which guarantees that the boundary-caused distortion does not appear in the synthesized view. In this paper, the data structure of the contour representation is re-designed. The new data structure (applying contour bypassing) can earn up to 60% bit rate reduction and keep the quality of the synthesized views. Since a specific layer-region data structure is adopted in this proposal, the parallel process becomes effective and relatively easy to implement. The parallelization design of our coding method is also illustrated in this paper.

The rest of this paper is arranged as follows: A general system introduction of the contour based depth map coding is given in Sect. 2. Sections 3 and 4 focus on contour bypassing issue and parallelization issue respectively. The conclusion is finally given in Sect. 5.

2   Contour Based Depth Map Coding

The framework of the proposed Contour Based Depth map Coding (CBDC) is illustrated in our previous work [7]. The flow chart of the coding process is shown in Fig. 1.

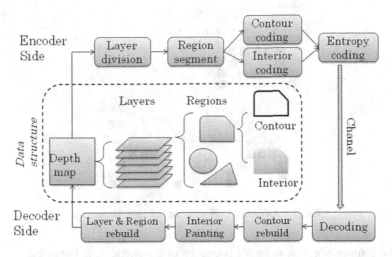

Fig. 1. Framework of CBDC

Some terms are also defined in that work [7], which are: Depth map, Layer, Region, Slice, Contour, Interior, Interior Paint. The contour coding is lossless. In the contour coding module, we adopt a vectorized coding strategy and retrieve straight lines as many as possible to achieve a high compression rate. The interior coding is lossy [10]. The interior reconstruction process in decoder is to paint the inside of the contour by a formula. The depth value of each pixel is a function of its coordinates [11] (Table 1).

Table 1. Straight-line-pixel Ratio*

Seq.	Balloons	Book_arrival	Lovebird1	Champagne Tower
SLP/ALL	92.7%	88.2%	81.8%	91.8%

* After applying the contour bypassing

3 Contour Bypassing

In our proposed CBDC, region is the basic coding unit and each region has an exclusive contour. On the one hand, the contour of each region must be represented precisely in decoder to guarantee the view synthesis quality. On the other hand, this data structure brings a large redundancy (see Fig. 2). In our previous works [7], since that is a preliminary framework, we code the entire contours for all regions, which is sufficient but not necessary. To distinguish different regions, only a single border line is necessary. It requires a strategy to determine which segment of contour can be bypassed when two regions are adjacent. The rules are as follows:

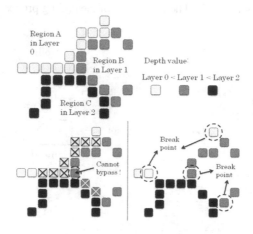

Fig. 2. An example of contour segment bypassing

1. If the contour segment is on the boundary of the frame, it is bypassed.
2. If a pixel on the contour has a neighbor pixel (8-connected) which is in a lower layer, it cannot be bypassed. Other pixels on the contour can be bypassed.

In the second rule, a lower layer denotes the layer in which the depth value is lower than it in the current layer. The rule means only the region in the higher layer should keep the contour for the adjacent layers. The contour segments which are bypassed are not coded in the encoder and not reconstructed in the decoder. An example is shown in Fig. 2. Assume we have 3 regions (A, B and C) which are adjacent to each other and they belongs to 3 different layers (0, 1 and 2) respectively. Region A in layer 0 has the lowest depth values and region C in layer 2 has the highest depth values. Our previous work just keeps the entire contour for all of them (upper left in Fig. 2). By applying the second rule, some pixels can be bypassed in region A and region B. Note that for the circled pixel in region B (lower left in Fig. 2), it cannot be bypassed. Although it is adjacent to a higher layer, it also has neighbor pixels from region A, which is in lower layer. The pixel which is next to the bypassed pixel is called break point (lower right in Fig. 2).

The statistics of the bit cost are illustrated in Table 2. Since half of the contour-pixels and all frame-boundary-pixels are bypassed, the bit cost of the contour is reduced by more than 50%. As the total system, the bit cost reduction rate (R.R.) is saved by more than 40%.

Table 2. Bit-cost per frame of CBDC (kbit)

*Seq.	w/o Bypass		w/Bypass			
	DIL8	DIL16	DIL8	R.R.%	DIL16	R.R.%
Book	682.2	336.0	291.8	57.2	164.4	43.6
Ball	471.1	222.9	189.0	59.9	112.3	40.6
Love	294.5	139.8	103.7	64.8	55.9	46.1
Tower	308.4	165.4	124.7	59.6	81.4	34.7
Avg.	439.0	216.0	177.3	59.6	103.5	41.6

* "Book" denotes to "Book arrival"; "Ball denotes to "Balloons"; "Love" denotes to "Lovebirds1"; "Tower" denotes to "Champagne tower"

We also evaluate the Bit-PSNR performance of the improved CBDC. The evaluating method refers to EL-Yamany's experiment [14]. For the bit cost, we calculate the sum of the bit cost of two related depth views. For the PSNR, we synthesize the reference virtual view with the original (uncoded) texture and depth views. Then we compare the reference virtual view with synthesized results (test virtual view) using compressed depth views and uncoded texture views. The results are shows in Tables 3 and 4.

Table 3. Bit cost per frame (kbit)

Seq.	CBDC				JM				
	DIL4	DIL8	DIL16	DIL32	QP16	QP20	QP24	QP28	QP32
Book8&10	366.43	291.78	164.44	93.54	600.95	403.65	281.24	195.68	126.98
Ball3&5	249.83	188.95	112.29	62.61	376.90	264.70	188.60	136.06	96.43
Love6&8	194.21	103.70	55.87	49.55	173.20	118.50	83.60	63.02	50.37
Tower37&39	169.73	124.66	81.41	52.00	408.50	318.20	249.00	195.13	146.87

Table 4. PSNR of the synthesized views (dB)

Seq.	CBDC				JM				
	DIL4	DIL8	DIL16	DIL32	QP16	QP20	QP24	QP28	QP32
Book9	54.891	51.756	47.490	43.239	54.5593	52.2364	50.296	48.2339	46.334
Ball4	56.483	51.117	47.277	44.044	52.380	50.946	49.711	48.351	47.026
Love7	52.459	48.398	44.873	44.306	54.000	52.073	50.646	49.595	48.037
Tower38	59.263	44.383	41.010	35.120	50.172	46.531	44.768	42.974	41.245

The comparison with Jager's work [12] is shown in Fig. 3. Since only the curves of "Book" and "Ball" are given in that paper, and no statistic data is given. In the experiment, we compare with the reference software of H.264/AVC (JM). CBDC can achieve higher PSNR in high bit cost scenarios. The coding efficiency of ours work is similar (in "Book") or higher (in "Ball") than the work in [12]. Since the DIL cannot be too large, our method cannot achieve lower bit rate. Compared with conventional block-based method (JM), CBDC achieves higher PSNR in most high bit cost scenarios. When the DIL is 4, our method is 3–15 dB better than JM in "Book", "Ball" and "Tower". The reason is that region size becomes small when DIL is short, and the interior can be fit by the model precisely. In the sequence "Love", depth distribution is unbalanced. The depth values are gathered in a small range (16 to 64).

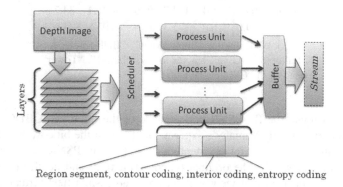

Region segment, contour coding, interior coding, entropy coding

Fig. 3. Layer-level parallel process

4   Parallelized Coding Structure

4.1   Layer-Region Structure

A significant advantage of the proposed method is the layer-region structure, which provides high parallelization capability to the algorithm design. Although the contour bypassing strategy brings data dependency to the decoder process, the contour reconstruction and interior painting can be parallelized without any data access conflict respectively. In the encoder side, because all regions can access the original depth map to bypass those contour segments simultaneously without any conflict, the parallelization is not influenced.

The depth is regarded as another dimension beside horizontal and vertical dimensions. The layer division's aim is to do the segmentation in the depth-dimension. Since applying different sampling rules, the physical surface of the objects is relatively smoother than the luminance or chrominance surfaces in mathematical expression. Thus dividing in the depth dimension maintains the continuity for the coding unit maximally. For the parallel process in block-based coding strategy (H.264/AVC) [15], data dependency between adjacent blocks is always the most critical issue. It is because as the basic coding unit, a block cannot represent an independent content segment.

When using block based coding method in depth map, the block-based segmentation breaks the good continuity and brings additional information entropy to the coding target.

In this work, the Depth Interval Length (DIL) of each layer is fixed to 4, 8, 16 or 32. The advantage of this setting is that the layer division is every fast. Although this fixed division strategy breaks the completeness of some physical objects, it makes a good tradeoff between the region continuity and computing complexity. Larger DIL will make the interior of each region contain more depth values and have more complex texture. Smaller DIL will increase the contours to be coded, which means higher bit cost and higher computation burden.

After the layer division, regions appear in each layer directly. On the one hand, compared with block-based division, pixels in the same region have much more correlation then the pixels in different regions and different layers. On the other hand, the DIL guarantees that the interior of the region is possible to be fit by a linear or quadrics model with low distortion.

In [12], the region is segmented by detect the large-depth-shifts. Although it has the advantage to maintain the completeness of objects, this segmentation process is with higher computing complexity and can hardly guarantee the closeness of the regions.

In the rest of this section, two kind of parallel strategies are proposed. They are adopted individually. To evaluate those strategies, not only total execution time, but also parallel efficiency are considered. The parallel efficiency in this paper is defined as follows (see Eq. 1).

$$Efficiency = \frac{PET}{N \times SET} \times 100\% \tag{1}$$

Where PET denotes to Parallel Execution Time, SET denotes to Serial Execution Time (only one PU is working), N denotes to the total number of PU.

4.2 Layer-Level Parallel

The data flow design of the parallel process in layer-level is unhindered. Each Process Unit (PU) can take the encoding task for one layer at one time. At the beginning of the task, each PU gets the depth interval of the layer and the entire global coding configuration. PU fetches the regions in the assigned layer, code the contour and interior for each region, and output the stream to the frame buffer.

All PU share a memory space (read only) which stores the original depth map, and access their exclusive memory (read and write) for other data. If the number of the PU is less than the number of layers, the scheduler will assign new task (layer) to the PU which finishes the last layer automatically. The frame buffer is employed to store and arrange the data output from all PU, and output the stream of one frame after all PU finish their tasks.

The workloads of the layers are different. Figure 3 illustrates the workload (execution time for single PU) of different layers in some test sequences.

Bars in Fig. 4 denote the execution time share of different layers. In one frame, the busy layer may cost as much as 10 times execution time than the light layer. The layer distributions of different sequences are also different. The workload in "Book" is much more even than it in "Love6". When only two PUs are working, the unbalanced workload does not influence the parallel efficiency very much. Because the scheduler always assigns the task to the idle PU, and the PU which is assigned with a busy layer can hardly get other tasks. When the PU number is increased, although total execution time is shortened, the parallel efficiency is lower. It is because the total waiting time is increased for the PUs which is assigned with light layers.

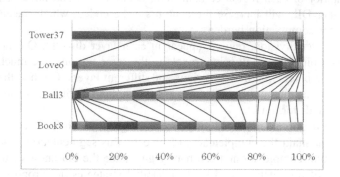

Fig. 4. Workload share for layers

4.3   Region-Level Parallel

In order to increase the parallel efficiency, tasks are always divided into smaller ones. In the proposed coding process, we can adopt a region-level parallel strategy to make a better workload balance and achieve a higher parallel efficiency. The region-level parallel flow chart is shown in Fig. 5.

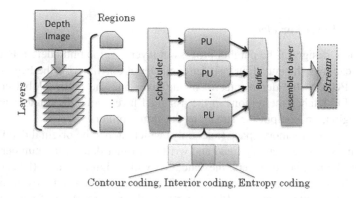

Fig. 5. Region-level parallel process

Before the region-level coding start, the region segmentation must be finished. Region segmentation is a labeling problem [13] and difficult to divide into parallel tasks. This solid part takes some limitation to the parallelization. The workload share

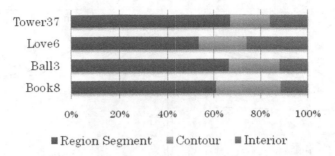

Fig. 6. Workload analysis in the layer process

for each step in the layer process is shown in Fig. 6. The region segmentation takes about 60% execution time on average in serial process. It indicates the maximum parallel-speedup for the region-level parallelization which is only about 167%.

Since the solid part takes most workload of the in-layer process, region-level parallelization is not effective. The number of the regions is large and the workload assigning overhead is relatively high for the process of each region. Thus region-level parallelization is not practical in CBDC.

4.4 Experimental Results of the Parallelization

We tested the speedup ratio and parallel efficiency of the layer-level parallelization. We implemented the CBDC based on Matlab and the parallelization is based on the Parallel Computing Toolbox. PU number is from 1 to 4. Tasks of layers are assigned to PU randomly. The average execution time is tested on our PC with Core i7 CPU (4 cores) and 16 GB memory.

Table 5. Speedup ratio and Parallel efficiency

Seq.	2PU		3PU		4PU	
	SR	EF	SR	EF	SR	EF
Book8	1.79	89.7%	2.27	75.7%	2.56	63.9%
Ball3	1.43	71.5%	2.03	67.8%	2.27	56.7%
Love6	0.98	48.8%	1.03	34.3%	1.08	27.0%
Tower37	1.81	90.4%	1.85	61.7%	2.23	55.8%

*SR: Speedup Ratio; EF: Parallel Efficiency
(see Eq. 1)

Speedup ratio and parallel efficiency are shown in Table 5 and Fig. 7. The speedup ratio increases linearly approximately when the number of the PU increases. It achieves up to 2.56 times accelerate when 4 PUs are employed. But the efficiency becomes lower when more PUs are employed. In Cheung's work [15], which is trying to use GPU (96 processors) to parallelize the intra-mode decision in H.264 and AVS, the speedup ratio is limited and the parallel efficiency (see Eq. 1) is low. Although the

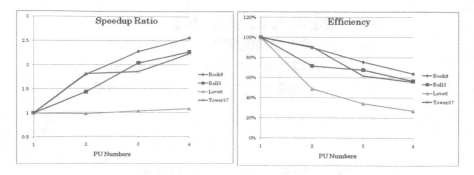

Fig. 7. Speedup ratio (upper) and parallel efficiency (lower)

coding workload is divided in to a large amount of small and balance tasks, the high data dependency among blocks limited the accelerating capacity.

In our proposal, the parallel efficiency is affected by the workload balance of the depth map. The efficiency is high in the sequence "Book" since the workload variance is relatively small. It is low in "Love" since the workloads are gathering in few layers.

5   Conclusion

A parallel processing of CBDC is proposed in this paper. Experimental results show that in most test sequences, CBDC can achieve high coding performance good parallel efficiency. The drawback of CBDC is the DIL is fixed after the coding process is started. It cannot divide the depth map into layers dynamically. For the depth map of which the depth value is gathered in a short range, such as "Love" in this paper, the interior painting module fails to fit the large size regions precisely. In the future work, we will develop a dynamical layer division strategy to fix this problem.

Acknowledgments. This research was supported by A Project [16ZA0131] which supported by Scientific Research Fund of Sichuan Provincial Education Department, [2018GZ0517] which supported by Sichuan Provincial Science and Technology Department, Science and Technology Planning Project of Guangdong Province [2017B010110007], [2018KF003] Supported by State Key Laboratory of ASIC & System.

References

1. Fehn, C.: Depth-image-based rendering (DIBR), compression and transmission for a new approach on 3D-TV. In: Proceedings of SPIE, Stereoscopic Displays and Virtual Reality Systems XI, vol. 5291, pp. 93–104 (2004)
2. Ugur, K., Liu, H., Lainema, J., Gabbouf, M., Li, H.: Parallel encoding - decoding operation for multiview video coding with high coding efficiency. In: 3DTV Conference 2007, 7–9 May 2007
3. Liu, D., Sun, X., Wu, F.: Intra prediction via edge-based inpainting. In: Data Compression Conference, DCC 2008, pp. 282–291 (2008)

4. Liu, D., Sun, X., Wu, F.: Edge-based inpainting and texture synthesis for image compression. In: IEEE International Conference on Multimedia and Expo, pp. 1443–1446 (2007)
5. Chaumont, M., Pateux, S., Nicolas, H.: Object-based video coding using a dynamic coding approach. In: International Conference on Image Processing, ICIP 2004, 24–27 October 2004
6. Liu, X., Chang, Y., Li, Z., Yuan, H.: A depth estimation method for edge precision improvement of depth map. In: International Conference on Computer and Communication Technologies in Agriculture Engineering (CCTAE), 12–13 June 2010
7. Wang, M., He, X., Jin, X., Goto, S.: Framework of contour based depth map coding system. In: The 2011 Pacific-Rim Conference on Multimedia (PCM 2011), Sydney, Australia, 20-22 December 2011
8. Yea, S., Vetro, A.: Multi-layered coding of depth for virtual view synthesis. In: Proceedings of Picture Coding Symposium (PCS 2009), Chicago, USA, May 2009
9. TrueType specification (Apple): http://developer.apple.com/fonts/TTRefMan/index.html
10. Wang, M., He, X., Jin, X., Goto, S.: Region-interior painting in contour based depth map coding system. In: 2011 IEEE 18th International Symposium on Intelligent Signal Processing and Communication Systems (ISPACS), ChiangMai, Thailand, 7–9 December 2011
11. Pinheiro, A., Ghanbari, M.: Piecewise approximation of contours through scale-space selection of dominant points. IEEE Trans. Image Process. **19**(6), 1442–1450 (2010)
12. Jager, F.: Contour-based segmentation and coding for depth map compression. In: Visual Communications and Image Processing, November 2011
13. Connected component labeling. http://en.wikipedia.org/wiki/Connected-component_labeling
14. El-Yamany, N.A., Ugur, K., Hannuksela, M.M., Gabbouj, M.: Evaluation of depth compression and view synthesis distortions in multiview-video-plus-depth coding systems. In: 3DTV-Conference: The True Vision - Capture, Transmission and Display of 3D Video (3DTV-CON), 7–9 June 2010
15. Cheung, N.M., Au, O.C., Kung, M.C., Wong, P.H.W., Liu, C.H.: Highly parallel rate-distortion optimized intra-mode decision on multicore graphics processors. IEEE Trans. Circuits Syst. Video Technol. **19**(11), 1692–1703 (2009)

Dual Subspaces with Adversarial Learning for Cross-Modal Retrieval

Yaxian Xia, Wenmin Wang[✉], and Liang Han

School of Electronic and Computer Engineering, Shenzhen Graduate School,
Peking University, Lishui Road 2199, Nanshan District, Shenzhen 518055, China
xiayaxian@pku.edu.cn,wangwm@ece.pku.edu.cn,1501213936@sz.pku.edu.cn

Abstract. Learning an effective subspace to calculate the correlation of items from different modalities is the core of cross-modal retrieval task, such as image, text or latent subspace. However, data in different modalities have imbalance and complementary relationships. Image contains abundant spatial information while text includes more background and context details. In this paper, we propose a model with dual parallel subspaces (visual and textual subspace) to better preserve modality-specific information. Triplet constraints are employed to minimize the semantic gap between items from different modalities with the same concept, while maximize that of concept-different image-text pair in corresponding subspace. Then we novelly combine adversarial learning with dual subspaces, which act as an interplay of two agents. The first agent, dual subspaces with similarity merging and concept prediction, aims to narrow the difference of data distributions from different modalities under the premise of concept invariance to fool the other agent, modality discriminator, which tries to distinguish image from text accurately. Extensive experiments on Wikipedia dataset and NUS-WIDE-10k dataset verify the effectiveness of our proposed model for cross-modal retrieval tasks, which outperforms the state-of-the-art methods.

Keywords: Dual subspaces · Similarity merging
Adversarial learning · Cross-modal retrieval

1 Introduction

Given an image, a person can find the corresponding text with similar meaning properly in a quick glance, and vice versa. But it can be quite difficult for a machine to understand the contents of images and sentences as well as the correlation between different modalities. The main challenge of cross-modal retrieval is that data from different modalities have inconsistent semantic distributions and learning the intrinsic correlation between them is quite elusive.

A large number of pioneering approaches have been proposed for breaking the heterogeneity gap between different modalities. The mainstream methods are

© Springer Nature Switzerland AG 2018
R. Hong et al. (Eds.): PCM 2018, LNCS 11164, pp. 654–663, 2018.
https://doi.org/10.1007/978-3-030-00776-8_60

learning the comparable representations in a single subspace. Most of the cross-modal retrieval methods learned a latent subspace to measure the correspondence of modality-specific items, such as CCA [14], Kernel-CCA [1], generalized multiview analysis (GMA)[7], etc. Some methods aligned multiple modalities in textual subspace [2] or visual subspace [4,9] only. Chen et al. [2] applied PLS to transfer the image features into text subspace. Word2VisualVec [4] learned to predict visual representations of textual items based on a DNN architecture.

However, data from different modalities contain diverse amount and types of information. For instance, image is more informative and contains abundant spatial details while natural language offers general descriptions as well as background and context information. Singly embedding modality-specific features visual subspace or textual subspace will lose plenty of details inevitably.

Different from methods above, dual parallel subspaces are constructed for better capturing domain-specific information in our proposed approach. In visual subspace, images are adjusted in the original feature space to adapt without losing too much image-specific details while texts are embedded into the visual subspace for follow-up alignment. In text subspace, texts are altered similarly for capturing more original textual information. Triplet constraints are employed to break the data distribution of the original space and to reconstruct that within the constraint of semantics, so that the modality-specific details are better preserved respectively.

In recent years, adversarial learning has developed rapidly. Generative adversarial networks(GANs) are a class of algorithms used unsupervised machine learning, in which the generative model aims to fool the discriminator to predict that synthetic images are real. GANs have been applied in a variety of fields of artificial intelligence since proposed by Goodfellow et al. [6], such as SRGAN [8], GAWWN [15], DCGANS [13], StackGAN [20], etc. Reed et al. [16] developed a GAN-based architecture to translate visual concepts from characters to pixels for automatic synthesis of realistic images from text. Peng et al. [10] proposed CM-GANs to identify the inter-modality and intra-modality correlation simultaneously for bridging the heterogeneity gap. For cross-modal retrieval task, adversarial learning can better learn the semantic distributions of input data, which is critical for eliminating the heterogeneity of different modalities. So in this paper, adversarial learning is utilized to optimize the cross-modal common representation learning.

Figure 1 illustrates the structure of our proposed model. The entire systems consists two training paths: multi-modal feature embedding in dual subspaces and adversarial learning. The first path maps different modality features into visual subspace and textual subspace respectively. Adversarial learning is applied for optimizing the distributions of modality-specific data in subspaces. The generator aims to minimize the heterogeneity gap and fool the discriminator to mistake image for text under the prerequisite of concept invariance while the discriminator is trained to distinguish image from text accurately.

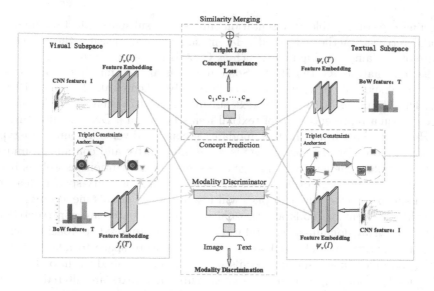

Fig. 1. Overview of proposed approach. The framework consists of two parts: Dual subspaces in parallel (visual and textual subspace) with triplet constraints. Adversarial learning acting as a minimax game: triplet loss and concept prediction aim at minimizing the differences of modality-specific distributions while ensuring the concept invariance after embedding. Modality discriminator is deployed to distinguish image from text accurately.

The contributions of this paper are as follows:

- We construct dual parallel subspaces (visual and textual subspace) to capture more domain-specific information. Then similarity merging is adopted to acquire the overall correlation of different media types, which can explore the complementary correlation between two semantic subspaces.
- Adversarial learning is employed to better capture semantic distributions and optimize common representations of modality-specific items. Our proposed model combine adversarial learning and dual subspaces creatively, which is proved to be effective by experimental results.

Extensive experiments on NUS-WIDE-10k and Wikipedia datasets have validated that our proposed approach outperforms the state-of-the-art methods. In this paper, Sect. 2 presents the architecture in details. Then we introduce the experiment settings and result analysis on two datasets in Sect. 3 while Sect. 4 concludes the paper.

2   Our Proposed Approach

In this section, we describe our proposed method with dual parallel subspaces based on adversarial learning for cross-modal retrieval in detail.

2.1   Visual Subspace

We attempt to find a 4096-dimension visual subspace where common representations of image and text are directly comparable.

Extracted image and text features are denoted as $I = \{i_1, i_2, ..., i_n\}$ and $T = \{t_1, t_2, ..., t_n\}$ respectively. $P_v^V = f_v(I)$ is utilized to adjust the semantic distributions of image in original 4096-dimension space without losing too much original image details while text features are embedded to visual subspace through $P_t^V = f_t(T)$ for further image-text alignment. f_v and f_t are embedding functions which are three-layer fully connected networks activated by ReLU.

Figure 2 depicts the training process of triplet loss for cross-modal retrieval task in visual and textual subspaces respectively, which aims to minimize the distance of anchor-positive pairs while maximizing that of anchor-negative pairs. In Fig. 2(a), image is considered as an anchor in visual space, triplet constraints can adapt the semantic distribution of texts by that of images. In this way, semantic gap between different media types can be minimized while original image-specific information are kept as much as possible.

(a) Illustration of triplet constraint in visual subspace   (b) Illustration of triplet constraint in textual subspace

Fig. 2. Triplet constraints for distribution optimization in dual subspaces.

In visual subspace, $\phi^V = (v_i, t_j^+, t_k^-)$ is the set of all possible triplets. The selected image and text with the same concept make a positive image/text pair (v_i, t_j^+), otherwise, a negative image/text pair (v_i, t_k^-). L2-norm is applied to calculated distances of embedded representation from different media types:

$$l_2^V(v, t) = \| f_v(v) - f_t(t) \|_2^2 \tag{1}$$

Triplet loss is adjusted to minimize $l_2^V(v_i, t_j^+)$ and maximize $l_2^V(v_i, t_k^-)$ while meeting an adding safety margin: $l_2^V(v_i, t_j^+) + \alpha < l_2^V(v_i, t_k^-), \alpha \in R^+$. The triplet loss in visual subspace is expressed as follows:

$$L_{tri}^V(\phi) = \sum_{\phi \in \Phi} max(0, l_2^V(v_i, t_j^+) - l_2^V(v_i, t_k^-) + \alpha) \tag{2}$$

It is worth noting that the negative pair is composed of a given query and a different modality item with different concept which has peak cosine similarity score with the query. Since the most similar item with different concept from the query can be distinguished, other items can obviously be discriminated.

2.2   Textual Subspace

For similar purpose, a textual subspace is constructed in parallel to visual subspace in our proposed method. The dimension of textual subspace is the same as that of text features T.

Text features are adjusted through $P_t^T = \psi_t(T)$ on the premise of dimension invariance while extracted image features are projected to textual subspace through $P_v^T = \psi_v(I)$. Three-layer fully connected networks activated by ReLU are applied for embedding.

We also obtained triplet constraints to optimize the distributions of the modality-specific items. In Fig. 2(b), text is regarded as an anchor, the semantic distribution of images are adapted for that of text. Thus semantic gap of image and text can be minimized without losing too much original text details.

Triples in textual subspace are represented as $\phi^T = (v_i, t_j^+, t_k^-)$. Distances of embedded representations of different media types are computed by L2-norm:

$$l_2^T(v,t) = \parallel \psi_v(v) - \psi_t(t) \parallel_2^2 \tag{3}$$

The overall triplet loss in textual subspace is calculated by the following equation:

$$L_{tri}^T(\phi) = \sum_{\phi \in \Phi} max(0, l_2^T(v_i, t_j^+) - l_2^T(v_i, t_k^-) + \alpha) \tag{4}$$

Where $\alpha \in R^+$ is the safety margin satisfying $l_2^T(v_i, t_j^+) + \alpha < l_2^T(v_i, t_k^-)$.

2.3   Similarity Merging

Inspired by [11], we merge the triplet loss functions in dual semantic subspaces to obtain the complementary and imbalance correlation of modality-specific data, which are fused with the following equation:

$$L_{tri} = \eta^V \cdot L_{tri}^V + \eta^T \cdot L_{tri}^T \tag{5}$$

Where η^V and η^T are hyperparameters representing weights of visual and text subspaces individually.

2.4   Concept Prediction

Concept prediction is deployed to guarantee the semantic invariance of items after feature embedding. Concepts are attached to instances in the set of image/text pairs correspondingly, which is represented as $c_i = \{c_{i1}, c_{i2}, ..., c_{im}\}$, where m is the number of semantic categories. If the ith instance belongs to the jth semantic category, $c_{ij} = 1$, otherwise $c_{ij} = 0$. The optimization objective is to minimize the differences between predicted concept p_c and the ground-truth concept c_i. L_c^V denotes the cross-entropy loss for concept prediction in visual subspaces:

$$L_c^V = -\frac{1}{N} \sum_{i=1}^{N} c_i \cdot (logp_c(f_v(I)) + logp_c(f_t(T))) \tag{6}$$

Where N is the number of instances in a batch. L_c^T is similar in textual subspace. The combination of concept invariance loss in dual subspaces is as follows:

$$L_c = L_c^V + L_c^T \tag{7}$$

2.5   Optimization: Adversarial Learning

In this subsection, we introduce the architecture of adversarial learning for optimizing the subspaces learning in detail.

Generative Model. The generator G is designed to learn the fitted joint distribution by modeling inter-modality correlation, which generates common representations (P^V and P^T mentioned in Sects. 2.1 and 2.2) in visual and text subspaces individually. The goal of G is to the minimize the differences between P^V and P^T while ensuring the concept invariance. In other words, the generator aims to minimize the heterogeneity gap between different modalities to fool the discriminator to mistake image for text, or in reverse. The generator is trained by descending its stochastic gradient with the following equation, while the discriminative model is fixed at this time:

$$L_G = \mu_c \cdot L_c + \mu_{tri} \cdot L_{tri} \tag{8}$$

where the hyper-parameters μ_c and μ_{tri} represent the contributions of the two terms.

Discriminative Model. As illustrated in Fig. 1, modality discriminator D is constructed to detect the modality of common representations as reliable as possible, which is divided to D^V for visual subspace and D^T for text subspaces. 3-layer feed-forward neural networks are applied for this modality detector:

$$L_{D^V} = -\frac{1}{N} \sum_{k=1}^{N} m_k \cdot (log D(f_v(I)) + log(1 - D(f_t(T)))) \tag{9}$$

Where m_k is the real modality distribution, which is expressed as a one-hot vector. $D(f_v(I))$ and $D(f_t(T))$ are the modality probability distributions of image and text detected by discriminators in visual subspace. L_{D^T} is similar for text subspaces. The overall modality discrimination aims to maximize the log-likelihood for correctly distinguishing image from text, by descending its stochastic gradient as follows, while the generator is fixed:

$$L_D = L_{D^V} + L_{D^T} \tag{10}$$

In our proposed method, generative model and discriminative model beat each other as a minimax game. Alternately training of Eqs. 9 and 10 can finally learn the optimal feature representation in dual subspaces.

3 Experiments

In this section, we describe our experimental methodology and results which validate the effectiveness of our model, compared with the existing state-of-the-art approaches.

3.1 Datasets

We report results on two widely-used datasets, Wikipedia and NUS-WIDE-10k.

(1) *NUS-WIDE-10k dataset*[1][3] : NUS-WIDE dataset is a web image dataset created by Lab for Media Search, which includes 269,648 images and the associated tags from Flickr and 81 concepts in alphabetical order. NUS-WIDE-10k is a subset of NUS-WIDE which is split into a training set with 8,000 images and a testing set with 2,000 images randomly.

(2) *Wikipedia dataset*[2][12]: Wikipedia dataset is generated from featured articles in Wikipedia, which consists of 2,866 image-text pairs with 10 most populated semantic categories, for instance, art and biology. The dataset is randomly split into a training set of 2,173 pairs and a testing set of 693 pairs.

3.2 Evaluation Scheme

For the evaluation of the results, mean Average Precision (mAP) is used on all returned results of the two datasets, which is applied to two retrieval tasks: retrieving text targets under a given image query *Img2Text* and retrieving image targets under a given text query *Text2Img*. The Average Precision (AP) of each query is defined as:

$$AP(q) = \frac{1}{R} \sum_{k=1}^{n} P(k)\delta_k \tag{11}$$

Where n is the number of instances in testing set and R is the number of instances relevant to the query in the retrieval set. $P(k)$ is the precision of the top k retrieved results. δ_k is set to be 1 if the returned item at rank k is relevant, otherwise, δ_k is set to be 0.

3.3 Experimental Results

The parameter α in Subsects. 2.1 and 2.2 is set to be 0.1 and the batch size is set to be 64. Since deep Convolutional Neural Network (CNN) is widely used as an effective image feature extractor, we apply CNN for feature extracting of input Image. The image features are 4096-dimension vectors extracted from the input image by the last fully connected layer (fc7) of VGGNet. Text instances are

---

[1] http://lms.comp.nus.edu.sg/research/NUS-WIDE.htm.
[2] http://www.svcl.ucsd.edu/projects/crossmodal/.

represented by Bag-of-Words (BoW) vectors with the TF-IDF weighting scheme. 3-layer fully-connected networks activated by ReLU are constructed to project extracted features from different modalities into visual and textual subspace individually. The learning rate is set as 0.001.

Table 1. The mAP score comparison on NUS-WIDE-10k and Wikipedia datasets

Methods	NUS-WIDE-10k dataset			Wikipedia dataset		
	Img2Text	*Text2Img*	*Average*	*Img2Text*	*Text2Img*	*Average*
CCA [14]	0.189	0.188	0.189	0.277	0.226	0.252
Corr-AE [5]	0.366	0.417	0.392	0.442	0.429	0.436
JRL [19]	0.426	0.376	0.401	0.479	0.428	0.454
Multimodal DBN [17]	0.201	0.259	0.230	0.204	0.183	0.194
ACMR [18]	0.544	0.538	0.541	0.509	0.431	0.470
MCSM [11]	–	–	–	0.516	0.458	0.487
Proposed Approach	**0.569**	**0.565**	**0.567**	**0.538**	**0.475**	**0.507**

Table 1 shows the mAP scores of our proposed approach as well as six compared approaches on NUS-WIDE-10k dataset and Wikipedia dataset. On the whole, our method outperforms both the traditional statistical correlation analysis methods (CCA [14], JRL [19]) and DNN-based methods (Corr-AE [5], ACMR [18] and MCSM [11]). This performance improvement clearly shows the advantage of our proposed method. While MSCM constructs two independent semantic spaces for each modality, the higher mAP scores of proposed method demonstrate the superiority of adversarial learning for cross-modal retrieval tasks. Our method is superior to ACMR based on GANs, which shows the effectiveness of constructing dual independent semantic subspaces for image and text respectively.

Table 2. Performance of cross-modal retrieval of our proposed method with dual subspaces, visual subspace only and textual space only

Methods	NUS-WIDE-10k dataset			Wikipedia dataset		
	Img2Text	*Text2Img*	*Average*	*Img2Text*	*Text2Img*	*Average*
Only visual subspace	0.572	0.516	0.544	0.527	0.465	0.496
Only textual subspace	**0.582**	0.494	0.538	0.531	0.465	0.498
Dual subspaces	0.569	**0.565**	**0.567**	**0.538**	**0.475**	**0.507**

Table 2 shows the performance of our proposed method with dual subspaces, visual subspace only and text subspace only. On Wikipedia dataset, our proposed method performs better than the method with only visual spaces or textual spaces. On NUS-WIDE-10k dataset, even though our proposed method with dual subspaces performs a little lower than our modal with only one subspace in *Img2Text* task, the average mAP score of proposed method with two subspaces is much higher than others.

Figure 3 shows two examples of *Text2Img* by our proposed method with dual subspaces. Row 1 and row 2 correspond to the samples from Wikipedia and NUS-WIDE-10k dataset. The first column represents the text queries and the second column contains the ground-truth images of queries. Column 3~7 are the top 5 retrieved images.

Fig. 3. Two examples of *Text2Img* by our proposed method on Wikipedia and NUS-WIDE-10k dataset.

From the results above, the separate contribution of each component in our proposed approach can be verified.

4   Conclusion

In this paper, we propose an approach which combines dual parallel semantic subspaces with adversarial learning for cross-modal retrieval. The retrieval results show that our proposed approach not only outperforms previous cross-modal retrieval works in a single subspace, but also exceeds state-of-the-art methods based on GANs. The combination of dual subspaces and adversarial learning turns out to be quite efficient. In the future work, we attempt to further improve the accuracy of our proposed method and generalize our method from image and text to more modalities.

Acknowledgement. This work is supported by Shenzhen Peacock Plan (20130408-183003656), Shenzhen Key Laboratory for Intelligent Multimedia and Virtual Reality (ZDSYS201703031405467) and National Natural Science Foundation of China (NSFC, No.U1613209).

References

1. Ballan, L., Uricchio, T., Seidenari, L., Bimbo, A.D.: A cross-media model for automatic image annotation. In: International Conference on Multimedia Retrieval, p. 73 (2014)
2. Chen, Y., Wang, L., Wang, W., Zhang, Z.: Continuum regression for cross-modal multimedia retrieval. In: IEEE International Conference on Image Processing, pp. 1949–1952 (2013)
3. Chua, T.S., Tang, J., Hong, R., Li, H., Luo, Z., Zheng, Y.T.: NUS-WIDE: a real-world web image database from National University of Singapore. In: Proceedings of ACM Conference on Image and Video Retrieval (CIVR 2009), Santorini, Greece, 8–10 July 2009
4. Dong, J., Li, X., Snoek, C.G.M.: Word2VisualVec: cross-media retrieval by visual feature prediction (2016)
5. Feng, F., Wang, X., Li, R.: Cross-modal retrieval with correspondence autoencoder, pp. 7–16 (2014)
6. Goodfellow, I.J., et al.: Generative adversarial nets. In: International Conference on Neural Information Processing Systems, pp. 2672–2680 (2014)
7. Jacobs, D.W., Daume, H., Kumar, A., Sharma, A.: Generalized multiview analysis: a discriminative latent space. In: IEEE Conference on Computer Vision and Pattern Recognition, pp. 2160–2167 (2012)
8. Ledig, C., et al.: Photo-realistic single image super-resolution using a generative adversarial network, pp. 105–114 (2016)
9. Norouzi, M., et al.: Zero-shot learning by convex combination of semantic embeddings. Eprint Arxiv (2013)
10. Peng, Y., Qi, J., Yuan, Y.: CM-GANs: cross-modal generative adversarial networks for common representation learning (2017)
11. Peng, Y., Qi, J., Yuan, Y.: Modality-specific cross-modal similarity measurement with recurrent attention network (2017)
12. Pereira, J.C., et al.: On the role of correlation and abstraction in cross-modal multimedia retrieval. IEEE Trans. Pattern Anal. Mach. Intell. 36(3), 521–35 (2014)
13. Radford, A., Metz, L., Chintala, S.: Unsupervised representation learning with deep convolutional generative adversarial networks. Comput. Sci. (2015)
14. Rasiwasia, N., et al.: A new approach to cross-modal multimedia retrieval. In: International Conference on Multimedia, pp. 251–260 (2010)
15. Reed, S., Akata, Z., Mohan, S., Tenka, S., Schiele, B., Lee, H.: Learning what and where to draw. New Republic (2016)
16. Reed, S., Akata, Z., Yan, X., Logeswaran, L., Schiele, B., Lee, H.: Generative adversarial text to image synthesis, pp. 1060–1069 (2016)
17. Srivastava, N., Salakhutdinov, R.: Learning representations for multimodal data with deep belief nets. In: ICML Workshop (2012)
18. Wang, B., Yang, Y., Xu, X., Hanjalic, A., Shen, H.T.: Adversarial cross-modal retrieval. In: ACM on Multimedia Conference, pp. 154–162 (2017)
19. Zhai, X., Peng, Y., Xiao, J.: Learning cross-media joint representation with sparse and semisupervised regularization. IEEE Trans. Circuits Syst. Video Technol. 24(6), 965–978 (2014)
20. Zhang, H., Xu, T., Li, H.: Stackgan: Text to photo-realistic image synthesis with stacked generative adversarial networks, pp. 5908–5916 (2016)

Special Session

Cross-Media Feature Learning Framework with Semi-supervised Graph Regularization

Tingting Qi[1,2], Hong Zhang[1,2(✉)], and Gang Dai[1,2]

[1] College of Computer Science and Technology, Wuhan University of Science and Technology, Wuhan, China
zhanghong_wust@163.com
[2] Hubei Province Key Laboratory of Intelligent Information Processing and Real-Time Industrial System, Wuhan, China

Abstract. With the development of multimedia data, cross-media retrieval has become increasingly important. It can provide the retrieval results with various types of media at the same time by submitting a query of any media type. In cross-media retrieval research, feature learning for different media types is a key challenge. In the existing graph-based methods, the similarity matrix denoting the affinities of data is usually constant matrix. Actually, calculating the similarity matrix based on the distances between the instances can more accurately represent the relevance of multimedia data. Furthermore, the dimensions of the original features are usually very high, which affects the computational time of algorithms. To address the above problems, we propose a novel feature learning algorithm for cross-media data, called cross-media feature learning frame-work with semi-supervised graph regularization (FLGR). FLGR calculates the similarity matrix based on the distances between the projected instances, which can not only accurately protect the relevance of multimedia data, but also effectively reduce the computational time of the algorithm. It explores the sparse and semi-supervised regularization for different media types, and integrates them into a unified optimization problem, which boosts the performance of the algorithm. Furthermore, FLGR studies the semantic information of the original data and further improve the retrieval accuracy. Compared with the current state-of-the-art methods on two datasets, i.e., Wikipedia, XMedia, the experimental results show the effectiveness of our proposed approach.

Keywords: Cross-media retrieval · Feature learning · The similarity matrix Semi-supervised

1 Introduction

With the development of multimedia data such as images, texts and video on the Internet, content-based multimedia retrieval has become increasingly important, and much research has been done for it [1–5]. Traditional content-based retrieval methods are generally focused on single-modality retrieval, such as image retrieval [6, 7], text retrieval [8, 9]. In this case, the query and retrieval results are of the same media type. However, single-modality retrieval cannot make full use of different media data. To

© Springer Nature Switzerland AG 2018
R. Hong et al. (Eds.): PCM 2018, LNCS 11164, pp. 667–677, 2018.
https://doi.org/10.1007/978-3-030-00776-8_61

handle this problem, cross-media retrieval has been proposed and become increasingly significant. It aims to take one type of data as query to retrieve relevant data objects of another type.

As for the cross-media feature learning of retrieval, correlation analysis among different media types is a key problem. Pairwise correlation can provide accurate relationship among objects of different media types. In the graph-based strategies, the similarity matrix plays a critical role, it can indicate the affinities of data. However, most existing works in the literature consider the similarity matrix is constant matrix [10–12]. In fact, calculating the similarity matrix based on the distances between the instances can more accurately represent the relevance of multimedia data. And others consider the similarity matrix is derived from the similarity between the raw data [10, 13], however, as we know, the dimensions of the original features are usually very high, which affects the computational time of algorithms.

To tackle above problems, in this paper, we propose a novel joint learning framework for the cross-media retrieval problem, which calculates the similarity matrix based on the distances between the projected data. On one hand, calculating the similarity matrix based on the distances between the data can accurately protect the relevance of multimedia data. On the other hand, the similarity matrix is derived from the similarity between the projected data, which reduce the dimension of the original features and thus our approach effectively reduces the computational time. FLGR also studies the semantic information focusing on high level abstraction of the original data, and jointing the correlation and semantic information [10] can further improve accuracy.

In summary, our approach has several distinct advantages.

(1) FLGR calculates the similarity matrix based on the distances between the projected data, which can more accurately protect the relevance of multimedia data and effectively reduces the computational time of the algorithm.
(2) FLGR explores the correlation and semantic information together. It can more accurate joint representation for cross-media data.
(3) FLGR explores both the labeled and unlabeled data of all the different media, unlabeled instances of different media types increase the diversity of training data and boost the performance of our approach.

The rest of this paper will be organized as follows. In Sect. 2, we introduce our learning framework. In Sect. 3, we present the objective function of FLGR, along with the optimization algorithm. Section 4 describes cross-media similarity measure in joint feature space. Experimental results are shown on two cross-media datasets in Sect. 5. Finally, we conclude the paper in Sect. 6.

2   Overview of Our Framework

We first present the formulation of problem definition for cross-media retrieval. The N labeled multimedia documents is denoted as $D = \left\{x_i^1, x_i^2, \ldots, x_i^S\right\}_{i=1}^{N}$, and each document contains data from S different modalities representing the same semantic. y_i represents the corresponding semantic label of x_i^p. $Y = [y_1, y_2, \ldots, y_N]^T$ denotes the corresponding the label matrix. $X_p = \left[x_1^p, x_2^p, \ldots, x_N^p\right] \in R^{d_p \times N}, p = 1, 2, \ldots, S$ represent the labeled data matrices from S modalities, respectively. $X_p^b = \left[x_1^p, x_2^p, \ldots, x_{N+E_p}^p\right] \in R^{d_p \times (N+E_p)}, p = 1, 2, \ldots, S$ represent the matrices of both labeled and unlabeled data, and the p-th modality has N labeled samples and E_p unlabeled samples embedded in the d_p dimensional space.

In the framework of our proposed algorithm, as shown in Fig. 1, first, we learn s transformation matrices M_p simultaneously for s original media types. Then all the original data x_i^p can be mapped into the common feature space as $f_i^p = M_p^T x_i^p$. Finally, to predict whether the query and retrieval results belong to the same semantic category, we will measure the similarity between two projected media objects $Sim\left(f_i^p, f_j^q\right)$ in the common space, using the KNN classifier and the Euclidian distances.

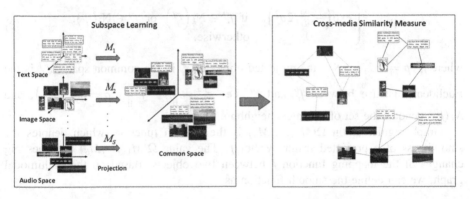

Fig. 1. Framework of our proposed method, which first projects the heterogeneous data of different modalities from their own feature spaces into a common space, then computes the cross-media similarity based on the nearest neighbors among all the media.

3   Subspace Learning

3.1   The Objective Function of FLGR

In existing cross-media retrieval methods [10, 14], correlation relationship between different types of media is widely used, i.e., if different media objects exist in the same document or are used jointly to represent for a given topic, so they should have the same semantic. And to explore jointly the correlation and semantic information, we further add the label consistent term into the loss function that is proposed by [10] as follows

$$loss(M_1, \ldots, M_S) = \sum_{p=1}^{S} \sum_{q=1}^{S} \left\| X_{ap}^T M_p - X_{aq}^T M_q \right\|_F^2 + \sum_{p=1}^{S} \left\| X_p^T M_p - Y_p \right\|_F^2 \quad (1)$$

Here, $\|A\|_F$ represents the Frobenius norm of matrix A, X_{ap} and X_{aq} denote two sets of media objects from p-th media and q-th media with the same labels.

If different modalities of data are related to the same content or topic, they should share similar semantics. This can be defined as the inter-modality similarity relationship. And within each single modality, the data objects with the neighborhood relationship should be close to each other in the common space. This can be defined as the intra-modality similarity relationship. We hope to preserve the inter-modality and intra-modality similarity relationship when learning the common space [13].

Graph regularization has been widely used to protect the similarity between multimedia data [10, 12, 13], where the similarity matrix expresses the similarity of multimedia data. We define the similarity matrix W_{ij} between the i-th modality and the j-th modality as follows

$$W_{ij} = \begin{cases} \exp\left(-z_{ij}^{pq}/2\sigma^2\right), & \text{if } f_i^p \in N_k\left(f_j^q\right) \text{ or } f_j^q \in N_k(f_i^p). \\ 0, & \text{otherwise.} \end{cases} \quad (2)$$

where $f_i^p, p = 1, 2, \ldots, S$ is the projected object of x_i^p in the common space, z_{ij}^{pq} is the Euclidean distance between f_i^p and f_j^q, i.e., $z_{ij}^{pq} = \left\| f_i^p - f_j^q \right\|^2$, $\sigma = max\left(z_{ij}^{pq}\right)$, and $N_k(f_i^p)$ denotes the set of k nearest neighbors of f_i^p.

Graph regularization $\Omega(M_1, \ldots, M_S)$ is the smooth function, which denotes the smoothness of a projected feature vector f. The value $\Omega(M_1, \ldots, M_S)$ punishes big changes of the mapping function f between two objects. Based on the multimodal graph, we can define the smooth function as

$$\Omega(M_1, \ldots, M_S) = \frac{1}{2} \sum_{i=1}^{\hat{N}} \sum_{j=1}^{\hat{N}} W_{ij} \left\| f_i^p - f_j^q \right\|^2 = Tr\left(FLF^T\right) \quad (3)$$

where $\hat{N}$ is the number of the total samples from all modalities, $F = \left(F_1^T, \ldots, F_S^T\right) = \left(M_1^T X_1^b, \ldots, M_S^T X_S^b\right)$ represents for all modalities of projected data in the common space. $L = D - W$ is the Laplacian matrix. Based on the above definition, Eq. (3) can be transformed into

$$\Omega(M_1, \ldots, M_S) = \sum_{p=1}^{S} \sum_{q=1}^{S} Tr\left(M_p^T X_p^b L_{pq} \left(X_q^b\right)^T M_q \right) \quad (4)$$

Based on the above, we obtain the objective function as follows

$$
\min_{M_1,\ldots,M_S} \alpha \sum_{p=1}^{S} \sum_{q=1}^{S} \left\| X_{ap}^T M_p - X_{aq}^T M_q \right\|_F^2 + \lambda_1 \sum_{p=1}^{S} \left\| M_p \right\|_{2,1}
$$
$$
+ \lambda_2 \sum_{p=1}^{S} \sum_{q=1}^{S} Tr\left(M_p^T X_p^b L_{pq} \left(X_q^b \right)^T M_q \right) + \lambda_3 \sum_{p=1}^{S} \left\| X_p^T M_p - Y_p \right\|_F^2
$$

(5)

3.2 Optimization of FLGR

Let $\varphi(M_p)$ denote the objective function in (5). Differentiating $\varphi(M_p)$ with respect to M_p and setting it to zero, we have the following equation

$$
\frac{\partial \varphi}{\partial M_p} = \alpha \sum_{q \neq p} X_p \left(X_p^T M_p - X_q^T M_q \right) + \lambda_1 R_p M_p
$$
$$
+ \lambda_2 \sum_{q \neq p} X_p^b L_{pq} \left(X_q^b \right)^T M_q + \lambda_2 X_p^b L_{pp} \left(X_p^b \right)^T M_p + \lambda_3 X_p \left(X_p^T M_p - Y_p \right) = 0
$$

(6)

where $R_p = Diag(r_p)$, where r_p represents an auxiliary vector of the $l_{2,1}$-norm [13]. Equation (6) can be rewritten as

$$
\left(\alpha \sum_{q \neq p} X_p X_p^T + \lambda_1 R_p + \lambda_2 X_p^b L_{pp} \left(X_p^b \right)^T + \lambda_3 X_p X_p^T \right) M_p
$$
$$
= \left(\alpha \sum_{q \neq p} X_p X_q^T - \lambda_2 \sum_{q \neq p} X_p^b L_{pq} \left(X_q^b \right)^T \right) M_q + \lambda_3 X_p Y_p
$$

(7)

Next, we propose an optimization approach to minimize the objective function (5) via solving the above linear system problem. The general procedure of this method is that we initialize M_p as identity matrix first, then, in each iteration, we calculate $\left\{ M_1^{t+1}, \ldots, M_p^{t+1} \right\}$ under the condition that $M_1^t, \ldots, M_p^t$ are given. Detailed steps will be shown in Algorithm 1.

Algorithm 1. Cross-Media Feature Learning Framework with Semi-Supervised Graph Regularization (FLGR)

Input: The matrix of both labeled and unlabeled data $X_p^b \in R^{d_p \times (N+E_p)}$; The matrix of labeled data $X_p \in R^{d_p \times N}$; The matrix of labels $Y \in R^{N \times c}$.

Output: The projection matrices $M_p \in R^{d_p \times c}, p = 1, 2, ..., S$.

Initialize $M_p^0, p = 1, 2, ..., S$ as identity matrix and set $t = 0$;

Repeat:

1. Compute the graph Laplacian matrix L^t according to $M_p^t \in R^{d_p \times c}, p = 1, 2, ..., S$;

2. Update M_p^{t+1} according to following equation:

$$M_p^{t+1} = \left(\alpha \sum_{q \neq p} X_p X_p^T + \lambda_1 R_p + \lambda_2 X_p^b L_{pp}^t \left(X_p^b \right)^T + \lambda_3 X_p X_p^T \right)^{-1}$$

$$\left(\left(\alpha \sum_{q \neq p} X_p X_q^T - \lambda_2 \sum_{q \neq p} X_p^b L_{pq}^t \left(X_q^b \right)^T \right) M_q^t + \lambda_3 X_p Y_p \right)$$

$$(8)$$

3. $t = t + 1$.

until Convergence

In the process of optimization approach, the iteration continues until convergence. In our experiment, the iteration stops when the ratio change between two iterations is less than 1%. In practice, the iteration only repeats several rounds before convergence.

4   Cross-Media Similarity Measure in Joint Feature Space

Until now, we have learned S projection matrices M_S for the original low-level features of multiple media types, and use them we can project all of the data point x_i^p into the common space $f_i^p = M_p^T x_i^p$. Then, we need to explore how to measure the cross-media similarity in the constructed common space. The cross-media similarity is defined as the marginal probability. The probability shows the semantic similarity of the two media objects regardless of their media types. We define the marginal probability of f_i^p and f_j^q that is proposed by [10] as

$$Sim\left(f_i^p, f_j^q \right) = P\left(y_i = y_j | f_i^p, f_j^q \right) = \sum_l p(y_i = l | f_i^p) p\left(y_j = l | f_j^q \right) \quad (9)$$

where $y_i (y_j)$ represents the label of $f_i^p \left(f_j^q \right)$, $p(y_i = l | f_i^p)$ represents the probability of f_i^p belonging to category l. $p(y_i = l | f_i^p)$ that is proposed by [10] is defined as

$$p(y_i = l|f_i^p) = \frac{\sum\limits_{f \in N_k(f_i^p) \wedge y=l} \sigma(\|f_i^p - f\|_2)}{\sum\limits_{f \in N_k(f_i^p)} \sigma(\|f_i^p - f\|_2)} \tag{10}$$

$N_k(f_i^p)$ represents the k-nearest neighbors of media f_i^p in training set, and y represents the label of media object f. $\sigma(z) = (1 + \exp(-z))^{-1}$ is the sigmoid function.

5 Experiments

5.1 Datasets and Parameter Setting

In this experiment, we used two real-world datasets to evaluate cross-media retrieval performance, i.e., Wikipedia and XMedia.

Wikipedia Dataset [15]: It is chosen from the Wikipedia's featured article. And it contains 2866 text-image pairs, which belong to ten different categories. In the experiment, we randomly select 2173 text-image pairs for training, the remaining 693 text-image pairs to test.

XMedia Dataset [10]: It contains five media types, i.e., 5000 texts, 5000 images, 500 videos, 1000 audios and 500 3D models. This dataset is divided into 20 categories, and each category has 600 media data. And this dataset is randomly divided into a training set of 9600 media data and a testing set of 2400 media data.

As we have seen the objective function in Eq. (5) involves four parameters $\lambda_1, \lambda_2, \lambda_3$ and α. λ_1 is the weighting parameter of the $l_{2,1}$-norms, λ_2 is the weighting parameter of the multimodal graph regularization, λ_3 is the parameter of the label consistent term, and α is the parameter of the loss function of cross-media regularization among multiple media. We tune them from $\{0.001, 0.01, 0.1, 1, 10, 100, 1000\}$ by cross validation. For the compared methods, we tune their parameters according to the corresponding literature.

5.2 Evaluation Metrics and Comparison Methods

In the experiment, we adopt mean average precision (MAP) [16] and precision-recall (PR) curve [16] to evaluate the performance of the algorithm. MAP for a set of queries is the mean of the average precision (AP) for each query. The larger the MAP, the better the performance. The calculation formula of AP is as follows

$$AP = \frac{1}{L} \sum_{r=1}^{N} P(r)\delta(r) \tag{11}$$

where L is the number of related items in the retrieved set; N is the total number of results returned by the query; $P(r)$ indicates the accuracy of the top r retrieved documents; $\delta(r) = 1$ if the result of the return is relevant and $\delta(r) = 0$ otherwise.

We compare our proposed FLGR algorithm with five state-of-the-art methods, which are summarized as follows.

(1) JGRHML [11]. It explores the heterogeneous metric, which can measure the content similarity between different media types.
(2) CMCP [17]. It is able to propagate the correlation between heterogeneous modalities, and deal with positive correlation and negative correlation simultaneously between media objects of different modalities.
(3) HSNN [18]. It could compute the similarity between media objects with different media types, which is further regarded as the weak ranker.
(4) JFSSL [13]. It can jointly deal with the measure of relevance and coupled feature selection in a joint learning framework.
(5) JRL [10]. It integrates the sparse and semisupervised regularization for different media types into one unified optimization problem,which is able to jointly exploit the pairwise correlation and semantic information.

5.3   Performance Comparisons to Rivals

In this section, we compare our proposed FLGR with five state-of-the-art methods for cross-media retrieval. It's important to note that A → B means that media A serves as the query and the results are media B in all experiments. Besides, note that each image feature is a 4,096-dimensional CNN feature vector, and each text feature is a 3,000-dimensional bag of words vector in our experiment. Particularly, in the XMedia dataset, each audio is denoted by a 29-dimensional MFCC feature vector, each video is denoted by a 4,096-dimensional CNN feature vector and each 3D model is denoted by the concatenated 4,700-dimensional vector of a set of Light Field descriptors. We conduct all methods in the experiment with same feature vectors for fair comparison. Due to the limitation of the pages, we only search for images, text, and video on the XMedia dataset.

Tables 1 and 2 show the performance of different methods based on the MAP scores of cross-media retrieval tasks on the Wikipedia dataset and on the XMedia dataset respectively. Compared with the state-of-the-art methods, our proposed FLGR improves the average MAP from 0.477 to 0.497 and from 0.641 to 0.667, respectively.

The corresponding precision-recall (PR) curve is plotted in Fig. 2 on the Wikipedia dataset and in Fig. 3 on the XMedia dataset. It can be seen that our proposed FLGR attains higher precision at most recall levels, outperforming those of compared methods.

Table 1. MAP comparison of different methods on the Wikipedia dataset

Dataset	Task	JGRHML	CMCP	HSNN	JFSSL	JRL	FLGR
Wikipedia dataset	Image→Text	0.250	0.388	0.450	0.493	0.513	**0.532**
	Text→Image	0.194	0.351	0.402	0.429	0.440	**0.461**
Average		0.222	0.370	0.426	0.461	0.477	**0.497**

Table 2. MAP comparison of different methods on the XMedia dataset

Dataset	Task	JGRHML	CMCP	HSNN	JFSSL	JRL	FLGR
XMedia dataset	Image→Text	0.458	0.727	0.742	0.876	0.901	**0.913**
	Image→Video	0.289	0.510	0.487	0.460	0.510	**0.542**
	Text→Image	0.367	0.723	0.756	0.882	0.909	**0.933**
	Text→Video	0.338	0.439	0.423	0.460	0.519	**0.567**
	Video→Image	0.251	0.485	0.486	0.436	0.500	**0.516**
	Video→Text	0.339	0.399	0.395	0.430	0.509	**0.532**
Average		0.340	0.547	0.548	0.591	0.641	**0.667**

As the MAP scores and the PR curves show, our proposed method has significant advantages because it calculates similarity matrix based on the distances between the projected data and thus can not only more precisely protect the relevance of multimedia data, but also effectively reduce the computational time of the algorithm. Furthermore, learning semantic information boosts the performance of the algorithm.

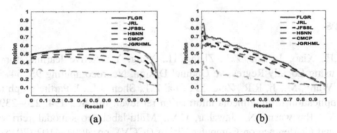

Fig. 2. Precision recall curves of cross-media retrieval on Wikipedia dataset. (a) Image→Text. (b) Text→Image.

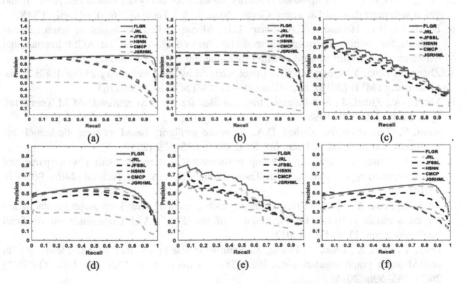

Fig. 3. Precision recall curves of cross-media retrieval on XMedia dataset (a) Image→Text. (b) Text→Image. (c) Image→Video. (d) Video→Image. (e) Text→Video (f) Video→Text.

6 Conclusions

In this paper, we have proposed a FLGR algorithm to explore the correlation between multimedia data. It learns the similarity matrix, which is calculated based on the distances between the projected data. And it explores the sparse and semi-supervised regularization for different media types, and integrates them into a unified optimization problem. Furthermore, FLGR studies the semantic information of the original data and further improve the retrieval accuracy. The experiments on two cross-media datasets show the effectiveness of our proposed approach, compared with the state-of-the-art methods. In the future, we intend to further optimize the graph regularization term to better protect the similarity of cross-media data, and we plan to apply FLGR to more applications.

Acknowledgment. This research is supported by the National Natural Science Foundation of China (No. 61373109).

References

1. Feng, Y.F., Xiao, J., Zha, Z.J., Zhang, H., Yang, Y.: Active learning for social image retrieval using Locally Regressive Optimal Design. Neurocomputing **95**, 54–59 (2012)
2. Gao, Y., Wang, M., Ji, R.R., Zha, Z.J., Zha, Z.J., Shen, J.L.: k-Partite graph reinforcement and its application in multimedia information retrieval. Inf. Sci. **194**, 224–239 (2012)
3. Ranjan, V., Rasiwasia, N., Jawahar, C.V.: Multi-label cross-modal retrieval. In: IEEE International Conference on Computer Vision (ICCV), pp. 4094–4102 (2015)
4. Hua, Y., Wang, S., Liu, S., Cai, A., Huang, Q.: Cross-modal correlation learning by adaptive hierarchical semantic aggregation. IEEE Trans. Multimed. (TMM) **18**(6), 1201–1216 (2016)
5. Peng, Y., Ngo, C.-W.: Clip-based similarity measure for query-dependent clip retrieval and video summarization. IEEE Trans. Circuits Syst. Video Technol. **16**(5), 612–627 (2006)
6. Escalante, H.J., Hérnadez, C.A., Sucar, L.E., Montes, M.: Late fusion of heterogeneous methods for multimedia image retrieval. In: Proceedings of the 1st ACM International Conference on Multimedia Information Retrieval, pp. 172–179 (2008)
7. Zhang, J., Peng, Y.: Query-adaptive image retrieval by deep weighted hashing. IEEE Trans. Multimed. (TMM) (2018). https://doi.org/10.1109/TMM.2018.2804763
8. Moffat, A., Zobel, J.: Self-indexing inverted files for fast text retrieval. ACM Trans. Inf. Syst. **14**(4), 349–379 (1996)
9. Amir, S., Tanasescu, A., Zighed, D.A.: Sentence similarity based on semantic kernels for intelligent text retrieval. J. Intell. Inf. Syst. **48**(3), 675–689 (2017)
10. Zhai, X., Peng, Y., Xiao, J.: Learning cross-media joint representation with sparse and semisupervised regularization. IEEE Trans. Circuits Syst. Video Technol. **24**(6), 965–978 (2014)
11. Zhai, X., Peng, Y., Xiao, J.: Heterogeneous metric learning with joint graph regularization for cross-media retrieval. In: Proceedings of the 27th AAAI Conference on Artificial Intelligence, pp. 1198–1204 (2013)
12. Peng, Y., Zhai, X., Zhao, Y., Huang, X.: Semi-supervised cross-media feature learning with unified patch graph regularization. IEEE Trans. Circuits Syst. Video Technol. (TCSVT) **26**(3), 583–596 (2016)

13. Wang, K., He, R., Wang, L., Wang, W., Tan, T.: Joint feature selection and subspace learning for cross-modal retrieval. IEEE Trans. Pattern Anal. Mach. Intell. **38**(10), 2010–2023 (2016)
14. Li, D., Dimitrova, N., Li, M., Sethi, I.: Multimedia content processing through cross-modal association. In: Proceedings of the ACM International Conference on Multimedia, pp. 604–611 (2003)
15. Rasiwasia, N., et al.: A new approach to cross-modal multimedia retrieval. In: Proceedings of the ACM International Conference on Multimedia (ACM-MM), pp. 251–260 (2010)
16. Rasiwasia, N., et al.: A new approach to cross-modal multimedia retrieval. In: Proceedings of the 18th ACM International Conference on Multimedia, pp. 251–260 (2010)
17. Zhai, X., Peng, Y., Xiao, J.: Cross-modality correlation propagation for cross-media retrieval. IEEE Int. Conf. Acoust. **22**(10), 2337–2340 (2012)
18. Zhai, X., Peng, Y., Xiao, J.: Effective heterogeneous similarity measure with nearest neighbors for cross-media retrieval. In: Schoeffmann, K., Merialdo, B., Hauptmann, A.G., Ngo, C.-W., Andreopoulos, Y., Breiteneder, C. (eds.) MMM 2012. LNCS, vol. 7131, pp. 312–322. Springer, Heidelberg (2012). https://doi.org/10.1007/978-3-642-27355-1_30

A Rapid Scene Depth Estimation Model Based on Underwater Light Attenuation Prior for Underwater Image Restoration

Wei Song[1], Yan Wang[1], Dongmei Huang[1(✉)],
and Dian Tjondronegoro[2(✉)]

[1] College of Information Technology, Shanghai Ocean University,
Shanghai, China
{wsong,dmhuang}@shou.edu.cn, yanwang9310@163.com
[2] Southern Cross University, Gold Coast, Australia
Dian.tjondronegoro@scu.edu.au

Abstract. Underwater images present blur and color cast, caused by light absorption and scattering in water medium. To restore underwater images through image formation model (IFM), the scene depth map is very important for the estimation of the transmission map and background light intensity. In this paper, we propose a rapid and effective scene depth estimation model based on underwater light attenuation prior (ULAP) for underwater images and train the model coefficients with learning-based supervised linear regression. With the correct depth map, the background light (BL) and transmission maps (TMs) for R-G-B light are easily estimated to recover the true scene radiance under the water. In order to evaluate the superiority of underwater image restoration using our estimated depth map, three assessment metrics demonstrate that our proposed method can enhance perceptual effect with less running time, compared to four state-of-the-art image restoration methods.

Keywords: Underwater image restoration · Underwater light attenuation prior
Scene depth estimation · Background light estimation
Transmission map estimation

1 Introduction

Underwater image restoration is challenging due to complex underwater environment where images are degraded by the influence of water turbidity and light attenuation [1]. Compared with green (G) and blue (B) lights, red (R) light with the longer wavelength is the most affected, thus underwater images appear blue-greenish tone. In our previous work [2], the model of underwater image optical imaging and the light selective attenuation are presented in detail.

An image restoration method recovers underwater images by considering the basic physics of light propagation in the water medium. The purpose of restoration is to deduce the parameters of the physical model and then recover the underwater images by reserved compensation processing. A simplified image formation model (IFM) [3–6] is

© Springer Nature Switzerland AG 2018
R. Hong et al. (Eds.): PCM 2018, LNCS 11164, pp. 678–688, 2018.
https://doi.org/10.1007/978-3-030-00776-8_62

often used to approximate the propagation equation of underwater Atmospheric scattering, can be shown as:

$$I_\lambda(x) = J_\lambda(x)t_\lambda(x) + (1 - t_\lambda(x))B_\lambda \qquad (1)$$

where x is a point in the image, λ represents RGB lights in this paper, $I_\lambda(x)$ and $J_\lambda(x)$ are the hazed image and the restored image, respectively, B_λ is regarded as the background light (BL), $t_\lambda(x)$ is the transmission map (TM), which is a function of both λ and the scene–camera distance $d(x)$ and can be expressed as:

$$t_\lambda(x) = e^{-\beta(x)d(x)} = Nrer(\lambda)^{d(x)} \qquad (2)$$

where $e^{-\beta(x)}$ can be represented as the normalized residual energy ratio $Nrer(\lambda)$, which depends on the wavelength of one channel and the water type in reference to [7].

In the IFM-based image restoration methods, a proper scene depth map is the key for background light and transmission map estimation. Since the hazing effect of underwater images caused by light scattering and color change is similar to the fog effect in the air, He's Dark Channel Prior (DCP) [3] and its variants [4, 8, 9] were used for depth estimation and restoration. Li et al. [10] estimated the background light by the map of the maximum intensity prior (MIP) to dehaze G-B channel and used Gray-World (G-W) theory to correct the R channel. However, Peng et al. [6] found the previous background light estimation methods were not robust for various underwater images, and they proposed a method based on the image blurriness and light absorption to estimate more accurate background light and scene depth to restore color underwater image precisely.

Learning-based methods for underwater image enhancement have been taken into consideration in recent years. Liu et al. [11] proposed the deep sparse non-negative matrix factorization (DSNMF) to estimate the image illumination to achieve image color constancy. Ding et al. [12] estimated the depth map using the Convolutional Neural Network (CNN) based on the balanced images that were produced by adaptive color correction. Although the above methods can obtain the scene depth map and enhance the underwater image, the deep learning is time consuming. A linear model to predict the scene depth of hazy images based on color attenuation prior is proposed by Zhu et al. [13]. This model trained by supervised learning method and its expression is as follows:

$$d(x) = \theta_0 + \theta_1 v(x) + \theta_2 s(x) + \epsilon(x) \qquad (3)$$

where $d(x)$ is the scene depth, $v(x)$ and $s(x)$ are the brightness and saturation components, respectively, $\epsilon(x)$ is a Gaussian function with zero mean and the standard deviation value σ. The model can successfully recover the scene depth map of outdoor hazed images. Unfortunately, it is not suitable for underwater images. Figure 1 shows an example image with the false depth map and invalid recovered image.

Fig. 1. An overview of Zhu et al.'s method. (a) Original underwater image; (b) Estimated depth map where the whiter the father; (c) Restored image.

In this paper, we reveal underwater light attenuation prior (ULAP) that the scene depth increases with the higher value of the difference between the maximum value of G and B lights and the value of the R light. On the basis of the ULAP and annotated scene depth data, we train a linear model of scene depth estimation. With the accurate scene depth map, BL and TM are easily estimated, and then underwater images are restored properly. The details of the scene depth learning model and the underwater image restoration process are presented in Sects. 2 and 3, respectively. In Sect. 4, we evaluate our results by comparing with four state-of-the-art image restoration methods from the perspectives of quantitative assessment and complexity. The conclusion is shown in Sect. 5.

2 Scene Depth Model Based on ULAP

2.1 Underwater Light Attenuation Prior

On account of little information about underwater scenes, restoring a hazed underwater image is a difficult task in computer vision. But, the human can quickly recognize the scene depth of the underwater image without any auxiliary information. When we explore a robust background light estimation, the farthest point in the depth map corresponding to the original underwater image is often considered as the background light candidate. With the light attenuation underwater, depending on the wavelength where the energy of red light is absorbed more than that of green and blue lights, the most intensity difference between R light and G-B light is used to estimate the background light [9, 10]. The rule motivates us to conduct experiments on different underwater images to discover an effective prior for single underwater image restoration.

After examining a large number of underwater images, we find the Underwater Light Attenuation Prior (ULAP), that is the difference between the maximum value of G-B intensity (simplified as MVGB) and the value of R intensity (simplified as VR) in one pixel of the underwater image is very strongly related to the change of the scene depth. Figure 2 gives an example with a typical underwater scene to show the MVGB, VR and the difference vary along with the change of the scene within different depths. As illustrated in Fig. 2(a), we select three regions as the test data from the close scene to the far scene and show the corresponding close-up patches in the right side. It is observed

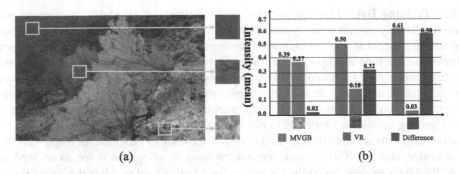

Fig. 2. The scene depth is positively correlated with the difference between MVGB and VR. (a) Original underwater image; (b) Three close-up patches of the close scene, moderately-close scene, relatively-far scene and their corresponding histograms, respectively.

in the left histogram of Fig. 2(b), in the close region, the MVGB and VR is relatively moderate and the difference is close to zero; In the middle histogram of Fig. 2(b), the MVGB in the moderately-close region increases while with the farther depth of the scene, the VR decreases at the same time, producing a higher value of the difference. Furthermore, in the utmost scene, the component of the red light remains nothing much due to the significant attenuation, the MVGB improves remarkably, and the difference is drastically higher than that in other patches in the right histogram of Fig. 2(b). In all, when the scene goes to a far region, the MVGB increase and the VR decreases, which leads to a positive correlation between the scene depth and the difference between MVGB and VR.

2.2 Scene Depth Estimation

Based on the ULAP, we define a linear model of the MVGB and VR for the depth map estimation as follows:

$$d(x) = \mu_0 + \mu_1 m(x) + \mu_2 v(x) \tag{4}$$

Where x represents a pixel, $d(x)$ is the underwater scene depth at point x, $m(x)$ is the MVGB, $v(x)$ is the VR.

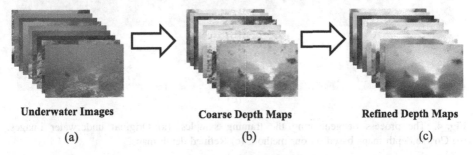

Underwater Images   **Coarse Depth Maps**   **Refined Depth Maps**

(a)   (b)   (c)

Fig. 3. The process of generating the training samples. (a) Original underwater images; (b) Coarse depth maps; (c) Refined depth maps by the guided filter [14].

The Training Data. In order to learn the coefficients μ_0, μ_1 and μ_2 accurately, we need relatively correct training data. In reference with the depth map estimation proposed by Peng et al. [6], we computed the depth maps of different underwater images based on the light absorption and image blurriness. 500 depth maps of underwater images were gained following by [6], and some maps exist obvious estimation errors (e.g., a close-up fish is white in the depth) were discarded. Hence 100 fully-proper depth maps are well-chosen by the final manual selection. The process of generating the training samples is illustrated in Fig. 3. Firstly, for the 500 underwater images, the depth map estimation method is used to obtain the corresponding depth maps with the same size, and then 100 accurate depth maps from the above maps are ascertained as the final training data. Secondly, the guided filter [14] is used to refine the coarse depth maps and the radius of the guided filter is set as 15 in this paper. Finally, 100 accurate depth maps verified by manually-operated selection according to the perception of the scene depths. The prepared dataset has a total of 24 million points of depth information, which are called reference depth maps (RDMs).

Coefficients Learning. Based on the reference depth maps (RDMs), Pearson correlation analysis for the MVGB and VR is firstly run. The range of Pearson correction coefficient (PCC) is $[-1, 1]$, and the PCC close to 1 or -1 indicates a perfect linear relationship and the value closes to 0 demonstrates no relations. The PCC values of the MVGB and VR are 0.41257 and -0.67181 ($\alpha \leq 0.001$), respectively. We can readily see that there is a tight correlation between RDMs and the MVGB, and between RDMs and the VR. Therefore, the proposed model in (4) is reasonable.

To train the model, we take the ratio of training and testing dataset as 7:3 and use 10-fold cross validation. The best learning result is $\mu_0 = 0.53214829$, $\mu_1 = 0.51309827$ and $\mu_2 = -0.91066194$. Once the values of the coefficients have been determined, this model will be used to generate the scene depths of single underwater images under different scenarios.

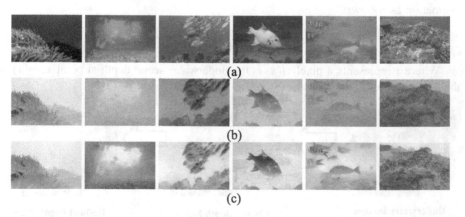

Fig. 4. The process of generating the training samples. (a) Original underwater images; (b) Coarse depth maps based on our method; (c) Refined depth maps.

Estimation of the Depth Map. As the relationship among the scene depth map $d(x)$, the MVGB and VR has been established and the coefficients have been learned, the depth maps of any underwater images can be obtained by Eq. (4). In order to check the validity of the assumption, we collected a large database of underwater images from several well-known photo websites, e.g., Google Images, Filck.com, and computed the underwater scene depth maps of each test images. Some of the results are shown in the Fig. 4. For different types of input underwater images, the corresponding estimated coarse depth maps and the refined depth maps are shown in the Fig. 4(a)–(c), respectively. As can be seen, the estimated depth maps have brighter color in the regions within the farther depth while having lighter color in the closer region as expected. After obtaining the correct depth map, the background light estimation and the transmission maps for RGB lights are rather simple and rapid.

3   Underwater Image Restoration

3.1   Background Light Estimation

The background light BL in the Eq. (1) is often estimated as the brightest pixel in an underwater image. However, the assumption is not correct in some situations, e.g., the foreground objects are brighter than the background light. The background light is selected from the farthest point of the input underwater image, i.e., the position of the maximum value in the refined depth map corresponding the input underwater image is the background light candidate value. But directly select the farthest point as the final background light, some suspended particles can interrupt the valid estimation result. After generating an accurate depth map, we firstly remove the effects of suspended particles via selecting the 0.1% farthest point, and then select the pixel with the highest intensity in the original underwater image. An example to illustrate the background light estimation method is shown in Fig. 5.

(a)   (b)   (c)

Fig. 5. An example to illustrate the global background light estimation algorithm. (a) Original image; (b) The result of searching for the 0.1% farthest pixels in the refined depth map; (c) The brightest intensity of the 0.1% farthest pixels corresponding to the original image.

3.2   Transmission Map Estimation for Respective R-G-B Channel

The relative depth map cannot be directly used to estimate the final TMs for R-G-B channel. To measure the distance from the camera to each scene point, the actual scene depth map d_a is defined as follows:

$$d_a(x) = D_\infty \times d(x) \tag{5}$$

where D_∞ is a scaling constant for transforming the relative distance to the real distance, and in this paper, the D_∞ is set as 10. With the estimated d_a, we can calculate the TM for the R-G-B channel as:

$$t_\lambda(x) = Nrer(\lambda)^{d_a(x)} \tag{6}$$

In approximately 98% of the world's clear oceanic or coastal water (ocean type I), the accredited ranges of $Nrer(\lambda)$ in red, green and red lights are 80%–85%, 93%–97%, and 95%–99%, respectively [7]. In this paper, we set the $Nrer(\lambda)$ for R-G-B light is 0.83, 0.95 and 0.97, respectively. Figure 6(c)–(e) gives an example of TMs for the RGB channels of a blue-greenish underwater image based on Eqs. (5)–(6). Figure 6(c)–(e) gives an example of TMs for the RGB channels of a blue-greenish underwater image based on Eqs. (5)–(6).

Now that we have the BL_λ and $t_\lambda(x)$ for the R-G-B channel, we can restore the underwater scene radiance J_λ with the Eq. (7). A lower bound and an upper bound for $t_\lambda(x)$ empirically set to 0.1 and 0.9, respectively.

$$J_\lambda(x) = \frac{1}{min(max(t_\lambda(x), 0.1), 0.9)}(I_\lambda(x) - BL_\lambda) + BL_\lambda \tag{7}$$

Figures 6(f) and 7(f) show some final results recovered by our proposed method.

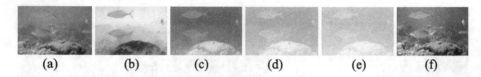

(a)   (b)   (c)   (d)   (e)   (f)

Fig. 6. The processing of underwater image restoration (a) Original image; (b) The refined depth map; (c) The estimated red transmission map; (d) The estimated green transmission map; (e) The estimated blue transmission map; (f) The restored image. (Color figure online)

4   Results and Discussion

In this part, we compare our proposed underwater image restoration method based on underwater light attenuation prior against with the typical image dehazing method by He et al. [3], the variant of the DCP (UDCP) by Drew et al. [8], the image restoration based on the dehazing of G-B channel and the correction of R channel [10], and Peng et al.'s method [6] based on both the image blurriness and light absorption. In order to

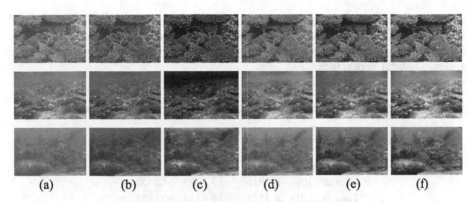

Fig. 7. (a) Original images with a size of 600×400 pixels; (b) He et al.'s results; (c) Drew et al.'s results; (d) Li et al.'s results; (e) Peng et al.'s results; (f) Our results.

demonstrate the outstanding performance of our proposed method, we present some examples of the restored images and introduce quantitative assessments and complexity comparison in this section.

Figure 7(a) shows four raw underwater images with different underwater characteristics in terms of color tones and scenes from our underwater images datasets. In Fig. 7(b), the DCP has no or little work on the test images due to incorrect depth map estimation. This indicates the direct application of the outdoor image dehazing is not suitable to the underwater image restoration. Figure 7(c) shows that the UDCP fails to recover the scene of the underwater images and even bring color distortion and error restoration. As shown in the Fig. 7(d)–(f), although all the methods can remove the haze of the input images, the color and contrast of Fig. 7(d–e) are not as good as those of Fig. 7(f) because the underwater light selection attenuation is ignored when estimating transmission maps or background light. As shown in Fig. 7(f), our image restoration method can effectively descatter and dehaze different underwater images, improves details and colorfulness of the input images and finally produce a natural underwater images.

4.1 Quantitative Assessment

Considering the fact that the clear underwater image presents better color, contrast and visual effect, we rely on three non-reference quantitative metrics: ENTROPY, underwater image quality measure (UIQM) [15] and the Blind/Referenceless Image Spatial Quality Evaluator (BRISQUE) [16] to assess the restored image quality. Table 1 lists the average scores of the three metrics on the recovered 100 low-quality underwater images and notes that the best results are in bold. Entropy represents the abundance of information. An image with higher entropy value prevents more valuable information. The UIQM is a linear combination of colorfulness, sharpness and contrast and a larger value represents higher image quality. The highest values of both ENTROPY and UIQM values mean that our proposed method can recover high-quality underwater images and reserve a lot of image information. The BRISQUE quantifies possible losses of naturalness in an image due to the presence of distortions. The BRISQUE value indicates the image quality from 0 (best) to 100 (worst). The lowest value of the BRISQUE obtained by our proposed method indicates the restored underwater image achieves natural appearance.

Table 1. Quantitative analysis in terms of ENTROPY, UIQM and BRISQUE.

Method	ENTROPY	UIQM	BRISQUE
He et al. [9]	6.1252	1.9296	33.5317
Drew et al. [8]	6.3714	2.6425	32.9704
Li et al. [10]	6.8031	3.3072	34.2450
Peng et al. [6]	6.7529	3.4514	30.1825
Ours	**7.2398**	**3.9884**	**28.9147**

4.2 Complexity Comparison

In this part, we compare running time of our proposed method with other methods (maps refined by the guided filter [14]), including He et al. [3], Drew et al. [8], Li et al. [10] and Peng et al. [6], processed on a Windows 7 PC with Intel(R) Core(TM) i7-4790U CPU@3.60 GHz, 8.00 GB Memory, running on Python3.6.5. When an underwater image with the size $m \times n$ is refined by the guided filter with the radius r, the complexity of our proposed underwater image restoration method is $O(m \times n \times r)$ after the linear model is used to estimate the depth map. In the Fig. 8, the running time

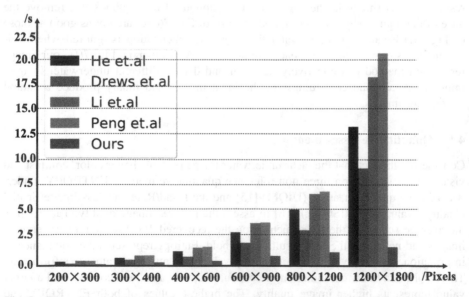

Fig. 8. The processing time (/s) of our proposed method (blue bar) with the different sizes of the input underwater images comparison with He et al. (brown bar), Drew et al. (green bar), Li et al. (red bar), Peng et al. (gray bar). (Color figure online)

(RT) is the average time (/s) of 50 underwater images with different sizes (/pixels). The RT of other compared methods improves significantly as the size test image becomes larger. Even though an underwater image is the size of 1200 × 1800 pixels, the RT of our method is lower than 2 s. In our method, the most time is used to obtain three transmission maps for respective RGB lights.

5 Conclusion

In this paper, we have explored the rapid and effective model of depth map estimation based on the underwater light attenuation prior within underwater images. After the linear model is created, the BL and the TMs for R-G-B light are smoothly estimated to restore the scene radiance simply. Our proposed method can achieve better quality of the restored underwater images, meanwhile it can save a mass of consumption time due to the depth map estimation using the linear model based on the ULAP, which further simplifies the deduction of the background light and transmission maps for R-G-B channel. The experimental results prove that our method can be well-suitable for underwater image restoration under different scenarios, faster and more effective to improve the quality of underwater images, according to the best objective evaluations and the lowest running time.

Acknowledgment. This work was supported by the National Natural Science Foundation of China (NSFC) Grant 61702323 and the Program for Professor of Special Appointment at Shanghai Institutions of Higher Learning (TP2016038). The first two authors contributed equally to this work.

References

1. Zhao, X., Jin, T., Qu, S.: Deriving inherent optical properties from background color and underwater image enhancement. Ocean Eng. **94**, 163–172 (2015)
2. Huang, D., Wang, Y., Song, W., Sequeira, J., Mavromatis, S.: Shallow-water image enhancement using relative global histogram stretching based on adaptive parameter acquisition. In: Schoeffmann, K., et al. (eds.) MMM 2018. LNCS, vol. 10704, pp. 453–465. Springer, Cham (2018). https://doi.org/10.1007/978-3-319-73603-7_37
3. He, K.M., Sun, J., Tang, X.O.: Single image haze removal using dark channel prior. IEEE Trans. Pattern Anal. Mach. Intell. **33**(12), 2341–2353 (2011)
4. Galdran, A., Pardo, D., Picón, A., Alvarez-Gila, A.: Automatic Red-Channel underwater image restoration. J. Vis. Commun. Image Represent. **26**, 132–145 (2015)
5. Wang, J.B., He, N., Zhang, L.L., Lu, K.: Single image dehazing with a physical model and dark channel prior. Neurocomputing **149**, 718–724 (2015)
6. Peng, Y.T., Cosman, P.C.: Underwater image restoration based on image blurriness and light absorption. IEEE Trans. Image Process. **26**(4), 1579–1594 (2017)
7. Chiang, J.Y., Chen, Y.C.: Underwater image enhancement by wavelength compensation and dehazing. IEEE Trans. Image Process. **21**(4), 1756–1769 (2012)
8. Drews, P., Nascimento, E., Moraes, F., Botelho, S., Campos, M.: Transmission Estimation in Underwater Single Images. In: Proceedings of IEEE ICCVW 2013, pp. 825–830 (2013)

9. Wen, H.C., Tian, Y.H., Huang, T.J., Guo, W.: Single underwater image enhancement with a new optical model. In: Proceedings of IEEE ISCAS 2013, pp. 753–756 (2013)
10. Li, C.Y., Quo, J.C., Pang, Y.W., Chen, S.J., Wang, J.: Single underwater image restoration by blue-green channels dehazing and red channel correction. In: ICASSP 2016, pp. 1731–1735 (2016)
11. Liu, X.P., Zhong, G.Q., Cong, L., Dong, J.Y.: Underwater image colour constancy based on DSNMF. IET. Image Process. 11(1), 38–43 (2017)
12. Ding, X.Y., Wang, Y.F., Zhang, J., Fu, X.P.: Underwater image dehaze using scene depth estimation with adaptive color correction. In: OCEAN 2017, pp. 1–5 (2017)
13. Zhu, Q.S., Mai, J.M., Shao, L.: A fast single image haze removal algorithm using color attenuation prior. IEEE Trans. Image Process. 24(11), 3522–3533 (2015)
14. He, K.M., Sun, J., Tang, X.O.: Guided image filtering. IEEE Trans. Pattern Anal. Mach. Intell. 35(6), 1397–1409 (2013)
15. Panetta, K., Gao, C., Agaian, S.: Human-visual-system-inspired underwater image quality measures. IEEE J. Ocean. Eng. 41(3), 541–551 (2016)
16. Mittal, A., Moorthy, A.K., Bovik, A.C.: No-reference image quality assessment in the spatial domain. IEEE Trans. Image Process. 21(12), 4695–4708 (2012)

Sequential Feature Fusion for Object Detection

Qiang Wang and Yahong Han[✉]

School of Computer Science and Technology, Tianjin University, Tianjin, China
{qiangw,yahong}@tju.edu.cn

Abstract. In an image, the category and the location of an object are related to global, spatial and contextual visual information of the object, which are all extremely important for accurate and efficient object detection. In this paper, we propose a region-based detector named Sequential Feature Fusion Network (SFFN) which simultaneously utilizes global, spatial and multi-scale contextual Region-of-Interest (RoI) features of an object and fuses them by a novel method. Specifically, we design a Feature Fusion Block (FFB) to fuse global and multi-scale contextual RoI features, which are extracted by RoI pooling layer. Then we apply the concatenation operation to integrate the fused feature with spatial RoI feature extracted by Positive-Sensitive RoI (PSRoI) pooling layer. The experimental results show that the performance of SFFN obtains significant improvements on both the PASCAL VOC 2007 and VOC 2012 datasets.

Keywords: Object detection · Region-of-Interest · Feature fusion

1 Introduction

Object detection is a challenging task which focuses on both object localization and classification. When humans recognize the location and the category of an object in an image, they would take all of the overall appearance, spatial structure and surrounding scene of the object into consideration to obtain global, spatial and contextual visual information of the object, respectively. As Fig. 1 shows, contextual visual information is helpful to predict the existence of objects and provides references for object identification; global visual information is used to identify the object according to object characteristics; and spatial visual information can refine object categories according to object structures in detail. Therefore, utilizing all of these visual information and considering the relationship among them benefit object detection.

Recently, there are two groups of object detectors based on Convolutional Neural Network (CNN) have been proposed: region-based detectors [2,5,12] and region-free detectors [9,13,14]. Due to the first stage of proposal generation, region-based detectors perform better than region-free detectors. The second important part in region-based detectors is Region-of-Interest (RoI) features

© Springer Nature Switzerland AG 2018
R. Hong et al. (Eds.): PCM 2018, LNCS 11164, pp. 689–699, 2018.
https://doi.org/10.1007/978-3-030-00776-8_63

Fig. 1. The relationship among global, spatial and contextual visual information of objects. Contextual visual information about the surrounding scene is helpful to predict the existence of objects. Global visual information about the object characteristic is used to identify the object. Spatial visual information about the object structure can refine the object category in detail.

extraction of objects in candidate proposals. To enrich visual information of an object, AC-CNN [21] explores multi-scale contextual RoI features in different receptive fields. But there is the drawback of losing spatial visual information of the object. CoupleNet [15] consists of a global branch for global and contextual RoI features and another local branch for spatial RoI features parallelly, then couples them in result level. However, the mergence in result level merely utilizes advantages of different visual information separately and then combines results, but ignores the essential complementarity relationship among them.

To solve deficiencies of methods mentioned above, we propose a novel region-based object detector named Sequential Feature Fusion Network (SFFN), which fuses global, spatial and multi-scale contextual visual information in feature level and considers the relationship among them. The overview is shown in Fig. 2. Specifically, we extract global and multi-scale contextual RoI features by RoI pooling layer [3], and extract spatial RoI features by Position-Sensitive RoI (PSRoI) pooling layer [12] for objects in candidate proposals. Then, we fuse these RoI features sequentially by a novel method. Specifically, We design FFB to fuse global and multi-scale contextual RoI features. Taking a feature in a specific receptive field and a fused feature of the last cycle as inputs, FFB generates a new fused RoI feature by filtering and selecting information of them. After all of global and multi-scale contextual RoI features are fused by applying FFB sequentially, the fused feature is integrated with spatial RoI feature by concatenation operation.

Our contributions are summarized as followed: We propose SFFN to jointly learn global, spatial and multi-scale contextual RoI features of objects, and design a novel sequential feature fusion method for them. Our method equipped with ResNet-101 achieves excellent performances of 82.3% mAP on PASCAL VOC 2007 and 79.7% mAP on VOC 2012 [1].

2   Related Work

Based on CNN, there are two groups of methods for object detection. The one is region-based detectors [2,5,12], which first generate candidate proposals from the input image, then output category classification and location regression for objects in candidate proposals according to corresponding RoI features. The other is region-free detectors [9,13,14], which just take object detection as a regression task and predict results from the input image directly. To a certain extent, region-based detectors are more accurate than region-free detectors because of the proposal generation stage.

As the most representative region-based detector, R-CNN [2] first proposes to utilize object features through CNN for category classification and location regression. Based on it, Fast/Faster R-CNN [3,5] applies RoI pooling layer to extract fixed-length RoI features of objects, which realizes sharing convolutional features between the whole image and objects inside it. To enhance the representation ability of RoI features, ResNet [8], DenseNet [10] and Deformable ConvNet [11] attempt to use deep and complex CNN structures for feature learning and achieve significant improvements in object detection. To enrich visual information of RoI features, HyperNet [7] and PVANet [16] utilize both the low-level image feature with high resolution and the high-level image feature with low resolution, which are beneficial for localization regression and category classification respectively. All of these methods focus on RoI features which are extracted by RoI pooling layer from a global perspective of objects. Besides, R-FCN [12] proposes PSRoI pooling layer for object detection, which pays more attention to spatial information of objects. To consider visual information from both global and spatial perspectives, Fan et al. [17] uses an extra objectness mask map to estimate existence probability of objects in pixel level, which is used to refine RoI features by element-wise production operation. In addition to global and spatial information of objects, contextual information around an object is also important. Li et al. [21] proposes AC-CNN to leverage contextual RoI features of multi-scale contextual proposals and combine them through Long Short-Term Memory (LSTM) and concatenation operation. Zeng et al. [20] proposes another method, named Gated Bi-directional CNN, to fuse multi-scale contextual RoI features. From above all, global, spatial and contextual information are all available, but these methods only utilize one or two kinds of information. Information combination has been applied widely on several tasks. Zhu et al. [22] utilizes visual features of multiple modalities for Mobile Landmark Search (MLS). Xie et al. [23] and Zhu et al. [24] proposes to use multi-view and context information for Image Retrieval, respectively. Besides, temporal, spatial and local information utilization is also considered as a kind of information combination in video tasks. Xu et al. [18] proposes Sequential Video VLAD for video representation with both spatial and temporal information. Yang et al. [19] utilizes global features and RoI features temporally for Video Captioning. Liu et al. [25] utilizes multi-scale CNN to improve feature extraction for Person Re-Identification. Liu et al. [26] proposes to add context information for Image Captioning.

Inspired by these methods, we propose SFFN, which utilizes all of global, spatial and multi-scale contextual RoI features simultaneously and fuses them by a novel fusion method according to the relationship among them.

3   Method

In this section, we introduce the architecture of the proposed SFFN. As shown in Fig. 2, our network takes an image as the input and outputs category classification and location regression for objects in the image. Similar to traditional region-based detectors, SFFN firstly extracts the image feature through CNN (e.g. ResNet-101) and applies RPN [5] to generate a set of candidate proposals by objectness prediction and location regression. Then, given the image feature and candidate proposals, it extracts global, spatial and multi-scale contextual RoI features and fuses them by the proposed method, which will be introduced in Sects. 3.1 and 3.2 respectively. Finally, the following fully convolutional layers are used to predict category probability and coordinate offset for objects according to the fused RoI features. For training, we apply the same multi-task loss as [5], including softmax cross-entropy loss for classification and Smooth-L1 loss for regression in both RPN sub-network and R-CNN sub-network. And the gradients are back propagated through all layers to perform end-to-end learning.

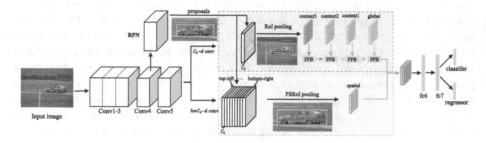

Fig. 2. The architecture of the proposed SFFN. (1) Generate the image feature through CNN. (2) Produce candidate proposals by RPN. (3) Extract global, spatial and multi-scale contextual RoI features and fuse them by proposed sequential feature fusion method. (4) Generate category classification and location regression based on fused RoI features of objects.

3.1   RoI Feature Extraction

Given the image feature and candidate proposals, we apply two methods to extract RoI features with different information. The one is RoI pooling layer [3] for global and multi-scale contextual RoI features; the other is PSRoI pooling layer [12] for spatial RoI feature. In order to fuse RoI features extracted by different methods, we standardize the channel number of RoI features to C_t.

To consider global and contextual visual information of objects, we first insert a dimension reduction layer which reduces the channel number of image feature to the unified channel number mentioned above. Given an image feature map represented by $I \in R^{H \times W \times C}$, a 1×1 convolutional layer is applied to change the channel number from C to C_t, and outputs a new image feature represented by $I' \in R^{H \times W \times C_t}$. Then we apply RoI pooling layer on I' to extract global and multi-scale contextual RoI features of the object according to corresponding candidate proposal and contextual proposals. Suppose that the shape of fixed-length features is $h \times w$, the shape of RoI features is $h \times w \times C_t$. As the essential difference between global and multi-scale contextual RoI features is the receptive field, we take global and multi-scale contextual RoI features as a whole contextual set with different receptive fields. Setting the length ratio of candidate proposals to 1.0, the set of RoI features is represented by $\{f_1, f_2, ..., f_n\}$ with length ratios of $\{s_1, s_2, ..., s_n\}$, and n is the length of the set.

In consideration of spatial visual information of objects, we aim to learn multiple sub-features for different parts of an object and integrate them to produce the spatial RoI feature. Based on the grid structure $h \times w$ for objects, we first insert a 1×1 convolutional layer on the original image feature I and produce a spatial feature represented by $I_s \in R^{H \times W \times hwC_t}$ for the whole image. Then, we use PSRoI pooling layer to extract spatial RoI feature of the object in each candidate proposal and the shape of the output is $h \times w \times C_t$.

3.2 Sequential Feature Fusion

Inspired by the principle of Recurrent Neural Network (RNN) which filters and memorizes sequential information, we design FFB to fuse global and multi-scale contextual RoI features. It takes an original feature in a specific receptive field and a fused feature of the last cycle as inputs, and produces a new fused feature. Specifically, there are two selection gates to filter valuable information of the input features. Besides, there is a mergence gate to fuse the two filtered features. The structure is shown in Fig. 3.

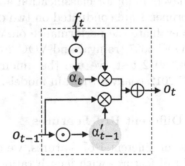

Fig. 3. The structure of Feature Fusion Block (FFB).

For the input of original feature f_t which is t-th feature in a specific fusion order, a selection gate computes an attention map α_t for it and multiplies the feature f_t with the attention map α_t to produce a filtered feature. For the input of fused feature O_{t-1} which filters visual information of all previous features until

f_{t-1} in last cycle, the other selection gate computes an attention map α'_{t-1} for it and multiplies the fused feature O_{t-1} with the attention map α'_{t-1} to produce a filtered feature. However, O_{t-1} is set to 0 when $t = 0$. After filtering information of inputs, the mergence gate sums up filtered features and the result O_t is used as an input for the next cycle:

$$\alpha'_{t-1} = \sigma(O_{t-1} \times W'_{t-1} + b'_{t-1})$$
$$\alpha_t = \sigma(f_t \times W_t + b_t) \tag{1}$$
$$O_t = (\alpha'_{t-1} \odot O_{t-1}) + (\alpha_t \odot f_t)$$

where σ is a sigmoidal non-linearity function; $\odot$ is the element-wise production operation; and $\times$ is the matrix multiplication operation. All weights denoted by W'_i or W_j and biases denoted by b'_i or b_j are trainable parameters.

To fuse global, spatial and multi-scale contextual RoI features, we propose a novel sequential feature fusion method. Firstly, we apply FFB to fuse global and multi-scale contextual RoI features, which are extracted by RoI pooling layer. After fusing them in descending order of the feature respective fields sequentially, the fused feature and spatial RoI feature are integrated by concatenation operation for following category classification and location regression.

4   Experiment

We implement the proposed SFFN based on Faster R-CNN [5] with OHEM [4] and without multi-scale training or testing in order to evaluate the effectiveness of proposed sequential feature fusion. Hyper-parameters for all experiments are remained same as [5]. And we use ResNet-101 as the basic CNN, which has the strong ability of feature learning. Besides, we apply fully convolutional layers to replace fully connection layers in [5] for classification and regression in order to reduce parameters. Experiments are conducted on two datasets: PASCAL VOC 2007 and VOC 2012. To evaluate the performance on VOC 2007 test, we train models on the union of VOC 2007 trainval and VOC 2012 trainval. To evaluate the performance on VOC 2012 test, we use the union of VOC 2007 trainval, VOC 2007 test and VOC 2012 trianval to train models.

4.1   Effectiveness of Different RoI Features

To study the effectiveness of different RoI features, we conduct experiments on a basic contextual set of RoI features with length ratios of $\{1.0, 2.0, H/h\}$ and spatial RoI features. Besides, we denote the spatial RoI features as SF. We denote the contextual RoI feature with length ratio of 1.0 as GF, which is the global RoI feature for object region essentially; The contextual RoI feature with length ratio of 2.0 and H/h are denoted by CF_1 and CF_2 respectively, which are RoI features for objects in expanded regions with different ratios.

As results on VOC 2007 test shown in Table 1, all of CF_1, CF_2 and SF improve the performance which illustrates the effectiveness of contextual and spatial RoI features. Besides, FFB works better for CF_1, CF_2 and GF while concatenation operation works better for SF and GF. It demonstrates that FFB is suitable for RoI features extracted by the same method, while concatenation operation works better for RoI features extracted by different methods.

Table 1. The experimental results of SFFN using different RoI features on VOC 2007 test.

Method	GF	SF	CF_1	CF_2	mAP (fused by concatenation)	mAP (fused by FFB)
baseline1	✓				80.4	-
baseline2		✓			80.4	-
SFFN	✓	✓			81.1	81.0
SFFN	✓		✓		80.9	81.9
SFFN	✓			✓	80.1	81.3

Figure 4 shows some examples of two baselines and SFFN on PASCAL VOC 2007 test with a score threshold of 0.5. Results of (a) only using global RoI features and results of (b) only utilizing spatial RoI features both ignore some small and unconspicuous objects. While results of (c) using all of global, spatial and multi-scale contextual RoI features can detect them. It also illustrates the perfect performance of SFFN for classification and regression.

Fig. 4. Examples of SFFN on the PASCAL VOC 2007 test with a score threshold of 0.5. (a) is results of baseline1 which only uses global RoI features of objects; (b) is results of baseline2 which only uses spatial RoI features of objects; (c) is results of SFFN which use all of global, spatial and multi-contextual RoI features of objects.

4.2 Effectiveness of Sequential Feature Fusion

To study the effectiveness of sequential feature fusion, we first reimplement several representative models. Then we fuse global and multi-scale contextual RoI features with length ratios of { 1.0, 2.0, H/h } by FFB sequentially, and replace the original RoI feature with the fused RoI feature.

Table 2. The experimental results of representative methods equipped with sequential feature fusion on VOC 2007 test.

Method	mAP (Reimplementation)	mAP (Sequential Feature Fusion)
PVANet [16]	71.5	73.9
Faster R-CNN [8]	80.4	82.2
R-FCN [12]	79.0	79.5
CoupleNet [15]	81.3	82.4

From experimental results in Table 2, models which use sequential feature fusion obtain further performance improvements more than 1.0 points for original PVANet [16], Faster R-CNN [8] and CoupleNet [15] except for R-FCN [12]. Moreover, R-FCN [12] only utilizes RoI features extracted by PSRoI pooling layer while others all use RoI features extracted by RoI pooling layer. It illustrates that sequential feature fusion for multi-scale contextual RoI features is more suitable for RoI features extracted by RoI pooling layer instead of PSRoI pooling layer.

4.3 Effectiveness of Different Contextual Sets

In experiments, we conduct experiments by only using global and multi-scale contextual RoI features on two contextual sets with different length ratios. One set consists of RoI features with length ratios of { 1.0, 2.0, H/h }, which considers the relationship between objects and the surrounding scene, and the relationship between objects and the whole image scene. The other set consists of RoI features with length ratios of {0.8, 1.2, 1.8, 2.7} as mentioned in [20], which considers smaller object regions to extract full object with less background and utilizes closer length ratios to reduce gap between adjacent RoI features.

The experimental results are shown in Table 3. Multi-scale contextual RoI features with length ratio of {0.8, 1.2, 1.8, 2.7} obtain higher performance than length ratio of {1.0, 2.0, H/h}. It illustrates the effectiveness of more and closer length ratio for contextual RoI features. To study the availability of SF, we conduct experiments by adding SF. Results shows that methods with SF works better, which illustrates the helpfulness of SF. Besides, the higher performance of methods using FFB also illustrates the effectiveness of FFB.

Table 3. The experimental results of different contextual sets on VOC 2007 test.

Method	{ s_1, s_2, ... }	SF	mAP (fused by concatenation)	mAP (fused by FFB)
SFFN	{1.0, 2.0, H/h}		81.1	81.1
SFFN	{0.8, 1.2, 1.8, 2.7}		81.0	82.2
SFFN	{1.0, 2.0, H/h}	✓	81.1	81.8
SFFN	{0.8, 1.2, 1.8, 2.7}	✓	81.5	**82.3**

4.4   Experiments on VOC 2007 and VOC 2012

Table 4 shows experimental results of SFFN on both VOC 2007 (82.3% mAP) and VOC 2012 (79.7% mAP), which obtain significant improvements compared with other CNN-based methods except for CoupleNet [15] on VOC 2012 (79.7% mAP). Compared with other CNN-based methods, the proposed SFFN and CoupleNet [15] both consider all of global, spatial and contextual visual information of objects. And the performance of them improve by more than 2.0 points on both VOC 2007 and VOC 2012, which illustrates the important of various information fusion. Moreover, SFFN, which considers fusion in feature level instead of in result level as CoupleNet [15], performs better on VOC 2007 but not on VOC 2012. The reason is that [15] shows mAP results with and without multi-scale training strategy on VOC 2007 test while only shows the mAP result with multi-scale training strategy on VOC 2012 test. And experiments of SFFN do not use multi-scale training strategy in order to purely evaluate the effectiveness of the proposed feature fusion.

Table 4. The experimental results on VOC 2007 test and VOC 2012 test.

Method	Network	mAP (VOC 2007 test)	mAP (VOC 2012 test)
YOLOv2(544) [9]	Darknet-19	78.6	73.4
SSD(512) [13]	VGG-16	76.8	74.9
DSOD(300) [14]	DS/64-192-48-1	77.7	76.3
AC-CNN [21]	VGG-16	72.0	70.6
Faster R-CNN [5]	VGG-16	73.2	70.4
ION [6]	VGG-16	79.2	76.4
HyperNet [7]	VGG-16	76.3	71.4
Faster R-CNN [8]	ResNet-101	76.4	73.8
R-FCN [12]	ResNet-101	79.5	77.6
CoupleNet [15]	ResNet-101	82.1	80.4
SFFN(Our method)	ResNet-101	**82.3**	**79.7**

Figure 5 shows some examples of SFFN on PASCAL VOC 2007 test with a score threshold of 0.5, which reflect perfect performances of SFFN for crowed and tiny objects, relational objects, and objects in special light.

Fig. 5. Examples of SFFN on the PASCAL VOC 2007 test (82.3 mAP) with a score threshold of 0.5. It shows the perfect performance for object detection in complex scene.

5   Conclusion

We propose a novel region-based detector SFFN. Based on the effective image feature, the proposed method simultaneously utilize global, spatial and multi-scale contextual RoI features and consider the relationship of them. In consideration of different methods to extract RoI features, we design a novel FFB to fuse global and contextual RoI features extracted by RoI pooling layer, then use concatenation operation to integrate the fused feature with spatial RoI features extracted by PSRoI pooling. Our approach achieves much improvement compared with previous methods and can be equipped with various CNNs or applied to other region-based object detectors to enrich RoI features.

Acknowledgments. This work is supported by the NSFC (under Grant U1509206, 61472276) and Tianjin Natural Science Foundation (no. 15JCYBJC15400).

References

1. Everingham, M., Van Gool, L., Williams, C.K., Winn, J., Zisserman, A.: The PASCAL Visual Object Classes (VOC) challenge. IJCV **88**(2), 303–338 (2010)
2. Girshick, R., Donahue, J., Darrell, T., Malik, J.: Rich feature hierarchies for accurate object detection and semantic segmentation. In: CVPR, pp. 580–587 (2014)
3. Girshick, R.: Fast R-CNN. In: CVPR, pp. 1440–1448 (2015)
4. Shrivastava, A., Gupta, A., Girshick, A.: Training regionbased object detectors with online hard example mining. In: CVPR, pp. 761–169 (2016)
5. Ren, S., He, K., Girshick, R., Sun, J.: Faster R-CNN: towards real-time object detection with region proposal networks. In: NIPS, pp. 91–99 (2015)
6. Bell, S., Zitnick, C.L., Bala, K., Girshick, R.: Inside-Outside net: detecting objects in context with skip pooling and Recurrent Neural Networks. In: CVPR, pp. 2874–2883 (2016)

7. Kong, T., Yao, A., Chen, Y., Sun, F.: Hypernet: towards accurate region proposal generation and joint object detection. In: CVPR, pp. 845–853 (2016)
8. He, K., Zhang, X., Ren, S., Sun., J.: Deep residual learning for image recognition. In: CVPR, pp. 770–778 (2016)
9. Redmon, J., Farhadi, A.: YOLO9000: Better, Faster, Stronger. arXiv preprint arXiv:1612.08242 (2016)
10. Huang, G., Liu, Z., Weinberger, K., van der Maaten, L.: Densely connected convolutional networks. In: CVPR (2017)
11. Dai, J., et al.: Deformable convolutional networks. In: ICCV, pp. 764–773 (2017)
12. Li, Y., He, K., Sun, J.: R-FCN: object detection via regionbased fully convolutional networks. In: NIPS, pp. 379–387 (2016)
13. Liu, W., et al.: SSD: single shot multibox detector. In: Leibe, B., Matas, J., Sebe, N., Welling, M. (eds.) ECCV 2016. LNCS, vol. 9905, pp. 21–37. Springer, Cham (2016). https://doi.org/10.1007/978-3-319-46448-0_2
14. Shen, Z., et al.: DSOD: learning deeply supervised object detectors from scratch. In: ICCV, pp. 1937–1945 (2017)
15. Zhu, Y., et al.: CoupleNet: coupling global structure with local parts for object detection. In: ICCV (2017)
16. Kim, K.-H., Hong, S., Roh, B., Cheon, Y., Park, M.: Pvanet: Deep but lightweight neural networks for real-time object detection. arXiv preprint arXiv:1608.08021 (2016)
17. Fan, X., Guo, H., Zheng, K., Feng, W., Wang, S.: Object Detection with Mask-based Feature Encoding. arXiv preprint arXiv:1802.03934 (2018)
18. Xu, Y., Han, Y., Tian, R.H.Q.: Sequential video VLAD: training the aggregation locally and temporally. IEEE Trans. Image Process. (IEEE TIP) 27(10), 4933–4944 (2018)
19. Yang, Z., Han, Y., Wang, Z.: Catching the temporal regions-of-interest for video captioning. In: ACM MM, pp. 146–153 (2017)
20. Zeng, X., Ouyang, W., Yang, B., Yan, J., Wang, X.: Gated Bi-directional CNN for object detection. In: Leibe, B., Matas, J., Sebe, N., Welling, M. (eds.) ECCV 2016. LNCS, vol. 9911, pp. 354–369. Springer, Cham (2016). https://doi.org/10.1007/978-3-319-46478-7_22
21. Li, J., et al.: Attentive contexts for object detection. IEEE Trans. Multimed. 19(5), 944–954 (2017)
22. Zhu, L., et al.: Discrete multimodal hashing with canonical views for robust mobile landmark search. IEEE Trans. Multimed. 19(9), 2066–2079 (2017)
23. Xie, L., Shen, J., Han, J., Zhu, L., Shao, L.: Dynamic multi-view hashing for online image retrieval. In: IJCAI, pp. 3133–3139 (2017)
24. Zhu, L., Huang, Z., Li, Z., Xie, L., Shen, H.T.: Exploring auxiliary context: discrete semantic transfer hashing for scalable image retrieval. IEEE Trans. Neural Netw. Learn. Syst. 99, 1–13 (2018). https://doi.org/10.1109/TNNLS.2018.2797248
25. Liu, J., et al.: Multi-scale triplet CNN for person re-identification. In: ACM MM, pp. 192–196 (2016)
26. Liu, D., Zha, Z.J., Zhang, H., Zhang, Y., Wu, F.: Context-aware visual policy network for sequence-level image captioning. In: ACM MM (2018)

Discriminative Correlation Quantization for Cross-Modal Similarity Retrieval

Jun Tang, XuanMeng Li, Nian Wang$^{(\boxtimes)}$, and Ming Zhu

Key Laboratory of Intelligent Computing and Signal Processing,
Ministry of Education, Anhui University, Hefei, China
tangjunahu@163.com, xmliahu@163.com, nwangahu@163.com, zhu_m@163.com

Abstract. Due to their superior performance on efficient search and discrete loss, quantization methods have attracted considerable attention for approximate nearest neighbor (ANN) search on large-scale multimedia data. In this paper, we aim to introduce quantization into cross-modal similarity search with a focus on learning discriminative binary codes. Different from existing cross-modal quantization algorithms that transform heterogeneous data to a common subspace by unsupervised approaches, we propose a novel cross-modal quantization method embedded in a supervised framework by exploring the discriminative property of label information. The proposed approach learns common semantic space from label information by linear classification, enabling the generated category-specific features to produce more discriminative quantization codes. Furthermore, the unified codebooks and quantization codes are adopted to preserve the correlation of similar inter-modal pairs in the learned semantic space. The overall optimization algorithm jointly learns the linear classifiers, category-specific features and the unified quantizer in an alternated strategy. Extensive comparative experiments on three benchmark datasets show the superiority of our approach over some state-of-the-art methods.

Keywords: Cross-modal · Quantization · Similarity retrieval

1 Introduction

With the rapid growth of Internet and social network, tremendous amounts of multimedia data (e.g., images, texts, audios and videos) have been stored and shared on websites. Consequently, scalable similarity search for multimedia data have attracted extensive attention. Nearest Neighbor (NN) search is one of the most frequently-used methods for similarity search. However, NN search is of linear time complexity so that it is intractable for massive data with high dimensionality. To improve search efficiency, some approximate nearest neighbor (ANN) search methods based on trees or graphs were proposed, including random multiple k-d trees [11], FLANN [8] and neighborhood graph search [16].

© Springer Nature Switzerland AG 2018
R. Hong et al. (Eds.): PCM 2018, LNCS 11164, pp. 700–710, 2018.
https://doi.org/10.1007/978-3-030-00776-8_64

Over past decades, hashing-based ANN search methods [1,17,21] have made remarkable progress arising from the advantage of compact binary representation and efficient Hamming distance computation. Although the advance is encouraging, the desirable quantization strategy remains a challenging problem. On one hand, hash codes are commonly produced by hard thresholding, which may lead to substantial information loss and unbalanced encoding. On the other hand, discrete optimization problems deduced by binary representation are often tackled by continuous relaxation, thus resulting in undesirable suboptimal hash codes.

Recently, quantization-based methods [5,9,12,19] have emerged as another promising solution to ANN search. One of the most representative methods is Product Quantization (PQ) [5], whose key idea is to decompose the original feature space into several disjoint subspaces and train codebooks for each subspace respectively. The resulting quantized vector is represented by the Cartesian product of codewords from each codebook. To speed up computation, the Euclidean distances between a query and codewords are computed in advance and stored in a distance table. Finally, the approximate distances between a query and database vectors can be efficiently obtained by looking up the distance table.

Quantization methods learn isomorphic binary codes in a more reasonable way, which have shown more powerful representation than hashing methods. With the boost of research interest on cross-modal similarity search, some cross-modal quantization methods were proposed. Rather than performing quantization directly in the original feature space, these methods embed heterogeneous data into a common space for the subsequent quantization. Compositional Correlation Quantization (CCQ) [7] optimally rotates the original feature space to a correlation-maximal latent space. Collaborative Quantization for Cross-Modal (CMCQ) [20] transforms semantic features to a joint abstract space using matrix factorization. However, label information, which is pivotal for producing discriminative binary codes, is not fully considered in quantization. Moreover, the preservation of correlation between relevant inter-modal pairs is not taken into account.

In this paper, we propose a quantization method named Discriminative Correlation Quantization (DCQ) for cross-modal retrieval. The main idea of DCQ is to learn both discriminative common space embedding and inter-modality correlation preservation in a joint optimization framework. We learn the common space by classification, enabling the proposed method to obtain category-specific features for quantization. In consideration of that the quantization representations for relevant inter-modal pairs should be similar, DCQ produces uniform codes to connect different modalities for cross-modal search. The contributions of our work are summarized as follows:

- We integrate the quantization method into the overall optimization framework, which is helpful to reduce discrete loss in the procedure of code generation.
- We learn isomorphic features by classification, consequently enhancing the discrimination of learned quantization codes.

– Extensive experiments on three public datasets demonstrate that the proposed method is competitive with state-of-the-art methods.

The rest of the paper is organized as follows. Section 2 introduces the related work on cross-modal hashing and quantization. Section 3 describes the detail of our method. Comparative experiments are presented in Sect. 4. Section 5 concludes this paper.

2   Related Work

As our work is tightly related to hashing methods and quantization methods, we review the related literature from these two aspects.

According to the utilization of label information, cross-modal hashing methods can be categorized into unsupervised ones and supervised ones.

Most unsupervised approaches exploit data distribution to learn hash functions. Inter-Media Hashing (IMH) [13] uses a linear regression model to learn hash codes for each modality, and both the inter-modal and the intra-modal consistency is considered in the objective function. The training phase of IMH is of cubic time complexity and therefore is not applicable for large-scale data. Linear Cross-Modal Hashing (LCMH) [23] represents data with small-sized anchors and consequently reduces the computational cost to linear time complexity. Collective Matrix Factorization Hashing (CMFH) [3] assumes that each view of instances has identical hash codes and learns a latent factor model from different views via collective matrix factorization. The query instance is converted into binary codes using the learned latent factor model. Latent Semantic Sparse Hashing (LSSH) [22] learns a high-level latent space based on sparse representations for image and semantic representations for text obtained by matrix factorization. LSSH also learns uniform hash codes for each view. Semantic Topic Multimodal Hashing (STMH) [15] explores semantic topics of texts within clustering patterns and models semantic concepts of images through robust matrix factorization. To match semantic concepts between different modalities, STMH learns a common latent space for topics and concepts for cross-modal retrieval.

Generally speaking, supervised methods are able to generate more discriminative codes than unsupervised ones by using label information. In [18], a semantic similarity matrix is constructed by measuring the cosine similarity between sample labels. SCM uses two distinct optimization strategies for learning hash functions. When orthogonal constraints are imposed, a closed-form solution is obtained by eigen-decomposition. Otherwise, hash codes are learned bit by bit using a sequential learning method. With the assumption that the sample distribution in Hamming space is similar to that in the original space, Semantics-Preserving Hashing (SePH) [6] uses the KL divergence to model the distribution difference. Learning hash functions is cast as a kernel-logistic regression problem. Supervised Matrix Factorization Hashing (SMFH) [14] can be viewed as a supervised variant of CMFH.

The basic idea of quantization methods derives from the K-Means algorithm, which groups a dataset into K clusters and assign each of them one codeword.

Accordingly, one database point can be further expressed as an indicator with only one entry being 1 and others being 0. The nonzero element indicates which codeword is chosen to quantize data point.

However, the K-Means algorithm is not applicable to large-scale datasets. To reduce storage and computation cost, an alternative choice is to use several codebooks for quantization rather than single codebook. PQ [5] divides the original feature space into M disjoint subspaces, which generates K^M codewords to encode data point with M codebooks. The quantization operation performs in different layers and assigns each subvector nearest cluster center at each subspace separably. The high-dimensional vector is finally denoted by a Cartesian product of a group of small cluster centers. Cartesian K-Means [9] multiplies a rotation matrix before quantization, which can be viewed as an extension of PQ. When the rotation matrix degrades to an identity matrix, Cartesian K-Means reduces to PQ. To accelerate query process, Composite Quantization (CQ) [19] introduces a constraint named constant inter-dictionary-element-product, which can also lead to smaller quantization errors.

Recently, vector quantization has been introduced into cross-modal similarity search to deal with discrete loss. Compositional Correlation Quantization (CCQ) [7] rotates the original space to an optimal quantization subspace using the Cartesian K-Means [9] quantizer. The overall framework of Collaborative Quantization for Cross-Modal (CMCQ) [20] is similar to LSSH [22]. The major difference is that the Composite Quantization [19] quantizer is employed to produce binary codes for each modality.

Enlightened by the success of the quantization methods and the significant improvement of supervised information in search accuracy, we make further efforts to exploit the quantization technique in a supervised learning framework.

3   Proposed Method

3.1   Problem Formulation

In this section, we describe the detail of the proposed approach. For simplicity, we study the cross-modal search problem over the objects with two modalities. Assume that we have N training objects denoted by $\mathbf{O} = \{o_i = (\mathbf{x}_i, \mathbf{y}_i)\}_{i=1}^{N}$, where $\mathbf{x}_i \in R^{d_x}$ and $\mathbf{y}_i \in R^{d_y}$ ($d_x \neq d_y$ in most cases) are the feature vectors of image and accompanying text, respectively. Without loss of generality, we assume that feature vectors are zero-centered, i.e., $\sum_{i=1}^{N} \mathbf{x}_i = 0$ and $\sum_{i=1}^{N} \mathbf{y}_i = 0$. Moreover, semantic labels $l_i \in R^c$ are available for training objects, where c is the number of categories. $l_{ij} = 1$ denotes that the i-th training object belongs to class j, otherwise $l_{ij} = 0$.

The target of DCQ is to learn identical codebooks $\{\mathbf{C}^m\}_{m=1}^{M}$ and quantization codes $\{\mathbf{b}_i^m\}_{i=1,m=1}^{N,M}$ for two modalities. $\mathbf{C}^m \in R^{d \times k}$ is the m-th codebook matrix; $\mathbf{b}_i^m \in \{0,1\}^k$ is the corresponding indicator for the i-th object; $\mathbf{M}$ is the number of quantization codebooks; $\mathbf{d}$ is the dimensionality of common space; $\mathbf{k}$ is the length of indicator. As each $\mathbf{b}_i^m$ is an indicator vector with only one nonzero element, hence the final quantization codes can be compressed into $\mathbf{H} = \mathbf{M}\log_2\mathbf{k}$ bits, where $\mathbf{H}$ is the length of codes.

We aim to learn modality-specific features for image data x_i and text data y_i by leveraging semantic labels directly. Intuitively, features are discriminative enough provided that they can be easily classified. Here, "discriminative" implies modality-specific features for different categories should be fully distinguished. Consequently, we import the linear classification form to get modality-specific features $\phi_X(x)$ and $\phi_Y(y)$. For image modality, the cost function is denoted by

$$\min_{W_x} \sum_{i=1}^{N} \|l_i - W_x \cdot \phi_X(x_i)\|_2^2 + \gamma R(W_x), \tag{1}$$

where $\mathbf{W}_x \in R^{c \times d}$ is the multi-class classifier parameter. γ is the regularization parameter and $R(\cdot) = \|\cdot\|_F^2$ is the regularization term to avoid over-fitting. Similarly, we can get discriminative text feature representation by analog with image classification form. Moreover, modality-specific features $\phi_X(x)$ and $\phi_Y(y)$ are also prepared to bridge the semantic gap between original image and text data. We suppose they are shared on common semantic space, where the uniform representation can be generated. Accordingly, We defined two linear projection P_x and P_y matrices as

$$\phi_X(x_i) = P_x \cdot x_i, \phi_Y(y_i) = P_y \cdot y_i, \tag{2}$$

that transform the original data to uniform representation.

To maximize the inter-modality correlation, we propose to share codebooks $\{\mathbf{C}^m\}_{m=1}^M$ for different modalities and share binary codes $\{\mathbf{b}_i^m\}_{i=1,m=1}^{N,M}$ for inter-modal pairs which are semantically relevant. So, the whole problem can be formulated by jointly learning the classifiers, category-specific features and the unified quantizer, as

$$\min_{W,P,C,B} \lambda \left(\sum_{i=1}^{N} \|l_i - W_X P_x x_i\|_2^2 + \mu \sum_{i=1}^{N} \left\| P_x x_i - \sum_{m=1}^{M} \mathbf{C}^m \cdot \mathbf{b}_i^m \right\|_2^2 \right)$$
$$+ (1-\lambda) \left(\sum_{i=1}^{N} \|l_i - W_Y P_y y_i\|_2^2 + \sum_{i=1}^{N} \left\| P_y y_i - \sum_{m=1}^{M} \mathbf{C}^m \cdot \mathbf{b}_i^m \right\|_2^2 \right) \tag{3}$$
$$s.t. \|\mathbf{b}_i^m\|_1 = 1, \mathbf{b}_i^m \in \{0,1\}^k, i \in \{1, \cdots, N\}, m \in \{1, \cdots, M\}.$$

where $\lambda \in (0,1)$ and μ are tradeoff parameters. We set $\lambda = 0.5$, $\gamma = 1e{-}3$, and $\mu = 100$, which are fine-tuned throughout this paper. And, following the settings in [7], we set $k = 256$.

3.2   Optimization

Our problem consists of six sets of variables, $\mathbf{W}_x$, $\mathbf{W}_y$, $\mathbf{P}_x$, $\mathbf{P}_y$, $\{\mathbf{C}^m\}_{m=1}^M$ and $\{\mathbf{b}_i^m\}_{i=1,m=1}^{N,M}$. We use the alternated optimization method to iteratively solve the problem by updating one set of variables with the rest variables fixed.

Update $\mathbf{W}_x$. With $\mathbf{W}_y$, $\mathbf{P}_x$, $\mathbf{P}_y$, $\{\mathbf{C}^m\}_{m=1}^M$ and $\{\mathbf{b}_i^m\}_{i=1,m=1}^{N,M}$ fixed, set the derivation of $O(\cdot)$ with respect to $\mathbf{W}_x$ to zero, and obtain

$$\mathbf{W}_x = \mathbf{LX}^T\mathbf{P}_x{}^T\left(\mathbf{P}_x\mathbf{XX}^T\mathbf{P}_x{}^T + \gamma\mathbf{I}\right)^{-1}, \tag{4}$$

where I is the identity matrix.

Update $\mathbf{W}_y$. With $\mathbf{W}_x$, $\mathbf{P}_x$, $\mathbf{P}_y$, $\{\mathbf{C}^m\}_{m=1}^M$ and $\{\mathbf{b}_i^m\}_{i=1,m=1}^{N,M}$ fixed, set the derivation of $O(\cdot)$ with respect to $\mathbf{W}_y$ to zero, and obtain

$$\mathbf{W}_y = \mathbf{LY}^T\mathbf{P}_y{}^T\left(\mathbf{P}_y\mathbf{YY}^T\mathbf{P}_y{}^T + \gamma\mathbf{I}\right)^{-1}. \tag{5}$$

Update $\mathbf{P}_x$. With $\mathbf{W}_x$, $\mathbf{W}_y$, $\mathbf{P}_y$, $\{\mathbf{C}^m\}_{m=1}^M$ and $\{\mathbf{b}_i^m\}_{i=1,m=1}^{N,M}$ fixed, set the derivation of $O(\cdot)$ with respect to $\mathbf{P}_x$ to zero, and obtain

$$\mathbf{P}_x = \left(\mathbf{W}_x{}^T\mathbf{W}_x + \mu\mathbf{I}\right)^{-1}\left(\mathbf{W}_x{}^T\mathbf{LX}^T + \mu\mathbf{CBX}^T\right)\left(\mathbf{XX}^T\right)^{-1}, \tag{6}$$

where $\mathbf{C} = \left[\mathbf{C}^1, \cdots, \mathbf{C}^M\right] \in R^{d \times (M \times k)}$, $\mathbf{B} = \begin{bmatrix} \mathbf{b}_1^1 & \cdots & \mathbf{b}_N^1 \\ \vdots & \ddots & \vdots \\ \mathbf{b}_1^M & \cdots & \mathbf{b}_N^M \end{bmatrix} \in R^{(M \times k) \times N}$.

Update $\mathbf{P}_y$. With $\mathbf{W}_x$, $\mathbf{W}_y$, $\mathbf{P}_x$, $\{\mathbf{C}^m\}_{m=1}^M$ and $\{\mathbf{b}_i^m\}_{i=1,m=1}^{N,M}$ fixed, set the derivate of $O(\cdot)$ with respect to P_y to zero, and obtain

$$\mathbf{P}_y = \left(\mathbf{W}_y{}^T\mathbf{W}_y + \mu\mathbf{I}\right)^{-1}\left(\mathbf{W}_y{}^T\mathbf{LY}^T + \mu\mathbf{CBY}^T\right)\left(\mathbf{YY}^T\right)^{-1}. \tag{7}$$

Update $\mathbf{C}$. With other variables fixed, we rewrite the objective function with respect to $\mathbf{C}$ in matrix formulation,

$$\min_C \lambda\left\|\mathbf{P}_x\mathbf{X} - \mathbf{CB}\right\|_F^2 + (1 - \lambda)\left\|\mathbf{P}_y\mathbf{Y} - \mathbf{CB}\right\|_F^2. \tag{8}$$

It is an unconstrained quadratic problem with analytic solution:

$$\mathbf{C} = \left(\lambda\mathbf{P}_x\mathbf{XB}^T + (1 - \lambda)\mathbf{P}_y\mathbf{YB}^T\right)\left(\mathbf{BB}^T\right)^{-1}. \tag{9}$$

Some algorithms such as L-BFGS can be used to accelerate computation.

Update B. As the binary vectors $\{\mathbf{b}_i\}_{i=1}^N$ are independent of each other, the optimization problem can be decomposed into N sub-problems,

$$\min_B \lambda \sum_{i=1}^N \left\|\mathbf{P}_x x_i - \sum_{m=1}^M \mathbf{C}^m \cdot \mathbf{b}_i^m\right\|_2^2 + (1 - \lambda) \sum_{i=1}^N \left\|\mathbf{P}_y y_i - \sum_{m=1}^M \mathbf{C}^m \cdot \mathbf{b}_i^m\right\|_2^2 \tag{10}$$
$$s.t.\|\mathbf{b}_i^m\|_1 = 1, \mathbf{b}_i^m \in \{0,1\}^k, i \in \{1, \cdots, N\}, m \in \{1, \cdots, M\}.$$

This optimization problem is NP-hard. Therefore, we turn to use an approximate strategy by greedily updating the $\mathbf{M}$ indicator vectors $\{\mathbf{b}_i^m\}_{m=1}^M$ alternately.

In more detail, with $\{\mathbf{b}_i^n\}_{n=1,n\neq m}^M$ fixed, $\mathbf{b}_i^m$ is updated by exhaustively checking all the codewords in codebook $\mathbf{C}^m$ such that Eq. (10) is minimized. Then, we set the corresponding entry of b_i^m to be 1 and all the others to be 0.

This algorithm terminates when reaching convergence or the maximal number of iterations. In the following experiments, we empirically set the maximal number of iterations to 300. The overall algorithm is summarized in Algorithm 1.

3.3   Search Process

Given a query q, we obtain its category-specific representation in the transformed space by $\tilde{q} = \mathbf{P}_x \cdot q$ or $\tilde{q} = \mathbf{P}_y \cdot q$. The approximate distance between a query and a database item is computed by

$$\left\| \tilde{q} - \sum_{m=1}^{M} \mathbf{C}^m \mathbf{b}_i^m \right\|_2^2 = -2 \sum_{m=1}^{M} \langle \tilde{q}, \mathbf{C}^m \mathbf{b}_i^m \rangle + \left\| \sum_{m=1}^{M} \mathbf{C}^m \mathbf{b}_i^m \right\|_2^2 + \|\tilde{q}\|_2^2 \qquad (11)$$

The first term in the right-hand side of Eq. (11) denotes the inner products between $\tilde{q}$ and codewords determined by b_i^m. For a given query, these inner products can be precomputed and stored in a $\mathbf{M} \times \mathbf{K}$ look-up table. The second and the third terms are both constants for all database samples.

Algorithm 1. Discriminative Correlation Quantization

Require:
    Semantic label matrix $\mathbf{L}$, data matrix $\mathbf{X}$, $\mathbf{Y}$; transformed dimension $\mathbf{d}$, parameters $\lambda, \mu, \gamma, \mathbf{M}$.

Ensure:
    Linear classifiers $\mathbf{W}_x$, $\mathbf{W}_y$, mappings $\mathbf{P}_x$, $\mathbf{P}_y$, codebooks $\{\mathbf{C}^m\}_{m=1}^{M}$ and binary indicators $\{\mathbf{b}_i^m\}_{i=1,m=1}^{N,M}$.

1: Initialize $\mathbf{W}_x$, $\mathbf{W}_y$, $\mathbf{P}_x$, $\mathbf{P}_y$ by random matrices, $\{\mathbf{C}^m\}_{m=1}^{M}$ by seeds, $\{\mathbf{b}_i^m\}_{i=1,m=1}^{N,M}$ by NN search.

2: **repeat**

3:     Fix $\mathbf{W}_y$, $\mathbf{P}_x$, $\mathbf{P}_y$, $\{\mathbf{C}^m\}_{m=1}^{M}$ and $\{\mathbf{b}_i^m\}_{i=1,m=1}^{N,M}$, update $\mathbf{W}_x$ by Eq.(4);

4:     Fix $\mathbf{W}_x$, $\mathbf{P}_x$, $\mathbf{P}_y$, $\{\mathbf{C}^m\}_{m=1}^{M}$ and $\{\mathbf{b}_i^m\}_{i=1,m=1}^{N,M}$, update $\mathbf{W}_y$ by Eq.(5);

5:     Fix $\mathbf{W}_x$, $\mathbf{W}_y$, $\mathbf{P}_y$, $\{\mathbf{C}^m\}_{m=1}^{M}$ and $\{\mathbf{b}_i^m\}_{i=1,m=1}^{N,M}$, update $\mathbf{P}_x$ by Eq.(6);

6:     Fix $\mathbf{W}_x$, $\mathbf{W}_y$, $\mathbf{P}_x$, $\{\mathbf{C}^m\}_{m=1}^{M}$ and $\{\mathbf{b}_i^m\}_{i=1,m=1}^{N,M}$, update $\mathbf{P}_y$ by Eq.(7);

7:     Fix $\mathbf{W}_x$, $\mathbf{W}_y$, $\mathbf{P}_x$, $\mathbf{P}_y$ and $\{\mathbf{b}_i^m\}_{i=1,m=1}^{N,M}$, update $\{\mathbf{C}^m\}_{m=1}^{M}$ by Eq.(9);

8:     **for** $i \to 1$ to N **do**

9:         Update $\{\mathbf{b}_i^m\}_{i=1,m=1}^{N,M}$ by greedy algorithm as Eq.(10);

10:    **end for**

11: **until** convergence.

---

4   Experiments

4.1   Experimental Setup

In this section, we evaluate the proposed approach on three benchmark datasets, *i.e.*, the [4] **MIRFlickr** dataset [2], the **NUS-WIDE** dataset and the **Wiki** dataset [10]. The details of three datasets are summarized in Table 1. We conduct two kinds of cross-modal retrieval task, *i.e.*, *Img to Txt* and *Txt to Img*. We employ mean Average Precision (mAP) as the evaluation metric and report the precision of the top 200 retrieved items. Besides, we report the results with respect to the *precision-recall* curve and the *topN-precision* curve. The *precision-recall* curve shows the precision varying at different recall and the *topN-precision* curve illustrates the precision varying at different numbers of retrieved samples. To validate the proposed approach, we comprehensively compare it with six state-of-the-art cross-modal search methods, including CMFH [3], LSSH [22], SCM [18], SePH [6], CCQ [7], and SMFH [14]. All the experiments are conducted on a workstation with Intel(R) Core(TM) CPU E5-2650@2.3 GHz and 64 GB RAM.

Table 1. The details of the evaluated datasets

Datasets	MIRFlickr	NUS-WIDE	Wiki
Database size	15902	186577	2866
Training size	5000	5000	2173
Query size	836	1866	693
No. of labels	24	10	10

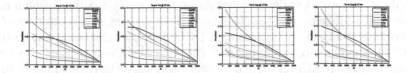

Fig. 1. *TopN-Precision* Curves on the MIRFlickr dataset varying code length.

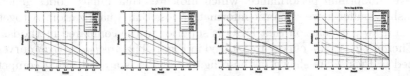

Fig. 2. *Precison-recall* Curves on the MIRFlickr dataset varying code length.

4.2   Experimental Results and Analysis

We report mAP results for DCQ and other compared methods in Table 2. From Table 2, we have the following three observations. Firstly, with the code length ranges varying from 16 bits to 64 bits, DCQ achieves the best performance on the MIRFlickr and NUS-WIDE datasets. Especially, DCQ outperforms all the compared methods on different kinds of retrieval tasks even when the code length is far shorter. For example, when conducting *Img to Txt* task on the MIRFlickr dataset with 16 bits code length, the mAP result of DCQ is 0.7186, which surpasses the best mAP results of all the compared methods by about 1%

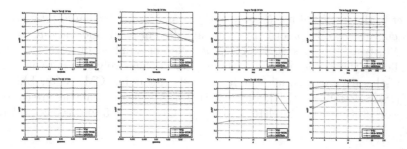

Fig. 3. Parameter sensitivity analysis.

with 64 bits code length. And when working on *Txt to Img* task in the same context, the mAP result of DCQ is 0.8562 and outperforms the best result of the compared methods by about 8% with 64 bits code length. Similar results can be observed with other cases. This indicates that our approach can capture much more semantic information with shorter code length when dealing with high-dimensional data. As for the Wiki dataset, the results are slightly different from those on the other two datasets. We consider that it may arise from that the Wiki is a relative low-dimensional dataset with only 128-dimensional image features and 10-dimensional text features, which can not play a positive role on the encoding ability of DCQ. Secondly, DCQ makes a substantial improvement over CCQ. For instance, DCQ outperforms CCQ by about 20% on the *Img to Txt* task and about 23% on the *Txt to Img* task, when the code length is 16 bits and the dataset is MIRFlickr. We attribute the better performance to its ability to explore the label information and hence produce more discriminative binary codes. Thirdly, SCM, SMFH and SePH are supervised hashing methods but achieve suboptimal performance, which indicates that binary codes generated by hashing suffer more semantic information loss than quantization. Altogether, we can draw a conclusion that DCQ is superior to the compared methods.

The *topN-precision* curves are plotted in Fig. 1. The *precision-recall* curves are plotted in Figs. 2. As shown in these figures, SePH is a challenging competitor for our approach. In most cases, DCQ produces better results on the top 200 retrieved instances and achieves higher precision than SePH when the recall level is less than 0.2. In practical retrieval systems, users have more interest in the top retrieved instances. Therefore, our approach is more suitable for real-world scenarios.

4.3 Parameter Sensitivity

Our approach has four parameters: λ, μ, γ, d. To validate the robustness of DCQ, we conduct parameter sensitivity study by varying the value of a parameter with other parameters fixed. We fix the code length at 16 bits, and use the mAP results to evaluate the performance. These experimental results are plotted in Fig. 3. We can learn that DCQ is able to yield satisfactory results in a wide range of parameter values.

Table 2. Overall comparison with mAP results of top 200 retrieved items on three datasets. Items in bold indicate the best performance.

Task	Methods	MIRFlickr				NUS-WIDE				Wiki			
		16bits	32bits	48bits	64bits	16bits	32bits	48bits	64bits	16bits	32bits	48bits	64bits
Img to Txt	LSSH	0.5999	0.6008	0.6124	0.6146	0.4434	0.4524	0.4586	0.4578	0.2166	0.2061	0.2188	0.2007
	CMFH	0.6194	0.6135	0.6110	0.6089	0.4359	0.4278	0.4287	0.4251	0.2224	0.2382	0.2360	0.2373
	CCQ	0.6285	0.6350	0.6319	0.6411	0.5179	0.5158	0.5114	0.5188	0.2151	0.2250	0.2359	0.2426
	SCM	0.6735	0.6797	0.6895	0.6975	0.5523	0.5689	0.5654	0.5697	0.2280	0.2243	0.2328	0.2278
	SePH	0.6481	0.6844	0.7074	0.7098	0.5648	0.5644	0.5649	0.5721	**0.2481**	**0.2716**	**0.2825**	**0.2987**
	SMFH	0.6637	0.6842	0.6838	0.6926	0.4785	0.4530	0.4320	0.4122	0.2312	0.2569	0.2476	0.2495
	DCQ	**0.7186**	**0.7209**	**0.7212**	**0.7250**	**0.6196**	**0.6269**	**0.6416**	**0.6433**	0.2379	0.2387	0.2489	0.2434
Txt to Img	LSSH	0.6542	0.6839	0.6898	0.7007	0.5365	0.5658	0.5734	0.5806	0.5752	0.5942	0.5925	0.5930
	CMFH	0.6657	0.6846	0.6920	0.7006	0.4212	0.4143	0.4122	0.4126	0.5617	0.5919	0.5938	0.5947
	CCQ	0.6273	0.6317	0.6396	0.6361	0.6033	0.6123	0.6101	0.6109	0.5667	0.5787	0.5873	0.5841
	SCM	0.6731	0.6854	0.6975	0.6985	0.5919	0.6199	0.6389	0.6343	0.3203	0.3613	0.3632	0.3514
	SePH	0.7134	0.7520	0.8009	0.7887	0.6962	0.7090	0.7027	0.7171	**0.6644**	**0.6805**	**0.6814**	**0.6829**
	SMFH	0.6982	0.7249	0.7317	0.7370	0.5419	0.5016	0.4607	0.4461	0.6273	0.6343	0.6348	0.6380
	DCQ	**0.8562**	**0.8570**	**0.8572**	**0.8610**	**0.7465**	**0.7532**	**0.7625**	**0.7648**	0.6219	0.6208	0.6240	0.6028

5 Conclusion

In this paper, we have proposed a novel supervised compact coding approach based on vector quantization, referred to Discriminative Correlation Quantization, for cross-modal similarity search. DCQ builds discriminative category-specific representation in terms of linear classification and jointly learns the unified quantizers for both modalities in quantization subspaces. Extensive experiments on several benchmark datasets have shown the superiority of the proposed approach.

Acknowledgments. This work was supported by the Natural Science Foundation of China under grants 61772032 and 61501003.

References

1. Chen, Z., Zhou, J.: Collaborative multiview hashing. Pattern Recognit. **75**, 149 160 (2017)
2. Chua, T.S., Tang, J., Hong, R., Li, H., Luo, Z., Zheng, Y.: NUS-wide: a real-world web image database from national university of Singapore. In: ACM International Conference on Image and Video Retrieval, p. 48 (2009)
3. Ding, G., Guo, Y., Zhou, J.: Collective matrix factorization hashing for multimodal data. In: IEEE Conference on Computer Vision and Pattern Recognition, pp. 2083–2090 (2014)
4. Huiskes, M.J., Lew, M.S.: The MIR FLICKR retrieval evaluation. In: ACM International Conference on Multimedia Information Retrieval, pp. 39–43 (2008)
5. Jegou, H., Douze, M., Schmid, C.: Product quantization for nearest neighbor search. IEEE Trans. Pattern Anal. Mach. Intell. **33**(1), 117–128 (2011)
6. Lin, Z., Ding, G., Hu, M., Wang, J.: Semantics-preserving hashing for cross-view retrieval, pp. 3864–3872 (2015)

7. Long, M., Cao, Y., Wang, J., Yu, P.S.: Composite correlation quantization for efficient multimodal retrieval. In: Computer Science, pp. 579–588 (2016)
8. Muja, M.: Fast approximate nearest neighbors with automatic algorithm configuration. In: International Conference on Computer Vision Theory and Application Vissapp, pp. 331–340 (2009)
9. Norouzi, M., Fleet, D.J.: Cartesian k-means. In: Computer Vision and Pattern Recognition, pp. 3017–3024 (2013)
10. Rasiwasia, N., Pereira, J.C., Coviello, E., Doyle, G., Lanckriet, G.R.G., Levy, R., Vasconcelos, N.: A new approach to cross-modal multimedia retrieval. In: International Conference on Multimedia, pp. 251–260 (2010)
11. Silpa-Anan, C., Hartley, R.: Optimised KD-trees for fast image descriptor matching. In: 2008 IEEE Conference on Computer Vision and Pattern Recognition, CVPR 2008, pp. 1–8 (2008)
12. Song, J., Gao, L., Liu, L., Zhu, X., Sebe, N.: Quantization-based hashing: a general framework for scalable image and video retrieval. Pattern Recognit. **75**, 175–187 (2017)
13. Song, J., Yang, Y., Yang, Y., Huang, Z., Shen, H.T.: Inter-media hashing for large-scale retrieval from heterogeneous data sources. In: ACM SIGMOD International Conference on Management of Data, pp. 785–796 (2013)
14. Tang, J., Wang, K., Shao, L.: Supervised matrix factorization hashing for cross-modal retrieval. IEEE Trans. Image Process. **25**(7), 3157–3166 (2016). A Publication of the IEEE Signal Processing Society
15. Wang, D., Gao, X., Wang, X., He, L.: Semantic topic multimodal hashing for cross-media retrieval. In: International Conference on Artificial Intelligence, pp. 3890–3896 (2015)
16. Wang, J., Li, S.: Query-driven iterated neighborhood graph search for large scale indexing. In: ACM International Conference on Multimedia, pp. 179–188 (2012)
17. Wang, K., Tang, J., Wang, N., Shao, L.: Semantic boosting cross-modal hashing for efficient multimedia retrieval. Inf. Sci. Int. J. **330**(C), 199–210 (2016)
18. Zhang, D., Li, W.J.: Large-scale supervised multimodal hashing with semantic correlation maximization. In: Twenty-Eighth AAAI Conference on Artificial Intelligence, pp. 2177–2183 (2014)
19. Zhang, T., Du, C., Wang, J.: Composite quantization for approximate nearest neighbor search. In: International Conference on Machine Learning, pp. 838–846 (2014)
20. Zhang, T., Wang, J.: Collaborative quantization for cross-modal similarity search. In: IEEE Conference on Computer Vision and Pattern Recognition, pp. 2036–2045 (2016)
21. Zhen, Y., Yeung, D.Y.: Co-regularized hashing for multimodal data. Adv. Neural Inf. Process. Syst. **2**, 1385–1393 (2013)
22. Zhou, J., Ding, G., Guo, Y.: Latent semantic sparse hashing for cross-modal similarity search, pp. 415–424 (2014)
23. Zhu, X., Huang, Z., Shen, H.T., Zhao, X.: Linear cross-modal hashing for efficient multimedia search, pp. 143–152 (2013)

GPU Assisted Towards Real-Time Reconstruction for Dual-Camera Compressive Hyperspectral Imaging

Shipeng Zhang[1], Lizhi Wang[2], Ying Fu[2], and Hua Huang[2(✉)]

[1] Xi'an Jiaotong University, Xi'an 710049, China
zsp6869123@stu.xjtu.edu.cn
[2] Beijing Institute of Technology, Beijing 100081, China
{lzwang,fuying,huahuang}@bit.edu.cn

Abstract. The dual-camera compressive hyperspectral imager (DCCHI) can capture 3D hyperspectral image (HSI) with a single snapshot. However, due to the high computation complexity of reconstruction methods, DCCHI cannot apply to the time-crucial applications. In this paper, we propose a GPU assisted towards real-time reconstruction framework for DCCHI. First, leveraging the fast convergence rate of the alternative direction multiplier method, we propose a reformative reconstruction algorithm which can achieve a fast convergence rate. Then, using the interpolation results of a low resolution reconstructed HSI as the warm start, we propose a fast reconstruction strategy to further reduce the computation burden. Last, a GPU parallel implementation is presented to achieve nearly real-time reconstruction. Evaluation experiments indicate our framework can obtain a significant promotion in reconstruction efficiency with a slight accuracy loss.

Keywords: Parallel optimization
Alternative direction multiplier method · Real-time reconstruction
Compressive hyperspectral imaging

1 Introduction

Compared with grayscale image or RGB color image, hyperspectral image (HSI) contains additional spectral information and thus owns a broader use. Nowadays, HSI has been applied to various fields such as medicine [10], agriculture [16], biology [4] and so on. In order to obtain HSI, the traditional hyperspectral imagers utilize the dispersive elements and the array detectors for continuous spectrum scanning [2,13]. However, such imagers have a significant limitation on scanning speed. Thanks to the development of compressive sensing (CS) theory, snapshot imaging in recent years has presented an alternative solution [6,9,15]. As the latest design, dual-camera compressive hyperspectral imaging (DCCHI) [19,20] incorporates a co-located grayscale camera to collect complementary information simultaneously with the coded aperture snapshot spectral imaging system

© Springer Nature Switzerland AG 2018
R. Hong et al. (Eds.): PCM 2018, LNCS 11164, pp. 711–720, 2018.
https://doi.org/10.1007/978-3-030-00776-8_65

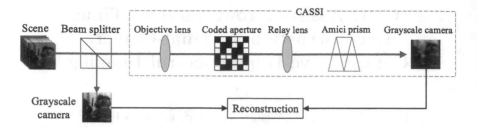

Fig. 1. The DCCHI system.

(CASSI) [1,17]. Compared with CASSI, DCCHI improves the imaging quality and obtains the potential to apply snapshot imaging into practice.

So far considerable reconstruction methods have been deployed on DCCHI. An approximate solution can be obtained by the two-step iterative shrinkage/thresholding (TwIST) [3] with L_1 regularization. Using the sparsity prior information, HSI can also be recovered by sparse representation. An adaptive dictionary derived from the panchromatic measurements was proposed in [20] for 4D hyperspectral video acquisition. Recently, an adaptive nonlocal sparse representation model is proposed to enhance the sparsity correlation among spectral-spatial dimension and achieved high-quality HSI reconstruction [18]. However, due to the high computation complexity, the reconstruction speed of all current algorithms are always slow. For example, TwIST takes 7 min to reconstruct a $512 \times 512 \times 31$ HSI for DCCHI on a platform of Inter i7 6800 K. It means that more than 3 h will be cost for reconstructing 1s hyperspectral video with 30 frames. Due to the heavy computational burden caused by dictionary learning and matrix inverse operation, sparse representation methods are more time-consuming than TwIST. Therefore, DCCHI cannot be featured as true real-time HSI acquisition system and has limitations for time-crucial applications.

In this paper, we propose a GPU assisted towards real-time reconstruction framework for DCCHI. The contributions of our work include: (1) Integrating the alternative direction multiplier method (ADMM) [5] with total variation (TV) regularization, a reformative reconstruction algorithm with a fast convergence rate is proposed. (2) Leveraging the interpolation results of the low resolution reconstructed HSI as a warm start, we propose a fast HSI reconstruction strategy for further reducing the computation burden. (3) A novel GPU implementation, including a series of parallel optimization, is presented to achieve nearly real-time HSI reconstruction.

2   Method Principles

2.1   Reformative TV Based Reconstruction Algorithm

Figure 1 shows the system diagram of DCCHI which consists of a CASSI system and a panchromatic camera (PanCam). Let $f(x, y, \lambda)$ denotes the original

3D HSI, where $1 \leq \lambda \leq \Omega$ indexes the spectral coordinate and $1 \leq x \leq M$, $1 \leq y \leq N$ index the spatial coordinate. After spatial modulation by the coded aperture and spectral dispersion by the Amici prism, the compressive measurement at position (x, y) on the camera plane of CASSI can be represented as

$$g_c(x, y) = 0.5 \sum_{\lambda=1}^{\Omega} \omega(\lambda) \varphi(x - \phi(\lambda), y) f(x - \phi(\lambda), y, \lambda) \tag{1}$$

where $\varphi(x, y)$ denotes the mask of the aperture code, $\phi(\lambda)$ denotes the dispersion relative distance introduced by Amici Prism and $\omega(\lambda)$ is the spectral response of the camera. In another branch, the grayscale camera directly obtains the uncoded panchromatic measurement, which can be represented as

$$g_p(x, y) = 0.5 \sum_{\lambda=1}^{\Omega} \omega(\lambda) f(x, y, \lambda) \tag{2}$$

For brevity, Eqs. 1 and 2 can be rewritten in an associated matrix form as

$$\mathbf{G} = \mathbf{HF} \tag{3}$$

where $\mathbf{G} = [\mathbf{G_c}^T, \mathbf{G_p}^T]^T$ is the vectorized representation of the dual measurements. $\mathbf{H} = [\mathbf{H_c}^T, \mathbf{H_p}^T]^T$ denotes the sparse matrix of the overall system forward response. $\mathbf{F}$ is the vectorized presentation of HSI. Under the assumption of piecewise smoothing in HSI, $\mathbf{F}$ can be reconstructed by solving

$$\mathbf{F} = \arg\min_{\mathbf{F}} \ 0.5 \|\mathbf{G} - \mathbf{HF}\|_2^2 + \alpha \|\mathbf{DF}\|_1 \tag{4}$$

where α is the regularization factor and $\|\mathbf{DF}\|_1$ is the spatial isotropy TV of HSI, which is defined as

$$\|\mathbf{DF}\|_1 = \sum_{x,y,\lambda} \sqrt{[f(x, y+1, \lambda) - f(x, y, \lambda)]^2 + [f(x+1, y, \lambda) - f(x, y, \lambda)]^2} \tag{5}$$

where $\mathbf{D}$ denotes the forward difference matrix and $\mathbf{DF} = [f(x, y+1) - f(x, y), f(x+1, y) - f(x, y)]^T$.

To achieve a fast convergence rate, we introduce ADMM to solve this problem. Specifically, Eq. 4 can be reformulated into

$$(\mathbf{F}, \mathbf{S}) = \arg\min_{\mathbf{F}, \mathbf{S}} \frac{1}{2} \|\mathbf{G} - \mathbf{HF}\|_2^2 + \alpha \|\mathbf{S}\|_1 \ s.t. \ \mathbf{S} = \mathbf{DF} \tag{6}$$

where $\mathbf{S}$ is am auxiliary variable. Then, the augmented Lagrangian function of Eq. 6 can be obtained

$$L_\mu(\mathbf{F}, \mathbf{S}, \mathbf{U}) = \frac{1}{2} \|\mathbf{G} - \mathbf{HF}\|_2^2 + \alpha \|\mathbf{S}\|_1 + \frac{\mu}{2} \left\| \mathbf{S} - \mathbf{DF} + \frac{\mathbf{U}}{\mu} \right\|_2^2 \tag{7}$$

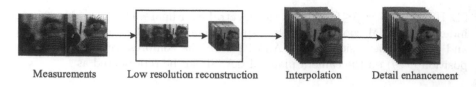

Measurements   Low resolution reconstruction   Interpolation   Detail enhancement

Fig. 2. Fast reconstruction strategy.

where $\mathbf{U}$ is a lagrangian multiplier and μ is a regularization factor. With the strategy of variable splitting and alternative updating, Eq. 6 can be solved iteratively

$$\mathbf{F}^{i+1} = (\mathbf{H}^T\mathbf{H} + \mu\mathbf{D}^T\mathbf{D})^{-1}(\mathbf{H}^T\mathbf{G} + \mathbf{D}^T(\mathbf{U}^i + \mu\mathbf{S}^i)) \tag{8}$$

$$\mathbf{S}^{i+1} = soft(\mathbf{D}\mathbf{F}^{i+1} - \frac{\mathbf{U}^i}{\mu}, \frac{\alpha}{\mu}) \tag{9}$$

$$\mathbf{U}^{i+1} = \mathbf{U}^i + \mu\left(\mathbf{S}^{i+1} - \mathbf{D}\mathbf{F}^{i+1}\right) \tag{10}$$

where $soft()$ is the soft thresholding function [7] and i denotes the iteration number. An approximate solution of Eq. 8 can be obtained by the conjugate gradient (CG) algorithm [11].

2.2 Fast Reconstruction Strategy

In this part, we present a HSI reconstruction strategy to further reduce computational cost. As shown in Fig. 2, the reconstruction is divided into three main stages: low resolution reconstruction, interpolation and detail enhancement. In the first stage, we use the proposed algorithm to reconstruct a low resolution HSI from the downsampling results of the measurements. The nearest interpolation and the average filtering are applied on the CASSI measurement and the PanCam measurement for downsampling, respectively. It should be noted that the downsamplings in spatial dimension and spectral dimension hold the same scale factor. Then the bilinear interpolation is adopted to obtain a warm start for the full-resolution HSI reconstruction. The proposed algorithm is utilized again in the detail enhancement to polish the initial value and guarantee the reconstruction fidelity.

Let β denotes the scale factor, the calculation cost in low resolution reconstruction is $\frac{1}{\beta^3}$ of that in detail enhancement. At the same time, the complexity of the interpolation operators is very low thus the first two stages can be carried out rapidly. Owing to the fast convergence rate of the reformative algorithm, an acceptable reconstruction result using the warm start can be achieved with a low iteration number. To sum up, this strategy can decrease the computation burden with a slight loss in reconstruction accuracy.

2.3   Parallel Implementation

In our work, the compute unified device architecture [8] is utilized for efficient parallel execution. The computation burden of the reconstruction is mainly determined by forward response, interpolation and difference operations. We will exploit the parallelism in these three aspects. Details are discussed as follows.

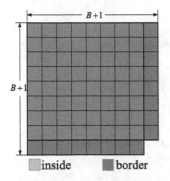

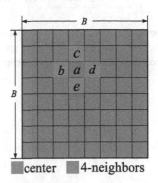

Fig. 3. The shared memory structure for forward difference optimization.

Fig. 4. CSM pattern.

Parallelism in Forward Response. A look-up table with the size of $2 \times (N + \Omega - 1)$ is created once for efficient execution of forward response operator, i.e., **HF**,

$$LUT(1,x) = \begin{cases} 1, & x \leq N \\ x - \Omega + 1 & \text{else} \end{cases}, \quad LUT(2,x) = \begin{cases} x, & x \leq \Omega \\ \Omega & \text{else} \end{cases} \quad (11)$$

Then Eq. 1 can be rewrite as:

$$g_c(x,y) = \sum_{m=LUT(1,x)}^{LUT(2,x)} H_c(x - m, y, m) \times f(x - m, y, m) \quad (12)$$

where $m = \varphi(\lambda)$ when $1 \leq x - \varphi(\lambda) \leq M$. Without the boundary judgments caused by the item $x - \varphi(\lambda)$ in Eq. 1, warps divergent in GPU can be validly ruled out. Because there is no dispersion in the PanCam branch, Eq. 2 can be directly computed. At the same time, a flexible shared memory for intermediate results can also been used for Eqs. 2 and 12.

Parallelism in Interpolation. The image interpolation can be treated as a linear combination of specific pixels. It means that a unified interpolation model can be represented as

$$d(j) = \sum_{m=1}^{K} s(I(j,m)) \times W(j,m) \quad (13)$$

Table 1. Performance comparison of the fast reconstruction strategy under different scale factors.

β	PSNR(dB)			SAM			Time(s)		
	2	3	4	2	3	4	2	3	4
$\sigma = 0.1$	36.45	35.81	35.72	0.19	0.20	0.20	16.08	15.62	15.60
$\sigma = 0.2$	34.86	34.48	34.43	0.24	0.24	0.25	16.19	15.38	15.16

Table 2. Performance of the fast reconstruction strategy. For compact tabulation, we use "w/FS" and "w/o FS" represent "with fast reconstruction strategy" and "without fast reconstruction strategy", respectively.

Noise variance	Loss		PSNR(dB)		Time(s)	
	w/FS	w/o FS	w/FS	w/o FS	w/FS	w/o FS
$\sigma = 0.1$	6.37E+03	6.19E+03	36.45	36.67	16.08	28.00
$\sigma = 0.2$	9.28E+03	8.98E+03	34.86	35.18	16.19	28.50

where the variable d and s denote the target image and the source image, respectively. j is the pixel index in the vectored image. K denotes the required number of pixels in the source. For example, $K = 1$ for the nearest interpolation and $K = 4$ for the bilinear interpolation. $I(j, m)$ and $W(j, m)$ denote the index matrix and the weight matrix, respectively. Once the resolution and scale factor are determined, we can generate I and W in advance as the lookup table, so as to implement the fast image interpolation.

Parallelism in Difference Operations. For the forward difference **DF**, we use a shared memory structure, as shown in Fig. 3, to optimize the data loading, where B denotes the side length of the square thread block layout adopted in our implementation. Two additional blocks are embedded for boundary access. Due to reduced redundant access to the global memory, a speed acceleration can be figured out.

For the inverse difference $\mathbf{D}^T\mathbf{D}_{\mathbf{hv}}$, where $\mathbf{D}_{\mathbf{hv}} = \mathbf{DF}$, we load all data directly from the global memory to avoid performance degradation caused by two extra thread synchronization in the shared memory.

We further simplify the calculation of $\mathbf{D}^T\mathbf{D}(\mathbf{F})$, which will be used in CG for solving Eq. 8. As shown in Fig. 4, the result of $\mathbf{D}^T\mathbf{D}(a)$ can be represented in one step

$$\mathbf{D}^T\mathbf{D}(a) = \mathbf{D}^T[\mathbf{D}(a)] = \mathbf{D}^T \begin{bmatrix} c - a \\ b - a \end{bmatrix} = 4a - b - c - d - e \tag{14}$$

A "Central Shared Memory support" (CSM) pattern shown in Fig. 4, is exploited for a further optimization. Specifically, the center pixel a is read from the shared memory, and the 4-neighbors b, c, d and e are read from the global memory directly. Compared with the "Full Shared Memory support" pattern introduced in [12], CSM can not only make use of the shared memory but also avoid divergent branches, thus promote the implementation performance.

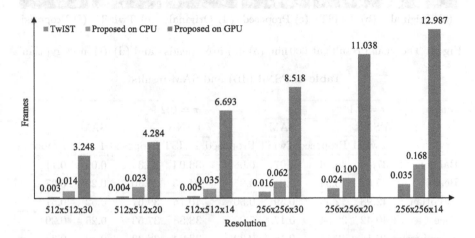

Fig. 5. Frame rates of different methods.

Table 3. Reconstruction time (in seconds) and parallel acceleration ratio

Spatial resolution		256×256			512×512			1024×1024		
Bands		14	20	30	14	20	30	14	20	30
TwIST		28.86	41.16	63.54	188.28	261.12	399.34	807.61	1177.81	1723.50
Proposed	On CPU	5.95	9.95	16.08	28.66	43.85	70.14	91.02	131.51	231.41
	On GPU	0.08	0.09	0.12	0.15	0.23	0.31	0.36	0.50	0.86
	Speedup	77.21	109.84	137.00	191.83	187.85	227.80	252.70	261.30	269.89

3 Experiments

In this section, we evaluate the experiment results of the proposed framework. The TwIST algorithm with TV regularization is used for comparison. The coded mask obeys a random Bernoulli distribution. White Gaussian noise is added in generating the measurements. Experiments are conducted on a platform of the Window 10 64-bit system with I7 6800 K and 64 GB RAM. The GPU device is a NVIDIA TITAN X. We test 10 representative HSIs from the CAVE dataset [21].

We set the parameters empirically. For the noise variance $\sigma = 0.1$, $\mu = 7.85$ and $\alpha = 0.25$. For $\sigma = 0.2$, $\mu = 8.70$ and $\alpha = 0.30$. The iteration numbers for both low resolution reconstruction and detail enhancement are set as 3. We evaluate the performance of our fast reconstruction strategy under different scale

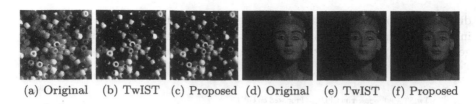

(a) Original   (b) TwIST   (c) Proposed   (d) Original   (e) TwIST   (f) Proposed

Fig. 6. The visual results at 600 nm. (a)–(c) are "beads" and (d)–(f) are "egyptian".

Table 4. PSNR (dB) and SAM results

Image	$\sigma = 0.1$				$\sigma = 0.2$			
	PSNR		SAM		PSNR		SAM	
	TwIST	Proposed	TwIST	Proposed	TwIST	Proposed	TwIST	Proposed
Balloons	37.56	36.76	0.07	0.08	36.04	35.34	0.09	0.11
Beads	25.82	25.64	0.25	0.25	24.65	25.30	0.29	0.27
Egyptian	42.98	42.33	0.39	0.50	40.75	39.67	0.52	0.57
Face	40.42	39.94	0.17	0.22	38.38	37.53	0.25	0.29
Strawberries	40.14	40.74	0.16	0.22	38.24	38.32	0.20	0.33
Feathers	35.63	35.38	0.12	0.13	34.14	34.24	0.14	0.16
Peppers	39.14	38.90	0.14	0.16	37.37	36.91	0.19	0.23
Sponges	35.79	34.46	0.04	0.04	34.21	33.48	0.05	0.05
Stuffed toys	33.59	33.23	0.12	0.14	32.19	32.15	0.15	0.17
Superballs	37.53	37.07	0.17	0.19	35.87	35.63	0.21	0.25
Average	**36.86**	**36.45**	**0.16**	**0.19**	**35.18**	**34.86**	**0.21**	**0.24**

factors, and the results are shown in Table. 1. With the increase of β, the reconstruction time gradually decreases. However, when β is larger than 2, there is a significant drop in PSNR. Therefore, We set β as 2 in our experiment.

We first verify the performance of our fast reconstruction strategy. Without the fast reconstruction strategy, we use the proposed reformative algorithm to recover HSIs directly as contrast. The results including the loss of objective function, PSNR and time cost are shown in Table 2. It indicates that our fast strategy can reduce the reconstruction time by 42% with a slight accuracy decrease.

Table 3 shows the average duration and the parallel acceleration ratio. It indicates that the reconstruction speed of the proposed method is at least 4 times faster than TwIST without parallel implementation. At the same time, our GPU optimization is very powerful and the speedup ratio rises with the increase of the underlying images resolution. Figure 5 shows the maximum frames that can be recovered within 1 s. It demonstrates that our parallel implementation is able to obtain 13 fps $256 \times 256 \times 14$ HSI reconstruction, which cannot be reached by the previous methods.

Table 4 shows the PSNR and spectral angle mapping (SAM) [14] results. A smaller value of SAM suggests a more accurate spectral fidelity. Compared with TwIST, our framework can obtain a comparable reconstruction result with a slight loss (0.32 dB—0.41dB) in PSNR. The reconstructed results of "beads" and "egyptian" at the wavelength of 600 nm when $\sigma = 0.1$ are shown in Fig. 6. we can see that our framework can obtain a similar visual results as TwIST.

4 Conclusion

In this paper, we propose a GPU assisted towards real-time reconstruction framework for the advanced DCCHI system. First, to improve the convergence rate, we propose a reformative TV based reconstruction algorithm. Then, we present a fast reconstruction strategy by interpolating the low resolution result as a warm start of the full-resolution HSI reconstruction, Last, an efficient parallel implementation is conducted for nearly real-time HSI reconstruction. Experiment results show the proposed framework is able to attain more than 12 fps $256 \times 256 \times 14$ HSI reconstruction with a slight loss in reconstruction fidelity.

Acknowledgements. This work was supported by the National Natural Science Foundation of China (Grant Nos. 61425013, 61701025, 61672096).

References

1. Arce, G.R., Brady, D.J., Carin, L., Arguello, H.: Compressive coded aperture spectral imaging: an introduction. IEEE Signal Process. Mag. **31**(1), 105–115 (2014)
2. Bao, J., Bawendi, M.G.: A colloidal quantum dot spectrometer. Sci. Found. China **523**(3), 5–5 (2015)
3. Bioucas-Dias, J.M., Figueiredo, M.A.T.: A new twist: two-step iterative shrinkage/thresholding algorithms for image restoration. IEEE Trans. Image Process. **16**(12), 2992–3004 (2007)
4. Borengasser, M., Hungate, W.S., Watkins, R.: Hyperspectral remote sensing: Principles and applications. Iasri.res.in **31**(12), 1249–1259 (2007)
5. Boyd, S., Parikh, N., Chu, E., Peleato, B., Eckstein, J.: Distributed optimization and statistical learning via the alternating direction method of multipliers. Found. Trends Mach. Learn. **3**(1), 1–122 (2011)
6. Cao, X., Du, H., Tong, X., Dai, Q., Lin, S.: A prism-mask system for multispectral video acquisition. IEEE Trans. Pattern Anal. Mach. Intell. **33**(12), 2423–2435 (2011)
7. Chen, S.S., Donoho, D.L., Saunders, M.A.: Atomic decomposition by basis pursuit. SIAM Rev. **43**(1), 129–159 (2001)
8. CUDA, C.: Programming guide v7. 0. NVIDIA, March 2015
9. Descour, M., Dereniak, E.: Computed-tomography imaging spectrometer: experimental calibration and reconstruction results. Appl. Opt. **34**(22), 4817–26 (1995)
10. Dicker, D.T., et al.: Differentiation of normal skin and melanoma using high resolution hyperspectral imaging. Cancer Biol. Ther. **5**(8), 1033–1038 (2006)
11. Hestenes, M.R., Stiefel, E.: Methods of conjugate gradients for solving linear systems. J. Res. Nat. Bur. Stand. **49**, 409–436 (1953). (1952)

12. Itu, L.M., Suciu, C., Moldoveanu, F., Postelnicu, A.: GPU optimized computation of stencil based algorithms. In: 2011 RoEduNet International Conference 10th Edition: Networking in Education and Research, pp. 1–6, June 2011
13. James, J.: Spectrograph Design Fundamentals. Cambridge University Press, Cambridge (2007)
14. Kruse, F., et al.: The spectral image processing system (sips)-interactive visualization and analysis of imaging spectrometer data. Remote. Sens. Environ. **44**(2), 145–163 (1993)
15. Okamoto, T., Yamaguchi, I.: Simultaneous acquisition of spectral image information. Opt. Lett. **16**(16), 1277 (1991)
16. Tilling, A.K.: Remote sensing to detect nitrogen and water stress in wheat. Reg. Inst. Ltd **104**(1–3), 77–85 (2006)
17. Wagadarikar, A., John, R., Willett, R., Brady, D.: Single disperser design for coded aperture snapshot spectral imaging. Appl. Opt. **47**(10), B44 (2008)
18. Wang, L., Xiong, Z., Shi, G., Wu, F., Zeng, W.: Adaptive nonlocal sparse representation for dual-camera compressive hyperspectral imaging. IEEE Trans. Pattern Anal. Mach. Intell. **39**(10), 2104–2111 (2017)
19. Wang, L., Xiong, Z., Gao, D., Shi, G., Wu, F.: Dual-camera design for coded aperture snapshot spectral imaging. Appl. Opt. **54**(4), 848–58 (2015)
20. Wang, L., Xiong, Z., Gao, D., Shi, G., Zeng, W., Wu, F.: High-speed hyperspectral video acquisition with a dual-camera architecture. In: Computer Vision and Pattern Recognition, pp. 4942–4950 (2015)
21. Yasuma, F., Mitsunaga, T., Iso, D., Nayar, S.K.: Generalized assorted pixel camera: postcapture control of resolution, dynamic range, and spectrum. IEEE Trans. Image Process. **19**(9), 2241–53 (2010)

Getting More from One Attractive Scene: Venue Retrieval in Micro-videos

Jie Guo[1], Xiushan Nie[2($\boxtimes$)], Chaoran Cui[2], Xiaoming Xi[2], Yuling Ma[1], and Yilong Yin[1($\boxtimes$)]

[1] School of Software, Shandong University, Jinan 250101, Shandong, China
ylyin@sdu.edu.cn
[2] Shandong University of Finance and Economics, Jinan 250014, Shandong, China
niexsh@sdufe.edu.cn

Abstract. Micro-videos, which contain interesting events occurring at a specific venue, are uploaded to social platforms with a high degree of subjectivity and arbitrariness. Because the duration of a micro-video is shorter than six seconds, the scene in the micro-video is always from a single venue. Therefore, the venue is an important location-related piece of information in micro-videos. In this preliminary research, we investigate micro-video venue retrieval. First we present the challenges of micro-video venue retrieval and then propose a new strategy based on a multi-layer neural network. Finally, we evaluate the proposed strategy on an actual micro-video dataset crawled from Vine. The experimental results show its superior performance.

Keywords: Micro-videos · Venue retrieval
Multi-layer neural network

1 Introduction

With the unprecedented growth of smart mobile devices, users can easily upload their short videos to social media websites. Therefore, micro-videos spread rapidly on various online video social platforms, such as Vine[1], Instagram[2], and Viddy[3], [16]. There are three characteristics of micro-videos that make them

This work is supported by the National Natural Science Foundation of China (61671274, 61573219, 61701281,61703234), China Postdoctoral Science Foundation (2016M592190,2018M632674), Shandong Provincial Key Research and Development Plan (2017CXGC1504), Shandong Provincial Natural Science Foundation (ZR2017QF009), Shandong Provincial High College Science and Technology Plan (J17KB161), Project of Shandong Province Higher Educational Science and Technology Program (J17KA065), and the Fostering Project of Dominant Discipline and Talent Team of Shandong Province Higher Education Institutions.

[1] https://vine.co/.
[2] https://www.instagram.com/.
[3] http://www.fullscreen.com/.

© Springer Nature Switzerland AG 2018
R. Hong et al. (Eds.): PCM 2018, LNCS 11164, pp. 721–733, 2018.
https://doi.org/10.1007/978-3-030-00776-8_66

different from traditional videos: (1) *Shortness.* A micro-video is always shorter than six seconds, which is much shorter than traditional videos; (2) *Randomness.* A micro-video is generated by mobile phone users, and is random and diverse; (3) *Social attributes.* A micro-video always has more social attributes, such as venue, loops, description, hashtag, and follower.

The numerous micro-videos has led to some new micro-video-related research, such as creative prediction [10], action recognition [6], tag prediction [6], and popularity prediction [3]. Generally, venue information is a latent and useful clue for these micro-video-related applications because the scene and event in the micro-video always occur in one single venue. Venue information is usually selected manually by users relying on a GPS enabled device, and each venue is automatically aligned with a venue category via the Foursquare API[4]. Venue information is beneficial scene recognition.

Venue retrieval, is a new topic in micro-video-related applications, e.g., for micro-video recommendation and video classification. However, there are, to our knowledge, no research regarding this issue because of the big challenges faced in micro-video venue retrieval.

Generally, venue information can be generated by the GPS device automatically. However, the venue information is always incomplete, because GPS can be turned-off by users. In addition, it is difficult to get the precise coordinates by GPS, especially for indoors. Although some researchers are trying to estimate the missing venue information, like location estimation and venue prediction [7,16], they can still not provide venues from all micro-videos. To address this issue, we try to utilize multi-modal features of the micro-video.

Multi-modal features are popularly used in traditional multimedia search and retrieval. Multi-view [17,18] and multi-feature [13] are all taken as multi-modal features. Most of these features are extracted by subspace learning, where a fusion of features in a common space is learned from multi-modal features, and used for retrieval. These methods are generally divided into two categories, i.e., unsupervised and supervised methods. The unsupervised methods include Canonical Correlation Analysis (CCA) [9], Partial Least Squares (PLS) [11], Bilinear Model (BLM) [14], and Deep Canonical Correlation Analysis (DCCA) [1], whereas the supervised methods include Generalized Multi-view Analysis (GMA) [12], Multi-view Discriminant Analysis (MvDA) [4,5], Multiple Feature Hashing (MFH) [13], and Semantic Correlation Maximization (SCM) [15].

In general, the subspace learning-based methods focus on exploring a common space to represent all features. However, when applied to micro-video venue retrieval, the performance is not acceptable due to the weak correlation between the venue label and multi-modal features. Therefore, in this study we try to investigate a new strategy that captures the complementary information rather than the common information.

---

[4] https://github.com/mLewisLogic/foursquare.

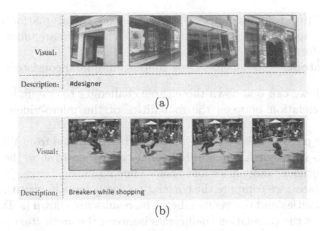

Fig. 1. Two micro-videos crawled from Vine, where "Visual" and "Description" are the visual and text modalities of the video, respectively.

Table 1. Correlation coefficient between original modality and venue labels

Modality	Correlation coefficient
Visual	0.6135
Audio	0.1755
Text	0.2827

The main contributions of this study are summarized as:

(1) We propose a new problem labeled venue retrieval in micro-videos, and we analyze in detail the differences and challenges compared to traditional videos.

(2) We propose a new strategy that explores discriminant venue features of micro-video from the complementary information among the multi-modal features. The remainder of the paper is organized as follows. Section 2 presents the topic of venue retrieval in micro-videos and analyzes its challenges, and Sect. 3 details our proposed strategy. Experimental results and analysis are reported in Sect. 4, followed by the conclusion and suggestions for future work in Sect. 5.

2   Challenges in Micro-video Venue Retrieval

As a first preliminary investigation, we explore venue retrieval to find more micro-videos that share venue information with the query. However, this is nontrivial due to the following challenges:

(1) **Incorrect or incomplete venue information**. The micro-video sometimes has venue information automatically added by GPS services. It will loss when GPS is turned-off or indoor situations.

(2) **Modal information loss.** Micro-videos on social platform are generated with subjectivity and arbitrariness. Therefore, many micro-videos lack modal information.

(3) **Inconsistent.** Some micro-videos content may be inconsistent with venue.

In general, we can boil down these three challenges to one underlying factor, i.e., weak correlation between the modalities of the micro-video and its true venue label.

To further prove this issue, we use a real dataset crawled from Vine, including 270,145 micro-videos distributed in 188 venue categories, to show the correlations between the original modalities and their venue label.

In this dataset, we compute the correlation coefficients between visual, audio, and text modalities and the venue label. The results are shown in Table 1, where we can see that the correlation coefficients between the audio/text modality and the venue label is very small. The correlation between the visual modality and venue has an acceptable value.

To further describe the weak correlation between the original modalities and venue label, we show two actual micro-videos crawled from Vine in Fig. 1, where both of these two micro-videos venue labels are "Mall" in ground truth. In Fig. 1(a), although we can easily recognize the venue label from the visual modality. However, another modality, text description, is "designer", which is weakly related to its venue "Mall". In contrast, as shown in Fig. 1(b), it is difficult to recognize the venue label from the visual modality because we only see a street dance show. However, the text description of this micro-video is "Breakers while shopping" which is related to the true venue label "Mall". From the example shown in Fig. 1, we can see that there are weak correlations between some original modalities and the true venue label. However, there still exist some complementary information between different modalities that are related to venue labels. A commonly used strategy in traditional video retrieval with multiple modalities is subspace learning. To verify whether the subspace learning is suitable for micro-video venue retrieval, we apply two popular subspace learning methods, i.e., CCA [9] and MvDA [5] to learn a common representation for all modalities of micro-videos. We then compute the correlation coefficients between the common representation of each modality and the venue label. The results are shown in Table 2. The correlations between the modalities and the venue label in the subspace are still weak. Furthermore, we compute the correlation coefficients between any pair of modalities in the subspace. We can see that the correlation coefficients are still not high from Table 3.

In general, utilizing the common representation of different modalities in a subspace to learn venue semantic is infeasible due to the weak correlation between the modalities and the venue label. Therefore, in this preliminary investigation, we try to explore venue semantic meaning of micro-videos based on the complementary information among multi-modal features.

Table 2. Correlation coefficients between modalities and venue labels in the subspace

Modality	Corrcoef_CCA	Corrcoef_MVDA
Visual	0.2502	0.2235
Audio	0.2074	0.0234
Text	0.2849	0.0964

Table 3. Correlation coefficient of modality pairs

Modality	Correlation coefficient
Visual & Audio	0.5036
Visual & Text	0.5069
Audio & Text	0.1217

3   Proposed Method

As we concluded in the previous section, it is difficult to extract common venue semantic features of a micro-video through the method of subspace learning. Therefore, we develop a multi-layer neural network to learn a discriminant venue representation for micro-video venue retrieval by exploiting both the complementary information and the nonlinear structure between different modalities.

Concatenating multiple features to a long vector is a strategy to combine all features together, and it can reflect the complementarity between multiple modalities. However, it contains a lot of redundant information. Therefore, directly utilizing concatenated multiple features is not suitable for venue retrieval. In this study, we first concatenate the features of different modalities, and then feed it to a multi-layer neural network, which cannot only capture the discriminant information for venue retrieval, but also remove the redundancy from different features. The framework of the proposed method is shown in Fig. 2, where we first train a multi-layer neural network to obtain a nonlinear transformation from the multiple modalities to its corresponding micro-video venue label, and then use the output of the last hidden layer as the discriminant feature for venue retrieval. The main notations used in the formulation are listed in Table 4.

3.1   Formulation

The objective is to learn a discriminant venue feature from multiple modalities of micro-videos. Given a concatenated feature from multiple modalities of the micro-video, the projected function is learned to produce a discriminant venue representation for its venue label $\mathbf{y}^i$. In this work, the form of the multi-layer neural network is adopted to generate the discriminant venue representation through multiple nonlinear transformations. Evidently, the multi-layer neural

Table 4. Notations

Notation	Description
$\mathbf{y}^i$	Venue label of the i_{th} micro-video
$\mathbf{p}^k$	Output vector of the k_{th} layer
$\mathbf{w}_j^k$	Weight vector of the j_{th} unit of the k_{th} layer
y_j^i	j_{th} element of venue label $\mathbf{y}^i$
t_j^k	Input of the j_{th} unit of the k_{th} layer
p_j^k	j_{th} element of the output vector $\mathbf{p}^k$
w_{jq}^k	q_{th} element of the weight vector of $\mathbf{w}_j^k$
c_j^k	Bias of the j_{th} unit at the k_{th} layer
$M^{(k)}$	Number of units in the k_{th} layer
N	Total number of micro-videos in each batch of the training set

network can better capture the complex relations between the modalities of the micro-video.

As shown in Fig. 2, we develop a network with K stacked layers of nonlinear transformations. Let $M^{(k)}$ be the number of units at the k_{th} layer, where $1 \leq k \leq K$. The input of the j_{th} unit at the k_{th} layer is computed iteratively as $t_j^k = \mathbf{w}_j^{k^T} \mathbf{p}^{k-1} + c_j^k$. The output of the j_{th} unit at the k_{th} layer is $p_j^k = S(t_j^k)$, where $S(\bullet)$ is the activation function. In this study, we use Relu and softmax as activation functions for the hidden and output layer, respectively.

The multi-layer neural network can be learned by solving an optimization problem to minimize the predefined loss measurement. In this study, we use cross entropy as the loss measurement. During the training, we divide the dataset into several batches. Assuming that N is the number of micro-videos in each batch, the loss function is as follows:

$$E(w_{jq}^k, \ c_j^k) = -\frac{1}{N} \sum_{i=1}^{N} \sum_{j=1}^{M^{(k)}} y_j^i \ln p_j^k + (1 - y_j^i) \ln(1 - p_j^k) \tag{1}$$

3.2 Optimization

The optimization problem can be solved by minimizing the loss function, we use the mini-batch gradient descent scheme to obtain the parameters. The gradients of the objective function $E(w_{jq}^k, \ c_j^k)$ with respect to the parameters $(w_{jq}^k, \ c_j^k)$ for the last output layer and hidden layer can be computed as follows:

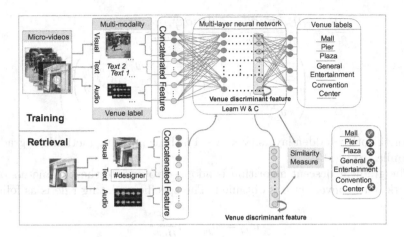

Fig. 2. The framework of the proposed method

For the last output layer (K_{th}), the gradients are

$$\frac{\partial E}{\partial w_{jq}^K} = \frac{\partial E}{\partial p_j^K} \frac{\partial p_j^K}{\partial t_j^K} \frac{\partial t_j^K}{\partial w_{jq}^K}$$

$$= -\frac{1}{N} \sum_{i=1}^{N} \sum_{j=1}^{M^{(K)}} \left(\begin{array}{c} (y_j^i \frac{1}{p_j^K} + (1 - y_j^i) \frac{-1}{(1-p_j^K)}) \\ * (e^{t_j^K} \sum_j e^{t_j^K} - e^{t_j^K} e^{t_j^K}) p_q^{K-1} \end{array} \right) \quad (2)$$

$$\frac{\partial E}{\partial c_j^K} = \frac{\partial E}{\partial p_j^K} \frac{\partial p_j^K}{\partial t_j^K} \frac{\partial t_j^K}{\partial c_j^K}$$

$$= -\frac{1}{N} \sum_{i=1}^{N} \sum_{j=1}^{M^{(K)}} \left(\begin{array}{c} (e^{t_j^K} \sum_j e^{t_j^K} - e^{t_j^K} e^{t_j^K}) \\ * (y_j^i \frac{1}{p_j^K} + (1 - y_j^i) \frac{-1}{(1-p_j^K)}) \end{array} \right) \quad (3)$$

For the last hidden layer ($k = K - 1$), the gradients are

$$\frac{\partial E}{\partial w_{jq}^k} = \frac{\partial E}{\partial p_j^k} \frac{\partial p_j^k}{\partial t_j^k} \frac{\partial t_j^k}{\partial w_{jq}^k} = \frac{\partial E}{\partial t_j^K} \frac{\partial t_j^K}{\partial p_j^k} \frac{\partial p_j^k}{\partial t_j^k} \frac{\partial t_j^k}{\partial w_{jq}^k}$$

$$= \left(\begin{array}{c} \frac{-1}{N} \sum_{i=1}^{N} \sum_{j=1}^{M^{(K)}} \left(\begin{array}{c} y_j^i \frac{1}{p_j^K} + (1 - y_j^i) \frac{-1}{(1-p_j^K)} \\ * (e^{t_j^K} \sum_j e^{t_j^K} - e^{t_j^K} e^{t_j^K}) \end{array} \right) \\ * w_{jq}^K * p_q^{k-1} \end{array} \right) \quad (4)$$

$$\frac{\partial E}{\partial c_j^k} = \frac{\partial E}{\partial p_j^k} \frac{\partial p_j^k}{\partial t_j^k} \frac{\partial t_j^k}{\partial c_j^k} = \frac{\partial E}{\partial t_j^K} \frac{\partial t_j^K}{\partial p_j^k} \frac{\partial p_j^k}{\partial t_j^k} \frac{\partial t_j^k}{\partial c_j^k}$$

$$= \left(\begin{array}{c} \frac{-1}{N} \sum_{i=1}^{N} \sum_{j=1}^{M^{(K)}} \left(\begin{array}{c} y_j^i \frac{1}{p_j^K} + (1 - y_j^i) \frac{-1}{(1 - p_j^K)} \\ * (e^{t_j^K} \sum_j e^{t_j^K} - e^{t_j^K} e^{t_j^K}) \end{array} \right) \\ * w_{jq}^K \end{array} \right) \tag{5}$$

For the other hidden layers $(k < K - 1)$, the computing methods of gradients are similar to Eq. (5).

The gradient descent algorithm is adopted to update the parameters of the network until convergence is obtained. The specific updating rule is as follows:

$$w_{jq}^k = w_{jq}^k - \eta \frac{\partial E}{\partial w_{jq}^k} \tag{6}$$

$$c_j^k = c_j^k - \eta \frac{\partial E}{\partial c_j^k} \tag{7}$$

where η is the learning rate.

4   Experiments

4.1   Experimental Settings

The data set we used in the experiment was crawled from Vine and has been made public on the website[5]. In the dataset, we map the venue information of micro-videos into Foursquare. The venue ID will be mapped into venue category in Foursquare. There are 270,145 micro-videos distributed in 188 venue categories. The venues are divided into 4 layers, with 10 non-leaf nodes, 341 leaf nodes, 312 leaf nodes, and 52 leaf nodes in the first, second, third, and fourth layer, respectively. For example, "Restaurant" is a non-leaf node and "Asia Restaurant" is a child node.

Generally, the number of samples in each category is extremely unbalanced. Some categories include dozens of samples and some categories even contain tens of thousands of samples. To simply the experiment, we select 11,500 micro-videos from 5 categories for balance. In the experiment, we use 100 micro-videos as queries, and the other micro-videos are used as training data.

We used three modalities, visual, audio, and text, in the experiment, extracted by deep convolutional neural networks, denoising autoencoder, and sentence2vector, respectively. The dimensions for these three modalities are 4096, 200, and 100, respectively.

Due to there are no suitable micro-venue retrieval works in literature, some unsupervised and supervised subspace learning methods presented in traditional video retrieval are used for comparison, which are described as follows:

---

[5] www.acmmm16.wixsite.com/mm16.

- **Linear Discriminant Analysis (LDA)** [2]. LDA tries to learn a linear transformation by maximizing between-class variation and minimizing within-class variation.
- **Canonical Correlation Analysis (CCA)** [9]. CCA is a typical unsupervised approach to obtain a common space, which attempts to learn two transforms, one for each view, to project the samples from the two views into a common subspace by maximizing the cross-correlation between the two views.
- **Multi-view Discriminant Analysis (MvDA)** [5]. MvDA seeks a single discriminant common space for multiple views in a non-pairwise manner by jointly learning multiple view-specific linear transforms.

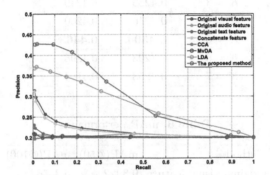

Fig. 3. PR curve of venue retrieval in micro-video

We also compare the proposed method with the methods that only use single modality or concatenating multiple features one by one.

4.2   Experimental Results and Analysis

In the experiments, the mean average precision (mAP) is adopted as the evaluation metric. The metric has been widely used in the literature [8,19].
For a given query, the average precision (AP) is calculated as

$$AP = \frac{1}{M} \sum_{r=1}^{R} pre(r)rel(r) \tag{8}$$

where R is the total number of retrieved micro-videos, M is the number of relevant micro-videos in the retrieved set, $pre(r)$ denotes the precision of top r retrieval micro-videos, which is defined as the ratio between the number of the relevant micro-videos and the number of retrieved micro-videos r. $rel(r)$ is an indicator function, with $= 1$ if the r_{th} micro-video is relevant to the query, and $= 0$ otherwise. mAP is defined as the average of the AP of all queries. A larger mAP means that the retrieval performance is better. In the experiments, we set R to 50 and 100 to collect results. Furthermore, the precision-recall curve is also reported to reflect the retrieval performance. The experimental results are shown in Table 6 and Fig. 3.

The important parameter in the proposed method is the number K of layers in the multi-layer neural network. We first carry out some experiments to choose K for the best performance. As shown in Table 5, the system achieves the best performance when $K = 5$. Therefore, we set $K = 5$ in the experiments.

Table 5. Different mAP performance with different number of layers of the neural network

Values of K	mAP(@50)	mAP(@100)
$K = 3$	0.4255	0.4240
$K = 4$	0.4246	0.4200
$K = 5$	**0.4504**	**0.4468**
$K = 6$	0.4293	0.4298
$K = 7$	0.4253	0.4164

Table 6. mAP performance

Method	mAP(@50)	mAP(@100)
Original visual feature	0.3922	0.3556
Original audio feature	0.2817	0.2481
Original text feature	0.3402	0.2918
Concatenating	0.3978	0.3583
LDA	0.4110	0.3929
CCA	0.2579	0.2338
MVDA	0.2824	0.2501
The proposed method	**0.4504**	**0.4468**

Table 6 shows the mAP performance of the different methods. According to the presented results, we clearly find that the proposed method has superior performance. Compared to the subspace learning-based method, such as CCA and MvDA, the proposed method achieves more than 50% improvement. This phenomenon is consistent with the analysis in Sect. 2, i.e., it is difficult to find a common space to represent micro-video venues. The performance of LDA is also better than the subspace learning-based method. One possible reason is that LDA tries to capture discriminant information among multiple features by maximizing between-class variations and minimizing with-class variations, which is similar to the proposed method. However, compared to the nonlinear transformation used in the proposed method, the LDA projects features to their labels by a linear transformation that is less effective in micro-video venue retrieval.

Table 7. Correlation coefficient between venue representation generated from different methods and the venue labels

Method	Correlation coefficient
CCA venue representation	0.3631
MvDA venue representation	0.2432
LDA venue representation	0.6392
The proposed venue representation	**0.8932**

In addition, the proposed method outperforms the single-modality cases from Table 6. This indicates the advantages of using the complementary assistance in multiple modalities.

Figure 3 shows the ROC (Receiver Operating Characteristic) curves (precision vs. recall) with respect to the comparison between the proposed method and other methods on the datasets. The figure shows that the performance of the proposed method is superior to those of the other methods.

As we describe in Sect. 3, we use the output of the last hidden layer as the discriminant feature to retrieval. To prove its effectiveness, we compute the correlation between the discriminant feature and the venue label of the corresponding micro-video. The experimental results are shown in Table 7, which shows that the discriminant feature used for venue retrieval has a higher correlation coefficient. This indicates that the proposed method has captured a suitable venue representation of the micro-videos with a strong correlation with the venue labels.

5   Conclusion

Micro-video venue retrieval is a new topic in the research field of social media, and it can be applied to many scenarios. In this study, as a preliminary investigation, we first explored the differences and challenges in micro-video venue retrieval compared to those in traditional videos. We then propose a new strategy to explore discriminant venue representation for micro-video retrieval. In the proposed method, multiple modalities are fed to a neural network for training a nonlinear transformation that can capture the complex relations between the modal information of micro-videos and their venue labels. The output of the last hidden layer is then obtained and adopted as the discriminant venue feature. Compared to the existing subspace learning-based methods, the proposed method achieves superior performance in micro-video venue retrieval.

However, owing to the special characteristics of micro-videos, there are still some open issues that should be addressed, such as the data imbalance and noisy labels in a real micro-video dataset. We will investigate these issues in future work.

References

1. Andrew, G., Arora, R., Bilmes, J., Livescu, K.: Deep canonical correlation analysis. In: Proceedings of the 2013 International Conference on Machine Learning, p. III-1247 (2013)
2. Belhumeur, P.N., Hespanha, J.P., Kriegman, D.J.: Eigenfaces vs. fisherfaces: recognition using class specific linear projection. IEEE Trans. Pattern Anal. Mach. Intell. **19**(7), 711–720 (1997)
3. Chen, J., Song, X., Nie, L., Wang, X., Zhang, H., Chua, T.S.: Micro tells macro: predicting the popularity of micro-videos via a transductive model. In: Proceedings of the 24th ACM International Conference on Multimedia, pp. 898–907. ACM (2016)
4. Kan, M., Shan, S., Zhang, H., Lao, S., Chen, X.: Multi-view discriminant analysis. In: Fitzgibbon, A., Lazebnik, S., Perona, P., Sato, Y., Schmid, C. (eds.) ECCV 2012. LNCS, vol. 7572, pp. 808–821. Springer, Heidelberg (2012). https://doi.org/10.1007/978-3-642-33718-5_58
5. Kan, M., Shan, S., Zhang, H., Lao, S., Chen, X.: Multi-view discriminant analysis. IEEE Trans. Pattern Anal. Mach. Intell. **38**(1), 188–194 (2016)
6. Nguyen, P.X., Rogez, G., Fowlkes, C., Ramamnan, D.: The open world of micro-videos (2016)
7. Nie, L., et al.: Enhancing micro-video understanding by harnessing external sounds. In: Proceedings of the 25th ACM International Conference on Multimedia, pp. 1192–1200. ACM (2017)
8. Nie, X., Yin, Y., Sun, J., Liu, J., Cui, C.: Comprehensive feature-based robust video fingerprinting using tensor model. IEEE Trans. Multimed. **19**(4), 785–796 (2017)
9. Rasiwasia, N., et al.: A new approach to cross-modal multimedia retrieval. In: Proceedings of the 18th ACM International Conference on Multimedia, pp. 251–260. ACM (2010)
10. Redi, M., Ohare, N., Schifanella, R., Trevisiol, M., Jaimes, A.: 6 seconds of sound and vision: creativity in micro-videos. In: Proceedings of the 2014 IEEE Conference on Computer Vision and Pattern Recognition, pp. 4272–4279. IEEE (2014)
11. Rosipal, R., Krämer, N.: Overview and recent advances in partial least squares. In: Saunders, C., Grobelnik, M., Gunn, S., Shawe-Taylor, J. (eds.) SLSFS 2005. LNCS, vol. 3940, pp. 34–51. Springer, Heidelberg (2006). https://doi.org/10.1007/11752790_2
12. Sharma, A., Kumar, A., Daume, H., Jacobs, D.W.: Generalized multiview analysis: a discriminative latent space. In: Proceedings of the 2012 IEEE Conference on Computer Vision and Pattern Recognition, pp. 2160–2167. IEEE (2012)
13. Song, J., Yang, Y., Huang, Z., Shen, H.T., Luo, J.: Effective multiple feature hashing for large-scale near-duplicate video retrieval. IEEE Trans. Multimed. **15**(8), 1997–2008 (2013)
14. Tenenbaum, J.B., Freeman, W.T.: Separating style and content with bilinear models. Neural Comput. **12**(6), 1247–1283 (2014)
15. Zhang, D., Li, W.J.: Large-scale supervised multimodal hashing with semantic correlation maximization. In: Proceedings of the 28th AAAI Conference on Artificial Intelligence, pp. 2177–2183. AAAI (2014)
16. Zhang, J., Nie, L., Wang, X., He, X., Huang, X., Chua, T.S.: Shorter-is-better: venue category estimation from micro-video. In: Proceedings of the 24th ACM International Conference on Multimedia, pp. 1415–1424. ACM (2016)

17. Zhu, L., Huang, Z., Chang, X., Song, J., Shen, H.T.: Exploring consistent preferences: discrete hashing with pair-exemplar for scalable landmark search. In: Proceedings of the 2017 ACM on Multimedia Conference, pp. 726–734. ACM (2017)
18. Zhu, L., Huang, Z., Liu, X., He, X., Sun, J., Zhou, X.: Discrete multimodal hashing with canonical views for robust mobile landmark search. IEEE Trans. Multimed. **19**(9), 2066–2079 (2017)
19. Zhu, L., Shen, J., Xie, L., Cheng, Z.: Unsupervised visual hashing with semantic assistant for content-based image retrieval. IEEE Trans. Knowl. Data Eng. **29**(2), 472–486 (2017)

Simulating Bokeh Effect with Kinect

Yang Yang$^{(\boxtimes)}$, Huiwen Bian, Yanhong Peng, Xiangjun Shen,
and Heping Song

School of Computer Science and Communication Engineering,
Jiangsu Univeristy, Zhenjiang, China
yyoung@ujs.edu.cn

Abstract. Bokeh effect is an artistic effect in photography, which is
the out-of-focus blur caused by the camera lens, typically, professional
cameras with wide aperture lenses are necessary to obtain the Bokeh
effect. In this paper, we propose a computational method to simulate
the Bokeh effect with the RGBD data captured by Kinect. The proposed method first refines the depth map by modeling it as an optimization problem with the guidance of the corresponding RGB image; the
size of the filter kernel is calculated from the refined depth map afterwards, finally, the RGB image with Bokeh effect is computed by filtering
with the varying-sized and varying-shaped kernels. We have conducted
experiment on 4 sets of data, and the experiment result suggests that
the proposed method is optimistic in computing natural Bokeh effect.

Keywords: Bokeh effect · Depth map refinement · Filter · Kinect

1 Introduction

Bokeh effect is widely explored by photographers in taking amazing photos,
especially in landscape and portrait photos. Bokeh effect can be used to highlight
the main object of the scene and blur the rest, which make it easier for people to
focus. In addition, the blur produced by natural Bokeh effect is quite pleasing.
The principle of Bokeh effect is quite straightforward, i.e., light rays from the
objects near the focal plane of the camera converges to the similar positions
on the camera sensor, forming sharply clear image, while light rays from the
scene that is farther away from the focal plane diverged to different positions on
the camera sensor, presenting a blur effect. In practice, a shallow depth of field
(DoF) created by a wide lens aperture is typically required in order to create
photos with Bokeh effect. However, such lens are typically available on high-end
cameras.

In this paper, we propose a method to computationally simulate the Bokeh
effect with the RGBD data captured by Kinect. Figure 1 shows the overview of
our proposed method. The initial depth map cannot be directly used, as it contains many empty and wrong depth information. The initial depth map is refined
by the guidance of the corresponding RGB image, so as to generate a high-quality

© Springer Nature Switzerland AG 2018
R. Hong et al. (Eds.): PCM 2018, LNCS 11164, pp. 734–743, 2018.
https://doi.org/10.1007/978-3-030-00776-8_67

depth map. The task is modeled as an optimization problem similar to [1], the spatial distance as well as the intensity difference are adapted in determining the coefficient of the system, the problem is then solved by existing solver. Based on the focal plane, aperture size as well as lens focal length specified by the users, the kernel size of each position on the RGB image can be calculated. Finally, by filtering the RGB image with the varying-sized and varying-shaped kernels, the image with Bokeh effect is finally rendered. An experiment is carried out to evaluate the proposed method. The dataset consists of 4 sets of RGBD pictures captured by the Kinect. The performance of the depth refinement algorithm, as well as the rendering algorithm are recorded. From the experiment result, it can be drawn that our proposed method is a practical to simulate natural and realistic Bokeh effect.

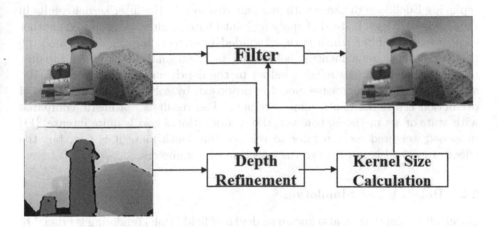

Fig. 1. Overview of the proposed method

2   Previous Work

2.1   Depth Map Refinement

In order to refine the depth map captured by Kinect, various methods have been proposed. Due to the strong correlations between the RGB images and corresponding depth map, almost all the depth map refinement methods are typically color guided. [2] first segments the color image, and interpolates the missing depth information within each segment. An early work [3] on joint bilateral filter (JBF) provides an elegant way to solve the depth map refinement problem.

By incorporating the intensity difference into the ranger filter, and spatial distance into the domain filter, the depth at a specific position is calculated as a weighted sum of depth at nearby positions. Joint bilateral filters could refine the depth map to some extent, but the quality of depth at depth discontinuities are quite low. To overcome the drawbacks of [3], [4] proposed a weighted joint bilateral filter, which reduces the weights of the edges, in addition, a post-processing technique to compensates slope depth is proposed to remove blur around object boundaries. However, in many cases, it will cause severe depth attening. The works in [5,6] extend [3] by proposing joint trilateral filter (JTF). In addition to spatial and range information, [5] proposed to use local gradient information of the depth map as guidance, [6] proposed to extract edges from the color image and use it as additional information to guide the refinement of depth map. It was reported that the edge discontinuities had been improved. [7] claimed that replacing Euclidean distance with geodesic distance in the filter kernel results in refined depth edges. Instead of applying joint bilateral filter, [8] proposed another edge preserving filter which is called guided filter to refine the depth map. [9] treats the depth refinement as hole filling problem, and adapt image inpainting algorithm to fill the missing values in the depth map. In the work of [10], a color guided autoregressive model is proposed, by solving an optimization of prediction error, the depth map is refined. The result is optimistic compared with state of art methods, however, the computational cost is quite intense. [11] proposed weighted median filter to recover the depth map, it is fast, but the effect of method is sensitive to the selection of parameters.

2.2   Bokeh Effect Simulation

Bokeh effect simulation, also known as depth of field (DoF) rendering is crucial to computer graphics applications. Many techniques for Bokeh effect are proposed which can be generally divided into two categories, namely physically-based approaches and image-based techniques. The former ones refer to the processing of light rays, while the latter ones deal with pixels on the image plane. For physically-based approaches, [12,13] used splatting techniques to achieve real-time rendering directly at each point, it renders the Bokeh for each pixel with a shaped spirit of the same size as the circle of confusion (CoC). Based on the samples of the light rays from the lens to the image plane, [14] explored Monte Carlo techniques to estimate the rendering equation which solves the problem of aliasing. Monte Carlo techniques were also adapted in [15] to express the intensity distribution in the CoC, and combines distributed ray tracing to achieve more precise effect than single view method. These methods could handle the intensity leakage and partial occlusion problem naturally, however, they are computationally intensive, which makes them unrealistic for complex scenarios. The image-based techniques are typically based on filtering with varying-sized kernel. [16] proposed to make use of the summed area table (SAT) for DoF rendering, each pixel of the rendered image can be calculated from 4 values in the SAT. This method is really fast, however, it can be difficult to apply it to Bokeh shapes other than rectangle. [17,18] proposed convolution on image pyramid which could enable the DoF rendering. [19] divided a 2D convolution into two 1D convolutions, the method does reduce the computational cost, however, it results in very soft blur. The researchers in the field of signal processing transform the convolutions in the spatial domain into the multiplication in the frequency domain [20], thus to enhance its performance for convolutions with large kernels. However, the images need to be segmented into areas with distinct depth values.

3   Proposed Method

3.1   Refining Depth Map

There is numerous missing information in the depth maps captured by Kinect. This is typically caused by the occlusions between the build-in infrared projectors and sensors. As we can see from the image in the top left corner of Fig. 2, the initial depth map contains a lot of holes on the image. The depth of neighboring pixels provides cues for the reconstruction of the missing depth information. In this paper, we model the problem as an optimization problem, the energy function is defined as the sum of the squared distance between the missing depth of the pixels and the weighted sum of the depth of their neighboring pixels. Notice that the depth of the neighboring pixels may also be missing here. Thus, we have our energy function $j(Z)$ as shown in Eq. (1).

$$j(Z) = \min \sum_{r \in M} \left(\left(Z(r) - \sum_{s \in N(r)} NW(s,r) * Z(s) \right)^2 \right) \tag{1}$$

Here, Z is the depth map, $Z(r)$ is the depth of pixel r, M is the set of pixels with missing depth, and $N(r)$ is the set of neighboring pixels around pixel r, $*$ is the operator for multiplication. NW is a normalized version of W which can be calculated from spatial and range kernel $W1$, $W2$ as shown in Eq. (2).

$$W(s,r) = W1(s,r) * W2(s,r) \tag{2}$$

The spatial kernel $W1$ measures the impact of pixel s on r with respect to the spatial distance. It is expected that if s is spatially closer to r, the depth of pixel r, thus pixel s will have greater weight in calculating the depth of pixel r, and vice versa. The spatial kernel $W1$ can be calculated by Eq. (3), $C(s)$ is the coordinate of pixel s, σ_1^2 is the standard deviation around the neighborhood in the spatial domain which is a parameter here.

$$W1(s,r) = e^{-\frac{(C(s)-C(r))^2}{2\sigma_1^2}} \tag{3}$$

The range kernel $W2$ refers to the impact of pixel s on r with respect to the color difference. It is natural that the pixel s with similar color would have more effect on the depth of pixel r, and vice versa. The formula to compute $W2$ is shown in Eq. (4), where $Y(s)$ is the color of pixel s, and σ_2^2 is user-defined parameter which is regarded as the standard deviation around the neighborhood in the range domain.

$$W2(s,r) = e^{-\frac{(Y(s)-Y(r))^2}{2\sigma_2^2}} \tag{4}$$

$$\begin{cases} Z(r) = \sum\limits_{s \in N(r)} NW(s,r) * Z(s) & \text{for all } r, \text{ where } Z(r) \text{ is missing} \\ Z(r) = d_r & \text{for all } r, \text{ where } Z(r) \text{ is not missing} \end{cases} \tag{5}$$

The optimization problem shown in Eq. (1) can be modeled as the simultaneous linear equations in Eq. (5), which can be efficiently solved with existing solver. If the depth of point r is missing, it should be weighted average of the depths of neighboring pixels; if the depth of point r is exists which is d_r, we have $Z(r) = d_r$. Notice that, due to the noise in the depth discontinuities, we adapted Canny edge detection algorithm to discover the edges in the depth map, and removed the depth information around the edges before formulating the optimization problem.

3.2   Simulating Bokeh Effect

In this paper, the Bokeh effect is rendered by filtering with the varying-sized and the varying-shaped kernels. Regarding a specific pixel in the scene, the kernel size determines to what extent the pixel is blurred. There are two factors which affects the kernel size, namely, aperture size and the distance to the focal plane. The aperture size controls the blurriness of all the pixels in the image, if the aperture is wider, the image gets more blurred. The distance controls the blurriness of a specific pixel, the larger the distance, the more the pixel gets blurred. The formula to calculate the radius of the kernel at pixel x is shown in Eq. (6), with the radius, the kernel size can be decided as $2 * rad(x) + 1$.

$$rad(x) = round(A * |disp(x) - disp(f)|) \tag{6}$$

Here, A is a coefficient which is proportional to the size of the aperture, which can be used to adjust the blurriness of the image. $disp(f)$ is the disparity of the focus plane, $disp(x)$ is the disparity of pixel x which can be calculated from depth with Eq. (7), notice that fl is focal length, T is baseline.

$$disp(x) = T * fl/Z(x) \tag{7}$$

The pixel x on the rendered image $R(x)$ can be derived by Eq. (8).

$$R(x) = \sum_{y \in \Omega(x)} I(y) * k^{(rad(x))}(y) \tag{8}$$

Here, $k^{(rad(x))}$ is the kernel at pixel x which has radius $rad(x)$. The kernel can be of different shapes, such as circular, star, etc. $\Omega(x)$ is the set of neighboring pixels of x which are within the range. I is the original image.

As there are limited different values for $rad(x)$, to speed up the calculation in Eq. (8), we propose to convolve the original image I with the kernels of different sizes, the final rendered image can be derived by combining these images as shown in Eq. (9).

$$R = \sum_{i=1}^{m} P^{(i)} \cdot * (I \otimes k^{(i)}) \tag{9}$$

Here, $P^{(i)}$ is the mask matrix, $P^{(i)}(x) = 1$ when $rad(x) = i$, $.*$ means element-wise multiplication, $\otimes$ is the operator for convolution. The high efficiency lies in that $I \otimes k^{(i)}$ can be done in the preprocessing step for all possible i, changing the focus would be simply summing the dot multiplication of the matrices, thus lead to linear time complexity.

4   Experiment Result

We have implemented our method in MATLB which runs on a Windows 10 operating system with an i5-6200U platform. We have selected 4 sets of data from Kinect 1.0, the resolution is 640 * 480.

Figure 2 shows the depth maps obtained by joint bilateral filtering (JBF), joint trilateral filtering (JTF) and our proposed method. Figure 2 shows the Bokeh effect with those methods. It can be seen that the edges are more sharp and accurate in the depth map retrieved by our method. This is important for the application in this paper, as otherwise the Bokeh effect would look unnatural and fake. From Fig. 3, we can see that if the depth map is not properly corrected (column 1 and column 2), there will be obvious artifacts in the rendered image. In addition, the proposed method could fill large holes in the depth map, although other methods could do the same with a larger neighborhood, however, it will cause further loss to the depth map.

Table 1. The running time to simulate Bokeh effect with different aperture sizes.

Image	A = 2	A = 4	A = 6
1	0.132788	0.158406	0.165642
2	0.138237	0.160106	0.165419
3	0.107712	0.133094	0.150457
4	0.108090	0.121217	0.166624

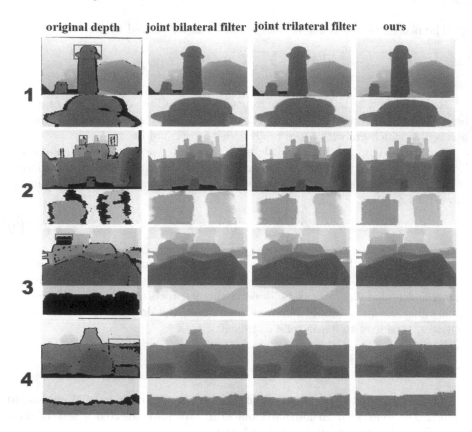

Fig. 2. The original depth map, depth map refined by JBF, JTF and our method.

The Bokeh effect could be rendered with different aperture shapes, such as, heart, diamond, star, or triangle. Altering the shapes of the aperture can make the pictures look different. As shown in Fig. 4, the black dots on the umbrella became hearts, diamonds, stars and triangles respectively. Figure 5 shows the Bokeh effect with different aperture sizes (A = 2, 4, 6), here we choose a circular aperture to illustrate the effect, and we can see circular black dots from the umbrella in the first image. Also, the aperture size affects the blurriness of the image, the larger the aperture, the more the image gets blurred.

Table 1 shows the running time (measured in seconds) to simulate the Bokeh effect with different aperture sizes. It can be seen that the calculations are quite efficient, although changing the size of the aperture would slightly increase the running time.

joint bilateral filter   **joint trilateral filter**   **ours**

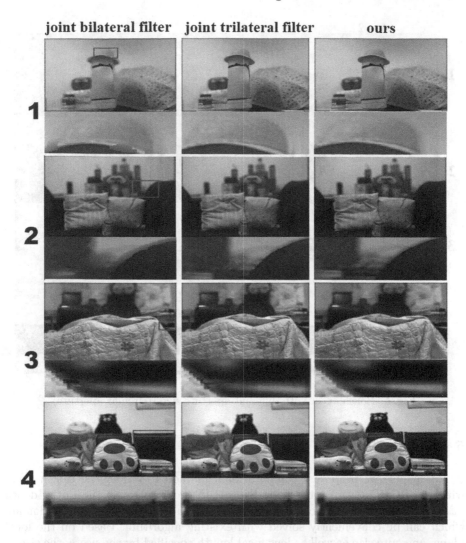

Fig. 3. Bokeh effect with the depth map refined by JBF, JTF and our method.

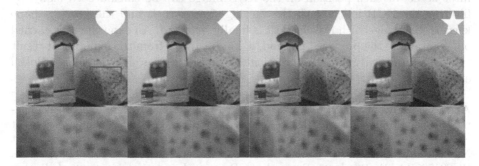

Fig. 4. Bokeh effect with different aperture shape

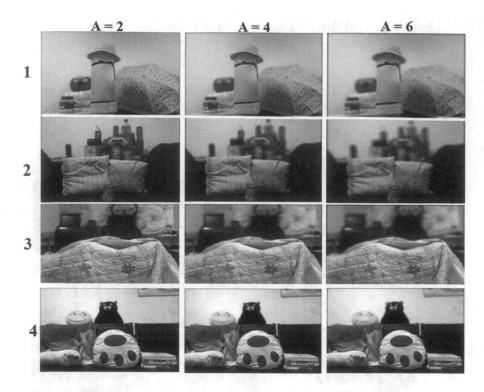

Fig. 5. Bokeh effect with different aperture sizes (circular aperture)

5   Conclusion

In this paper, we propose a computational method to render the Bokeh effect with the RGBD data captured by Kinect. In order to refine the initial depth map, we modeled it as an optimization problem in the range and spatial domain which can be conveniently solved with existing algorithm. Based on the focal plane, aperture size as well as lens focal length specified by the users, the size of filter kernel at each position on the RGB image can be calculated. Finally, by filtering the RGB image with the varying-sized kernels and the varying-shape kernels, the image with Bokeh effect is finally rendered. An experiment is carried out to evaluate the proposed method, it can be drawn from the experiment result that our proposed method is practical to simulate natural and realistic Bokeh effect.

Acknowledgment. This work was supported by National Natural Science Foundation of China (Grant No. 61402205), China Postdoctoral Science Foundation (Grant No. 2015-M571688), University Science Research Project of Jiangsu Province (Grant No. 16KJB520008), Natural Science Foundation of Jiangsu Province (Grant No. BK20170558), Postgraduate Research & Practice Innovation Program of Jiangsu Province (Grant No. SJCX17_0575).

References

1. Levin, A., Lischinski, D., Weiss, Y.: Colorization using optimization. In: ACM SIGGRAPH, pp. 689–694. ACM (2004)
2. Garro, V., Zanuttigh, P., Cortelazzo, G.M.: A new super resolution technique for range data (2010)
3. Kopf, J., Cohen, M.F., Lischinski, D.: Joint bilateral upsampling. ACM Trans. Graph. (TOG) 26(6), 96 (2007)
4. Matsuo, T., Fukushima, N., Ishibashi, Y.: Weighted joint bilateral filter with slope depth compensation filter for depth map refinement. In: International Conference on Computer Vision Theory and Applications (2015)
5. Lo, K.H., Wang, Y.F., Hua, K.L.: Edge-preserving depth map upsampling by joint trilateral filter. IEEE Trans. Cybern. 48(1), 371–384 (2018)
6. Zhang, S., Zhong, W., Ye, L., Zhang, Q.: A modified joint trilateral filter for depth image super resolution. In: Yang, X., Zhai, G. (eds.) IFTC 2016. CCIS, vol. 685, pp. 53–62. Springer, Singapore (2017). https://doi.org/10.1007/978-981-10-4211-9_6
7. Liu, M.Y., Tuzel, O., Taguchi, Y.: Joint geodesic upsampling of depth images. In: Computer Vision and Pattern Recognition, pp. 169–176 (2013)
8. He, K., Sun, J., Tang, X.: Guided image filtering. IEEE Trans. Pattern Anal. Mach. Intell. 35(6), 1397–1409 (2013)
9. Ho, Y.S., Lee, S.B.: Joint multilateral filtering for stereo image generation using depth camera (2013)
10. Yang, J., Ye, X., Li, K.: Color-guided depth recovery from RGB-D data using an adaptive autoregressive model. IEEE Trans. Image Process. 23(8), 3443–3458 (2014)
11. Ma, Z., He, K., Wei, Y.: Constant time weighted median filtering for stereo matching and beyond. In: IEEE International Conference on Computer Vision, pp. 49–56 (2014)
12. Krivanek, J., Zara, J., Bouatouch, K.: Fast depth of field rendering with surface splatting. In: Computer Graphics International, pp. 196–201 (2003)
13. Lee, S., Kim, G.J., Choi, S.: Real-time depth-of-field rendering using point splatting on per-pixel layers. Comput. Graph. Forum 27(7), 1955–1962 (2008)
14. Cook, R.L.: Stochastic sampling in computer graphics. ACM Trans. Graph. 5(1), 51–72 (1986)
15. Buhler, J., Wexler, D.: A phenomenological model for Bokeh rendering. In: Computer Graphics Proceedings, Annual Conference Series, ACM SIGGRAPH Abstracts and Applications, p. 142 (2002)
16. Justin, H., Thorsten, S., Greg, C., et al.: Fast summed-area table generation and its applications. Comput. Graph. Forum 24(3), 547–555 (2010)
17. Kraus, M.: Using opaque image blur for real-time depth-of-field rendering. In: GRAPP 2011 - Proceedings of the International Conference on Computer Graphics Theory and Application, pp. 153–159 (2011)
18. Rokita, P.: Fast generation of depth of field effects in computer graphics. Comput. Graph. 17(5), 593–595 (1993)
19. Zhou, T., Chen, J.X., Pullen, M.: Accurate depth of field simulation in real time. In: Computer Graphics Forum, pp. 15–23 (2007)
20. Gonzalez, R.C., Woods, R.E.: Digital Image Processing, vol. 28(4), pp. 484–486. Prentice Hall International (1977)

LFSF: Latent Factor-Based Similarity Framework and Its Application for Collaborative Recommendation

Liangliang He[1], Zhenhua Tan[1,2(✉)], Guibing Guo[1], Qiuyun Chang[1], and Danke Wu[1]

[1] Software College, Northeastern University, Shenyang 110819, China
tanzh@mail.neu.edu.cn
[2] Academy of Information Technology, Northeastern University, Shenyang 110819, China

Abstract. Similarity computation is the critical component of collaborative filtering recommendation. Traditional similarity measures are usually calculated on user-item rating matrix directly, and couldn't well describe users' internal similar relations and personalization of rating habits. This paper proposes a novel similarity framework based on latent factor model, named LFSF (Latent Factor-based Similarity Framework). Different from the traditional similarity calculation based on rating matrix, the LFSF uses latent factor vectors instead of rating vectors to compute similarity. Since the latent factors are learned from the user feedback by the LFMs, the proposed LFSF appropriately merges advantages of the LFM into the similarity calculation, which improves the predictive accuracy of similarity measures. We illustrate the proposed framework in detail and propose related application instance of LFSF in this paper. We find that latent factors indicate the user preferences and reflect the item attributes more specifically, and the value of latent factors are less affected by user rating habits and item popularity. Experiments based on the real datasets show that the proposed LFSF is more rational and effective than traditional rating-based similarity.

Keywords: Collaborative filtering · Latent factor · Recommendation
Similarity measure

1 Introduction

With the rapid development of the Internet, recommender technologies have been applied to a variety of Internet-based systems. Recommendation techniques mainly include content-based recommendation, collaborative filtering (CF) recommendation and hybrid recommendation. Collaborative filtering techniques are more frequently implemented and often result in better predictive accuracy [1, 2]. These techniques recommend items based on the opinions of other like-minded users or identify items that are similar to those previously rated by the target user, and mainly include item-based CF, which associates an item with nearest neighbors, and user-based CF, which associates a set of nearest neighbors with each user [3–5].

© Springer Nature Switzerland AG 2018
R. Hong et al. (Eds.): PCM 2018, LNCS 11164, pp. 744–754, 2018.
https://doi.org/10.1007/978-3-030-00776-8_68

The similarity measure is a critical component of CF recommendation for computing similarities between users' past behaviors [6–8]. It provides a direct recommendation pattern since users with similar past preferences have similar opinions on items. Traditional similarity is usually calculated based on a user-item rating matrix, where each row is a rating vector evaluated by a related user and each column includes ratings of a specific item given by users. The similarity measure models that are commonly used for CF recommender systems are mainly Cosine Similarity (COS) and Pearson's Correlation Coefficient (PCC) [9–11]. PCC calculates similarity as the covariance of two users' preferences (ratings) divided by their standard deviations based on co-related items. The formula is:

$$PCC(u,v) = \frac{\sum_{i \in I_{u,v}} (r_{u,i} - \bar{r}_u) \cdot (r_{v,i} - \bar{r}_v)}{\sqrt{\sum_{i \in I_{u,v}} (r_{u,i} - \bar{r}_u)^2} \sqrt{\sum_{i \in I_{u,v}} (r_{v,i} - \bar{r}_v)^2}}, \tag{1}$$

where $I_{u,v}$ is the set of co-related items of both users u and v, $r_{u,i}$ is the rating of item i by user u, and $\bar{r}_u$ is the average rating of user u for all the correlated items. COS calculates the similarity between two users by measuring the value of the cosine angle between the two vectors of ratings; a smaller angle indicates greater similarity. The COS formula is:

$$COS(u,v) = \frac{\sum_{i \in I_{u,v}} r_{u,i} \cdot r_{v,i}}{\sqrt{\sum_{i \in I_{u,v}} r_{u,i}^2} \sqrt{\sum_{i \in I_{u,v}} r_{v,i}^2}}, \tag{2}$$

where $I_{u,v}$ and $r_{u,i}$ have the same meaning as in Eq. (1).

However, the two traditional similarity measures have inherent limitations. (1) With the continuous increasing of new users and items updated in the system [12], the prediction efficiency of traditional similarity based on rating matrix would be decreased. (2) Some inaccurate ratings result in that the similarity based on rating matrix couldn't reflect the relationships between users or items. (3) Personalization of user rating habits [2] and difference of item popularities [8] result in that the similarity based on rating matrix couldn't illustrate preference of a user to an item. (4) In addition, CF-based recommendation systems always suffer from the cold-start problem, where users do not have enough co-related ratings for prediction from the sparse rating matrix.

To address these observed problems of traditional similarity measures, this paper proposes a novel similarity calculation framework based on latent factors, named LFSF (latent factor-based similarity framework). The initiative opinion is to calculate similarity based on latent factor matrix other than on the original rating matrix. The LFMs [13–15] can decompose the user-item rating matrix R into users' latent factor matrix P and items' latent factor matrix Q, where P reflects the weight of users' preference on each latent factor and Q reflects the weight of items' characteristics. Flexibility is the best advantage of proposed LFSF, existing LFMs can be used to get latent factors, such as SVD++ [13] and TrustSVD [15]; and existing similarity model can be applied to measure users' or items' similarity in LFSF, such as Cosine and PCC. The proposed

LFSF consists of two critical components, one is LFM and other is similarity measure. This work makes the following main contributions.

(1) Different from the traditional similarity calculation based on rating matrix, the LFSF uses latent factor vectors instead of rating vectors to compute similarity. Since the latent factors are learned from the user feedback by the LFMs, the proposed LFSF appropriately merges advantages of the LFM into the similarity calculation, which improves the predictive accuracy of similarity measures.

(2) The latent factors are usually fixed by specific LFM, and the number of factors won't increase with the increasing of users or items. Thus, the proposed LFSF can improve the calculation efficiency of similarity, including user-based and item-based. Since the latent factor matrix is dense, the similarity measure in LFSF can alleviate the shortcomings of sparseness and cold-start to some extent.

(3) We also propose an application instance of LFSF, using SVD++ to get latent factors and Cosine to be similarity model. Experiments based on real datasets prove the efficiency and predictive accuracy. Based on the proposed instance, readers can implement more.

The rest of this article is organized as follows. We introduce the proposed LFSF and its the application instance is proposed in Sect. 2. Experiments and analysis is in Sect. 3, with conclusions afterwards in Sect. 4.

2 Proposed Latent Factor-Based Similarity Framework

2.1 LFSF: Latent Factor Based Similarity Framework

Different traditional similarity calculation based on user-item rating matrix directly, the proposed latent factor based similarity framework (LFSF) firstly decomposes the user-item rating matrix R by LFM into latent factor matrices P and Q, which denotes users' latent factors and items' latent factors respectively. Then perform the user-based similarity on latent factor matrix P, and item-based similarity on latent factor matrix Q. The framework is illustrated in Fig. 1.

There are two important components, LFM and latent factor-based Similarity. We formalized the LFSF as $LFSF(LFM, LFSim)$. A LFM is used to learn the latent factor matrices of users and items from the user feedbacks which include the rating information and additional implicit attributes from rating matrix (other additional information), such as social trust data of users. User latent factor matrix and item latent factor matrix are used to replace the row vector matrix and column vector matrix of the user-item ratings matrix, respectively, as shown in Fig. 2. The core decomposition in LFM is $\hat{r}_{ui} = q_i^T p_u$ where q_i^T represents the transpose of q_i. Existing SVD, SVD++, TrustSVD could be used as LFM in LFSF. Latent Factor-based similarity evaluates users' or items' similarities. The user-based similarity on LFSF can be denoted as $LFSim(u_i, u_j) = f(R, LFM, P^T)$ based on the P^T, and item-based similarity on LFSF can be denoted as $LFSim(I_i, I_j) = f(R, LFM, Q^T)$ based on Q^T. Traditional Cosine, PCC could be used to calculate the $LFSim$ based on latent factors.

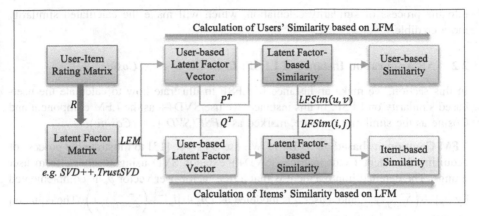

Fig. 1. Proposed latent factor-based similarity framework.

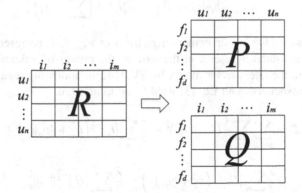

Fig. 2. Basic matrix decomposition model by LFM.

In terms of the LFMs, the number of latent factors is fixed artificially, such as 10, 20 and 50, and it doesn't increase as the total amount of data in the systems increases. Compared to the number of the users and the items in the system, the number of the latent factors is a very small order of magnitude and the *Latent Factor Matrix* can be obtained during the preprocessing stage, so the LFSF can greatly improve the computational efficiency of similarity. Since the data in the latent factor spaces are equivalent to users' weights or items' weights on each latent factor, these values, different from the ratings, are equivalent to ratings in a higher dimension and they are specific to the characteristics of the items. In the case of the film, they are actually the equivalent of the ratings to the film genres, the actor or actress of a film or the director of a film, etc., so compared to the rating spaces, the latent factor spaces can better indicate the preferences of users and well explain the strength of the various characteristics the items have. Because of the scalability of the LFMs, other additional information sources, besides the rating data, will have the opportunity to be incorporated into the process of generating the latent factor spaces. Therefore, another advantage of the LFSS is that the more information sources can be easily incorporated into the process of similarity calculation, which will make the calculated similarity more credible.

2.2   An Application Instance of LFSF: $LFSF(SVD++, Cosine)$

In this section, we make an instance of LFSF to illustrate how to calculate the user-based similarity on LFSF. In this instance, we use SVD++ as the LFM component and Cosine as the similarity measure, marked as $LFSF(SVD++, Cosine)$.

LFM Component based on SVD++. We use SVD++ [13] to illustrate the process of acquiring the latent factor matrices. SVD++ extends SVD using feedback from user ratings. The detailed improvement is that a free factor-user vector p_u is complemented by $|I_u|^{-\frac{1}{2}} \cdot \left(\sum_{j \in I_u} y_j \right)$, and a user u is modeled as $\left(p_u + |I_u|^{-\frac{1}{2}} \cdot \left(\sum_{j \in I_u} y_j \right) \right)$. The equation of SVD++ is:

$$\hat{r}_{ui} = b_u + b_i + \mu + q_i^{\mathsf{T}} \left(p_u + |I_u|^{-\frac{1}{2}} \cdot \left(\sum_{j \in I_u} y_j \right) \right) \tag{3}$$

where I_u represents the set of item ratings by user u; $y_j \in Y$ represents the implicit influence of items rated by user u in the past on the ratings of unknown items in the future. To obtain the latent factor matrix by SVD++, the following regularized squared error function associated with Eq. (3) need to be minimized:

$$\mathcal{L} = \sum_{u \in U_i} \sum_{i \in I_u} \frac{1}{2} (\hat{r}_{ui} - r_{ui})^2 + \frac{\lambda}{2} \sum_u |I_u|^{-\frac{1}{2}} \left(b_u^2 + \|p_u\|_F^2 \right) +$$

$$\frac{\lambda}{2} \sum_i |U_i|^{-\frac{1}{2}} \left(b_i^2 + \|q_i\|_F^2 \right) + \frac{\lambda}{2} \sum_j |U_j|^{-\frac{1}{2}} \|y_j\|_F^2 \tag{4}$$

where U_i, U_j are the set of users who rate items j and i, respectively; $\|\cdot\|$ denotes the Frobenius norm, and λ is to alleviate operation complexity and avoid over-fitting. To obtain a local minimization of the objection function $\mathcal{L}$, we can perform the following gradient descents on b_u, b_i, p_u, q_i and y_j for all the users and items in a given training dataset:

$$\left. \begin{array}{l} \frac{\partial \mathcal{L}}{\partial b_u} = \sum_{i \in I_u} e_{ui} + \lambda |I_u|^{-\frac{1}{2}} \cdot b_u \\[4pt] \frac{\partial \mathcal{L}}{\partial b_i} = \sum_{u \in U_i} e_{ui} + \lambda |U_i|^{-\frac{1}{2}} \cdot b_i \\[4pt] \frac{\partial \mathcal{L}}{\partial p_u} = \sum_{i \in I_u} e_{u,i} q_i + \lambda |I_u|^{-\frac{1}{2}} p_u \\[4pt] \frac{\partial \mathcal{L}}{\partial q_i} = \sum_{u \in U_i} e_{u,i} \left(p_u + |I_u|^{-\frac{1}{2}} \sum_{j \in I_u} y_j \right) + \lambda |U_i|^{-\frac{1}{2}} q_i \\[4pt] \forall j \in I_u, \frac{\partial \mathcal{L}}{\partial y_j} = \sum_{i \in I_u} e_{u,i} |I_u|^{-\frac{1}{2}} q_i + \lambda |U_j|^{-\frac{1}{2}} y_j \end{array} \right\} \tag{5}$$

where $e_{ui} = \hat{r}_{ui} - r_{ui}$. Obviously, user and item latent factor matrices, P and Q, will be output when the error function $\mathcal{L}$ reaches a local minimization.

Similarity Measure Based on Cosine We simply use cosine to measure the similarity between user vectors based on user latent factor matrix P, marked as *LFCOS*. That is,

$$LFCOS(u,v) = \frac{\sum_{f=1}^{D} r_{u,f} \cdot r_{v,f}}{\sqrt{\sum_{f=1}^{D} f_{u,f}^2} \sqrt{\sum_{f=1}^{D} r_{v,f}^2}} \tag{6}$$

where $r_{u,f} \in p_u$ and $r_{v,f} \in p_v$ denote the rating value of user u and user v to factor f, respectively.

Algorithm. Based on the above, we propose a user-based similarity algorithm associated to SVD++ and COS on the LFSF, as shown in Algorithm 1. The similarity between any two users in the system can be returned based on this algorithm.

Algorithm 1. User-based similarity on LFSF(SVD + +, Cosine)

Input: rating record, number of factors d, number of iterations T, trade-off parameter λ, learning rate γ, Convergence condition parameter φ;

Output: cosine similarity between user u and v, i.e., $LFCOS^{SVD++}(u,v)$

(1) set $\gamma \leftarrow 0.02$; initialize vectors B_u and B_i, matrix P and Y with random values in (0,1);

(2) **for** t=1,..., T **do**

(3)   **for** l=1,..., $|\mathcal{R}|$ **do**

(4)   calculate gradients according to Eq.(5);

(5)   Randomly pick up a rating record $(x, i, r_{x,y})$ from rating dataset.

(6) $b_x \leftarrow b_x - \gamma \frac{\partial \mathcal{L}}{\partial b_x}$;

(7) $b_i \leftarrow b_i - \gamma \frac{\partial \mathcal{L}}{\partial b_i}$;

(8) $p_x \leftarrow p_x - \gamma \frac{\partial \mathcal{L}}{\partial p_x}$;

(9) $q_i \leftarrow q_i - \gamma \frac{\partial \mathcal{L}}{\partial q_i}$;

(10) $\forall j \ni I_u, y_j \leftarrow y_j - \gamma \frac{\partial \mathcal{L}}{\partial y_j}$;

(11)   **end**

(12)   **if** $(\partial \mathcal{L}(t) - \partial \mathcal{L}(t-1)) < \varphi$ **end**

(13) **end**

(14) set $sum_{uv} \leftarrow 0, sum_u \leftarrow 0, sum_v \leftarrow 0$;

(15) **for** f=1..D **do**

(16) $sum_{uv} = r_{u,f} \cdot r_{v,f}$;// $r_{u,f}$ and $r_{v,f}$ all belong to factor-user matrix P

(17) $sum_u = r_{u,f}^2$;

(18) $sum_v = r_{v,f}^2$;

(19) **end**

(20) **return** $LFCOS^{SVD++}(u,v)$ by Eq.(6);

Complexity Analysis. In terms of space complexity, it is obvious that the growth on Algorithm (1) almost cannot be ignored compared with COS similarity based on RSS, because bias vectors B_u and B_i, factor-user or factor-item matrix P or Q and so on need to be stored by using additional space. And in terms of time complexity, Algorithm (1) mainly relies on the SVD++. The time complexity of the SVD++ model is mainly from the loss function $\mathcal{L}$ (in Eq. (4)) and its gradients. In terms of $\mathcal{L}$, the time complexity is $O(d \cdot |R|)$, where d represents the matrix dimensionality, and $|R|$ represents the number of the observed ratings. And in terms of the gradients, the time complexities of $\frac{\partial \mathcal{L}}{\partial b_u}, \frac{\partial \mathcal{L}}{\partial b_i}, \frac{\partial \mathcal{L}}{\partial p_u}$, and $\frac{\partial \mathcal{L}}{\partial q_i}$ are $O(d \cdot |R|)$, and the time complexity of $\frac{\partial \mathcal{L}}{\partial y_j}$ is

$O(k \cdot d \cdot |R|)$, where k represents the average of the number a user marks or an item receives, respectively. Obviously $O(k \cdot d \cdot |R|)$ is considerably larger than $O(|I_{u,v}| \cdot O(1))$ which is the time complexities of COS similarity based on RSS. However, for a given dataset, factor-user or factor-item matrix P or Q can be pre-learned and pre-stored before they are used directly in similarity calculations, so the time complexities of COS similarity based on LFSS can be reduced to $O(D \cdot O(1))$ which is considerably less than $O(|I_{u,v}| \cdot O(1))$, because the scale of ratings is considerably larger than factors.

3   Experiments and Analysis

In order to illustrate the rationality and effectiveness of the LFSF, we perform experiments on two real datasets, ml-100 k and FilmTrust (shown in Table 1).

Table 1. Experimental datasets.

Datasets	Users	Movies	Ratings	Density	Rating ranges	Trust ratings
FilmTrust	1508	2071	35497	1.14%	[0.5,4]	1853
ml-100 k	943	1682	100000	6.30%	[1, 5]	null

We implement two LFSF instances to calculate user-based similarity on datasets, including $LFSF(SVD++, Cosine)$ and $LFSF(TrustSVD, Cosine)$, similar to the proposed method in Sect. 2.2. Since no trust information exists in dataset ml-100k, the $LFSF(TrustSVD, Cosine)$ is only performed on dataset FilmTrust in experiments. The user-based similarity from $LFSF(SVD++, Cosine)$ is denoted as $LFCOS^{SVD++}(u, v)$, while the user-based similarity from $LFSF(TrustSVD, Cosine)$ is denoted as $LFCOS^{TrustSVD}(u, v)$ in the following description. The prediction is calculated by following equation:

$$p(u, i) = \bar{r}_u + \frac{\sum_{v \in U(i)} (sim(u, v) \cdot (r_{vi} - \bar{r}_v))}{\sum_{v \in U(i)} |sim(u, v)|}, \tag{7}$$

where $sim(u, v)$ is the similarity measure, such as $LFCOS^{SVD++}(u, v)$, $LFCOS^{TrustSVD}(u, v)$, and Cosine. The evaluation performance of predictive accuracy is measured by MAE (Mean Absolute Error), RMSE (Root Mean Square Error) and RE (Relative Error), respectively, where a lower value indicates better predictive accuracy. And

$$MAE = \frac{\sum_{(u,i) \in \tau} |r_{ui} - \hat{r}_{ui}|}{N},$$

$$RMSE = \sqrt{\frac{\sum_{(u,i)\in\tau}(r_{ui} - \hat{r}_{ui})^2}{N}}, \quad RE = \sqrt{\frac{\sum_{(u,i)\in\tau}(r_{ui} - \hat{r}_{ui})^2}{\sum_{(u,i)\in\tau}(r_{ui})^2}},$$

where $\tau = \{(u,i)|\exists_r((u,i,r)\in R)\}$, N is the number of observed ratings.

Experiment (1): Performance of User-based Similarity on LFSF with Different Latent Factors.

This experiment is to estimate the performance of $LFCOS^{SVD++}(u,v)$ and $LFCOS^{TrustSVD}(u,v)$ on ml-100 k and FilmTrust, with different number of latent factors among [1, 20]. Each dataset is divided into two parts, the first part (80%) is used for training, the other one (20%) for prediction. The traditional Cosine measure is used for comparison. Figure 3 shows the results.

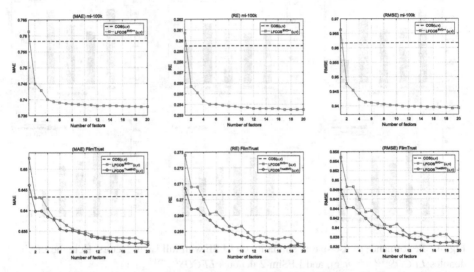

Fig. 3. Performance of user-based similarity on LFSF with different factors. Latent factors Scale: [1, 20]; LFSF Instances: $LFCOS^{SVD++}(u,v)$ and $LFCOS^{TrustSVD}(u,v)$.

It can be seen from the results that the number of factors increases gradually from 1 to 20, and the values of the three evaluation metrics are getting smaller gradually. It proves that the predictive accuracies of the similarities of $LFCOS^{SVD++}(u,v)$ and $LFCOS^{TrustSVD}(u,v)$ get better with the increasing number of latent factors, while traditional Cosine remains same all the time. On dataset ml-100K, the performance of MAE, RE and RMSE of Cosine is 0.7583, 0.2595 and 0.9617, respectively. The performance of proposed $LFCOS^{SVD++}(u,v)$ arrives to 0.7377 (MAE), 0.2535(RE) and 0.9393 (RMSE), all are better than traditional Cosine similarity. On dataset FilmTrust, performances of $LFCOS^{SVD++}(u,v)$ and $LFCOS^{TrustSVD}(u,v)$ are also superior to traditional Cosine. It proves the proposed LFSF can efficiently improve the predictive accuracy, and proves the rightness of the design of similarity based on latent factors in LFSF.

Experiment (2): Performance of User-Based Similarity on LFSF with Full Latent Factors.

This experiment is to estimate the predictive accuracies of LFSF instances on full latent factor which number is fixed to 50. We take original rating matrix based similarities of COS, PCC, PIP [8], and BS [6] as comparison objects. Results on datasets ml-100 k and FilmTrust are shown in Fig. 4, and each dataset is divided into two parts, the first part (80%) is used for training, the other one (20%) for prediction. In the figure, *LFSim-1* and *LFSim-2* denote $LFCOS^{SVD++}(u, v)$ and $LFCOS^{TrustSVD}(u, v)$ respectively, since the original names of them too long to show in one figure together. It can be seen from the results that both of the two LFSF instances have superior performances on both datasets.

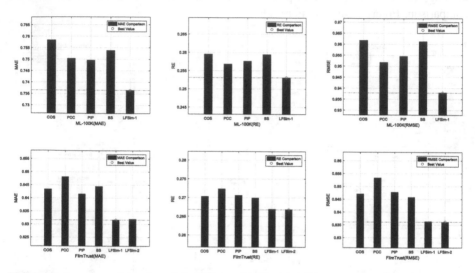

Fig. 4. Performance of user-based similarity on LFSF with full latent factors. Here the LFSim-1 denotes $LFCOS^{SVD++}(u, v)$, and LFSim-2 denotes $LFCOS^{TrustSVD}(u, v)$.

On dataset ml-100K, the predictive accuracy of traditional Cosine remains 0.7583 (MAE), 0.2595(RE) and 0.9617 (RMSE), while the proposed $LFCOS^{SVD++}(u, v)$ arrives to 0.7365(MAE), 0.2531 (RE) and 0.9379 (RMSE). Compared with Cosine, $LFCOS^{SVD++}(u, v)$ improves the predictive accuracy to 2.87% (MAE), 2.466% (RE), and 2.474% (RMSE), respectively. Compared with current efficient similarity such PIP and BS, the proposed LFSF instances are also superior and efficient. On dataset FilmTrust, both $LFCOS^{SVD++}(u, v)$ and $LFCOS^{TrustSVD}(u, v)$ are efficient and their predictive accuracy rank in the first two, comparing with Cosine, PCC, PIP [8] and BS [6]. The performances of $LFCOS^{SVD++}(u, v)$ are 0.6316 (MAE), 0.2669 (RE) and 0.8363 (RMSE), and $LFCOS^{TrustSVD}(u, v)$ arrives at 0.6319 (MAE), 0.2668 (RE) and 0.8361 (RMSE). And we can observed that the performance of $LFCOS^{TrustSVD}(u, v)$ has slight advantage to $LFCOS^{SVD++}(u, v)$. Results of this experiment prove similarity based on full latent factors are great efficient, and illustrate that the effectiveness and efficiency of proposed LSFS.

4 Conclusions

In recommender systems, similarity measures for users or items are usually calculated on original user-item rating matrix which is usually sparse and updated dynamically. Personalization of rating habits and the popularity difference of the items lead to the similarities based on rating matrix do not efficiently reflect personalized similarity relationships between users. Latent factor models can address these shortcomings. Therefore, we propose a novel latent factor-based similarity framework, where the similarity is calculated on latent factor matrices learned from original rating matrix. We introduce the proposed LFSF and propose instances of LFSF, experiments on real datasets prove the rightness and effectiveness of LFSF, and prove the efficiency of proposed instances on LFSF.

Acknowledgement. This work is supported by the National Natural Science Foundation of China under Grants No. 61772125, No. 61702084, No.61702090 and No. 61402097; and the Fundamental Research Funds for the Central Universities under Grant No. N151708005.

References

1. Candillier, L., Meyer, F., Fessant, F.: Designing specific weighted similarity measures to improve collaborative filtering systems. In: Proceedings of ICDM, Leipzig, Germany, pp. 242–255(2008)
2. Tan, Z., He, L., Li, H., Wang, X.: Rating personalization improves accuracy: a proportion-based baseline estimate model for collaborative recommendation. In: Proceedings of CollaborateCOM, Beijing, China, pp. 104–114 (2016)
3. Cao, G., Kuang, L.: Identifying core users based on trust relationships and interest similarity in recommender system. In: Proceeding of ICWS, San Francisco, CA, USA, pp. 284–291 (2016)
4. Tan, Z., He, L.: An efficient similarity measure for user-based collaborative filtering recommender systems inspired by the physical resonance principle. IEEE Access $5(1)$, 27211–27228 (2017)
5. Li, W., Xu, H., Ji, M., Xu, Z., Fang, H.: A hierarchy weighting similarity measure to improve user-based collaborative filtering algorithm. In: Proceedings of ICCC, Chengdu, China, pp. 843–846 (2016)
6. Guo, G., Zhang, J., Yorke-Smith, N.: A novel bayesian similarity measure for recommender systems. In: Proceedings of IJCAI, Beijing China, pp. 2619–2625 (2013)
7. Guo, G., Zhang, J., Yorke-Smith, N.: A novel evidence-based Bayesian similarity measure for recommender systems. ACM Trans. Web $10(2)$, 1–30 (2016)
8. Ahn, H.J.: A new similarity measure for collaborative filtering to alleviate the new user cold-starting problem. Inf. Sci. $178(1)$, 37–51 (2008)
9. Breese, J.S., Heckerman, D., Kadie, C.: Empirical analysis of predictive algorithms for collaborative filtering. In: Proceedings of UAI, Madison, Wisconsin, pp. 43–52 (1998)
10. Lathia, N., Hailes, S., Capra, L.: The effect of correlation coefficients on communities of recommenders. In: Proceedings of SAC, Fortaleza, Ceara, Brazil, pp. 2000–2005 (2008)
11. Schwarz, M., Lobur, M., Stekh, Y.: Analysis of the effectiveness of similarity measures for recommender systems. In: Proceedings of CADSM, Lviv, Ukraine, pp. 275 –277 (2017)

12. Sarwar, B., Karypis, G., Konstan, J., Riedl, J.: Item-based collaborative filtering recommendation algorithms. In: Proceedings of the 10th International Conference on World Wide Web, pp. 285–295. ACM (2001)
13. Koren, Y.: Factorization meets the neighborhood: a multifaceted collaborative filtering model. In: Proceedings of the 14th ACM SIGKDD International Conference on Knowledge Discovery and Data Mining, pp. 426–434. ACM (2008)
14. Kumar, R., Verma, B.K., Rastogi, S.S.: Social popularity based SVD++ recommender system. International Journal of Computer Applications, 87(14) (2014)
15. Guo, G., Zhang, J., Yorke-Smith, N.: TrustSVD: collaborative filtering with both the explicit and implicit influence of user trust and of item ratings. In: Proceedings of AAAI, pp. 123–129. (2015)

Image Aesthetics Assessment Based on User Social Behavior

Huihui Liu[1], Chaoran Cui[2(✉)], Yuling Ma[1], Cheng Shi[1], Yongchao Xu[1], and Yilong Yin[3(✉)]

[1] School of Computer Science and Technology, Shandong University, Jinan, China
[2] School of Computer Science and Technology, Shandong University of Finance and Economics, Jinan, China
crcui@sdufe.edu.cn
[3] School of Software Engineering, Shandong University, Jinan, China
ylyin@sdu.edu.cn

Abstract. Automatically assessing image quality from an aesthetic perspective is emerging as a promising research topic due to its potential in numerous applications. Generally, existing methods perform aesthetics assessment purely based on image visual content. However, aesthetic perceiving is essentially a human cognitive activity, and it is necessary to consider user cognitive information when judging the image aesthetic quality. In this paper, inspired by the observation that human cognition and behavior influence each other, we propose to sense users' cognition to images from their social behavior, and further integrate this knowledge into image aesthetics assessment. To alleviate the uncertainty of social behavior, we merge different types of raw social behavior into clusters, and represent each image using a social distribution over different clusters. We borrow the idea of transfer learning to establish a social behavior detector with social images, but apply it to extract the user cognitive features of web images. In this manner, our approach is generalized to common web images, for which user social behavior is not visible. Finally, the user cognitive information and image visual content are effectively fused to enhance image aesthetics assessment. Extensive experiments on two benchmark datasets have well verified the promise of our approach.

Keywords: Image aesthetics assessment · User cognitive modeling Social behavior sensing

1 Introduction

The task of image aesthetics assessment is to automatically measure whether an image looks beautiful from human's perception. It has attracted more and more attention due to its potential in lots of applications [2, 26–29]. For example, modern image search engines are expected to rank results not only by topical relevance but also by aesthetic quality [3].

© Springer Nature Switzerland AG 2018
R. Hong et al. (Eds.): PCM 2018, LNCS 11164, pp. 755–766, 2018.
https://doi.org/10.1007/978-3-030-00776-8_69

Image aesthetics assessment is typically cast as a classification problem. Recent research advances stem from elaborately designing handcrafted features [14,16], and evolve into systematically learning deep representations for visual aesthetics [17,18]. Overall, existing studies performed aesthetics assessment purely based on image visual content. However, we argue that aesthetic perceiving is essentially a human cognitive activity, and it is necessary to consider users' cognition to images when judging their aesthetic qualities.

Despite extensive research in recent decades, how to understand and model user cognition still poses a tremendous challenge. Today, with the emergence of social media platforms (e.g., Flickr), billions of users proactively interact with huge volumes of social images, including bookmarking images as favorite, organizing images into interest groups, and adding tags to images. In social psychology, it has proven that human cognition and behavior influence each other [1]. Motivated by this, we propose to sense users' cognition to images from their behavior in social media platforms, and further integrate this knowledge into image aesthetics assessment.

However, this idea faces the following fundamental problems:

- **The Uncertainty of Social Behavior.** In social media platforms, the motivation of user behavior is rather casual and complicated, resulting in the fact that the social behavioral information accompanied by images is uncertain [4]. How to make the right use of such inaccurate resources to capture user cognition is a critical issue.
- **The Lack of Social Behavior for Web Images.** Most web images do not have social interactions with users, since they are not produced or shared in social media platforms. How to generalize our approach to common web images emerges as a great challenge.
- **The Fusion of User Cognitive Information and Image Visual Content.** Our ultimate vision is to enhance image aesthetics assessment by jointly taking user cognitive information as well as image visual content into account. How to effectively fuse the two important factors plays a crucial role in our problem.

To address the above challenges, we propose a novel framework of Social-sensed Image Aesthetics Assessment (SIAA). In our approach, we resolve the uncertainty problem by merging the raw social behavior into clusters, and represent each image using a social distribution over different clusters. To avoid being limited to social images, we realize an end-to-end social behavior detector from social images using a deep convolutional neural network, and apply it to predict different types of social distribution for general web images. In this process, the user cognitive features of web images are extracted from their mid-level representations in the network. Lastly, a fusion sub-network is designed to effectively balance the cognitive features and visual features for image aesthetics assessment.

The main contributions can be summarized as follows:

- We propose to sense users' cognition to images from their social behavior, and further integrate this knowledge into image aesthetics assessment. To the best of our knowledge, our study is the first attempt to consider user cognitive information when judging the image aesthetic quality.

- We introduce the idea of transfer learning [23] to generalize our approach to common web images. A social behavior detector is learned with social images, but applied to extract the user cognitive features of web images.
- We evaluate our approach on two benchmark datasets. The experimental results demonstrate the potential of our approach in comparison with the state-of-the-art methods, and highlight the benefits of user cognitive information for image aesthetics assessment.

2   Related Work

Image aesthetics assessment has driven considerable research attention owing to its wide range of applications. It is predominantly cast as a classification problem, in which the glaring challenge is how to extract effective image features [8]. Many early studies designed handcrafted features based on the intuitions about human aesthetic perception or photographic rules. For example, Datta et al. [5] designed 56 features including the indictors of colorfulness, rule of thirds, and depth of field. Ke et al. [12] developed seven kinds of features to represent images through simplicity, contrast, brightness, etc. Dhar et al. [9] learned several high-level describable attributes of images from the perspectives of layout, content, and illumination. Tang et al. [24] argued that different types of images are related to different aesthetic evaluation criteria, and designed features in different ways based on the variety of image content. Lo et al. [16] proposed aesthetic features with high efficiency to compute. Unlike the above efforts extracting features from the whole image, Luo et al. [19] first extracted the subject region from an image, and then formulated a number of high-level semantic features based on this subject and background division. Furthermore, generic image descriptors, like Bag-of-Visual-Words and the Fisher Vector, were also used to assess image aesthetic quality [21]. Despite the significant progress in image aesthetics assessment, developing handcrafted features still requires a lot of engineering skill and domain expertise.

Deep neural networks have also been used to learn aesthetic features, and obtained excellent performance. Lu et al. [17] proposed a double-column convolutional neural network to capture both global and local characteristics of images, and employed the style and semantic attributes of images to boost the aesthetics categorization performance. By extending this idea, they further proposed a deep multi-patch aggregation network [18], which simultaneously accepts multiple patches from a single image, and aggregates the features of individual patches to predict the aesthetic quality of that image. Kao et al. [13] proposed to use a related task, the semantic recognition, to assist the aesthetics assessment through a multi-task deep network. To circumvent the fixed-size constraint of convolutional neural networks, Mai et al. [20] presented a composition-preserving method that directly learns aesthetic features from the original inputs without any image transformations. Similarly, Fang et al. [10] introduced a fully convolutional network, enabling inputs of varying sizes.

Overall, existing methods extract aesthetic features purely based on image visual content. They overlook the fact that aesthetics is a human cognitive activity, and do not take user cognitive information into account. In contrast, inspired by the observation that human cognition and behavior influence each other, we seek to sense users' cognition to images from their social behavior, and further make the best use of this knowledge for image aesthetics assessment.

3   Framework

In this section, we introduce a novel framework of Social-sensed Image Aesthetics Assessment (SIAA). To formulate our problem, we use bold letters (e.g., $\mathbf{x}$), non-bold letters (e.g., x), and calligraphic capital letters (e.g., $\mathcal{X}$) to denote vectors, scalars, and sets, respectively. For a vector $\mathbf{x}$, x_i denotes the i-th element of $\mathbf{x}$.

SIAA consists of three main components: (1) social distribution representation, (2) user cognition extraction, and (3) cognitive and visual feature fusion. In the following, we elaborate on each component and describe the implementation details.

3.1   Social Distribution Representation

In social media platforms, the typical social behavior associated with images includes user favoring, group sharing, and user tagging. Following the idea of bag-of-visual-words, an image can thus be represented by $i = \{\mathcal{U}_i, \mathcal{G}_i, \mathcal{T}_i\}$, where $\mathcal{U}_i$, $\mathcal{G}_i$, and $\mathcal{T}_i$ denote the set of users who favor the image i, groups where i is shared, and tags annotated to i, respectively. However, as mentioned previously, the social behavior information is uncertain. To address this issue, we propose to merge the raw social behavior into clusters, and represent each image using a social distribution over different clusters.

Taking users as an example, if an image is favored by two users, it can be considered that the interests of the two users are partially similar. Based on the assumption, we first establish a similarity graph among users. Let $\mathcal{I}(u)$ be the set of favorite images of a user u. For two users u and v, we estimate their similarity by measuring how many images co-favored by u and v:

$$sim(u, v) = \frac{|\mathcal{I}(u) \cap \mathcal{I}(v)|}{|\mathcal{I}(u) \cup \mathcal{I}(v)|}. \tag{1}$$

Then, we conduct the affine propagation clustering [11] over the user similarity graph, resulting in k user clusters with $\mathcal{C}_j$ denoting the j-th cluster. Note that due to the uncertainty of user favoring behavior, we cannot judge that users in the same cluster must be similar, and vise versa. In light of this, we further assign each user a reliability score indicating the confidence that the user is tied to the correct cluster. Specifically, for the cluster $\mathcal{C}_j$, we denote by $\mathbf{r}_j$ the vector of reliability scores of the users in $\mathcal{C}_j$. Similar to [15], $\mathbf{r}_j$ is computed by the PageRank algorithm and iteratively updated by:

$$\mathbf{r}_j(t) = d\mathbf{P}_j\mathbf{r}_j(t-1) + (1-d)\mathbf{e}, \tag{2}$$

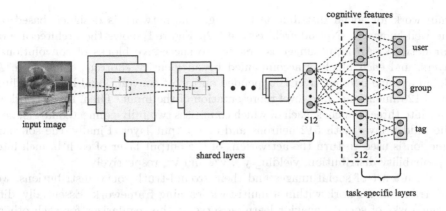

Fig. 1. The architecture of the network for social distribution learning.

where $\mathbf{r}_j(t)$ is the vector of reliability scores in the t-th iteration. $\mathbf{P}_j$ is the transaction matrix normalized from the pairwise similarity matrix between the users in $\mathcal{C}_j$, $\mathbf{e}$ is a normalized vector whose element are all $1/|\mathcal{C}_j|$, and d is the damping factor. Intuitively, the reliability score r_{ju} of a user $u \in \mathcal{C}_j$ reflects its centrality in $\mathcal{C}_j$.

Finally, we map the image i into a k-dimensional vector $\mathbf{y}_i^u$, whose element y_{ij}^u is defined by:

$$y_{ij}^u = \begin{cases} \frac{\sum_{u \in \mathcal{U}_i \cap \mathcal{C}_j} r_{ju}}{|\mathcal{U}_i \cap \mathcal{C}_j|}, & \mathcal{U}_i \cap \mathcal{C}_j \neq \emptyset; \\ 0, & \text{otherwise.} \end{cases} \tag{3}$$

$\mathbf{y}_i^u$ is subsequently normalized to be a distribution with y_{ij}^u indicating the probability of i belonging to $\mathcal{C}_j$. Note that such a distribution offers two additional advantages: (1) It is much lower-dimensional than the representation based on the set of individual users; (2) It is stable in dimension when a new user comes. In analogous ways, we also represent each image using its shared groups and annotated tags. For the image i, $\mathbf{y}_i^u$, $\mathbf{y}_i^g$, and $\mathbf{y}_i^t$ denote the social distribution of i over user clusters, group clusters, and tag clusters, respectively. In the following, we shall omit the subscript i for notational simplicity.

3.2 User Cognition Extraction

In our study, we propose to sense users' cognition to images from their social behavior. To avoid being limited to social images, our approach borrows the idea of transfer learning, where a social behavior detector is learned in advance from social images, and further applied to extract the user cognitive features of general web images.

To be specific, we seek to predict the social distribution of an image over user clusters, group clusters, and tag clusters, i.e., $\widehat{\mathbf{y}}^u$, $\widehat{\mathbf{y}}^g$, and $\widehat{\mathbf{y}}^t$, respectively. The task is termed *social distribution learning*. Inspired by the recent success of deep learning, we utilize a deep convolutional network to realize an end-to-end framework for social distribution learning. The network is designed based on the well-known ResNet-50 architecture [12]. Figure 1 shows the architecture of our network. An input image is first fed to successive blocks of convolutional layers, and each block is accompanied by an identity shortcut connection. A global average pooling layer is appended on top of convolutional layers, leading to a 512-dimensional mid-level representation of the input. Then, the network is split into three branches, each of which comprises two fully-connected layers, i.e., one hidden layer with 512 neurons and one output layer. Finally, the softmax function is used to turn the activations of the output layer of each branch into a probability distribution, yielding $\widehat{\mathbf{y}}^u$, $\widehat{\mathbf{y}}^g$, and $\widehat{\mathbf{y}}^t$, respectively.

With a set of social images and their ground-truth social distributions, we can train our network within a multi-task learning framework. Essentially, different tasks of social behavior learning serve as the regularizer for each other. This may lead to a model with better generalization capability. We choose the KL divergence as the loss function to penalize the deviation of the predicted distribution from the ground-truth distribution, i.e.,

$$l = \sum_j y_j^u \ln \frac{y_j^u}{\widehat{y}_j^u} + \sum_j y_j^g \ln \frac{y_j^g}{\widehat{y}_j^g} + \sum_j y_j^t \ln \frac{y_j^t}{\widehat{y}_j^t}. \tag{4}$$

The weights of the network are determined by minimizing the loss for all social images. We resort to mini-batch Stochastic Gradient Descent (SGD) for the optimization problem.

Once the training is completed, we run a forward pass through the network for a web image, and extract the activations from the penultimate fully connected layer of each branch. As aforementioned, users' social behavior is reckoned to be the reflection of their cognition to images. So we regard the three 512-dimensional activations as three types of user cognitive features of the input web image.

3.3   Cognitive and Visual Feature Fusion

In our study, we leverage both user cognitive information and image visual content to facilitate image aesthetics assessment. The user cognitive features of images are obtained following the procedures as described in the above subsection. As for the visual features, we use a ResNet-50 model pre-trained on ImageNet [6], and take the 512-dimensional outputs for each image from the global average pooling layer.

A high-level sub-network is elaborately designed to fuse the cognitive features and visual features. Figure 2 displays the overall structure of the fusion sub-network, which is composed of two hidden layers and one output layer. Specifically, each type of cognitive features are first fed into a fully connected layer with 256 neurons, respectively. The first hidden layer has the effect of dimension reduction, helping us preserve key information from each type of cognitive features and reduce the subsequent computational burden. Then, the outputs of the first hidden layer across different types of cognitive features are fused by a common fully-connected layer, yielding a unified representation with 256

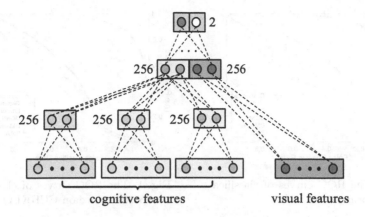

Fig. 2. The overall structure of the fusion sub-network.

neurons. At the same time, the visual features also go through a fully-connected layer and are transformed into a 256-dimensional hidden representation. Lastly, the cognitive and visual hidden representations are concatenated and mapped into a binary aesthetic label (i.e., high-aesthetic or low-aesthetic) via a softmax classifier layer.

3.4 Implementation Details

For the dumping factor d in Eq. (2), we set $d = 0.8$. To tackle the overfitting problem, we randomly sampled 224×224 crop from each image or its horizontal flip. We also subtracted each pixel by the mean value computed over the training set. For the mini-batch SGD algorithm, we set the batch size to 64. The initial learning rate was 0.001 for all layers, and was annealed by a factor of 0.1 every time the validation loss plateaus. We used weight decay of 10^{-4} and momentum of 0.9. The training phase would be early stopped when the learning rate dropped to 10^{-7}. At test time, we made predictions on ten 224×224 random crops of each test image and took the average as the final prediction result.

4 Experiments

4.1 Datasets

We randomly selected 50,000 images from the Yahoo Flickr Creative Commons dataset [25], and crawled the social behavior information associated with these images, including the users who favor them, the groups that they belong to, as well as the tags annotated to them. Since the information is usually noisy, we left out those users/groups/tags occurring less than 50 or more than 2,000 times. This filtering process leads to 5,776 users, 2,861 groups, and 2,210 tags, respectively. As described previously, we further applied the affine propagation

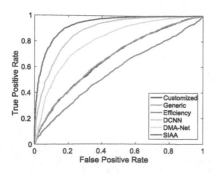
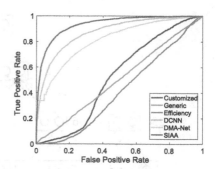

Fig. 3. The ROC curves of classification performance on AVA.

Fig. 4. The ROC curves of classiffication performance on CUHKPQ.

clustering on the raw social behavior information, and finally identified 240 user clusters, 327 group clusters, and 399 tag clusters, respectively.

Our experiments on image aesthetics assessment were conducted on two benchmark datasets, namely, AVA [22] and CUHKPQ [24]. The AVA dataset contains about 255,530 images. Each image receives an average of 210 aesthetic ratings ranging from 1 to 10. We followed the same procedure as the previous studies [17,18,22] to assign a binary aesthetic label to each image. Specifically, images with mean ratings smaller than $5-\delta$ are labeled as low quality, those with mean ratings greater than $5+\delta$ are labeled as high quality, and the others are considered ambiguous and discarded. In our case, we chose $\delta = 1$. We randomly picked out 70% of the images for training, 10% for validation, and the remaining for testing.

The CUHKPQ dataset consists of 17,690 images. Each image has already been labeled as either high-aesthetic or low-aesthetic by at least 8 out of 10 independent viewers. To verify the generalization ability of different methods, we performed a cross-set evaluation. That is, we trained a model only on AVA, but tested it on AVA as well as CUHKPQ simultaneously. In other words, CUHKPQ was used merely for testing.

4.2   Evaluation Metrics

In our study, we adopted classification accuracy to evaluate the performance on image aesthetics assessment. The accuracy is obtained by binarizing the outputs of a method with the threshold value of 0.5 and comparing the results to the ground-truth binary aesthetic labels. Moreover, on account of the imbalance between the positive and negative test images, we plotted the Receiver Operating Characteristic (ROC) curve of classification performance. To quantitatively compare different methods, we also computed the metric of Area Under the ROC Curve (AUC).

Table 1. Aesthetic classification performance on AVA in terms of accuracy and AUC, respectively.

Mereics	Customized	Generic	Efficiency	DCNN	DMA-Net	SIAA
Accuracy	0.866	0.802	0.817	0.906	0.870	**0.930**
AUC	0.669	0.671	0.567	0.880	0.801	**0.938**

Table 2. Aesthetic classification performance on CUHKPQ. Note that all methods are trained on AVA but tested on CUHKPQ.

Mereics	Customized	Generic	Efficiency	DCNN	DMA-Net	SIAA
Accuracy	0.256	0.265	0.334	0.598	0.507	**0.730**
AUC	0.524	0.507	0.418	0.875	0.815	**0.934**

4.3 Comparison with State-of-the-Art

We compared our approach **SIAA** against several state-of-the-art methods for image aesthetics assessment, e.g. **Customized** [14], **Generic** [21], **Efficiency** [16], **DCNN** [17], and **DMA-Net** [18].

Figure 3 plots the ROC curves of different methods on AVA, and Table 1 lists the performance in terms of accuracy and AUC. As can be seen, SIAA outperforms the other competitors in different metrics. A possible reason is that SIAA introduces users' cognitive information sensed from their social behavior, which plays a critical role in the process of image aesthetics assessment.

The comparison results on CUHKPQ are displayed in Fig. 4 and Table 2, respectively. As expected, SIAA still achieves the best performance. One thing worth noting is that all methods are trained on AVA but tested on CUHKPQ. In this circumstance, the other competitors experiences a sharp degradation in terms of accuracy, while SIAA maintains relatively stable performance. The results indicate the superior generalization ability of our approach for image aesthetics assessment.

4.4 Benefits of Cognitive Information

In our study, SIAA leverages three types of user cognitive information sensed from the social behaviors of user favoring (**F-Cognition**), group sharing (**G-Cognition**), and user tagging (**T-Cognition**), respectively. To investigate the contribution of user cognitive information to image aesthetics assessment, we implement three variant methods using only one type of information, and compare them against our original framework SIAA. Moreover, a method (**Visual**) that only considers the visual features of images is also introduced to the comparison.

Due to space limitation, we only report the comparison results of different methods on AVA in terms of accuracy and AUC, respectively. As shown in

Table 3. Comparison between the methods using different types of user cognitive information.

Mereics	Visual	F-Cognition	G-Cognition	T-Cognition	SIAA
Accuracy	0.909	0.918	0.924	0.922	**0.930**
AUC	0.889	0.915	0.929	0.926	**0.938**

Table 4. Comparison between our approach and the method with raw user social behavior.

Mereics	Raw	SIAA
Accuracy	0.922	**0.930**
AUC	0.921	**0.938**

Table 3, F-Cognition, G-Cognition, and T-Cognition all offer better performance over Visual. Meanwhile, SIAA slightly outperforms any of these variant methods. To sum up, these results imply that each type of user cognitive information indeed contributes to the performance improvement.

4.5   Effect of Social Behavior Uncertainty

In our approach, since the raw user social behavior is uncertain, we merge them into clusters and represent each image over the resulting clusters. To verify the necessity of this step, we also experiment our approach with the raw user social behavior. Specially, each social image is represented based on the set of individual social behavior, and the user cognitive features are extracted in the process of predicting the occurrence of individual social behavior for images. Table 4 shows the comparison results between SIAA and this implementation (**Raw**). It can be seen that SIAA is superior to Raw in both metrics. This finding underlines the importance of addressing the uncertainty issue in our framework.

5   Conclusions

In this paper, we have investigated image aesthetics assessment from a new perspective of introducing user cognitive information. Given the fact that human cognition and behavior influence each other, we have proposed to sense users' cognition to images from their social behavior. To generalize our approach to common web images without user social behavior, we have learned a social behavior detector with social images, and applied it to extract the user cognitive features of web images. The experimental results on two benchmark datasets have verified the effectiveness of our approach, and highlighted the benefits of user cognitive information for image aesthetics assessment.

In the future, we plan to realize personalized image aesthetics assessment that provides different assessments regarding the same image for users with different aesthetic preferences [7]. For this problem, how to understand an individual users aesthetic preference poses a formidable challenge.

Acknowledgements. This work is supported by the National Natural Science Foundation of China (61573219, 61701281), Shandong Provincial Natural Science Foundation (ZR2017QF009), and the Fostering Project of Dominant Discipline and Talent Team of Shandong Province Higher Education Institutions.

References

1. Albarracin, D., Wyer Jr., R.S.: The cognitive impact of past behavior: influences on beliefs, attitudes, and future behavioral decisions. J. Pers. Soc. Psychol. **79**(1), 5 (2000)
2. Cheng, Z., Shen, J., Zhu, L., Kankanhalli, M., Nie, L.: Exploiting music play sequence for music recommendation. In: Twenty-Sixth International Joint Conference on Artificial Intelligence, pp. 3654–3660 (2017)
3. Cui, C., Fang, H., Deng, X., Nie, X., Dai, H., Yin, Y.: Distribution-oriented aesthetics assessment for image search. In: Proceedings of the 40th International ACM SIGIR Conference on Research and Development in Information Retrieval, pp. 1013–1016 (2017)
4. Cui, C., Shen, J., Nie, L., Hong, R., Ma, J.: Augmented collaborative filtering for sparseness reduction in personalized poi recommendation. ACM Trans. Intell. Syst. Technol. **8**(5), 71 (2017)
5. Datta, R., Joshi, D., Li, J., Wang, J.Z.: Studying aesthetics in photographic images using a computational approach. In: Leonardis, A., Bischof, H., Pinz, A. (eds.) ECCV 2006. LNCS, vol. 3953, pp. 288–301. Springer, Heidelberg (2006). https://doi.org/10.1007/11744078_23
6. Deng, J., Dong, W., Socher, R., Li, L.J., Li, K., Fei-Fei, L.: Imagenet: a large-scale hierarchical image database. In: Proceedings of the IEEE Conference on Computer Vision and Pattern Recognition, pp. 248–255 (2009)
7. Deng, X., Cui, C., Fang, H., Nie, X., Yin, Y.: Personalized image aesthetics assessment. In: Proceedings of the 2017 ACM on Conference on Information and Knowledge Management, pp. 2043–2046 (2017)
8. Deng, Y., Loy, C.C., Tang, X.: Image aesthetic assessment: an experimental survey. IEEE Sig. Proc. Mag. **34**(4), 80–106 (2017)
9. Dhar, S., Ordonez, V., Berg, T.L.: High level describable attributes for predicting aesthetics and interestingness. In: Proceedings of the IEEE Conference on Computer Vision and Pattern Recognition, pp. 1657–1664 (2011)
10. Fang, H., Cui, C., Deng, X., Nie, X., Jian, M., Yin, Y.: Image aesthetic distribution prediction with fully convolutional network. In: Proceedings of the 24th International Conference on Multimedia Modeling, pp. 267–278 (2018)
11. Frey, B.J., Dueck, D.: Clustering by passing messages between data points. Science **315**(5814), 972–976 (2007)
12. He, K., Zhang, X., Ren, S., Sun, J.: Deep residual learning for image recognition. In: Proceedings of the IEEE Conference on Computer Vision and Pattern Recognition, pp. 770–778 (2016)
13. Kao, Y., He, R., Huang, K.: Deep aesthetic quality assessment with semantic information. IEEE Trans. Image Process. **26**(3), 1482–1495 (2017)

14. Ke, Y., Tang, X., Jing, F.: The design of high-level features for photo quality assessment. In: Proceedings of the IEEE Conference on Computer Vision and Pattern Recognition, pp. 419–426 (2006)
15. Liu, S., Cui, P., Zhu, W., Yang, S.: Learning socially embedded visual representation from scratch. In: Proceedings of the 23rd Annual ACM Conference on Multimedia Conference, pp. 109–118. ACM (2015)
16. Lo, K.Y., Liu, K.H., Chen, C.S.: Assessment of photo aesthetics with efficiency. In: Proceedings of the 21st International Conference on Pattern Recognition, pp. 2186–2189 (2012)
17. Lu, X., Lin, Z., Jin, H., Yang, J., Wang, J.Z.: Rating image aesthetics using deep learning. IEEE Trans. Multimedia **17**(11), 2021–2034 (2015)
18. Lu, X., Lin, Z., Shen, X., Mech, R., Wang, J.Z.: Deep multi-patch aggregation network for image style, aesthetics, and quality estimation. In: Proceedings of the IEEE International Conference on Computer Vision, pp. 990–998 (2015)
19. Luo, Y., Tang, X.: Photo and video quality evaluation: focusing on the subject. In: Forsyth, D., Torr, P., Zisserman, A. (eds.) ECCV 2008. LNCS, vol. 5304, pp. 386–399. Springer, Heidelberg (2008). https://doi.org/10.1007/978-3-540-88690-7_29
20. Mai, L., Jin, H., Liu, F.: Composition-preserving deep photo aesthetics assessment. In: Proceedings of the IEEE Conference on Computer Vision and Pattern Recognition, pp. 497–506 (2016)
21. Marchesotti, L., Perronnin, F., Larlus, D., Csurka, G.: Assessing the aesthetic quality of photographs using generic image descriptors. In: Proceedings of the IEEE International Conference on Computer Vision, pp. 1784–1791 (2011)
22. Murray, N., Marchesotti, L., Perronnin, F.: Ava: A large-scale database for aesthetic visual analysis. In: Proceedings of the IEEE Conference on Computer Vision and Pattern Recognition, pp. 2408–2415 (2012)
23. Pan, S.J., Yang, Q.: A survey on transfer learning. IEEE Trans. Knowl. Data Eng. **22**(10), 1345–1359 (2010)
24. Tang, X., Luo, W., Wang, X.: Content-based photo quality assessment. IEEE Trans. Multimedia **15**(8), 1930–1943 (2013)
25. Thomee, B., Shamma, D.A., Friedland, G., Elizalde, B., Ni, K., Poland, D., Borth, D., Li, L.J.: Yfcc100m: the new data in multimedia research. Commun. ACM **59**(2), 64–73 (2016)
26. Xie, L., Shen, J., Han, J., Zhu, L., Shao, L.: Dynamic multi-view hashing for online image retrieval. In: Twenty-Sixth International Joint Conference on Artificial Intelligence, pp. 3133–3139 (2017)
27. Zhu, L., Huang, Z., Chang, X., Song, J., Shen, H.T.: Exploring consistent preferences: discrete hashing with pair-exemplar for scalable landmark search. In: Proceedings of the 2017 ACM on Multimedia Conference, pp. 726–734 (2017)
28. Zhu, L., Huang, Z., Li, Z., Xie, L., Shen, H.T.: Exploring auxiliary context: discrete semantic transfer hashing for scalable image retrieval. IEEE Trans. Neural Networks Learn. Syst. **99**, 1–13 (2018). https://doi.org/10.1109/TNNLS.2018.2797248
29. Zhu, L., Huang, Z., Liu, X., He, X., Sun, J., Zhou, X.: Discrete multimodal hashing with canonical views for robust mobile landmark search. IEEE Trans. Multimed. **19**(9), 2066–2079 (2017)

Adaptive STBC Scheme for Soft Video Transmission with Multiple Antennas

Ya Guo[1] ⓘ, Anhong Wang[1](✉), Haidong Wang[1], Suyue Li[1],
and Jie Liang[2]

[1] Institute of Digital Multimedia and Communication,
Taiyuan University of Science and Technology, Taiyuan 030024, China
18734166761@163.com, wah_ty@163.com
[2] Engineering Science, Simon Fraser University,
Burnaby, BC V5A 1S6, Canada
jiel@sfu.ca

Abstract. In this paper, we report our study of SoftCast-based soft video transmission with multiple antennas, which provides high spectrum efficiency and a high transmission rate and overcomes the cliff effect of the digital scheme. However, current soft video transmission with multiple antennas cannot achieve better video performance because it uses only a single space time block code (STBC). In this paper, we propose an adaptive STBC (ASTBC) scheme for soft video transmission with multiple antennas. When the symbol rate is given, under different channel conditions, the closed-form expression is derived for the rearrangement of packets to be transmitted. Orthogonal STBC (OSTBC) is used to transmit some of the most important packets at a 1/2 transmission rate while discarding the same number of the least important packets; the other packets are transmitted using quasi-orthogonal STBC (QSTBC). The experimental results show that, in a 4×1 system, our scheme provides much higher quality than using pure OSTBC or pure QSTBC.

Keywords: SoftCast · Adaptive STBC · Multiple antennas

1 Introduction

Recently, wireless video transmission with multiple antennas has become very popular [1]. A wireless video soft multicast scheme, SoftCast, was proposed in [2–4], in which a sender broadcasts a single stream, and each receiver watches a video with quality that matches its channel quality. It overcomes the cliff effect of the traditional digital scheme. In addition, due to the relative simplicity and the feasibility of having multiple antennas at the base station in a multi-antenna system, transmit diversity has been studied extensively as a method of combating the detrimental effects in wireless fading channels [5]. One attractive approach to transmit diversity is space time block code (STBC) [6–8], in which full diversity is achieved by a very simple, maximum-likelihood decoding algorithm at the decoder. However, current soft video transmission with multiple antennas cannot achieve better video performance because it only uses a single space time block code (STBC). The orthogonal space time block code (OSTBC)

© Springer Nature Switzerland AG 2018
R. Hong et al. (Eds.): PCM 2018, LNCS 11164, pp. 767–775, 2018.
https://doi.org/10.1007/978-3-030-00776-8_70

of complex symbols, such as Alamouti's code [6], with a full transmission rate is not possible for more than two antennas. When the symbol rate is given, only half of all symbols is used for OSTBC in a 4×1 system for 1/2 transmission rate, although it provides an obvious transmit diversity gain. The quasi-orthogonal space time block code (QSTBC) proposed by Jafarkhani [9] provides higher transmission rates, but it sacrifices full diversity. Thus, when the symbol rate is given, combining OSTBC with QSTBC is the best choice.

In this paper, we propose an adaptive STBC (ASTBC) scheme for soft video transmission with multiple antennas to obtain optimal video quality. When the symbol rate is given, under different channel conditions, a closed-form expression is derived for the rearrangement of packets to be transmitted. The OSTBC is used to transmit some of the most important packets at 1/2 transmission rate while discarding the same number of the least important packets, and QSTBC is used to transmit the other packets. At the decoder, maximum ratio combining (MRC) is used to decode OSTBC, and pairs of transmitted symbols can be decoded separately for QSTBC. We reconstruct the original signal using the linear least square estimator (LLSE), so users obtain the reconstructed video with optimal quality.

2 Related Work

2.1 SoftCast

SoftCast mainly includes four parts, i.e., decorrelation, power allocation, whitening, and decoding of the linear least square estimator (LLSE) [4]. The first step in encoding the frame is to remove correlation across pixels in the same block. SoftCast uses a two-dimensional Discrete Cosine Transform (DCT) of pixel luminance in each block. Then, power allocation is used to minimize the mean squared reconstruction error. Next, whitening is used to offer protection from packet loss, which ensures that these packets have equal power to transmit in the channel with additive white Gaussian noise. Then LLSE is chosen to remove noise and estimate the original signal.

2.2 STBC

The OSTBC codeword in a 4×1 system is designed as:

$$
\mathbf{X} = \begin{bmatrix}
x_1 & x_2 & x_3 & x_4 \\
-x_2 & x_1 & -x_4 & x_3 \\
-x_3 & x_4 & x_1 & -x_2 \\
-x_4 & -x_3 & x_2 & x_1 \\
x_1^* & x_2^* & x_3^* & x_4^* \\
-x_2^* & x_1^* & -x_4^* & x_3^* \\
-x_3^* & x_4^* & x_1^* & -x_2^* \\
-x_4^* & -x_3^* & x_2^* & x_1^*
\end{bmatrix}. \tag{1}
$$

Where * denotes the complex conjugate. The column vector corresponds to antenna order, while the row vector corresponds to the slot order. The method can transmit four symbols in eight time slots and allow a receiver with one antenna to decode.

The QSTBC codeword in a 4×1 system is designed as:

$$\mathbf{X} = \begin{bmatrix} x_1 & x_2 & x_3 & x_4 \\ -x_2^* & x_1^* & -x_4^* & x_3^* \\ -x_3^* & -x_4^* & x_1^* & x_2^* \\ x_4 & -x_3 & -x_2 & x_1 \end{bmatrix}. \tag{2}$$

Equation (2) uses four slots to transmit four symbols, so it has a higher rate, but it does not have full diversity.

3 ASTBC Scheme Design

In order to obtain optimal video quality for soft video transmission with multiple antennas, the proposed ASTBC scheme is shown in Fig. 1, where the BS transmitter is equipped with four antennas, and the terminal receiver has one antenna.

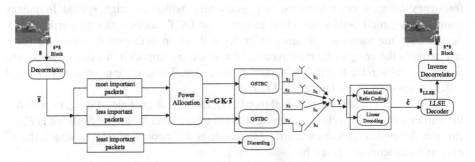

Fig. 1. Framework of our proposed scheme

A frame is divided into 8×8 blocks. Each block is transformed via a 2-dimensional DCT and a zig-zag order to a 64×1 column vector of spatial frequencies. The vectors corresponding to different blocks are then tiled together to form a matrix. Each row vector of the matrix forms one packet, so there are 64 packets for each frame. The packets are divided into three classes, i.e., those that are most important, those that are less important, and those that are least important. Power allocation is used for the first two classes, OSTBC is used for the most important and QSTBC is used for the less important packets, and the least important packets are discarded. The symbol rate is measured in symbols per second. When the symbol rate, R, is given, which is an arbitrary constant, $R^{(O)}$ is the symbol rate of taking OSTBC, and $R^{(Q)}$ is the symbol rate of taking QSTBC. Let J be the number of most important packets that will be transmitted by OSTBC. Then, the J least significant packets are discarded, and the other $64 - 2J$ packets are coded by QSTBC. We also derive a formula to characterize the rearrangement of packets to be transmitted for different channel conditions. Thus, transmitting proper J packets using OSTBC can obtain optimal quality of the reconstructed video, i.e., the error between a decoded frame $\hat{s}$ and its original version s has its minimum value. The number of columns and the number of rows in a frame are represented by n and m, respectively, and the optimization problem can be formulated as:

$$\text{argmin}_{J,R^{(O)},R^{(Q)}} \frac{1}{mn} \sum_{u=0}^{m-1} \sum_{v=0}^{n-1} |\hat{s}(u,v) - s(u,v)|. \tag{3}$$

$$\text{s.t. } 0 \leq J \leq 32, 2R^{(O)} + R^{(Q)} = R \left(0 \leq R^{(O)} \leq R/2, 0 \leq R^{(Q)} \leq R\right)$$

Note that, J that depends on the video contents and channel conditions is an adaptive value. However, it is difficult to find the closed-form solution for this problem, so we use a grid search method to find an approximate solution of the optimal $\{J, R^{(O)}, R^{(Q)}\}$. More precisely, we assume the channel SNR is known, and use a search step size of 4 for J and 1 for $R^{(O)}$ and 2 for $R^{(Q)}$.

The original signal, s, must take DCT to obtain a decorrelated signal, $\tilde{s}$. We assume that $\tilde{s}$ is the packet signal and that the ith row of $\tilde{s}$ corresponds to the ith packet. Since the images are relatively smooth, most of their energy is concentrated in the low spatial frequency components (slow-changing gradients), while the high spatial frequency components (small details) are close to zero. The DCT has two nice properties: (1) it decorrelates the video signal since it projects it on an orthogonal basis, and (2) it redistributes the energy (the information) in a block to compact it in a few components, which typically refer to the low spatial frequencies. Thus, we can obtain packets of different importance by taking 2D-DCT.

Thus, we can obtain the diagonal matrix Λ_s, the ith diagonal element, λ_i, of which is the variance of the ith packet. Assuming that the total power budget of SoftCast is P, we can get the power allocation matrix G, which is a diagonal matrix in which the ith diagonal elements are given by:

$$g_i = \sqrt{\frac{P}{\sqrt{\lambda_i} \sum_{u=1}^{64-J} \sqrt{\lambda_u}}}. \tag{4}$$

The J least significant packets are discarded for 1/2 transmission rate of taking OSTBC, and the packet vector, $\tilde{c}$, can be formulated as:

$$\tilde{c} = G \cdot K \cdot \tilde{s}, \tag{5}$$

where is $\tilde{c}$ the powered signal, and $\tilde{c}$ and $\mathbf{K}$ can be written as:

$$\tilde{c} = \begin{bmatrix} \tilde{c}_1 \\ \vdots \\ \tilde{c}_J \\ \tilde{c}_{J+1} \\ \vdots \\ \tilde{c}_{64-J} \\ \tilde{c}_{64-J+1} \\ \vdots \\ \tilde{c}_{64} \end{bmatrix}, \ and \ \mathbf{K}^T = \begin{bmatrix} k_1 \\ \vdots \\ k_J \\ k_{J+1} \\ \vdots \\ k_{64-J} \\ k_{64-J+1} \\ \vdots \\ k_{64} \end{bmatrix}, \tag{6}$$

where $\{k_1 = k_2 = \cdots = k_{64-J} = 1, k_{64-J+1} = \cdots = k_{64} = 0\}$, which means that the last J packets are discarded. After power allocation, we use an adaptive STBC scheme to transmit these packets in a 4×1 system, and we assume that Rayleigh channels are used. In order to obtain optimal reconstructed video quality, when the symbol rate is given the OSTBC is used to transmit some of the most important packets $\{\tilde{c}_1, \cdots, \tilde{c}_J\}$ at 1/2 transmission rate, the same number of the least important packets $\{\tilde{c}_{64-J+1}, \cdots, \tilde{c}_{64}\}$ are discarded, and QSTBC is used to transmit other packets $\{\tilde{c}_{J+1}, \cdots, \tilde{c}_{64-J}\}$.

Let $\{x_1, x_2, x_3, x_4, \cdots\} \subseteq \{\tilde{c}_i, i = 1 \cdots 64-J\}$ be coded by OSTBC and QSTBC. The OSTBC codeword in Eq. (1) is transmitted, and the QSTBC codeword in Eq. (2) is transmitted.

For the most important packets on the receiver side, the codeword of OSTBC in Eq. (1) is transmitted in the eight time slots, and the 1-antenna receiver receives eight wireless symbols, denoted by $\mathbf{Y}$, in the eight slots. Thus, we have:

$$\mathbf{Y} = \mathbf{X} \cdot \mathbf{H} + \mathbf{N}, \tag{7}$$

Where $\mathbf{Y} = [y_1, y_2, y_3, y_4, y_5, y_6, y_7, y_8]^T$, $\mathbf{H} = [h_1, h_2, h_3, h_4]^T$; h_1, h_2, h_3 and h_4 are channel coefficients which are assumed to be unchanged in the eight slots; and the equivalent Gaussian noise, $\mathbf{N}$, is given by $\mathbf{N} = [n_1, n_2, n_3, n_4, n_5, n_6, n_7, n_8]^T$. The maximum ratio combining (MRC) is used to decode OSTBC, and Eq. (7) can be rewritten as:

$$\mathbf{Y}_1 = [y_1, y_2, y_3, y_4, y_5^*, y_6^*, y_7^*, y_8^*]^T = \mathbf{\Omega} \cdot \mathbf{X}_1 + \mathbf{N}_1, \tag{8}$$

where

$$\Omega = \begin{bmatrix} h_1 & h_2 & h_3 & h_4 \\ h_2 & -h_1 & h_4 & -h_3 \\ h_3 & -h_4 & -h_1 & h_2 \\ h_4 & h_3 & -h_2 & -h_1 \\ h_1^* & h_2^* & h_3^* & h_4^* \\ h_2^* & -h_1^* & h_4^* & -h_3^* \\ h_3^* & -h_4^* & -h_1^* & h_2^* \\ h_4^* & h_3^* & -h_2^* & -h_1^* \end{bmatrix}, \tag{9}$$

$\mathbf{X}_1 = [x_1, x_2, x_3, x_4]^T$ and $\mathbf{N}_1 = [n_1, n_2, n_3, n_4, n_5^*, n_6^*, n_7^*, n_8^*]^T$. Left-multiplying Ω^H to each item in Eq. (8) and letting $\mathbf{Y}' = \Omega^H \cdot \mathbf{Y}_1$ and $\mathbf{N}' = \Omega^H \cdot \mathbf{N}_1$, we get:

$$\mathbf{Y}' = \mathbf{H}' \cdot \mathbf{X}_1 + \mathbf{N}', \tag{10}$$

where $\mathbf{H}'$ is a 4×1 diagonal matrix with $2\sum_{i=1}^{4} \|h_i\|^2$ on the diagonal. We obtain:

$$\tilde{\mathbf{X}}_1 = \frac{\mathbf{Y}'}{2\sum_{i=1}^{4} \|h_i\|^2}, \tag{11}$$

For the less important packets, the codeword of QSTBC in Eq. (2) is transmitted in four time slots, and the 1-antenna receiver receives four wireless symbols, denoted by Eq. (7), and $\mathbf{Y} = [y_1, y_2, y_3, y_4]^T$, $\mathbf{N} = [n_1, n_2, n_3, n_4]^T$. Linear decoding is used to decode QSTBC, so Eq. (7) can be rewritten as:

$$\mathbf{Y}_1 = [y_1, y_2^*, y_3^*, y_4]^T = \Omega \cdot \mathbf{X}_1 + \mathbf{N}_1, \tag{12}$$

Equation (9) can be rewritten as:

$$\Omega = \begin{bmatrix} h_1 & h_2 & h_3 & h_4 \\ h_2^* & -h_1^* & h_4^* & -h_3^* \\ h_3^* & h_4^* & -h_1^* & -h_2^* \\ h_4 & -h_3 & -h_2 & h_1 \end{bmatrix}, \tag{13}$$

and $\mathbf{N}_1 = [n_1, n_2^*, n_3^*, n_4]^T$. Left-multiplying Ω^H to each item in Eq. (12), we get:

$$\mathbf{Y}' = \Omega^H \cdot \mathbf{Y}_1 = \begin{bmatrix} a & 0 & 0 & b \\ 0 & a & -b & 0 \\ 0 & -b & a & 0 \\ b & 0 & 0 & a \end{bmatrix} \cdot \mathbf{X}_1 + \Omega^H \cdot \mathbf{N}_1, \tag{14}$$

where $a = \sum_{i=1}^{4} \|h_i\|^2$, and $b = 2\text{Re}(h_1 \cdot h_4^* - h_2 \cdot h_3^*)$. By left-multiplying $\mathbf{W}$ to each item in Eq. (14) [10], it can be written as:

$$\mathbf{Y}'' = \mathbf{W} \cdot \mathbf{Y}' = \left(\frac{a^2 - b^2}{a}\right) \begin{bmatrix} 1 & 0 & 0 & 0 \\ 0 & 1 & 0 & 0 \\ 0 & 0 & 1 & 0 \\ b & 0 & 0 & 1 \end{bmatrix} \cdot \mathbf{X}_1 + \mathbf{W} \cdot \boldsymbol{\Omega}^H \cdot \mathbf{N}_1, \qquad (15)$$

where

$$\mathbf{W} = \begin{bmatrix} 1 & 0 & 0 & -\frac{b}{a} \\ 0 & 1 & \frac{b}{a} & 0 \\ 0 & \frac{b}{a} & 1 & 0 \\ -\frac{b}{a} & 0 & 0 & 1 \end{bmatrix}, \qquad (16)$$

finally, we obtain:

$$\tilde{\mathbf{X}}_1 = [\tilde{x}_1, \tilde{x}_2, \tilde{x}_3, \tilde{x}_4] = \frac{a}{a^2 - b^2} \cdot \mathbf{Y}''. \qquad (17)$$

The decoded symbols $\{\tilde{x}_1, \tilde{x}_2, \tilde{x}_3, \tilde{x}_4, \cdots\} \subseteq \{\hat{c}_i, i = 1 \cdots 64 - J\}$ are combined to form the packet vector $\hat{c}$ and the packet signal can be reconstructed by an LLSE decoder, and the reconstructed signal is:

$$s_{LLSE} = \Lambda_s \cdot \mathbf{G}^T \cdot (\mathbf{G}\Lambda_s\mathbf{G}^T + \boldsymbol{\Sigma})^{-1} \cdot \hat{c}, \qquad (18)$$

where s_{LLSE} refers to the LLSE estimation of packet signal $\tilde{s}$, and $\boldsymbol{\Sigma}$ is a diagonal matrix with the noise power of each packet in the diagonal element. Then, the reconstructed packet signal with LLSE, i.e., s_{LLSE}, can be obtained. Reorganize the reconstructed signal s_{LLSE} into an 8×8 DCT block by undoing the zigzag scanning, multiplying by the inverse DCT matrix to generate the original video block, and combining the blocks into a video frame. The decoded frame $\hat{s}$ is obtained.

4 Simulation

Our experiments are conducted based on four standard video test sequences, i.e., *foreman*, *football*, *news*, and *akiyo*, with a resolution of CIF@ 352×288. These videos have different motion characteristics, background textures, and energy distributions. In our simulations, the DCT coefficients are divided into 64 chunks, and the channel coefficients are unchanged in a frame. Because the reconstruction error in our optimization formula is inversely proportional to the peak signal-to-noise ratio (PSNR), the quality of the reconstructed video is assessed by the objective metric PSNR (in dB). Since the proposed formula is not a simple convex function, we are unable to obtain the most suitable packet allocation factor, J, directly with minimum error. Thus, we use the numerical method to plot the experimental results for the four video sequences.

Figure 2 shows the average PSNR of *football* by different Js and schemes in a 4×1 system. Figure 2(a) shows the performance of different Js using OSTBC in a 4×1 system. When $J = 0$, it indicates that all of the packets are transmitted by QSTBC, and $J = 4$ indicates that we use OSTBC to transmit the four most important packets and discard the four least important packets; also, QSTBC is used to transmit the 56 less important packets, ..., and J = 32 indicates that half packets are transmitted using OSTBC. When the conditions of the channel are bad, i.e., snr = 0 and 5 dB, no matter how many packets are transmitted using OSTBC, the average PSNR always is better than using pure QSTBC, and it is no different from using pure OSTBC. With the improvement of the condition of the channel, i.e., snr = 10, 15, or 20 dB, using pure QSTBC or pure OSTBC does not obtain optimal video performance, we can obtain the best average PSNR by choosing a suitable value for J.

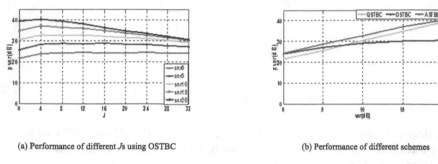

(a) Performance of different Js using OSTBC   (b) Performance of different schemes

Fig. 2. Average PSNR of *football* by different Js and schemes in a 4 × 1 system

When the symbol rate is given, we make comparisons with two reference schemes, i.e., (1) using pure QSTBC and using pure OSTBC, as shown in Fig. 2(b). We draw the simulation results of our scheme by selecting appropriate J value that achieves optimal video performance. Because *football* includes large areas of texture and complex motions, the performance of our proposed scheme is in the range of 0.9–3 dB better than using pure QSTBC, and it is in the range of 0.2–9 dB better than using pure OSTBC.

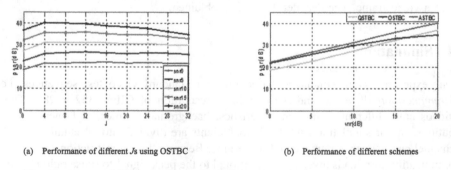

(a)   Performance of different Js using OSTBC   (b)   Performance of different schemes

Fig. 3. Average PSNR of *foreman* by different Js and schemes in a 4 × 1 system

Figure 3 shows the average PSNR of *foreman* by different Js and schemes in a 4 × 1 system. Figure 3(a) shows simulation results similar to those in Fig. 2(a). Because *foreman* contains an active closed up object and panning background, Fig. 3(b) shows that the performance of our proposed scheme is about 3–4 dB better than using pure QSTBC, and it is in the range of 0.4–5 dB better than using pure OSTBC. The other test video sequences produce similar simulation results.

5   Conclusion

In this paper, we proposed an adaptive STBC scheme for soft video transmission with multiple antennas. In a 4×1 system, the OSTBC is used to transmit some of the most important packets at 1/2 transmission rate to obtain significant gain in the diversity of the transmission with the same number of the least important packets discarded. QSTBC is used to transmit the other packets. The experimental results show that, in a 4×1 system, our scheme provided much higher quality than using pure OSTBC or pure QSTBC.

Acknowledgements. This work has been supported in part by the National Natural Science Foundation of China (No. 61672373, No. 61501315), Scientific and Technological Innovation Team of Shanxi Province (No. 201705D131025), Key Innovation Team of Shanxi 1331 Project (2017015), Collaborative Innovation Center of Internet+3D Printing in Shanxi Province (201708), The Program of "One hundred Talented People" of Shanxi Province.

References

1. Cui, H., Luo, C., Chen, C.W., et al.: Scalable video multicast for MU-MIMO systems with antenna heterogeneity. IEEE Trans. Circuits Syst. Video Technol. **26**(5), 992–1003 (2016). https://doi.org/10.1109/TCSVT.2015.2430651
2. Katabi, D., Rahul, H., Jakubczak, S.: Softcast: one video to serve all wireless receivers. CSAIL Technical reports (2009)
3. Jakubczak, S., Katabi, D.: A cross-layer design for scalable mobile video. In: Proceedings of ACM Mobicom 2011, pp. 289–300. ACM, New York (2011). https://doi.org/10.1145/2030613.2030646
4. Jakubczak, S., Katabi, D.: Softcast: one-size-fits-all wireless video. In: SIGCOMM 2010 – Proceedings of SIGCOMM 2010 Conference, New Delhi, India, pp. 449–450 (2010). https://doi.org/10.1145/1851182.1851257
5. Tarokh, V., Seshadri, N., Calderbank, A.R.: Space-time codes for high data rate wireless communication: performance analysis and code construction. IEEE Trans. Inform. Theory **44**, 744–765 (1998). https://doi.org/10.1109/18.661517
6. Alamouti, S.M.: A simple transmit diversity technique for wireless communications. IEEE J. Sel. Areas Commun. **16**(8), 1451–1458 (1998). https://doi.org/10.1109/9780470546543.ch2
7. Tarokh, V., Jafarkhani, H., Calderbank, A.R.: Space-time block coding from orthogonal designs. IEEE Trans. Inform. Theory **45**, 1456–1467 (1999)
8. Tarokh, V., Jafarkhani, H., Calderbank, A.R.: Space-time block coding for wireless communications: performance results. IEEE J. Select. Areas Commun. **17**, 451–460 (1999). https://doi.org/10.1109/49.753730
9. Jafarkhani, H.: A quasi-orthogonal space-time block code. IEEE Trans. Commun. **49**(1), 1–4 (2001). https://doi.org/10.1109/WCNC.2000.904597
10. He, L., Ge, H.: Reduced complexity maximum likelihood detection for V-BLAST systems. In: IEEE Military Communications Conference, pp. 1386–1391. IEEE (2003). https://doi.org/10.1109/MILCOM.2003.1290429

Hypergraph-Based Discrete Hashing Learning for Cross-Modal Retrieval

Dianjuan Tang, Hui Cui, Dan Shi, and Hua Ji$^{(\boxtimes)}$

School of Information Science and Engineering,
Shandong Normal University, Jinan 250014, China
jihua_sdnu@hotmail.com

Abstract. Hashing has drawn increasing attention in cross-modal retrieval due to its high computation efficiency and low storage cost. However, there is a certain lack in the previous cross-modal hashing methods that they can not effectively represent the correlations between paired multi-modal instances. In this paper, we propose a novel Hypergraph-based Discrete Hashing (BGDH) to solve the limitation. We formulate a unified unsupervised hashing framework which simultaneously performs hypergraph learning and hash codes learning. Hypergraph learning can effectively preserve the intra-media similarity consistency. Furthermore, we propose an efficient discrete hash optimization method to directly learn the hash codes without quantization information loss. Extensive experiments on three benchmark datasets demonstrate the superior performance of the proposed approach, compared with state-of-the-art cross-modal hashing techniques.

Keywords: Hypergraph · Unsupervised · Hashing
Cross-modal retrieval

1 Introduction

As the major component of big data, multimedia data including image, text, video and audio have emerged on the Internet rapidly. It is now imperative to exploit the correlations among multimedia data. Therefore, cross-modal retrieval technique has attracted increasing public attention that it just needs one modality of a query and can retrieve nearest neighbors in different modalities. The main objective of cross-modal retrieval is to effectively model the correlations among the multimedia data from different modalities.

As an advanced Approximate Nearest Neighbor (ANN) search technique, hashing has attracted considerable attention due to its high computation efficiency and low storage cost. Hashing-based methods play a significant role in many important applications, such as information retrieval, data mining, and computer vision [11,14,16]. Hashing methods aim to learn the hash functions for converting high-dimensional data into low-dimensional compact binary codes

© Springer Nature Switzerland AG 2018
R. Hong et al. (Eds.): PCM 2018, LNCS 11164, pp. 776–786, 2018.
https://doi.org/10.1007/978-3-030-00776-8_71

while retaining the relationship of original data as much as possible. With hashing, the time-consuming ANN search can be efficiently implemented by simple Hamming distance computation, and thus it can be performed for cross-modal retrieval. Various cross-modal hashing methods have been proposed recently. Multimodal Latent Binary Embedding (MLBE) employs a binary latent factor with a probabilistic model to learn hash codes [12]. Cross View Hashing (CVH) learns hash functions by minmizing the weighted average Hamming distance of the codewords for the training objects over all the views [5]. Inter-Media Hashing (IMH) simultaneously preserves intra-modality similarities and correlates heterogeneous modalities to learn hash codes [10]. Semantics-Preserving Hashing (SePH) transforms the supervised information into a probability distribution and learns hash codes by minimizing KL-divergence [7]. Furthermore, Collective Matrix Factorization Hashing (CMFH) learns unified hash codes by collective matrix factorization with latent factor model from different modalities of one object [3]. However, these hashing methods still have the limitation in correlation analysis for different types of media data.

In this paper, we propose a novel cross-modal hashing approach, dubbed as Hypergraph-based Discrete Hashing (BGDH). The core idea is to take advantage of a hypergraph which can enhance the semantic correlations among instances more effectively. Furthermore, we propose an efficient discrete hash optimization method to iteratively learn the hash codes without relaxing quantization information loss. The main contributions of our work are as follows:

1. We develop a unified based on hypergraph hash learning framework to preserve the inter-media and intra-media similarity consistency effectively. It utilizes a hypergraph to capture the high-order relations among instances and matrix factorization to model inter-media correlation for cross-modal retrieval. To the best of our knowledge, there is still no similar work.
2. In this paper, we propose an efficient discrete hash optimization technique to directly solve the hash codes and reduce the quantization information loss.
3. Extensive experiments on three benchmark datasets demonstrate the state-of-the-art performance of our method and validate the desirable advantage on boosting cross-modal retrieval performance.

The rest of the paper is organized as follows. In Sect. 2, we detail the proposed method. In Sect. 3, we introduce the experimental results. Finally, we conclude the paper in Sect. 4.

2   The Proposed Method

2.1   Notations and Definitions

We suppose that there are two modalities including image modality $X = [x_1, ..., x_N] \in R^{d_1 \times N}$ and text modality $Y = [y_1, ..., y_N] \in R^{d_2 \times N}$, where N is the number of instances, d_1 and d_2 are the feature dimensions of data in each modality respectively. Our goal is to learn unified hash codes $B = [b_1, ..., b_N] \in R^{K \times N}$ to preserve the similarity among instances, where K is the length of the learned hash codes.

2.2 Objective Formulation

The objective function of BGDH consists of three parts: intra-media semantic similarity preservation, inter-media semantic similarity preservation and hash function learning.

Intra-media Semantic Similarity Preservation. In this subsection, we introduce a hypergraph that effectively preserves intra-media semantic similarity. Unlike a simple graph, where two related vertices are linked on one edge, a hypergraph has a specific structure that one edge can connect more than two vertices. These edges are referred to hyperedges. With its specific structure, a hypergraph can capture the complex and high-order relations among vertices. Therefore, we preserve intra-media semantic similarity by a hypergraph. $G(V, E, w)$ denotes a hypergraph. V is the vertex set. Each vertex in the hypergraph represents an instance, thus there are N vertices in the generated hypergraph. E is the hyperedges set, $E = E_1 + E_2$. E_1 is the image hyperedges set, E_2 is the text hyperedges set. We exploit K-means clustering to create hyperedges. Therefore, the number of hyperedges is decided by the number of clusters in each modality. W denotes the diagonal matrix of the hyperedge weights with each entry that $w(e)$ is the weight of hyperedge. We set the weight of each edge as 1 in this paper. The vertex degrees and the hyperedge degrees are represented by the diagonal matrices D_v and D_e respectively.

D_v with each diagonal element is defined as

$$d(v) = \sum_{e \in E} w(e)h(v, e). \tag{1}$$

As aforementioned, we set the weight of each edge as 1. Thus the vertex degree becomes

$$d(v) = \sum_{e \in E} h(v, e). \tag{2}$$

D_e with each diagonal element is defined as

$$\delta(e) = \sum_{v \in V} h(v, e). \tag{3}$$

The hypergraph $G(V, E, w)$ can be denoted by a $|V| \times |E|$ incidence matrix H with each entry defined as follows

$$h(v, e) = \begin{cases} \exp(-(Dis(v, e))/\sigma), & \text{if } v \in e \\ 0, & otherwise \end{cases} \tag{4}$$

where $Dis(\cdot)$ defines the Euclidean distance, σ is the bandwith parameter. According to [4], we construct the hypergraph laplacian L as

$$L = I - D_v^{-\frac{1}{2}} H W D_e^{-1} H^T D_v^{-\frac{1}{2}} \tag{5}$$

where I is the identity matrix. In order to perform the intra-media relations well, L is used to constrain hash code B in the overall formulation.

Inter-media Semantic Similarity Preservation. Inspired by Collective Matrix Factorization Hashing (CMFH) [3], we can learn latent semantic feature from multi-modal data by matrix factorization. Furthermore, matrix factorization is a well-established technique in recommender system [2]. Supposed that the interlinked data should have the same semantic representation, we detect the latent semantic structure of images and text by matrix factorization as follows

$$\min_{Z,U_X,U_Y} ||X - U_X Z||_F^2 + ||Y - U_Y Z||_F^2 \tag{6}$$

where U_X is basis feature matrix of images datasets and U_Y represents the basis feature matrix of text. In addition, Z involves latent semantic topic of image and text. If we consider each hashing bit as latent semantic topic, hash codes can be understood as semantic topic distribution. Thus, we directly define the distribution of Z as hash codes B in hashing learning.

Hash Function Learning. In our formulation, we employ linear projection to learn hash functions, which is similar to previous work [3,5]. The objective function of hash learning is

$$\min_{B,W_X,W_Y} ||B - W_X X||_F^2 + ||B - W_Y Y||_F^2 \tag{7}$$

where W_X and W_Y denote the projection matrix of image datasets and text datasets respectively.

Overall Formulation. As aforementioned, the intra-media similarity, the inter-media consistency and the hash learning have been comprehensively considered. The overall objective formulation of BGDH is

$$\begin{aligned}
\min_{B,W_X,W_Y,U_X,U_Y} & ||X - U_X B||_F^2 + ||Y - U_Y B||_F^2 \\
& + \delta(||B - W_X X||_F^2 + ||B - W_Y Y||_F^2) \\
& + \gamma(||U_X||_F^2 + ||U_Y||_F^2 + ||W_X||_F^2 + ||W_Y||_F^2) \\
& + \beta tr(B(I - D_v^{-\frac{1}{2}} H W D_e^{-1} H^T D_v^{-\frac{1}{2}})B^T) \\
s.t. \quad & B \in \{-1,1\}^{K \times N}
\end{aligned} \tag{8}$$

where γ is defined to avoid overfitting. δ and β balance for the regularization terms.

2.3 Efficient Discrete Optimization Algorithm

Equation (8) is in actually NP-hard and difficult to solve with the existence of discrete constraints. Most existing hashing method generally exploit a two-step relaxing+rounding to solve hash codes, which may lead to significant information loss. Inspired by [15], we adopt a novel optimization framework to directly solve the discrete constraint. We introduce an auxiliary variable A to represent the binary codes B. Thus the Eq. (8) can be transformed into

$$\min_{B,W_X,W_Y,U_X,U_Y,A} ||X - U_X B||_F^2 + ||Y - U_Y B||_F^2$$
$$+ \delta(||B - W_X X||_F^2 + ||B - W_Y Y||_F^2)$$
$$+ \gamma(||U_X||_F^2 + ||U_Y||_F^2 + ||W_X||_F^2 + ||W_Y||_F^2)$$
$$+ \beta tr(B(I - D_v^{-\frac{1}{2}} H W D_e^{-1} H^T D_v^{-\frac{1}{2}})A^T) + \frac{\mu}{2}||B - A + \frac{D}{\mu}||_F^2$$
$$s.t. \quad B \in \{-1,1\}^{K \times N}$$
$$(9)$$

where D illustrates the deviation between B and A, μ is used to balance the target value and auxiliary variable value.

With the auxiliary variable, the objective function becomes a tractable problem. There are $B, W_X, W_Y, U_X, U_Y, A, D$ that need to be solved. We can adopt an iterative process to solve the discrete hash codes optimization problem. The steps are listed as follows.

Update U_X and U_Y. With B, W_X, W_Y, A, D fixed, the optimization formulas for U_X, U_Y are

$$\min_{U_X} ||X - U_X||_F^2 + \gamma||U_X||_F^2$$
$$\min_{U_Y} ||Y - U_Y||_F^2 + \gamma||U_Y||_F^2.$$
$$(10)$$

We set the derivation of the objective function with respective to U_X, U_Y to 0. Then we gain that

$$U_X = XB^T(BB^T + \gamma I)$$
$$U_Y = YB^T(BB^T + \gamma I).$$
$$(11)$$

Update W_X and W_Y. By fixing other variables, the optimization formulas for W_X, W_Y are

$$\min_{W_X} ||B - W_X X||_F^2 + \frac{\gamma}{\delta}||W_X||_F^2$$
$$\min_{W_Y} ||B - W_Y Y||_F^2 + \frac{\gamma}{\delta}||W_Y||_F^2.$$
$$(12)$$

The solution of W_X and W_Y can be regarded as the regularized least squares problem. Thus we obtain that

$$W_X = BX^T(XX^T + \frac{\gamma}{\delta}I)$$
$$W_Y = BY^T(YY^T + \frac{\gamma}{\delta}I).$$
$$(13)$$

Update A. The optimization formula for A is as follows while other variables fixed

$$\min_A \beta tr(B(I - L_n)A^T) + \frac{u}{2}||B - A + \frac{D}{u}||_F^2$$
$$(14)$$

where $L_n = D_v^{-\frac{1}{2}} H W D_e^{-1} H^T D_v^{-\frac{1}{2}}$. The formula of Eq. (14) can be simplified into

$$\min_A Tr((\beta B(I - L_n) - uB - D)A^T)$$
$$= \min_A ||A - (uB + D - \beta B(I - L_n))||_F^2.$$
$$(15)$$

Finally, the updating result of A is as follows

$$A = uB + D - \beta B(I - L_n). \tag{16}$$

Update B. With fixed other variables, the optimization function of B is represented as

$$\min_B ||X - U_X B||_F^2 + ||Y - U_Y B||_F^2 + \delta(||B - W_X X||_F^2 + ||B - W_Y Y||_F^2)$$

$$+ \beta tr(B(I - D_v^{-\frac{1}{2}} HWD_e^{-1}H^T D_v^{-\frac{1}{2}})A^T) + \frac{\mu}{2}||B - A + \frac{D}{\mu}||_F^2 \tag{17}$$

$$s.t. \quad B \in \{-1, 1\}^{K \times N}.$$

We simplify the function Eq. (17) for B as

$$\min_B - Tr(B^T U_X X) - Tr(B^T U_Y Y) - \delta(Tr(B^T W_X X) + Tr(B^T W_Y Y))$$

$$+ \beta Tr(B(I - D_v^{-\frac{1}{2}} HWD_e^{-1}H^T D_v^{-\frac{1}{2}})A^T) - \mu Tr(B^T A - \frac{B^T D}{\mu}) \tag{18}$$

$$= \min_B -Tr(B^T M)$$

$$s.t. \quad B \in \{-1, 1\}^{K \times N}$$

where

$$M = \mu A - D - \beta A(I - D_v^{-\frac{1}{2}} HWD_e^{-1}H^T D_v^{-\frac{1}{2}})$$

$$+ U_X X + U_Y Y + \delta W_X X + \delta W_Y Y. \tag{19}$$

Thus the discrete solution of B can be directly calculation as

$$B = sign(M) \tag{20}$$

where $sign(\cdot)$ is the signum function which equals -1 if $x < 0$, 1 if $x \geq 0$.

Update D. With fixed other variables, the optimization function of D can be represented as follows

$$D = \mu(A - B). \tag{21}$$

The complete optimization algorithm to solve the problem Eq. (8) is detailed in Algorithm 1 below.

Convergence. During the iterative process, the updating of variables retains decreasing towards the lower bound of objective function in Eq. (8). Inspired by ALM optimization theory [6], the optimization will converge with the increasing iterations.

Complexity Analysis. The hypergraph construction includes hyperedge generation and distance computation between vertices and hyperedges. The time complexity of this process is $O(N|E|)$, where $|E|$ is the number of hyperedges. Solving discrete hash codes is conducted in an iterative process, the computation complexity is $O(t(dNK + NK^2 + K^3 + KNd + Nd^2 + d^3))$, where t denotes the number of iterations, and $d = \max\{d_1, d_2\}$. Considering $N \gg d_1(d_2) > K$, this process is linear to N. The time complexity of calculating hash codes is $O(N)$, which indicates the desirable scalability of the proposed method.

Algorithm 1. Hypergraph-based Discrete Hashing

Input: Image matrix $X \in R^{d_1 \times N}$, Text matrix $Y \in R^{d_2 \times N}$, parameter γ, δ, β, μ;
Output: B, W_X, W_Y;
1: Randomly initialize B, W_X, W_Y, U_X, U_Y;
2: Construct hypergraph laplacian L according to Eq. (5);
3: **while** not convergence **do**
4:    Update U_X, U_Y according to Eq. (11);
5:    Update W_X, W_Y according to Eq. (13);
6:    Update A according to Eq. (16);
7:    Update B according to Eq. (19) and Eq. (20);
8:    Update D according to Eq. (21);
9: **end while**
10: Return B, W_X, W_Y

---

3   Experiments

3.1   Experimental Datasets

We conduct experiment on three benchmark datasets including WiKi, MIR-Flickr, NUS-WIDE.

WiKi dataset contains $2,866$ instances collected from featured Wikipedia articles. Each instance consists of image-text pairs which belong to 10 semantic categories. Each image is represented by 128 dimensional SIFT histogram [9] and the text in each document is represented by a 10 dimensional feature vector generated by latent Dirichlet allocation (LDA) [1]. **MIRFlickr** contains $25,000$ samples annotated by 38 unique semantic labels. Each instance is an image with its associated contextual tags. Images are described by $1,000$ dimensional dense SIFT histogram. The contextual text feature is described by 457 dimensional binary vector. **NUS-WIDE** is a large-scale dataset. It consists of $269,648$ images with tags which collected from the Web. There are 81 semantic categories but some of them are limited. As a consequence, we select $195,834$ pairs which are correspond with 10 most common concepts in our experience. And each image is represented by a 500 dimensional SIFT histograms. We extract $1,000$ dimensional binary vector to describe the corresponding text.

On Wiki and MIRFlikr datasets, we select 10% of samples as the query set, and the remaining 90% as database. From the database, 30% of instances are randomly chosen as training dataset. For NUS-WIDE dataset, we take 1% of the dataset as the query set and the rest as database. Then, we randomly sample 5,000 instances from database as training set. The instances are considered to be relevant if they share at least one label.

3.2   Comparison Algorithms and Evaluation Metrics

We compare our method with several unsupervised state-of-the-art cross-modal hashing methods including Cross-View Similarity Search (CVH) [5], Inter-Media

Hashing (IMH) [10], Linear Cross-Modal Hashing (LCMH) [17], Collective Matrix Factorization Hashing (CMFH) [3], Latent Semantic Sparse Hashing (LSSH) [13] and Cross-Modal Discrete Hashing (CMDH) [8].

We adopt two standard evaluation metrics to measure the performance of cross-modality retrieval, i.e., mean Average Precision (mAP) [3] and Precision-Scope curve [13].

Table 1. mAP of all approaches on three datasets for Cross-modal Retrieval. The best result in each column is marked with bold.

Task	Method	WiKi				MIRFlickr				NUS-WIDE			
		16	32	64	128	16	32	64	128	16	32	64	128
Image to Text	CVH	0.1769	0.1438	0.1438	0.1438	0.5995	0.5923	0.5875	0.5846	0.4023	0.3917	0.3869	0.3802
	IMH	0.1755	0.1640	0.1345	0.1273	0.5983	0.5952	0.5882	0.5856	0.3982	0.3898	0.3839	0.3809
	LCMH	0.1748	0.1910	0.1791	0.1727	0.6020	0.5964	0.6012	0.6041	0.4282	0.4447	0.4367	0.4480
	CMFH	0.1990	0.1877	0.1923	0.2010	0.5730	0.5806	0.5771	0.5761	0.3629	0.3684	0.3655	0.3658
	LSSH	0.1573	0.2022	0.1760	0.1824	0.5559	0.5581	0.5692	0.5656	0.2826	0.3536	0.3643	0.3765
	CMDH	0.2166	0.2121	0.2092	0.2278	0.5675	0.5787	0.5871	0.6019	0.3921	0.4109	0.4075	0.4369
	BGDH	**0.2251**	**0.2429**	**0.2188**	**0.2279**	**0.6154**	**0.6116**	**0.6174**	**0.6309**	**0.4324**	**0.4457**	**0.4667**	**0.4818**
Text to Image	CVH	0.2248	0.2019	0.1707	0.1723	0.5964	0.5917	0.6095	0.6087	0.3906	0.3888	0.3829	0.3870
	IMH	0.2297	0.1983	0.1724	0.1574	0.5950	0.5971	0.5895	0.5959	0.3973	0.3887	0.3871	0.3898
	LCMH	0.2238	0.2312	0.2690	0.2933	0.5886	0.6023	0.6025	0.6071	0.3929	0.4045	0.4146	0.4281
	CMFH	0.3026	0.3311	0.3540	0.3580	0.5718	0.5833	0.5944	0.5728	0.3671	0.3597	0.3671	0.3721
	LSSH	0.2390	0.2737	0.2437	0.2799	0.5540	0.5736	0.5998	0.5823	0.3688	0.4042	0.4147	0.4230
	CMDH	0.2398	0.2866	0.3009	0.3066	0.5741	0.5839	0.6020	0.6197	0.3825	0.3909	0.3966	0.4130
	BGDH	**0.3340**	**0.3915**	**0.3994**	**0.4141**	**0.6192**	**0.6337**	**0.6459**	**0.6555**	**0.4443**	**0.4628**	**0.4963**	**0.5054**

3.3 Comparison Results

Table 1 illustrates mAP values of all cross-modal hashing methods on three datasets. Code length on all datasets is varied in the range of $\{16, 32, 64, 128\}$. Figure 1 demonstrates the Precision-Scope curves with code length fixed 64. For WiKi, the search scope is ranged from 100 to 1,000 with stepsize 100. For MIRFLickr and NUS-WIDE, the search scope is ranged from 500 to 5,000 with stepsize 500. The presented results clearly demonstrate that BGDH consistently outperforms the comparison methods on all datasets and code lengths. Specifically, on MIRFLickr and NUS-WIDE, BGDH outperforms the second best performance by more than 3% when a text querying relevant images. The promising performance of BGDH is mainly attributed to the effective inter-media and intra-media similarity preservation of hash codes.

3.4 Effects of Hypergraph and Discrete Optimization

Our approach employs a hypergraph to enhance the intra-media similarity relationship among instances and adopts the discrete optimization to directly learn hash codes simultaneously. We compare BGDH with two variant approaches to

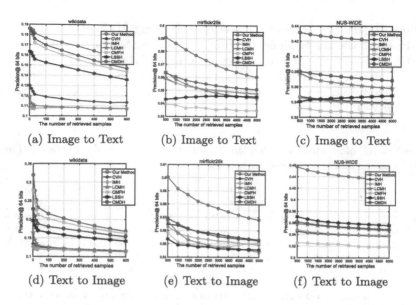

(a) Image to Text   (b) Image to Text   (c) Image to Text

(d) Text to Image   (e) Text to Image   (f) Text to Image

Fig. 1. *Precision-Scope* curves on three datasets **@64 bit**.

validate the effects of the hypergraph and the discrete optimization method. We denote BGDH-I as the variant approach with removing the hypergraph and BGDH-II as the variant approach with relaxing the discrete constraint. The comparison results are summarized in Table 2. From the results, we can clearly observe that BGDH outperforms the variants BGDH-I and BGDH-II on all code lengths and datasets for cross-modal retrieval. The results validate the effects of the hypergraph and discrete optimization.

Table 2. Effects of hypergraph and discrete optimization

Task	Method	WiKi				MIRFlickr				NUS-WIDE			
		16	32	64	128	16	32	64	128	16	32	64	128
	BGDH-I	0.2177	0.2285	0.2165	0.2197	0.5740	0.5820	0.5907	0.5876	0.4202	0.4414	0.4614	0.4784
Image to Text	BGDH-II	0.2116	0.2067	0.2066	0.2146	0.5664	0.5701	0.5706	0.5728	0.3584	0.3549	0.3550	0.3545
	BGDH	**0.2251**	**0.2429**	**0.2188**	**0.2279**	**0.6154**	**0.6116**	**0.6174**	**0.6309**	**0.4324**	**0.4457**	**0.4667**	**0.4818**
	BGDH-I	0.2812	0.3239	0.3378	0.3574	0.6002	0.6295	0.6402	0.6348	0.4011	0.4042	0.4162	0.4240
Text to Image	BGDH-II	0.2682	0.2929	0.2792	0.2766	0.5974	0.5963	0.5675	0.5630	0.3619	0.3666	0.3677	0.3675
	BGDH	**0.3340**	**0.3915**	**0.3994**	**0.4141**	**0.6192**	**0.6337**	**0.6459**	**0.6555**	**0.4443**	**0.4628**	**0.4963**	**0.5054**

4   Conclusions

In this paper, we propose a novel based on hypergraph hashing method for the large-scale cross-modal retrieval. Our approach enhances the intra-media semantic relationship by constructing a hypergraph and preserves the inter-media semantic relationship by matrix factorization simultaneously. In addition, an efficient discrete hash optimization method is developed to directly solve the hash codes and avoid the quantization loss. Experimental results on three benchmark datasets demonstrate the superiority of BGDH for cross-modal retrieval.

References

1. Blei, D.M., Ng, A.Y., Jordan, M.I.: Latent dirichlet allocation. J. Mach. Learn. Res. Arch. **3**, 993–1022 (2003)
2. Cheng, Z., Shen, J., Zhu, L., Kankanhalli, M.S., Nie, L.: Exploiting music play sequence for music recommendation. In: Proceedings of the Joint Conference on Artificial Intelligence (IJCAI), pp. 3654–3660 (2017). https://doi.org/10.24963/ijcai.2017/511
3. Ding, G., Guo, Y., Zhou, J.: Collective matrix factorization hashing for multimodal data. In: Proceedings of the IEEE International Conference on Computer Vision and Pattern Recognition (CVPR), pp. 2083–2090 (2014). https://doi.org/10.1109/CVPR.2014.267
4. Gao, Y., Wang, M., Zha, Z., Shen, J., Li, X., Wu, X.: Visual-textual joint relevance learning for tag-based social image search. IEEE Trans. Image Process. **22**(1), 363–376 (2013). https://doi.org/10.1109/TIP.2012.2202676
5. Kumar, S., Udupa, R.: Learning hash functions for cross-view similarity search. In: Proceedings of the Joint Conference on Artificial Intelligence (IJCAI), pp. 1360–1365 (2011). https://doi.org/10.5591/978-1-57735-516-8/IJCAI11-230
6. Lin, Z., Chen, M., Ma, Y.: The augmented lagrange multiplier method for exact recovery of corrupted low-rank matrices. CoRR 1009.5055 (2010)
7. Lin, Z., Ding, G., Hu, M., Wang, J.: Semantics-preserving hashing for cross-view retrieval. In: Proceedings of the IEEE International Conference on Computer Vision and Pattern Recognition (CVPR), pp. 3864–3872 (2015). https://doi.org/10.1109/CVPR.2015.7299011
8. Liong, V.E., Lu, J., Tan, Y.: Cross-modal discrete hashing. Pattern Recognit. **79**, 114–129 (2018). https://doi.org/10.1016/j.patcog.2018.02.002
9. Lowe, D.G.: Distinctive image features from scale-invariant keypoints. Int. J. Comput. Vis. **60**(2), 91–110 (2004)
10. Song, J., Yang, Y., Yang, Y., Huang, Z., Shen, H.T.: Inter-media hashing for large-scale retrieval from heterogeneous data sources. In: Proceedings of the ACM International Conference on Management of Data (SIGMOD), pp. 785–796 (2013). https://doi.org/10.1145/2463676.2465274
11. Xie, L., Shen, J., Han, J., Zhu, L., Shao, L.: Dynamic multi-view hashing for online image retrieval. In: Proceedings of the Joint Conference Artificial Intelligence (IJCAI), pp. 3133–3139 (2017). https://doi.org/10.24963/ijcai.2017/437
12. Zhen, Y., Yeung, D.: A probabilistic model for multimodal hash function learning. In: Proceedings of the ACM International Conference on Knowledge Discovery and Data Mining (KDD), pp. 940–948 (2012). https://doi.org/10.1145/2339530.2339678
13. Zhou, J., Ding, G., Guo, Y.: Latent semantic sparse hashing for cross-modal similarity search. In: Proceedings of the ACM International Conference on Information Retrieval (SIGIR), pp. 415–424 (2014). https://doi.org/10.1145/2600428.2609610

14. Zhu, L., Huang, Z., Chang, X., Song, J., Shen, H.T.: Exploring consistent preferences: discrete hashing with pair-exemplar for scalable landmark search. In: Proceedings of the ACM International Conference on Multimedia (MM), pp. 726–734 (2017). https://doi.org/10.1145/3123266.3123301
15. Zhu, L., Huang, Z., Li, Z., Xie, L., Shen, H.T.: Exploring auxiliary context: discrete semantic transfer hashing for scalable image retrieval. IEEE Trans. Neural Netw. Learn. Syst. **99**, 1–13 (2018). https://doi.org/10.1109/TNNLS.2018.2797248
16. Zhu, L., Huang, Z., Liu, X., He, X., Sun, J., Zhou, X.: Discrete multimodal hashing with canonical views for robust mobile landmark search. IEEE Trans. Multimed. **19**(9), 2066–2079 (2017). https://doi.org/10.1109/TMM.2017.2729025
17. Zhu, X., Huang, Z., Shen, H.T., Zhao, X.: Linear cross-modal hashing for efficient multimedia search. In: Proceedings of the ACM International Conference on Multimedia (MM), pp. 143–152 (2013). https://doi.org/10.1145/2502081.2502107

Improve Predictive Accuracy by Identifying Collusions in P2P Recommender Systems

Qiuyun Chang, Zhenhua Tan[(⊠)], and Guangming Yang

Software College, Northeastern University, Shenyang 110819, China
tanzh@mail.neu.edu.cn

Abstract. According to the malicious collusion behavior of nodes in P2P networks, a method of collusion identification is proposed based on the idea of clustering. This method considers the behavior characteristics of collusion, screens the rating nodes of the target node by three attributes of the rating extremes, rating time and historical similarity. According to the size of the suspected degree of collusion, some sets of malicious collusive nodes are selected. Experiments on a real data set show that the accuracy of recommendation has been significantly improved after excluding the identified collusive nodes, which proves the effectiveness of the method proposed in this paper.

Keywords: Collusion · Clustering · The characteristics of behavior

1 Introduction

With the wide application of P2P network technology, the characteristics of openness, anonymity and loose coupling between nodes lead to some hidden security risks in the network, which facilitates the development of malicious nodes. As a result, a large number of trust models and anti-malicious mechanisms have emerged for preventing this phenomenon. Some models use the global trust to evaluate the trust value of the node, such as [2, 5, 7, 8, 10]. Some measure the trustworthiness of a node by its local trust value, such as [1, 6, 9, 12, 13]. And they control the malicious behavior of the node to some extent. In addition, [7, 9] considered the impact factors of participating in trust assessment from multiple dimensions. They can avoid the problem of a single measure dimension effectively. [3, 4] proposed a corresponding identification method by analyzing the characteristics of the behaviors of collusive nodes. They improved the collusion problem in P2P network environment effectively.

Malicious behavior in a P2P network can be roughly divided into two types: malicious behavior of single nodes and collusive behavior of some groups. Malicious behavior of single nodes includes spreading malicious data, issuing false evaluation and abusing resources of network, etc. Compared with the malicious behavior of single nodes, the collusive behavior is more complicated. On the one hand, these nodes show malicious behavior when they participate in collusion. On the other hand, they may behave normally when they as single nodes perform some activities respectively. Thus, [3] proposed a collusion identification method based on fuzzy logic, which calculates the behavioral similarity between abnormal nodes through the three similarities of

© Springer Nature Switzerland AG 2018
R. Hong et al. (Eds.): PCM 2018, LNCS 11164, pp. 787–796, 2018.
https://doi.org/10.1007/978-3-030-00776-8_72

rating similarity, time similarity and rating deviation between nodes. However, there are still some limitations to identify the collusive groups with the complex behavior. Although the fuzzy method reflects the uncertainty between objective things and subjective evaluation, its standard of fuzzy judgment is subjective. [4] proposed a collusion recognition model based on behavioral similarity. Through the trust management nodes to monitor the abnormal behavior of other nodes, the behavioral similarity between the monitored abnormal nodes is calculated to identify the collusive groups. However, this method will identify some malicious individuals with special behaviors into the collusive groups. And it also has weak recognition ability for those complicating collusive groups.

Aiming at the malicious collusion in P2P networks, this paper adopts the step-by-step clustering and screening method on the basis of the behavioral characteristics of collusion. The malicious collaborating groups in the network are identified through three aspects: the rating and the rating time of a node, and the evaluation history that includes the rating and the rating time of a node. The main contributions of the collusion identification method proposed in this paper are as follows:

(1) The model adopts a step-by-step screening method to identify the collaborating groups in this paper, which reduces the computational complexity of operating on all nodes in the entire network;
(2) Considering that the collusive node is colluding on the current target node, it may also have collaborated with other nodes in the past, the evaluation historical factor is introduced to effectively avoid the problem that some single malicious nodes are misidentified into a collusion group;
(3) The concept of suspected degree of collusion is introduced.

The remainder of the paper is organized as follows: Sect. 2 details the identification method of collusion. Section 3 describes the relevant experiments. And Sect. 4 conducts the full text to sum up.

2   Collusion Identification Model

2.1   Screening Nodes with Extreme Rating

Definition 1: The $r_{u_j,i}$ represents the rating of node j on target node i; C_i represents the set of nodes evaluating node i.

For any node i, traverses all the nodes that it interacts with, and stores these nodes in the collection C_i according to the order of transaction time.

$$C_i = \{u_1, u_2, u_3, u_4, \ldots\}$$

As the nodes of collusive group conduct collusion (that is, maliciously raising or reducing the rating of a certain node), the more extreme the rating is, the more significant the effect is. Therefore, this paper only deals with extreme rating nodes when identifying collusion community nodes.

Definition 2: The r_{min} indicates the minimum value of the predefined rating in the system, and r_{max} indicates the maximum value of the predefined rating in the system (for example, if the rating range in a system is [1, 5], the minimum rating is 1 and the maximum rating is 5.)

When $r_{max} - 1 \leq r_{u_j,i} \leq r_{max}$, node j is stored in set A_i with extremely high rating;

$$A_i = \{u_1, u_2, u_3, u_4, \ldots\}$$

When $r_{min} \leq r_{u_j,i} \leq r_{min} + 1$, node j is stored in set B_i with extremely low rating;

$$B_i = \{u_5, u_6, u_7, u_8, \ldots\}$$

2.2   Time Clustering of Nodes

As the transaction time of the collusive group is more concentrated, the collusion attack is better. Therefore, for each node group obtained by the extremes of rating in all nodes that rate the target node, the nodes are clustered by a bottom-up hierarchical clustering method according to their rating time. Here we only keep the set of nodes whose number of nodes is greater than 10.

Definition 3: Δt denotes the difference in rating time between node m_i and node n_i; φ denotes the threshold for the difference of the rating time.

The process of the node clustering:

(1)   Each node in the collection (A_i, B_i) is considered a class $(A_{i1}, A_{i2}, \ldots, B_{i1}, B_{i2}, \ldots)$, and each node is the center node of the class;
(2)   The difference between the rating time of the center point of each class and the rating time by other nodes is used as a clustering basis ($\Delta t = t_{m_i} - t_{n_i}$). In the first clustering, other nodes closest to the center node in each class are aggregated. After this aggregation, the center node of each class is calculated again;
(3)   The process of the second step is repeated until the difference of rating time between the center node of each class and the nodes outside the class is greater than a certain threshold (That is, $\Delta t > \varphi$). And when the nodes in each class tend to be stable and no longer changing, the aggregation will stop.

A set of nodes with extremely high rating and more concentrated rating time:

$$A_{i1} = \{u_1, u_2, u_3, \ldots\}$$

A set of nodes with extremely low rating and more concentrated rating time:

$$B_{i1} = \{u_5, u_6, \ldots\}$$

2.3   The Similarity of the Evaluation History of Nodes

As the nodes of the collaborating groups perform the collusion, their goals and the purpose are same. And these groups are likely to perform collusion more than once for multiple target nodes. Therefore, this paper will calculate the similarity of the evaluation history between the nodes for filtering the collusive nodes (The similarity is the composite values in two aspects of the rating and rating time).

Definition 4: The $\text{HisSim}(u_m, u_n)$ indicates the similarity of the evaluation history between the node u_m and the node u_n; $\text{RtSim}(u_m, u_n)$ indicates the rating similarity between the node u_m and the node u_n.$\text{RtSim}(u_m, u_n)$ indicates the similarity of the rating time between the node u_m and the node u_n.

(1) The calculation of rating similarity between two nodes in a set will be implemented by PIP similarity [11]:

$$\text{RSim}(u_m, u_n) = \sum_{k \in C_{m,n}} PIP(r_{mk}, r_{nk}) \tag{1}$$

where the $C_{m,n}$ represents the set of nodes rated by both node m and node n,and r_{mk} and r_{nk} represent the rating of node k by node m and node n, respectively. The PIP calculation method for any two ratings is as follows:

$$PIP(r_1, r_2) = proximity(r_1, r_2) * impact(r_1, r_2) * popularity(r_1, r_2) \tag{2}$$

Where the proximity, impact and popularity are three factors for the calculation of PIP similarity.

(2) The similarity of the rating time between two nodes in a collection is calculated by a coarse-grained method. That is, when the difference of the rating time between two nodes $(\Delta t = t_m - t_n)$ is within a certain specified interval, the degree of similarity between the rating times of the two nodes is a specific value. The specific settings are as follows:

$$\text{RtSim}(u_m, u_n) = \begin{cases} 1, & 0 \leq \Delta t \leq 1 \\ 0.8, & 1 < \Delta t \leq 2 \\ 0.6, & 2 < \Delta t \leq 3 \\ 0.4, & 3 < \Delta t \leq 4 \\ 0.2, & 4 < \Delta t \leq 5 \\ 0, & \Delta t > 5 \end{cases} \tag{3}$$

(Here the unit of time is "day")

The similarity of the evaluation history between two nodes in the collection is as follows:

$$\text{HisSim}(u_m, u_n) = \text{RSim}(u_m, u_n) * \text{RtSim}(u_m, u_n) \tag{4}$$

The similarity of the evaluation history between two nodes is regarded as the weight value, and the maximum tree clustering algorithm is used to reassemble the nodes in each set after rating time clustering. The specific methods are as follows. First, the nodes in the set are connected into a tree according to the similarity of the evaluation history. Then, the edge of the tree with the weights (that is, the similarity of the evaluation history) less than γ and the similarity of the rating time less than 0.5 will be cut off. If there are at least one connected edges among the remaining nodes, the connectable nodes belong to the same class. After that, the sets are filtered by the number of nodes. Here, we only preserve the node categories whose number of nodes is not less than 10.

2.4   The Suspected Degree of Collusion

The sets of the nodes who rated the target node have been selected according to the three attributes of the rating extremes, the similarity of rating time and the similarity of the evaluation history, which are collected to calculate the suspected degree of the collusion. The calculation of the suspected degree of collusion includes three aspects: the number of nodes in the group, the rating deviation of the nodes in the group to the current target node, and the rating variance of nodes in the group.

Definition 5: The DSC indicates the suspected degree of collusive communities that have been screened out.

(1) Calculating the suspected degree of collusion, the formula is as follows:

$$\text{DSC} = \frac{d}{\sigma^2 + 1 + d} * \frac{|S_n|}{5 * \log(10)} \tag{5}$$

where d denotes the rating deviation generated by the nodes of the group for the target node, δ^2 denotes the rating variance of the nodes of the group, and the $|S_n|$ denotes the number of nodes in the nth sets filtered out in step four.

The rating deviation d is calculated as follows:

$$d = \left| \overline{r_{S_n,i}} - \overline{r_{N,i}} \right| \tag{6}$$

$$\overline{r_{N,i}} = \frac{1}{n} \sum_{j \in N} r_{j,i} \tag{7}$$

where $\overline{r_{S_n,i}}$ denotes the average rating of the nodes in the set S_n for the node i, $\overline{r_{N,i}}$ represents the average rating of the remaining relatively normal nodes outside the suspected collusion screened by step four for the target node i, and $r_{j,i}$ represents the average rating of the node j in the relatively normal node set to the target node i.

The rating variance is calculated as follows:

$$\sigma^2 = \frac{1}{n}\sum_{j\in S_n}\left(r_{j,i} - \bar{r}\right)^2 \tag{8}$$

where $r_{j,i}$ denotes the rating of node j in the set S_n for the node i, and $\bar{r}$ denotes the average rating of all nodes in the set S_n for the current target node i.

(2) The sets of rating nodes are sorted according to the suspected degree of collusion;
(3) Set the threshold of the suspected degree of collusion θ, and consider the sets whose suspicion is not less than the threshold (i.e.,DSC ≥ θ) is regarded as a collusive group.

The algorithm of collusive group identification is as follows:

Input: the rating record, the threshold of rating time φ, the similarity threshold of evaluation history γ, the threshold of colluding suspected degree θ

Output: the collusion node sets $(A_{i2}', B_{i2}', ...)$
(1)   **for** $(\forall I_i \in I)$ **do**
(2)   Query the rating records of each target item I_i;
(3)   **for** $(\forall r_{u_i,i} \in I_i)$ **do**
(4)   Query the extremely high and low raring sets (A_i, B_i) respectively;
(5)   **end**
(6)   **for** $(\forall t_{u_i,i} \in A_i)$ **do**
(7)   Cluster the closest nodes for each rating set according to the formula $(\Delta t = t_{u_m,i} - t_{u_n,i})$;
(8)   **if** $(\Delta t > \varphi)$ **end**
(9)   **end**
(10)   Obtain the sets of nodes with concentrated rating tiame: $(A_{i1}, A_{i2}, ..., B_{i1}, B_{i2}, ...)$
(11)   **for** $(\forall u_i \in A_{i1})$ **do**
(12)   Calculate the similarity of evaluation history HisSim(u_m, u_n) according to the Eq.(1-4);
(13)   Filter the nodes with the similar evaluation history by the maximum tree clustering algorithm;
(14)   **if** (HisSim$(u_m, u_n) > \gamma$ && RtSim$(u_m, u_n) \geq 0.5$)
(15)   Obtain the sets of nodes with concentrated rating time: $(A_{i1}', A_{i2}', ..., B_{i1}', B_{i2}', ...)$ **end**
(16)   **end**
(17)   Calculate the suspected degree of each node group by the Eq.(5-8);
(18)   **if** (DSC ≥ θ)
(19)   Output the collusive groups $(A_{i2}', B_{i2}', ...)$ **end**
(20) **end**

3 Experiments and Analysis

This paper uses the Eclipse development environment and the Java programming language to design and implement the proposed method for the identification of collusive groups. The experiment will verify the effectiveness of the proposed collusion identification method on the actual data set Movielens-1 M by calculating the average absolute error (MAE).

The Movielens-1 M dataset in this paper is from Movielens. The specification of this dataset is: This dataset contains 6040 users, 3706 movies, and a total of 1000209 rating data. This paper will take 50 movies, 100 movies, and 200 movies in the Movielens-1 M data set for experiments.

3.1 The Evaluation Index of Experimental Performance

The mean absolute error (MAE) is to measure the accuracy of the overall recommendation by calculating the average value of the absolute deviation between the predicted rating and the real rating. And the formula is as follows:

$$MAE = \frac{\sum_{(u,i)\in T}|\hat{r}_{u,i} - r_{u,i}|}{|T|} \tag{9}$$

where T is a rating matrix for (user, item), $|T|$ is the number of elements in the set T, $\hat{r}_{ui}$ represents the predicted rating of the user u on the item i, and r_{ui} is the true rating of the user u on the item i.

The calculation of the predicted rating of the user u on the item i is as follows:

$$\hat{r}_{u,i} = \bar{r}_u + \frac{\sum_{v\in N}(r_{v,i} - \bar{r}_v) \cdot S_{u,v}}{\sum_{v\in N}|S_{u,v}|} \tag{10}$$

where $r_{v,i}$ denotes the rating of the user u on the item i, $\bar{r}_v$ denotes the average rating of the user v, and $S_{u,v}$ denotes the similarity between the target user u and the user v that is in the neighbors of user u. Here, the set of neighbors refers to all the sets of users whose similarity with the target user u is not less than the threshold ω among all the users who rate the target item.

3.2 Experimental Results and Discuss

The Effectiveness of Collusive Identification Method Under Different Data Sizes. The settings of experiment-related parameters are shown in the following Table 1:

(1) When the mae values are calculated by different similarities under different data sizes, the performance of the collusion method is verified.

Table 1. Parameters for the first experiment.

The parameters	Settings
The threshold of rating time (φ)	10 days
The similarity threshold of evaluation history under 50 Movies (γ)	5000
The similarity threshold of evaluation history under 100 Movies (γ)	8000
The similarity threshold of evaluation history under 200 Movies (γ)	9000
The threshold of colluding suspected degree in 50 movies (θ)	0.1
The threshold of colluding suspected degree in 100 movies (θ)	0.1
The threshold of colluding suspected degree in 200 movies (θ)	0.1

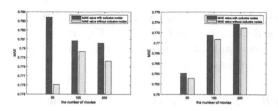

Fig. 1. The mae is calculated by cosine similarity and PCC similarity

Figure 1 shows the performance of the collusive identification method under different data scales when mae values are calculated by different similarities. As shown above, the MAE is smaller when the identified collusive nodes are removed. The recommended accuracy is higher.

The Effectiveness of the Collusion Identification Method When the Similarity Thresholds of Evaluation History Are Different.

The settings of experiment-related parameters are shown in the following Table 2:

Table 2. Parameters for the second experiment

The parameters	Settings
The threshold of rating time (φ)	10(unit: day)
The threshold of colluding suspected degree in 50 movies (θ)	0.1
The threshold of colluding suspected degree in 100 movies (θ)	0.1
The threshold of colluding suspected degree in 200 movies (θ)	0.1
The similarity selected in the MAE calculation	PCC

(1) The effectiveness of collusion recognition method is evaluated with different similarity thresholds of the evaluation history,when the number of movies is 50, 100, 200 respectively.

As shown in Fig. 2, when the number of movies is 50, 100, 200, the threshold of the evaluating historical similarity is taken as 5,000, 8000, 9000 respectively, the calculated MAE value is the smallest. That is, the recommendation accuracy is higher than other values.

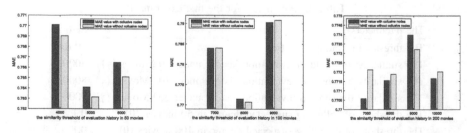

Fig. 2. The selection of thresholds for the evaluating historical similarity in 50, 100, 200 movies, respectively

The Effectiveness of the Collusion Identification Method With Different Suspicion Thresholds.

The settings of experiment-related parameters are shown in the following Table 3:

Table 3. Parameters for the third experiment

The parameters	Settings
The threshold of rating time (φ)	10 days
The similarity threshold for evaluating history under 50 Movies (γ)	5000
The similarity threshold for evaluating history under 100 Movies (γ)	8000
The similarity threshold for evaluating history under 200 Movies (γ)	9000
The similarity selected in the MAE calculation	PCC

(1) The effectiveness of collusion recognition method is evaluated with the different suspicion thresholds when the number of movies is 50, 100, 200 respectively.

As shown in Fig. 3 when the suspicion threshold is taken as 0.1 and the number of movies is 50 and 100, the MAE value is the smallest, and the collusion identification method is most effective. The accuracy of the recommendation is higher than other values. In addition, when the suspicion threshold is taken more than 0.1, the collusion identification method is effective. But when the suspicion threshold is taken as 0.12, the collusion identification method is most effective, and the MAE value is smallest.

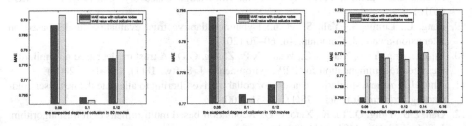

Fig. 3. The selection of suspicion thresholds in 50, 100, 200 movies respectively

In conclusion, the collusion identification method presented in this paper shows a certain degree of effectiveness on different data scales. However, the selection of the corresponding parameters in the method requires artificial adjustment.

4   Conclusion

The method of recognizing collusion proposed in this paper is to select the nodes of the collusive groups from three aspects: the rating of the node, the rating time and the evaluation history. The performance of effectiveness has been verified under different data scales. However, the method still has some limitations on the identification of complexly collusive nodes (such as spy nodes, etc.), which is also the focus of our next work on the improvement of this recognition method.

Acknowledgement. This work is supported by the National Natural Science Foundation of China under Grants No. 61772125, No. 61702084, No. 61702090 and No. 61402097; and the Fundamental Research Funds for the Central Universities under Grant No. N151708005.

References

1. Yichun, L., Yinghong, L.: Dynamic P2P trust model based on context factors. J. Commun. **37**(8), 34–45 (2016)
2. Tan, Z., Xingwei, W., Xueyi, W.: A novel iterative and dynamic trust computing model for large scaled P2P networks. Mobile Inf. Syst. **2016**, 1–12 (2016). Article No.: 3610157
3. Miao, G., Feng, D., Su, P.: A collusion detector based on fuzzy logic in P2P trust model. J. Comput. Res. Develop. **48**(12), 2187–2200 (2011)
4. Miao, G., Feng, D., Su, P.: Colluding clique detector based on activity similarity in P2P trust model. J. Commun. **30**(8), 9–20 (2009)
5. Kamvar, S.D., Schlosser, M.T., Garcia-Molina, H.: The Eigentrust algorithm for reputation management in P2P networks. In: International Conference on World Wide Web, pp. 640–651. ACM (2003)
6. Zhang, L., Rao, K., Wang, R.: Dynamic trust evaluation model based on evaluation credibility in cloud computing. J. Commun. (s1), 31–37 (2013)
7. Zhenhua, T., Xingwei, W., Wei, C., et al.: A distributed trust model for peer-to-peer networks based on multi-dimension-history vector. Chin. J. Comput. **33**(9), 1725–1735 (2010)
8. Ma, Y., Wang, D.: A novel trust model for P2P networks. In: International Conference on Natural Computation, Fuzzy Systems and Knowledge Discovery, pp. 1969–1973. IEEE (2016)
9. Feng, L., Limin, S., Yali, S., et al.: Dynamic adaptive trust evaluation model based on interaction-aware. J. Commun. **10**, 60–70 (2012)
10. Li, J.T., Jing, Y.N., Xiao, X.C., Wang, X.P., Zhang, G.D.: A trust model based on similarity-weighted recommendation for P2P environments. J. Softw. **18**(1), 157–167 (2007)
11. Ahn, H.J.: A new similarity measure for collaborative filtering to alleviate the new user cold-starting problem. Inf. Sci. **178**(1), 37–51 (2008)
12. Gan, Z.B., Ding, Q., Li, K., Xiao, G.Q.: Reputation-based multi-dimensional trust algorithm. J. Softw. **22**(10), 2401–2411 (2011)
13. Li, X.Y., Gui, X.L.: Cognitive model of dynamic trust forecasting. J. Softw. **21**(1), 163–176 (2010)

Multi-graph Regularized Deep Auto-Encoders for Multi-view Image Representation

Jiaying Fang, Yongzhao Zhan, and Xiangjun Shen[✉]

School of Computer Science and Telecommunication Engineering,
Jiangsu University, Zhenjiang, China
xjshen@ujs.edu.cn

Abstract. Deep auto-encoders combined with the manifold construction have attracted much attention as they can preserve local manifolds when the encoding function is learned. This paper proposes a novel framework which is named multi-graph regularized deep auto-encoders (MGDAE). Different from the previous work of graph regularized auto-encoders, our proposed framework incorporates multiple manifolds to well preserve multiple localities in multi-view datasets. With this framework, the low multi-view dimensional features can be obtained due to diverse graph constructions, which vary smoothly along the geodesics of multi-view data manifolds and facilitate the deep auto-encoders. Extensive experimental results on COIL20 and CIFAR-10 for image classification demonstrate our proposed framework outperforms other auto-encoders proposed in deep learning literature, such as AE and LAE.

Keywords: Auto-encoders · Manifold learning
Multi-view image representation · Image classification

1 Introduction

As deep learning can automatically extract low dimensional features from large volume of data in many deep hidden layers, it has been successfully applied to image recognition, speech recognition, face recognition and other fields. Among popular techniques in deep learning, auto-encoder (AE) is a technique that can reconstruct the input data points in low dimensional representations through designing encoder-decoder hidden layers.

In the literature, many methods [1, 15–17] have been proposed. For example, Rumelhart [15] put forward the concept of auto-encoders in 1986, which aims to learn compressed feature representations and use for processing the high-dimensional and complex data. Hinton [16] proposed a denoising auto-encoder that takes a partially corrupted input while training to recover the original undistorted input. It competed the pre-training of the hidden layer with greedy layer-wise training algorithm and used back propagation algorithm to optimize the

© Springer Nature Switzerland AG 2018
R. Hong et al. (Eds.): PCM 2018, LNCS 11164, pp. 797–807, 2018.
https://doi.org/10.1007/978-3-030-00776-8_73

parameter of the neural network system. Bengio [1] proposed the concept of sparse auto-encoders, which can automatically learn features from unlabeled data and give better descriptions than the origin data. Vincent et al. [14] proposed a stacked denoising auto-encoder to stack multiple DAE to form a deep neural network for extracting useful feature representations that produce a significant improvement over the stacking of ordinary auto-encoders. In 2015, an unsupervised manifold learning method termed Laplacian Auto-Encoders (LAE) [10] was proposed from the manifold learning perspective. LAE regularized the training of auto-encoders with the learned encoding function having the locality preserving property for data points on the manifold. Liao et al. proposed a local invariant deep nonlinear mapping algorithm called graph regularized auto-encoder (GAE) [11] that preserves the local connectivity from the original space to the representation space, which provides powerful expressive capacity and extends the conventional manifold learning algorithms into the context of deep architecture.

Multimedia data such as image, video or audio, has multi-view feature representations [7–9]. However, current methods are not applicable for modeling such multi-view datasets. In fact, many manifold algorithms have been proposed for dimensionality reduction, such as Locality Preserving Projections (LPP) [6], Neighborhood Preserving Embedding (NPE) [5], Sparse Preserving Projection (SPP) [13], Locality Sensitive Discriminant Analysis (LSDA) [3], to name but a few [4]. It is limited to only use a single graph to the deep auto-encoders [12,18,19] if we want to preserve multi-view representations. Thus how to build a general framework of AE to incorporate multiple graph manifolds together for modeling multi-view dataset representations is an important issue.

Based on above discussions, we propose a framework that embedding multigraph with auto-encoders, which is called multi-graph deep auto-encoders, motivated by the advances in auto-encoders and manifold learning. To summarize, the main contributions of this paper are highlighted as follows:

- With multiple graphs regularizations being imposed to the deep autoencoders, the MGDAE can preserve local neighborhood structures of data points and learn multi-view representations. The application of multiple manifold learning methods in MGDAE is meaningful for the feature representation of multi-view.
- Multiple graph constructions are correlated in our framework and a constraint term is formulated into the objective function of auto-encoders in the proposed framework. The constraint term of multi-graph is designed for optimizing the weight of MGDAEs, which enhances the discriminative ability of the network.
- Experimental results are achieved on different image classification datasets, such as COIL-20 and CIFAR-10 for object recognition. The proposed MGDAE demonstrates superior performance and generalized capability compared with other auto-encoders and manifold learning algorithms.

The rest of this paper is organized as follows. Section 2 overviews related works of auto-encoders and manifold learning. The proposed framework and the multi-graph regularized auto-encoders are presented in Sect. 3. Section 4 shows the experimental results. A conclusion and future work are provided in Sect. 5.

2   Related Work

In this section, a concise review of literature in terms of auto-encoders will be given. To understand multi-graph constructions, we will represent some relevant dimensionality reduction methods.

2.1   Auto-Encoders

Auto-encoder (AE) is a kind of neural networks applying back-propagation algorithm, which consists of 3 layers including the input layer, hidden layer and output layer. Given a dataset $X = \{x_1, x_2, ..., x_n\} \in \mathbb{R}^{m \times n}$ containing n samples, and m is the dimension of each column. Auto-encoders learn to encode the inputs into low dimensional representation and decode the hidden layer back into input space, which can be defined as follows:

$$
\begin{aligned}
H = f(X) = s(WX + b_h) \\
Y = g(H) = s(W^T H + b_y)
\end{aligned}
\tag{1}
$$

where H and Y denote the hidden representation and the reconstruction respectively, W denote the weight matrix, b_h and b_y denotes the bias matrix. The sigmoid function $s(x) = 1/(1 + e^{-x})$ is the activation of encoder and decoder. The process of training the AE [2] is to find the minimum reconstruction error of the parameters $\theta = \{W, b_h, b_y\}$. The objective function is:

$$
J = \arg\min \sum_{x \in X} L(x, g(f(x)))
\tag{2}
$$

Generally, the reconstruction error function L can be measured by a mean-square error function or a cross-entropy loss function.

2.2   Graph Construction Methods

Neighborhood Preserving Embedding (NPE). Neighborhood preserving embedding is a linear dimensionality reduction algorithm [5]. Given a set of points $X = \{x_1, x_2, ..., x_n\}$ that the dimension of a sample is m, find a transformation matrix T that maps the dataset X to a low dimensional space $Y = \{y_1, y_2, ..., y_n\}$. NPE reconstructs each data point with K-nearest neighbor algorithm so that it can preserve the local structure between neighboring points. Let $y_{ij}(j = 1, 2, ..., k)$ represent the k-th neighbor of points and a_{ij} represent the weight matrix between y_i and y_j. The corresponding calculation method of adjacency graph is k-nearest neighbors. Thus the objective function of NPE is as follow:

$$
\min J(Y) = \sum_{i=1}^{n} \left| y_i - \sum_{j=1}^{k} a_{ij} y_{ij} \right|^2 = Y^T M Y
\tag{3}
$$

where $M = (I - A)^T (I - A)$.

Locality Sensitive Discriminant Analysis (LSDA). Locality sensitive discriminant analysis is a supervised dimension reduction exploiting the geometry of the data manifold for studying the class relationship between samples [3]. This method optimally preserves the local neighbor information and discriminant information by construct graphs. Given a set of points $X = \{x_1, x_2, ..., x_n\}$ that the dimension of a sample is m, LSDA can build a nearest neighbor graph G to model the local geometrical structure of samples. For each sample x_i, let $l(x_i)$ be the label of x_i, $N(x_i) = \{x_i^1, ..., x_i^k\}$ be the set of its k nearest neighbors. $N(x_i)$ is consist of $N_w(x_i)$ and $N_b(x_i)$. $N_w(x_i)$ is the set of neighbors sharing the same labels with x_i. The nearest neighbor graph G is split into within-class graph (G_w) and between-class graph (G_b) by class labels in order to discover geometrical and discriminant structure of the data manifold. The graph G is a combination of G_w and G_b. The $A_{w,ij}$ and $A_{b,ij}$ are weight matrices of G_w and G_b. The objective function is defined as follow:

$$\arg\max Y^T(\alpha L_b + (1 - \alpha)A_w)Y \tag{4}$$

where $Y = (y_1, y_2, ..., y_n)^T$ is mapped from X, $L_b = D_b - A_b$ and $D_{b,ii} = \sum_j A_{b,ij}$ is the Laplacian matrix of G_b.

3   Multi-graph Regularized Deep Auto-Encoders

3.1   The Framework and Multi-graph Regularization

Multi-graph auto-encoder (MGAE) is a method embedding different graph constructions into auto-encoders. MGAE is composed of one encoder with multi-graph embedding and one decoder. The encoder that combing multiple graph constructions transforms the given training samples into low-dimensional representation. These graph constructions, such as NPE, LSDA and the Laplacian graph, find multi-view low-dimensional representations of samples and preserve local neighborhood structures after encoding. The concrete graph constructions are described above. Thus the representation can discover the latent structure of the original data from the manifold learning perspective. Multi-graph deep auto-encoder is a framework that stacks multi-graph auto-encoders as shown in Fig. 1. The objective function of MGAE includes two parts: a reconstruction error of training samples, and a multi-graph regularization. The solution is defined as follow:

$$J_{MGAE}(f, g) = \sum_{x \in X} E(x, g(f(x))) + \frac{1}{2}\alpha \sum_{k=1}^{K} r_k Tr(H^T GH) \tag{5}$$

where $E(x, y)$ is the mean square error loss, α is the weight coefficient of graph regularizations, and K is the number of graph constructions. The r_k is a regularization term that controls the trade-off among different graph constructions

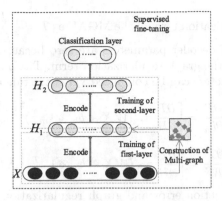

Fig. 1. Framework of our proposed multi-graph auto-encoders

and the sum of r_k is one. Then the r_k is calculated as follow:

$$r_k = \frac{\left(\frac{1}{tr(HGH^T)}\right)^{\frac{1}{d-1}}}{\sum\limits_{k=1}^{K}\left(\frac{1}{tr(HGH^T)}\right)^{\frac{1}{d-1}}} \tag{6}$$

where d represents the control parameter for the weights of graph constructions. Meanwhile the $G = [G_1, G_2, ..., G_L]^T$ is a normalized combination of graph constructions from multiply local embedding methods, such as LPP, NPE, LSDA. The form of G is $G = \sum\limits_{i=1}^{K} G_i$. Thus, with the proposed framework multiple graphs are combined for preserving the local multiple manifold structures. Furthermore, the MGDAEs are capable of achieving a balance in terms of local structures constructed by multiple graphs. By applying the chain rule to derive the partial derivatives of $E(x,y)$ w.r.t decoder parameters, we have:

$$\frac{\partial E}{\partial b_y} = \frac{\partial E}{\partial Y}\frac{\partial Y}{\partial b_y} = \left[\frac{\partial E}{\partial Y} \circ s'(W^T H + b_y \otimes 1_m{}^T)\right] 1_m \tag{7}$$

$$\frac{\partial E}{\partial W^T} = \frac{\partial E}{\partial Y}\frac{\partial Y}{\partial W^T} = \left[\frac{\partial E}{\partial Y} \circ s'(W^T H + b_y \otimes 1_m{}^T)\right] H^T \tag{8}$$

where $\circ$ is Hadamard product, $\otimes$ is outer product, and 1_m denotes a column vector of length m with all entry values of 1. The partial derivatives of $E(x,y)$ w.r.t encoder parameters using backward recurrence are calculated as follows:

$$\frac{\partial E}{\partial b_h} = \frac{\partial E}{\partial H}\frac{\partial H}{\partial b_h} = \left[\frac{\partial E}{\partial H} \circ s'(WX + b_h \otimes 1_m{}^T)\right] 1_m \tag{9}$$

$$\frac{\partial E^X}{\partial W} = \frac{\partial E^X}{\partial H}\frac{\partial H}{\partial W} = \left[\frac{\partial E^X}{\partial H} \circ s'(WX + b_h \otimes 1_m{}^T)\right] X^T \tag{10}$$

Denote graph regularization term of the MGAE as $T = \frac{1}{2} \sum_{k=1}^{K} r_k Tr(H^T G H)$, the partial derivatives of decoder parameters are zero because there is no decoder function involved in the graph regularization term. The partial derivatives of T w.r.t. encoder parameters can be computed via chain rule as:

$$\frac{\partial T}{\partial b_h} = \left[\frac{\partial T}{\partial H} \circ s'(WX + b_h \otimes 1_m{}^T) \right] 1_m \tag{11}$$

$$\frac{\partial T}{\partial W} = \left[\frac{\partial T}{\partial H} \circ s'(WX + b_h \otimes 1_m{}^T) \right] X^T \tag{12}$$

Considering reconstruction error and graph regularization terms, we can compute partial derivatives by averaging over the n training samples in the mini-batch and get partial derivatives of the objective function J_{MGAE} w.r.t. the parameters of MGAE as follows:

$$\Delta W := \frac{1}{2} \left[\frac{\partial J_{MGAE}}{\partial W} + \left(\frac{\partial J_{MGAE}}{\partial W^T} \right)^T \right]$$
$$= \frac{1}{2} \left[\frac{1}{n}(\frac{\partial E^X}{\partial W} + \lambda \frac{\partial T}{\partial W}) + \left(\frac{1}{n}(\frac{\partial E}{\partial W^T} + \lambda \frac{\partial T}{\partial W^T}) \right)^T \right] \tag{13}$$

$$\Delta b_y := \frac{\partial J_{MGAE}}{\partial b_y} = \frac{1}{n}(\frac{\partial E}{\partial b_y} + \lambda \frac{\partial T}{\partial b_y}) \tag{14}$$

$$\Delta b_h := \frac{\partial J_{MGAE}}{\partial b_h} = \frac{1}{n}(\frac{\partial E}{\partial b_h} + \lambda \frac{\partial T}{\partial b_h}) \tag{15}$$

3.2 Deep Network Training

Multi-graph deep auto-encoders attempt to extract high-level semantic feature representations using deep architectures composed of multiple MGAEs. Multi-graph deep auto-encoders consist of two stages: unsupervised pre-training and supervised fine-tuning. In the pre-training stage, unlabeled training data is used to optimize the parameters of MGAEs in a greedy layer-wise manner. For pre-training a single-layer MGAE, the purpose is to solve the optimization problem in (5). The process of training is shown in Table 1. Then the fine-tuning stage uses the labeled data to adjust the weight parameters of the MGAEs. Finally, a classifier layer is built on the top of multi-graph deep auto-encoders for classification tasks.

4   Experimental Results

In this section, the experiments are carried out in classification tasks to verify the performance of multi-graph auto-encoders.

Table 1. The process of deep network training.

Algorithm 1. Stochastic Gradient Descent for Pre-Training Single Layer of the Multi-

Graph Auto-Encoders (MGAE)

---

1   **Input:** Input dataset X , randomly initialized $\theta = \{W, b_h, b_y\}$

2   **Output:** $\theta = \{W, b_h, b_y\}$

3   **While** not stopping criterion **do**

4   Obtain G_{NPE}, G_{LAE}, G_{LSDA} according to corresponding methods

5   Obtain J_{MGAE} according to eqn.(5)

6   Calculate $\Delta W, \Delta b_h, \Delta b_y$ according to sample x_i

7   Update W, b_h, b_y according to $\Delta W, \Delta b_h, \Delta b_y$

8   **End**

Table 2. Descriptions of datasets

Dataset	Size: train, test	Dimensions	Classes
CIFAR-10	60000:50000,10000	3072	10
COIL-20	1440:960,480	1024	20

4.1 Evaluation Datasets

The summary of datasets is shown in Table 2. The CIFAR-10 dataset consists of 60000 32×32 color images in 10 classes, with 6000 images per class. The dataset is divided into five training batches and one test batch, each with 10000 images. The test batch contains exactly 1000 randomly-selected images from each class. The training batches contain the remaining images in random order, but some training batches may contain more images from one class than another. Between them, the training batches contain exactly 5000 images from each class. Columbia Object Image Library (COIL-20) is a database of 32×32 gray-scale images of 20 objects. It is composed of 1440 images with varying angles. Here we divided the dataset into train set and test set.

4.2 Image Classifications on CIFAR-10 and COIL-20

We evaluate the classification results by comparing the class labels of each image with its ground truth label provided by datasets. The standard metric accuracy (ACC) is used to measure the classification performance. Given a data sample

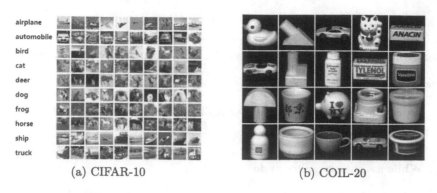

(a) CIFAR-10   (b) COIL-20

Fig. 2. Sample images of (a) CIFAR-10 and (b) COIL-20

x_i with classification label c_i and ground truth label g_i, the accuracy is defined as follows:

$$ACC = \frac{1}{n} \sum_{i=1}^{n} \delta(g_i, map(c_i)) \tag{16}$$

where n is the number of samples, $\delta(a, b)$ is a delta function, which equals 1 when $a = b$ and equals 0 when $a \neq b$. The function $map(c_i)$ is the permutation mapping function that maps each classification label to the best label c_i from the dataset.

The filters are illustrated learned by auto-encoder, sparse auto-encoder and multi-graph auto-encoder in Fig. 3, where 100 filters are visualized. The results demonstrated that the filters of AE learn meaningless shapes because no graph regularization is applied to constrain hidden layers. The filters of SAE and LAE achieve better results than the AE since additional constraints were imposed on hidden layers. For the filters learned by MGDAE, the shape reflects multi-view features that lead to good generalization performance.

We perform image classification tasks on COIL-20 and CIFAR-10 to evaluate the proposed multi-graph deep auto-encoder. The same network architecture is used for all methods. Table 3 reports the results of different methods, which suggest that the proposed MGAE achieves best performance among all methods on both CIFAR-10 and COIL-20. Compared MGAE with auto-encoder and single graph auto-encoder, we can find that all of these results demonstrate competitive results with consideration of multi-view preserving. We note that the results of AE are the worst because it only calculates the reconstruction error of the model. The results of SAE, LAE and AE combining with single-graph construction such as NPE and LSDA are better than the results of AE due to the sparse term constraint or graph embedding. MGAE imposes multi-graph constructions into auto-encoders so that it achieves best results and it is effective in learning multi-view feature representation. Here we divided the dataset into train set and test set. The sample images of datasets are shown in Fig. 2.

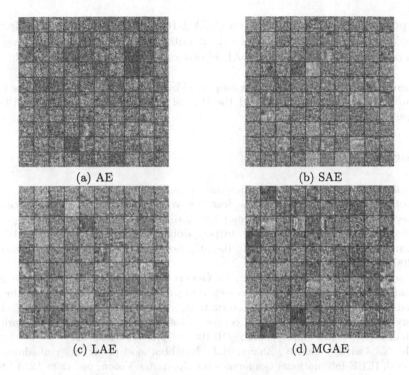

(a) AE

(b) SAE

(c) LAE

(d) MGAE

Fig. 3. Visualization filters learned by different methods on COIL-20

Table 3. Classification results (%) of comparison methods on COIL-20 and CIFAR-10

Models	COIL-20	CIFAR-10
AE	79.99 ± 0.70	51.36 ± 0.87
SAE	81.98 ± 0.91	51.58 ± 1.40
LAE	80.67 ± 0.45	51.53 ± 0.90
AE+NPE	81.11 ± 0.38	51.75 ± 0.99
AE+LSDA	81.78 ± 0.47	52.65 ± 1.07
MGDAE	**82.22 ± 0.89**	**53.58 ± 0.94**

5   Conclusion

In this paper, we propose a novel framework incorporating multiple graph constructions to auto-encoders, which can learn multi-view feature representation from data points. The purpose of combing various manifold learning algorithms is to find different views of low dimensional representation that preserves the local geometric properties of high dimensional observations. The advantages of multi-graph deep auto-encoders are revealed that it can preserve multiple localities in datasets by carrying out the theoretical analysis. Visualization results are presented to demonstrate that MGDAEs can extract the multi-view feature from data points. Experimental results on CIFAR-10 and COIL-20 show the effectiveness of the multi-graph regularized deep auto-encoders. In the future, we will focus on the applications of MGDAE about convolutional neural network.

Acknowledgments. This work was supported by the National Natural Science Foundation of China (No. 61672268) and the Primary Research & Development Plan of Jiangsu Province (No. BE2015137).

References

1. Bengio, Y.: Learning Deep Architecture for AI (2009)
2. Bottou, L.: Large-scale machine learning with stochastic gradient descent. In: Lechevallier, Y., Saporta, G. (eds.) Proceedings of COMPSTAT 2010, pp. 177–186. Springer, Heidelberg (2010). https://doi.org/10.1007/978-3-7908-2604-3_16
3. Cai, D., He, X., Zhou, K., Han, J., Bao, H.: Locality sensitive discriminant analysis **2007**, 708–713 (2007)
4. Hao, Y., Han, C., Shao, G., Guo, T.: Generalized graph regularized non-negative matrix factorization for data representation. In: Lu, W., Cai, G., Liu, W., Xing, W. (eds.) Proceedings of the 2012 International Conference on Information Technology and Software Engineering. Lecture Notes in Electrical Engineering. Springer, Heidelberg (2013). https://doi.org/10.1007/978-3-642-34528-9_1
5. He, X., Cai, D., Yan, S., Zhang, H.J.: Neighborhood preserving embedding. In: Tenth IEEE International Conference on Computer Vision, pp. 1208–1213 (2005)
6. He, X., Niyogi, P.: Locality Preserving Projections (lPP). Adv. Neural Inf. Process. Syst. **16**(1), 186–197 (2002)
7. Hong, R., Zhang, L., Tao, D.: Unified photo enhancement by discovering aesthetic communities from flickr. IEEE Trans. Image Process. **25**(3), 1124–1135 (2016)
8. Hong, R., Hu, Z., Wang, R., Wang, M., Tao, D.: Multi-view object retrieval via multi-scale topic models. IEEE Trans. Image Process. **25**(12), 5814–5827 (2016)
9. Hong, R., Zhang, L., Zhang, C., Zimmermann, R.: Flickr circles: aesthetic tendency discovery by multi-view regularized topic modeling. IEEE Trans. Multimed. **18**(8), 1555–1567 (2016)
10. Jia, K., Sun, L., Gao, S., Song, Z., Shi, B.E.: Laplacian auto-encoders: an explicit learning of nonlinear data manifold. Neurocomputing **160**, 250–260 (2015)
11. Liao, Y., Wang, Y., Liu, Y.: Graph regularized auto-encoders for image representation. IEEE Trans. Image Process. **26**(6), 2839–2852 (2017)
12. Lu, S., Liu, H., Li, C.: Manifold regularized stacked autoencoder for feature learning. In: IEEE International Conference on Systems, Man, and Cybernetics, pp. 2950–2955 (2016)
13. Qiao, L., Chen, S., Tan, X.: Sparsity preserving projections with applications to face recognition. Pattern Recogn. **43**(1), 331–341 (2010)
14. Rifai, S., Vincent, P., Muller, X., Glorot, X., Bengio, Y.: Contractive auto-encoders: explicit invariance during feature extraction. In: ICML (2011)
15. Rumelhart, D.E., Hinton, G.E., Williams, R.J.: Learning representations by back-propagating errors **323**(6088), 399–421 (1986)
16. Schölkopf, B., Platt, J., Hofmann, T.: Greedy layer-wise training of deep networks. In: International Conference on Neural Information Processing Systems, pp. 153–160 (2006)

17. Vincent, P., Larochelle, H., Lajoie, I., Bengio, Y., Manzagol, P.A.: Stacked denoising autoencoders: learning useful representations in a deep network with a local denoising criterion. J. Mach. Learn. Res. **11**(12), 3371–3408 (2010)
18. Yang, S., Li, L., Wang, S., Zhang, W., Huang, Q.: A graph regularized deep neural network for unsupervised image representation learning. In: Computer Vision and Pattern Recognition, pp. 7053–7061 (2017)
19. Yu, W., Zeng, G., Luo, P., Zhuang, F., He, Q., Shi, Z.: Embedding with autoencoder regularization. In: Blockeel, H., Kersting, K., Nijssen, S., Železný, F. (eds.) ECML PKDD 2013. LNCS (LNAI), vol. 8190, pp. 208–223. Springer, Heidelberg (2013). https://doi.org/10.1007/978-3-642-40994-3_14

Discrete Semi-supervised Multi-label Learning for Image Classification

Liang Xie[✉], Lang He, Haohao Shu, and Shengyuan Hu

Department of Mathematics, Wuhan University of Technology, Wuhan, China
whutxl@hotmail.com, {helang,215738,ihuhiccup}@whut.edu.cn

Abstract. Multi-label image classification is a critical problem in semantic based image processing. Traditional semi-supervised multi-label learning methods usually learn classification functions in continuous label space. And the ignorance of discrete constraint of semantic labels impedes the classification performance. In this paper, we specifically consider the discrete constraint and propose Discrete Semi-supervised Multi-label Learning (DSML) for image classification. In DSML, we propose a semi-supervised framework with discrete constraint. Then we introduce anchor graph learning to improve the scalability, and derive an ADMM based alternating optimization process to solve the framework. Experimental results demonstrate the superiorly of DSML compared with several advanced semi-supervised methods.

Keywords: Discrete learning · Multi-label learning
Image classification

1 Introduction

With the advance of multimedia technology, learning semantics of multimedia content is becoming important. Multi-label image classification is a critical semantic learning approach for images, it can assign multiple semantic labels to an image. Multi-label image classification can be applied to many image/multimedia applications, such as image retrieval, image annotation, and etc.

In semi-supervised learning, training set contains both unlabeled and labeled data. Thus semi-supervised learning is suitable for the real-world application where the semantic labels are very limited. General semi-supervised learning methods simultaneously learn the label vectors of unlabeled data and the classification functions which map image features into the label space [1]. Many semi-supervised multi-label methods focus on learning continuous label vectors from unlabeled and labeled data [14,19], then image features are mapped into this continuous label space. Obviously, these methods ignore the special characteristic of semantic labels that they are essentially binary representations of the image. Learning classification functions from continuous label vectors makes these methods be more like the regression model but not the classification model.

© Springer Nature Switzerland AG 2018
R. Hong et al. (Eds.): PCM 2018, LNCS 11164, pp. 808–818, 2018.
https://doi.org/10.1007/978-3-030-00776-8_74

Since semantic labels are intrinsically discrete, the learned continuous vectors of labels should be transferred to discrete space, which will obviously cause quite a lot of quantization loss. As a result, although some well designed semi-supervised methods can effectively learn the semantic information from unlabeled data, they cannot effectively preserve semantic information in final results.

The traditional continuous semi-supervised learning approach cannot effectively obtain sufficient semantic information in the learning process, which impedes their performance in multi-label classification. In order to deal with this problem, discrete learning can be used to replace traditional continuous learning. Discrete learning has shown promising performance in unsupervised learning [18,21]. Generally, semi-supervised learning frameworks are based on the assumption of unsupervised learning, thus we can also expect the promising performance of discrete learning in semi-supervised learning.

In this paper, we directly solve the learning problem of discrete labels. And we propose an effective discrete learning based method, which is dubbed as Discrete Semi-supervised Multi-label Learning (DSML), for image classification. Unlike previous methods which treat the label matrix as continuous. We preserve the discrete constraint of labels and propose a discrete learning framework to solve this constraint. To improve the efficiency and scalability of our method, we introduce anchor graph learning to our discrete learning framework. In the optimization, an ADMM based alternating process is derived to learn binary label matrix and mapping matrix of classification function. The experimental results on MIR Flickr and NUS-WIDE show the promising performance of the proposed method.

The rest of this paper is organized as follows. Section 2 reviews the related work. Details about the proposed methodology are presented in Sect. 3. In Sect. 4, we introduce our experiments. Section 5 concludes this paper.

2   Related Works

2.1   Semi-supervised Multi-label Learning

Semi-supervised multi-label learning has gained much attention in recent years. In [20], semi-supervised learning is formulated as a Non-negative Matrix Factorization (NMF) problem. Graph learning is a prevalent technology in semi-supervised learning [1], and some methods based on graph learning consider the label correlation to improve classification performance. [14,16,19] deploy multi-task shared subspace learning to utilize the label correlation. In [11], a semi-supervised singular value decomposition (SVD) is proposed for exploiting correlations between labels and making up for the lack of labeled data or even missing labeled data. Most of the above methods consider the label space of unlabeled data as continuous, thus they cannot satisfy the discrete characteristic of semantic labels.

Some semi-supervised methods also improve the advanced discriminative method to deal with unlabeled data, and some of them can learn discrete labels.

Transductive SVMs exploits specific iterative algorithms with a transductive process that incorporates both labeled and unlabeled samples in the training phase [3]. Cost interval semi-supervised large margin distribution machine (cisLDM) optimizes the margin distribution on both labeled and unlabeled data [22]. However, most of these methods solve single-label (or two class) problem, which may be not suitable for multi-label problem.

2.2 Discrete Learning

Discrete learning is first proposed for the hashing task [13,15,25]. Unlike continuous learning which relaxes the constraint of discrete variables [17], discrete learning methods [23,24] specifically consider solving the discrete constraint. Recently, it also becomes popular in other multimedia applications. Discrete collaborative filtering is proposed in [21] for recommendation system, and it shows that the introduce of discrete learning will significantly improve the recommendation performance. Recently, discrete learning also shows the superiority in unsupervised task, such as clustering [18].

3   The Proposed Methodology

3.1   Notations and Problem Description

In this paper, we focus on multi-label image classification task which is different to single-label or multi-class image classification. And we also only concentrate on the semi-supervised problem. Suppose we have the training data $X = [X_L; X_U] \in \mathbb{R}^{N \times m}$, where $X_L \in \mathbb{R}^{N_L \times m}$ denotes the labeled data, and $X_U \in \mathbb{R}^{N_U \times m}$ denotes the unlabeled data. N_L is the number of labeled data, N_U is the number of unlabeled data, $N = N_L + N_U$ is the number of all training data, m is the dimension of image features. The labeled data have the corresponding labeled matrix $Y_L \in \{-1, 1\}^{N_L \times k}$, where $Y_{ij} = 1$ denotes that i_{th} image is labeled by j_{th} label, and $Y_{ij} = -1$ denotes that i_{th} image is not labeled by j_{th} label. k is the number of labels, and unlabeled data have no label matrix. The goal of our method is to learn the following classification function which can predict the labels of new images

$$y = \text{sgn}(xW) \tag{1}$$

where x is the feature vector of new image, y is the predicted label vector, and $W \in \mathbb{R}^{m \times k}$ is the mapping matrix. Since we will introduce anchor graph in our framework, the feature vector x can be replaced by truncate similarity vector z, which will be discussed in later subsection.

3.2   Basic Formulation

The main purpose of this paper is to improve multi-label image classification by discrete learning. So we construct the basic semi-supervised framework based on the success of previous state-of-the-art methods [19]. However, our formulation is very different to the previous methods, in that we have the important discrete constraint in our framework, which is formulated as

$$\min \frac{1}{2N} Tr(B^T LB) + \alpha \|B_L - Y_L\|_F^2$$
$$s.t \qquad B \in \{1, -1\}^{N \times k} \qquad (2)$$

where B is the learned binary label matrix for all training data. α is a hyper-parameter which is used to ensure consistence of the predicted label matrix and true label matrix of labeled data. L is the Laplacian matrix, we will discuss it in details later.

The above objective function (2) has several difficulties to be directly optimized. At first, the discrete constraint make the optimization be a NP-hard problem. Secondly, computing Laplacian matrix L is time-consuming when the training data is large. We will solve these two difficulties in the following sub-sections.

3.3 Discrete Learning

Traditional semi-supervised learning relax the discrete constraint and treat label matrix as continuous variables. Therefore, they will loss many semantic information in the learning process. In this paper, we propose an optimization approach based on alternating direction method of multipliers (ADMM) [2,9] to cope with the discrete constraint. Our idea is to introduce an auxiliary variable to transform the objective function to an equivalent one that is tractable. By introducing an intermediate matrix F to represent B and applying the ADMM, the basic formulation can be transformed to

$$\min \frac{1}{2N} Tr(F^T LF) + \frac{\alpha}{2} \|F_L - Y_L\|_F^2 + \frac{\alpha}{2} \|B_L - Y_L\|_F^2$$
$$+ \langle \Lambda, B - F \rangle + \frac{\rho}{2} \|B - F\|_F^2$$
$$s.t \qquad B \in \{1, -1\}^{N \times k} \qquad (3)$$

where Λ is the dual variable, ρ is the hyperparameter of the corresponding term.

The formulation (3) can be optimized by the following alternating sub-optimization process

$$F^{t+1} = \arg\min \frac{1}{2N} Tr(F^T LF) + \frac{\alpha}{2} \|F_L - Y_L\|_F^2 + \frac{\rho}{2} \left\| B - F + \frac{\Lambda}{\rho} \right\|_F^2 \qquad (4)$$

$$B = \arg\min \frac{\alpha}{2} \|B_L - Y_L\|_F^2 + \frac{\rho}{2} \left\| B - F + \frac{\Lambda}{\rho} \right\|_F^2 \qquad (5)$$

$$\Lambda = \Lambda + \rho(B - F) \qquad (6)$$

The above sub-optimization steps are tractable, and the whole optimization can be solved by the alternating process of these steps.

3.4   Anchor Graph Learning

The optimization of (3) is still difficult, in that it is very time-consuming. The time complexity of constructing Laplacian matrix is about $O(mN^2)$, and the optimization of F will be not less than $O(N^2)$. Thus we introduce the framework of anchor graph learning [12] to improve the time efficiency of our method.

We first randomly select N_A anchors $X_A \in R^{N_A \times m}$ from training data. Then the truncated similarities Z_{ij} is computed between all N training data and N_A anchors

$$Z_{ij} = \frac{\exp\left(-dist\left(x_i, x_j^{(A)}\right)^2 \Big/ \delta\right)}{\sum_j \exp\left(-dist\left(x_i, x_j^{(A)}\right)^2 \Big/ \delta\right)} \tag{7}$$

where $x_j^{(A)}$ is the feature vector of $j_t h$ anchor. Then we can generate the truncate similarity matrix $Z \in \mathbb{R}^{N \times N_A}$ of all training data. $dist(\cdot)$ is the L2 distance, and δ is the mean of all distances of training data.

The adjacency matrix can be efficiently approximated by $A = ZD^{-1}Z^T$, where $D = diag(Z^T 1\} \in \mathbb{R}^{N_A \times N_A}$ and 1 is the vector with all ones. Then the Laplacian matrix can be approximately computed by $L = I - ZD^{-1}Z^T$.

Since the size of L remains $N \times N$, directly solving F will also cost $O(N^2)$. Therefore, we further consider the approximation of F. F is used to represent B, and it can be regarded as the continuous version of predicted label matrix. So we use the label function (1) to factorize F as $F = ZW$, and we use Z to replace X. Then the objective function (3) becomes

$$
\begin{aligned}
min \qquad & \frac{1}{2N} Tr(W^T Z^T Z D^{-1} Z^T Z W) + \frac{\alpha}{2} \|Z_L W - Y_L\|_F^2 \\
& + \frac{\alpha}{2} \|B_L - Y_L\|_F^2 + \langle \Lambda, B - ZW \rangle + \frac{\rho}{2} \left\| B - ZW + \frac{\Lambda}{\rho} \right\|_F^2 + \frac{\lambda}{2} \|W\|_F^2 \\
s.t. \qquad & B \in \{1, -1\}
\end{aligned} \tag{8}
$$

where $Z_L \in \mathbb{R}^{N_L \times N_A}$ is the truncate similarity matrix of labeled data, λ is the hyperparameter of the regularization term of W.

3.5   Optimization Process

Based on the discrete learning and anchor graph learning, the optimization process can become tractable and efficient. In this subsection we discuss the detailed alternating optimization process of (4) which is based on ADMM. In each step of alternating process, we only optimize one variable and fix other variables.

Update W. By fixing other variables, the sub-optimization problem of W becomes

$$min \ \frac{1}{2N}Tr(W^T Z^T Z D^{-1} Z^T ZW) + \frac{\alpha}{2} \|Z_L W - Y_L\|_F^2$$

$$+\frac{\rho}{2} \left\| B - ZW + \frac{\Lambda}{\rho} \right\|_F^2 + \frac{\lambda}{2} \|W\|_F^2 \tag{9}$$

The above objective function (10) can be simplified as following formulation

$$min \ \frac{1}{2N}Tr(W^T Z^T Z D^{-1} Z^T ZW) + \frac{1}{2}Tr((ZW - Y)^T A(ZW - Y))$$

$$+\frac{\rho}{2} \left\| B - ZW + \frac{\Lambda}{\rho} \right\|_F^2 + \frac{\lambda}{2} \|W\|_F^2 \tag{10}$$

where A is a diagonal matrix, $A_{ii} = \alpha$ if i_{th} image is labeled, otherwise $A_{ii} = 0$, $Y = [Y_L; Y_U]$ and $Y_U \in R^{N_U \times k}$ is an all zero matrix.

By solving (10), we can update W via

$$W = G(H + \rho Z^T B + Z^T \Lambda) \tag{11}$$

where

$$H = Z^T AY$$

$$G = \left(\frac{1}{N} Z^T Z D^{-1} Z^T ZW + Z^T AZ + \rho Z^T Z + \lambda I \right)^{-1} \tag{12}$$

Update B. By fixing other variables, the sub-optimization problem of B becomes:

$$min \frac{\alpha}{2} \|B_L - Y_L\|_F^2 + \frac{\rho}{2} \left\| B - ZW + \frac{\Lambda}{\rho} \right\|_F^2$$

$$s.t. \qquad B \in \{1, -1\} \tag{13}$$

The above formulation can be transformed to

$$min \ \frac{\alpha}{2} \|B_L - Y_L\|_F^2 + \frac{\rho}{2} \left\| B_L - Z_L W + \frac{\Lambda}{\rho} \right\|_F^2 + \frac{\rho}{2} \left\| B_U - Z_U W + \frac{\Lambda}{\rho} \right\|_F^2$$

$$s.t. \ B_L, B_U \in \{1, -1\} \tag{14}$$

where B_U is the predicted label matrix of unlabeled data, $B = [B_L; B_U]$, and Z_U is the truncate matrix of unlabeled data. From (14), we can update B by

$$B_L = sgn\,(\alpha Y_L + \rho Z_L W - \Lambda)$$

$$B_U = sgn\left(Z_U W - \frac{\Lambda}{\rho} \right) \tag{15}$$

Update Λ. Based on (4), we can update Λ by

$$\Lambda = \Lambda + \rho(B - ZW) \tag{16}$$

The overall optimization process is summarized in Algorithm 1. We can find that the time complexities of updating process (11), (15) and (16) are all $O(N)$. Thus the whole process of Algorithm 1 also cost $O(N)$, which means its time complexity is linear to the size of training data.

Algorithm 1. Optimization process of DSML

Require:
   $X_L, X_U, Y_L,\ \alpha, \lambda, \rho$
Ensure:
   W, B
1: Initialize B;
2: Compute G, H according to (12);
3: **while** Not Converge **do**
4:     Update W according to (11);
5:     Update B according to (15);
6:     Update Λ according to (16);
7: **end while**

---

4   Experiment

4.1   Datesets and Feature Preprocessing

In this paper, two real world image datasets: MIR Flickr [10] and NUS-WIDE [5] are used for evaluation. The statistics of these two datasets are summarized in Table 1.

Table 1. Statitics of MIR Flickr and NUS-WIDE

Dataset	MIR Flickr	NUS-WIDE
Training data size	12,500	161,789
Labeled data size	1,250	15,000
Test size	12,500	78,370
Label number	38	20

MIR Flickr dataset consists of 25,000 images downloaded from the Flickr through its public API, and each image is associated with textual tags. Images in MIR Flickr are annotated for 24 concepts, including object categories such as clouds, tree or dog, as well as more general categories such as sky, water or people. For 14 of the 24 concepts a second, stricter annotation was made: for each concept a subset of the positive images is selected where the concept is salient in the image. These more strictly annotated classes are denoted by using ∗ as a suffix. In total there are 38 categories for this dataset. For MIR Flickr, 15 visual features [8] are used for image modality, including 4 types of SIFT BOVW, 4 types of Hue BOVW, 4 types of color histograms and GIST.

NUS-WIDE dataset contains 269,648 image downloaded from Flickr. Each image is also associated with textual tags, and all images are labeled by 81 concepts. On NUS-WIDE dataset we directly use features provided by [5]: image features include 64-D color histogram, 144-D color correlogram, 73-D edge direction histogram, 128-D wavelet texture, 225-D block-wise color moments and 500-D BOVW based on SIFT descriptions.

In our experiment, MIR Flickr is split to 12,500 training images and 12,500 test images [8], where 1,250 of the training images are chosen as labeled, and the rest are regarded as unlabeled. NUS-WIDE is split to 161,789 training images and 107,859 test images. On NUS-WIDE, 1,250 training imges is chosen as labeled, and the rest training images are assumed to be unlabeled. In NUS-WIDE, some concepts contain few images, so we select top 20 concepts with largest number of images. Then we remove test images with no concepts and our test size becomes 78,370. We preserve all 161,789 training images, and select 15,000 of them as labeled data.

Images in these two datasets are represented by multiple visual features, and our method is only suited to single feature. Therefore, we should preprocess visual features. In this work, we use kernel PCA (KPCA) [16] to transform multiple visual features to a single feature. Finally we obtain a 500-D visual feature for each image on MIR Flickr, and a 100-D visual feature for each image on NUS-WIDE.

4.2 Evaluation Metrics

In our experiment, we compute the average precision (AP) for each label, and also using the mean AP (mAP) of all concepts to measure performance. We choose the AP score used in PASCAL VOC challenge evaluation [6], it is defined as:

$$AP = \frac{1}{11} \sum_r P(r) \tag{17}$$

where $P(r)$ is the maximum precision over all recalls which are larger than $r \in \{0, 0.1, 0.2, \ldots, 1\}$, a larger AP score means a better performance. We also use precision, recall and F1-score to measure the performance, and they are widely used for image annotation task [7] which can be seem as equivalent to multi-label learning.

4.3 Experimental Results

Our method uses linear classification functions for prediction. In order to assure the fairness and show the effect of discretely learning, we only compare our method with linear semi-supervised methods. We compare our method with three state-of-the-art semi-supervised multi-label learning methods, manifold regularized multitask learning (MRMTL) [14], label correlation mining with relaxed graph embedding (LMGE) [19] and cost interval semi-supervised large margin distribution machine (cisLDM) [22]. MRMTL and LMGE both output continuous label scores, thus we have to transform them to discrete labels. For each test image, we choose 3–5 labels with highest scores.

In the implementation of all methods, we choose appropriate hyperparameters to report their best experimental results. In our method, we set $\{\alpha, \rho, \lambda\} =$

Table 2. Comparison of performance on MIR Flickr.

Method	mAP	Precision	Recall	F1-score
MRMTL [14]	0.2546	0.4621	0.2446	0.3199
LMGE [19]	0.2449	0.3422	0.2688	0.3011
cisLDM [22]	0.2584	0.4170	0.2542	0.3158
DSML	**0.2959**	**0.5505**	**0.2699**	**0.3622**

Table 3. Comparison of performance on NUS-WIDE.

Method	mAP	Precision	Recall	F1-score
MRMTL [14]	0.1920	0.2835	0.3609	0.3175
LMGE [19]	0.2226	0.3068	0.3884	0.3428
cisLDM [22]	0.2191	0.2834	0.3649	0.3190
DSML	**0.2870**	**0.4645**	**0.4537**	**0.4590**

$\{100, 0.0001, 1000\}$ on MIR Flickr, and set $\{\alpha, \rho, \lambda\} = \{100, 0.001, 1\}$ on NUS-WIDE.

Table 2 shows the performance of all methods on MIR Flickr in terms of mAP, precision, recall and F1-score. From Table 2 we can observe that DSML outperforms other compared methods on four evaluation metrics. The superiority of DSML on recall is not significant, the reason is that some labels in MIR Flickr contain very few images, and they are always not used for labeling. The other scores of DSML are much larger than compared methods, which approves the advantage of discrete learning in semi-supervised multi-label image classification.

In the experiment of NUS-WIDE, since the training data is very large, we slightly change the algorithm of MRMTL and LMGE by using anchor graphs to construct Laplacian matrices. Base on this improvement, the optimization processes of MRMTL and LMGE become feasible. Table 3 shows the performance of all methods on NUS-WIDE. The results of Table 3 demonstrate the superiority of DSML on all four types of scores, and DSML outperforms the second best method more than 10%. We have purified the NUS-WIDE dataset, thus the superiority of DSML on NUS-WIDE is more significant than MIR Flickr.

5   Conclusion

In this work, we propose Discrete Semi-supervised Multi-label Learning (DSML) for image classification in the case of limited labels. In DSML we construct a semi-supervised framework with discrete constraint, and introduce anchor graph learning to improve scalability. Then an ADMM based optimization algorithm is derived to directly solve the discrete learning problem of DSML. Experimental results show the superiority of DSML. Moreover, DSML can be improved in future. For example, our DSML uses linear function for prediction, and advanced nonlinear discriminative functions such as CNN and other deep networks can be integrated into DSML. Besides, based on the promising performance of discrete learning, DSML can be applied for other image tasks such as image retrieval and multimedia recommendation [4].

Acknowledgements. This work was supported by the National Natural Science Foundation of China (No. 61702388) and the Fundamental Research Funds for the Central Universities (WUT: 2018IVB021).

References

1. Belkin, M., Niyogi, P., Sindhwani, V.: Manifold regularization: a geometric framework for learning from labeled and unlabeled examples (2006). JMLR.org
2. Boyd, S., Parikh, N., Chu, E., Peleato, B., Eckstein, J.: Distributed optimization and statistical learning via the alternating direction method of multipliers. Found. Trends Mach. Learn. **3**(1), 1–122 (2010)
3. Bruzzone, L., Chi, M., Marconcini, M.: A novel transductive SVM for semisupervised classification of remote-sensing images. IEEE Trans. Geosci. Remote Sens. **44**(11), 3363–3373 (2006)
4. Cheng, Z., Shen, J., Zhu, L., Kankanhalli, M., Nie, L.: Exploiting music play sequence for music recommendation. In: Twenty-Sixth International Joint Conference on Artificial Intelligence, pp. 3654–3660 (2017)
5. Chua, T.S., Tang, J., Hong, R., Li, H., Luo, Z., Zheng, Y.: NUS-WIDE: a real-world web image database from National University of Singapore. In: Proceedings of the ACM International Conference on Image and Video Retrieval, p. 48. ACM (2009)
6. Everingham, M., Van Gool, L., Williams, C., Winn, J., Zisserman, A.: The pascal visual object classes challenge 2007 (voc 2007) results (2007). http://www.pascal-network.org/challenges/VOC/voc2007/workshop/index.html (2008)
7. Guillaumin, M., Mensink, T., Verbeek, J., Schmid, C.: TagProp: discriminative metric learning in nearest neighbor models for image auto-annotation. In: IEEE International Conference on Computer Vision, pp. 309–316 (2009)
8. Guillaumin, M., Verbeek, J., Schmid, C.: Multimodal semi-supervised learning for image classification. In: 2010 IEEE Conference on Computer Vision and Pattern Recognition (CVPR), pp. 902–909. IEEE (2010)
9. Hajinezhad, D., Chang, T.H., Wang, X., Shi, Q., Hong, M.: Nonnegative matrix factorization using ADMM: algorithm and convergence analysis. In: IEEE International Conference on Acoustics, Speech and Signal Processing, pp. 4742–4746 (2016)
10. Huiskes, M.J., Lew, M.S.: The MIR Flickr retrieval evaluation. In: Proceedings of the 1st ACM International Conference on Multimedia Information Retrieval, pp. 39–43. ACM (2008)
11. Jing, L., Shen, C., Yang, L., Yu, J., Ng, M.K.: Multi-label classification by semi-supervised singular value decomposition. IEEE Trans. Image Process. **26**(10), 4612–4625 (2017)
12. Liu, W.: Hashing with graphs. In: Proceedings of International Conference on Machine Learning, pp. 1–8, June 2011
13. Liu, W., Mu, C., Kumar, S., Chang, S.F.: Discrete graph hashing. In: Advances in Neural Information Processing Systems, pp. 3419–3427 (2014)

14. Luo, Y., Tao, D., Geng, B., Xu, C., Maybank, S.J.: Manifold regularized multitask learning for semi-supervised multilabel image classification. IEEE Trans. Image Process. **22**(2), 523–536 (2013)

15. Shen, F., Shen, C., Liu, W., Shen, H.T.: Supervised discrete hashing. In: Computer Vision and Pattern Recognition, pp. 37–45 (2015)

16. Xie, L., Pan, P., Lu, Y., Wang, S.: A cross-modal multi-task learning framework for image annotation. In: ACM International Conference on Conference on Information and Knowledge Management, pp. 431–440 (2014)

17. Xie, L., Shen, J., Han, J., Zhu, L., Shao, L.: Dynamic multi-view hashing for online image retrieval. In: Twenty-Sixth International Joint Conference on Artificial Intelligence, pp. 3133–3139 (2017)

18. Yang, Y., Shen, F., Huang, Z., Shen, H.T., Li, X.: Discrete nonnegative spectral clustering. IEEE Trans. Knowl. Data Eng. **PP**(99), 1 (2017)

19. Yang, Y., Wu, F., Nie, F., Shen, H.T., Zhuang, Y., Hauptmann, A.G.: Web and personal image annotation by mining label correlation with relaxed visual graph embedding. IEEE Trans. Image Process. A Publ. IEEE Signal Process. Soc. **21**(3), 1339–51 (2012)

20. Yi, L., Rong, J., Liu, Y.: Semi-supervised multi-label learning by constrained nonnegative matrix factorization. In: National Conference on Artificial Intelligence and the Eighteenth Innovative Applications of Artificial Intelligence Conference, 16–20 July 2006, Boston, Massachusetts, USA, pp. 421–426 (2006)

21. Zhang, H., Shen, F., Liu, W., He, X., Luan, H., Chua, T.S.: Discrete collaborative filtering. In: International ACM SIGIR Conference on Research and Development in Information Retrieval, pp. 325–334 (2016)

22. Zhou, Y.H., Zhou, Z.H.: Large margin distribution learning with cost interval and unlabeled data. IEEE Trans. Knowl. Data Eng. **28**(7), 1749–1763 (2016)

23. Zhu, L., Huang, Z., Chang, X., Song, J., Shen, H.T.: Exploring consistent preferences: discrete hashing with pair-exemplar for scalable landmark search, pp. 726–734 (2017)

24. Zhu, L., Huang, Z., Li, Z., Xie, L., Shen, H.T.: Exploring auxiliary context: discrete semantic transfer hashing for scalable image retrieval. IEEE Trans. Neural Netw. Learn. Syst. **PP**(99), 1–13 (2018). https://doi.org/10.1109/TNNLS.2018.2797248

25. Zhu, L., Huang, Z., Liu, X., He, X., Sun, J., Zhou, X.: Discrete multimodal hashing with canonical views for robust mobile landmark search. IEEE Trans. Multimedia **19**(9), 2066–2079 (2017)

Multispectral Foreground Detection via Robust Cross-Modal Low-Rank Decomposition

Aihua Zheng, Yumiao Zhao, Chenglong Li$^{(\boxtimes)}$, Jin Tang, and Bin Luo

School of Computer Science and Technology, Anhui University, Hefei 230601, China
{ahzheng214,tj,luobin}@ahu.edu.cn, ymiaozhao@foxmail.com,
lcl1314@foxmail.com

Abstract. In this paper, we propose a novel approach which pursues cross-modal low-rank decomposition for robust multi-spectral foreground detection. For each spectrum, we employ the idea of low-rank and sparse decomposition to detect sparse moving objects against background with low-rank structure for its robustness to noises. Unlike simply combining multi-modal detecting results or compulsively enforcing a shared foreground mask in existing methods, we propose to pursue the cross modality consistency among heterogeneous modalities by introducing a soft cross-modality consistent constraint to the multi-modal low-rank decomposition model. Extensive experiments on the benchmark dataset GTFD suggest that our approach achieves superior performance over the state-of-the-art algorithms.

Keywords: Cross-modality consistency · Foreground detection
Low-rank decomposition

1 Introduction

Foreground detection is a fundamental research topic in computer vision and essential in many related scenarios, such as video surveillance [25], behavior analysis [5], visual tracking [15] and object retrieval [10] et al. Despite of the great progress in the past decades, it is still a challenging task due to the complex factors, such as background clutter, illumination, bad weather, et al.

Extensive methods have been proposed for single-modality foreground detection over the past decades. The representative methods include Gaussian Mixture Models (GMM) [24], non-parameter algorithms [1], multiple features based methods [23], low-rank decomposition models [13,32] and convolutional neural network methods [4,17]. However, single visual sensor suffers from the aforementioned challenging scenarios. Recently, some literatures integrated the complementary thermal infrared sensor to effectively boost the performance in challenging scenarios. Han et al. [9] proposed a hierarchical scheme to automatically align synchronous grayscale and thermal frames, and probabilistically combined

© Springer Nature Switzerland AG 2018
R. Hong et al. (Eds.): PCM 2018, LNCS 11164, pp. 819–829, 2018.
https://doi.org/10.1007/978-3-030-00776-8_75

cues from registered grayscale-thermal frames for human silhouette detection. Davis et al. [6] proposed a new background-subtraction technique fusing contours from thermal and grayscale videos for object detection in urban settings. Zhao et al. [31] integrated the infrared and visible images with different strategies on salient and non-salient regions, then employed GMM to achieve the background subtraction.

Recently, some literatures focused on the multi-modal foreground detection on low-rank decomposition framework for its robustness to noises [14,28]. Li et al. [14] proposed a weighted low-rank decomposition method by learning the shared foreground mask matrix for different modalities to achieve adaptive fusion of different source data. Yang et al. [28] proposed a fast grayscale-thermal foreground detection via collaboratively separating and integrating the foregrounds from different modalities in low-rank decomposition framework. However, the hard consistency [14] with shared foreground among different modalities may be overstrict. Inspired by the fact that different images can be perceived from multi-view features [11], we argue that the different modalities are heterogeneous with different properties as shown in Fig. 1. Furthermore, the independency between modalities [28] ignored the complementary benefits from different modalities as shown in Fig. 1. Where the visible spectrum disturbed by low illumination benefits from thermal source in Fig. 1(a), and the thermal one disturbed by glass and thermal crossover benefits from visible one in Fig. 1(b). Therefore, we argue to pursuing the cross modality consistency among the heterogeneous modalities to capture these benefits.

Fig. 1. Sample of the multi-modal image pairs from GTFD dataset, the grayscale and thermal modalities are heterogeneous with different properties.

Based on above discussion, we propose a novel and robust multispectral foreground detection approach to capture cross modality consistency among the heterogeneous modalities in a unified low-rank decomposition framework, we first accumulate sequential frames as two input matrices from the grayscale and thermal videos. The underlying background images are linearly correlated in each modality when ignoring the sparse and heterogeneous foregrounds and outliers. After introducing the appearance consistency and spatial compactness constraint among the neighborhood in each heterogeneous modality, we propose to construct the cross-modal graph to pursue the cross-modality consistency among the heterogeneous foregrounds into a unified low-rank decomposition framework. Finally, we jointly optimize the proposed multi-modal low-rank decomposition to generate the heterogeneous background models and the foreground masks simultaneously.

2   Our Algorithm

Given a grayscale-thermal video pair, we solve the multi-modal foreground detection based on the low-rank decomposition framework in a batch manner.

2.1   Model Formulation

Given the k-th modal video, we accumulate n frames into a matrix by reshaping each frame into a column vector, $i.e.$, $\mathbf{D}^k = [\mathbf{d}_1^k, \mathbf{d}_2^k, ..., \mathbf{d}_n^k] \in R^{m \times n}$, with $k = 1, \cdots, K$ and m is the number of pixel on each frame. Herein, the grayscale-thermal data in this paper is the special case with $K = 2$. First, we assume that the underlying background images are linearly correlated in each modality video and the foregrounds are sparse and contiguous. This assumption has been successfully applied in background modeling [7,32].

Heterogeneous Decomposition. As we discussed above, the different modalities are heterogeneous with different properties. Therefore, we decompose the input matrices into heterogeneous foreground/background for each modality as: $\mathbf{D}^k = \mathbf{B}^k + \mathbf{S}^k$, where $\mathbf{B}^k \in R^{m \times n}$ is the low-rank background matrix, and $\mathbf{S}^k \in R^{m \times n}$ denotes the sparse heterogeneous foreground matrix of the k-th modality, which can be formulated as:

$$\min_{\mathbf{B}^k, \mathbf{S}^k} \frac{1}{2} \|f_{\mathbf{S}_\perp^k}(\mathbf{D}^k - \mathbf{B}^k)\|_F^2 + \beta \|vec(\mathbf{S}^k)\|_0, \tag{1}$$
$$s.t.\ rank(\mathbf{B}^k) \leq r^k,\ k = 1, 2, ..., K,$$

where β is a balance parameter. $vec(\cdot)$ is a vectorize operator on a matrix. $\| \cdot \|_F$ and $\| \cdot \|_0$ indicate the Frobenius norm of a matrix and the l_0 norm of a vector, respectively. r^k is a constant that suppresses the complexity of the background model in each modality. $f_{\mathbf{S}}(\mathbf{X})$ represents the orthogonal projection of a matrix $\mathbf{X}$ onto the linear space of matrices supported by $\mathbf{S}$:

$$f_{\mathbf{S}}(\mathbf{X})(i,j) = \begin{cases} 0, & \mathbf{S}_{ij} = 0, \\ \mathbf{X}_{ij}, & \mathbf{S}_{ij} = 1. \end{cases} \tag{2}$$

and $f_{\mathbf{S}_\perp}(\mathbf{X})$ is its complementary projection, $i.e.$, $f_{\mathbf{S}}(\mathbf{X}) + f_{\mathbf{S}_\perp}(\mathbf{X}) = \mathbf{X}$.

Appearance Consistency and Spatial Compactness. We observe that the neighbouring pixels have high probability with similarity appearance, which has been successfully applied in foreground detection [26,27]. Based on this consideration, we encourage the appearance consistency by constructing adaptive weights $w^k_{ij,pq}$ into the spatial compactness constraint:

$$||\mathbf{C}^k\ vec(\mathbf{S}^k)||_1 = \sum_{(ij,kl)\in\varepsilon^k} w^k_{ij,pq}\ |\mathbf{S}^k_{ij} - \mathbf{S}^k_{pq}|;$$

$$w^k_{ij,pq} = \exp\frac{-||d^k_{ij} - d^k_{pq}||_2^2}{2\theta^2}. \tag{3}$$

where, $||\mathbf{X}||_1 = \sum_{ij}|\mathbf{X}_{ij}|$ denotes the l_1-norm, ε^k denotes the edge set connecting spatially neighboring pixels in the k-th modality. $\mathbf{C}^k$ is the node-edge incidence matrix denoting the connecting relationship among pixels in the k-th modality, d^k_{ij} and d^k_{pq} represent the intensity of pixel ij and pq in the k-th modality respectively and θ is a tunning parameter.

Cross-Modality Consistency. Different from the existing multispectral foreground detection methods that consider the information from individual modality are independent, we further propose to enforce the cross-modality consistency among the multispectral data. Meanwhile, to deal with occasional perturbation or malfunction of individual sources, we construct the cross-modality graph among the quad on one modality (thermal image) for each pixel from the other modality (grayscale image). This constraint is defined as:

$$\sum_{k=2,(ij,mn)\in\mathcal{F}}^{K} ||\mathbf{S}^k_{ij} - \mathbf{S}^{k-1}_{mn}||_F^2, \tag{4}$$

where $\mathcal{F}$ denotes edge set connecting spatially cross-modality pixels in the k-th modality (as shown in Fig. 2). Equation (4) encourages the pixel $\mathbf{S}^{k-1}_{ij}$ and its quad neighbors on the other modality $[\mathbf{S}^k_{ij}, \mathbf{S}^k_{(i+1)j}, \mathbf{S}^k_{(i+1)(j+1)}, \mathbf{S}^k_{i(j+1)}]$ belonging to the same pattern. Therefore, our model can be rewritten as:

$$\min_{\mathbf{B}^k,\mathbf{S}^k} \frac{1}{2}||f_{\mathbf{S}^k_\perp}(\mathbf{D}^k - \mathbf{B}^k)||_F^2 + \beta||vec(\mathbf{S}^k)||_0 + \mu||\mathbf{C}^k\ vec(\mathbf{S}^k)||_1$$

$$+ \gamma \sum_{k=2,(ij,mn)\in\mathcal{F}}^{K} ||\mathbf{S}^k_{ij} - \mathbf{S}^{k-1}_{mn}||_F^2,\ \ s.t.\ rank(\mathbf{B}^k) \leq r^k,\ k = 1,2,...,K, \tag{5}$$

Equation (5) is a NP-hard problem, to make Eq. (5) tractable, we relax the rank operator on $\mathbf{B}^k$ with the nuclear norm, which has proven to be an effective convex surrogate of the rank operator [22]. Meanwhile, we impose the low-rank constraints on the joint background matrix that concatenates all matrices of different modalities together to optimize them collaboratively. The formulation of collaborative low-rank representation model is proposed as follows:

Algorithm 1. Optimization Procedure to Eq. (6)

Require: $\mathbf{D}^k$, $(k = 1, ..., K)$.
    Set $\mathbf{B}^k = \mathbf{D}^k$ $(k = 1, 2, ..., K)$, $\mathbf{S}^k = \mathbf{0}$, $maxIter = 20$, $\epsilon = 1e - 4$.
Ensure: $\mathbf{B}^k$, $\mathbf{S}^k$, $(k = 1, 2, ..., K)$.
1: **for** $i = 1 : maxIter$ **do**
2:    Parallelly update $\mathbf{B}^k$ by Eq. (7);
3:    **if** $rank(\hat{\mathbf{B}}^k) \leq r^k$ **then**
4:        tuning parameters λ , return to step 2.
5:    **end if**
6:    Update $\{\mathbf{S}^k\}$ by Eq. (9);
7:    Check the convergence condition: if the maximum objective change between two consecutive iterations is less than ϵ, then terminate the loop.
8: **end for**

$$\min_{\mathbf{B},\mathbf{S}^k} \sum_{k=1}^{K} \frac{1}{2}||f_{\mathbf{S}^k_\perp}(\mathbf{D}^k - \mathbf{B}^k)||_F^2 + \beta||vec(\mathbf{S}^k)||_0 + \mu||\mathbf{C}^k \ vec(\mathbf{S}^k)||_1$$

$$+ \lambda||\mathbf{B}^k||_* + \gamma \sum_{k=2,(ij,mn)\in\mathcal{F}}^{K} ||\mathbf{S}^k_{ij} - \mathbf{S}^{k-1}_{mn}||_F^2, \tag{6}$$

where γ and λ are balance parameters, $|| \cdot ||_*$ denotes the nuclear norm of a matrix.

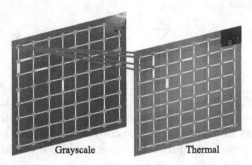

Fig. 2. The graph construction of cross-modality consistency, each pixels in grayscale image are connected to the corresponding four neighborhoods in thermal image.

2.2 Optimization

Equation (6) can be efficiently solved by the alternating optimization algorithm.

B–*subproblem.* Given an current estimate of the foreground mask $\hat{\mathbf{S}}^k$, estimating $\mathbf{B}^k$ by minimizing Eq. (6) turns to be the matrix completion problem:

$$\min_{\mathbf{B}^k} \sum_{k=1}^{K} \frac{1}{2}||f_{\hat{\mathbf{S}}^k_\perp}(\mathbf{D}^k - \mathbf{B}^k)||_F^2 + \lambda \ || \ \mathbf{B}^k \ ||_*, \tag{7}$$

This is to learn a low-rank background matrix from partial observations, which can be computed via SOFT-IMPUTE [19] by iteratively using Eq. (8):

$$\hat{\mathbf{B}}^k \longleftarrow \Theta_\lambda(P_{\hat{\mathbf{S}}_\perp}(\mathbf{D}^k) + P_{\hat{\mathbf{S}}^k}(\hat{\mathbf{B}}^k)), \tag{8}$$

S–subproblem. Given an current estimate of the background position matrix $\hat{\mathbf{B}}^k$, Eq. (6) can be transferred into following optimization function:

$$\min_{\mathbf{S}^k} \sum_{k=1}^{K} \frac{1}{2} \|f_{\mathbf{S}^k_\perp}(\mathbf{D}^k - \hat{\mathbf{B}}^k)\|_F^2 + \beta\|vec(\mathbf{S}^k)\|_0 + \mu\|\mathbf{C}^k \ vec(\mathbf{S}^k)\|_1$$
$$+ \gamma \sum_{k=2,(ij,mn)\in\mathcal{F}}^{K} \|\mathbf{S}^k_{ij} - \mathbf{S}^{k-1}_{mn}\|_F^2 \tag{9}$$

The energy function Eq. (9) can be rewritten in line with the standard form of a first-order Markov Random Fields [16] as:

$$\min_{\mathbf{S}^k} \frac{1}{2} \sum_{k=1}^{K} \sum_{ij} (\mathbf{D}^k_{ij} - \hat{\mathbf{B}}^k_{ij})^2 (1 - \mathbf{S}^k{}_{ij}) + \beta \sum_{ij} \mathbf{S}^k_{ij} + \mu\|\mathbf{C}^k \ vec(\mathbf{S}^k)\|_1$$
$$+ \gamma \sum_{k=2,(ij,mn)\in\mathcal{F}}^{K} \|\mathbf{S}^k_{ij} - \mathbf{S}^{k-1}_{mn}\|_F^2 = \min_{\mathbf{S}^k} \sum_{ij} [\beta - \frac{1}{2} \sum_{k=1}^{K} (\mathbf{D}^k_{ij} - \hat{\mathbf{B}}^k_{ij})^2] \tag{10}$$
$$+ \mu\|\mathbf{C}^k \ vec(\mathbf{S}^k)\|_1 + \gamma \sum_{k=2,(ij,mn)\in\mathcal{F}}^{K} \|\mathbf{S}^k_{ij} - \mathbf{S}^{k-1}_{mn}\|_F^2 + \mathscr{C}$$

where $\mathscr{C} = \frac{1}{2}\sum_{k=1}^{K}\sum_{ij}(\mathbf{D}^k_{ij} - \hat{\mathbf{B}}^k_{ij})^2$ is a constant with respect to $\mathbf{S}^k$. The Eq. (10) can be efficiently solved by graph cut algorithm [2,12].

A sub-optimal solution can be obtained by alternating optimization to $\{\mathbf{B}^k\}$, $\{\mathbf{S}^k\}$ as summarized in Algorithm 1. The convergence of our model can be guaranteed obviously, as each sub-problem converges to a optimal solution.

3   Experiments

We evaluate our method against the state-of-the-arts on the public challenging GTFD dataset [14]. It consists of 25 video sequence pairs with grayscale and thermal modalities captured from fifteen different scenes, including laboratory rooms, campus roads, playgrounds and water pools, etc. The main challenges include intermittent motion, low illumination, bad weather, intense shadow, dynamic scene, background clutter.

3.1 Parameters

There are five parameters in our method, we adjust one parameter while fixing other parameters and then obtain better performance for our approach. The parameter β controls the sparsity of the foreground masks. We typically set $\beta = 4.5\sigma^2$, where σ is estimated online by the mean variance of $\{\mathbf{D}^k - \mathbf{B}^k\}$. The parameter μ controls the spatial smoothness to punish the neighboring pixels with different labels. The parameter γ controls the cross-modality consistency of the foreground masks to promote the pixels with same label from different modality. The parameter r constrains the complexity of the background model. The parameter θ is the tunning parameter for the appearance consistency. The final parameters are empirically set as $\{\beta, \mu, \gamma, r, \theta\} = \{4.5\sigma^2, 0.5\beta, 0.5\beta, \sqrt{n}, 10\}$, where n is the total number of pixels.

3.2 Comparison Results

We compare our approach with some state-of-the-art foreground detection algorithms, including grayscale, thermal and grayscale-thermal detection methods. Following the protocols in [14, 28], we choose the detection result under grayscale scenarios as the final foreground.

Quantitative Results. Figure 3 demonstrates several detected results from GTFD dataset. From which we can see, the cross-modality consistent constraints can better preserve the foreground structures from both modalities, and achieve promising performance in both grayscale and thermal modalities even if there are misalignment among the image pairs. Furthermore, our method can produce more compact structured foregrounds.

Qualitative Results. Table 1 reports comparison results on precision, recall, F-measure together with the running speed on public GTFD dataset. We can conclude that: (1) Our method substantially outperforms other grayscale-thermal methods in precision, recall and F-measure, verifying the contribution of the proposed cross-modality consistent constraints. (2) Although WELD [14] and CLoD [28] achieve satisfying performance after fusing the grayscale and thermal results, but perform much worse in each single modality than ours. (3) The running speed of our method is lower than CLoD [28], but with much higher precision, recall and F-measure. Therefore, our method keeps a good balance between the efficiency and accuracy.

3.3 Component Analysis

To justify the component contributions of the proposed approach, we evaluate several variants of our model and report the results in Table 2, where Ours: the proposed model; Ours-I: our model without cross-modality consistency by setting γ to 0 in Eq. (6); Ours-II: our model without appearance consistency by setting adaptive weighting factor $w^k_{ij,kl}$ to 1 in Eq. (6); Ours-III: our model without spatial smoothness and appearance consistency by setting μ to 0.

Table 1. Average Precision (P), Recall (R), and F-measure (F) of our method against state-of-the-arts. The bold fonts of results indicate the best performance.

Algorithm	Grayscale			Thermal			Grayscale-Thermal			Code type	FPS
	P	R	F	P	R	F	P	R	F		
ASOM [18]	0.18	0.07	0.06	0.16	0.07	0.08	–	–	–	C++	111.11
FCFT [30]	0.39	0.20	0.22	0.25	0.22	0.20	–	–	–	C++	38.46
APKV [20]	0.38	0.42	0.36	0.42	0.20	0.24	–	–	–	Matlab, C++	0.03
ViBe [1]	0.41	0.49	0.41	0.41	0.47	0.39	–	–	–	C++	318.47
TTD [21]	0.59	0.29	0.32	0.58	0.38	0.40	–	–	–	Matlab	0.07
PCP [3]	0.28	0.18	0.21	0.49	0.40	0.43	–	–	–	Matlab	20.42
GMM [24]	0.48	0.65	0.52	0.48	0.65	0.50	–	–	–	C++	93.37
SAC [7]	0.42	0.74	0.41	0.47	0.71	0.53	–	–	–	Matlab	1.15
DECOLOR [32]	0.54	0.84	0.59	0.52	0.82	0.59	–	–	–	Matlab, C++	1.98
MAMR [29]	0.57	0.67	0.60	0.59	0.63	0.59	–	–	–	Matlab, C++	3.37
GMM-GT [24]	–	–	–	–	–	–	0.53	0.60	0.53	C++	34.04
JSC [8]	–	–	–	–	–	–	0.17	0.43	0.18	Matlab	10.21
WELD [14]	0.58	0.80	0.64	0.50	0.63	0.50	**0.64**	0.81	0.67	Matlab, C++	2.43
CLoD [28]	0.53	0.71	0.55	**0.63**	0.62	0.57	0.62	0.80	0.66	Matlab, C++	45.66
Ours	**0.66**	**0.86**	**0.71**	0.65	**0.85**	**0.70**	0.66	**0.86**	**0.71**	Matlab,C++	3.51

Table 2. Average Precision (P), Recall (R), and F-measure (F) of our method and its variants. The bold fonts of results indicate the best performance.

Algorithm	Grayscale			Thermal		
	P	R	F	P	R	F
Ours	**0.66**	**0.86**	**0.71**	**0.65**	**0.85**	**0.70**
Ours-I	0.59	0.73	0.60	0.60	0.70	0.61
Ours-II	0.64	**0.86**	0.70	0.63	**0.85**	0.69
Ours-III	0.62	0.85	0.68	0.61	**0.85**	0.67

The evaluation results demonstrate that: (1) Each component plays important roles in our model. (2) The cross-modality consistency contributes most by comparing Ours-I to Ours, which consequentially verifies the significance of the proposed model. Note that the higher recall in Ours-II and Ours-III results from the coarse boundary of the detected foregrounds.

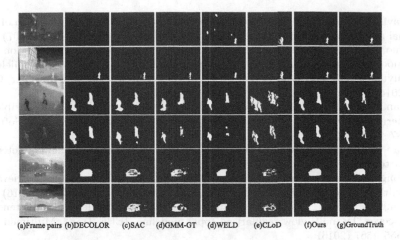

(a)Frame pairs (b)DECOLOR   (c)SAC   (d)GMM-GT   (d)WELD   (e)CLoD   (f)Ours   (g)GroundTruth

Fig. 3. Sample results of our method against other methods. The odd rows indicate the grayscale frames and the corresponding detection results generated by grayscale methods, and the even rows denote the thermal frames and the corresponding detection results generated by thermal methods.

4   Conclusion

In this paper, we have proposed novel multispectral foreground detection approach by exploring the cross-modality consistency in the low-rank and sparse decomposition framework. Extensive experiments on the GTFD dataset suggest that our approach achieved superior performance against other state-of-the-art approaches. In future work, we will develop prior models on foreground or background into our framework to further improve the robustness, and extend our algorithm into a streaming or an online fashion.

Acknowledgment. This work was partially supported by the National Natural Science Foundation of China (61502006, 61702002, 61472002 and 61671018) and the Natural Science Foundation of Anhui Higher Education Institutions of China (KJ2017A017).

References

1. Barnich, O., Droogenbroeck, M.V.: ViBe: a universal background subtraction algorithm for video sequences. IEEE Trans. Image Process. **20**, 1709–1724 (2011)
2. Boykov, Y., Veksler, O., Zabih, R.: Fast approximate energy minimization via graph cuts. IEEE Trans. Pattern Anal. Mach. Intell. **23**, 1222–1239 (2001)
3. Candes, E., Li, X., Ma, Y., Wright, J.: Robust principal component analysis? J. ACM **58**, 175–181 (2011)
4. Chen, Y., Wang, J., Zhu, B., Tang, M., Lu, H.: Pixel-wise deep sequence learning for moving object detection. IEEE Trans. Circuits Syst. Video Technol., 1 (2017)
5. Cho, S.H., Kang, H.B.: Abnormal behavior detection using hybrid agents in crowded scenes. Pattern Recogn. Lett. **44**, 64–70 (2014)

6. Davis, J., Sharma, V.: Background-subtraction using contour-based fusion of thermal and visible imagery. Comput. Vis. Image Underst. **106**, 162–182 (2007)
7. Guo, X., Wang, X., Yang, L., Cao, X., Ma, Y.: Robust foreground detection using smoothness and arbitrariness constraints. In: Fleet, D., Pajdla, T., Schiele, B., Tuytelaars, T. (eds.) ECCV 2014. LNCS, vol. 8695, pp. 535–550. Springer, Cham (2014). https://doi.org/10.1007/978-3-319-10584-0_35
8. Han, G., Cai, X., Wang, J.: Object detection based on combination of visible and thermal videos using a joint sample consensus background model. J. Softw. **8**, 987–994 (2013)
9. Han, J., Bhanu, B.: Fusion of color and infrared video for moving human detection. Pattern Recogn. **40**, 1771–1784 (2007)
10. Hong, R., Hu, Z., Wang, R., Wang, M., Tao, D.: Multi-view object retrieval via multi-scale topic models. IEEE Trans. Image Process. **25**, 5814–5827 (2016)
11. Hong, R., Zhang, L., Zhang, C., Zimmermann, R.: Flickr circles: aesthetic tendency discovery by multi-view regularized topic modeling. IEEE Trans. Multimedia **18**, 1555–1567 (2016)
12. Kolmogorov, V., Zabin, R.: What energy functions can be minimized via graph cuts? IEEE Trans. Pattern Anal. Mach. Intell. **2**, 147–159 (2004)
13. Li, C., Lin, L., Zuo, W., Wang, W., Tang, J.: An approach to streaming video segmentation with sub-optimal low-rank decomposition. IEEE Trans. Image Process. **25**, 1947–1960 (2016)
14. Li, C., Wang, X., Zhang, L., Tang, J., Wu, H., Lin, L.: Weighted low-rank decomposition for robust grayscale-thermal foreground detection. IEEE Trans. Circuits Syst. Video Technol. **27**, 725–738 (2017)
15. Li, C., Wu, X., Bao, Z., Tang, J.: ReGLe: spatially regularized graph learning for visual tracking. In: ACM, pp. 252–260 (2017)
16. Li, S.Z.: Markov Random Field Modeling in Image Analysis. Springer, London (2009). https://doi.org/10.1007/978-1-84800-279-1
17. Lim, L.A., Keles, H.Y.: Foreground segmentation using a triplet convolutional neural network for multiscale feature encoding. arXiv preprint arXiv:1801.02225 (2018)
18. Lucia, M., Alfredo, P.: A self-organizing approach to background subtraction for visual surveillance applications. IEEE Trans. Image Process. **17**, 1168–1177 (2008)
19. Mazumder, R., Hastie, T., Tibshirani, R.: Spectral regularization algorithms for learning large incomplete matrices. J. Mach. Learn. Res. **11**, 2287–2322 (2010)
20. Narayana, M., Hanson, A., Learned-miller, E.: Background modeling using adaptive pixelwise kernel variances in a hybrid feature space. In: Proceedings of the IEEE Conference on Computer Vision and Pattern Recognition (2012)
21. Oreifej, O., Li, X., Shah, M.: Simultaneous video stabilization and moving object detection in turbulence. IEEE Trans. Pattern Anal. Mach. Intell. **35**, 450–462 (2013)
22. Recht, B., Fazel, M., Parrilo, P.A.: Guaranteed minimum-rank solutions of linear matrix equations via nuclear norm minimization. SIAM Rev. **52**, 471–501 (2010)
23. St-Charles, P.L., Bilodeau, G.A., Bergevin, R.: Subsense: a universal change detection method with local adaptive sensitivity. IEEE Trans. Image Process. **24**, 359–373 (2014)
24. Stauffer, C., Grimson, W.E.L.: Adaptive background mixture models for real-time tracking. In: 1999 Proceedings of the IEEE International Conference on Computer Vision, pp. 246–252 (1999)
25. Wang, X.: Intelligent multi-camera video surveillance: a review. Pattern Recogn. Lett. **34**, 3–19 (2013)

26. Xin, B., Tian, Y., Wang, Y., Gao, W.: Background subtraction via generalized fused lasso foreground modeling. arXiv preprint, pp. 4676–4684 (2015)
27. Xu, M., Li, C., Shi, H., Tang, J., Zheng, A.: Moving object detection via integrating spatial compactness and appearance consistency in the low-rank representation. In: Yang, J., et al. (eds.) CCCV 2017. CCIS, vol. 773, pp. 50–60. Springer, Singapore (2017). https://doi.org/10.1007/978-981-10-7305-2_5
28. Yang, S., Luo, B., Li, C., Wang, G., Tang, J.: Fast grayscale-thermal foreground detection with collaborative low-rank decomposition. IEEE Trans. Circuits Syst. Video Technol., 1 (2017)
29. Ye, X., Yang, J., Sun, X., Li, K., Hou, C., Wang, Y.: Foreground-background separation from video clips via motion-assisted matrix restoration. IEEE Trans. Circuits Syst. Video Technol. **25**, 1721–1734 (2015)
30. Zhang, H., Xu, D.: Fusing color and texture features for background model. In: Wang, L., Jiao, L., Shi, G., Li, X., Liu, J. (eds.) FSKD 2006. LNCS (LNAI), vol. 4223, pp. 887–893. Springer, Heidelberg (2006). https://doi.org/10.1007/11881599_110
31. Zhao, B., Li, Z., Liu, M., Cao, W., Liu, H.: Infrared and visible imagery fusion based on region saliency detection for 24-hours-surveillance systems. In: Proceeding of the IEEE International Conference on Robotics and Biomimetics (2013)
32. Zhou, X., Yang, C., Yu, W.: Moving object detection by detecting contiguous outliers in the low-rank representation. IEEE Trans. Pattern Anal. Mach. Intell. **35**, 597–610 (2013)

Leveraging User Personality
and Tag Information for One Class
Collaborative Filtering

Jianshan Sun$^{(\boxtimes)}$, Deyuan Ren, and Dong Xu

School of Management, Hefei University of Technology,
Hefei 230009, Anhui, China
sunjs9413@hfut.edu.cn

Abstract. Recommender systems have been great tools for electronic market companies to satisfy customer. A large volume of information is generated by online users and how to appropriately provide personalized content is becoming more challenging. Traditional recommendation models are overly dependent on preference ratings and often suffer from the problem of "data sparsity". The one class collaborative filtering (OCCF) method is more applicable in the electronic market scenario yet it is insufficient for item recommendation. In this study, we develop a novel personality-tag-aware item recommendation framework, referred to as PT_OCCF, in order to tackle the above challenges. We leverage user personality and tag information and OCCF models to improve recommendation performance. We conduct comprehensive experiments on a public dataset to verify the effectiveness of the proposed framework and methods. The results show that the proposed methods are effective in improving the performance of the baseline OCCF methods.

Keywords: Electronic market · Recommender system · Personality
Tag · OCCF

1 Introduction

In the OCCF setup, there are only limited data given by users as positive examples, which lead to data sparse and class unbalance. To address this issue, some researchers have exploited rich user generated content as supplementary resources to build models [1,2]. Recently, researchers have paid attention to personality traits to improve personalized services [3]. Previous research indicated that a user's personality traits had a substantial influence on preferences [4,5]. People with similar personality characteristics are more likely to have similar

This work is supported by the National Natural Science Foundation of China (71501057), the Fundamental Research Funds for the Central Universities (JZ2017HGTB0185) and Anhui Provincial Natural Science Foundation (1608085QG166).

© Springer Nature Switzerland AG 2018
R. Hong et al. (Eds.): PCM 2018, LNCS 11164, pp. 830–840, 2018.
https://doi.org/10.1007/978-3-030-00776-8_76

interests and preferences. The majority of these studies only combined personality information into the basic neighborhood-based collaborative filtering models. Advanced one-class collaborative filtering models were not explored. Therefore, the exploitation of user personality and tag information in a recommender system is limited and more extensive experimental work is required.

In this paper, we develop a personality-tag-aware one-class collaborative filtering recommendation framework, referred to as PT_OCCF, to tackle the key challenges highlighted above. The rest of the paper is organized as follows. In Sect. 2, we review related work on personality, tag and recommender systems. The details of the personality-tag-aware recommendation framework are introduced in Sect. 3, while Sect. 4 presents the design and methodology used in the experiments. The results are analyzed in Sect. 5. We conclude the work in Sect. 6.

2 Related Work

In the OCCF scene, there are large parts of unlabelled examples and we only have positive examples, which provoke an extremely sparse data in the system. Previous research on OCCF considered the unlabelled examples as starting point and they focused on distinguishing positive and negative examples from unlabelled examples. The traditional method to solve these unlabelled examples is treating all the missing examples as negative examples.

Personality is a measurable and quite useful factor that informs human behaviors. The big five factor model (FFM) [6] is widely used to build user personality profiles. The FFM model defined independent five dimensions (factors or traits) to describe an individual's personality: openness to experience (OPE), conscientiousness (CON), extraversion (EXT), agreeableness (AGR) and neuroticism (NEU) [7]. Previous studies have revealed that people who have similar personality factors are likely to have similar interests [11], and demonstrated that the user personality has close links to user behaviour and preference. In this paper, we will mine this relationships and leverage personality information to design recommender system. Tag information also is one of main research directions to enhance recommendations. Tag information implies the user's interest and preference. When introduce tag information, researchers usually build distinctive three dimensional relationships among users, tags and items and improve the relationships [12,13]. Many tag-based collaborative filtering methods have been applied to practical scenarios [14,15]. In addition, in recent years, the aesthetic tendency can also serve as user information [8], and can be found from user's photo information [9,10]. The aesthetic tendency is also related to user's personality.

In the case of electronic market websites, it is typical that only positive implicit feedback is available. For a user preference matrix, the portion of unlabeled examples is large and the matrix becomes extremely sparse. Due to this limitation, Hu and Pu [16] showed that the personality could help overcoming the cold start problem. Some personality-based recommendation methods employing the FFM was proposed in [3,17]. Previous research only employed basic OCCF models (UCF or ICF methods) to fuse personality information. In our work, we will place a heavier reliance on advanced OCCF models.

3   Personality-Tag-Aware One-Class Collaborative Filtering Recommender System

In this paper we propose an one-class collaborative filtering methods to provide recommendations in electronic markets. The principle in our methods is to incorporate personality and tag information into OCCF models appropriately. We define the problem of item recommendation and then proceed to introduce representative models. In this paper, we are interested in the recommendation of electronic market items to online users by leveraging user personality and tag information. For the sake of illustration, we introduce the key notations. For a given user u, the task of item recommendation is to provide a sorted list of top N items, L, based on the available user-item preference matrix R and user-personality matrix P. U and I respectively represent the set of users and items. p is the set of user personality, $p = (ope, con, ext, agr, neu)$ and $p_{\{ope,con,ext,agr,neu\}} \in [1,5]$. u_i and i_j respectively represent the index of user and item. tag^U and tag^I respectively represent the set of user tagged tags and item's tags. r_{ui} is the rating of user u on item i. $\left(R_{ij}\right)_{|U| \times |I|}$ is user-item preference matrix, $R_{ij} \in \{0,1\}$. $\left(P_{ij}\right)_{|U| \times |p|}$ is user-personality matrix.

We first introduce baseline OCCF models which employ user preference feedback (user-item matrix such as purchase log or listening logs) and then present personality-tag-aware OCCF models which combine personality and tag information with preference feedback.

3.1   Baseline OCCF Recommendation Models

In memory-based CF methods, the recommendation score is computed based on the entire the user-item preference matrix. The similarity computation between users or items is the main task of these methods. Neighborhood-based CF is one of memory-based CF models for recommender systems. The method typically includes the following steps: the similarity computation between pair of users or items, then finding the k most similar users or items and assembling the k's neighbour ratings to generate top N items. The user-based CF (UCF) is one popular model of neighbourhood-based CF models. The UCF can be formally defined as Eq. (1), which provides recommendations by generating the most similar neighbours for a given user then aggregating their rating.

$$P(u,i) = \bar{r}_u + \frac{\sum_{v \in Nr_u} sw(u,v)(r_{vi} - \bar{r}_v)}{\sum_{v \in Nr_u} sw(u,v)} \tag{1}$$

where $P(u,i)$ is the predicted rating of the target user u to an item i which her never rated, r_{vi} is the rating of user v for item i, $\bar{r}_u$ and $\bar{r}_v$ are the mean ratings of u and v, $Nr_u \in U$ is the set of the nearest neighbours of the target user u and $sw(u, v)$ is the similarity weight between user u and user v. The similarity weight $sw(u, v)$ can be calculated by a function such as Cosine similarity Eq. (2) or Pearson similarity Eq. (3).

$$sw(u, v) = \frac{\sum_{i=1}^{|I|} r_{ui} * r_{vi}}{\sqrt{\sum_{i=1}^{|I|} r_{ui}^2 \sum_{i=1}^{|I|} r_{vi}^2}} \tag{2}$$

$$swp(u, v) = \frac{N \sum r_{ui} r_{vi} - \sum r_{ui} r_{vi}}{\sqrt{N \sum r_{ui}^2 - (\sum r_{ui})^2} \sqrt{N \sum r_{vi}^2 - (\sum r_{vi})^2}} \tag{3}$$

Therefore, we have chosen the UCF model as our baseline model in the memory-based methods. In model-based CF methods, recommendation system attempt to learn more complex patterns based on the training data (for example, preference matrix) and provide recommendations based on the learned models. Matrix factorization (MF) techniques are typical model-based CF methods and have been widely and successfully applied in movie recommendation context. Many studies have empirically suggested that MF methods perform better than neighbourhood-based methods in rating predication tasks. The idea behind a MF-based method is to predict the rating r_{ui} by the learned latent user factors $x_u \in R^f$ and latent item factors $y_i \in R^f$. In order to learn these latent factors, recent studies have suggested direct modelling on the observed ratings and use of the Tikhonov regularization to avoid over-fitting problems. Therefore, the learning process is equivalent to solving the unconstrained optimization problem:

$$Arg \sum_{u,i} \left(\left\| r_{ui} - x_u^T y_i \right\|^2 + \lambda \left(\left\| x_u \right\|^2 - \left\| y_i \right\|^2 \right) \right). \tag{4}$$

Where $\|\sim\|$ is the of a matrix and λ is used to regularize the model to avoid over-fitting. This model has already demonstrated its prediction excellence for the Netflix datasets. In the OCCF setting, the above model treat all missing values in R as negative examples (AMAN) where the response variables only take the value 0 or 1. Hereafter, we call this model AMAN_MF. However, this strategy is obviously limited since it does not cover the real world case that missing examples are likely to be positive ones. A better approach proposed by Pan et al. (Pan et al. 2008) treat all missing values as weighted negative examples (wAMAN). The weights are used to control their relative contribution to the loss function:

$$Arg \sum_{u,i} w_{ui} \left(\left\| r_{ui} - x_u^T y_i \right\|^2 + \lambda \left(\left\| x_u \right\|^2 - \left\| y_i \right\|^2 \right) \right). \tag{5}$$

Minimization of the function (5) allows the parameters to be learnt using Alternating Least Squares (ALS). The ALS algorithm is an effective and efficient iterative method to solve the optimization problem. In the case of the wAMAN model, a weighted ALS (wALS) was proposed in reference [18] and used to solve the above low rank approximation problem mentioned above. In this paper, we decide to use the AMAN_MF and wAMAN_MF methods as our baselines in the model-based methods. In the case of the wAMAN_MF method, we use the global weighting method to assign the weights. In Sect. 3.3, we will incorporate user personality and tag information into the wAMAN method to enhance the basic models.

3.2 Calculate Probability of User's Preference

In the OOCF framework for recommend system, abundant user information and product content information can alleviate the problem of data sparsity in the user-item preference matrix R and find the positive or negative examples from the missing data.

For items, their tags indicate some attributes of items, showing essential characteristics of them. User usually likes a class of items, which have similar attributes or tags. The probability of user's preference to item can be measured with the intersection of user's tags and item's tags. Meanwhile, users with similar personalities have similar preferences. The probability also can be calculated by users' similarity and the rating for user to item. In this work, we will use methods to get the probability of user's preference to item, using tags and personality information respectively.

As to user's tags, we need to build the user tags library firstly. In our model, user's tags are affected by the following two dimensions: (1) User's tag of implicit preference. In the music social website named LastFm, users tend to listen to a series of songs of similar styles. These songs may have the same wind or come from similar singers, even though songs don't have the obvious classifications in the user interface. Indeed, these songs has tags in backstage database. These tags are called user's tag of implicit preference. (2) User's tag of explicit preference. When a user has heard a song, the user labeled the song according to his own cognition. These tags is described as user's tag of explicit preference.

In Fig. 1, it shows the process of building a user's preference tags based on user's implicit and explicit preference. First, it need to find the set I^n of all the items of the user n through the user preference matrix. Then, depending on the item tag set tag^i, the user n's all tags of implicit preference $tag_n^{implicit}$ will be matched, as well as the number of each coincident tag. At last, the user tags library tag_n^{pre} will be obtained according to the explicit preference tags $tag_n^{explicit}$ labeled by the user n and the implicit preference tags $tag_n^{implicit}$ of the user. The user tags library about the preference of items tag_n^{pre} is defined as:

$$tag_n^{pre} = tag_n^{implicit} \bigcap tag_n^{explicit}. \tag{6}$$

The user tags library explains that user's preference for items in a specific domain is affected by certain attributes of the item. Now, The probability $pre_{u,i}^{tag}$ of user u's preference to the item i will be calculated based on user tags library. $pre_{u,i}^{tag}$ can be written as:

$$pre_{u,i}^{tag} = \sum_{k=1}^{K} \frac{count\left(t_k^{ui}\right)}{sum\left(tag^u\right)}. \tag{7}$$

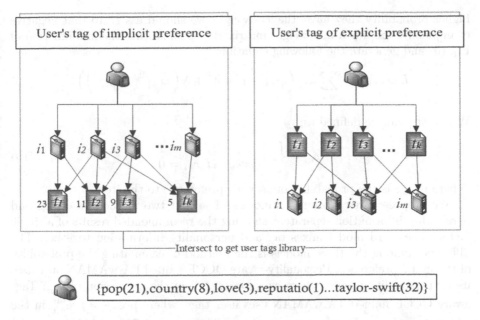

Fig. 1. The build progress of user tags library.

Where, $count\left(t_k^{ui}\right)$ represents the number of tags, matching the k-th tag of the item i, in the user u's tags library. $sum\left(tag^u\right)$ is defined as the total number of tags in the user u's tags library.

Relative to the tags, the process of using personality information is relatively simple. The probability of user u's preference to the item i, $pre_{u,i}^{sim}$ is calculated by personality information as follows:

$$pre_{u,i}^{sim} = \sum_{q=1}^{Q} sim\left(u, u_q\right) * r_{u_q i}. \tag{8}$$

Where $sim\left(u, u_q\right)$ is computed by a function such as Cosine similarity Eq. (2) or Pearson similarity Eq. (3) via personality information. u_q represents the user q of user u's nearest neighbours. $r_{u_q i}$ is the rating for user q to the item i.

3.3 Personality-Tag-Aware OCCF Recommendation Models

For the MF framework, AMAN_MF and wAMAN_MF are two baseline methods for item recommendation. We leverage user personality and tag information in order to enhance the wAMAN_MF method using a different weighting scheme. We also consider missing examples as negative examples, we assign the weight for each negative example by leveraging user related information instead of using a global weighting scheme, a better way for assigning the weight for each negative example is to look at the probability of the user's preference to the item: the higher probability they have, the less weight we should assign to that negative example. We employ a new weight matrix W' to substitute the loss function of Eq. (5) and generate the following equation:

$$L(x_u, y_i) = \sum_{u,i} w'_{ui} \left(\left\| r_{ui} - x_u^T y_i \right\|^2 + \lambda \left(\|x_u\|^2 - \|y_i\|^2 \right) \right). \tag{9}$$

Where w'_{ui} can be defined as:

$$w'_{ui} = \begin{cases} 1 & if\ r_{ui} = 1 \\ 1 - pre_{u,i} & if\ r_{ui} = 0 \end{cases} \tag{10}$$

Where $pre_{u,i}$ is the probability of user u's preference to the item i.

We propose three recommend models. The first two methods use tags and personality information separately, showing the recommended results of a single factor. The third model mixes tag and personality information together. The difference among the three models is the method of calculating the probability of the user's preference. Personality-aware OCCF named PA_wAMAN only uses user's personality profile, where $pre_{u,i} = pre_{u,i}^{sim}$ in the expression of w'_{ui}. Tag-aware OCCF named TA_wAMAN uses user tags, where $pre_{u,i} = pre_{u,i}^{tag}$ in the expression of w'_{ui}. Personality-tag-aware OCCF named PT_wAMAN uses the user personality profile and tags, where $pre_{u,i} = \delta pre_{u,i}^{tag} + (1 - \delta) pre_{u,i}^{sim}$ in the expression of w'_{ui}. In addition, the loss function can be optimized by the wALS algorithm [18].

4   Experimental Design

4.1   Dataset and Evaluation Metrics

In order to evaluate these proposed recommendation frameworks, we investigate the Mypersonality dataset. MyPersonality is a Facebook application which allows users to take real psychometric tests and records their psychological and Facebook profiles. Due to the size and complexity of the database, in this paper we perform experimental tests on the Last.fm music listening data kindly provided Liam McNamara in the Mypersonality project [19]. From this data, we extracted 1618 Last.fm users with related information about personality tests and listening histories. The dataset consists of 1107468 records from 1618 users on 38952 artists. The number of tags is 13646. Data sparsity is 98.23%. Each user has a personality record which is represented by a five-dimension personality vectors (ope, con, ext, arg, neu). The value of dimension is from 0 to 5. The distribution of user Big-Five personality score is that most users have high scores in the dimension of "openness to experience", middle scores in the dimensions of "conscientiousness, extraversion, and agreeableness", and low scores in the dimension of "neuroticism".

In the case of item recommendation, we treat it as a content retrieval system that recommends electronic market items to online users. The evaluation metrics, Precision@K (Prec@K), Mean Average Precision (MAP), Mean Reciprocal

Rank (MRR) (Manning, Raghavan 2008), Recall@K and Normalized Discounted Cumulative Gain (NDCG@L) are employed to evaluate the recommendation accuracy of different methods. NDCG@L measure the ability for guaranteeing highly relevant items appearing earlier in result list. The Prec@K measure only evaluates the ability to return overall relevant items. However, the MAP and MRR measures consider the rank information of relevant items in the recommendation list.

4.2 Experimental Procedure

To minimize the influence of variations in the training set, 5-fold cross validation is performed. For each user, the preference records are partitioned into five subsets with similar sizes and distributions. The union of 4 subsets is then used as the training set while the remaining subset is used as the test set, which is repeated five times such that every subset has been used as the test set once. The average test result is regarded as the result of the 5-fold cross validation. The above process is repeated 5 times with random partitions of the 5 subsets, and the average result of these different partitions is recorded. For each user, one record is randomly selected as training candidate and the remaining records are considered testing candidates.

We implement the baseline OCCF models in the literature and proposed methods for the performance comparison. They are listed as follows: Refer to Eq. (3), UCF focus on users' similarities computation from the implicit feedback matrix. Refer to Eq. (4), AMAN focus on the matrix factorization technique for the implicit feedback matrix. Refer to Eq. (5), wAMAN focus on the proposed weighted matrix factorization technique for the implicit feedback matrix. PA_wAMAN focus on user's personality information. TA_wAMAN focus on user's tag information. PT_wAMAN we proposed focus on user's similarities computation from both user tag and personality information.

5 Experiment Results

In this section, we present the results of the six recommendation models utilized on the Mypersonality dataset (Table 1). Overall, we see that in all cases personality-tag-aware methods significantly outperforms baseline methods (Wilcoxon signed-rank test, $p < 0.05$). It can be observed that personality and tag information is highly beneficial and it is good supplement to the preference information.

In the case of the Neighborhood-based methods, it can be observed that the result of UCF is worst without any extra information. For MF-based methods, AWAN_MF performs worst since treating all missing examples as negative examples without weighting does not work well in the case of the highly sparse OCCF problem. In our work, there is little difference between PA_wAMAN and TA_wAMAN. Meanwhile, personality-aware method and tag-aware method are a little worse than personality-tag-aware methods. As additional information, personality information and tag information have same effect in reducing data sparsity.

Table 1. Performance comparison of models on LastFm dataset

Methods	Prec@5	Prec@10	Recall@5	Recall@10	MAP	MRR	NDCG
UCF	0.1059	0.1061	0.0108	0.0178	0.0493	0.1671	0.4471
AMAN	0.2094	0.1986	0.0171	0.0312	0.0775	0.2312	0.5085
wAMAN	0.332	0.2971	0.0229	0.039	0.0971	0.2823	0.5325
TA_wAMAN	0.4487	0.4047	0.0267	0.0438	0.1231	0.286	0.5562
PA_wAMAN	0.4533	0.4044	0.027	0.0463	0.1239	0.2886	0.5573
PT_wAMAN	0.4638	0.4189	0.0279	0.047	0.1312	0.2924	0.5639

Furthermore, the MAP improvements (Table 2) for PT-enhanced MF method (PT_wAMAN) over other models are obvious (35.11% over wAMAN, 6.58% over TA_wAMAN, 5.89% over PA_wAMAN). The personality and tag information demonstrates its effectiveness as additive evidence to the baseline model. Using personality and tag to determine the weights of the wAMAN_MF model, our proposed PT_wAMAN model is able to get the best results.

Table 2. Performance comparison of models

Methods	Prec@5	Prec@10	MAP	MAP improvement		
				wAMAN	TA_wAMAN	PA_wAMAN
wAMAN	0.332	0.2971	0.0971	-	-	-
TA_wAMAN	0.4487	0.4047	0.1231	26.77%	-	-
PA_wAMAN	0.4533	0.4044	0.1239	27.61%	-	-
PT_wAMAN	0.4638	0.4189	0.1312	35.11%	6.58%	5.89%

6   Conclusion

In this paper, we proposed a personality-tag-aware recommender system for user in electronic market situation or social media websites. The proposed method leveraged personality information for users to improve OCCF performance. Comprehensive experiments were conducted on a public dataset in order to evaluate the effectiveness of the proposed models. After careful analysis of the results, we have arrived at the following conclusions: (1) Personality and tag information are very effective to overcome the sparsity of OCCF. (2) Neighborhood-based models work robustly when the data matrix is very sparse, while MF-based models perform better when the data matrix is less sparse. (3) Personality-tag-aware OCCF models have better performance than the model with single external information.

References

1. Zhang, Z.-K., Zhou, T., Zhang, Y.-C.: Tag-aware recommender systems: a state-of-the-art survey. J. Comput. Sci. Technol. **26**(5), 767 (2011)
2. Li, Y., Zhai, C.X., Chen, Y.: Exploiting rich user information for one-class collaborative filtering. Knowl. Inf. Syst. **38**(2), 277–301 (2014)
3. Nunes, M.A.S.N., Hu, R.: Personality-based recommender systems: an overview. In: ACM Conference on Recommender Systems, pp. 5–6. ACM, Dublin (2012)
4. Kosinski, M., Bachrach, Y., Kohli, P., Stillwell, D., Graepel, T.: Manifestations of user personality in website choice and behaviour on online social networks. Mach. Learn. **95**(3), 357–380 (2014)
5. Greenberg, D.M., Baron-Cohen, S., Stillwell, D.J., Kosinski, M., Rentfrow, P.J.: Musical preferences are linked to cognitive styles. PloS one **10**(7), e0131151 (2015)
6. Costa, P.T., McCrae, R.R.: The revised neo personality inventory (neo-pi-r). In: The SAGE Handbook of Personality Theory and Assessment, vol. 2(2), pp. 179–198 (2008)
7. Fernández-Tobías, I., Braunhofer, M., Elahi, M., Ricci, F., Cantador, I.: Alleviating the new user problem in collaborative filtering by exploiting personality information. User Model. User-Adapt. Interact. **26**(2–3), 221–255 (2016)
8. Hong, R., Zhang, L., Zhang, C., Zimmermann, R.: Flickr circles: aesthetic tendency discovery by multi-view regularized topic modeling. IEEE Trans. Multimed. **18**(8), 1555–1567 (2016)
9. Hong, R., Zhang, L., Tao, D.: Unified photo enhancement by discovering aesthetic communities from Flickr. IEEE Trans. Image Process. **25**(3), 1124–1135 (2016)
10. Hong, R., Hu, Z., Wang, R., Wang, M., Tao, D.: Multi-view object retrieval via multi-scale topic models. IEEE Trans. Image Process. **25**(12), 5814–5827 (2016)
11. Cantador, I., Fernandez-Tobias, I., Bellogin, A., Kosinski, M., Stillwell, D.J.: Relating personality types with user preferences multiple entertainment domains. In: The Workshop on Emotions & Personality in Personalized Services. UMAP, Aalborg (2013)
12. Liang, H., Xu, Y., Li, Y., Nayak, R.: Collaborative filtering recommender systems using tag information. In: IEEE/WIC/ACM International Conference on Web Intelligence and Intelligent Agent Technology, vol. 03, pp. 59–62. IEEE Computer Society, Sydney (2008)
13. Tso-Sutter, K.H.L., Marinho, L.B., Schmidt-Thieme, L.: Tag-aware recommender systems by fusion of collaborative filtering algorithms. In: ACM Symposium on Applied Computing, pp. 1995–1999. ACM, Vila Galé (2008)
14. Zhen, Y., Li, W.J., Yeung, D.Y.: TagiCoFi: tag informed collaborative filtering. In: ACM Conference on Recommender Systems, pp. 69–76. ACM, New York (2009)
15. Reyn, N., Shinsuke, N., Jun, M., Shunsuke, U.: Tag-based contextual collaborative filtering. IAENG Int. J. Comput. Sci. **34**(2), 214–219 (2007)
16. Hu, R., Pu, P.: Enhancing collaborative filtering systems with personality information. In. Proceedings of the Fifth ACM conference on Recommender systems, pp. 197–204. ACM, Chicago (2011)
17. Tkalcic, M., Kunaver, M., Tasic, J., Košir, A.: Personality based user similarity measure for a collaborative recommender system. In: Proceedings of the 5th Workshop on Emotion in Human-Computer Interaction-Real world challenges, pp. 30–37. Fraunhofer, Cambridge (2009)

18. Pan, R., Zhou, Y., Cao, B., Liu, N.N., Lukose, R., Scholz, M., et al.: One-class collaborative filtering. In: Eighth IEEE International Conference on Data Mining, pp. 502–511. IEEE Computer Society, Pisa (2008)
19. Mcnamara, L., Mascolo, C., Capra, L.: Media sharing based on colocation prediction in urban transport. In: International Conference on Mobile Computing and Networking, pp. 58–69. ACM SIGMOBILE, San Francisco (2008)

Enhanced Linear Discriminant Canonical Correlation Analysis for Cross-modal Fusion Recognition

Chengnian Yu, Huabin Wang(✉) ⓘ, Xin Liu, and Liang Tao

School of Computer Science and Technology, Anhui University,
230601 Hefei, China
wanghuabin@ahu.edu.cn

Abstract. Based on discriminant canonical correlation analysis of LDA, a new method of multimodal information analysis and fusion is proposed in this paper. We process data from two perspectives, single modality and cross-modal. More specifically, firstly, LDA is utilised to obtain the best projection matrix, this way, the data in each within-modal can be as centralized as possible. Secondly, the improved DCCA is used to process the output of first step in order to maximize within-class correlation and minimize between-class correlation. The above two steps prove beneficial to obtain the feature with higher discriminating ability which is essential for the average fusion recognition accuracy improvement. We show state-of-art results or better than state-of-art on widely used USM benchmarks against all existing results include CCA, LDA, DCCA, GCCA and KCCA.

Keywords: Feature level fusion · Multimodal analysis
Canonical correlation analysis · Linear discriminant analysis

1 Introduction

Single modality biometric identification technology, such as face recognition [1], the gesture recognition [2], the iris recognition [3], has plagued researchers with a variety of formidable challenges over the years. However, multimodal fusion recognition [4, 5] is a technology that combines multiple biological features and uses fusion algorithms to recognition. Most recently, the concept has attracted increasing attention for feature set matching in [6, 7]. Wang et al. [8] proposed dynamic fusion methods that corresponding weights were assigned to each modality. It has strong robustness. Liu et al. [9] proposed an adaptive multi-feature fusion algorithm. This refer to as infrared object tracking method based on adaptive multi-feature fusion and Mean Shift (MS), which can achieve target tracking in complex scene. Liu et al. [10] proposed a new learning framework for projection dictionary was established to solve the problem of weak

Huabin Wang: The research work is supported by the National Natural Science Foundation of China (grant no.61372137).

R. Hong et al. (Eds.): PCM 2018, LNCS 11164, pp. 841–853, 2018.
https://doi.org/10.1007/978-3-030-00776-8_77

matching and multimodal information fusion. Zeng et al. [11], from the viewpoint of fusing appearance statistical features, proposed human target recognition algorithm based on appearance statistics feature fusion.

According to the fusion of different information, the multimodal biometric fusion technology can be divided into sensor level fusion [12], feature level fusion [13], matching-score level fusion [14], and decision level fusion [15]. Among them, feature level fusion has great advantages. It not only retains more information of original samples, but also eliminates redundant information between different features. Theoretically, the superior fusion performance can be obtained. However, feature level fusion currently met with many challenges, such as the Curse of dimensionality, the incompatibility of the feature space, poor correlation between features, and how to design effective fusion strategy. We hope to solve above problems, so the goal for mutual utilization and supplement between different features can be achieved. At present, two well-known and typical feature fusion methods are: serial feature fusion [16] and parallel feature fusion [17, 18].

Recently, Sun et al. [19] used canonical correlation analysis (CCA) to achieve feature fusion by maximizing the correlation of two sets of variables in the projection space. A discriminative canonical correlation analysis (DCCA) algorithm, proposed by Kim et al. [20], this method can simultaneously maximize the within-class correlation and minimize the between-class correlation. Generalized canonical correlation analysis (GCCA) [21] made full use of class label information by minimizing and Constraint the within-class scatter matrix, so as to improve the discriminating ability of features. Kernel canonical correlation analysis (KCCA) [22], an extension method of CCA, cast a light on nonlinear problem, in this way, samples were mapped to kernel space through kernel functions and extracted features in kernel space. Haghighat et al. [23] proposed the discriminant correlation analysis (DCA), this work removed the correlation between feature sets of different classes through between-class scatter matrix, thus realizing the purpose of using class information.

CCA and derivative algorithm had solved some specific problems and gained exceeding recognition performance. The above performance, however, is achieved at the cost of not taking into account solving the problem of high dimension and similarity between different features at the same time. Additionally, it ignored the between-class relationship of the same modality feature sample. In this paper, we propose an enhanced fusion algorithm for linear discriminant canonical correlation analysis. The algorithm is described as follows: Firstly, each modality is treated widthwise, and LDA is applied to process each modality dataset respectively. In this way, data in each modality will be more centralized as much as possible. Secondly, the improved DCCA is used to process the outputs of first step, for the purpose of simultaneously maximize the within-class correlation and minimize the between-class correlation. The establishment and solution of the objective function is based on (DMCCA) [24]. The benefits of the proposed paper are multi-fold: 1, dimension curse problem of feature fusion is successfully solved: 2, Feature sets of each modality and different modalities have been well processed. The proposed algorithm simultaneously maximizes the within-class correlation and minimize the between-class correlation. More importantly, the within-class feature is more centralized in each modality.

2   Related Knowledge-Canonical Correlation Analysis

The canonical correlation analysis was proposed by Hotelling et al. [25] in 1936. The object of CCA is to capture the correlations between two sets of variables. The methods are based on singular value decomposition for both representations. The details are as follows: let $x \in R^{p \times 1}$, $y \in R^{q \times 1}$ be two sets of zero-mean random variables, the CCA is to find a pair of projection directions w_x and w_y, such that the linear correlations between the projections onto these basis vectors are mutually maximized: $X_1 = w_x^T x$, $Y1 = w_y^T y$, then X_1, Y_1 is the first pair of canonical variables. It needs to satisfy the canonical property that the first projection is uncorrelated with the second projection, etc. All the correlation features of x and y are extracted. The criterion function is defined as follows:

$$\arg\max_{w_x, w_y} w_x^T * C_{xy} * w_y \quad s.t. w_x^T C_{xx} w_x = 1, w_y^T C_{yy} w_y = 1 \tag{1}$$

Where C_{xx}, C_{yy} are the within-sets covariance matrices of X and Y respectively, C_{xy} represents the between-set covariance matrix (note that $C_{yx} = C_{xy}^T$). The way to solve these problems can be referred to [26] for detail. In the aspect of fusion, according to the theory of canonical correlation analysis of Sun et al. [27], the method of concatenation or summation is put forward to fuse the feature vectors. The method is called FFS-1(F1) and FFS-2(F2), respectively.

Feature layer fusion strategy FFS-1: $Z_1 = \begin{pmatrix} X^* \\ Y^* \end{pmatrix} = \begin{pmatrix} W_x^T X \\ W_y^T Y \end{pmatrix}$

Feature layer fusion strategy FFS-2: $Z_2 = X^* + Y^* = W_x^T X + W_y^T Y$

3   Fusion Algorithm

In order to tackle the problem that covariance irreversibility caused by the number of samples is smaller than the number of dimensions. The traditional feature fusion method such as proposed in [28], which adopt method in two stages way, PCA + CCA. However, this method ignores that PCA may cause the loss of feature information, and ignores the discriminative information between samples as well. Motivated by the ideas of this combination. By this, algorithm is based on the dimension reduction method of LDA [29] to find the projection of the best separation class. In this paper, not only the problem of covariance irreversible is solved, but also achieves better results for each modality and cross-modal data sets.

3.1   Linear Discriminate Analysis

The main contribution of LDA is the idea that by finding a better projection vector space for high-dimensional data sets, so the distances within-class become smaller and the distances between-classes become larger in the projected space. Let's assume that the samples can be divided into C separate groups, where n_i columns belong to the ith

class $\left(N = \sum_{i=1}^{C} n_i\right)$. Suppose that $X = \{X_1, X_2 \cdots X_N\}$ denote a matrix, contains N training feature vectors. Let $X_i = R^d$ denote a feature vector. $\bar{u}_i$ is the means of the X_i vectors in the ith class, $\bar{u}$ is the global mean of the entire sample set.
The within-class scatter matrix is defined as

$$S_W = \sum_{i=1}^{C} S_i = \sum_{i=1}^{C} \sum_{X_i \in W_i} (X_i - \bar{u}_i)(X_i - \bar{u}_i)^T \tag{2}$$

The between-class scatter matrix is defined as

$$S_B = \sum_{i=1}^{C} n_i(\bar{u}_i - \bar{u})(\bar{u}_i - \bar{u})^T \tag{3}$$

Therefore, a criterion function can be established:

$$J(W) = \frac{W^T S_B W}{W^T S_W W} \tag{4}$$

By constraining the denominator $W^T S_W W = 1$, the objective function is maximized to find the optimal value. The Lagrangian multiplier method is employed and then Eq. (4) is converted to find the eigenvalue of $S_W^{-1} S_B W = \lambda_i W$ to solve the problem.

3.2   Linear Discrimination Canonical Correlation Analysis

By the 3.1 algorithm, thereby achieving between-class is centralized and the effects of within-class are decentralized. Due to the difference in the dimension of the two modal samples, the two Training sets are normalized. Next, the algorithm of 3.2 is to establish correlations between the samples of the two modalities. The purpose is to maximize the similarities of a pairs of sets of within-class while minimizing the correlations between-class. $X = (X_1, X_2, \cdots X_C), Y = (Y_1, Y_2 \cdots Y_C)$ are training sets of two modalities, where $X_i \in R^{p \times ni}, Y_i \in R^{q \times ni}$, each class has ni samples, N represents the total number of training samples, X and Y have $C = (W_1, W_2, \ldots, W_c)$ class samples, and subsamples of each class can be represented as $Y_i = (y_1, y_2, \cdots, y_{ni}) X_i = (x_1, x_2, \cdots, x_{ni})$, $x_i \in R^{p \times 1}, y_i \in R^{q \times 1}$. The correlation of category W_i in sample space X and sample space Y can be expressed as:

$$C_{WXY,i} = X_i(Y_i)^T \tag{5}$$

$$C_{BXY,i,j} = X_i(Y_j)^T \tag{6}$$

The within-class correlation matrix is expressed as:

$$C_{WXY} = \sum_{i=1}^{C} C_{WXY,i} = \sum_{i=1}^{C} X_i(Y_i)^T \tag{7}$$

The between-class correlation matrix is expressed as:

$$
\begin{aligned}
C_{BXY} &= \sum_{i=1}^{C}\sum_{j=1,j\neq i}^{C} C_{BXY,i,j} = \sum_{i=1}^{C}\sum_{j=1,j\neq i}^{C} X_i(Y_j)^T \\
&= \sum_{i=1}^{C}\sum_{j=1}^{C} X_i(Y_j)^T - \sum_{i=1}^{C} X_i(Y_i)^T \\
&= \left(\sum_{i=1}^{C} X_i\right)\left(\sum_{j=1}^{C} Y_i\right)^T - \sum_{i=1}^{C} X_i(Y_i)^T
\end{aligned}
\tag{8}
$$

The criterion function model can be established as:

$$J(W, V) = \max_{W,V} \frac{W^T \tilde{C}_{XY} V}{\sqrt{W^T C_{XX} W}\sqrt{V^T C_{YY} V}} \quad s.t. W^T C_{XX} W = 1, V^T C_{YY} V = 1 \tag{9}$$

Among them, C_{xx}, C_{yy} are the within-sets covariance matrixes of X and Y, respectively, where $\tilde{C}_{XY} = C_{Wxy} - \eta C_{Bxy}, (\eta > 0)$. From the above equation, by adjusting the size of η, so that $\tilde{C}_{XY}$ makes certain trade-offs between within-class and between-class. The above problem becomes that under the constraint condition, the maximum projection vector W, V of the criterion function is obtained.

$$L(W, V) = W^T \tilde{C}_{XY} V - \lambda_1(W^T C_{XX} W - 1) - \lambda_2(V^T C_{YY} V - 1) \tag{10}$$

Where λ_1, λ_2 is a Lagrangian multiplier and the derivation of W and V respectively, and then obtained:

$$\frac{\partial L}{\partial W} = \tilde{C}_{XY} V - \lambda_1 C_{XX} W = 0 \tag{11}$$

$$\frac{\partial L}{\partial V} = \tilde{C}_{XY}^T W - \lambda_2 C_{YY} V = 0 \tag{12}$$

Multiply W^T, V^T by (11), (12) above, and then convert (11), (12) to (13):

$$\left.\begin{aligned}
W^T \tilde{C}_{XY} V &= \lambda_1 W^T C_{XX} W = \lambda_1 \\
\left(W^T \tilde{C}_{XY} V\right)^T &= \lambda_2 V^T C_{YY} V = \lambda_2
\end{aligned}\right\} \tag{13}$$

From the constraints, $\lambda_1 = \lambda_2 = \lambda$, $\tilde{C}_{YX} = \tilde{C}_{XY}^T$ can be obtained, C_{XX}, C_{YY} is reversible, then (13) can be converted to:

$$\left.\begin{array}{c}\tilde{C}_{XY}V = \lambda C_{XX}W\\ \tilde{C}_{XY}^T W = \lambda C_{YY}V\end{array}\right\} \Rightarrow \left.\begin{array}{c}\left(C_{XX}^{-1/2}\tilde{C}_{XY}C_{YY}^{-1}\tilde{C}_{YX}C_{XX}^{-1/2}\right)\left(C_{XX}^{1/2}W\right) = \lambda^2\left(C_{XX}^{1/2}W\right)\\ \left(C_{YY}^{-1/2}\tilde{C}_{YX}C_{XX}^{-1}\tilde{C}_{XY}C_{YY}^{-1/2}\right)\left(C_{YY}^{1/2}W\right) = \lambda^2\left(C_{YY}^{1/2}W\right)\end{array}\right\} \quad (14)$$

At this point, finding a maximizing projection vector for a criterion function is converted to solve generalized characteristic equations of formula (14). The matrix required to solve the eigenvalue decomposition (EVD) is a square matrix and singular value decomposition (SVD) [30] is a decomposition method applicable to any matrix, so this paper uses SVD decomposition for Eq. (15). Let $H = C_{XX}^{-1/2}\tilde{C}_{XY}C_{YY}^{-1/2}, \bar{W} = C_{XX}^{1/2}W, \bar{V} = C_{YY}^{1/2}V$, then (14) becomes the following form:

$$\begin{array}{c}HH^T\bar{W} = \lambda^2\bar{W}\\ H^T H\bar{V} = \lambda^2\bar{V}\end{array} \quad (15)$$

It can be seen that $\bar{W} = C_{XX}^{1/2}W$ and $\bar{V} = C_{YY}^{1/2}V$ are the feature vectors of HH^T and $H^T H$, respectively, and singular value decomposition $H = P\Lambda Q^T$ is performed on H, where $\Lambda = diag(\lambda_1, \lambda_2, \cdots, \lambda_r)$, $P = (p_1, p_2, \cdots, p_r)$, $Q = (q_1, q_2, \cdots q_r)$, $r = rank(\tilde{C}_{XY})$. HH^T and $H^T H$ have common non-zero eigenvalues, the column vector of P is the eigenvector corresponding to the eigenvalue λ^2 of HH^T, and the column vector of Q is the eigenvector corresponding to the eigenvalue λ^2 of $H^T H$. Then, the maximum solution of the criterion function model is obtained:

$$\left.\begin{array}{c}W = C_{XX}^{-1/2}\bar{W}\\ V = C_{YY}^{-1/2}\bar{V}\end{array}\right\} \quad (16)$$

The resulting W, V is the standard function projection vector. Therefore, the feature set X^*, Y^* is obtained, where X', Y' are the feature sets obtained in the first time, and the projection matrix W_x, W_y is as follows:

$$\left.\begin{array}{c}X^* = W^T X' = W^T W_{x'}^T X = W_X X\\ Y^* = V^T Y' = V^T W_{y'}^T Y = W_Y Y\end{array}\right\} \quad (17)$$

3.3   The Flow of the Algorithm

(1) Enter the two sets of centered training feature sets X and Y, and the category label information.
(2) According to the algorithm in Sect. 3.1, the LDA algorithm is used for the two feature sets, the projection matrix $W_{x'}, W_{y'}$ is obtained.

(3) According to the algorithm in Sect. 3.2, similarity processing is performed on the processed data sets of the two modalities. The projection matrix W, V is obtained from the criterion function in 3.2.

(4) According to the fusion strategy of Part 2, the test samples are classified and identified by using the nearest neighbor classifier.

4 Simulation Experiment

To verify the validity of algorithm in this paper, the database used for the USM database is the finger vein database, the open palm print database and the finger database of Hong Kong Polytechnic University, and the multi-feature handwritten data sets. Among them, the mean LBP method was used to extract finger veins and palm veins features. The finger feature is calculated by histogram of gradients (HOG) for detecting the contour image of finger vein. In this paper, we utilized the mean classification accuracy (MCA) [31] to evaluate the performance of the fusion algorithm, which is defined as follows:

Let's assume that the N samples as the testing subsets, while truth class labels are $\{\bar{y}_1, \bar{y}_2, \ldots, y_N\}$ and the predicted class labels are $\{\bar{y}_1, \bar{y}_2, \ldots, y_N\}$.

$$MCA = \frac{N_1}{N} \tag{18}$$

Where N_i denotes the number of samples with an error not greater than k between the predicted category label and the real category label, parameter k is set to 1.

4.1 Experiment 1

Multi-feature hand-written data sets are multimodal datasets in the UCI dataset, including 0 to 9 total 10 hand-written digital features, 200 samples in each category, 2000 samples in total. Each sample contains 6 features, which are morphological features (mfeat_mor), Zernike moment features (mfeat_zer), KL expansion coefficients (mfeat_kar), Fourier coefficients (mfeat_fou), contour correlation features (mfeat_fac), and pixel averaging (mfeat_pix). If you select any two features as fusion feature sets, there are 15 combinations. Randomly select 100 samples for each class as the training sets. The remaining samples are used as the testing sets. The nearest neighbor method is used for classification. The testing process repeats 20 cycles independently, and the average of 20 results is used as the final performance measure. Table 1 provides the recognition rate in single modality. Table 2 is based on the FFS-1(denote F1) and FFS-2 (denote F2) feature fusion strategies.

Table 1. Average classification accuracy in single modality

Single modality	fac	fou	kar	mor	pix	zer
MCA	0.944	0.823	0.967	0.422	0.967	0.809

Table 2. Average classification accuracy of the F1 and F2 under cross- modal

Multi-feature fusion	MCA											
	CCA		DCCA		GCCA		DCA		KCCA		ELDCCA	
	F1	F2	F1	F2	F1	F2	F1	F2	F1	F2	F1	F2
fac + fou	0.85	0.89	0.84	0.91	0.97	0.97	0.94	0.90	0.70	0.69	0.98	0.98
fac + kar	0.91	0.93	0.92	0.93	0.97	0.95	0.96	0.96	0.48	0.47	0.98	0.98
fac + mor	0.89	0.93	0.91	0.94	0.95	0.96	0.92	0.86	0.41	0.49	0.98	0.98
fac + pix	0.83	0.84	0.81	0.85	0.96	0.96	0.96	0.94	0.65	0.67	0.98	0.98
fac + zer	0.92	0.94	0.92	0.93	0.97	0.95	0.90	0.88	0.60	0.62	0.98	0.97
fou + kar	0.88	0.87	0.87	0.88	0.96	0.96	0.94	0.90	0.72	0.78	0.97	0.96
fou + mor	0.75	0.77	0.74	0.75	0.51	0.49	0.76	0.56	0.52	0.47	0.85	0.84
fou + pix	0.70	0.75	0.69	0.79	0.93	0.93	0.95	0.89	0.81	0.86	0.97	0.96
fou + zer	0.79	0.81	0.80	0.80	0.84	0.84	0.84	0.81	0.69	0.64	0.86	0.85
kar + mor	0.93	0.95	0.93	0.95	0.92	0.90	0.81	0.48	0.68	0.64	0.97	0.97
kar + pix	0.77	0.86	0.79	0.85	0.96	0.90	0.95	0.94	0.70	0.68	0.94	0.94
kar + zer	0.94	0.93	0.94	0.95	0.96	0.93	0.76	0.83	0.67	0.66	0.96	0.95
mor + pix	0.69	0.77	0.67	0.77	0.91	0.90	0.93	0.93	0.27	0.26	0.97	0.97
mor + zer	0.81	0.81	0.80	0.81	0.74	0.74	0.73	0.66	0.36	0.30	0.84	0.82
pix + zer	0.79	0.87	0.76	0.84	0.94	0.87	0.88	0.96	0.80	0.73	0.96	0.94

From Tables 1 and 2, it can be seen that under the double modality, the MCA of the proposed method is higher than that in the single modality. Under the combinations of 13 features situation, we show that our approach performs better than single modality. However, the fusion recognition accuracy is only slightly lower than single modality in combinations of two features situation. Firstly, the processing of features makes the two modal data more centralized. Secondly, cross modal maximizes the within-class sample correlation between two modalities. The above two steps prove beneficial to the improvement of recognition accuracy. From Table 2, it can be seen that the recognition performance is better than the existing algorithms whether in series or parallel way.

From the analysis of the algorithm in Table 3 in Sects. 3.1 and 3.2, this illustrates the reasonableness of the 3.1 algorithm that make the data more centralized in advance. A large of experiments show that put the 3.1 algorithm into the standard function of algorithm 3.2 directly can't achieve double standard effect at the same time. Therefore, in Table 3, the recognition rate of the combination algorithm in the above manner is also better than the recognition rate of single algorithm. The Sect. 3.1 algorithm improves the performance of the Sect. 3.2. The combination of the two algorithms shows the rationality and effectiveness of the proposed algorithm.

4.2   Experiment 2

The datasets we use are finger vein library provided by the USM database, palm print database and finger database of Hong Kong Polytech University. However, there are no multimodal databases with multiple biometric features from the same user. Taking into

Table 3. Uses the F1 and F2 to combine the average classification accuracy of the proposed algorithm

Multi-feature fusion	MCA					
	Algorithm 3.1		Algorithm 3.2		ELDCCA	
	F1	F2	F1	F2	F1	F2
fac + fou	0.973	0.973	0.924	0.944	0.983	0.984
fac + kar	0.974	0.971	0.921	0.931	0.983	0.980
fac + mor	0.975	0.975	0.915	0.952	0.986	0.986
fac + pix	0.979	0.961	0.834	0.852	0.982	0.981
fac + zer	0.978	0.974	0.930	0.956	0.983	0.979
fou + kar	0.957	0.956	0.946	0.957	0.975	0.968
fou + mor	0.932	0.814	0.829	0.829	0.851	0.842
fou + pix	0.933	0.937	0.700	0.805	0.979	0.965
fou + zer	0.814	0.818	0.849	0.854	0.869	0.850
kar + mor	0.960	0.955	0.937	0.952	0.979	0.970
kar + pix	0.949	0.916	0.782	0.855	0.948	0.947
kar + zer	0.958	0.924	0.949	0.949	0.968	0.957
mor + pix	0.941	0.936	0.715	0.776	0.970	0.970
mor + zer	0.789	0.780	0.808	0.806	0.841	0.829
pix + zer	0.958	0.935	0.806	0.861	0.965	0.945

account the independence of biometric feature, so we choose the single modal feature of each user in compromise.64 users were selected in total, each user's single modality image consist of 6 samples, 3 training samples and the remaining 3 as testing samples. The nearest neighbor method is leveraged to feature classification. The test process was repeated 20 times independently and the average results were taken as the final performance metric.

Figures 1 and 2 refers to the combination of vein and finger shape. Serial and parallel strategies are adopted respectively. The recognition rate of each algorithm is obtained by setting different feature dimensions, and then the recognition results are compared with the existing algorithms. Figures 3 and 4 is the combination of finger veins and palmprint. In the same spirit, do as operations of Figs. 1 and 2 did. The conclusions can be drawn from Figs. 1 to 4: the algorithm proposed in this paper is more robust than the existing method. According to the following figure, the recognition rate of some algorithms improved with the increase of the feature dimension, while others fluctuate greatly. The reason for the decline is that redundant information is added to the feature dimension, which does harm to the acquisition of information, worst of all, and the recognition accuracy. In this paper, the recognition accuracy improves steadily.

Tables 4 and 5 shows the comparison between our algorithm and CCA and their improved methods in hand features. It can be seen from the above table that our algorithm has shown outstanding results on hand biometrics. This also fully demonstrates that a projection matrix can't simultaneously achieve data within-modal more

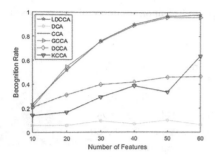

Fig. 1. Average classification accuracy of finger veins and finger outline using the FFST-1

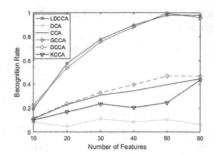

Fig. 2. Average classification accuracy of finger veins and finger outline using the FFST-2

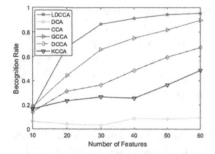

Fig. 3. Average classification accuracy of finger veins and palm prints using the FFST-1

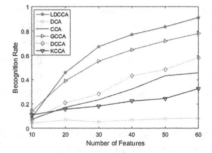

Fig. 4. Average classification accuracy of finger veins and palm prints using the FFST-2

Table 4. The average classification accuracy of the F1 and F2 under cross-modal

Multi-feature fusion	MCA											
	CCA		DCCA		GCCA		DCA		KCCA		ELDCCA	
	F1	F2	F1	F2	F1	F2	F1	F2	F1	F2	F1	F2
Finger Vein + Contour	0.77	0.65	0.77	0.77	0.97	0.97	0.27	0.29	0.76	0.61	0.99	0.99
Finger Vein + Palm	0.93	0.90	0.93	0.91	0.26	0.25	0.93	0.93	0.94	0.91	0.97	0.98

centralized and maximum correlation between cross-modal. Table 5 show the results obtained from different feature combinations, using serial and parallel feature fusion strategies respectively. The original feature and the double standard effect are preserved. We can see that the recognition performance has been greatly improved. More importantly, the connection between single modal data has not been destroyed. This combination has achieved outstanding classification performance.

Table 5. Uses the F1 and F2 to combine the average classification accuracy of the proposed algorithm

Multi-feature fusion	MCA					
	Algorithm 3.1		Algorithm 3.2		ELDCCA	
	F1	F2	F1	F2	F1	F2
Finger Vein + Contour	0.994	0.994	0.993	0.994	0.997	0.998
Finger Vein + Palm	0.947	0.963	0.848	0.947	0.973	0.984

5   Conclusions

In this paper, a canonical correlation analysis algorithm based on linear discriminant analysis is proposed. In order to improve the effect of feature layer fusion, firstly, this article replaces the correlation analysis of image set classes in DCCA with the correlation analysis between samples. It can simultaneously maximize the within-class correlation and minimize the between-class correlation. At the same time, in order to extract feature sets with higher discriminative ability, LDA is adopted to make Samples in the same class for each modality as concentrat as possible. Results demonstrated, that in the USM database, the palmprint open database, the finger database and the multi feature handwritten data set of Hong Kong Polytech University, the algorithm is superior to other internationally popular comparison methods.

References

1. Li, L., Jun, Z., Fei, J., et al.: An incremental face recognition system based on deep learning. In: 2017 Fifteenth IAPR International Conference on Machine Vision Applications (MVA), pp. 238–241. IEEE (2017). https://doi.org/10.23919/mva.2017.7986845
2. Sun, Z., Zhang, H., Tan, T., et al.: Iris image classification based on hierarchical visual codebook. IEEE Trans. Pattern Anal. Mach. Intell. **36**(6), 1120–1133 (2014). https://doi.org/10.1109/tpami.2013.234
3. Liu, S., Liu, Y., Yu, J., et al.: Hierarchical static hand gesture recognition by combining finger detection and HOG features. J. Image Graph (2015)
4. Seng, K., Ang, L.M., Ooi, C.: A combined rule-based and machine learning audio-visual emotion recognition approach. IEEE Trans. Affect. Comput. (2016). https://doi.org/10.1109/TAFFC.2016.2588488
5. Ahlawat, M., Kant, C.: An introduction to multimodal biometric system: an overview. Int. J. Sci. Res. Dev. **3**(02), 2321–0613 (2015)
6. Zheng, H., Geng, X.: A multi-task model for simultaneous face identification and facial expression recognition. Neurocomputing **171**, 515–523 (2016). https://doi.org/10.1016/j.neucom.20-15.06.079
7. Wen, H., Liu, Y., Rekik, I., et al.: Multi-modal multiple kernel learning for accurate identification of tourette syndrome children. Pattern Recogn. **63**, 601–611 (2017). https://doi.org/10.1016/j.patcog.2016.09.039
8. Wang, S., Zhang, J., Zong, C.: Learning multimodal word representation via dynamic fusion methods. arXiv preprint arXiv:1801.00532 (2018)

9. Liu, Q., Tang, L., Zhao, B.-J., et al.: Infrared target tracking based on adaptive multiple features fusion and mean shift. J. Electron. Inf. Technol. **34**(5), 1137–1141 (2012). https://doi.org/10.3724/SP.J.1146.2011.01077

10. Liu, H., Wu, Y., Sun, F., et al.: Weakly paired multimodal fusion for object recognition. IEEE Trans. Autom. Sci. Eng. **15**(2), 784–795 (2018). https://doi.org/10.1109/TASE.2017.2692271

11. Zeng, M., Wu, Z., Tian, C., et al.: Fusing appearance statistical features for person re-identification. J. Electron. Inf. Technol. **36**(8), 1845–1851 (2014). https://doi.org/10.3724/SP.J.1146.2013.01389

12. Meng, W., Wong, D.S., Furnell, S., et al.: Surveying the development of biometric user authentication on mobile phones. IEEE Commun. Surv. Tutorials **17**(3), 1268–1293 (2015). https://doi.org/10.1109/COMST.2014.2386915

13. Haghighat, M., Abdel-Mottaleb, M., Alhalabi, W.: Discriminant correlation analysis for feature level fusion with application to multimodal biometrics. In: IEEE International Conference on Acoustics, Speech and Signal Processing, pp. 1866–1870 (2016). https://doi.org/10.1109/icassp.2016.7472000

14. He, M., Horng, S.J., Fan, P., et al.: Performance evaluation of score level fusion in multimodal biometric systems. Pattern Recogn. **43**(5), 1789–1800 (2010). https://doi.org/10.1016/j.patcog.2009.11.018

15. Liu, H., Li, S.: Decision fusion of sparse representation and support vector machine for SAR image target recognition. Neurocomputing **113**, 97–104 (2013). https://doi.org/10.1016/j.neucom.2013.01.033

16. Liu, C., Wechsler, H.: A shape-and texture-based enhanced fisher classifier for face recognition. IEEE Trans. Image Process. **10**(4), 598–608 (2001). https://doi.org/10.1109/83.913594

17. Yang, J., Yang, J., Zhang, D., et al.: Feature fusion: parallel strategy vs. serial strategy. Pattern Recogn. **36**(6), 1369–1381 (2003). https://doi.org/10.1016/S0031-3203(02)00262-5

18. Sun, Q., Zeng, S., Yang, M., et al.: A new method of feature fusion and its application inimage recognition. Pattern Recogn. **38**(12), 2437–2448 (2005). https://doi.org/10.1016/j.patcog.2004.12.013

19. Sun, Q., Zeng, S., Yang, M., et al.: Combined feature extraction based on canonical correlation analysis and face recognition. J. Comput. Res. Dev. **42**(4), 614–621 (2005). https://doi.org/10.1360/crad20050413

20. Kim, T.K., Kittler, J., Cipolla, R.: Discriminative learning and recognition of image set classes using canonical correlations. IEEE Trans. Pattern Anal. Mach. Intell. **29**(6), 1005–1018 (2007). https://doi.org/10.1109/TPAMI.2007.1037

21. Tenenhaus, A., Philippe, C., Guillemot, V., et al.: Variable selection for generalized canonical correlation analysis. Biostatistics. **15**(3), 569–583 (2014). https://doi.org/10.1093/biostatistics/kxu001

22. Jia, Z.: Multi-feature combination face recognition based on kernel canonical correlation analysis. Int. J. Signal Process. Image Process. Pattern Recogn. **9**(7), 221–230 (2016)

23. Haghighat, M., Abdel-Mottaleb, M., Alhalabi, W.: Discriminant correlation analysis: real-time feature level fusion for multimodal biometric recognition. IEEE Trans. Inf. Forensics Secur. **11**(9), 1984–1996 (2016). https://doi.org/10.1109/TIFS.2016.2569061

24. Gao, L., Qi, L., Chen, E., et al.: Discriminative multiple canonical correlation analysis for information fusion. IEEE Trans. Image Process. **27**(4), 1951–1965 (2018). https://doi.org/10.1109/TIP.2017.2765820

25. Hotelling, H.: Relations between two sets of variates. Biometrika **28**(3–4), 321–377 (1992). https://doi.org/10.2307/2333955

26. Weenink, D.: Canonical correlation analysis. In: Proceedings of the Institute of Phonetic Sciences of the University of Amsterdam, vol. 25, pp. 81–99. University of Amsterdam (2003)
27. Sun, Q., Zeng, S., Yang, M., et al.: A new method of feature fusion and its application in image recognition. Pattern Recogn. **38**(12), 2437–2448 (2005). https://doi.org/10.1016/j.patcog.2004.12.013
28. Correa, N.M., Adali, T., Li, Y.-O., Calhoun, V.D.: Canonicalcorrelation analysis for data fusion and group inferences. IEEE Signal Process. Mag. **27**(4), 39–50 (2010). https://doi.org/10.1109/34.598228
29. Belhumeur, P.N., Hespanha, J.P., Kriegman, D.: Eigenfaces vs. fisherfaces: recognition using class specific linear projection. IEEE Trans. Pattern Anal. Mach. Intell. **19**(7), 711–720 (1997). https://doi.org/10.1109/34.598228
30. Dan, K.: A singularly valuable decomposition: the SVD of a matrix. Coll. Math. J. **27**(1), 2–23 (1996). https://doi.org/10.2307/2687269
31. Zhou, H., Chen, S.: Ordinal discriminative canonical correlation analysis. J. Softw. **25**(9), 2018–2025 (2014). https://doi.org/10.13328/j.cnki.jos.004649

Spatial-Temporal Attention for Action Recognition

Dengdi Sun[1], Hanqing Wu[1], Zhuanlian Ding[1(✉)], Bin Luo[1,2], and Jin Tang[1,2]

[1] School of Computer Science and Technology, Anhui University,
Hefei 230601, China
{sundengdi,wuhanqing26,dingzhuanlian}@163.com, {luobin,tj}@ahu.edu.cn
[2] Key Lab of Industrial Image Processing & Analysis of Anhui Province,
Hefei, China

Abstract. We present a new framework for end-to-end classifying actions in video called Spatial-Temporal Attention Model (STAM) which integrates ideas from attention mechanism to better explore spatial and temporal information, simultaneously, in videos. Specifically, spatial attention is used to locate the action-relevant parts within each video frame, temporal attention is used to measure the discrimination of each time step, and jointly learn the video-level representation for final prediction. Experimental results show our model is comparable to state-of-arts. We also visualize the learned models, which makes our model more interpretable and demonstrates the effectiveness of Spatial-Temporal Attention Model.

Keywords: Action recognition · Spatial attention
Temporal attention · ConvGRU

1 Introduction

Action recognition aims to analyze human behavior in videos and give a unique label. It has been applied to numerous problems including video surveillance, video indexing, and human-computer interaction (HCI). Inspired by image classification, the traditional approach train a classifier based on encoded appearance features to recognize actions frame by frame, but lacks the representation of motion information hidden in videos. To overcome the defect, it is necessary to extract trajectory information based on optical flow and spatiotemporal information, and various methods have been proposed consequently.

Recently, motivated by the success of convolutional neural network (CNN) on image classification and object detection tasks, CNNs also have been applied to understand the human actions. To incorporate appearance and motion information, a two-stream CNN framework was proposed in [12] firstly, which learn appearance representation from frames and motion representation from optical flows simultaneously. Then, 3D convolutional kernels was employed to learn those

© Springer Nature Switzerland AG 2018
R. Hong et al. (Eds.): PCM 2018, LNCS 11164, pp. 854–864, 2018.
https://doi.org/10.1007/978-3-030-00776-8_78

representations in [14]. Moreover, due to the advantages in processing sequential data, recurrent neural network (RNN) also utilized in video understanding. Donahue et al. [1] feed the CNN features into RNN in order to model temporal dynamics in video.

Furthermore, it has been noted that humans focus sequentially on different parts of the scene based on attention mechanism, which can also potentially infer the action in videos by concentrating only on the significant relevant information in frames and time intervals. According to this principle, Sharma et al. [11] firstly integrate soft-attention with LSTM to extract the relevant part in each frame. Kar et al. [5] propose to select fewer and task-related frames with a temporal attention mechanism.

In this paper, we introduce a Spatial-Temporal Attention Model (STAM) for action recognition which utilizes attention both spatial and temporal (see overview in Fig. 1). STAM not only focus on the salient observations which are important to classify actions and discard irrelevant parts in each frame, but also more tend to choose those frames that better distinguish actions among all video frames.

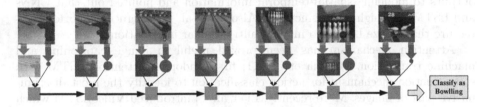

Fig. 1. Our model first utilize spatial attention mechanism to focus on the salient areas of current frame (the green arrow represents spatial attention processor), and temporal attention mechanism determine the discriminative ability of the current frame (the blue arrow represents temporal attention processor and the stronger the ability, the bigger the red circle.) Orange squares represent video-level representation until current frame. Classification is based on the final representation. (Color figure online)

In summary, our STAM model has following advantages for video action recognition.

1. We choose Gated Recurrent Unit (GRU) to implement recurrent steps which has fewer parameters and leads to faster training rather than LSTM.
2. Spatial attention module is able to locate more informative part in each frame automatically.
3. Temporal attention module can select relevant segments during the entire action stage.
4. STAM is an end-to-end framework that can be trained with traditional cross-entropy loss function.

2 Related Work

CNN is a popular deep learning framework in many computer vision tasks. Simonyan et al. [12] utilized two separated ConvNet to model appearance information and motion information respectively. More specifically, appearance information are just video frames and push into spatial CNN, motion information consist of optical flows between adjacent frames and feed into a temporal CNN. Feichtenhofer et al. [2] further improved two-stream ConvNets by using convolutional layers rather than averaging softmax scores to better model videos.

Ji et al. [4] firstly attempted use 3D convolutional kernels and operated on stacked video frames to learn spatial-temporal at the same time. Then, Tran et al. [14] trained a 3D neural network on massive sport video datasets and yield better performance.

RNN models have shown its powerful ability to model sequential data in tasks like machine translation, language modeling and image captioning which also make it popular in action recognition. Donahue et al. [1] proposed to apply two-layer LSTM on top of ConvNet features and do classification at every time steps. Wu et al. [17] presented a framework that fusing three independent LSTM outputs to modeling spatial-temporal information and pointed out that CNNs and LSTMs are high complementary. Also, Li et al. [7] designed a hybrid architecture that utilize LSTM to model multi-granular information.

Attention mechanism has been proved useful in image captioning and machine translation. Sharma et al. [11] firstly adopted attention LSTM with soft-attention mechanism for action classification to identify the most discriminative part of images. More recently, Li et al. [8] introduced VideoLSTM which add convolutional operation to LSTM to preserve spatial structure in recurrent steps. Zhu et al. [18] proposed a key volume mining mechanism to select task related video frames and its a kind of temporal attention mechanism. Kar et al. [5] also designed a frame selection framework to make better use of temporal information.

3 The Proposed Approach

We propose an end-to-end video classification framework with spatial and temporal attention mechanisms. In the following, we describe major components of our framework - CNN feature extraction, Convolutional GRU, Spatial Attention Module and Temporal Attention Module.

3.1 Convolutional Features

Unlike previous study [11] that extract last fully connected layer outputs as input feature, we choose the last convolutional layer's feature map as inputs, because it preserves spatial structure of video frames in some extent. Specifically, we choose BNInception [3] as our base ConvNet, the last convolutional layer has D feature maps and each one have size $K \times K$. For a video composed of T frames, after apply feature extraction we can get a representation $X = \{X_1, X_2, ..., X_T\}$ $\in \mathbb{R}^{T \times D \times K \times K}$.

In temporal attention module, we only need the relationship between past and future frames, spatial information is redundant for our model, so we choose last fully connected layer's feature as temporal attention module inputs.

3.2 Convolutional GRU

Video always contains a lot of temporal information that represents the process of the movement, so recurrent neural networks is needed to model dynamic information. But traditional GRU is fully connected in which no spatial structure is encoded, so it is inefficient to model video that contains spatial-temporal information. In order to preserve the spatial information over time, we can replace vector multiplication operations in the state-to-state and input-to-state transitions with convolutional operations, the main equations of ConvGRU is shown below:

$$Z_t = \sigma(W_z * \widetilde{X_t} + U_z * H_{t-1} + b_z) \tag{1}$$

$$R_t = \sigma(W_r * \widetilde{X_t} + U_r * H_{t-1} + b_r) \tag{2}$$

$$G_t = tanh(W_g * \widetilde{X_t} + U_g * (R_t \odot H_{t-1}) \tag{3}$$

$$H_t = Z_t * H_{t-1} + (1 - Z_t) * G_t \tag{4}$$

where $*$ denotes convolutional operations, $\odot$ denotes the Hadamard product. $W_\sim$, $U_\sim$, are 2D convolutional kernels, $b_\sim$ are bias term, those learnable parameters are shared over time. Two gates Z_t and R_t, candidate hidden state G_t and hidden state H_t are 3D tensors, $\widetilde{X_t}$ is spatial attention processed input, we will discuss it in the next section. We also add zero-padding before convolutional operations to ensure hidden states have same spatial size over time.

3.3 Spatial Attention Module

To generate spatial attention map for current input X_t, we adopt a small ConvNet with two convolution layer, and a 2D softmax is used to compute the normalized spatial attention map, formally:

$$S_t = U_s * tanh(W_{xs} * X_t + W_{hs} * H_{t-1} + b_s) + b_{us} \tag{5}$$

where X_t is current input, H_{t-1} is previous hidden state of ConvGRU, U_s, W_{xs}, W_{hs} are convolutional kernels with size 1×1. Then we apply 2D softmax on S_t to get the spatial attention map of current input.

$$A_t^{ij} = \frac{exp(S_t^{ij})}{\sum_k \sum_l exp(S_t^{kl})} \tag{6}$$

where A_t^{ij} is the attention weight for position (i, j).

Now, we have spatial attention map for current input X_t, and ConvGRU input $\widetilde{X_t}$ can easily calculate by element-wise product between attention map and each channel of feature maps

$$\widetilde{X_t} = A_t \odot X_t \tag{7}$$

3.4   Temporal Attention Module

The reason for joining the temporal attention mechanism is that different frames in videos contain different information, those informative and discriminative frames should be given more attention, those non-discriminative frames should be discard.

In addition, in order to determine the attention value for each time step, we not only need the information of the current moment, but also consider information from the adjacent moments in both directions - past and future. So we using a bi-directional GRU to model this adjacent information, and temporal attention value is calculated as below:

$$\beta_t = f(x_t, \overrightarrow{h_t}, \overleftarrow{h_t}) \tag{8}$$

which depends on the current input x_t, hidden states of bi-directional GRU model $\overrightarrow{h_t}$ and $\overleftarrow{h_t}$. Here x_t is ConvNets' last fully connected outputs. Three fully connected layers are used to implement the function $f(\cdot)$, and use *sigmoid* as activation function to squash output between $[0, 1]$. The graphical structure of temporal attention model is illustrated in Fig. 2.

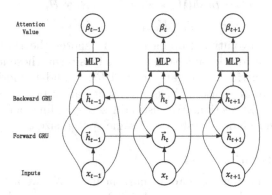

Fig. 2. This graphic shows our temporal attention module. The bottom is inputs, middle is bi-directional GRU, MLP calculates the temporal attention value based on input and hidden states at current time step.

3.5   Joint Spatial and Temporal Attention

Now, we integrate all the above modules together to represent a video. Let Γ_t represents video feature until time t, γ_t is the information at time t and is calculates as:

$$\gamma_t = ReLU(H_t) \tag{9}$$

H_t is the hidden state of ConvGRU at time t, video-level feature transition operation is similar to ConvGRU hidden state transition operation and is formulated as:

$$\Gamma_t = (1 - \beta_t) * \Gamma_{t-1} + \beta_t * \gamma_t \tag{10}$$

where β_t is the temporal attention value at time t, finally we do action classification base on Γ_T.

Also, for end-to-end training the model, our loss function using cross-entropy coupled with penalty on temporal attention module, making the full objective function as:

$$L = -\sum_{i=1}^{C} y_i log \hat{y}_i - \lambda \sum_{t=1}^{T} \frac{e^{\beta_t}}{N} log(\frac{e^{\beta_t}}{N}) \tag{11}$$

$$\lambda \geq 0, N = \sum_{t=1}^{T} e^{\beta_t} \tag{12}$$

λ is the temporal attention penalty coefficient.

4   Experiments

4.1   Datasets

We perform our experiments on two public benchmark datasets, namely UCF101 and HMDB51. Both datasets contain short trimmed videos and single label.

UCF101 [13]. The UCF101 dataset contains 101 action classes and is composed of 13320 realistic video clips. This dataset gives the largest diversity in term of actions and with the presence of large variations in camera motion, viewpoint cluttered background, and object/context, so action classification in this dataset is challenging. There are three training/testing splits for evaluation scheme, we only report classification accuracy on split 1.

HMDB51 [6]. Human Motion Database dataset also provides three different training/testing splits each consisting of 6766 video clips from various sources, and has 51 action classes. Many of the videos in this dataset are of low quality and the total amount of this dataset is not enough to effectively train a deep network. We report classification accuracy for 51 classes on split 1.

4.2   Implementation Details

ConvNet Features. With the development of deep learning, the network has become deeper and deeper, and lots of work points out that deeper network has better performance, so we choose the BN-Inception [3] pretrained on ImageNet as our feature extractor, due to its efficiency and accuracy. Like two-stream ConvNets, the spatial stream ConvNet takes a single RGB frame as inputs, and the motion stream ConvNet operates on a stack of consecutive optical flow fields. Also we use cross-modality pretraining like [16] to initialize motion stream ConvNet, data augmentation is similar to [16].

Spatial Attention Module. We use the last inception's outputs as feature maps with size $7 \times 7 \times 1024$. Convolutional kernels size for ConvGRU are 3×3, and 1×1 convolutions are used to generate attention map. All the channels in ConvGRU and spatial attention module are set to 1024.

Temporal Attention Module. Hidden size of bi-directional GRU is 1024, and use three layer MLP to generate temporal attention value, the first two activation function are *tanh*, the last activation function are *sigmoid* to compress the output to $[0, 1]$.

In training phase, a mini-batch of 10 videos randomly sampled from training set, for each sample, 25 frames or stacked optical fields are randomly selected with data augmentation. Note that, there are about 50 videos less than 25 frames in HMDB51, we do padding with last frame or optical flow. Regularization term λ set to 10, dropout is also used before classification layer with ratio 0.8 for RGB inputs, and 0.7 for optical flow inputs. We use SGD as our optimizer and the initial learning rates set to 1e-6 for base ConvNet architecture, 1e-2 for Spatial-Temporal Module and decrease to its $\frac{1}{10}$ every 20 epochs, the maximum epochs is set to 100. We implement our network using the PyTorch toolkit.

In testing phase, like [12], 25 frames is selected with uniform sampling. For each frames, we obtain 10 inputs by cropping and flipping four corners and the center. Classification result is obtained by averaging 10 inputs' class scores.

4.3   Visualization of Spatial Attention

Figure 3 shows some examples of our spatial attention module on UCF101 dataset. In Fig. 3a, spatial attention model focus on different parts of frame. First, the relevant area is people, then people and ball, and move to basketball and hoop, finally focus on hoop and people, while correctly classifying the action as "Basketball". Similary, in Fig. 3b, the model attends to the bicycle and recognizes the activity as "Biking", but more interesting, in this case, model even ignores the human body.

(a) Classified as "Basketball", correct   (b) Classified as "Biking", correct

Fig. 3. Spatial attention over time. Module learns to focus on the relevant parts of frames.

In Fig. 4a, a man is playing golf, our spatial model gives a rough result which related to the whole person, but should focus more on hands and golf clubs. In Fig. 4b, spatial attention doesn't capture the right person until the last moment and classify as "JavelinThrow" which should be "LongJump", maybe the scene is too confusing and leads to wrong answer.

(a) Classified as "SoccerJuggling", incor- (b) Classified as "JavelinThrow", incorrect   rect

Fig. 4. Spatial attention over time but focus on irrelevant area, resulting wrong answers.

4.4 Comparison Temporal Attention with Mean Pooling

To verify the effectiveness of our temporal attention module, we conduct a comparison between temporal attention and mean-pooling. For mean-pooling, we omit temporal attention module of STAM, and in order to obtain video-level feature, every time steps γ_t (in Eq. (9)) is averaged, classification is based on the averaged representation. Comparison experiments based on RGB images on UCF101 and HMDB51.

Table 1. Comparison results of mean-pooling with temporal attention.

Dataset	Mean-pooling	Temporal-attention
UCF101	84.0%	84.9%
HMDB51	44.7%	45.2%

Table 1 shows comparison results based on RGB images on UCF101 and HMDB51 datasets. We observe that temporal attention consistently performs betters on both datasets. The improvement on UCF101 dataset is slightly better than HMDB51, this gap is reasonable, mainly because HMDB51 is mostly made up of film clips which is difficult to judge whether a frame or a fragment is important when compared with sport actions, and UCF101 dataset has action related to sports.

4.5 Comparison with State-of-the-Art

Our proposed model achieves competitive performance with current state-of-the-art methods (see Table 2). Also, STAM performs better than those methods employing recurrent neural networks, e.g., 91.6% for STAM and 89.2% 89.4% for [8], [5] on UCF101 which demonstrating that our model is capable of extract video-level representations. When combined with C3D method, classification accuracy is future improved.

Table 2. Comparison with state-of-art.

Model	UCF101	HMDB51
IDT+FV [15]	57.2%	85.9%
IDT+HSV [10]	87.9%	61.1%
Two Stream [12]	88.0%	59.4%
Two Stream + LSTM [9]	88.6%	-
C3D [14]	85.2%	-
Soft Attention + LSTM [11]	85.0%	41.3%
VideoLSTM [8]	89.2%	56.4%
AdaScan [5]	89.4%	54.9%
Conv Two Stream [2]	92.5%	65.4%
Key Volume [18]	93.1%	67.2%
STAM	91.6%	59.6%
STAM + C3D	93.3%	65.6%

(a) "HighJump"

(b) "IceDancing"

Fig. 5. Spatial and temporal attention exhibition. In each picture, top row is original video frames, the bottom row is spatial processed frames. Temporal attention value at each time step is in the lower left corner.

Figure 5 displays the result of our proposed framework. As we can see, the temporal attention trend of two actions are different. HighJump's temporal attention value is jittering over time, but IceDancing's temporal attention value is decreasing over time. This is what we expected, because HighJump and IceDancing are two different type of actions which is changing over time or repeating periodically. So in order to recognize IceDancing, a small number of image frames is enough, in contrast, when classifying HighJump, we have to pay different temporal attention to different sub-stage. This figure exhibits the interpretability of our model and the ability to represent videos.

5 Conclusion

In this work, we presented the Spatial-Temporal Attention Model (STAM), an end-to-end framework for classifying actions in video. Our model is motivated by attention mechanism from machine translation and image captioning, then we extended this idea to each video frame and video sequences. Experimental results on two widely used action recognition tasks demonstrate that our model is able to extract video-level features and outline the practical and theoretical merits of STAM.

Acknowledgment. This work was supported by the National Basic Research Program (973 Program) of China (2015CB351705), National Natural Science Foundation of China (61402002, 61671018, and 61472002), Natural Science Foundation of Anhui Higher Education Institutions of China (KJ2018A0023 and KJ2016A040), Key Research and Development Program of Anhui Province (1804a09020101), and Science and Technology Planning Project of Guangdong Province (2017B010110011).

References

1. Donahue, J., et al.: Long-term recurrent convolutional networks for visual recognition and description. In: Proceedings of the IEEE Conference on Computer Vision and Pattern Recognition, pp. 2625–2634 (2015)
2. Feichtenhofer, C., Pinz, A., Zisserman, A.: Convolutional two-stream network fusion for video action recognition. In: Proceedings of the IEEE Conference on Computer Vision and Pattern Recognition, pp. 1933–1941 (2016)
3. Ioffe, S., Szegedy, C.: Batch normalization: accelerating deep network training by reducing internal covariate shift. arXiv preprint arXiv:1502.03167 (2015)
4. Ji, S., Xu, W., Yang, M., Yu, K.: 3d convolutional neural networks for human action recognition. IEEE Trans. Pattern Anal. Mach. Intell. **35**(1), 221–231 (2013)
5. Kar, A., Rai, N., Sikka, K., Sharma, G.: Adascan: adaptive scan pooling in deep convolutional neural networks for human action recognition in videos. In: Proceedings of the IEEE Conference on Computer Vision and Pattern Recognition, pp. 3376–3385 (2017)
6. Kuehne, H., Jhuang, H., Stiefelhagen, R., Serre, T.: Hmdb51: a large video database for human motion recognition. In: Nagel, W., Kröner, D., Resch, M. (eds.) High performance computing in science and engineering, vol. 12, pp. 571–582. Springer, Heidelberg (2013). https://doi.org/10.1007/978-3-642-33374-3_41

7. Li, Q., Qiu, Z., Yao, T., Mei, T., Rui, Y., Luo, J.: Action recognition by learning deep multi-granular spatio-temporal video representation. In: Proceedings of the 2016 ACM on International Conference on Multimedia Retrieval, pp. 159–166. ACM (2016)

8. Li, Z., Gavrilyuk, K., Gavves, E., Jain, M., Snoek, C.G.: Videolstm convolves, attends and flows for action recognition. Comput. Vis. Image Underst. **166**, 41–50 (2018)

9. Ng, J.Y.H., Hausknecht, M., Vijayanarasimhan, S., Vinyals, O., Monga, R., Toderici, G.: Beyond short snippets: deep networks for video classification. In: Proceedings of the IEEE Conference on Computer Vision and Pattern Recognition, pp. 4694–4702. IEEE (2015)

10. Peng, X., Wang, L., Wang, X., Qiao, Y.: Bag of visual words and fusion methods for action recognition: comprehensive study and good practice. Computer Vis. Image Underst. **150**, 109–125 (2016)

11. Sharma, S., Kiros, R., Salakhutdinov, R.: Action recognition using visual attention. arXiv preprint arXiv:1511.04119 (2015)

12. Simonyan, K., Zisserman, A.: Two-stream convolutional networks for action recognition in videos. In: Advances in neural information processing systems, pp. 568–576 (2014)

13. Soomro, K., Zamir, A.R., Shah, M.: Ucf101: a dataset of 101 human actions classes from videos in the wild. arXiv preprint arXiv:1212.0402 (2012)

14. Tran, D., Bourdev, L., Fergus, R., Torresani, L., Paluri, M.: Learning spatiotemporal features with 3d convolutional networks. In: Proceedings of the IEEE International Conference on Computer Vision, pp. 4489–4497. IEEE (2015)

15. Wang, H., Schmid, C.: Lear-inria submission for the thumos workshop. In: ICCV Workshop on Action Recognition with a Large Number of Classes, vol. 2, p. 8 (2013)

16. Wang, L., et al.: Temporal segment networks: towards good practices for deep action recognition. In: Leibe, Bastian, Matas, Jiri, Sebe, Nicu, Welling, Max (eds.) ECCV 2016. LNCS, vol. 9912, pp. 20–36. Springer, Cham (2016). https://doi.org/10.1007/978-3-319-46484-8_2

17. Wu, Z., Wang, X., Jiang, Y.G., Ye, H., Xue, X.: Modeling spatial-temporal clues in a hybrid deep learning framework for video classification. In: Proceedings of the 23rd ACM International Conference on Multimedia, pp. 461–470. ACM (2015)

18. Zhu, W., Hu, J., Sun, G., Cao, X., Qiao, Y.: A key volume mining deep framework for action recognition. In: Proceedings of the IEEE Conference on Computer Vision and Pattern Recognition, pp. 1991–1999. IEEE (2016)

Prediction Method of Parking Space Based on Genetic Algorithm and RNN

Jilun Qiu[1,4], Jianrong Tian[2], Haipeng Chen[3,4(✉)], and Xuwang Lu[1,4]

[1] College of Software, Jilin University, Changchun 130012, China
[2] The School of Chinese Language, Jilin University, Changchun 130012, China
[3] College of Computer Science and Technology, Jilin University, Changchun
130012, China
Chenhp@jlu.edu.cn
[4] Key Laboratory of Symbolic Computation and Knowledge Engineering
of Ministry of Education, Jilin University,
Changchun 130012, China

Abstract. With respect to the prediction of short-term unoccupied parking space of parking guidance and information system (PGIS),a prediction method using genetic algorithm combined with recurrent neural network (RNN) is proposed. First, set the parameters of the RNN population, including the search space of the neural network's hidden layers, neuron number, and neuron type. Then by setting the parameters of the genetic algorithm to drive and control the RNN training process, and using the RMSE value of the prediction result as the fitness function of the genetic algorithm to perform the individual evaluation index of the RNN. Finally, the RMSE values of the predicted results of all RNN individuals on the experimental dataset are compared through two different scenarios of prediction examples to obtain the best prediction model. The results of experiments show that this method has excellent prediction accuracy and wide applicability for the prediction of short-term and parking spaces in parking guidance information systems.

Keywords: Parking guidance information system
Short-term unoccupied parking space · Genetic algorithm · RNN

1 Introduction

The problem of "hard parking" is a common problem in the central areas of major cities in the world. And even in some areas it has reached the point where "one space is hard to find". In order to alleviate the contradiction between the supply and demand of motor vehicle spaces in the central urban area, the parking guidance information system (PGIS) emerge as the times require. It can predict the parking space characteristics in real time. Reasonable parking induction can not only effectively reduce the driver's time to find a parking space, but also reduce the traffic congestion and environmental pollution. The accurate real-time parking space prediction is an important technical content of the parking guidance information system. Under realistic scenarios, the choice of parking plan is affected by many factors, such as parking price, traffic

© Springer Nature Switzerland AG 2018
R. Hong et al. (Eds.): PCM 2018, LNCS 11164, pp. 865–876, 2018.
https://doi.org/10.1007/978-3-030-00776-8_79

conditions, and arrival time. According to different research directions, domestic and foreign scholars have more research results at a different angle.

Firstly, in detecting the parking status of parking spaces, Choeychuen [1] proposed a parking space exploration algorithm with automatic threshold. However, the cost of using image processing method is expensive. It is more practical to use hardware solutions. Zheng [2] proposed a mechanism to predict the state of parking spaces using three feature sets with selected parameters. The forecasting method is carried out in an ideal parking environment. It ignores the diversification of factors that affect the parking prediction in the actual situation. Keat [3] used the bayesian rules to extract the parking position from data obtained from laser scanning. The above prediction methods based on the status of the parking spaces are all carried out in an ideal parking environment, which ignore the diversification of factors that affect the parking space prediction in actual situations.

Followed by researches on parking spaces. Pattanaik et al. [4] proposed a cluster-based analysis of traffic conditions to avoid congestion. Kotb et al. [5] proposed a novel intelligent parking system based on dynamic resource allocation and pricing. This system establishes a mixed-integer linear programming mathematical model. Real-time parking spaces are used to dynamically perform system decisions and provide parking services for users. It can reduce the time for the driver to search for parking spaces, minimize the cost and ensure the maximum utilization of parking resources. Benson [6] proposed a wireless sensor network solution, which combines parking management and routing protocols to improve the reliability of the transmission. The method is only to propose to manage parking lots from the perspective of a single parking lot. All of the above researches have related to the problem of poor information sharing capability in smart parking services. At the same time, the above methods are all focused on the research of a certain aspect of the parking management system. They neglect the service needs of users on other aspects and fail to integrate these functions on a platform.

In recent years, the artificial intelligence, with its ability to handle non-linear problems, has achieved remarkable applications in the field of prediction. In view of the superiority of neural network in solving complex problems, Yan-jie et al. [7, 8] proposed a parking space prediction model based on wavelet neural network, wavelet transform and particle swarm wavelet neural network. Haipeng et al. [9] used the combination of the wavelet transform and the ELM neural network to predict the short-space idling parking spaces, which can improve the training speed of traditional neural networks while maintaining the original prediction accuracy. The above-mentioned research methods based on deep learning often have difficulty in training neural networks and it is difficult to obtain a global optimal solution.

From the above three types of research methods, we can see there are following three problems:

The cost of inducing system construction is high and the accuracy of forecasting is not good.

The applicability of the method is not high. The prediction method is often aiming at a certain parking lot, but the prediction effect for other parking lots is not ideal.

Model is difficult to train and the optimal solution is often difficult to obtain.

We propose a parking space prediction method based on genetic algorithm and RNN to remedy the shortages in the above prediction methods. The method combines the advantages of the genetic algorithm in the global optimal solution search and the advantage of the neural network in dealing with non-linear problems. It not only reduces the difficulty of model training but also improves the accuracy of the prediction. Moreover, it can adapt different parking scenes by adjusting the parameters of the genetic algorithm. This greatly improves the forecasting applicability of the method. It is a powerful aid for parking guidance system. It has certain research value.

2 The Prediction Method

2.1 Genetic Algorithm

The genetic algorithm was first proposed by Professor J. Holland in 1975 [10]. Its discovery research is based on the theory of evolution and Mendel's genetic theory. It uses computer as a tool to realize the imitation of biological evolution mechanism, and explores the discovery of random global search and optimization methods in the process of imitation and inquiry. Unlike traditional search algorithms, a genetic algorithm starts with a set of randomly generated initial solutions, called populations. Each individual in the population is a solution to the problem and is called a chromosome. These chromosomes evolved in subsequent iterations, called genetic. Genetic algorithms are mainly implemented through crossover, mutation, and selection operations. Crossover or mutation operations generate the next generation of chromosomes, called descendants. The quality of chromosomes is measured by the result of fitness function. According to the value of the fitness, we select a certain number of individuals from the previous generation as the next generation and continue to evolve. After a few generations, the algorithm converges on the best chromosome. It is likely to be the optimal solution to the problem or suboptimal solution.

2.2 RNN (Recurrent Neural Network)

A recurrent neural network or RNN is a class of neural networks used to process sequence data [11]. In recent years, RNN has rapidly been widely used in speech, image recognition, and natural language processing. The common neural network has many disadvantages. For example, when predicting the next word in a sentence, it must be predicted based on the previous word and its sentence structure, and the word to be predicted cannot be directly and independently predicted. This is because in traditional neural network model, the nodes between the input layer, hidden layer, and output layer are not connected. Only the layers are fully connected. Therefore, we need to find a way to improve the traditional neural network so that it can solve more problems. Because the traditional neural network only has a short-term memory, and the gradient disappears and the gradient explodes easily during the training process. RNN not only has the advantage of long-term memory, but its variants such as LSTM and GRU completely circumvent such hidden dangers. Due to its powerful function, it can be said that this network is also the preferred network for solving time series.

2.3   Prediction Method Combining Genetic Algorithm with RNN

In traditional neural network model, data flows from the input layer to the hidden layer to the output layer. The adjacent layers are fully connected and the nodes in the layer are connectionless. However, this common neural network is incapable of many problems. For example, if you want to predict what the next word of a sentence is, you generally need to use the previous word because the context in a sentence is important. RNN is a recurrent neural network, that is, the current output of a sequence is also related to the previous output. The specific manifestation of RNN is that the network will memorize the previous information and apply it to the calculation of the current output. That is, the nodes between hidden layers are connected, and the input of the hidden layer includes not only the output of the input layer but also the output of the hidden layer at the previous moment. In theory, the RNN can process any length of sequence data. Therefore, RNN is the preferred neural network for dealing with time series problems. Using the advantage of genetic algorithm for efficient search of problem optimal solutions, control the training process of RNN to obtain the optimal network structure, which not only reduces the difficulty of RNN training, but also greatly enhances the prediction accuracy. Specifically, according to parameters of different functions, set the parameters in the RNN training process, including hidden layer search space, the number of neuron search space and the network type search space (GRU, LSTM, or a mixture of both). In order to filter the optimal RNN, the optimizer used for RNN iterative learning is undetermined, and an optimizer search space is set. Then using the idea of the heuristic search genetic algorithm (GA), the driving algorithm performs an optimal solution search in its search space, and finally outputs the prediction result. The following Fig. 1 shows the specific steps of the predicting method.

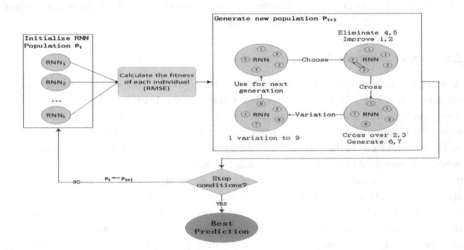

Fig. 1. Algorithm flowchart

The specific steps of the prediction method are as follows:

Step 1: Initialize the RNN population and randomly generate various individuals with different parameters in the population. These different parameters are the chromosomes of the genetic algorithm.

Step 2: The newly randomly generated individuals in the population are evaluated fitness individually. This evaluation criteria selected is the Root Mean Square Error (RMSE) of each individual prediction result generated after the raw data is trained.

Step 3: Perform genetic operations to preserve individuals with better fitness through selection, crossover, and mutation to generate new generations and generate new populations.

Step 4: Determine whether to meet the conditions of stop. If it is not satisfied, replace the original population with a new population and repeat steps 2 and 3. If the stopping condition is satisfied, the genetic operation is terminated and the best individual in the output population is the optimal solution. The stopping condition is to reach the maximum genetic generation and population size.

Step 5: The best individual in the RNN population is learnt to obtain the final prediction result.

2.4   Evaluation Standard of Prediction

Time series prediction uses the sample values observed in the sequence to estimate the value of a certain time in the future [12]. The error between the predicted value and the true value is usually expressed as following formula (1) to formula (5).

$$SSE(Sum\ of\ square\ error) = \sum_{i=1}^{n} w_i(y_i - \hat{y}_j)^2 \tag{1}$$

$$MSE(Mean\ square\ error) = \frac{SSE}{n} = \frac{1}{n}\sum_{i=1}^{n} w_i(y_i - \hat{y}_j)^2 \tag{2}$$

$$RMSE(root\ mean\ square\ error) = \sqrt{MSE} = \sqrt{\frac{SSE}{n}} = \sqrt{\frac{1}{n}\sum_{i=1}^{n} w_i(y_i - \hat{y}_j)^2} \tag{3}$$

$$MAE(Mean\ absolute\ error) = \frac{1}{n}\sum_{i=1}^{n} |(y_i - \hat{y}_j)| \tag{4}$$

$$MAPE(Mean\ absolute\ percentage\ error) = \sum_{i=1}^{n} \left|\frac{y_i - \hat{y}_j}{y_i}\right| \times \frac{100}{n} \tag{5}$$

Where y_i represents the predicted value and $\hat{y}_j$ represents the true value.

From the definitions of MSE, RMSE, and SSE, it can be seen that MSE and RMSE are both variants of SSE. Considering the unity of error magnitude, RMSE is chosen to describe the error between the original data and the predicted value. The smaller the value, the smaller the error and the higher the accuracy. Since the original parking lot data is full of parking spaces, the corresponding number of free parking spaces may be 0. At this time, the denominator in the MAPE formula is 0, and the statistics are inconvenient. Therefore, the average absolute error is selected to represent the system error.

3 Experiment Analysis

3.1 The Data of Experiments

In order to test the practicability and wide applicability of the algorithm, this experiment selected parking data from two cities as experimental data to test the algorithm. Select historical data of Beijing shopping mall parking lot and a hotel parking lot in Shanghai for a period of 7 days from April 3 to April 9, 2017 as algorithm training data to predict parking data at each time on April 10th. The number of parking spaces in the shopping mall parking lot is 136. Since the mall is only open during the period from 9:00 to 21:00, we split the original data for the time interval of 10 min and count the number of free parking spaces in each period as a record. A total of 584 historical data were obtained, including 511 training data and 73 test data. However, the hotel is open for 24 h and has 130 parking spaces. The same time intervals are 10 min. A total of 1152 data are obtained, of which 1,008 are training data and 144 are test data. The partial data of the two data sets are shown in Figs. 2 and 3.

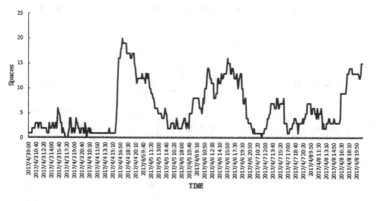

Fig. 2. Free parking spaces of shopping mall parking in Beijing

3.2 Parameter Setting of Genetic Algorithm

In this paper, the parameters of the genetic algorithm are set according to the ideas. Imagine the solution space as a geometrical space. If we want to get better solutions, we need increase the number of population, therefore, the density of points can be increased, but the calculation is increasing. So it is necessary to select the appropriate density population. It is also very important to obtain the optimal solution. For the distribution of these solutions, first of all cannot be too concentrated, if the probability

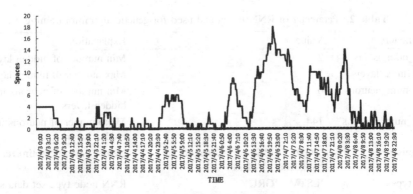

Fig. 3. Free parking spaces of hotel parking lot in Shanghai

of the mutation will increase, the probability of crossover will be reduced, so it is appropriate to disperse, and dispersion is relatively, in the vicinity of local solutions can be relatively concentrated. However, if the aggregation area is not significant, the mutation probability needs to be reduced.

In summary, the final genetic algorithm parameters are set as Table 1.

Table 1. Genetic algorithm parameters

Parameters	Value	Explanation
mutation_rate	0.1	Genetic algorithm mutation rate
min_mutation_momentum	0.0001	Min mutation momentum
max_mutation_momentum	0.1	Max mutation momentum
min_population	6	Min population for GA
max_population	72	Max population for GA
num_Iterations	3	Number of iterations to evaluate GA

3.3 Parameter Setting of RNN

The design of RNN population, including four chromosomes, is used to describe the different aspects of neural network, which are hidden layer number, neuron number, optimization function set, and neuron structure set. The above chromosome design provides the corresponding search space for the RNN of the optimal structure. Table 2 is the parameter setting of individual RNNs for genetic algorithm training.

3.4 The Results of Experiments

We use the Parking space prediction method proposed in this paper to perform prediction experiments in two parking lots, and get the corresponding individuals of the best RNN population. First, we analyze the results of the experiment. Figure 4 shows the forecast results of the short-term parking spaces of a shopping mall in Beijing.

Table 2. Parameter of RNN individual used for genetic algorithm training

Parameters	Value	Explanation
min_num_layers	2	Min number of hidden layers
max_num_layers	5	Max number of hidden layers
min_num_neurons	6	Min number of neurons in hidden layers
max_num_neurons	144	Max number of neurons in hidden layers
optimisers	'SGD', 'RMSprop', 'Adagrad', 'Adadelta', 'Adam'	Neural network optimizer set
rnn_types	'LSTM', 'GRU'	RNN node type set data set
rnn_epochs	3	Epochs for RNN

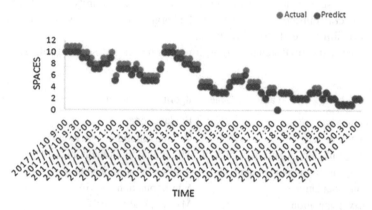

Fig. 4. Experimental results of shopping mall parking lot in Beijing

Figure 5 shows the forecast results of the short-term free parking spaces of a hotel in Shanghai. It can be seen from the experimental results that the predicted results of the two scenes are very similar to the actual results, indicating that the accuracy of the method is excellent.

Next, we analyze the best individual parameters of the method in two experiments. Figure 6 shows the best RNN individual parameters obtained in the experiment of Beijing shopping mall, in which the hidden layer consists of the dense layer, LSTM layer, and Dense. Each layer has a corresponding input and output dimension. The input dimension represents the dimension that accepts the input from the previous network layer, and the output dimension indicates that after the operation of the layer, the data is exported to the next level of data dimension.

Figure 7 shows the best RNN individual parameters obtained in the experiment of the Shanghai hotel. Different from the previous experiment, the hidden layer consists of the dense layer, the Simple RNN layer, and Dense. Moreover, the input and output

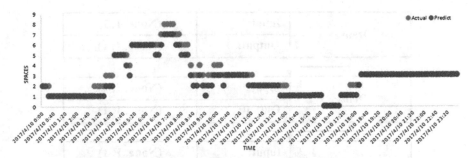

Fig. 5. Experimental results of hotel parking lot in Shanghai

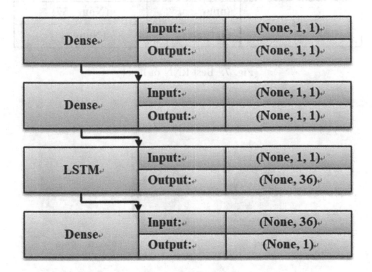

Fig. 6. Best RNN of mall

dimensions of the Simple RNN layer and the LSTM layer also differ, which shows that the method automatically adjusts the parameters in different applications, to ensure the best predictive effect. Thus, this method has a wide range of applicability.

Finally, in order to comprehensively prove the excellent performance of this method, we compare this method with algorithms currently used in the field of parking prediction, specifically to analyze the RMSE value of the experimental results of each algorithm. In addition to the introduction of the BP [13] algorithm based on neural networks and the combination of wavelet transform and particle swarm wavelet neural network [8], we also introduced universally used and statistically significant ARIMA (autoregressive integral sliding average model). Models [14] algorithm to compare algorithmic results. The experimental results of two application scenarios of Beijing Mall and Shanghai Hotel are shown in Fig. 8 and Fig. 9 respectively.

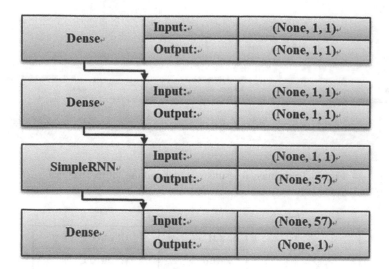

Fig. 7. Best RNN of hotel

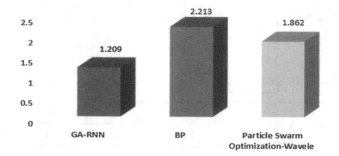

Fig. 8. Best RNN of hotel

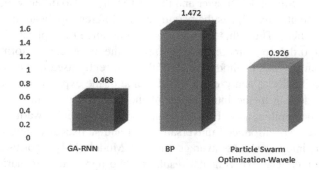

Fig. 9. Best RNN of hotel

The experimental results show that under the two parking lots, the RMSE value obtained by the proposed method is the smallest, which means that the method has less error and excellent accuracy.

4 Conclusion

In this paper, the combination of genetic algorithm and recurrent neural network (RNN) is used to predict the short-term parking space in parking guidance system. The paper first discusses the principle and implementation process of the method, then demonstrates and analyzes it through two prediction examples in different regions and different scenarios. The results show that this method is effective in the prediction of shortterm parking space berths in parking guidance systems. With a very high accuracy of prediction, and for different scenes, it can have excellent applicability by automatically adjusting training parameters.

Acknowledgment. This research is supported by the National Natural Science Foundation of China (61672259, 61602203), Key Projects of Jilin Province Science and Technology Development Plan (20180201064SF), and Outstanding Young Talent Foundation of Jilin Province (20170520064JH, 20180520020JH).

References

1. Choeychuen, K.: Automatic parking lot mapping for available parking space detection. In: 5th International Conference on Knowledge and Smart Technology (KST), pp. 117–121. IEEE (2013)
2. Zheng, Y., Rajasegarar, S., Leckie, C.: Parking availability prediction for sensor-enabled car parks in smart cities. In: IEEE Tenth International Conference on Intelligent Sensors, Sensor Networks and Information Processing (ISSNIP), pp. 1–6. IEEE (2015)
3. Keat, C.T.M., Pradalier, C., Laugier, C.: Vehicle detection and car park mapping using laser scanner. In: IEEE/RSJ International Conference on Intelligent Robots and Systems (IROS), pp. 2054–2060. IEEE (2005)
4. Pattanaik, V., Singh, M., Gupta, P.K., et al.: Smart real-time traffic congestion estimation and clustering technique for urban vehicular roads. In: Region 10 Conference (TENCON), pp. 3420–3423. IEEE (2016)
5. Kotb, A.O., Shen, Y.C., Zhu, X., et al.: iParker—a new smart car-parking system based on dynamic resource allocation and pricing. IEEE Trans. Intell. Transp. Syst. **99**, 1–11 (2016)
6. Benson, J.P., O'Donovan, T., O'Sullivan, P., et al.: Car-park management using wireless sensor networks. In: 31st IEEE Conference on Local Computer Networks, pp. 588–595. IEEE (2006)
7. Yan-jie, J., Dou-nan, T., Blythe, P., et al.: Short-term forecasting of available parking space using wavelet neural network model. IET Intel. Transp. Syst. **9**(2), 202–209 (2015)
8. Ji, Y.J., Chen, X.S., Wang, W., et al.: Short-term forecasting of parking space using particle swarm optimization-wavelet neural network model. J. Jilin Univ. Eng. Ed. **46**(2), 399–405 (2016)
9. Haipeng, C., Xiaohang, T., Yu, W., et al.: Short-term parking space prediction based on wavelet-ELM neural networks. J. Jilin Univ. Sci. Ed. **55**(2), 388–392 (2017)

10. Sampson, J.R.: Adaptation in natural and artificial systems (John H. Holland). SIAM Rev. **18**(3), 53 (1976)
11. Rumelhart, D.E., Hinton, G.E., et al.: Learning representations by back-propagating errors. Nature **323**(6088), 533–536 (1986)
12. Wang, Y.: Application Time Series Analysis. China Renmin University Press, Beijing (2012)
13. Gao, G.Y., Ding, Y., Jiang, F., Li, C.: Prediction of parking guidance space based on BP neural networks. Comput. Syst. Appl. **26**(1), 236–239 (2017)
14. Zhang, G.P.: Time series forecasting using a hybrid ARIMA and neural network model. Neurocomputing **50**(1), 159–175 (2003)

Perceptual Image Dehazing Based on Generative Adversarial Learning

Fangfang Wu[1], Yifan Li[2], Jianwen Han[1], Weisheng Dong[1(✉)],
and Guangming Shi[1]

[1] School of Artificial Intelligence, Xidian University, Xi'an, China
wsdong@mail.xidian.edu.cn
[2] Faculty of Science and Technology, University of Macau, Macau, China

Abstract. Convolutional Neural Networks (CNN) based single image dehazing methods have recently gained much attention. However, as they heavily rely on synthetic haze images, existing CNN-based dehazing methods have limitations in achieving visually pleasant results, especially for real haze images. Inspired by the recent advances in generative adversarial networks (GAN), this paper proposes a novel end-to-end image dehazing network for image dehazing. Different from the existing CNN-based dehazing methods that were trained with paired hazy and hazy-free images, the proposed network was trained with paired and unpaired hazy datasets. To this end, the perception loss expressing high-level semantic information has been proposed. Experimental results show that the proposed method achieve substantial improvements over current state-of-the-art dehazing methods.

Keywords: Image dehazing · Generative adversarial networks
Weakly supervised · Perceptual loss

1 Introduction

High-quality images are desired in many image processing and computer vision applications. However, outdoor images are often degraded due to the bad weather such as fog and haze. Recovering a clear images from a hazy image is a challenging ill-posed inverse problem. The image degradation model [16] describing the relationship between the hazy image and the original clear image can be expressed as

$$I(x) = J(x)t(x) + A(1 - t(x)), \tag{1}$$

where I is the captured hazy image, J is the original haze-free image, A denotes the global atmosphere light, t denotes the transmission map describing the light that directly reaches the camera without scattering and x denotes the pixel index.

F. Wu, Y. Li and G. Shi—Contributed equally to this paper. This work was supported in part by the Natural Science Foundation of China under Grant 61622210, Grant 61471281, Grant 61632019, Grant 61621005, and Grant 61390512.

To solve the ill-posed problem, conventional dehazing methods usually estimated the transmission map and atmosphere light based on some assumptions, e.g., the local max contrast prior [20] and dark channel prior (DCP) [9]. In recent years, some handcrafted priors have also been proposed for image dehazing [21,25]. Most recently, convolutional neural networks (CNNs) have also been proposed to estimate the transmission map based on massive synthetic training data [5, 14,19], and then the dehazed image can be recovered based on the atmospheric scattering model Eq. 1.

Despite the remarkable progress of those learning-based methods, current learning-based methods still have some limitations. First, most of the existing methods follow the conventional dehazing framework, i.e., first estimating the atmospheric light and transmission map and then solving for the clear images by inverting the hazy model. Obviously, inaccurate estimation of these parameters will degrade the dehazing performance significantly. Second, the mean square error (MSE) loss cannot insure visually pleasant results. Third, current learning-based methods relied on synthetic training datasets and often fail to generate visually pleasant results for real hazy images. Inspired by the powerful image generation ability [11–13,24] of generative adversarial networks (GAN) [7], we propose a GAN-based image dehazing method. Different from existing CNN-based dehazing methods that relied on paired synthetic hazy and haze-free images, we train the proposed network on both paired image dataset and unpaired real haze and haze-free image dataset. Moreover, perceptual loss has also been adopted to further improve the visual quality of the dehazed images. Experimental results show that the proposed method outperform existing dehazing methods on both synthetic and real haze images.

2   Related Work

Most traditional dehazing methods focused on the estimation of the accurate transmission map based on some handcrafted priors. Fattal et al. [6] assumed that the transmission and albedo were uncorrelated and used independent component analysis to estimate the transmission map. Tan et al. [20] proposed a contrast maximization based dehazing method under the assumptions that clear images have higher contrast. He et al. [9] proposed a dark channel prior (DCP) for haze removal, assuming that the dark channel of clear images usually remains dark or low intensities. Some improved DCP-based dehazing methods have also been proposed to improve the efficiency and effectiveness [8,9,18,22]. However, DCP-based methods often failed when there exists large sky area. Meng et al. [17] proposed to estimate the transmission map through bounded constraint and contextual regularization. Berman et al. [4] proposed a global prior for haze removal, assuming that clear images can be approximated by a few hundreds of colors. However, the method become less effective when the atmospheric light is strong. A fusion strategy blending the white balanced and the contrast enhanced hazy image has also been proposed [2]. All of these method mentioned above used handcrafted priors, which cannot accurately describe the real hazy images.

In addition to the handcrafted priors, data-driven dehazing methods have also been developed. Tang et al. [21] proposed to combine different handcraft priors using the random forest technique. Zhu et al. [25] proposed a Color Attention Prior (CAP) and used a linear model to predict the depth of hazy images. Recently convolutional neural network (CNN) based method has also been proposed. Cai et al. [5] proposed a DehazeNet utilizing the BReLU unit and Maxout function to estimate the transmission maps. A coarse-to-fine network has also been proposed to estimate the transmission [19]. Note that these methods only attempted to predict the transmission maps and still needed to estimate the atmosphere light using conventional methods. To address this issue, Li et al. [14] proposed the AOD-Net to predict a new variable which integrates the transmission and atmosphere light through a linear transformation. However, this still falls into the physical model.

3   Proposed Method

Instead of estimating the intermediate transmission maps for dehazing, in this paper we direct reconstruct the dehazed images through end-to-end learning. Different from previous CNN-based methods that relied on synthetic hazy images, we train the deep neural network using both synthetic hazy images and real hazy images. To this end, we proposed a weakly supervised generative adversarial network (WSGAN) for image dehazing, whose architecture is shown in Fig. 1. Two types of image dataset are used to train this network, i.e., the paired synthetic hazy and clear image and the unpaired real hazy and clear images. As shown in Fig. 1, the generator G and the discriminator D_{pair} is used for supervised adversarial learning on paired synthetic dataset, while the other discriminator D_{unpair} is used for weakly supervised adversarial learning on unpaired dataset. The add-haze network R is used for dual constraint in the weakly supervised module. Thus, the visual quality of the dehazed image can be improved through the adversarial training using real high quality natural images. To optimize the proposed network, we propose hybrid loss functions, i.e., the content loss, adversarial loss, consistence loss, and the perceptual loss. All the sub-networks and the loss functions will be described in the following subsections.

3.1   Network Architecture

Generator Network. The generator G, whose architecture is shown in Fig. 2, aims to generate a haze-free image from a hazy image. It consists of three parts, i.e., the encoder, transformer and decoder. As shown in Fig. 2, the encoder containing 3 convolutional layers encodes the visual feature of the input image, while the transformer containing 6 Resblocks further transforms the visual features. The decoder that contains 3 deconvolutional layers reconstruct the haze-free image from the transformed visual features. To avoid vanishing gradient problem, skip-connections from the input of the convolutional layer to the output of the deconvolutional layers are adopted. Each convolutional layer is followed by

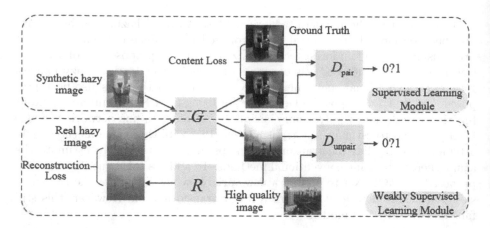

Fig. 1. The architecture of the proposed dehaze method.

the instance normalization [23] and the leaky ReLU activation function, except the final layer where the Tanh activation is adopted for better reconstruction performance. All the layers use the 3 × 3 kernels and generate 32 feature maps, except the first convolutional layer that uses 5 × 5 kernels. Note that the add-haze subnetwork R has same architecture as that of generator G.

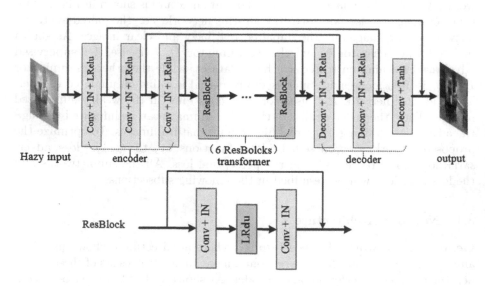

Fig. 2. The structure of generator network.

Discriminator Network. The Discriminator networks D_{pair} and D_{unpair} have the same architecture, as shown in Fig. 3. The input of D_{pair} is the dehazed results of the synthetic hazy images (with label 0) or the corresponding original images (with label 1), whereas the input of D_{unpair} is the dehazed results of the real hazy images (with label 0) or the original high quality images (with label 1). The output of each discriminators is a 70×70 image which denotes the accurate probability. As shown in Fig. 2, the first 4 convolutional layers are followed by the instance normalization [23] and the leaky ReLU activation function, while the final layer is followed by the sigmoid activation function for calculation of the probability. All the layers use the 3×3 kernels. The convolutional strides of the first three layers are set to 2 to reduce the spatial resolution of the feature maps, and the strides of the last two layers are set to 1 to remain the same resolution. The number of the feature maps of the 5 layers are 64, 128, 256, 512 and 512, respectively.

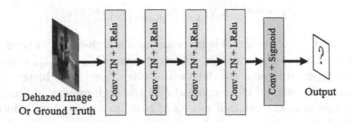

Fig. 3. The structure of discriminator network.

3.2 Loss Function

As illustrated in Fig. 1, content loss and supervised adversarial loss are used for supervised leaning, while weakly supervised adversarial loss and consistence loss are employed for weakly supervised learning. All the loss functions are described as follows.

Content Loss. The content loss contains MSE loss and perceptual loss. Given a pair of images $\{I^{Sh}, I^{Sl}\}$, I^{Sh} is the synthetic hazy image and I^{Sl} is the corresponding original image. MSE loss is defined as

$$L_{MSE}^{pair} = \frac{1}{CWH} \sum_{c=1}^{C} \sum_{w=1}^{W} \sum_{h=1}^{H} \left\| I^{Sl} - G\left(I^{Sh}\right) \right\|^2 \qquad (2)$$

where $C \times W \times H$ denotes the size of input image. Minimizing MSE loss function help to obtaining high signal-to-noise ratio (PSNR). However, this kind of pixel loss struggles to recover high-frequency image details. To overcome this limitation, we introduce the perceptual loss developed in [13] as

$$L_p^{pair} = \frac{1}{C_f H_f W_f} \sum_{j=1}^{3} \left\| \phi_j\left(I^{Sl}\right) - \phi_j\left(G\left(I^{Sh}\right)\right) \right\|_2^2 \qquad (3)$$

where ϕ_j denotes the activation of jth layer of a pre-trained deep CNN. Here we choose the "ReLu3","ReLu4" and "ReLu5" layers of the pre-trained VGG19 to compute the activation. $C_f \times W_f \times H_f$ denotes the size of active feature maps.

Adversarial Loss. The supervised adversarial loss for adversarial learning on paired dataset is formulated as

$$L_a^{pair} = \log\left(1 - D_{pair}\left(G\left(I^{Sh}\right)\right)\right) \tag{4}$$

By minimizing the adversarial loss, we enforce the dehazed result of the simulated hazy images should be much similar to the corresponding original images and try to fool D_{pair}. In addition to the supervised adversarial loss, the following weakly supervised adversarial loss has also be adopted, expressed as

$$L_a^{unpair} = \log\left(1 - D_{unpair}\left(G\left(I^{Rh}\right)\right)\right) \tag{5}$$

where I^{Rh} and I^{Rc} deonte a real hazy image and a high-quality image randomly selected from the unpaired dataset. Different from D_{pair}, D_{unpair} aims to distinguish two types of inputs: the dehazed results of real hazy images (for which we don't have the original clear images) and the high quality images. Through minimizing this loss, the dehazed results of real hazy images are enforced to be similar with high-quality natural images in terms of contrast, brightness and colors.

Consistence Loss. Similar with [10], to ensure the stability of weakly supervised adversarial learning, a dual learning is also employed to add haze to the dehazed images of real hazy images. The recovered images should be close to the original real hazy images. This can be enforced by imposing the following consistence loss

$$L_{MSE}^{cons} = \frac{1}{CWH} \sum_{c=1}^{C} \sum_{w=1}^{W} \sum_{h=1}^{H} \left\| I^{Rh} - R\left(G\left(I^{Rh}\right)\right)\right\|^2 \tag{6}$$

Total Loss. The whole loss function for optimizing the generator G can be formulated as

$$L_G = L_{MSE}^{pair} + L_p^{pair} + L_{MSE}^{cons} + 0.001(L_a^{pair} + L_a^{unpair}), \tag{7}$$

To make the adversarial learning work, discriminators D_{pair} and D_{unpair} should be trained adequately to ensure their ability to distinguish different types of input. The loss functions for optimizing D_{pair} and D_{unpair} are expressed as follows:

$$L_{D_{pair}} = -\log\left(D_{pair}\left(I^{Sl}\right)\right) - \log\left(1 - D_{pair}\left(G\left(I^{Sh}\right)\right)\right), \tag{8}$$

$$L_{D_{unpair}} = -\log\left(D_{unpair}\left(I^{Rc}\right)\right) - \log\left(1 - D_{unpair}\left(G\left(I^{Rh}\right)\right)\right). \tag{9}$$

4   Experiments

4.1   Training Dataset

Synthetic Dataset. The recently developed RESIDE dataset [15], which contains 8971 clear outdoor images and their corresponding depth images, is used as the synthetic dataset. We randomly selected 8000 clear images and the depth images from this dataset and simulated the corresponding hazy images based on the atmospheric scattering model (1) where the atmospheric lights A is randomly set in the range of $[0.7, 10]$ and the scattering coefficient β is randomly set in the range of $[0.6, 1.8]$. The clear images and the simulated hazy images were croped and resized to be size of 256×256.

Real Hazy Dataset. In addition to the simulated hazy images, we also generate a set of unpaired hazy and clear images. We downloaded a set of real hazy images from the internet and also collect a set of high-quality images from the internet and the NTIRE challenge dataset [1]. Note that there are no one-to-one correspondences between the hazy images and the clear images. All the images were cropped and resized to be size of 256×256. Image argumentation by rotations and flips were also adopted to generate unpaired 8000 real hazy images and high-quality images.

4.2   Implementation Details

The generator G, the discriminators D_{pair}, D_{unpair}, and the add-haze subnetwork R were alternatively trained by minimizing the loss functions of Eqs. (7), (8), (9), and (6), respectively. Specifically, G was first trained by fixing others, followed by training D_{pair}, D_{unpair}, and R sequentially with other fixed. The ADAM optimizer are used to train all the networks with $\beta_1 = 0.5, \beta_2 = 0.999$. The initial learning rate is set to 0.0002, and we linearly decreased it to zero after 50 epochs. We implemented all of our networks using TensorFlow and the training was performed on a single NVIDIA 1080Ti GPU. Since the generator network is based on full convolutions, arbitrary size of image can be processed directly.

To verify the effectiveness of the proposed weakly supervised learning GAN for dehazing (denoted as WSGAN-DH), we also proposed a variant of the proposed method by removing the discriminator D_{unpair} and the add-haze subnetwork R, where the generator G was trained only on the synthetic dataset. We denote this variant as GAN-DH. To verify the performance of the proposed method, the standard synthetic D-HAZY Middlebury dataset [3], which contains 23 standard synthetic hazy images with atmospheric lights $A = 1$ and scattering coefficient $\beta = 1$ were used in the comparison study. Some real hazy images were used as for verify the dehazing performance of the proposed methods. We compared our methods with several leading dehazing methods, including DCP [9], CAP [25], Dehaze-Net [5], MSCNN [19] and AOD-Net [14]. In addition to the PSNR and SSIM metrics, the entropy was also used to evaluate the competing methods.

4.3   Results

Results on Synthetic Hazy Images. Table 1 shows the average results of the test methods on the D-HAZY Middlebury dataset [3]. As shown in Table 1, our methods perform better than other methods, except the CAP method. While the proposed method is slightly worse than the CAP method in terms of PSNR and SSIM, the proposed method outperforms CAP method in terms of entropy metric, indicating that the proposed method can recovering more image details.

Table 1. Quantitative results on D-HAZY Middlebury dataset.

Metrics	PSNR/dB	SSIM	Entropy
DCP [9]	12.43	0.7906	6.9921
CAP [25]	**14.61**	**0.7993**	7.2866
DehazeNet [5]	13.94	0.7948	7.1764
MSCNN [19]	13.72	0.7795	7.2477
AODNet [14]	13.36	0.7785	7.0913
DHGAN	14.51	0.7941	7.3096
WDHGAN	14.01	0.7806	**7.3198**

Figure 4 shows parts of the dehazed images by the test methods on the D-HAZY Middlebury dataset [3]. It can be seen that the image produced by DCP [9] were over enhanced. The CAP [25] method can better preserve the original image colors. However, some remaining haze can be observed in the dehazed images. Similarly, haze can still be observed in the dehazed images by the DehazeNet [5], MSCNN [19]and AOD [14] methods. In contrast, the images dehazed by proposed methods are more clear than other competing methods. By

Fig. 4. Subjective results on D-HAZY Middlebury dataset.

Table 2. Quantitative entropy results on real hazy images.

Tests	DCP [9]	CAP [25]	DehazeNet [5]	MSCNN [19]	AOD [14]	DHGAN	WDHGAN
Mountain	6.7719	7.2815	7.4203	7.0345	7.3279	7.3950	**7.4676**
Cloud	7.4114	7.4702	7.3784	**7.5033**	6.5704	7.1223	7.2247
Village	6.9595	6.9551	6.9389	7.0790	6.5367	7.1957	**7.5172**
Waterfall	6.6015	7.4161	7.3955	7.0799	7.1944	7.4798	**7.6934**
Gym	7.2564	7.3802	7.2971	7.2564	7.0190	7.6553	**7.7888**
House	7.3089	7.4520	7.4551	**7.5458**	7.2591	7.3966	7.4777
City	7.1973	7.2369	7.1425	7.4483	6.7325	7.3982	**7.5778**
Traffic	6.5632	6.8432	6.6961	6.9151	6.2384	7.3587	**7.5230**
Train	6.1305	6.6384	6.7991	6.8412	6.9233	7.0888	**7.4104**
Plant	7.0157	7.2651	7.3339	7.1660	6.9689	7.6417	**7.6553**
Average	6.9216	7.1939	7.1857	7.1869	6.8771	7.3732	**7.5336**

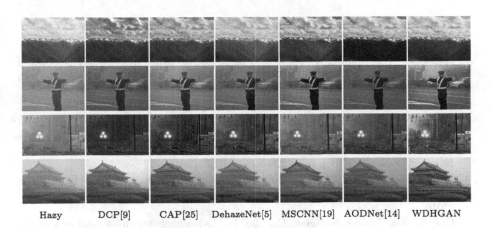

Hazy   DCP[9]   CAP[25]   DehazeNet[5]   MSCNN[19]   AODNet[14]   WDHGAN

Fig. 5. Subjective results on real hazy images.

employing the weakly supervised learning the proposed WSGAN-DH methods can further improve the visual quality of the dehazed images.

Results on Real Hazy Images. In addition to the synthetic haze images, we also evaluate the test methods on some real haze images used in [9, 19, 25]. Table 2 shows the entropy results on the 10 real hazy images. We can be see that the proposed methods perform better than other methods on most of the test images. It can also been seen that the proposed WSGAN-DH method consistently outperforms it counterpart GAN-DH on this scenario. Some dehazed images by the test methods are shown in Fig. 5. It can be seen that the dehazed images by other methods contain many artifacts, e.g., color deviations, color over saturation. Compared to other methods, the proposed methods generated more clear images. With weakly supervised learning, the WSGAN-DH further improves the visual quality of the dehazed images.

5   Conclusions

In this paper, we proposed a weakly supervised generative adversarial network for image dehazing. Different from current deep learning based methods that rely on strong supervised learning on synthetic dataset, the proposed method train the proposed network is trained by both supervised learning and weakly supervised learning. To this end, two kinds of dataset, i.e., the synthetic dataset containing paired synthetic haze image and clear images and the real dataset consisting of unpaired real hazy and clear images, are built. To the proposed model, diverse loss functions, including the perception loss, supervised adversarial loss, MSE loss and the consistence loss are adopted. Experimental results on both synthetic haze images and real haze images show the the effectiveness of the proposed method.

References

1. Agustsson, E., Timofte, R.: Ntire 2017 challenge on single image super-resolution: dataset and study. In: The IEEE Conference on Computer Vision and Pattern Recognition (CVPR) Workshops, vol. 3, p. 2 (2017)
2. Ancuti, C.O., Ancuti, C.: Single image dehazing by multi-scale fusion. IEEE Trans. Image Process. **22**(8), 3271–3282 (2013)
3. Ancuti, C., Ancuti, C.O., De Vleeschouwer, C.: D-hazy: a dataset to evaluate quantitatively dehazing algorithms. In: 2016 IEEE International Conference on Image Processing (ICIP), pp. 2226–2230. IEEE (2016)
4. Berman, D., Avidan, S., et al.: Non-local image dehazing. In: Proceedings of the IEEE Conference on Computer Vision and Pattern Recognition, pp. 1674–1682 (2016)
5. Cai, B., Xu, X., Jia, K., Qing, C., Tao, D.: Dehazenet: an end-to-end system for single image haze removal. IEEE Trans. Image Process. **25**(11), 5187–5198 (2016)
6. Fattal, R.: Single image dehazing. ACM Trans. Graph. (TOG) **27**(3), 72 (2008)
7. Goodfellow, I., et al.: Generative adversarial nets. In: Advances in Neural Information Processing Systems, pp. 2672–2680 (2014)
8. He, K., Sun, J., Tang, X.: Guided image filtering. In: Daniilidis, K., Maragos, P., Paragios, N. (eds.) ECCV 2010. LNCS, vol. 6311, pp. 1–14. Springer, Heidelberg (2010). https://doi.org/10.1007/978-3-642-15549-9_1
9. He, K., Sun, J., Tang, X.: Single image haze removal using dark channel prior. IEEE Trans. Pattern Anal. Mach. Intell. **33**(12), 2341–2353 (2011)
10. Ignatov, A., Kobyshev, N., Timofte, R., Vanhoey, K., Van Gool, L.: Wespe: weakly supervised photo enhancer for digital cameras. arXiv preprint arXiv:1709.01118 (2017)
11. Isola, P., Zhu, J.Y., Zhou, T., Efros, A.A.: Image-to-image translation with conditional adversarial networks. arXiv preprint (2017)
12. Kim, T., Cha, M., Kim, H., Lee, J., Kim, J.: Learning to discover cross-domain relations with generative adversarial networks. arXiv preprint arXiv:1703.05192 (2017)
13. Ledig, C., et al.: Photo-realistic single image super-resolution using a generative adversarial network. arXiv preprint (2016)

14. Li, B., Peng, X., Wang, Z., Xu, J., Feng, D.: AOD-Net: all-in-one dehazing network. In: Proceedings of the IEEE International Conference on Computer Vision, pp. 4770–4778 (2017)
15. Li, B., et al.: Benchmarking single image dehazing and beyond (2018)
16. McCartney, E.J.: Optics of the Atmosphere: Scattering by Molecules and Particles, 421 p. Wiley, New York, Inc. (1976)
17. Meng, G., Wang, Y., Duan, J., Xiang, S., Pan, C.: Efficient image dehazing with boundary constraint and contextual regularization. In: 2013 IEEE International Conference on Computer Vision (ICCV), pp. 617–624. IEEE (2013)
18. Nishino, K., Kratz, L., Lombardi, S.: Bayesian defogging. Int. J. Comput. Vision 98(3), 263–278 (2012)
19. Ren, W., Liu, S., Zhang, H., Pan, J., Cao, X., Yang, M.-H.: Single image dehazing via multi-scale convolutional neural networks. In: Leibe, B., Matas, J., Sebe, N., Welling, M. (eds.) ECCV 2016. LNCS, vol. 9906, pp. 154–169. Springer, Cham (2016). https://doi.org/10.1007/978-3-319-46475-6_10
20. Tan, R.T.: Visibility in bad weather from a single image. In: IEEE Conference on Computer Vision and Pattern Recognition, CVPR 2008, pp. 1–8. IEEE (2008)
21. Tang, K., Yang, J., Wang, J.: Investigating haze-relevant features in a learning framework for image dehazing. In: Proceedings of the IEEE Conference on Computer Vision and Pattern Recognition, pp. 2995–3000 (2014)
22. Tarel, J.P., Hautiere, N.: Fast visibility restoration from a single color or gray level image. In: 2009 IEEE 12th International Conference on Computer Vision, pp. 2201–2208. IEEE (2009)
23. Ulyanov, D., Vedaldi, A., Lempitsky, V.: Instance normalization: The missing ingredient for fast stylization (2016)
24. Zhu, J.Y., Park, T., Isola, P., Efros, A.A.: Unpaired image-to-image translation using cycle-consistent adversarial networks. arXiv preprint arXiv:1703.10593 (2017)
25. Zhu, Q., Mai, J., Shao, L.: A fast single image haze removal algorithm using color attenuation prior. IEEE Trans. Image Process. 24(11), 3522–3533 (2015)

Author Index

Printed in the United States
By Bookmasters